Plant Physiology

Volume IX: Water and Solutes in Plants

PLANT PHYSIOLOGY

A TREATISE

Plant Physiology

A TREATISE

EDITED BY

F. C. STEWARD

Professor Emeritus
Cornell University
Ithaca, New York

Volume IX: Water and Solutes in Plants

Coedited by

F. C. STEWARD
Charlottesville, Virginia

JAMES F. SUTCLIFFE
The University of Sussex
Falmer, Brighton, England

JOHN E. DALE
Botany Department
University of Edinburgh
Edinburgh, Scotland

1986

ACADEMIC PRESS, INC.

Harcourt Brace Jovanovich, Publishers

Orlando San Diego New York Austin
Boston London Sydney Tokyo Toronto

ACADEMIC PRESS, INC.
Orlando, Florida 32887

United Kingdom Edition published by
ACADEMIC PRESS INC. (LONDON) LTD.
24–28 Oval Road, London NW1 7DX

Library of Congress Cataloging in Publication Data
(Revised for volume 9)

Plant physiology.

Includes bibliographies and indexes.
Contents: v. 1. A cellular organization and respiration. B. Photosynthesis and chemosynthesis. 2 v. – v. 2. Plants in relation to water and solutes – v. 3. Inorganic nutrition of plants – – [etc.] – v. 9. Water and solutes in plants.
1. Plant physiology–Collected works. I. Steward, F. C. (Frederick Campion)
QK711.P58 1959 581.1 59-7689
ISBN 0–12–668609–2 (v. 9)

PRINTED IN THE UNITED STATES OF AMERICA

86 87 88 89 9 8 7 6 5 4 3 2 1

Contents

CHAPTER FOUR

CHAPTER FIVE

CHAPTER SIX

CHAPTER SEVEN

Contributors to Volume IX

Numbers in parentheses indicate the pages on which the authors' contributions begin.

JOHN E. DALE (1, 455), *Department of Botany, University of Edinburgh, Edinburgh EH9 3JH, Scotland*

W. J. DAVIES (49), *Department of Biological Sciences, University of Lancaster, Lancaster, England*

D. H. JENNINGS (225), *Department of Botany, The University, Liverpool L69 3BX, England*

T. A. MANSFIELD (155), *Department of Biological Sciences, University of Lancaster, Lancaster, England*

F. C. STEWARD (551), *Charlottesville, Virginia 22901*

JAMES F. SUTCLIFFE[1] (1, 381, 455), *The University of Sussex, Falmer, Brighton BN1 9QG, England*

[1] Deceased, June 1983.

Preface to Volume IX

Volume VII dealt with the organization of plant cells, the metabolism of their carbon compounds, and the ways in which they mobilize energy for useful purposes. Volume VIII dealt specifically with the use by plants of compounds of nitrogen and the transformations of inorganic nitrogen into the great range of its organic compounds that are either simple and soluble or highly complex. The continually expanding picture from these and earlier volumes is one of interlocking gene-regulated enzyme systems which, at membranes in organelles and in organs, constitute the machinery of metabolism and keep its wheels turning.

Volume IX is organized in three parts. The first three chapters redirect attention to cells and plants in relation to water (cf. Chapters 2 and 3 of Volume II) and the need to recognize changes in terminology that have occurred in the interim. Chapters 4, 5, and 6 dwell on the problems of solutes (organic and inorganic) that comprise the osmotica of cells (cf. Chapter 4 of Volume II) and which may be translocated (cf. Chapters 5 and 6 of Volume II). The final chapter, Chapter 7, addresses the problems of solutes in cells inasmuch as they are inherent in cells and organs as they grow and develop. To this extent it is a transition to and anticipates some of the problems of growth and development which are to follow.

Volume II brought the distinctive problems of water and solutes in plant cells inherited from the classical period of plant physiology (which had yielded the equilibrium concepts of osmotic pressure, the colligative properties of solutes in solutions, the applicability of the gas laws to solutes in solution, and an osmotic view of vacuolated plant cells) to an essentially modern outlook on these questions. After twenty-five years of further intensified research, information has proliferated. As the chapters of Volume IX unfold, they review a wealth of detail from rigorously selected areas.

However, the problems of the water and solutes of plants as they grow extend also to our fields and forests, and thus they have far-reaching consequences for climates, the economy, and human welfare.

Finally, the editors are indebted to Dr. W. J. Dress for the Index to Plant Names and for his oversight of the names used in the text. This volume is the first in the treatise without an Author Index. Such indexes, when rigorously compiled, became more voluminous than their usefulness warranted. All other relevant information can be found in the alphabetized reference lists at the ends of the chapters. For assistance with the Subject Index we are indebted to Mrs. Shirley Bidwell. Again it is a pleasure to express our appreciation to the staff of Academic Press for their understanding of and help with the special problems Volume IX presented.

The Editors

An Explanatory Note

The decision to resume publication of supplementary volumes to this treatise was made in conjunction with three prospective coeditors. In view of his early association with Volume II and his continued contact with the editor, Dr. James F. Sutcliffe was one of them. His principal role was to coedit what is now Volume IX. While Dr. Sutcliffe had participated in the early planning of this volume and had recruited and encouraged its authors, his untimely death in June 1983 occurred before any manuscripts were finally cleared for publication. Moreover, Dr. Sutcliffe had elected to contribute chapters to this volume for which he would be an author.

Dr. John E. Dale, a long-time friend and collaborator of Dr. Sutcliffe, agreed to serve as an assistant editor to complete the editorial work on this volume. However, Dr. Sutcliffe's name is being retained as coeditor, essentially as an In Memoriam tribute to him. It would be difficult to overstate the debt which the editors and the authors now owe to Dr. Dale for his generous gestures. Dr. Dale also agreed to complete the chapter on Phloem Transport that he and Dr. Sutcliffe had planned (it is now Chapter 6). An incomplete chapter on Salt Relations of Intact Plants (now Chapter 5) needed to be revised and completed, and this was also done by Dr. Dale. The chapter on Water Relations of Plant Cells (now Chapter 1) was written by Dr. Dale drawing on his personal knowledge of the subject matter and certain notes and guidelines found in Dr. Sutcliffe's papers.

The plan to conclude Volume IX with Chapter 7 was made while Dr. Sutcliffe was still active as a coeditor, though he never saw the final text. This essentially personal account is, therefore, presented in memory of Dr. Sutcliffe since in his capacity as editor of the *Annals of Botany* he would have been familiar with both its style and content. However, without the contributions of the various collaborators named in the published papers this overall survey could not have been made.

F. C. Steward

Note on the Use of Plant Names

The policy has been to identify by its scientific name, whenever possible, any plant mentioned by a vernacular name by the contributors to this work. In general, this has been done on the first occasion in each chapter when a vernacular name has been used. Particular care was taken to ensure the correct designation of plants mentioned in tables and figures which record actual observations. Sometimes, when reference has been made by an author to work done by others, it has not been possible to ascertain the exact identity of the plant material originally used because the original workers did not identify their material except by generic or common name.

It should be unnecessary to state that the precise identification of plant material used in experimental work is as important for the enduring value of the work as the precise definition of any other variables in the work. "Warm" or "cold" would not usually be considered an acceptable substitute for a precisely stated temperature, nor could a general designation of "sugar" take the place of the precise molecular configuration of the substance used; "sunflower" and "*Helianthus*" are no more acceptable as plant names, considering how many diverse species are covered by either designation. Plant physiologists are becoming increasingly aware that different species of one genus (even different varieties or cultivars of one species) may differ in their physiological responses as well as in their external morphology and that experimental plants should therefore be identified as precisely as possible if the observations made are to be verified by others.

On the assumption that such common names as lettuce and bean are well understood, it may appear pedantic to append the scientific names to them—but such an assumption cannot safely be made. Workers in the United States who use the unmodified word "bean" almost invariably are referring to some form of *Phaseolus vulgaris;* whereas in Britain *Vicia faba*, a plant of another genus entirely, might be implied. "Artichoke" is another such name that comes to mind, sometimes used for *Helianthus tuberosus* (properly, the Jerusalem artichoke), though the true artichoke is *Cynara scolymus*.

By the frequent interpolation of scientific names, consideration has also been given to the difficulties that any vernacular English name alone may present to a reader whose native tongue is not English. Even some American and most British botanists would be led into a misinterpretation of the identity of "yellow poplar," for instance, if this vernacular American name were not supplemented by its scientific equivalent *Liriodendron tulipifera,* for this is not a species of *Populus* as might be expected, but a member of the quite unrelated magnolia family.

When reference has been made to the work of another investigator who, in his published papers, has used a plant name not now accepted by the nomenclature authorities followed in the present work, that name ordinarily has been included in parentheses, as a synonym, immediately after the accepted name. In a few instances, when it seemed expedient to employ a plant name as it was used by an original author, even though that name is not now recognized as the valid one, the valid name, preceded by the sign =, has been supplied in parentheses: e.g., *Betula verrucosa* (= *B. pendula*). Synonyms have occasionally been added elsewhere also, as in the case of a plant known and frequently reported on in the literature under more than one name: e.g., *Pseudotsuga menziesii* (*P. taxifolia*); species of *Elodea* (*Anacharis*).

Having adopted these conventions, their implementation rested first with each contributor to this work; but all outstanding problems of nomenclature have been referred to Dr. W. J. Dress of the L. H. Bailey Hortorium, Cornell University. The authority for the nomenclature now employed in this work has been the Bailey Hortorium's "Hortus Third." For bacteria Bergey's "Manual of Determinative Bacteriology" and Skerman, McGowan, and Sneath's "Approved List of Bacterial Names" and for fungi Ainsworth and Bisbee's "Dictionary of the Fungi" have been used as reference sources; other names have been checked where necessary against Engler's "Syllabus der Pflanzenfamilien." Recent taxonomic monographs and floras have been consulted where necessary. Dr. Dress's work in ensuring consistency and accuracy in the use of plant names is deeply appreciated.

The Editors

Plant Physiology

Volume IX: Water and Solutes in Plants

CHAPTER ONE

Water Relations of Plant Cells

J. E. DALE AND J. F. SUTCLIFFE

I. Introduction

"The study of water economy of plants is a major preoccupation of physiologists and ecologists, and as water is the chief constituent of most tissues, every aspect of cell physiology is involved" (1). These opening words of Bennet-Clark's lucid and still useful article in Volume II of this treatise remain appropriate today. Since 1959 new research utilizing powerful techniques and linked to a growing appreciation of basic concepts has contributed much to our understanding of plant cell water relations. In aiming to identify and discuss the major advances that have occurred over the last 25 years or so, we shall build on and complement Bennet-Clark's chapters, which will remain essential reading for those interested in the older work and in treatments of osmosis and plasmolysis.

When Bennet-Clark wrote his article he was fully aware of the many terms used by plant physiologists to describe the water relations of plant cells. These included *Saugkraft* and *Saugdruck* (translated from the German as suction force and suction pressure), osmotic equivalent, and diffusion pressure deficit. While these terms are now of historical interest only, the use by Bennet-Clark of osmotic pressure and turgor pressure is certainly familiar to present-day plant physiologists.

Bennet-Clark recognized that "the passage of water by diffusion from one phase to another depends upon a free-energy difference," and soil physicists had for some years been attempting to introduce a consistent thermodynamic terminology to describe soil–water relationships. In a seminal paper in 1960 Slatyer and Taylor (85) introduced the term *water potential* and summarized the views of an interdisciplinary discussion group of biological and physical scientists which had led to its formulation (94). It was the objective to produce a consistent, unified, and unifying terminology, based on sound thermodynamic considerations, which would have equal relevance and application to all aspects of soil–plant–atmosphere water relations. To a large extent this has been achieved, and water potential is widely used by plant physiologists and ecologists. However, the terminology has been criticized by Weatherley (115) on grounds of convenience and by Oertli (72), Spanner (86), and Zimmermann and Steudle (131) as being imprecise and lacking in rigor; significantly, the lengthy treatment of cell water by House (51) does not mention water potential at all. Nor, as will be seen in Section II, is there a uniform presentation by those who use the term.

Despite these shortcomings, it seems likely that the water potential terminology will continue to be used by plant physiologists for some time to come, in view of its convenience for describing the water relations of plant communities, individual plants and their organs, and plant cells alike. Without prescribing a correct usage, we shall try to present a unified and consistent treatment in keeping with the aims of the 1960 conference (94), while at the same time recognizing that terminology of itself is less important than understanding the processes involved. The terms, units, and abbreviations used in this chapter are summarized in Table I.

Hand in hand with the development of the concept of water potential and its components has been the use of the theory of the thermodynamics of irreversible processes, pioneered by Kedem and Katchalsky (61) and Dainty (19). This approach examines the relationships between the flow or flux of matter or energy and driving forces responsible for such flows. The theory, described in a number of articles (61, 70, 102), is implicit in much that has been published since 1963. This article is no exception, for although we do not set out to develop the theoretical background, many of the important concepts and conclusions deriving from it are dealt with.

Bennet-Clark (1) surveyed the main methods available for studying the water relations of plant cells and tissues. These include vapor pressure lowering methods, cryoscopy, plasmolytic methods, and use of the pressure chamber, first devised by Dixon (28) and later rediscovered by Scholander and colleagues (82). Commercially available psychrometers for vapor pressure determinations, osmometers, and pressure chambers have ensured that these techniques remain in use today. A number of critical assessments of the available methods have appeared and that by Turner (100) gives useful practical detail on many of the techniques, and we do not propose to cover this ground again. Here, however, an exception must be made in the case of the pressure chamber, which has proved to be a versatile instrument yielding important results, anticipated by Dixon (28), but only recently appreciated more widely (48, 105, 107). Another extremely important development has been the use of the pressure probe (55, 88, 91, 99) to examine directly, and in considerable detail, the water relations of individual cells. We shall discuss the techniques currently in use and some of the important data obtained by using them.

Research on plant cell water relations has long made use of the giant cells of certain algae, especially *Nitella, Chara,* and *Valonia.* The historical background to such work, going back more than 200 years, has been given by Hope and Walker (50), who also survey much of the recent work on the physiology of these organisms. In describing the work of Kamiya and Tazawa (58) on transcellular osmosis in *Nitella,* Bennet-Clark (1) recognized the great convenience of using these giant cells for the measurement of parameters that are much more difficult to obtain for individual cells of higher plants, in which we usually have to make do with average or "bulk" values. More recent work on the water relations of algal cells has led to a renewed interest in cell turgor and the recognition that plant cells may regulate their osmotic composition. An introduction to the ideas of turgor regulation is given in Section V and the topic is also considered in Chapters 2, 4, and 5.

Traditionally, cell water relations have been concerned with the fully mature vacuolate plant cell with an elastic wall and a thin layer of parietal cytoplasm. Over the last 25 years there have been a number of theoretical and experimental studies on the dynamics of growth of expanding plant cells. These have brought out the importance of cell wall parameters and turgor, and we shall end by discussing this topic.

Inevitably, given the great and continuing complexity of the subject, treatment here must be selective and far from comprehensive. Those who wish to take this fascinating topic further are referred to the books by Briggs (7), House (51), Kramer (63), Milburn (69), and Slatyer (84); important reviews by Dainty (21), Tyree and Jarvis (107), Weatherley (115), and Zimmermann and Steudle (131) will be referred to frequently. A recent book by Franks (33) presents a concise account of the chemistry of water.

II. Water Potential Terminology

A. FREE ENERGY, CHEMICAL POTENTIAL, AND WATER POTENTIAL

If water diffuses into or out of a plant cell, the free energy of that cell will change also; under the special conditions where neither temperature nor pressure changes, spontaneous influx of water will be accompanied by a loss of free energy from the external medium and a gain in free energy by the cell. The direction of water movement will be along a free-energy gradient from high to low, and given the absence of any constraining factors, movement will continue until the gradient disappears.

The use of chemical potential is a way of describing the free energy of a system and is central to our understanding of water potential. Suppose that a drop of a dyestuff, dissolved in water, is carefully placed in a beaker containing pure water. The dye will diffuse into the surrounding bulk water so that the drop will spread. At the same time there will be diffusion of water molecules from the bulk solution, where they are more concentrated, to the dye solution. Net diffusion will cease when water and dye molecules are uniformly distributed throughout the solution. Put in thermodynamic terms, the chemical potential of the dyestuff is higher in the initial drop than in the bulk water, where it is zero, and the chemical potential of water is initially higher in the bulk solution than in the dye droplet. In this model diffusion of both water and dye occurs along gradients of chemical potential from high to low.

Expressed formally, the chemical potential of water μ_w can be defined as the partial molal Gibbs free energy of water

$$\mu_w = (\partial G/\partial n_w)P,\ T,\ n_j \qquad (1)$$

where G is the Gibbs free energy; n_w is the number of moles of water in the system; and P, T, n_j are the conditions of pressure, temperature, and number of moles of all species in the system held constant for the differentiation.

The units of μ_w are energy per mole. Gibbs free energy is expressed in relative terms since absolute values cannot be measured (70). Nor is it possible to measure absolute values of chemical potential, but by applying Raoult's law it is possible to estimate the difference in chemical potential of water in a system (μ_w) from that of pure water at the same temperature and pressure of 1 atm (μ_w^0) (see Table I) as

$$\mu_w - \mu_w^0 = RT\ \ln(e/e^0) \qquad (2)$$

TABLE I

SYMBOLS AND UNITS USED IN THIS CHAPTER[a]

Symbol	Meaning	Units[b]
a_j	Chemical activity of species j	Dimensionless
A	Area	m^2
D	Diffusivity, diffusion coefficient	$m^2\ s^{-1}$
e	Vapor pressure of water	Pa
G	Gibbs free energy	J
J_i	Flux of species i	$mol\ m^{-2}\ s^{-1}$
L_p	Hydraulic conductance	$m\ Pa^{-1}\ s^{-1}$
L	Hydraulic conductivity	$m^2\ Pa^{-1}\ s^{-1}$
N_j	Number of moles of species j	Dimensionless
P	Turgor pressure of cell	Pa
R	Gas constant	$8.314\ J\ mol^{-1}\ {}^\circ T^{-1}$
t	Time	s
T	Temperature	°C
V	Volume	m^3
W	Weight	g
Y	Wall yield threshold	Pa
ϵ	Elastic modulus	Pa
μ	Chemical potential	$J\ mol^{-1}$
π	Osmotic potential	Pa
Π	Osmotic pressure	Pa
σ	Reflection coefficient	Dimensionless
τ	Matric potential	Pa
ϕ	Wall extensibility	$Pa\ s^{-1}$
ψ	Water potential	Pa

[a] SI units are used throughout and where necessary in later tables and figures original data have been converted from other units.

[b] The SI unit of pressure is the pascal (Pa); $1\ Pa = 1\ N\ m^{-2} = 1\ kg\ m^{-1}\ s^{-2}$. To convert older units to their SI equivalent: 1 bar = 0.987 atm = 0.1 MPa.

where R is the gas constant (Table I), and e and e^0 are the equilibrium vapor pressures of water in the system and pure water at the same temperature and atmospheric pressure, respectively. The vapor pressure terms can be measured using psychrometry so that derivation of $\mu_w - \mu_w^0$ is possible. When combined with the partial molal volume of water, which in the context of the cell is approximately the volume occupied by a mole of water, V_w, we have a definition of water potential ψ,

$$\psi \simeq (\mu_w - \mu_w^0)/V_w \tag{3}$$

This equation, although approximate, is a convenient working definition, and the inclusion of V_w means that it has the dimensions of energy/volume equals pressure, so that the units used (Table I) are consistent with those used in older literature. By convention μ_w^0 is set equal to zero and water potential can be defined in words as the chemical potential per unit partial molar volume. In fact, there has been considerable debate over the use of partial molar volume in the definition of water potential (71, 72, 86), particularly as to whether it is more appropriate to use the partial molar volume of water in the system or the molar volume of pure water. Numerically there is little difference, and in disregarding the controversy, we accept that water potential, as defined, is a useful practical measure of the free-energy state of water in a plant cell system. Since the chemical potential of pure water μ_w^0 is set at zero, it follows that under similar conditions of temperature and pressure $\mu_w < 0$, since the chemical potential of pure water must always be greater than that of an aqueous solution, and thus water potential is also equal to, or less than, zero. It may be noted that Meyer and Anderson (68) defined what they termed diffusion pressure deficit, given the symbol S in Bennet-Clark's chapter, as

$$S = (\mu_w^0 - \mu_w)/V_w$$

This definition, which is not the usual thermodynamic one, results in S being positive since $\mu_w^0 > \mu_w$. Diffusion pressure deficit and water potential are numerically equal but of opposite sign, i.e., $\psi = -S$.

B. THE COMPONENTS OF WATER POTENTIAL

Conceptually, it is easy to see that the chemical potential of water in a solution can be lowered by increasing the proportion of solute to water molecules or raised by decreasing the proportion or by increasing the pressure. The partitioning of water potential into components has led to controversy and confusion, particularly with respect to the variety of symbols and expressions that have been used to describe the water relations of the vacuolated plant cells (see Table I). It is generally agreed that the components include one ascribable to solutes present in the system, which tend to lower the water potential, and another ascribable to the hydrostatic pressure which can exist in plant cells bounded by a more or less rigid cell wall, which tends to raise it. But the form of the equation used to describe the relationship of water potential to its components varies, as the following examples show:

Bennet-Clark (1):

$$S = P - W$$

where P is "osmotic pressure of a solution" and W "excess hydrostatic pressure or turgor pressure, or wall pressure" (but see p. 10).

Dainty (19):

$$S = \Pi - P$$

where Π is "osmotic pressure of vacuolar contents" and P is "the hydrostatic pressure of the vacuolar contents, and is called the turgor pressure or wall pressure" (but see p. 10).

Dainty (21):

$$\psi = P - \pi - \tau \tag{4}$$

where P and π have the same definitions as in the preceding equation and τ is a matrix component.

Tyree and Jarvis (107):

$$\psi = P + \pi + \tau \tag{5}$$

where π is defined as osmotic potential rather than osmotic pressure.

As will be shown, the difference between the last two equations is readily resolved, but the existence of several forms of the same basic equation can hardly be welcome to those who sought a unified terminology in the 1960s.

1. *Osmotic Pressure and Osmotic Potential*

Bennet-Clark (1) pointed out that "the osmotic pressure of a solution is most rigidly defined as the excess hydrostatic pressure which must be applied to the solution in order to make the chemical potential (activity) of the solvent in the solution equal to that of the pure solvent at the same temperature. This, in effect, is nearly the same as stating that it is the pressure that the solute would exert if it were a gas occupying the volume of the solution. As so defined, osmotic pressure is independent of the presence of any membrane—ideal or otherwise." This quotation gives not only a useful and unambiguous definition, but also a reminder of the similar relationship between pressure and volume for ideal gases and ideal solutions. Thus, for a dilute solution, we have

$$RT \ln a_w = -V_w \Pi$$

or

$$\Pi = (RT \ln a_w)/V_w \tag{6}$$

where R is the gas constant (Table I), Π is the osmotic pressure, T is the absolute temperature, and a_w is the chemical activity of water.

The van't Hoff relationship, embodied in Eq. (6), applies to ideal dilute solutions and the inclusion of a_w makes allowance for the interactions between solute molecules that occur in nonideal solutions. In very dilute aqueous solutions the presence of solutes tends to decrease the activity of water, the concentration of which is diluted. Since for pure water $a_w = 1$, and the presence of solutes reduces it, ln a_w becomes negative, and osmotic pressure Π is positive, consistent with the verbal definition in the quotation given earlier. The symbol Π will be used henceforth only for osmotic pressure.

Although Bennet-Clark argued against the need for the term osmotic potential, it is now widely used. It originates from the reasonable view that the solutes in a solution will tend to lower the water potential and that this effect will be concentration, or more realistically activity dependent. Since water potential is equal to or less than zero, osmotic potential, also referred to as solute potential by some authors, must be negative. To avoid confusion with osmotic pressure, osmotic potential is here given the symbol π; to indicate that it is a component of water potential some writers use the symbols ψ_π or ψ_s, but the single symbol is preferred despite the possibility of confusion with Π. Numerically Π and π are identical, differing only in sign so that

$$-\Pi = \pi$$

It is important to realize that an increase in solute concentration *raises* the osmotic pressure of a solution but *lowers* the osmotic potential. Put another way, a solution having an osmotic pressure of say 1.0 MPa has a lower osmotic potential than one having an osmotic pressure of 0.5 MPa. The differences between Eqs. (4) and (5) lie in the fact that the one uses osmotic pressure, the other osmotic potential. That confusion is so easily engendered should make all workers in the field especially careful about the terms and symbols they use.

2. *The Matric Potential*

In their treatment of the water relations of soils, Taylor and Slatyer (94) were concerned with the fact that there could be physical interactions between water and the solid matrix of the soil, so that the chemical potential of water would be lowered, not by osmotic effects, but because of these interactions. To allow for this they included a matric potential term τ in their equations for water potential. There has been continuing debate over the nature and importance of the matric potential in the water relations of plant cells (73, 74, 107, 114).

Matric forces can be considered to be those which occur between liquid and solid phases in a soil or a cell. They are short-range, involving hydrogen bonding and van der Waals forces. The force of attraction between the surface of the solid phase, especially if it is charged, and the water molecules can lead to an ordering of the latter, reducing their free energy relative to that of molecules outside the force field. Also, as Dainty (21) points out, where the interface is charged, as most surfaces in cells are, the presence of an electrical double layer will lead to local accumulation of ions of the opposite charge, with significant local osmotic effects. However, these effects are not sharply defined and in any case fall off rapidly as distance from the charged surface increases (Fig. 1) (107, 108).

Cell walls and membrane systems are sites at which solid–liquid interactions can occur and as such are clearly components of the matrix. It is much less clear what size a molecule or macromolecule has to reach to be classed as matrix rather than as a contributor to the osmotic component. For example, water in the hydration shells of cations will have a different free energy from that of "free" water. At the level of bulk water relations of cells, such local effects will be included in the osmotic component. Furthermore, because the interactions between solid and liquid phases occur only over short distances of the order of a few nanometers (107), it is reasonable to assume that the contribution of the matrix term is negligibly small, at least when considering mature vacuolated cells in which vacuolar volume

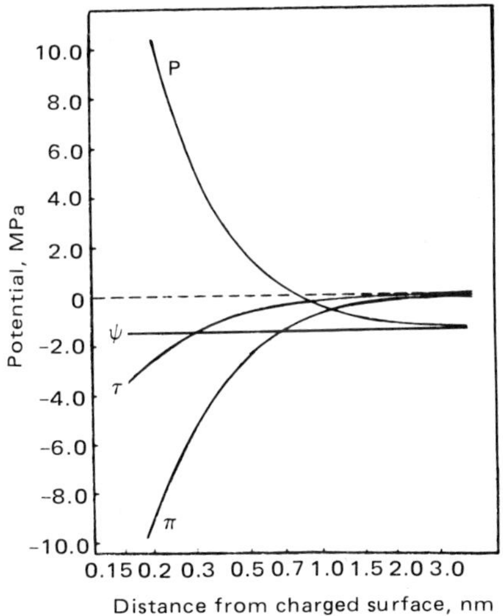

FIG. 1. The calculated effects of distance from a charged surface on the components of water potential. In this example the univalent ion concentration outside the electrical double layer is taken as 10 mM and the surface charge as -4×10^{-5} C cm^{-2}. Proximity to the charged surface affects the components only when the distance is less than about 10 nm in this example. After Tyree and Karamanos (108).

is a major component of total cell volume. Such an assumption may be unjustified for growing cells, although neglecting it is unlikely to lead to seriously erroneous conclusions.

3. The Pressure Term

Application of pressure to an aqueous solution will bring water molecules slightly closer together, making the water more concentrated and thus tending to increase the chemical potential and to raise the water potential. For a plant cell, the difference in hydrostatic pressure between the inside of the cell and its surroundings is termed the turgor pressure P. As Bennet-Clark (1) defined it, "the turgor pressure is the outwardly directed *excess* hydrostatic pressure which is equal and opposite to the inwardly directed reaction due to the structured cell wall," which he called wall pressure (98). Provided that the cell is turgid and not plasmolyzed, P will be positive; at incipient plasmolysis it falls to zero and remains thus during plasmolysis. A point often forgotten is that the turgor of plant cells can be several times that of an inflated car tire.

Whether turgor pressure can be negative in plant cells has been considered (103, 131), with the conclusion reached that this is not likely and that states of tension do not normally exist in plant cells. (An exception, described by Bennet-Clark (1), is the establishment of tension in the annulus cells of the fern sporangium as part of the mechanisms of dehiscence.) For intracellular tensions to occur three conditions have to be met. These are (1) that the cell wall is very rigid and will not buckle under tension, (2) that there is continuity of the liquid phase in the cell, and (3) that there is complete absence of nuclei around which gas bubbles could form. If these points are not met, there will be rapid formation of gas vacuoles and cavitation in the cell or tissue (131).

The maintenance of turgor is important if the plant body is not to become flaccid and wilt. At the cellular level, the regulation of turgor may involve a variety of mechanisms; these are discussed in Section V. Methods for the measurement of turgor are discussed in the following section.

4. The Components of Water Potential in Cells and Tissues

From the preceding discussion water potential can be partitioned into components such that

$$\text{Water potential} = \text{turgor pressure} + \text{osmotic potential} + \text{matric potential}$$

[cf. Eq. (5)]. The turgor pressure will be zero or positive; all others will be zero or negative.

In a cell in which there is no net influx or efflux of water, the water potential of all parts will be the same. But as Weatherley (115) has pointed out, the components of water potential may differ greatly between, say, the cell wall, in which the matric component may be important, and the vacuole, in which the presence of solutes will lower the osmotic potential. The position is even more complicated if one considers adjacent cells in a tissue which may well show substantial differences in the osmotic potential of their vacuolar contents, with compensating differences in turgor pressure to bring ψ to similar values for each cell. Heterogeneity between and within cells means that almost invariably we deal with, and measure, average values for the components. That values for organs such as leaves or stems are "bulk" values is widely recognized; that values for cells are also averaged values is less so.

III. Recent Developments in Techniques

A. THE PRESSURE CHAMBER AND PRESSURE–VOLUME CURVES

In its modern form the pressure chamber, or "bomb," consists of a thick-walled metal vessel sealed with a rubber gasket and connected to a pressure gauge and a supply of nitrogen or compressed air (Fig. 2). Leaves, branches, or stem pieces are placed inside the chamber with a portion of the petiole or stem protruding through the sealing gasket. Pressure in the chamber is increased until sap is forced out of the tissue and just appears at the cut surface. Many workers use a microscope or lens to detect the point at which solution just appears. The pressure when this occurs is noted and is sometimes referred to as the balance pressure. It is usually regarded as equivalent to the water potential of the tissue ψ_{tissue}, although strictly it is a measure of the pressure that has to be applied to force water out of the lumina of the xylem vessels, i.e., to expel water from the apoplast. In equating this pressure with the water potential *of the tissue,* it is assumed that ψ_{apoplast} and ψ_{symplast} are the same and in equilibrium. This assumption is reasonable, but the position is rather more complicated when the apparatus is used to obtain so-called pressure–volume curves (21, 107).

In this technique, the leaf or stem is allowed to rehydrate by standing in water for several hours before being sealed into the chamber, usually enclosed in a polyethylene bag to minimize evaporational losses. The balance pressure is obtained in the usual way; if rehydration is complete, the value of tissue is close to zero. [Hellkvist *et al.* (48) reported values of -0.05 to -0.1 MPa for shoots of Sitka spruce.] The pressure is now increased

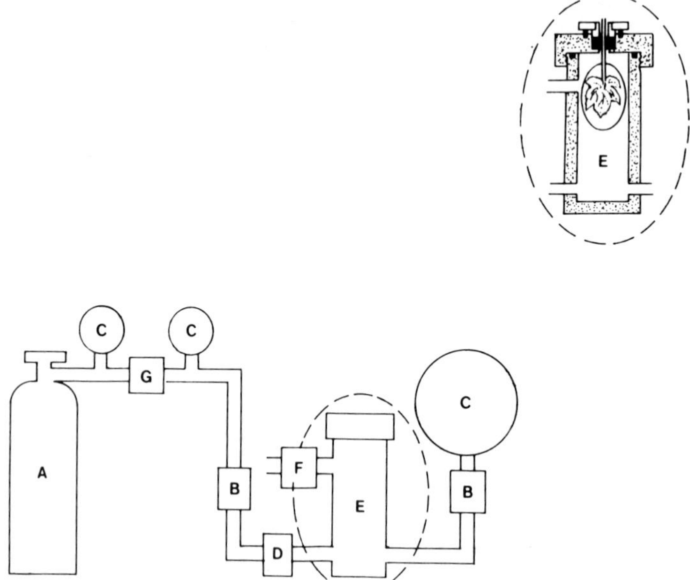

FIG. 2. Diagrammatic representation of the pressure chamber circuit. The insert shows the construction of the chamber itself with a leaf in position enclosed in a polyethylene bag to reduce evaporative losses of water; the petiole just protrudes through the sealing gasket of the chamber and can be viewed under a microscope. In operation, nitrogen or compressed air from cylinder A is passed through a reduction valve G and on/off and time control valves (B and D) to the chamber. Pressure is recorded at the gauges C. After measurement gas is vented from the chamber through the valve F. From Turner (100).

beyond the balance point so that sap is expressed; this is collected and weighed for a known amount of excess pressure. The pressure is then reduced and a new balance pressure obtained. Because water has been lost, ψ_{tissue} will be lower than it was initially. The operation is repeated to obtain a number of points which enable the relationship between balance pressure and volume of sap expressed to be determined. An example of such a pressure–volume curve is shown in Fig. 3. When plotted this way, i.e., as the inverse of the balance pressure, important additional information can be obtained.

The theory of pressure–volume curves has been examined in detail by Tyree and colleagues (104–107). The form of Fig. 3 shows an initial curvilinear portion, followed by a linear phase. This is interpreted as follows: the pressure applied to the system will counteract turgor, which will eventually fall to zero, at which point the curve becomes linear. At this

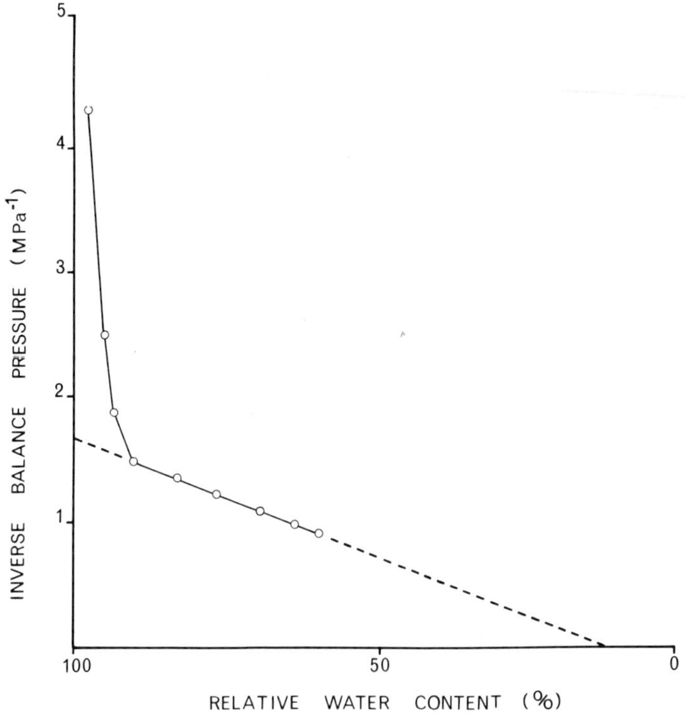

FIG. 3. A pressure–volume curve obtained for a young primary leaf of *Phaseolus vulgaris*. In this example inverse balance pressure is plotted against relative water content; in other published examples (e.g., 21, 100) the plot is against volume of sap expressed. Extrapolation of the linear portion of the curve backward to the ordinate gives the inverse of osmotic potential at full turgor (i.e., $\psi = 0$). The inverse balance pressure at the point at which curved and linear portions of the graph meet is the inverse of osmotic potential at limiting plasmolysis (i.e., $P = 0$). Extrapolation to the baseline enables apoplastic water content to be determined from the distance between the intercept and zero relative water content. Unpublished data of Stuart Milligan.

point also, since $P = 0$, the bulk osmotic potential is equal to the water potential of the tissue ψ_{tissue}. It is thus possible to obtain value of $\bar{\pi}$ for the conditions of incipient plasmolysis (i.e., $P = 0$), even though separation of protoplast and wall does not occur. At this stage, the relationship between the applied pressure in the chamber P_c and the volumes of water in the tissue V_0 and expressed V_e is given by

$$\frac{1}{P_c} = \frac{V_0 - V_e}{RTN_s} \tag{7}$$

where N_s is the number of moles of solute in the leaf. Extrapolation of the linear portion of the curve back to the ordinate enables the bulk osmotic potential to be obtained for the tissue at the start of the run, i.e., at or near full turgor. The nonlinear portion of the curve represents the net effect of falling turgor and falling osmotic potential on tissue ψ. If we ignore the matric potential, it is possible, having derived ψ and $\bar{\pi}$, to calculate P, the turgor component averaged for all the cells in the tissue as

$$\bar{P} = \psi + \bar{\pi} \quad (8)$$

If the linear portion of the pressure–volume curve is extrapolated downward to the abscissa (i.e., to the point at which $1/P_c = 0$ and the applied pressure is infinite), the intersect gives the maximum volume of water in the symplast, V_s, where $V_s > V_0$ and $V_s = V_0 + V_e$. The total amount of water in apoplast and symplast together V_{max} can be obtained from the fresh weight of tissue less the dry weight. If it is assumed that pressure does not cause a reduction in the volume of apoplastic water V_a, this can be estimated as

$$V_a = V_{max} - V_s$$

The validity of the underlying assumption has been questioned by Tyree and Richter (109), and values of V_a obtained by the pressure chamber method may be underestimates if significant loss of water occurs from xylem vessels and tracheids as they shrink. [The physiological significance of apoplastic water, estimates of which range from 5 to 40%, especially in the *water relations* of large trees is discussed by Tyree and Jarvis (107).] Irrespective of the accuracy of estimates of V_a, knowledge of V_{max} and of V_e means that changes in water potential and its components can be set against relative water content, this being given as $[(V_{max} - V_e)/V_{max}] \times 100$.

The relationships among the various parameters derivable from pressure–volume curves are shown in Fig. 4 [see also (80)]. Traditionally, plant cell water relations have been expressed diagrammatically in what is usually called a Höfler diagram (49) but perhaps should more correctly be named after Thoday (97). Over the years the form of such diagrams has changed and Fig. 4 is a modern version which takes account of current terminology. The derivation of this diagram from Fig. 3 should be obvious.

B. The Pressure Probe

It must again be stressed that the pressure chamber method gives data which are average values for the tissue. The method cannot yield data for individual cells. In contrast, the pressure probe apparatus described by

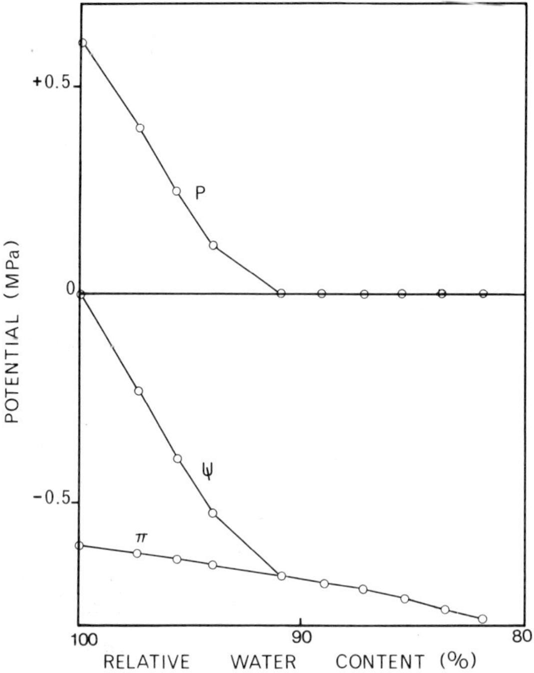

FIG. 4. A Höfler diagram derived from the pressure–volume curve shown in Fig. 3.

Zimmermann, Steudle, and colleagues (29, 55, 88–92, 124–131) can give values for single cells.

Early attempts to measure cell turgor directly were made by Green (36, 39), who inserted small microcapillary manometers into cells of *Nitella*. The manometers contained small air bubbles, and turgor was determined from the change in dimension of these when sealed into the cell. This approach has a number of shortcomings, including the fact that the gas in the bubble could dissolve in the cell and that the cell volume could change due to the compressibility of the air. In the pressure probe these disadvantages have been overcome. In its original form this apparatus (Fig. 5) consists of a screw micrometer connected to a thin plunger sealed into a small chamber containing silicone oil. This chamber is sealed to a glass micropipette at one end and to a very sensitive pressure transducer which transforms pressure to a proportional voltage. In use, the micropipette is inserted into a plant cell and pressure in the system is measured directly. It is also possible to inject a small volume of oil ΔV into the cell and to measure the instantaneous rise in hydrostatic pressure ΔP. Changes in P with time can be measured and from such values important data on water fluxes and on the elastic modulus of the cell can be derived (see Section IV).

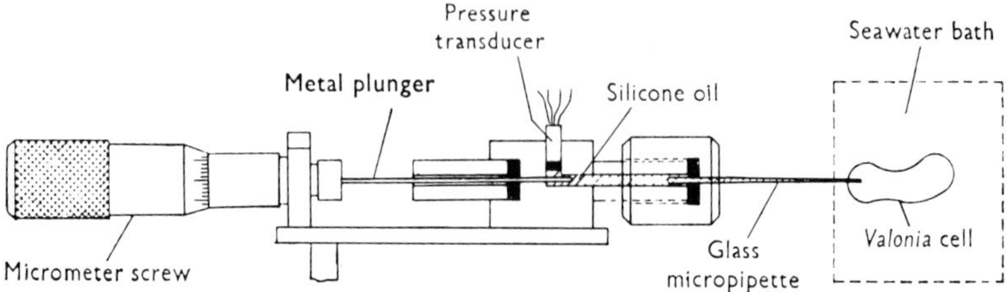

FIG. 5. The original pressure probe apparatus designed to be used for studies on giant algal cells. Using a micromanipulator (not shown) the micropipette probe is inserted into the cell. Pressure inside the cell is transmitted through the small silicone oil reservoir and recorded by the pressure transducer. Pressure in the system can be varied (see, e.g., Fig. 8) by moving the micrometer screw. From Zimmermann (124).

The limitations of this form of pressure probe come from the relative volumes of cell and silicone oil and from the compressibility of the latter (55, 125). For these reasons it has been used mainly to study the water relations of cells of the giant algae *Nitella* and *Valonia,* although Meidner and Edwards (67) made unsuccessful attempts to measure the turgor of stomatal guard cells of *Commelina* with a probe working on similar principles.

More recently, Hüsken *et al.* (55) designed a miniaturized version of the pressure probe which, because it has a very small compressible volume of silicone oil (<0.1 nl), can be used to measure cell volume changes in cells of higher plants with diameters of 50 μm or less. The modified design is shown in Fig. 6. The apparatus is both expensive and difficult to use, but a number of studies have been made on a variety of higher plant cells including epidermal cells of *Mesembryanthemum* (88), epidermal and mesophyll cells of the CAM plant *Kalanchoe daigremontiana* (89), and epidermal cells of *Tradescantia virginiana* (99) and of *Rhoeo spathacea (R. discolor)* (29). Unfortunately, this approach has not yet been used to study the water relations of growing cells (see Section Vc).

IV. Elastic Properties and Hydraulic Conductivity

A. THE ELASTIC MODULUS

In simple terms a vacuolated plant cell can be regarded as a bag of fluid surrounded by a thin layer of cytoplasm bounded by membranes and usually pressing on a wall which is more or less elastic. The elastic nature of the cell wall is important in turgor maintenance at least over the short

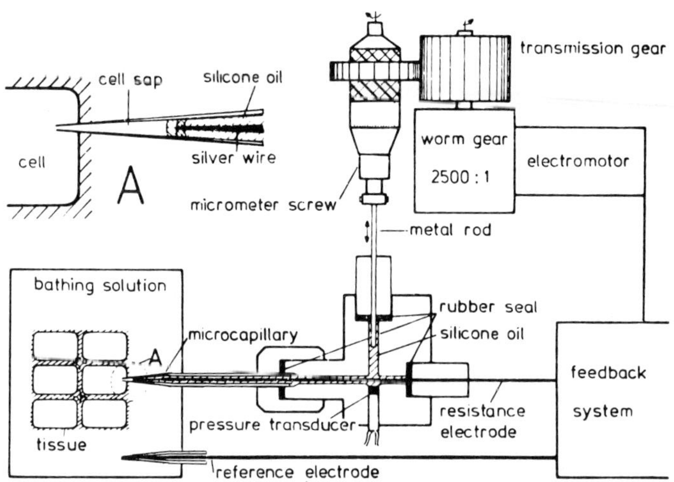

FIG. 6. The modified pressure probe scaled down for studies on cells of higher plants. The microcapillary probe (insert A), which is substantially finer than that used in the original apparatus (Fig. 5), is inserted into a cell and the pressure inside is transmitted by the silicone oil in a very small volume reservoir and monitored by a pressure transducer as before. Results can be affected by temperature fluctuations in the system and by any slight compression of the oil. These problems are overcome by accurate adjustment of the cell sap–oil boundary in the probe tip, using a feedback system employing the measured resistance between the silver electrode in the probe and the reference electrode. From Hüsken *et al.* (55).

term. If the cell wall were inelastic, reduction in the water content of the cell would reduce turgor to zero and either the protoplast would separate from the wall, or if this did not occur the wall would be pulled inward with the shrinking protoplast and would buckle or rupture. At the other extreme the elastic properties enable the wall to accommodate increases in turgor and volume without rupture outward.

During growth of the cell, the wall is subjected to stretching by *stress,* which causes deformation or *strain.* When deformation is permanent, the strain is said to be plastic, and when it is completely reversible, it is said to be elastic. In growth, both elastic and plastic properties are important, but even walls of mature cells can be permanently strained by an appropriate stress. However, as mentioned earlier, it is the elastic properties of the wall that are intimately connected with turgor. These properties are described by a coefficient ϵ, the volumetric elastic modulus, which relates the change in cell volume dV in response to a small change in the pressure inside the cell dP:

$$\epsilon = (dP/dV) \times V \tag{9}$$

where ϵ has the units of pressure (see Table I).

Although it is the case that the more elastic the cell wall, the smaller is the value of the elastic modulus, the position is rather more complicated than this simple statement suggests. Because of its detailed architecture, the cell wall (20) is anisotropic and can often be stretched more easily in one plane than in another. For a cylindrical cell, linear stress will produce some circumferential strain as well as stretching in the linear plane. Conversely, circumferential stress may affect linear strain. Detailed analyses of shearing and extensional stresses are complicated mathematically (75, 113), but simplifying approximations have been made (25). These have enabled estimates of the elastic modulus to be derived from simple measurements of the swelling and shrinkage of cells of *Chara* and *Nitella* exposed to solutions of polyethylene glycol and other nonpermeating osmotica.

Using the pressure probe it is possible to estimate the elastic modulus since the effect of small instantaneous changes in pressure and volume can be determined; ϵ is then calculated from Eq. (9). Values obtained by this method are given in Table II.

Strictly speaking, the definition of the elastic modulus in Eq. (9) applies only to isolated cells which are not subject to pressure interactions with other cells, such as are found in tissues. It is nevertheless possible to obtain

TABLE II

EFFECTS OF TURGOR PRESSURE P ON THE ELASTIC MODULUS ϵ OR BULK ELASTIC MODULUS $\bar{\epsilon}$ OF A VARIETY OF PLANT CELLS AND TISSUES

Species	P (MPa)	ϵ or $\bar{\epsilon}$	MPa	Reference
Nitella flexilis	Low	ϵ	6.0	(129)
	High		20.0	
Valonia utricularis	0	ϵ	3.5	(128)
	0.3		18.0	
Mesembryanthemum crystallinum	0	ϵ	0.5	(88)
Bladder cells	0.3–0.4		5.0–11.0	
Capsicum annuum	0	ϵ	0.2	(55)
Cells of fruit	0.25		0.8–2.5	
Helianthus annuus	0.34	$\bar{\epsilon}$	1.4	(34)
	0.34		4.7	
Gossypium hirsutum	0.2	$\bar{\epsilon}$	1.5	(34)
	0.2		6.0	
Pernettya mucronata	0	$\bar{\epsilon}$	1.0	(106)
	2.3		30.0	
Picea glauca	0	$\bar{\epsilon}$	0.2	(106)
	7.5		7.5	
Podocarpus nubigena	0	$\bar{\epsilon}$	0	(106)
	1.0		12.5	

an average value $\bar{\epsilon}$, the bulk elastic modulus, for tissue as a whole from analysis of pressure–volume curves obtained using the pressure chamber. Tyree and Jarvis (107) point out that for a tissue

$$\bar{\epsilon} = (d\bar{P}/dW) \times W \tag{10}$$

where $\bar{P}$ is the average value of P for the cells of the tissue and W is the weight of water in the symplast. Thus $\bar{\epsilon}$ can be derived from the slope of the plot of P against W, at any particular value of W, provided, of course, that $P > 0$. That is to say, only values on the nonlinear part of the curve (see Fig. 3) are used. It also follows that the elastic modulus is not constant but varies with turgor, being much smaller when P is low and larger when P is large. This is reflected in the published values (Table II and Fig. 7), and means that as turgor rises, resistance to elastic stretching of the cell wall becomes progressively greater.

The variation in ϵ (and $\bar{\epsilon}$) and its dependence upon water content, and by implication on cell volume (92), is significant for another reason, as

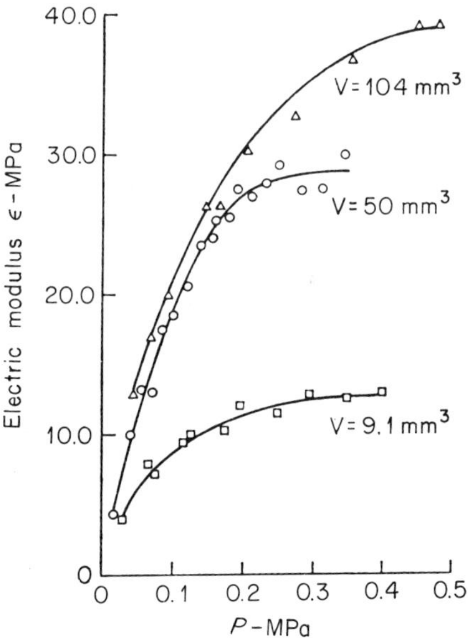

FIG. 7. The relationship between elastic modulus of the cell wall and turgor pressure for three cells of *Valonia utricularis* with different volumes. The data were obtained using the pressure probe and, as well as indicating a positive nonlinear relationship between ϵ and P, show the relationship to be dependent on cell volume. After Zimmermann and Steudle (129).

Dainty (21) has pointed out. Rearranging Eq. (9) gives

$$dP = \epsilon \times dV/V \tag{9a}$$

But osmotic potential is also a function of cell volume so that

$$d\pi = -\pi \times dV/V \tag{9b}$$

assuming that πV is constant, from the Boyle–van't Hoff relationship. On combining Eqs. (9a) and (9b) and expressing in terms of ψ,

$$\begin{aligned} d\psi &= dP - d\pi \\ &= (\epsilon + \pi)dV/V \end{aligned} \tag{11}$$

Numerically the elastic modulus is often larger than osmotic potential (see Table II) and this means that the relationship between water potential and V will be influenced more by ϵ than by π, especially where values of P are high and ϵ is high also. The problem of osmotic adaptation and turgor regulation is considered in Section V.

Properties of the wall important in cell growth are discussed further in Section V.

B. Hydraulic Conductivity and Hydraulic Conductance

If the chemical potential of water inside a vacuolated cell is the same as that of the external medium surrounding it, the system is in equilibrium and there will be not net flow of water into, or out of, the cell. Water potential inside ψ_i and outside the cell, ψ_i and ψ_o, respectively, will be the same. Now suppose that equilibrium no longer obtains and that $\psi_o \neq \psi_i$. Water will move from the region of high water potential to one of low ψ. Hence, if $\psi_o > \psi_i$, water will move into the cell, and if $\psi_i > \psi_o$, water will leave the cell. Since at atmospheric pressure both ψ_o and ψ_i are less than or equal to zero, as we have already seen, the lowest water potential will be that with the largest numerical value, but, of course, with negative sign.

The flow of water J_w, usually termed the volume flux, with units of meters per second, is proportional to the difference in water potential $\Delta\psi$, the proportionality coefficient being denoted by L_p. Formally,

$$J_w = L_p\, \Delta\psi \tag{12}$$

where $\Delta\psi = \psi_o - \psi_i$. The units of L_p are meters per pascal second.

Flux of water into a cell is often considered in terms of an analogy with Ohm's law whereby

Flow = Driving force/Resistance

Here the driving force is the difference in water potential and L_p is the reciprocal of the resistance to flow, termed the hydraulic conductance. When the length x of the path over which water moves is taken into account, we have $L_p = L/x$, L being the hydraulic conductivity (units of square meters per pascal second).

Hydraulic conductance can be regarded as a measure of the permeability of the cell to water. Wall, membranes, and cytoplasm may have different water permeabilities, but it is usually considered that the main resistance to water movement is at the cell membrane and that the conductivity of the wall and the cytoplasm is high (but see Table III).

Bennet-Clark (1) reviewed the methods available for measuring hydraulic conductance (he called it K). Of these, the osmotic methods (54) based on measurement of the shrinkage and swelling of plasmolyzed protoplasts, have been severely criticized (19, 21, 64, 87), largely because other methods have indicated that values of L_p may depend upon turgor, which, of course, is zero in plasmolyzed cells. A more acceptable method, used by Gutknecht (42) and colleagues among others, is to measure the time course of swelling and shrinkage of nonplasmolyzed cells in which the water potential of the external solution is varied. The pressure probe can also be used to induce rapid and reproducible changes in turgor in cells and to monitor water flux into or out of the cell in terms of changes in P and V. Figure 8 shows the time course of pressure changes of a cell of *Valonia*

TABLE III

VALUES OF HYDRAULIC CONDUCTANCE L_p FOR CELLS OF GIANT ALGAE AT DIFFERENT TURGORS OBTAINED BY VARIOUS METHODS

Species	L_p ($m\ s^{-1}\ MPa^{-1}$)	P (MPa)	Method	Reference
Chara corallina	1×10^{-5}	—	Transcellular osmosis	(23)
Chara fragilis	1×10^{-5}	0.4	Pressure probe	(130)
	2×10^{-5}	0	Pressure probe	(130)
Chara intermedia	7×10^{-6}	0.4	Pressure probe	(130)
	2×10^{-5}	0	Pressure probe	(130)
Nitella flexilis	$1-3 \times 10^{-4}$[a]		Transcellular osmosis	(58)
	2×10^{-5}	0.2	Pressure probe	(91)
	5×10^{-5}	0	Pressure probe	(129)
N. flexilis cell wall	7×10^{-5}	—	Perfusion	(130)
Valonia utricularis	5×10^{-7}	0.15	Pressure probe	(129)
	3×10^{-6}	0	Pressure probe	(128)
Valonia ventricosa	2×10^{-7}	—	Perfusion	(42)

[a] The lower value was obtained by exosmosis, the higher by endosmosis; see text.

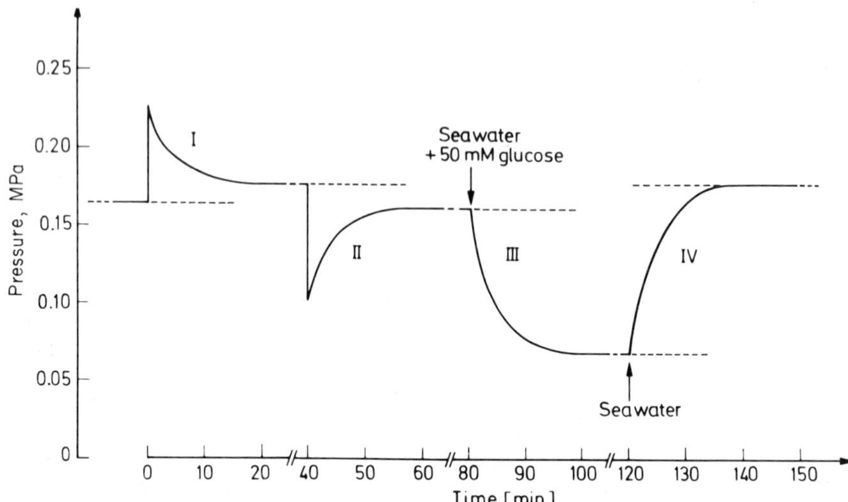

FIG. 8. The use of the pressure probe to alter pressures in a cell of *Valonia utricularis* (curves I and II); pressures were also altered by changing the water potential of the external medium (curves III and IV). Hydraulic conductivity L_p can be calculated from the kinetics of pressure change [see text and Eq. (13)]. After Steudle and Zimmerman (90).

treated both with the pressure probe and by an alteration in the water potential of the external medium.

The time course of swelling and shrinkage in experiments of the kind indicated is exponential, and it can be shown (21) that for a given volume V

$$L_p = t_c V / A(\epsilon + \pi_i) \tag{13}$$

where t_c is a time constant. Overall, A is the area of the cell across which water moves and π_i is the osmotic potential of the cell contents. It is often impossible to determine A accurately, and when this is the case, a value of "apparent conductance" L_p' is sometimes quoted, the area term being omitted so that the units are cubic meters per pascal second (112).

The pressure probe has also been used to measure hydraulic conductance for cells of higher plants. From measurements on 50 cells in the epidermis of *T. virginiana*, Tomos *et al.* (99) found substantial variations in both the volume elastic modulus and L_p. For the latter, values ranged from 0.2 to 11×10^{-5} m MPa^{-1} s^{-1}, and the authors consider that part of this variation is attributable to the difficulty of determining the effective area of the cell over which water exchange occurs. In other studies (29) with cells of *Rhoeo*, treatment with the growth regulator abscisic acid (ABA) has been found to increase values of apparent hydraulic conductivity by a factor of about two (Fig. 9).

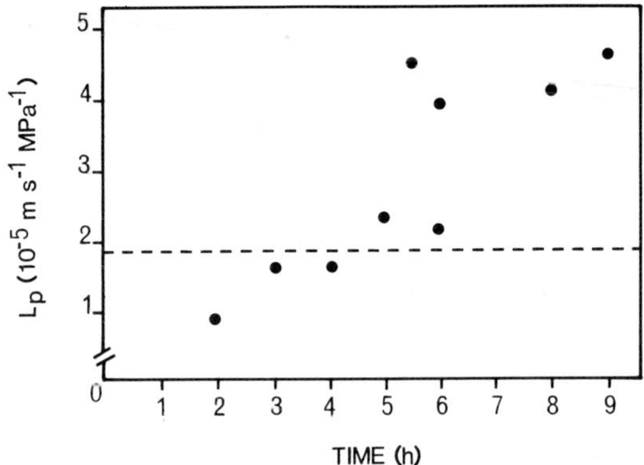

Fig. 9. Changes in L_p with time following immersion of an entire leaf of *Rhoeo spathacea* in a solution containing 100 μM abscisic acid. The broken line gives the mean value of L_p for an entire leaf kept throughout in air. Redrawn from Eamus and Tomos (29).

An alternative approach for the determination of hydraulic conductance, touched on by Bennet-Clark (1), involves measurement of exchange of labeled water, DHO or THO, between plant tissues and the surrounding medium. Although calculation of L_p is straightforward, use of this isotope-exchange method for maize root tissues (52, 117) has given very low values which are up to two orders of magnitude less than those obtained by the pressure probe. There is thus some doubt about the validity of this approach (21).

Bennet-Clark (1) also described the method of Kamiya and Tazawa (58), who used transcellular osmosis in *Nitella* to determine L_p (see Fig. 10). This work has been repeated and extended by several workers (25, 62, 95, 96). Among the interesting results obtained by Kamiya and Tazawa was that values of L_p for water efflux and influx were different, permeability being much greater for influx than for outflow (Table III). This polarity with respect to water permeability, first shown with *Nitella flexilis,* has been confirmed with other giant algal cells, including another species of *Nitella* (23, 24). It has been suggested (62) that the effect may involve differences in hydration of the cytoplasm between the ends of the cell exposed or not to the external osmoticum and that these affect the membrane properties on which L_p largely depends.

A selection of published values of L_p are shown in Tables III and IV, which also includes the method used for measurement. The data include values from pressure probe studies on giant algal cells which clearly indicate that L_p is dependent upon cell turgor. In cases in which turgor falls to less than 0.2 MPa and the cell nears the point of incipient plasmolysis, values of L_p increase by one or more orders of magnitude. One conse-

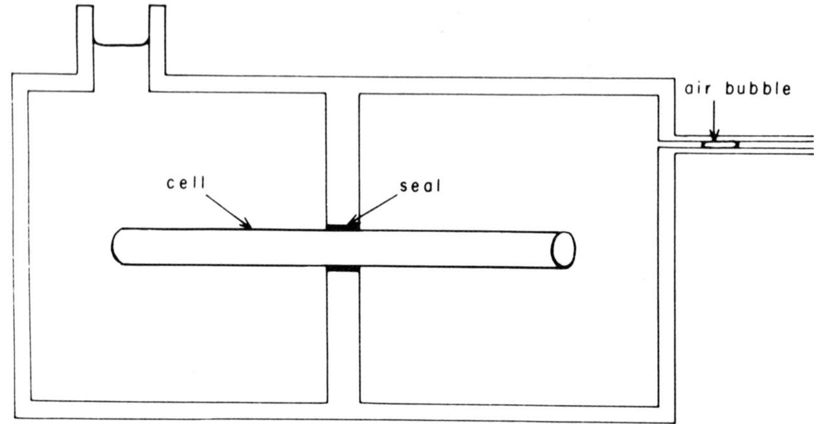

FIG. 10. Apparatus used for transcellular osmosis in Characean cells. The cell is sealed between two chambers, and water flux across the cell, induced by changes in water potential of the solution in the open chamber, is measured by movement of the air bubble in the capillary. After Dainty (21).

quence of this is that the conductivity *of the cell* to water approaches that of the cell wall alone, which could become rate limiting to water flow at low values of *P*. The very high permeability of giant algal cells to water was noted by Bennet-Clark (1), and the remarkably high values of L_p for these cells at low turgor remain unexplained. It would clearly seem to be an

TABLE IV

VALUES OF HYDRAULIC CONDUCTANCE L_p FOR CELLS FROM HIGHER PLANT TISSUES OBTAINED USING THE PRESSURE PROBE

Species and cell type	L_p (10^5 m MPa^{-1} s^{-1})	Reference
Rhoeo spathacea (R. discolor)		
Intact leaf epidermis	1.82 (±0.97)	(29)
Leaf infiltrated with DMSO	0.99 (±0.46)	
Tradescantia virginiana		
Leaf epidermis	0.2–11	(99)
Mesembryanthemum crystallinum		
Epidermal bladder cell	2	(88)
Kalanchoe daigremontiana		
Mesophyll cells	6.9 ± 4.6 (range 1.5–16)	(89)
Capsicum annuum		
Fruit cells	0.5 ("first approximation")	(55)

adaptational advantage in that L_p would not limit water flux in cases in which incipient plasmolysis is approached. It is not known whether hydraulic conductance is similarly pressure dependent for cells in tissues of higher plants, although with the highly specialized epidermal bladder cells of *Mesembryanthemum crystallinum*, pressure probe data suggest that L_p does not change over a range of P values from 0 to 0.5 MPa (55). However, the substantial variations in L_p recorded for cells of *Tradescantia* epidermis (99) prevent an unequivocal answer on this point.

C. THE REFLECTION COEFFICIENT

So far it has been assumed that only water moves into or out of cells. This assumption is unrealistic for solutes are likely to move as well, either by diffusion or by other transport mechanisms (see Chapter 4). Consequently, the solute composition of a cell and the surrounding medium cannot be regarded as constant. It follows that

$$J_v \neq L_p(\Delta P + \Delta\pi) \tag{14}$$

where J_v is the total volume flux of water *and solute*. [Compare with Eq. (12).] Instead, the relationship becomes

$$J_v = L_p(\Delta P + \sigma\,\Delta\pi) \tag{15}$$

where σ is a dimensionless constant characteristic of each particular solute, such that $0 < \sigma < 1$. If the membrane is permeable to water only and not to solutes, then $J_v = L_p\,\Delta\psi$ as before [cf. Eq. (12)] and $\sigma = 1$. When the membrane is permeable to solute as well as to water, $\sigma < 1$. We may think of σ as being a measure of the extent to which solutes are reflected at the membrane, and in fact σ is known as the reflection coefficient.

Values of the reflection coefficient obtained using the pressure probe σ are listed in Table V. It may be noted that cells of *N. flexilis* are almost completely impermeable to sucrose and glucose, but that membrane permeability to a variety of alcohols is considerable. For tissues of higher plants, it is widely believed that the reflection coefficient for sucrose is significantly less than 1, although reliable data are scarce; mannitol is relatively impermeant with σ close to 1 (83).

Determinations of the reflection coefficient can give considerable information about cell membranes and in particular the passage of water and solute across them. Details of the approach and the theory behind it may be consulted elsewhere (19, 21, 22, 50, 84), but results for *Nitella* (23, 24) suggest that here the plasmalemma may contain water-filled pores. Against this, for *Valonia ventricosa* Gutknecht (42) could find no evidence

TABLE V

REFLECTION COEFFICIENT σ FOR VARIOUS SOLUTES MEASURED FOR CELLS OF *Nitella flexilis*, USING THE PRESSURE PROBE[a]

Solute	σ
Sucrose	0.97
Glucose	0.96
Ethylene glycol	0.94
Acetamide	0.91
Urea	0.91
Glycerol	0.80
Ethanol	0.34
Methanol	0.31
N-Propanol	0.17

[a] Data from Steudle and Zimmermann (91).

for bulk flow of water across pores in the membrane, although for a related species, *Valonia utricularis*, the data suggest that bulk flow may occur! The lack of agreement, not uncommon in plant physiology, requires more detailed examination.

V. The Control of Turgor

A. THE GIANT ALGAL CELL AS A MODEL

Mention has already been made of some of the important results that have been obtained from studies of giant algal cells. What has emerged over recent years is that these cells do not behave as though they are simple osmometers. Such parameters as hydraulic conductivity and the eleastic modulus are not constant but change with turgor or with cell volume, and there is good evidence that these cells show properties whereby turgor pressure is controlled. These aspects are considered here, but first some attention must be given to terminology. The term osmoregulation has long been used by animal physiologists; and with the discovery that osmotic potential could change in plant cells in response to variations in the external environment, the term has been used by plant physiologists [e.g. (47)] as well. Cram (18) and Reed (79) have criticized the use of the term, considering turgor regulation to be more appropriate since it is turgor rather than osmotic potential that is the regulated output resulting from osmotic adjustments of walled cells. This view is accepted and followed here notwithstanding the widespread use of the term osmoregulation.

Two reasons have been suggested why turgor regulation is important in giant cells (44). The first is that their small surface : volume ratio may result in a slow adaptation to any change in osmotic potential of the external medium. This would be important for algae living in regions such as tidal zones, rock pools, and estuarine habitats in which there may be marked and rapid changes in salinity. The second reason is that such cells have low turgor (values of P of less than 0.2 MPa are not unusual) and are thus vulnerable to plasmolysis if salinity in the medium rises; conversely, because of their geometry and the small radius of curvature of such cells, there would be a danger of bursting if salinity falls. The maintenance of turgor within relatively narrow limits is thus important, and this is achieved by variation in the elastic modulus in the short term and by ionic adjustment over longer periods.

Evidence for turgor regulation in cells of the algae *Valonia macrophysa, Codium decorticatum,* and *Halicystis parvula* has come from the work of Gutknecht and collaborators (3, 35, 41–44, 46), and some of their results are shown in Fig. 11. Cells of the three species were allowed to equilibrate slowly in media whose osmotic potential ranged from −1.5 to about −4.5 MPa. Turgor P was determined by knowing the osmotic potential of the medium and the cell contents. The most noteworthy point is that the values of P varied remarkably little despite large variations in the external osmotic potential. Judged by the slope of the lines fitted to the data of Fig. 11, the effectiveness of turgor regulation was estimated to be 100% for

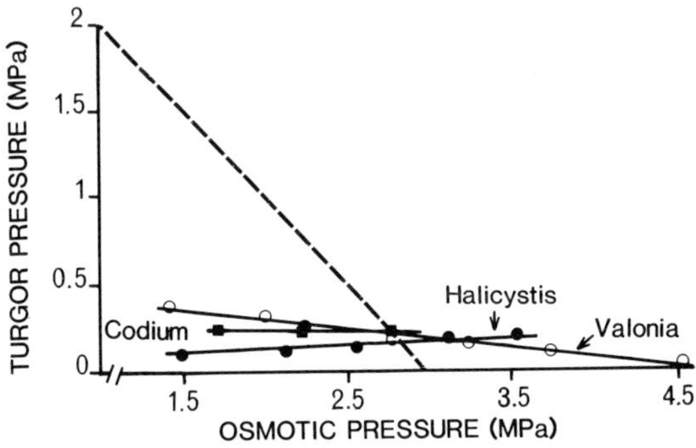

FIG. 11. The effect of varying external osmotic potential (equivalent to ψ) on turgor of cells of three species of marine algae. A slope of zero would indicate that turgor regulation is 100% effective; a slope of −1.0 (broken line) would indicate that internal osmotic potential is unaffected by the external solution. Redrawn from Gutknecht and Bisson (43).

Codium and *Halicystis* and 90% for *Valonia* (43). Similar regulation of turgor in *Valonia* has also been found using the pressure probe (124), and here it should be noted that the adjustment is comparatively slow, taking nearly 20 h under the experimental conditions used. Short-term fluctuations in the osmotic environment are unlikely to be adequately compensated for by ionic adjustment, and variations in the elastic modulus are probably of more immediate practical importance to the cell. Since such changes are passive and do not involve an active response of the cell to the external stimulus of changing osmotic potential, they are not regulatory in the strict sense.

The mechanisms involved in the control of intracellular osmotic pressure have been discussed by Cram (18) against a background of control theory. He points out that while in many cases the main component of osmotic pressure is potassium, this is not always the case (Table VI; see also Chapter 7 by Steward in this volume).

The nature of the regulating mechanisms involved has also been considered by Gutknecht *et al.* (44), whose interpretation of a turgor-regulation system based on transport of solutes and water is summarized in Fig. 12. The system is envisaged to compensate for fluctuation in the osmotic potential of the external environment. As π_o changes, the associated water flux will cause turgor to change and depart from an optimum or "desired"

TABLE VI

IONIC CONCENTRATIONS IN A VARIETY OF ALGAE AND THE EXTENT TO WHICH THE LISTED IONS ACCOUNT FOR π[a]

	Ionic concentration (mol m^{-3})				
Species	K^+	Na^+	Cl^-	Other	% Accounted for
Freshwater species					
Chlorella pyrenoidosa	110	1	1		40
Hydrodictyon pateraeforme	75	3	60	$Ca^{2+} + Mg^{2+}$: 4 SO_4^{2-}	100
Nitella clavata	54	10	91	$Ca^{2+} + Mg^{2+}$: 28	85
Nitella flexilis	75	11	134	Ca^{2+}: 9	90
Marine species					
Acetabularia mediterranea	335	65	480	Oxalate: 55 Mg^{2+}: 50	90
Codium fragile	16	475	495	SO_4^{2-}: 116	100
Enteromorpha intestinalis	450	260	370		50
Halicystis osterhoutii	6	557	603		100
Valonia ventricosa	550	75	625		98

[a] From data accumulated by Cram (18).

value. This difference ΔP is perceived, transduced, and passed as an error signal to a control mechanism which amplifies it to alter the rate of solute transport into or out of the cell. As a consequence, π_i will be adjusted, and this in turn will lead to turgor adjustment, changes in ΔP, and a reduction in the error signal. The system thus shows negative feedback.

In mechanistic terms the model system in Fig. 12 leaves a number of important questions unanswered. For example, what determines the optimum or "desired" value of P? And what is the nature of the turgor transducer and how does it operate? It also excludes passive turgor control brought about by rapid swelling or shrinkage, which is possible where the elastic modulus is small and the wall is readily and reversibly stretched. Significantly, ϵ is highly sensitive to pressure in these giant cells (see Table II), and short-term adjustments by this mechanism may well be very important.

The nature of the turgor-transducing mechanism has been the subject of much speculation. Cram (18) speculates that the primary detector of osmotic pressure may be the volume of an organelle. It seems logical that in walled cells the transducer would be located at the point at which turgor forces operate, i.e., at the interface between plasmalemma and wall. There are a number of possibilities for such a mechanism which are not necessar-

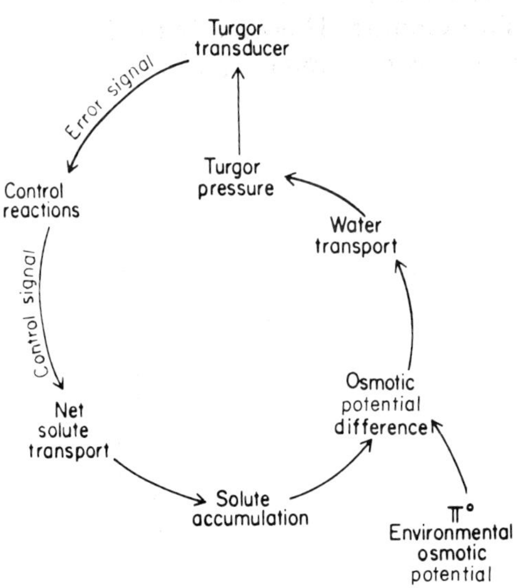

Fig. 12. Gutknecht's scheme (44) for turgor regulation in algal cells based on the accumulation and transport of solutes and water. After Gutknecht *et al.* (44).

ily mutually exclusive: (1) that pressure of the wall microfibrils on the membrane blocks or restricts ion channels or pumping sites; (2) that as pressure on the membrane changes, charged groups on it are moved farther apart or closer together, with resultant changes in electrostatic forces changing the membrane properties; (3) that changes in pressure on the membrane change its characteristics directly, without necessarily involving electrostatic effects as in (2); (4) that the membrane is normally stretched or folded around the wall microfibrils (Fig. 13) and that as turgor falls and the membrane relaxes, a chemical signal is released from it which stimulates inward pumping of ions. This last possibility is favored by Gutknecht *et al.* (44). There is some evidence for turgor-induced appression of membranes against cell wall microfibrils from freeze-fracture studies with electron microscopy (Fig. 14), but even if this is accepted, there remains the problem of explaining how membrane activity is affected by compression or stretching. Berezin *et al.* (2) studied the activity of chymotrypsinogen embedded in polyacrylamide gel and reported that activity of the enzyme was increased more than 20-fold when the gel was compressed mechanically. Whether deformation of a membrane *in vivo,* which has been shown to be possible (116), has similar effects, especially on the ion pumps involved in turgor regulation, is not known. If membrane activity is turgor dependent, then the value of many *in vitro* studies on isolated membrane preparations must inevitably be questioned.

Turgor regulation by control of π seems to be a common feature of all the marine algae so far examined. However, in the freshwater genus *Nitella,* which does not usually encounter significant variation in external

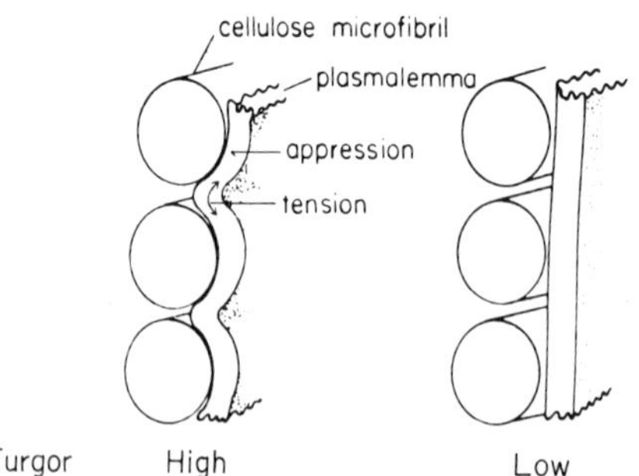

FIG. 13. Hypothetical turgor-induced membrane (right) stretching in *Valonia.* In the relaxed state the plasmalemma makes limited contact with the innermost wall microfibrils, but as turgor increases (left) the contact becomes closer, leading to stretching and appression of the membrane against the wall. From Gutknecht *et al.* (44).

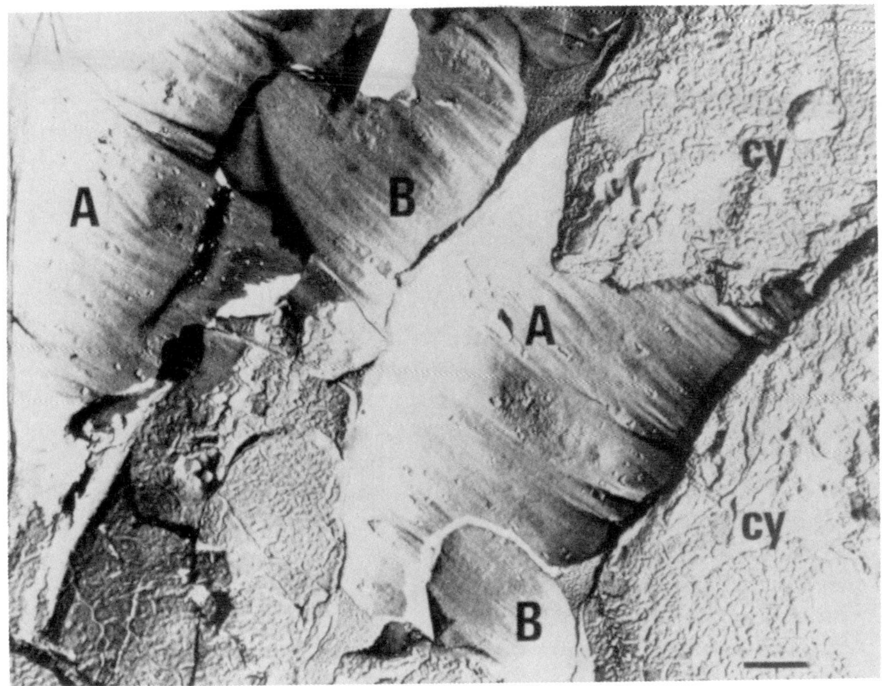

FIG. 14. Freeze-fractured root tip cells of *Phaseolus vulgaris* showing intimate contact between wall and plasmalemma. A and B are fracture faces showing undistorted plasmalemma pressed tightly against the walls so that the interface topography is seen as impressions. The cytoplasm of ruptured cells is also seen (cy). Bar line 1 μm. From Willison (116).

osmotic pressure, regulation seems to be absent. Instead, short-term adjustments of volume depending on ϵ are found.

The wall-less alga *Ochromonas malhamensis* appears to show volume regulation (59, 60) in which the contractile vacuole plays a significant role by secretion of water to balance the inwardly directed water flux when the alga is in fresh water. When exposed to medium or lower osmotic potential, the cells shrink rapidly and volume regulation is achieved by adjusting the number of solute molecules within the cell. This is achieved, at least in part, by synthesis of α-galactosylglycerol, also known as isofloridoside (4, 59).

B. TURGOR REGULATION AND TURGOR SENSING IN HIGHER PLANTS

There is a large and growing literature showing that turgor regulation also occurs in the cells and tisues of higher plants [e.g., (18, 30, 31, 40, 47,

53, 57, 76, 93, 101, 118–120, 122)]. The subject is also discussed in the chapters by Davies and by Sutcliffe in this volume, and because of this only two general aspects are dealt with here, namely, the reason for turgor regulation and the nature of the solutes involved.

Although it is often convenient and attractive to consider the behavior of giant algal cells as models for the behavior of other cells, there are dangers in this approach. For example, the cells of higher plants have a much higher surface : volume ratio than do giant algal cells, and values of turgor tend also to be higher. The reason suggested for turgor regulation in the giant cells (Section V,A) are unlikely to be important for the smaller cells of higher plants. Instead, it has been argued [e.g. (53, 101)] that small changes in *P* are the most likely means by which changes in water status affect metabolism; in other words, turgor and metabolic activity are related so that changing the former also changes the latter. How can this be brought about? A major theme of Chapter 7 by Steward is that cells accumulate ions during growth as an osmotic necessity and that it is the total solute content that is important in this context, even though there are temporal variations in the patterns of uptake of individual ions in different cells. As will be considered in Section V,C, turgor and growth are intimately associated; the osmotic necessity referred to earlier is a reflection of the need to maintain turgor during growth by increasing the solute content of the cell. If, as has been suggested on p. 30, membrane structure and function depend on turgor, then the passage of materials into and out of the cell, during growth and subsequently, will be affected by the water relations of the cell. The plasmalemma (16, 17), and perhaps the tonoplast, is seen as the site of transduction of the signal of water status into a change in metabolite flux and hence in metabolism. There is an urgent need for much more sound data on the effects of small measured turgor changes on the metabolism of cells of higher plants as they expand and once they attain full size. The technical difficulties of such work are obvious.

Because the vacuole is often the largest component of the cells of higher plants, it is often considered to be the most affected when the cell experiences water stress. Undoubtedly, shrinkage of the vacuole will be an important contribution to falling turgor, but effects on the cytoplasm itself are also important, sufficiently so for Schobert (81) to argue that water retention by binding to a protein–solute complex is of major significance under such conditions. Without developing this argument further, acceptance of the more widely held view of turgor regulation requires that solute accumulation in both vacuole and cytoplasm must be involved. Potassium, sodium, and chloride are important osmotica for vacuolate cells, and there is some scope for altering the concentration of these electrolytes without affecting metabolic activity. However, KCl and NaCl are relatively toxic to

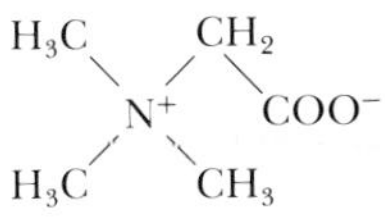

$(H_3C)_2N^+(CH_2COO^-)CH_3$

Glycinebetaine

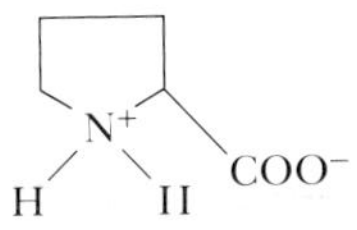

Proline

a number of enzymes, including malate dehydrogenase (32). Effectively this restricts the extent to which salts accumulated from the soil can contribute to the cytoplasmic solute content. Instead, in a number of species the synthesis and accumulation of nontoxic osmotica, known also as compatible solutes (118, 121), such as glycinebetaine and the imino acid proline serves to lower the osmotic potential of the cytoplasm. Table VII, adapted from one prepared by Cram (18), shows the principal solutes contributing to the osmotic potential of a range of tissues from higher plants. No distinction is made for cytoplasmic and vacuolar solutes and it is likely that solutes from the vacuole, because of its relative size, dominate the analyses. The available data are scanty when compared with those considered by Steward in Chapter 7. They also show substantial within-species variation as must be inevitable for plants grown under widely differing conditions. (Compare, for example, values for cotton, *Gossypium hirsutum,* and tomato, *Lycopersicon esculentum.*) Nevertheless, it is clear that, except in the halophytic species, potassium is a major component of π and that sugars are of little importance; sugars are much more significant in the carrot tissue cultures grown heterotrophically on a sugar source, while sodium is a major contributor to π in the tissues of halophytes.

TABLE VII

IONIC CONCENTRATIONS IN LEAVES AND SHOOTS OF A VARIETY OF HIGHER PLANT SPECIES GROWN UNDER WIDELY DIFFERING CONDITIONS[a]

	Ionic concentration (mol m^{-3})							
Species	K^+	Na^+	Ca^{2+}	Mg^{2+}	Cl^-	SO_4^{2-}	Organic acid	Sugars
Hordeum vulgare	228	22	18	18	72	8	?	15
Sorghum bicolor (S. vulgare)	179	1	32	37	26	11	?	—
Lycopersicon esculentum	161	6	58	42	25	75	?	—
Lycopersicon esculentum	58	19	40	15	12	11	180	—
Gossypium hirsutum	228	18	126	54	18	92	?	—
Gossypium hirsutum	77	8	99	23	11	36	250	10
Beta vulgaris	167	229	1	51	44	20	?	—
Beta vulgaris	200	68	22	57	23	10	274	17

[a] From data accumulated by Cram (18).

C. TURGOR AND CELL GROWTH

1. *General Considerations*

So far this chapter has been concerned with the water relations of fully expanded and vacuolated plant cells. Much less attention has been given to the water relations of enlarging cells, although even so there has been controversy over interpretation of both theoretical and experimental studies (8, 9, 77).

Essentially, growth consists of an irreversible increase in cell volume which can be both substantial and rapid (see Fig. 15). It involves the uptake of water and the plastic deformation of the wall and incorporation of new wall material. It would be expected therefore that cell growth rate dV/dt is controlled both by the inwardly directed water flux and by those properties of the wall which govern wall extensibility. The relationship between growth rate and water influx is

$$dV/dt = L'_{\mathrm{p}}\, \Delta\psi \qquad (16)$$

where L'_{p} is the apparent hydraulic conductance (see p. 22) over the entire

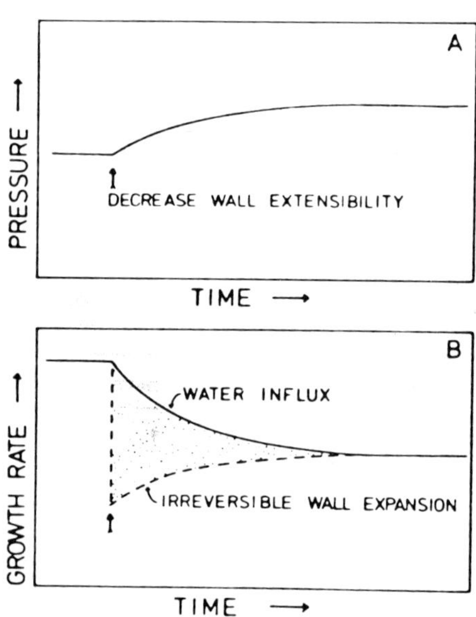

FIG. 15. The effects of an instantaneous fall in wall extensibility on cell turgor (A) and on growth rate measured in terms of water uptake and irreversible wall expansion (B). See text for further details. From Cosgrove (15).

cell membrane (cubic meters per pascal second) [cf. Eq. (12)] and $\Delta\psi$ is the difference of water potential between the cell and its surroundings. Because the membrane is not truly semipermeable $\Delta\psi = \sigma\pi - P$. Clearly, if the permeability of the membrane to water is low, cell growth could be limited just as much as if $\Delta\psi$ were low. The cell's water potential can be lowered (i.e., $\Delta\psi$ increased) by reducing either osmotic potential or turgor. Because of the volume increase osmotic potential will tend to rise (become less negative) during cell expansion unless there is an increase in total cell solutes to balance or exceed this dilution effect (see Chapter 7 by Steward).

In an important paper Lockhart (65, 66) derived a mathematical analysis of irreversible cell growth which combined a consideration of water flux [as in Eq. (16)] with a treatment of the inevitable and essential changes in the cell wall which occur during cell expansion. The equation he used for irreversible wall expansion can be modified and simplified to

$$dV/dt = \phi(P - Y) \tag{17}$$

where ϕ is the extensibility of the wall, i.e., the extent to which it can be stretched in growth, with units of square meters per pascal second; and Y is the threshold value of turgor below which growth will not occur, also known as the wall yield stress. Thus irreversible cell extension is governed by wall extensibility and the excess of turgor over a minimum threshold value. If wall extensibility is constant, or nearly so, cell extension is governed by $(P - Y)$, and if $P < Y$, then $dV/dt = 0$. Lockhart's approach has been taken up and extended by a number of workers (13, 15, 53, 78, 112), many of whom have, following Lockhart, combined Eqs. (16) and (17) to give the general equation

$$\frac{dV}{dt} = \frac{L'_p\phi[\Delta\psi + (P - Y)]}{\phi + L'_p} \tag{18}$$

Since $\Delta\psi = \pi_e - \pi - P$, an alternative form of Eq. (18) is

$$\frac{dV}{dt} = \frac{L'_p\phi(\pi_o - \pi - Y)}{\phi + L'_p} \tag{19}$$

Taking into account the fact that the membranes of the growing cells are unlikely to be truly semipermeable and including a reflection coefficient term σ, we have

$$\frac{dV}{dt} = \frac{L'_p\phi(\sigma\,\Delta\pi - Y)}{L'_p + \phi} \tag{20}$$

The growth of a cell, and so far the treatment used relates only to single cells, is determined by four parameters (five if the reflection coefficient term is included), two of which, ϕ and Y, relate to the wall; one, L'_p, to membrane characteristics; and the remaining one to osmotic potential.

Here L'_p can be converted to hydraulic conductivity per cell, designated L, by taking account of cell surface area A and volume V so that $L = L'_p A / V$. This is convenient because L has the units of pressure over time, i.e., square meters per pascal second, which are the same as those for ϕ. Inclusion of L instead of L'_p in the term outside the brackets in Eqs. (18)–(20) makes it obvious that where $L < \phi$ growth is limited by wall extensibility, and vice versa. In growing cells of *Nitella* (38), ϕ has been calculated to be about two orders of magnitude less than L, so that wall extensibility rather than hydraulic conductance would seem to be the major limiting factor to cell growth.

Equations (18)–(20) relate, strictly speaking, to steady-state conditions. Where this is not the case, e.g., if wall extensibility were to change suddenly, Eqs. (16) and (17) are not equal to each other and the subsequent combining is invalid. Cosgrove (15) has examined the consequences of sudden changes in parameters such as wall extensibility. Interested readers are referred to his paper for the derivations of equations used to calculate effects of departure from the steady state. An illustration of the effect of a sudden perturbation of wall extensibility on turgor and cell growth rate is shown in Fig. 15. An instantaneous fall in ϕ is followed by a gradual increase in P to a new steady-state value. At the same time, water influx declines as P increases, again to reach a new, lower, steady-state value as P stabilizes. Growth rate, considered as irreversible wall expansion, falls sharply as wall extensibility is decreased and then begins to rise, equaling the rate of growth seen in terms of water flux at the time that P also stabilizes. Prior to the attainment of a new steady state, the cell has two "growth rates," that based on water influx being higher than that based on wall expansion; the difference is shown by the shaded area in Fig. 15 and is accommodated by elastic changes in the cell wall, depending, of course, on the elastic modulus. The extent to which such elastic changes are necessary in normally growing cells is uncertain. Large and sudden changes in ϕ seem unlikely except for cells exposed to the rigors of experimental treatments. Consequently, lack of equality between Eqs. (16) and (17) may well be small.

The approach used by Lockhart and developed by later workers applies properly only to single cells, and not to tissues, although a number of workers have applied the method to multicellular material. Cosgrove (15) has also extended the analysis to cover the complications that arise in growing tissues when the resistance to water flow is located throughout the growing system, and not only at its boundary as is the case for a single cell.

A consequence of this is that a term to account for tissue diffusivity of water D (in square meters per second) has to be introduced to replace L_p' (or L). In a series of model calculations, Cosgrove showed that for whole tissues changes in D and ϕ have effects similar to those of changes in L_p' and ϕ for single cells, ϕ wall extensibility controlling growth when it has low values but not when its value is large.

This conclusion is reassuring, for a number of workers have applied the approach for single cells to a variety of tissues and organs, e.g., coleoptiles of rye *(Secale cereale)* and oat (10, 37); hypocotyls of soybean (6); and leaves of *Zea,* sorghum, and *Phaseolus vulgaris* (53, 110–112).

2. *Cell Expansion in Primary Leaves of* Phaseolus vulgaris

The later stages of lamina expansion of primary leaves of *Phaseolus* do not involve cell divison but only cell expansion (26, 27, 110) (Fig. 16). Because of this, these leaves have been a convenient model for a number of studies of cell expansion in rapidly growing multicellular systems.

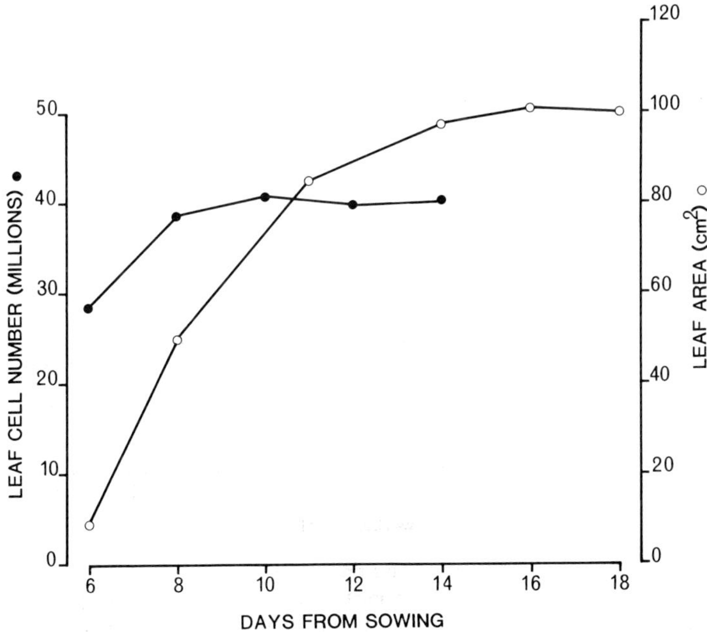

Fig. 16. The time course for increase in cell number and in leaf area for a primary leaf of *Phaseolus vulgaris* grown in 12-h days with photon flux density of 200 μmol m^{-2} s^{-1}.

It has been known for some time that cell expansion in *Phaseolus* leaves is dependent on high-energy light (27), and this suggested to Van Volkenburgh and Cleland (112) that light must affect one or more of the parameters in Eq. (20). They therefore attempted to measure each of these factors by using sections of expanding leaf tissue in which cell division had ceased. The water potential of the leaf tissue and the osmotic potential were determined by thermocouple psychrometry (5), and from these data turgor potential was calculated by difference. Hydraulic conductance was calculated from the relation $dV/dt = L'_p \, \Delta\psi$ [see Eq. (16)]. Wall yield stress Y was determined by incubating leaf pieces in solutions of polyethylene glycol and measuring growth in length. When growth is plotted against the concentration of PEG, two-phase curves of the sort shown in Fig. 17 result. The intersection between the curves, or break point, indicates the minimum osmotic concentration in which growth does not occur. In these experiments (112) the concentration of PEG required to prevent growth was about 100 mOsm, equivalent to 0.25 MPa. Since the osmotic concen-

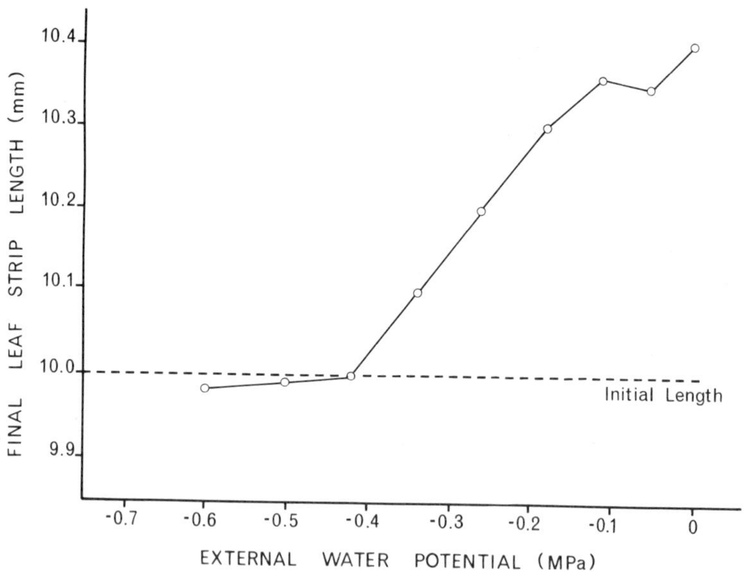

FIG. 17. Break point curves obtained for strips of young expanding primary leaves of *Phaseolus vulgaris*. In this example the break point, i.e., the water potential of the incubating solution of polyethylene glycol in which no change in strip dimensions occurred, is -0.4 MPa. Osmotic potential of the tissue was measured as -0.7 MPa. Wall yield stress is given by the difference between the two values, i.e., $Y = 0.3$ MPa. Unpublished data of Stuart Milligan.

tration of the leaf sap, obtained separately by psychrometry, was about 280 mOsm (0.7 MPa), the wall yield stress was about 180 mOsm, or 0.45 MPa. Finally, wall extensibility was measured directly using a linear stress extension meter or an Instron apparatus (Fig. 18) (11, 12, 14a).

The results obtained by Van Volkenburgh and Cleland (112) are summarized in Table VIII. In their experiments leaves of 10-day-old plants were used, incubated in dim red light, in which cell expansion is slight, or in white light, in which leaf growth is much more rapid. Variation in the estimates of water potential and its components suggests that these were not affected by treatment. Any downward trend in ψ in the growing tissue

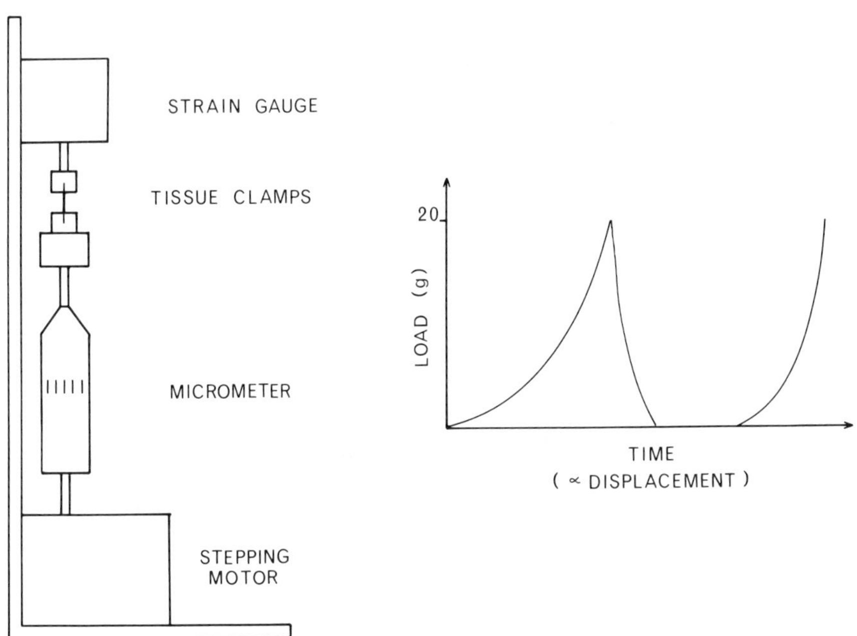

FIG. 18. Diagram of apparatus used for determination of extensibility of cell walls and the graphical output. A variety of forms of the apparatus exist, including the Instron stress–strain analyzer, which is commercially available from Instron Corporation, Canton, Massachusetts, and versions following the design of Van Volkenburgh *et al.* (112a). In operation, pieces of tissue, e.g., leaf lamina, are cut to a standard size and killed by boiling in methanol. They are then inserted between the clamps of the tensiometer and extended twice to a load of 20*g*. The force–extension relationships obtained by recording the output from the strain gauge against time are analyzed to give, first, total (i.e., plastic and elastic) extensibilities of the tissue and, second, elastic extensibility only. The difference between the two is plastic extensibility, equated with wall extensibility ϕ. A critical review of the method has been published by Cleland (14a).

TABLE VIII

EFFECT OF LIGHT TREATMENT ON CELL GROWTH VARIABLES FOR LEAF TISSUE OF *Phaseolus vulgaris*[a]

Parameter	Dark control	Red light	White light
Leaf growth rate ($m^3\ h^{-1} \times 10^6$)	1.30 ± 0.47	0.92 ± 0.23	3.38 ± 0.69
Tissue water potential (MPa)	-0.22 ± 0.09	-0.16 ± 0.09	-0.36 ± 0.07
Tissue osmotic potential (MPa)	-1.04 ± 0.04	-0.94 ± 0.16	-0.86 ± 0.12
Tissue turgor potential (MPa)	0.82 ± 0.08	0.78 ± 0.09	0.50 ± 0.19
Apparent hydraulic conductance ($m^2\ MPa^{-1}\ s^{-1}$)	1.9×10^{-11}	2.0×10^{-11}	2.6×10^{-11}
Wall yield threshold (MPa)		-0.45	-0.45
Wall extensibility (measured as % plastic extension per 10*g* load)		2.8	3.8

[a] Data adapted from Van Volkenburgh and Cleland (112).

was associated with a reduction in *P* which more than offset a slight increase in osmotic potential. The estimates of apparent hydraulic conductance indicate little difference due to treatment, with higher values for the growing tissue if anything. There was no effect on wall yield stress (see Fig. 17). In contrast, there was a substantial increase in wall extensibility of leaf tissue treated with white light. Following 8h in white light after initial culture in red light, wall extensibility rose so that a 10*g* load on a standard piece of leaf tissue gave a 3.5–4% extension compared with an extension of less than 3% in tissue cultured throughout in red light. In these experiments, therefore, it was concluded that the effect of white light was exerted mainly on wall extensibility, with only minor effects on the other parameters. Once again a note of caution should be sounded because the values of wall extensibility are bulk or average values for all cells including, for example, palisade and epidermal cells which normally show their maximum extension growth in planes at right angles to each other!

In further investigations of the effect of the light treatment it was shown that exposure to white light stimulates proton secretion by bean leaf tissue (111). It was argued, therefore, that in line with the acid growth theory proposed for other systems (see 11, 45) proton secretion may be responsible for wall loosening in the cells of *Phaseolus* leaves; the exact nature of the mechanisms involved has been discussed elsewhere (14, 123).

A correlation between the failure of leaf tissue to expand and a fall in wall extensibility has also been shown for leaves of *Phaseolus* plants subjected to root cooling or partial root removal (Fig. 19). It is of interest that these data come from long-term experiments over several days, in contrast

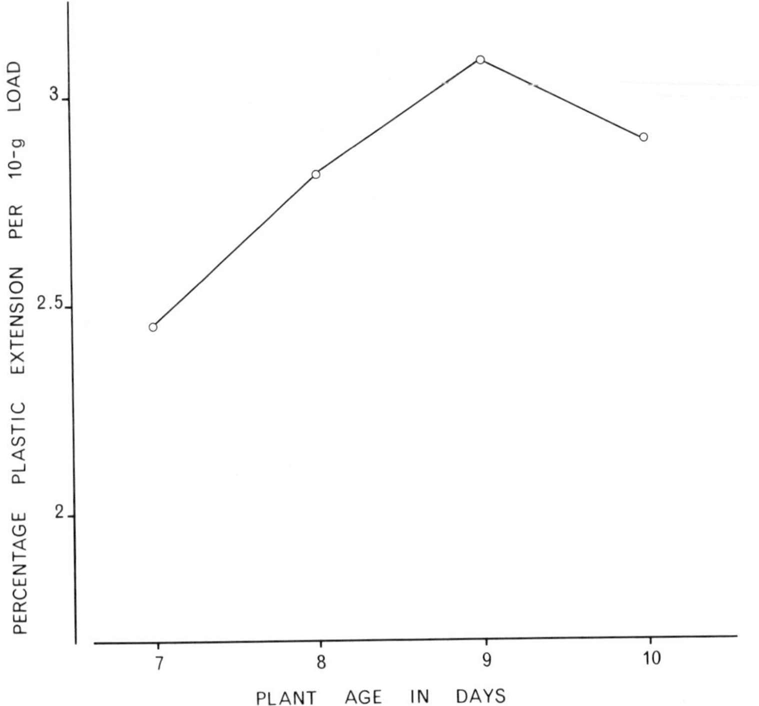

FIG. 19. Changes in wall extensibility for samples of primary leaf tissue of *Phaseolus vulgaris* of various ages. The leaves emerge into the light and unfold on day 7 coincident with a rapid increase in wall extensibility. Unpublished data of Stuart Milligan.

to the short-term experiments on light effects. Whether in development over the longer term there are significant changes in other parameters in the Lockhart equation, particularly in the wall yield stress term, remains at present uncertain.

VI. Future Developments

When Bennet-Clark's article appeared in 1959 in Volume II of this treatise, it was already apparent that changes in the terminology of plant water relations were imminent and that these together with the development of the ideas and concepts of irreversible thermodynamics were likely to have great significance for the cell physiologist concerned with fluxes of water and ions. What are likely to be the major developments over the next

25 years? Gazing into the crystal ball is a notoriously risky business, but already the outlines of three areas of advance are appearing: greater availability of key data for individual cells, increased understanding of the significance of turgor regulation, and developments in the understanding of the growth processes of expanding cells.

The main features of the water relations of vacuolate cells have been understood in outline for many years, based often on measurements of bulk values rather than values for individual cells. Development of the pressure probe has meant that detailed measurements on individual cells, including those in tissues, are now possible. Parameters such as elastic modulus and hydraulic conductivity, which hitherto have proved difficult to measure, can now be determined with considerable accuracy. Already it is clear that there is considerable variation from cell to cell in the key parameters of cell growth and water relations; we may expect the reasons for this and its significance to be understood. At the same time, the increasing precision of the available techniques will mean that more exact mathematical analyses and definitions of the relationships among cell parameters will evolve.

The acquisition of more accurate and precise data is likely to be of considerable importance to our understanding of turgor regulation. It is increasingly recognized that the maintenance of turgor is a major activity of plant cells which impinges on a large number of other cellular activities. To understand this will require a greater appreciation than we have at present of pressure-sensing mechanisms and transduction, together with better knowledge of the function and control of the ion pumps and biochemical mechanisms involved in regulation of solute content.

Third, and of no less importance, we can expect to see continued attacks on the relationship between turgor and cell growth. In particular, the approach initiated by Lockhart may have to be modified as more knowledge accrues on the ways in which wall extensibility and wall yield threshold change as a cell grows. Here progress will certainly require information on the biochemistry of wall growth and deposition. At the same time, our knowledge of the formation and growth of vacuoles is likely to be extended.

It is hoped that this article reinforces the view that cell water relations are not and should not be regarded as a topic isolated from other facets of cell and plant physiology. Much of the intellectual excitement in this area now comes from a growing awareness that it is possible to integrate knowledge from classical water relations, salt relations, membrane physiology, and cell wall studies into coherent models of the growing differentiating cell. The synthesis of these cognate aspects into a clearer understanding of cell development is a major task which will take us well into the twenty-first century.

References

1. Bennet-Clark, T. A. (1959). Water relations of cells. *In* "Plant Physiology: A Treatise" (F. C. Steward, ed.), Vol. 2, pp. 105–191. Academic Press, New York.
2. Berezin, I. V., Kubanov, A. M., and Martinek, K. (1974). The mechanochemistry of immobilised enzymes. How to steer a chemical process at the molecular level by a mechanical device. *Biochim. Biophys. Acta* **364,** 193–199.
3. Bisson, M. A., and Gutknecht, J. (1975). Osmotic regulation in the marine alga, *Codium decorticatum* 1. Regulation of turgor pressure by control of ion composition. *J. Membr. Biol.* **24,** 183–200.
4. Borowitzka, L. J., and Brown, A. D. (1974). The salt relations of marine and halophilic species of the unicellular green alga *Dunaliella.* The role of glycerol as a compatible solute. *Arch. Microbiol.* **96,** 37–52.
5. Boyer, J. S., and Knipling, E. B. (1965). Isopiestic technique for measuring leaf water potentials with a thermocouple psychrometer. *Proc. Natl. Acad. Sci. U.S.A.* **54,** 1044–1045.
6. Boyer, J. S., and Wu, G. (1978). Auxin increases the hydraulic conductivity of auxin-sensitive hypocotyl tissue. *Planta* **139,** 227–237.
7. Briggs, G. E. (1967). "Movement of Water in Plants." Blackwell, Oxford.
8. Burstrom, H. G. (1971). Wishful thinking on turgor. *Nature (London)* **234,** 488.
9. Burstrom, H. G. (1979). In search of a plant growth paradigm. *Am. J. Bot.* **66,** 98–104.
10. Cleland, R. E. (1959). Effect of osmotic concentration on auxin action and an irreversible and reversible expansion of *Avena* coleoptile. *Physiol. Plant.* **12,** 809–825.
11. Cleland, R. E. (1967). Extensibility of isolated cell walls: Measurement and changes during cell elongation. *Planta* **74,** 142–152.
12. Cleland, R. E. (1971). Cell wall extension. *Annu. Rev. Plant Physiol.* **22,** 197–222.
13. Cleland, R. E. (1977). The control of cell enlargement. *Symp. Soc. Exp. Biol.* **31,** 101–115.
14. Cleland, R. E. (1983). The mechanism of auxin-induced proton efflux. *In* "Plant Growth Substances 1982" (P. F. Wareing, ed.), pp. 23–31. Academic Press, London.
14a. Cleland, R. E. (1984). The Instron technique as a measure of immediate-past wall extensibility. *Planta* **160,** 514–520.
15. Cosgrove, D. J. (1981). Analysis of the dynamic and steady-state responses of growth rate and turgor pressure to changes in cell parameters. *Plant Physiol.* **68,** 1439–1446.
16. Cosgrove, D. J., and Steudle, E. (1981). Water relations of growing pea epicotyl segments. *Planta* **153,** 343–350.
17. Coster, H. G. L., Steudle, E., and Zimmermann, U. (1976). Turgor pressure sensing in plant cell membranes. *Plant Physiol.* **58,** 636–643.
18. Cram, W. J. (1976). Negative feedback regulation of transport in cells. The maintenance of turgor, volume and nutrient supply. *In* "Encyclopedia of Plant Physiology, New Series" (U. Lüttge and M. G. Pitman, eds.), Vol. 2, Part A, pp. 284–316. Springer-Verlag, Berlin and New York.
19. Dainty, J. (1963). Water relations of plant cells. *Adv. Bot. Res.* **1,** 279–326.
20. Dainty, J. (1972). Plant cell-water relations: The elasticity of the cell wall. *Proc.—R. Soc. Edinburgh, Sect. A: Math. Phys. Sci.* **70,** 89–93.
21. Dainty, J. (1976). Water relations of plant cells. *In* "Encyclopedia of Plant Physiology, New Series" (U. Lüttge and M. G. Pitman, eds.), Vol. 2, Part A, pp. 12–35. Springer-Verlag, Berlin and New York.
22. Dainty, J., and Ginzburg, B. Z. (1963). Irreversible themodynamics and frictional models of membrane processes with particular reference to the cell membrane. *J. Theor. Biol.* **5,** 256–265.

23. Dainty, J., and Ginzburg, B. Z. (1964). The measurement of hydraulic conductivity (osmotic permeability to water) of internodal characean cells by means of transcellular osmosis. *Biochim. Biophys. Acta* **79,** 102–111.
24. Dainty, J., and Ginzburg, B. Z. (1964). The reflection coefficient of plant cell membranes for certain solutes. *Biochim. Biophys. Acta* **79,** 129–137.
25. Dainty, J., Vinters, H., and Tyree, M. T. (1974). A study of transcellular osmosis and the kinetics of swelling and shrinking in cells of *Chara corallina. In* "Membrane Transport in Plants" (U. Zimmermann and J. Dainty, eds.), pp. 59–63. Springer-Verlag, Berlin and New York.
26. Dale, J. E. (1976). Cell division in leaves. *In* "Cell Division in Higher Plants" (M. M. Yeoman, ed.), pp. 315–345. Academic Press, London.
27. Dale, J. E., and Murray, D. (1968). Photomorphogenesis, photosynthesis and early growth of primary leaves of *Phaseolus vulgaris. Ann. Bot. (London)* [N.S.] **32,** 767–780.
28. Dixon, H. H. (1914). "Transpiration and the Ascent of Sap in Plants." Macmillan, London.
29. Eamus, D., and Tomos, A. D. (1983). The influence of abscisic acid on the water relation of leaf epidermal cells of *Rhoeo discolor. Plant Sci. Lett.* **31,** 253–259.
30. Flowers, T. J., Hall, J. L., and Ward, M. E. (1976). Salt tolerance in the halophyte *Suaeda maritima.* Further properties of the enzyme malate dehydrogenase. *Phytochemistry* **15,** 1231–1234.
31. Flowers, T. J., Troke, P. F., and Yeo, A. R. (1977). The mechanism of salt tolerance in halophytes. *Annu. Rev. Plant Physiol.* **28,** 89–121.
32. Flowers, T. J., Ward, M. E., and Hall, J. L. (1976). Salt tolerance in the halphyte *Suaeda maritima.* Some properties of malate dehydrogenase. *Philos. Trans. R. Soc London, Ser. B* **273,** 523–540.
33. Franks, F. (1983). "Water." Royal Society of Chemistry, London.
34. Gardner, W. R., and Ehlig, C. F. (1965). Physical aspects of the internal water relations of plant leaves. *Plant Physiol.* **40,** 705–710.
35. Graves, J. S., and Gutknecht, J. (1976). Ion transport studies and determination of the cell wall elastic modulus in the marine alga *Halicystic parvula. J. Gen. Physiol.* **67,** 579–597.
36. Green, P. B. (1968). Growth physics in *Nitella:* A method for continuous *in vivo* analysis of extensibility based on a micromanometer technique for turgor pressure. *Plant Physiol.* **68,** 1169–1184.
37. Green, P. B., and Cummins, W. R. (1974). Growth rate and turgor pressure. Auxin effect studies with an automated apparatus for coleoptiles. *Plant Physiol.* **54,** 863–870.
38. Green, P. B., Erickson, R. O., and Buggy, J. (1971). Metabolic and physical control of cell elongation rate. *In vivo* studies in *Nitella. Plant Physiol.* **47,** 423–430.
39. Green, P. B., and Stanton, F. W. (1967). Turgor pressure: Direct manometric measurement in single cells of *Nitella. Science* **155,** 1675–1676.
40. Greenway, H., and Munns, R. A. (1980). Mechanisms of salt tolerance in non-halophytes. *Annu. Rev. Plant Physiol.* **31,** 149–190.
41. Guggino, S., and Gutknecht, J. (1982). Turgor regulation in *Valonia macrophysa* following osmotic shock. *J. Membr. Biol.* **67,** 155–164.
42. Gutknecht, B. (1968). Permeability of *Valonia* to water and solutes: Apparent absence of aqueous membrane pores. *Biochim. Biophys. Acta* **163,** 20–29.
43. Gutknecht, J., and Bisson, M. A. (1977). Ion transport and osmotic regulation in giant algal cells. *In* "Water Relations in Membrane Transport in Plants and Animals" (A. M. Jungreis, T. K. Hodges, A. Kleinzeller, and S. G. Schultz, eds.), pp. 3–14. Academic Press, New York.

44. Gutknecht, J., Hastings, D. F., and Bisson, M. A. (1978). Ion transport and turgor regulation in giant algal cells. *In* "Membrane Transport in Biology" (G. Giebisch, D. C. Tosteson, and H. H. Ussing, eds.), Vol. 3, pp. 125–174. Springer-Verlag, Berlin and New York.
45. Hager, A., Menzel, C., and Krauss, A. (1971). Versuche und Hypothese zur Primarwirkung des Auxins beim Streckungswachstum. *Planta* **100,** 47–75.
46. Hastings, D. F., and Gutknecht, J. (1976). Ionic relations and the regulation of turgor pressure in the marine alga *Valonia macrophysa*. *J. Membr. Biol.* **28,** 263–275.
47. Hellebust, J. A. (1976). Osmoregulation. *Annu. Rev. Plant Physiol.* **27,** 485–507.
48. Hellkvist, J., Richards, G. P., and Jarvis, P. G. (1974). Vertical gradients of water potential and tissue water relations in Sitka spruce trees measured with the pressure chamber. *J. Appl. Ecol.* **11,** 637–667.
49. Höfler, K. (1920). Ein Schema für die osmotische Leistung der Pflanzenzelle. *Ber. Dtsch. Bot. Ges.* **38,** 288–298.
50. Hope, A. B., and Walker, N. A. (1975). "The Physiology of Giant Algal Cells." Cambridge Univ. Press, London and New York.
51. House, C. R. (1974). "Water Transport in Cells and Tissues." Edward Arnold, London.
52. House, C. R., and Jarvis, P. (1968). Effect of temperature on the radial exchange of labelled water ion maize roots. *J. Exp. Bot.* **19,** 31–40.
53. Hsioa, T. C., Acevedo, E., Fereres, E., and Henderson, D. W. (1976). Stress metabolism. Water stress, growth and osmotic adjustment. *Philos. Trans. R. Soc. London, Ser. B* **273,** 479–500.
54. Huber, B., and Höfler, K. (1930). Die Wasserpermeabilität des Protoplasmas. *Jahrb. Wiss. Bot.* **73,** 351–511.
55. Hüsken, D., Steudle, E., and Zimmermann, U. (1978). Pressure probe technique for measuring water relations of cells in higher plants. *Plant Physiol.* **61,** 158–163.
56. Jarvis, P., and House, C. R. (1967). The radial exchange of labelled water in maize roots. *J. Exp. Bot.* **18,** 695–706.
57. Jefferies, R. L. (1980). The role of organic solutes in osmo-regulation in halphytic higher plants. *In* "Genetic Engineering of Osmoregulation. Impact on Plant Productivity for Food, Chemicals and Energy" (D. W. Rains, R. C. Valentine, and A. Hollaender, eds.), pp. 133–154. Plenum, New York.
58. Kamiya, N., and Tazawa, M. (1956). Studies of water permeability of a single plant cell by means of transcellular osmosis. *Protoplasma* **46,** 394–422.
59. Kauss, H. (1973). Turnover of galactosylglycerol and osmotic balance in *Ochromonas*. *Plant Physiol.* **52,** 613–615.
60. Kauss, H. (1978). Osmotic regulation in algae. *Prog. Phytochem.* **5,** 1–27.
61. Kedem, O., and Katchalsky, A. (1958). Thermodynamic analysis of the permeability of biological membranes to non-electrolytes. *Biochim. Biophys. Acta* **27,** 229–246.
62. Kiyosawa, K., and Tazawa, M. (1973). Rectification characteristics of *Nitella* membranes in respect to water permeability. *Protoplasma* **78,** 203–214.
63. Kramer, P. J. (1983). "Water Relations of Plants." Academic Press, New York.
64. Kuhn, P. G., and Dainty, J. (1966). The measurement of permeability to water in disks of storage tissue. *J. Exp. Bot.* **17,** 809–821.
65. Lockhart, J. A. (1965). An analysis of irreversible plant cell elongation. *J. Theor. Biol.* **8,** 264–275.
66. Lockhart, J. A. (1965). Cell extension. *In* "Plant Biochemistry" (J. Bonner and J. E. Varner, eds.), 2nd ed. pp. 826–849. Academic Press, New York.
67. Meidner, H., and Edwards, M. (1975). Direct measurement of turgor potential of guard cells. *J. Exp. Bot.* **26,** 319–330.

68. Meyer, B. S., and Anderson, D. B. (1952). "Plant Physiology." Van Nostrand-Reinhold, Princeton, New Jersey.
69. Milburn, J. A. (1979). "Water Flow in Plants." Longmans, Green, London and New York.
70. Nobel, P. S. (1983). "Biophysical Plant Physiology and Ecology." Freeman, San Francisco, California.
71. Noy-Meir, I., and Ginzburg, B. Z. (1967). An analysis of the water potential isotherm in plant tissue. 1. The theory. *Aust. J. Biol. Sci.* **20,** 695–621.
72. Oertli, J. J. (1969). Terminology of plant-water energy relations. *Z. Pflanzenphysiol.* **61,** 264–265.
73. Passioura, J. B. (1980). The meaning of matric potential. *J. Exp. Bot.* **31,** 1161–1169.
74. Philip, J. R. (1958). The osmotic cell, solute diffusibility and the plant water economy. *Plant Physiol.* **33,** 264–271.
75. Probine, M. C. (1963). Cell growth and the structure and mechanical properties of the wall in internodal cells of *Nitella opaca*. III. Spiral growth and cell wall structure. *J. Exp. Bot.* **14,** 101–113.
76. Rains, D. W., Valentine, R. C., and Hollaender, A. (1980). "Genetic Engineering of Osmoregulation: Impact on Food, Chemicals and Energy." Plenum, New York.
77. Ray, P. M., Green, P. B., and Cleland, R. E. (1972). Role of turgor in plant cell growth. *Nature (London)* **239,** 163–164.
78. Rayle, D. L., and Cleland, R. E. (1970). Enhancement of wall loosening and elongation by acid solutions. *Plant Physiol.* **46,** 250–253.
79. Reed, R. H. (1984). Use and abuse of osmo-terminology. *Plant, Cell Environ.* **7,** 165–170.
80. Richter, H. (1978). A diagram for the description of water relations in plant cells and organs. *J. Exp. Bot.* **29,** 1197–1203.
81. Schobert, B. (1977). Is there an osmotic regulatory mechanism in algae and higher plants? *J. Theor. Biol.* **68,** 17–26.
82. Scholander, P. F., Hammel, H. T., Bradstreet, E. D., and Hemmingsen, E. A. (1965). Sap pressure in vascular plants. *Science* **148,** 339–346.
83. Slatyer, R. O. (1966). An underlying cause of measurement discrepancies in determinations of osmotic characteristics in plant cells and tissues. *Protoplasma* **62,** 34–43.
84. Slatyer, R. O. (1967). "Plant-Water Relationships." Academic Press, New York.
85. Slatyer, R. O., and Taylor, S. A. (1960). Terminology in plant and soil water relationships. *Nature (London)* **187,** 922–924.
86. Spanner, D. C. (1973). The components of the water potential in plants and soils. *J. Exp. Bot.* **24,** 816–819.
87. Stadelmann, E. (1966). Evaluation of turgidity, plasmolysis and deplasmolysis of plant cells. *In* "Methods in Cell Biology" (D. M. Prescott, ed.), Vol. 2, pp. 143–216. Academic Press, New York.
88. Steudle, E., Lüttge, U., and Zimmermann, U. (1975). Water relations of the epidermal bladder cells of the halophytic species *Mesembryanthemum crystallinum:* Direct measurements of hydrostatic pressure and hydraulic conductivity. *Planta* **126,** 229–246.
89. Steudle, E., Smith, J. A. C., and Lüttge, U. (1980). Water-relation parameters of individual mesophyll cells of the crassulacean acid metabolism plant *Kalanchoe diagremontiana. Plant Physiol.* **66,** 1155–1163.
90. Steudle, E., and Zimmermann, U. (1971). Hydraulic conductivity of *Valonia utricularis. Z. Naturforsch., B: Anorg. Chem., Org. Chem., Biochem., Biophys., Biol.* **26B,** 1302–1311.
91. Steudle, E., and Zimmermann, U. (1974). Determination of the hydraulic conductivity and of reflection coefficients in *Nitella flexilis* by means of direct cell-turgor pressure measurements. *Biochim. Biophys. Acta* **332,** 399–412.

92. Steudle, E., Zimmermann, U., and Lüttge, U. (1977). Effect of turgor pressure and cell size on the wall elasticity of plant cells. *Plant Physiol.* **59,** 285–289.
93. Stevenson, T. T., and Cleland, R. E. (1981). Osmoregulation in the *Avena* coleoptile in relation to auxin and growth. *Plant Physiol.* **67,** 749–753.
94. Taylor, S. A., and Slatyer, R. O. (1961). Proposals for a unified terminology in studies of plant-soil-water relationships. *In* Plant-Water Relationships in Acid and Semi-acid Conditions," pp. 339–349. UNESCO Paris.
95. Tazawa, M., and Kamiya, N. (1966). Water permeability of a Characean internodal cell with special reference to its polarity. *Aust. J. Biol. Sci.* **19,** 399–419.
96. Tazawa, M., and Kiyosawa, K. (1973). Analysis of transcellular water movements in *Nitella:* A new procedure to determine the inward and outward water permeabilities of membranes. *Protoplasma* **78,** 349–364.
97. Thoday, D. (1918). On turgescence and the absorption of water by the cells of plants. *New Phytol.* **17,** 108–113.
98. Thoday, D. (1950). On the water relations of plant cells. *Ann. Bot. (London)* [N.S.] **14,** 1–6.
99. Tomos, A. D., Steudle, E., Zimmermann, U., and Schulze, E.-D. (1981). Water relations of leaf epidermal cells of *Tradescantia virginiana. Plant Physiol.* **68,** 1135–1143.
100. Turner, N. C. (1981). Techniques and experimental approaches for the measurement of plant water status. *Plant Soil* **58,** 339–366.
101. Turner, N. C., and Jones, M. M. (1980). Turgor maintenance by osmotic adjustment: A review and evaluation. *In* "Adaptation of Plants to Water and High Temperature Stress" (N. C. Turner and P. J. Kramer, eds.), pp. 87–103. Wiley, New York.
102. Tyree, M. T. (1970). The symplast concept. A general theory of symplastic transport according to the thermodynamics of irreversible processes. *J. Theor. Biol.* **26,** 181–214.
103. Tyree, M. T. (1976). Negative turgor pressure in plant cells: fact or fallacy? *Can. J. Bot.* **54,** 2738–2746.
104. Tyree, M. T., and Dainty, J. (1973). The water relations of hemlock *(Tsuga canadensis).* II. Kinetics of water exchange between the symplast and apoplast. *Can. J. Bot.* **51,** 1481–1489.
105. Tyree, M. T., Dainty, J., and Benis, M. (1973). The water relations of hemlock *(Tsuga canadensis).* I. Some equilibrium water relations as measured by the pressure bomb technique. *Can. J. Bot.* **51,** 1471–1480.
106. Tyree, M. T., and Hammel, H. T. (1972). The measurement of the turgor pressure and the water relations of plants by the pressure bomb technique. *J. Exp. Bot.* **23,** 267–282.
107. Tyree, M. T., and Jarvis, P. G. (1982). Water in tissues and cells. *In* "Encyclopedia of Plant Physiology, New Series" (O. L. Lange, P. S. Nobel, C. B. Osmond, and H. Ziegler, eds.), Vol. 12B, pp. 35–77. Springer-Verlag, Berlin and New York.
108. Tyree, M. T., and Karamanos, A. J. (1980). Water stress as an ecological factor. *In* "Plants and Their Atmospheric Environment" (J. Grace, E. D. Ford, and P. G. Jarvis, eds.), pp. 237–261. Blackwell, Oxford.
109. Tyree, M. T., and Richter, H. (1982). Alternate methods of analysing water potential isotherms: Some cautions and clarifications. II. Curvilinearity and water potential isotherms. *Can. J. Bot.* **60,** 911–916.
110. Van Volkenburgh, E., and Cleland, R. E. (1979). Separation of cell enlargement and division in bean leaves. *Planta* **146,** 245–247.
111. Van Volkenburgh, E., and Cleland, R. E. (1980). Proton excretion and cell expansion in bean leaves. *Planta* **148,** 273–278.
112. Van Volkenburgh, E., and Cleland, R. E. (1981). Control of light-induced bean leaf expansion: Role of osmotic potential, wall yield stress and hydraulic conductivity. *Planta* **153,** 572–577.

112a. Van Volkenburgh, E., Hunt, S., and Davies, W. J. (1983). A simple instrument for measuring cell wall extensibility. *Ann. Bot. (London)* [N.S.] **51,** 669–672.
113. Vinters, H., Dainty, J., and Tyree, M. T. (1977). Cell wall elastic properties of *Chara corallina*. *Can. J. Bot.* **55,** 1933–1939.
114. Warren Wilson, J. (1967). The components of leaf water potential. 1. Osmotic and matric potentials. *Aust. J. Biol. Sci.* **20,** 329–347.
115. Weatherley, P. E. (1970). Some aspects of water relations. *Adv. Bot. Res.* **3,** 171–206.
116. Willison, J. H. M. (1976). An examination of the relationship between freeze-fractured plasmalemma and cell-wall microfibrils. *Protoplasma* **88,** 187–200.
117. Woolley, J. T. (1965). Radial exchange of labelled water in intact maize roots. *Plant Physiol.* **40,** 711–717.
118. Wyn Jones, R. G. (1980). An assessment of quaternary ammonium and related compounds as osmotic effectors in crop plants. *In* "Genetic Engineering of Osmoregulation: Impact on Plant Productivity for Food, Chemicals and Energy" (D. W. Rains, R. C. Valentine, and A. Hollaender, eds.), pp. 155–170. Plenum, New York.
119. Wyn Jones, R. G., Brady, C. J., and Speirs, J. (1979). Ionic and osmotic relations in plant cells. *In* "Recent Advances in the Biochemistry of Cereals" (D. L. Laidman and R. G. Wyn Jones, eds.), pp. 63–118. Academic Press, London.
120. Wyn Jones, R. G., and Gorham, J. (1983). Osmoregulation. *In* "Encyclopedia of Plant Physiology, New Series" (O. L. Lange, P. S. Nobel, C. B. Osmond, and H. Ziegler, eds.), Vol. 12C, pp. 35–58. Springer-Verlag, Berlin and New York.
121. Wyn Jones, R. G., and Storey, R. (1982). Betaines. *In* "The Physiology and Biochemistry of Drought Resistance in Plants" (L. G. Paleg and D. Aspinall, eds.), pp. 171–204. Academic Press, Sydney.
122. Yeo, A. R. (1983). Salinity resistance: Physiologies and prices. *Physiol. Plant.* **58,** 214–222.
123. Zeroni, M., and Hall, M. A. (1980). Molecular effects of hormone treatment on tissue. *In* "Encyclopedia of Plant Physiology, New Series" (J. Macmillan, ed.), Vol. 9, pp. 511–586. Springer-Verlag, Berlin and New York.
124. Zimmermann, U. (1977). Cell turgor pressure regulation and turgor pressure-mediated transport processes. *Symp. Soc. Exp. Biol.* **31,** 117–154.
125. Zimmermann, U. (1978). Physics of turgor and osmoregulation. *Annu. Rev. Plant Physiol.* **29,** 121–148.
126. Zimmermann, U., Hüsken, D., and Schulze, E.-D. (1980). Direct turgor pressure measurements in individual leaf cells of *Tradescantia virginiana*. *Planta* **149,** 445–453.
127. Zimmermann, U., Räde, H., and Steudle, E. (1969). Kontinuierliche Druckmessung in Pflanzenzellen. *Naturwissenschaften* **56,** 634.
128. Zimmermann, U., and Steudle, E. (1974). The pressure-dependence of the hydraulic conductivity, the membrane resistance and membrane potential during turgor pressure regulation in *Valonia utricularis*. *J. Membr. Biol.* **16,** 331–352.
129. Zimmermann, U., and Steudle, E. (1974). Hydraulic conductivity and volumetric elastic modulus in giant algal cells: Pressure and volume dependence. *In* "Membrane Transport in Plants" (U. Zimmermann and J. Dainty, eds.), pp. 64–71. Springer-Verlag, Berlin and New York.
130. Zimmermann, U., and Steudle, E. (1975). The hydraulic conductivity and volumetric elastic modulus of cells and isolated cell walls of *Nitella* and *Chara* spp. Pressure and volume effects. *Aust. J. Plant Physiol.* **2,** 1–13.
131. Zimmermann, U., and Steudle, E. (1978). Physical aspects of water relations of plant cells. *Adv. Bot. Res.* **6,** 45–117.

CHAPTER TWO

Transpiration and the Water Balance of Plants

W. J. DAVIES

I. Introduction

The distribution of vegetation over the earth is influenced more by the availability of water than by any other environmental factor except perhaps temperature. This is because the normal functioning of a plant depends fundamentally on its water status. The colonization of the land has meant that plants exposed to the desiccating influence of the air are constantly faced with the problem of maintaining an aqueous environment within their leaves. The first plants which colonized land were small and their cells could be adequately supplied with water. As plants of greater size evolved, they developed an efficient system allowing rapid movement of water from the absorbing surfaces to the transpiring surfaces, while at

Plant Physiology
A Treatise
Vol. IX: Water and Solutes in Plants

the same time maintaining some restriction of water loss. The successful land plant can maintain a favorable water status by continually balancing water loss and intake. This chapter is concerned with this balance and with the factors influencing the movement of water into, through, and out of the plant.

Plant water relationships were comprehensively reviewed by Kramer in Volume II of this treatise (166). Since 1959 the emphasis in water relations research has shifted somewhat and there is now a considerable interest in the *movement* of water through the soil, the plant, and into the air. The tissue water balance is after all a dynamic one, a continual balance of loss and gain.

Of all the advances made in the past 25 years, perhaps the most notable has been the improvement in the terminology of plant water relations (see Chapter 1 of this volume). As Kramer (168) points out, physiologists have been slow to accept the idea that water movement is controlled by gradients in chemical potential, and it was not until the early 1960s that the term *water potential* began to be accepted (306). Taylor and Slatyer (321) originally defined water potential ψ as $\Delta\mu_w$, the difference in chemical potential of water between the test system and the standard state (pure free water). This definition proved to be unpopular, for it was subsequently changed by dividing $\Delta\mu_w$, which has units of joules per mole, by the partial molal volume of water $\bar{V}_w$, which has units of cubic meter per mole, to give pressure units, i.e., joules per cubic meter. Although this approach has been used widely, it is unsatisfactory on theoretical grounds (236) since while the definition of equilibrium in an isothermal system uninfluenced by external forces is that μ_w is contant, $\bar{V}_w$ is not necessarily constant and so ψ may vary. In practice this variation may be small, but in very precise experiments, the use of water potential defined in mechanical rather than thermodynamic terms must cause some problems (e.g., 368).

With the acceptance of the thermodynamic concept that the availability of soil water is best represented by its free energy status, the concept of available water was drastically revised (301, 302), and our understanding of soil–plant water relations has improved greatly. Recent research has suggested that very small soil water deficits, of the order of 0.01 to 0.02 MPa, may markedly reduce plant growth (1). With the work of Monteith and Owen (221) and Richards and Ogata (273), who developed the thermocouple psychrometer, and that of Scholander and co-workers (281), who developed Dixon's (75) early ideas on the pressure chamber, convenient methods to measure plant and soil water potential have become available. Psychrometers, hygrometers, and osmometers are now used routinely to measure the solute potential of plant material (13, 17, 41), while turgor can now be measured directly by means of the pressure probe (111, 310, 369). It is now clear that in many situations turgor is a better indicator of plant water deficit than is plant water potential (see Section V,B). Routine measurement of leaf solute potential has highlighted the importance of solute accumulation in leaves as a means of lowering water potential while keeping turgor, the driving force of growth, at a high level (132, 290, 328).

Passioura (249) has discussed a second problem of thermodynamic treatment which can arise when water potential is split into its component parts—turgor, osmotic potential, and matric potential. This is that ψ_m, the matric potential, defined as the difference in ψ between the test system and its equilibrium dialyzate (at the same height, temperature, and pressure), is supposed to account for the influence of solids on water potential. If the pressure component is defined consistently, however, as the hydrostatic pressure in the liquid phase (no matter whether this be in the vacuole, the cytoplasm, or the apoplast), and if the counterions that accompany any charged solid are accounted for under the solute component, then the matric component is zero. This is the case not because solids have no influence on water potential, but because any such influence is accounted for by the pressure and osmotic components. Passioura (247) has also questioned the "erroneous pedantry" of the use of the terms osmotic potential and pressure potential rather than osmotic pressure and hydrostatic pressure and recommends avoidance of the word potential (but see also Chapter 1).

In the past 25 years much has been added to our understanding of the role of stomata in the control of gas exchange and the responses of stomata to the environment. This followed the publication by Gaastra (97) of a means of quantifying diffusion resistance by simple measurements. In addition, techniques for monitoring stomatal behavior have become more sophisticated. The use of resistances and driving forces to describe flow through the soil–plant–atmosphere continuum (108, 126), has been extended, and new, ideas on the pathways of water movements in roots and leaves are gaining acceptance (29, 335).

As a result of their many and varied effects throughout the plant, growth regulators can greatly modify the water relations of plants. Some of these influences are described in this chapter and by Mansfield (Chapter 3). It is noteworthy that many of the recent developments cited involve a reassessment and development of concepts which have existed for 60 years or more. Kramer's chapter (166) remains an authoritative account of much of the early work in the field.

Excellent reviews that relate to this chapter are found in contributions by Bradford and Hsiao (32), Cowan (52), Kozlowski (157–164), Kramer (167), Lange *et al.* (172), Meidner and Sheriff (207), Milburn (216), Passioura (249), Slatyer (303), Turner and Kramer (328a), Paleg and Aspinall (241), and Tyree and Jarvis (334), Weatherley (352).

II. Soil-Plant-Atmosphere Continuum

A. THE TRANSPORT EQUATION

A close similarity among equations describing transport of water, carbon dioxide, heat, momentum, and electric charge [see, for example, Jones (149)] has resulted in these being referred to as examples of the general transport equation

$$\text{Flux density} = \text{driving force} \times \text{proportionality constant} \quad (1)$$

It is often convenient to measure the driving force for transport (in the case of water this is the vapor concentration difference) at two positions in a system, and thus the transport equation is commonly applied in an integrated form. The flux density J_i of an entity i is given by Fick's first law of diffusion as

$$J_i = -D_i(\partial c_1/\partial x) \quad (2)$$

where J_i is the flux density of an entity i per unit area, $\partial c_i/\partial x$ is the connection gradient across a plane, and D_i is the diffusion coefficient. Integration of this equation between planes at x_1 and x_2, a distance of l apart, gives

$$J_i = D_i(C_i1 - C_i2)/l \quad (3)$$

where C_i1 and C_i2 are the concentration of i at x_1 and x_2 in the diffusion pathway.

In Eq. (3) the driving force is the concentration difference across the pathway, and therefore D_i/l is the proportionality constant, which in plant physiology is commonly called a conductance (designated g). As we shall see later, it is sometimes more convenient to convert conductance to its reciprocal, namely, a resistance (designated r), although this can be misleading in systems with one dominant resistance.

The analogies between several different transport processes are summarized in Table I, in which it is clear that the units for conductance depend on what is chosen for the driving force. The dominant practice in plant physiology has been to express mass and heat transfer resistances in units of seconds per meter. These arise if the flux is expressed as a mass flux density (e.g., kilograms per square meter second) and the driving force is a concentration difference (kilograms per cubic meter). It is becoming increasingly common, however, to express flux as a molar flux density (moles per square meter second). If concentrations are expressed as dimensionless partial pressures or mole fractions, conductance will have the same units as flux density (Table I). This can be confusing, but this definition of

TABLE I

ANALOGIES BETWEEN DIFFERENT MOLECULAR TRANSFER PROCESSES

General transport equation	Flux density	=(Apparent) driving force	×Conductance
Ohm's law	/(flux)	$=V$	$\times 1/R$
(electric charge)	(A)	(W A^{-1})	(A^2 W^{-1})
Fick's law	J_i	$=\Delta c_i$	$\times D_i/l\ (=g_i)$
(mass transfer)	(kg m^{-2} s^{-1})	(kg m^{-3})	(m s^{-1})
	J_i^m	$=\Delta x_i$	$\times PD_i/lRT\ (=g_i^m)$
	(mol m^{-2} s^{-1})	(dimensionless)	(mol m^{-2} s^{-1})
	J_i^m	$=(P/RT)\Delta x_i$	$\times D_i/l\ (=g_i)$
	(mol m^{-2} s^{-1})	(mol m^{-3})	(m s^{-1})
Fourier's law	C	$=\Delta T$	$\times k/l$
(heat conduction)	(J m^{-2} s^{-1})	(K)	(W m^{-2} K^{-1})
	C	$=\rho c_p\,\Delta T\ (=\Delta c_H)$	$\times D_H/l\ (=g_H)$
	(J m^{-2} s^{-1})	(J m^{-3})	(m s^{-1})
Poiseuille's law	J_v	$=\Delta P$	$\times r^2/8ln\ (=L_p)$
	(m s^{-1})	(kg m^{-1} s^{-2})	(m^2 s kg^{-1})

conductance has some advantages, particularly where the system is non-isothermal.

Mass flow of water can be described by the transport equation, and for flow in porous media and in capillaries, the appropriate driving force is the hydrostatic pressure gradient ($\partial P/\partial x$) and so

$$J_v = -L(\partial P/\partial x) \tag{3a}$$

where J_v is the volume flux density (cubic meters per square meter second) and L is a hydraulic conductivity coefficient (square meters per pascal second), a property of the material through which flow varies and corresponding to the diffusion coefficient in Eq. (2). When flow from cell to cell through semipermeable membranes is described, the driving force becomes total water potential. Equation (3a) is known as Darcy's law and can be used to describe flow in water-saturated soils. If the equation is applied in an integrated form, path length l is included in the equation to give

$$J_v = L\,\Delta P/l = L_p\,\Delta P \tag{3b}$$

where L_p is a hydraulic conductance (meters per second pascal), which for a uniform path is given by L/l (analogous to g and depending on path length).

For hydraulic flow through xylem conduits, the hydraulic conductance can be expressed in terms of tube radius and fluid viscocity (see Table I). Because of the complexity of the flow pathway, however, steady-state flow is often analyzed in terms of a simple resistance model linking flow in each segment of the pathway described by the general transport equation.

In the past 30 years particularly, the continuity of water movement through the soil, the plant, and the atmosphere has been emphasized. The Gradmann (108)–van den Honert (126) principle can be represented by the equation

$$F = \frac{\psi_s - \psi_r}{R_s} = \frac{\psi_r - \psi_l}{R_p} = \frac{\psi_l - \psi_a}{R_a} \tag{4}$$

where ψ_s, ψ_r, ψ_l, and ψ_a are the water potentials of the soil at some distance from the root, of the root surface, of the evaporating surfaces of the leaf, and of the bulk air, respectively, comprising the components of the driving forces for water movement. Here R_s, R_p, and R_a are the resistances of the soil, plant, and vapor pathways, respectively. Apart from emphasizing the links in water transport through the soil, plant air, continuum, Eq. (4) has been used to illustrate the plant's control over total water flux. For example, while the drop in water potential in the two liquid phases of the pathway is normally no greater than 3 MPa, the drop in water potential between the leaf and the air ($\psi_l - \psi_a$) can be 100 MPa or greater. It is clear that the stomata are situated in the segment of the soil–plant–air continuum where the largest drop in potential occurs and in the most advantageous position to control the flow of water through the system. However, caution is needed when attempting to extrapolate from Eq. (4). Flow through the soil, plant, and atmosphere may be considered as two coupled but separate systems: the uptake and redistribution of water from the soil and the loss of water vapor from the plant to the atmosphere. These are different types of transport (liquid and vapor) and the effective resistances to water movement differ considerably in magnitude. There is a natural discontinuity at the leaf surface because of the different characteristics of liquid and vapor transfer. The movement of liquid water in the plant, as well as in saturated soil, is approximately proportional to the difference in vapor concentration or any linearly related parameter such as vapor pressure. Since the relationship between water potential and vapor pressure is logarithmic [see, for example, Kramer (167)], it follows that vapor flux cannot be proportional to a water potential gradient and therefore that the vapor and liquid phase resistances are not comparable. A calculation using gradient of water potential as the driving force for vapor movement will lead to a substantial overestimation of the magnitude of the resistance of the vapor phase. Rawlins (269) demonstrated that vapor flow can be described in terms of water potential only if the resistance is made a function of vapor concentration. Such a manipulation is most inconvenient and is seldom used.

There is a second type of effective discontinuity at the leaf surface. Over the normal range of leaf water potentials, the vapor concentration at the surface is very nearly constant at the saturation value (207). This means that the vapor concentration gradient, and therefore transpiration rate, are not directly affected by changes in leaf water potential. This is equivalent to saying that the vapor phase dominates the flow. This situation changes when the liquid phase resistance begins to increase rapidly and the surface dries severely, a situation commonly referred to as incipient drying. Since the stomata generally close well before this point, it follows that leaf water status affects transpiration only if it changes the resistance to vapor diffusion through the stomata.

For most purposes, the van den Honert model, with a catena of resistances from soil to leaf, is too simple. It assumes a steady-state situation, something which seldom exists in the plant. In addition, Jarvis (142) has suggested that the tissues of plants with significant water storage capacity (see Section III,D) can be regarded as a number of alternative sources of water linked in parallel with each other and with the soil. As an alternative simplified model of the plant system, Powell and Thorpe (260) propose a network of water flux in the liquid phase (Fig. 1) with two storage elements, C_1 and C_2, representing bark and leaves connected to the main transpiration pathway via resistances R_{st_1} and R_{st_2}. The flux of water from a given storage element depends on the potential difference between it and the main pathway and the resistance to flow between the storage element and the pathway. The consequence of this is that at any instant, the water uptake from the soil is unlikely to be equal to water loss from the leaves. Variable resistances to water movement in the soil and in the plant will further complicate the van den Honert analysis, and it is best to regard this as useful in a limited number of situations but generally rather misleading.

B. HYDRAULIC LINKS BETWEEN ROOTS AND SHOOTS

In a living plant, continuous columns of water are maintained in the xylem conduits during growth and development of leaves and shoots. The plant can be thought of as a single hydraulic system, all functioning parts of which have a continuous liquid water connection between them.

Renner (272) pointed out over 70 years ago that the cohesion theory for the ascent of sap is the only hypothesis which explains how absorption of water and transpiration are effectively coupled together. Despite this, recent texts continue to devote considerable space to its defense. Of the essential features of the cohesion theory, the most contentious has been the necessity to demonstrate high cohesive forces between water mole-

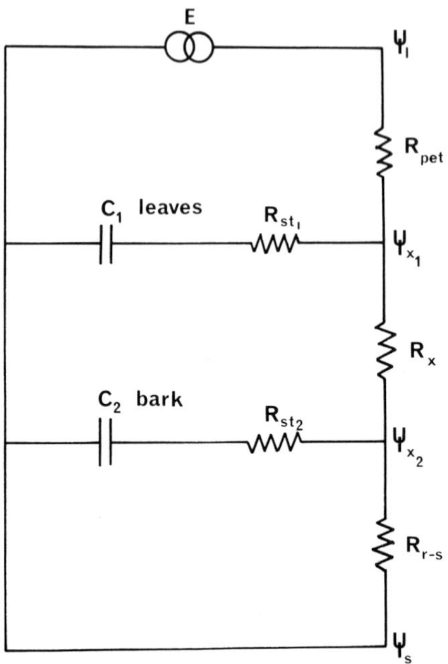

FIG. 1. Network model for the pathways of water movement through a tree. The current generator E represents the transpiration rate; ψ_s soil water potential; ψ_{x_1} and ψ_{x_2} xylem water potentials; ψ_l leaf water potential; R_{r-s}, R_x, and R_{pet} resistances to flow in root–soil, xylem, and petiole, respectively. C_1 and C_2 are storage elements in the leaves and bark, and R_{st_1} and R_{st_2} are resistances to flow between the storage elements and the main pathway of water flow. After Powell and Thorpe (260).

cules. Such forces must occur if water columns are not to break under tension. That the water columns have the necessary tensile strength variously estimated at 10 to 30 MPa was demonstrated first by Askensay (9) and Dixon and Joly (76) and may easily be demonstrated using the pressure chamber apparatus.

Raschke (265, 266) showed that stomata of detached maize leaves responded almost instantaneously to a change in the hydrostatic pressure of the water supply to the leaf base. This is a simple demonstration of the hydraulic connections throughout the leaf. Almost immediate resumption in leaf elongation by unwatered maize *(Zea mays)* plants following rewatering has been demonstrated by Hsiao *et al.* (131) and Acevedo *et al.* (1). Such a response can only be explained by transmission of the change in water potential of the root medium, via a change in xylem tension, to the leaf cells, in which an increase in hydrostatic pressure causes a resumption in leaf expansion. The speed of response to such a pressure change will depend upon the distance from the site of the original change in pressure, friction in the water path, and the elasticity of the tissues, but responses are generally recorded within seconds.

III. Water Uptake and Movement through Plants

In our consideration of water movement through plants we will start with a consideration of the pathway or pathways taken. Many details of the pathway are well established, but there are several areas of uncertainty, in particular the actual route taken through nonvascular tissues.

A. THE PATHWAY OF WATER MOVEMENT THROUGH PLANTS

1. Movement across the Root

Water taken up by the root moves radially inward across the root until it reaches the xylem, often a distance of less than 1 mm. There may be considerable resistance to this movement despite the short distances involved. To reach the xylem in a young root, water must cross the epidermis, the cortex, the endodermis, and the pericycle. Possible pathways for water movement are shown in Fig. 2. If water follows pathway 1, it will move from vacuole to vacuole down a gradient of water potential. When entering each cell, water must cross the paradermal cell wall, the plasma-

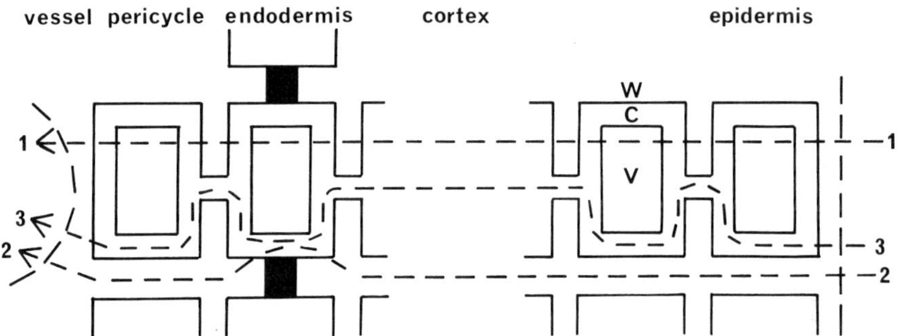

FIG. 2. Diagram of part of transverse section of root showing possible pathways for water movement. C, cytoplasm; V, vacuole; W, wall. After Newman (231).

lemma, the cytoplasm, and the tonoplast and traverse the same structures again on leaving. In this pathway the membranes of the cells are assumed to provide the main resistance to movement.

Pathway 2 is the free space – endodermis pathway. Here water moves in the cell walls in response to a hydrostatic pressure gradient. At the endodermal cylinder the cell wall pathway is interrupted by the Casparian strips (137) and the water must enter endodermal cells and move through the cytoplasm or the vacuole or both. In this pathway the membranes of the endodermal cells would provide the main resistance to movement. Once past the endodermis, water can again move in the cell wall system of the stele.

Pathway 3 is the symplastic pathway. Here water enters the cytoplasm through the plasmalemmae of the epidermal and cortical cells and moves via the plasmodesmata through the whole cortex and stele. Inside the endodermis, water will leave the symplasm and enter the xylem conduits either directly or via the stelar free space. If the cell walls are pathways highly resistant to water movement, most water will enter the symplasm in the epidermis and leave near the xylem conduits. If cell walls have a lower resistance to water movement, water may enter the symplasm anywhere across the cortex and leave anywhere inside the endodermal cylinder. Resistances to movement will be provided by membranes and by plasmodesmata.

With the three possible pathways arranged in parallel, most of the water will flow along the path of least resistance. When the roots of an intact plant are immersed in a solution of a dye or a substance that can be viewed with the electron microscope, root sections reveal the presence of the tracer in all the walls of the epidermal and cortical cells. This is strong evidence that exterior to the endodermis water penetrates the cell walls, but this is not necessarily proof of major water flow since tracers can move in this pathway by diffusion. Certainly the symplast and membranes are also, to some degree, permeable to water. One approach to determining the path of major flow of water is to calculate the resistances of the alternate pathways by using estimates of their dimensions and permeabilities of their component parts (34, 103, 154). As Newman (231) points out, the disagreement over the results of these calculations may be attributed to an uncertainty over what to take as realistic values for the permeabilities of membranes, wall material, and plasmodesmata. By choosing suitable values for permeability, it is possible to make any of the pathways in Fig. 2 the one of lowest resistance. Permeabilities of whole root systems can, however, be measured reliably (224). Taking each pathway in turn, Newman (231) has estimated permeabilities of membranes, cell walls, and plasmodesmata which would be required to compile a realistic value for total permeability of a root of known dimensions. The required values may then be compared with the measured values and the likelihood of any one pathway being the major pathway for water flow can be assessed.

Using this technique, Newman has concluded that the required membrane permeability for the vacuolar pathway to function is higher than any value reported for vascular plants by a factor of at least 10. It is therefore unlikely that the vacuoles comprise a major pathway for water movement. According to Newman's calculations, if the free space–endodermis pathway is to function effectively, the permeability of the wall material must also be 10 times higher than any value reported in the literature. This conclusion must be somewhat doubtful, since the permeability to flow through, rather than across, the walls of parenchyma cells is difficult to measure. Newman's conclusions on the free space–endodermis pathway rest on the assumption that the apoplastic resistance is a minor part of the total resistance. His argument that this must be the case when total root resistance is sensitive to the environment is not convincing. In addition, many of the studies describing metabolic sensitivity have been performed at low water flow rates, where ion uptake may well be influenced. Uptake of salts at low flow rates can have major effects on the apparent resistance and is discussed later (see Section III,C).

Newman's calculations for the symplastic pathway provide root permeabilities which agree well with published values. Overall, his conclusions favor this pathway but obviously rest on the assumption that the permeability values reported in the literature are realistic. Ginsburg and Ginzburg (103) have also suggested that the symplastic pathway may be of importance. Other authors (230, 309, 331, 332, 350) have discounted the importance of the vacuolar pathway but like Newman (231) have not been able to decide unequivocally between the free space–endodermis and symplastic pathway. The actual pathway must be a combination of all three possibilities, with the quantitative importance of each varying with species, root age, and environmental conditions. Tyree and co-workers (335) have considered the possibility that a significant proportion of the water moving through the leaf to the evaporation sites may move through the symplast. Clearly these considerations (discussed further in Section III,A,3) are relevant here.

Water flow through the many intercellular spaces in the cortex must also be taken into account (81). Passioura (248) notes that if there are water films partly occupying these, then the permeability of the cortex could be increased appreciably. The intercellular spaces of the cortex of wheat and maize are 3–10 μm in diameter and capillary theory predicts that at a matric potential of -0.1 MPa, hydrophilic pores of a chamber up to 3 μm in diameter will be filled with water. Many of the grooves between cells may contain water even if the whole of the intercellular space is not filled (248). Intercellular spaces are arranged predominantly in columns (235) and radial connections may be rare. Passioura makes several assumptions about the frequency and size of such connections and calculates a permeability 100 times larger than the value calculated by Newman (231). Even if this calculation is based on inaccurate assumptions, the point is made that flow along cell walls may be very much easier than flow through them.

A plant root system consists of roots in various stages of differentiation. Fully mature roots may be enclosed in a layer of suberized tissue, having lost their surface layer and cortex. Variations in root structure obviously suggest wide variations in permeability which in fact can change markedly within a few centimeters of the root tip (4, 35). It is interesting to note that the effect of root age on root permeability differs depending on the rate of water uptake by the plant (35). Generally, as roots get older their permeability declines. There are large differences among species (Table II) in root permeability, which can presumably be explained in part by differences in root structure and possibly in root age. Up to now we have considered the role of young, unsuberized roots in absorption. Considerable absorption of water must occur through older suberized roots (167), especially in woody perennials which possess only a comparatively small percentage of unsuberized roots (169). Presumably, water enters suberized roots via lenticels and wounds. Graham *et al.* (109) have shown substantial water uptake by roots which possess a suberized endodermis, while Clarkson *et al.* (44) have suggested that water may move through a suberized endodermis via plasmodesmata, which can be numerous (274). In other plants, water movement through the suberized endodermis may occur via passage

TABLE II

PERMEABILITY PER UNIT SURFACE AREA OF WHOLE ROOT SYSTEMS OF FIVE SPECIES GROWN UNDER THE SAME CONDITIONS[a]

Plant	Permeability ($nm\ s^{-1}\ MPa^{-1}$)
Broad bean	5.4
Dwarf bean	5.6
Sunflower	7.1
Maize	2.2
Tomato	61.0

[a] After Newman (229).

cells (36). Having crossed the endodermis, water enters the xylem and may then move rapidly throughout the plant.

2. *Longitudinal Water Movement through the Plant*

a. Structure and Hydraulic Conductivity. The longitudinal path of water movement through the plant may be easily followed by the injection of a dye or a radioactive isotope (165). The bulk of the water naturally follows the path of least resistance and moves by mass flow along gradients of hydrostatic pressure, mainly in the xylem conduits. The xylem forms a continuous system running from near the tips of the roots, through the stem, and into the leaves. Profuse branching of xylem in the leaves means that in some plants vein endings may only be 50 μm, or one or two cells, apart (364).

The xylem cells principally involved in the conduction of water in angiosperms are the vessels and the tracheids. Tracheids may be only a few millimeters in length (262), while single vessels, composed of many vessel elements, can have minimum lengths of the order of centimeters or even meters (117), so that water moving in vessels need only pass through walls infrequently. Flow of water in the lumen of conduits is laminar (342) and obeys the Hagen–Poiseuille law (74). Thus, in ideal capillaries

$$\text{Flow rate of sap} = (P_2 - P_1)(\pi r^4/8L\eta) \tag{5}$$

where η is the viscosity of the sap (approximately that of water), $(P_2 - P_1)$ is the pressure gradient along the capillary, L is the length of the pathway, and r is the radius of the capillary. It is apparent from this equation that differences in the radii of conduits will have a pronounced effect on flow rate. For example, if the radius is doubled, the flow rate will increase 16-fold. Clearly a much greater quantity of water will flow through wide vessels than through narrow vessels and the much narrower tracheids (see, for example, 366b).

From the Hagen–Poiseuille equation, frictional resistances to flow can be calculated. Xylem elements do not behave as ideal capillaries. However, Dimond (74), working with tomato, obtained reasonable agreement between determinations of conductivity of stem segments and calculations based on studies of vascular anatomy. Dimond's data suggest that in the vessels of tomato the main resistance to flow is in the lumen and that the vessel end walls offer little resistance. These results would, however, seem to be exceptional. The general finding is that due to end walls, wall roughness, and cross-wall remnants, the measured resistance is at least twice the calculated resistance. It is possible that in tomato there are fewer end walls and cross-wall remnants than in other plants.

The development of xylem in woody stems varies greatly from species to species. In ring-porous trees (some oaks, elms, etc.) water conduction takes place primarily through the large early-wood vessels of the current year (134). Larger vessels in older wood are mostly gas filled. Early-wood vessels are produced in the spring as leaves expand, and thus some trees can produce an efficient water-conducting system as it is needed. These plants can, therefore, afford to lose continuity in the water columns of the previous year's wood. The outermost annual ring might be expected to function as the chief path for sap flow because it is most directly connected to the new leaves and there has been less time for the conducting elements to become blocked. In conifers and diffuse porous trees, however, more than one growth ring functions in conduction (165, 316).

In any study of water flow through plants the vascular structure of the stem and the phyllotaxy must be considered. Fiscus *et al.* (94) have shown that when modeling water flow, the plant stem cannot be treated as a hollow pipe in which all leaves have equal access to the contents of any given level. In tobacco, the transvascular resistance is significant when compared to the axial resistance; i.e., the total resistance to flow is not simply a function of distance between any two leaves under consideration.

Plant diseases can, in some circumstances, produce long-term irreversible increases in resistance. Pits will block with fungal or bacterial toxins or other large molecules (275, 337) and resistance to flow may also be increased by gums and tyloses produced by the plant (10, 319). In trees which rely on the latest growth ring for conduction, an attack of a disease such as Dutch Elm disease or Chestnut blight can lead to the blockage of the whole of the water-conducting system, with very serious consequences.

b. Cavitation, Collapse of Xylem Conduits, and Large Axial Resistances. Data on xylem permeabilities in stem sections are collected under conditions where all vessels and tracheids are filled with water. In the intact plant this is not always the case. As transpiration increases, leaf water potential falls, causing an increased tension in the xylem. Although water columns can withstand tensions of at least 30 MPa, there is evidence that in stems and petioles air bubbles will form when xylem water is under tension (115). These may act as flaws in the water column. When water columns are disrupted, the process is referred to as cavitation. When cavitation occurs, the ensuing shock wave produces vibrations, which, with the aid of sensitive transducers, may be heard as clicks (213–215). A vascular strand with a newly broken water column probably contains mainly water vapor at a pressure determined by the water potential and temperature of the tissue (261). Later, air may come out of solution and fill the conduit.

Water columns may be broken by jarring of the tissue, by mechanical injury, or by winter freezing. Although the shock wave of cavitation may cause cavitation in adjacent conduits, under normal circumstances an air embolism will remain confined to the tracheid or vessel in which it has occurred. In vessels, the spread of an air embolism is limited by end walls, while in conifers, where pit perforations are large, aspiration (Fig. 3) of bordered pits will act to restrict the spread of an embolism (367). When a pit aspirates, the thickened center of the pit membrane (the torus) is pulled sideways against the pit borders, effectively blocking the pit perforation. The pit membrane and torus can therefore act as a valve.

Byrne *et al.* (39) have reported that the larger xylem elements in a section of lateral root from a stressed cotton plant, viewed under reflected light, appeared to be empty of water. Their experiments suggested that root resistance to water flow increased rapidly when water potential fell below −2 MPa, and they concluded that cavitation in the root xylem may be partly or wholly responsible for this phenomenon. It is unlikely, however, that all the xylem conduits in a thickened root or stem would cavitate at the same time, and even if the large xylem elements were blocked, it seems probable that flow would increase in the unblocked vessels. Lateral transport from conduit to conduit means that cavitated areas in the xylem

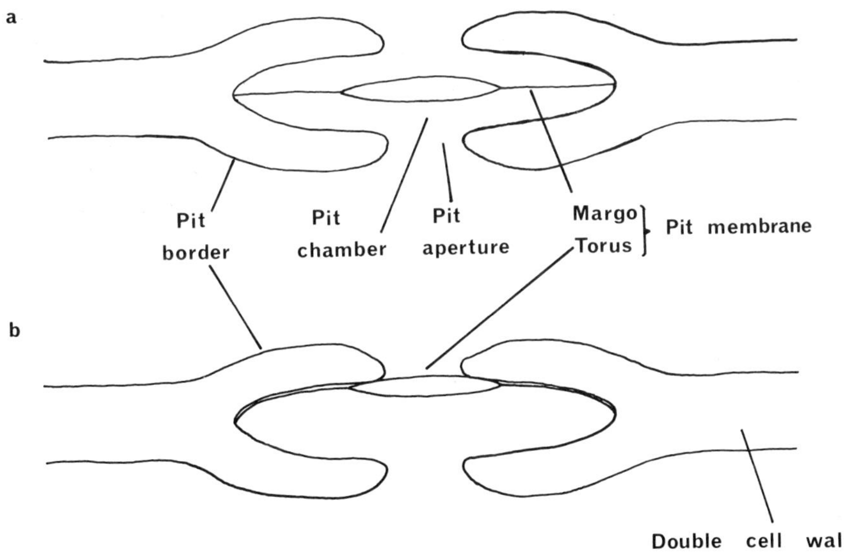

FIG. 3. Diagrammatic representation of (a) open and (b) aspirated, bordered pit of a conifer.

may be bypassed, and although a measurable increase in resistance may result from cavitation, a large increase in water deficit need not occur. Cavitation is discussed at length by Zimmermann (366b).

A simple experiment in which overlapping saw cuts are made in the trunk of a tree (257), while it does not indicate the quantity or rate of water movement, demonstrates some lateral transport of water in the stem and suggests that many plants possess an excess of vascular tissue which can cope with such an emergency. Passioura (248) has noted, however, that particularly in the roots of monocotyledons there may not be an excess of vascular tissue and a high axial resistance may have a large effect on leaf water relationships (128). For example, wheat plants without nodal roots may rely on three seminal axes per plant. Each axis typically contains only one large xylem vessel. Passioura suggests that the tissues of the stele of unhardened plants may not be able to withstand the large pressure gradients which form in the xylem. Transections of the axes of severely water-stressed wheat plants (248) show that the main xylem vessels may collapse and therefore presumably offer such a resistance to axial flow that the plant always exhibits water deficit, even after rewatering.

Passioura (248) has also described a possible water transport problem in both *Sorghum* and blue grama *(Bouteloua gracilis)* seedlings. In both species, drought may prevent the formation of nodal roots and thus water must move to the leaves through single seminal roots which may provide a huge axial resistance and prevent adequate water uptake.

c. Reformation of Broken Water Columns. In some plants, disrupted water columns are able to reform following an increase in the water content of the stem. Refilling will occur partly as a result of the partial vacuum in the empty conduit pulling the two parts of the broken water column together. In conifers, such a process may cause aspirated pits to reopen, so that cavitated conduits can refill from adjacent noncavitated conduits. Zimmermann and Brown (367) have suggested that root pressure (see Section III,B,1) could cause a refilling of empty conduits. Interestingly, diffuse porous species such as birches produce considerable root pressure in early spring, which may explain why conduits in several growth rings conduct water in a given season. Presumably, if some water can be absorbed by leaves during wet or humid periods (313), a more favorable leaf water balance may enable some conduits to be refilled from the top of the plant (216).

3. Water Movement through the Leaf to the Evaporating Sites

The observation that most leaves survive the cutting of major veins suggests that they have excess conductive capacity (253). The larger veins are important as structural elements and as water mains through which the smaller veins are supplied. The lowest resistant pathway for water movement in leaves is probably through the vascular tissue, with even the minor veins offering only a relatively small resistance to water movement (28, 208). However, Tyree *et al.* (335) have noted that contrary to early indications (27, 29) the xylem resistance in sunflower leaves may be substantial and may account for at least 50% of the total leaf resistance to water movement. There is some uncertainty about this conclusion since xylem resistances in leaves are often determined by measuring the resistance to water flow through the veins via the cut margins of leaves. This technique may be of doubtful accuracy, perhaps overestimating xylem resistance as a result of plugging of veins by cellular debris (335). Using this technique, values of R_{xylem}/R_{leaf} for sunflower leaves are of varying magnitude but certainly not consistently less than 0.1, which would be necessary before it would be safe to say that xylem resistance is negligible (335).

Regardless of the resistance of the xylem, water must move through several mesophyll, and possibly several epidermal, cells (364) before reaching the evaporating sites.

Where Are the Evaporating Sites in the Leaf? Although direct measurements of the hydraulic conductivity of cell wall material are variable and few and far between, the traditional view is that after leaving the vascular tissue, water moving through the leaf travels almost exclusively through the cell walls (27, 57, 252, 348). In an analysis of these conclusions, Tyree *et al.* (335) modeled water movement through leaves and compared the results with data obtained from rehydrating sunflower leaves. Models applied were (a) a series model in which it is proposed that water moved serially through the cells by a transcellular pathway (cell wall resistance larger than the membrane resistance) and (b) a parallel model in which it is proposed that water travels through the tissue via the low resistance of the cell wall pathway and enters the cells through the relatively high resistance of the cell membranes. Both models fit the data equally well, and the authors were unable to draw firm conclusions concerning the pathway of water movement in mesophyll and epidermal tissues.

As we have seen, Newman (231) concluded that the cell walls in the root provide too high a resistance to be a major pathway for water movement. Interestingly, Boyer (29) has reached a similar conclusion for water movement through leaf cell walls but rather than suggesting that water moves through the symplast proposes a low-resistance pathway for water movement away from the xylem by liquid flow for only a short distance from the vein before evaporation takes place. Boyer bases his hypothesis on an experiment comparing the kinetics of water loss from single mesophyll cells and from an intact leaf in a pressure bomb. The half-time for water loss from the whole leaf is substantially longer than that for water loss from single cells. He argues that if cell walls had a low resistance to water movement, water should flow reasonably freely to all the cells in the leaf and they should hydrate virtually in unison, with kinetics similar to the kinetics of single cells. Boyer's single cells were obtained from homogenized tissue and the differences in hydration times might be explained if the homogenization treatment increased the permeability of the cell membranes. Nevertheless, Boyer's hypothesis is supported by evidence that the exposed walls of the leaf mesophyll cells, the guard cells, and the epidermis in the vicinity of the stomata are covered with a nonwettable, hydrophobic lipid layer (96, 179, 234, 287, 288) (Fig. 4). Meidner (205) has pointed out, however, that these mesophyll cell walls can be wetted from within (i.e., via other mesophyll cells), although they are hydrophobic with respect to water placed on their surface. This situation applies even if a cuticle is present on the cell wall. Classically, the mesophyll cell walls have been considered to be the source of most transpirational water loss, with water escaping from the walls to the intercellular spaces and then through the stomatal pores into the atmosphere. Recent work has suggested that the importance of the epidermal cells as a source of transpirational water may have been underestimated and provides good evidence that the cells adjacent to the guard cells and the guard cells themselves can lose appreciable quantities of water.

Meidner (202) notes that the classical view of leaf water loss does not take into account several observations concerning the path of water movement in the leaf. Using fluorescent dyes (315), tritiated water (185), gold sols (98), and lead chelates (320) added to the transpiration stream, several workers have proposed that the epidermal cell walls are an important path for water movement through the leaf to the evaporating sites. Considering this and other evidence, Meidner (202) has proposed that the most active evaporation per unit area of cell wall may occur in the vicinity of stomatal pores. He suggests that the paths of least resistance for diffusion of water vapor to the stomatal pore are not from the mesophyll cells walls, but from internal epidermal cell walls adjacent to the stomatal pores and from the guard cells themselves. Figure 4 emphasizes the prominent role of the guard cell and subsidiary cell walls in forming the boundary to the substomatal cavity in *Lolium*. An important series of papers by Maier-Maercker (186–191) has also emphasized the importance of evaporation sites on the epidermis of a number of species. Recent evidence (6, 7, 144) suggests that a considerable amount of water may be lost from the outer surfaces of the guard cells even when stomatal pores are closed, and this may have important implications for maintenance of guard cell turgor and

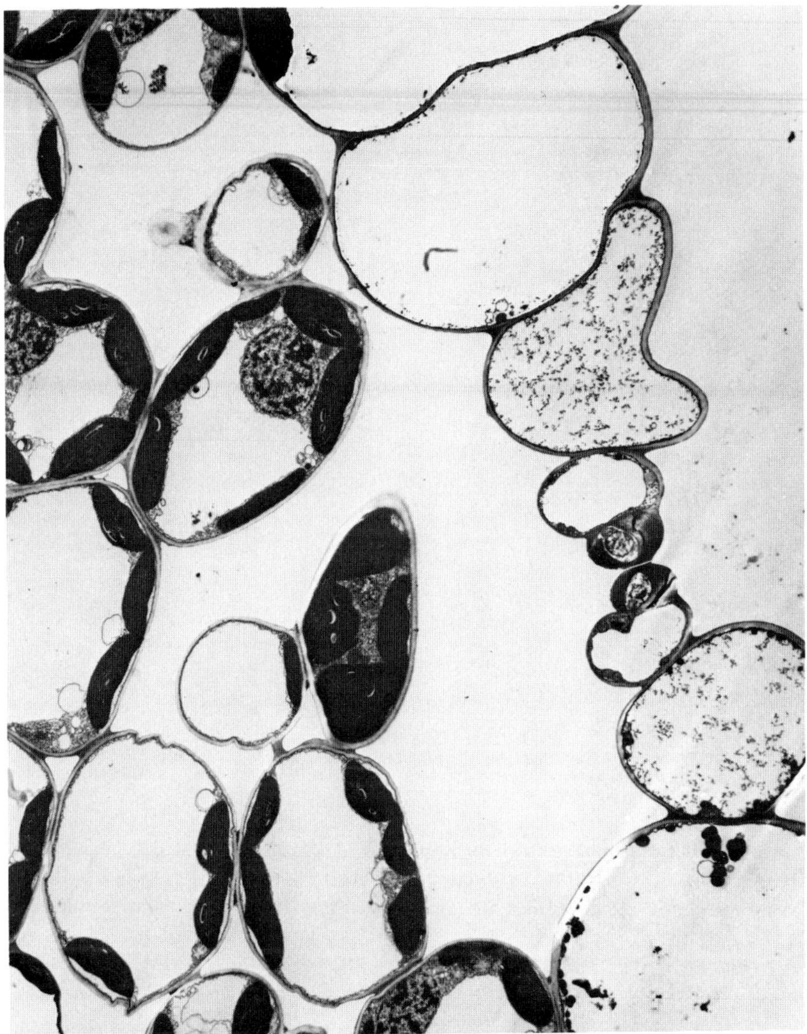

FIG. 4. Cross section through guard cells of *Lolium perenne* L. (×12,000). D. E. Molyneaux and W. J. Davies (unpublished).

for stomatal humidity responses (see Section IV,B and Mansfield, Chapter 3 in this volume) (Fig. 5).

Support for the hypothesis that the epidermis is a major route for water movement comes from an observation that the inner epidermal walls alone can sustain a reasonable evaporation rate into a moist atmosphere (202). In

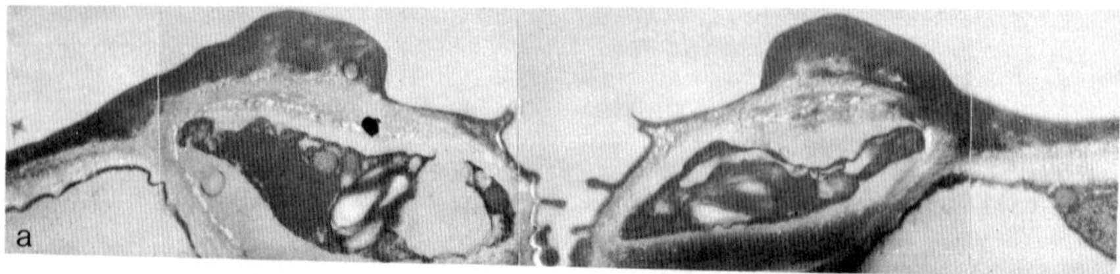

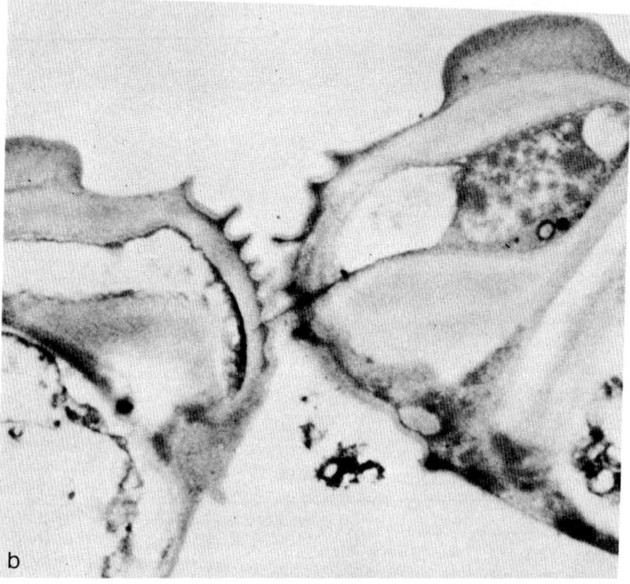

FIG. 5. (a) Transmission electron micrograph of a cross section of a *Quercus* leaf showing cuticular development on guard cells and epidermal cells (×5045). (b) Transmission electron micrograph of a cross section of a *Quercus* leaf showing lanthanum staining of the ventral guard cell walls (×6560). After Appleby and Davies (7).

these experiments, the epidermal strip was prevented from drying completely by placing its end in a small well containing water. A similar experiment (203) showed that rates of vapor loss through open stomata of cuticle-covered epidermal tissue with the mesophyll removed compared in magnitude with rates of water loss from intact leaves. Working with a scale model of a substomatal cavity, Meidner (204) showed that the rates of evaporation inside the cavity were greater from areas near the pore than from areas some distance away. In many plants, however, the internal exposed surface of the mesophyll is 10 to 15 times greater than the epidermal surface. The amount of epidermal surface exposed to air is further reduced by the mesophyll cells which are in contact with it. Therefore, although the rate of water loss per unit area of epidermis may be high and can be of considerable physiological significance, the total water loss from the exposed mesophyll cells will be substantial.

Sheriff (296) has suggested that under certain conditions, water may be distilled from the mesophyll cell surfaces to the internal surfaces of the epidermal cells. The reason for this is the relatively high saturation vapor pressure of the water in the mesophyll cell walls caused by the heating of these cells by absorption of radiant energy. The epidermal cells, which are relatively colorless, can lose heat by convection and by radiation. Under certain conditions it seems possible that the inner walls of the epidermis may lose a negligible amount of water or even show a net gain in water.

B. THE DRIVING FORCE FOR WATER MOVEMENT

1. Free Energy Flow

The volume flow J_v between two regions in a hydraulic system is described by the following simple extension of Eq. (3b)

$$J_v = L_p(\Delta P + \sigma\,\Delta\pi + \Delta\tau) \tag{6}$$

where L_p is the hydraulic conductivity and σ the reflection coefficient of the channel through which the flow takes place. The components of the water potential gradient between the two regions are ΔP, $\Delta\pi$, and $\Delta\tau$, the differences in pressure and osmotic and matric potential, respectively (see Dale and Sutcliffe, Chapter 1 in this volume). In a path which does not discriminate between components of a solution, e.g., the xylem, the reflection coefficient is zero. In addition, matric potentials can be ignored in the xylem, in vacuoles, and in saturated cell walls and so the driving force in such a situation is the pressure or hydrostatic potential.

If the matric potential gradient is negligible, flow through cell membranes, e.g., radial flow across roots, will be influenced by osmotic and hydrostatic forces. If the pathway is perfectly semipermeable ($\sigma = 1$) (103), Eq. (6) becomes

$$J_v = L_p(\Delta P + \Delta\pi) \tag{7}$$

It is often considered that a distinction between osmotic- and pressure-induced flow is not particularly useful since the nature of water movement is similar whether induced by a hydrostatic pressure gradient or by solutes. The driving force for movement is the difference in water potential between the root surface and the xylem conduits, and if the root is behaving like a perfect osmometer, it makes no difference whether the water potential in the xylem is lowered by the hydrostatic pressure gradient or by the accumulation of solutes. It has been suggested, however, that interactions between osmotic- and pressure-induced water flow in plant roots may explain the apparent changes in root resistance which take place under the influence of a hydrostatic pressure gradient (89, 91, 93) (see Section III,C). If this is the case, it is not sufficient to know only the magnitude of the water potential gradient. Values for ΔP and $\Delta \pi$ are also required. In the case of a plant that is transpiring very slowly or in a detopped plant, ΔP is small, while $\Delta \pi$ may be large as a result of solute accumulation in the xylem. Under these conditions osmotically induced water flow will occur and the plant will develop root pressure. When transpiration is rapid, ΔP is large, and under these conditions, the uptake of water may be so great that the solute concentration in the xylem conduits is reduced even below that in the soil solution.

2. *Variation in Leaf Water Potential*

The driving force for water movement against the frictional resistances in the pathway is derived from the evaporation of water from the leaf. As a result of this loss of water, the water potential of the leaf falls; in other words, the water deficit of the leaf increases. Water deficit is the inevitable consequence of the flow of water along a pathway in which frictional resistance and gravitational potential have to be overcome. It is often stated that water deficits arise when transpiration exceeds absorption. This is, of course, true, but this discrepancy can only occur because of the contribution of the water content of the tissues of the plant to the transpiration stream. If the pathway were completely rigid and the storage capacity of the plant could contribute no water to the transpiration stream, a water potential gradient would still be necessary to drive flow against frictional resistance and gravitational potential even if the absorption and transpiration rates were equal.

The causes of water deficit in a transpiring plant can be understood with reference to Fig. 6 (349). Two porous pots represent the leaf and root, respectively. They are filled with water and are connected by a tube of low resistance representing the transpiration pathway. The manometer represents the cells adjacent to the pathway and is separated from it by a resistance to flow. In the steady state the rate of water uptake will equal the rate of water loss, while the manometer in the side arm of the apparatus will register the reduction in pressure arising in the pathway, an indication of the water deficit or reduction in water potential within the cells. If transpiration increases from the steady rate, the manometer will register an abstraction of water from the reservoir into the transpiration stream such that

$$F = f_a + f_c \tag{8}$$

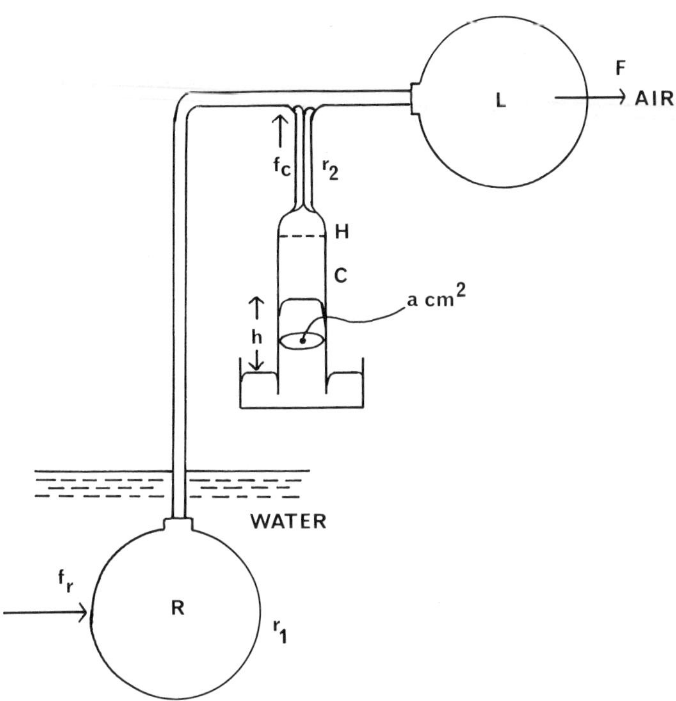

FIG. 6. Model of the transpiring plant illustrating the relationship between the cells and the pathway of water movement through the cell walls. After Weatherley (349).

where F is the rate of transpiration, f_a the rate of absorption through the roots, and f_c the adjustment in the water content of the cells. This last term will decline to zero when the adjustment ceases and the rate of transpiration will again equal the rate of absorption. The water deficit in the plant will have increased. The size of this deficit will depend on the moisture characteristics of the cells (the relationship between water content and water potential; see Section III,D), the resistances in the flow pathway and the rate of flow of water through the plant.

In a tall plant the leaf water potential will be reduced in relation to its height above the point of water uptake (0.01 MPa m^{-1}). Clearly the value of soil water potential ψ_{soil} will influence the leaf water potential whether the plant is transpiring or not. Finally, the reduction of water potential will be dependent on the flow of water F through the resistances of the plant r_p. This flow obeys a typical transport-type law [Eq. (1)] so that $\Delta\psi = Fr_p$. Thus, leaf water potential ψ_{leaf} is determined as

$$\psi_{leaf} = \psi_{soil} - Fr_p - 0.1h \quad (9)$$

where h is the height of the leaf in meters. With a small plant, $0.1h$ is negligible and

$$\psi_{\text{leaf}} = \psi_{\text{soil}} - Fr_{\text{p}} \tag{9a}$$

In some herbaceous and woody plants, leaf water potential is a linear function of transpiration rate (Fig. 7). From Eq. (9a) the slope of line A in Fig. 7 gives the effective total plant resistance to water transfer and the intercept on the ordinate gives the source water potential. Relationships between leaf water potential and transpiration rate, however, are not always linear. Hysteresis is sometimes evident in the relationship because leaf water potential shows a tendency to fall to a minimum later in the day than when the transpiration peak is reached. Jarvis (143) has suggested that this may result from an increase in resistance to flow through a plant during the course of a day, possibly as a result of cavitation in the xylem. In addition, the hysteresis may be caused by a change in the sources of available stored water in the plant. Following an increase in transpiration, it is expected that water will move to the evaporating sites in the leaf from sources in the plant and the soil. Initially, most water will move from sources where water is most readily available, but late in the day most water transpired must either be drawn from soil or drawn progressively from sites within the plant at which water is less readily available. Consequently, later in the day, a particular transpiration rate will be likely to result in a lower water potential.

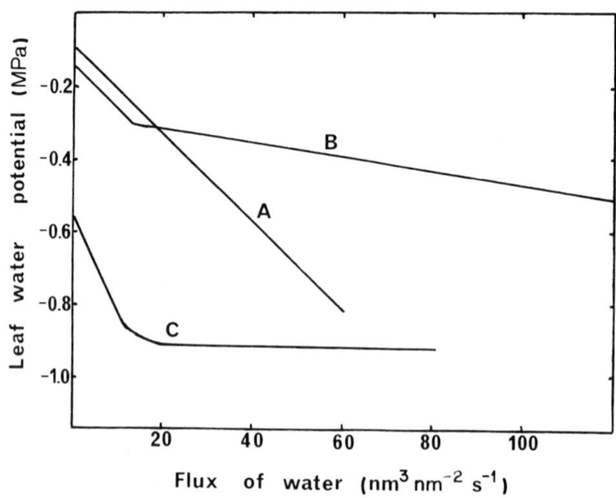

Fig. 7. Variation of leaf water potential due to variation of rate of evaporation. A, after Barrs (15); B, after Boyer (27); C, after Stoker and Weatherley (312).

Several workers have cited instances in which the resistance to flow through the plant is apparently dependent upon the rate of flow, and thus the force–flux relationship is not linear (Fig. 7) (9, 27, 312). Many workers have found that for some plants leaf water potential is essentially constant over a wide range of transpiration rates. Workers in different laboratories have obtained different force–flux relationships, in some cases for the same species. Several hypotheses to explain apparent changes in resistance with changes in flux are discussed in Section III,C.

Jarvis (142) has noted that despite variation in plant size, leaf water potentials in the majority of well-watered plants fall within the range of −0.5 to −2.0 MPa. It might be expected that in larger plants, in which large volumes of water are transported great distances, greater water potential differences would be necessary to drive water movement, but this seems not to be the case.

When leaf turgor falls to a critical level, stomatal closure will result in an increase in plant turgor and leaf water potential. If this were the only control of leaf water potential, stomata of large plants would be closed for long periods of time, but this is not the case. Reductions in plant resistance with increasing flow (Fig. 7) (351) and relative constancy of Fr_p [Eq. (9a)] (142) will help to keep leaf water potential within narrow limits during the course of a day, without the necessity of stomatal closure. Linear correlations between foliage mass and cross-sectional area of sapwood have been found in a large number of plants (e.g., 114). Since the cross-sectional area of the conducting tissue must increase in proportion to the flow if the leaf water potential is not to decrease, such correlations must result in relative constancy of Fr_p [Eq. (9a)].

Leaf water potential may vary rhythmically with the period of the rhythm ranging from minutes to hours. Such oscillations in leaf water balance often result from cycling in stomatal aperture (14) (see Section IV,B,3). Stomatal oscillations, which often have a period of 10 to 40 min, are commonly brought about by rapid water loss from guard cells or cells adjacent to guard cells. The resulting stomatal closure is followed by rehydration of the epidermis and reopening of stomata (Mansfield, Chapter 3 in this volume). Oscillations may be initiated by changes in the external environment or by changes in the hydraulic properties of the plants, e.g., changes in the apparent resistance to water uptake by the plant as a function of time of day (246).

In small plants, stomatal apertures in different leaves tend to oscillate in phase, partly because all the leaves are subjected to similar environmental conditions (293) but also because the pressure of water in the xylem of the plant exerts a unifying influence (51, 56, 171). In large plants, stomata of different leaves may oscillate out of phase (127, 322). In these plants, leaves will have significantly different microenvironments, and in addition the resistance, capacitance, and inductance of a large hydraulic system will cause appreciable time lags in the transmission of pulses.

Fluctuations in leaf water balance with a period of 1 to 10 min occur in a number of plant species (11, 292, 294, 295, 299). On occasion, when transpiration rate is high and the leaf is not suffering a water deficit, these fluctuations show no correlation with changes in stomatal behavior (Fig. 8). Sheriff (294, 295) has suggested that these fluctuations occur as a result of time lags in transmission of changes in flow rate, as explained above, combined with changes in the resistance to water movement in the apoplast of the leaf. Such changes in resistance may be caused by changes in leaf water content or flow rate (254–256). Fluctuations of this kind are probably not great enough to cause water deficit in leaves.

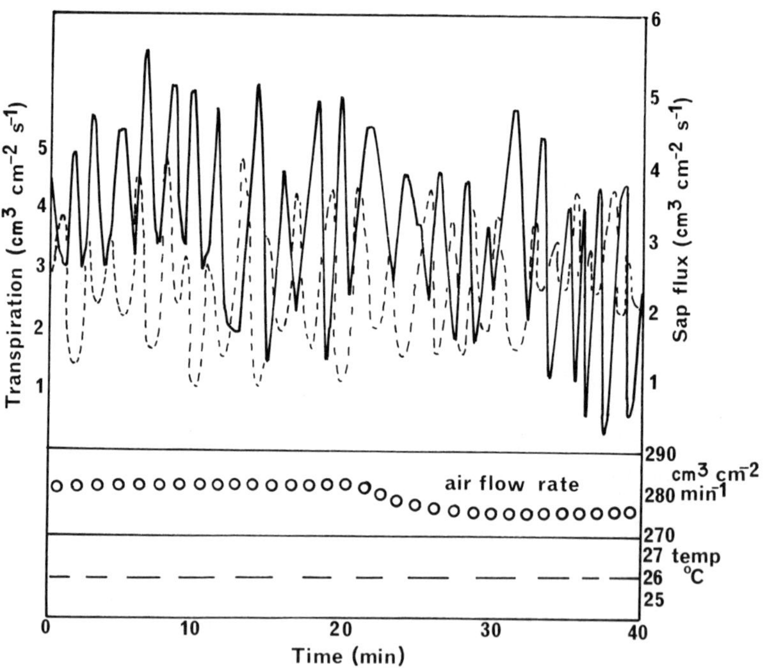

FIG. 8. Fluctuations in water uptake into, and water vapor loss from, a leaf of *Erythrina variegata (E. indica)* fastened in a potometer while air was pressed through the leaf under pressure. The viscous flow resistance of the leaf (stomatal aperture) OOO and the air temperature --- did not vary significantly. After Sheriff (296).

C. RESISTANCE TO WATER MOVEMENT

1. *Plant Resistances*

In many experiments with different species the largest resistance to liquid water flow in the plant has been shown to be in the roots. Root resistance is generally from 2 to 10 times the magnitude of the flow resistances provided by the stem or the leaves (Table III). However, high stem resistances have been found in Sitka spruce (122), and high leaf resistances have been shown to occur in a number of species (27, 170). We have seen that resistances to flow are apparently a function of the rate of water loss by the plant and thus it seems pointless to quote figures for resistances in absolute terms (351). Similarly, the ratio of resistances throughout the plant cannot be constant if one or more of the components is dependent on flux. Several general relationships between leaf water potential and flux have been described. These are shown in Fig. 7. Differences in intercept and slope are generally ascribed to variation in plant resistance. Relationship B in Fig. 7 suggests that the variable (root?) resistance is large and that the other (xylem?) resistances are negligible, while relationship A suggests that for these plants the resistance of the whole plant is constant and that any variable resistance is only small.

The shape of these relationships indicates that at least when the soil is moist, soil resistances do not increase with increasing flux (see Section III,C,2) and that cavitation in the xylem (see Section III,A,2,b), which may occur when the plant suffers increasing water deficit, does not influence the movement of water at "normal" transpiration rates. It seems possible that physiological conditioning or varietal differences might account for different force–flux responses within the same species (142), but several other explanations have been put forward to explain intra- and interspecific differences.

TABLE III

AVERAGE RESISTANCES AND RESISTANCE RATIOS FOR WATER MOVEMENT IN YOUNG APPLE *(Malus)* TREES[a]

Time (GMT)	Total plant resistance r_p (MPa s m^{-3} × 10^{-9})	Xylem resistance r_x	Root resistance r_r	r_r/r_p (%)
0600–0800	8.2	4.3	3.9	48
0800–1000	12.9	3.4	9.5	74
1000–1200	10.0	2.6	7.4	74
1200–1400	11.8	4.2	7.6	64
1400–1600	10.8	3.5	7.3	68

[a] After Landsberg *et al.* (170).

a. Apparent Changes in Plant Resistance. Because the root resistance is generally considered to be the major resistance to liquid flow and because nonlinearity of the force–flux relationship can be demonstrated with decapitated root systems, a variable hydraulic resistance is thought to be located in the root. In an attempt to explain the variation in the force–flux relationships which have been demonstrated in some plants, Fiscus (89) has presented a general model for coupled solute and water flow through roots based on the thermodynamics of irreversible processes. In his model, the root has a single inner compartment separated from the external compartment by a single "membrane." The membrane is defined as the structure which separates the two points of the system to which we have ready access, i.e., the external root medium and the xylem exudate from detopped plants. The exact nature and structure of the membrane are treated as unknowns—a black box with simple input and outputs. The rate of water uptake across the membrane is described by Eq. (7). Substituting into this equation Fiscus obtained, for a situation in which the reflection coefficient is unity,

$$J_v = L_p(\Delta P - RT[C^o - (J_s^*/J_v)]) \tag{10}$$

where R is the gas constant, T is the absolute temperature, C^o is the external solute concentration in moles per cubic centimeter, and J_s^* is the active solute flux in moles per square centimeter second.

Solving this equation for ΔP

$$\Delta P = (J_v/L_p) + RT[C^o - (J_s^*/J_v)] \tag{11}$$

The apparent resistance R^a to pressure flow is the slope of the force–flux curve when the force is given as the dependent variable and is therefore the first derivative of ΔP with respect to J_v,

$$R^a = \frac{d\,\Delta P}{dJ_v} = \frac{1}{L_p} + \frac{RTJ_s^*}{J_v^2} \tag{12}$$

The second part of the resistance term in this equation decreases with the inverse square of the flow rate. The effect of this is shown in Fig. 9, where the insert shows data collected by Lopushinsky (182, 183). It is clear from Fig. 9 and Eq. (11) that the nonlinear water flow with respect to pressure, described by Lopushinsky, is explained by the decrease in one driving force ($\Delta\pi$), as a result of sweeping away or a dilution effect, and an increase in the other driving force (ΔP). The system is, however, still linear with respect to the total inwardly directed driving force ($\Delta P - \Delta\pi$), and therefore this model will not explain the results of Mees and Weatherley (201), who observed nonlinear flow with respect to pressure, but also with respect to the total inwardly directed driving force.

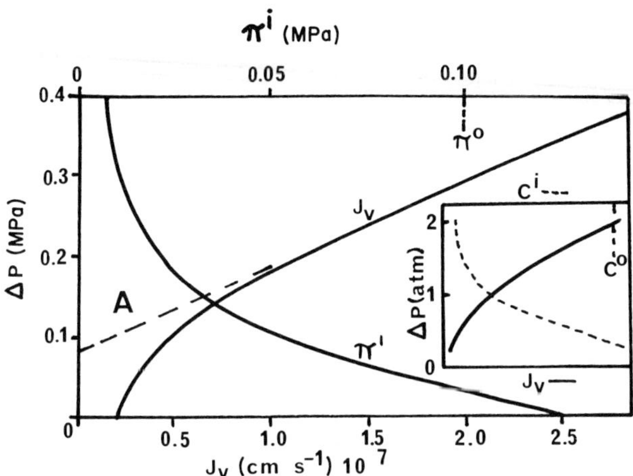

FIG. 9. Relationship among applied pressure, volume flow rate, and internal osmotic pressure for the model system proposed by Fiscus (89). $J_s^* = 0.1\ (10^{-11})$ mol cm^{-2} s^{-1}; $L_p = 0.1\ (10^{-5})$ cm s^{-1} MPa^{-1}; $\pi^0 = 0.1$ MPa. The inset is redrawn from Lopushinsky (183). Line A is the tangent to the curve J_v against ΔP [Newman (232)]. After Fiscus (89).

Fiscus (89) has developed his model to account for solute movement through leaky membranes (reflection coefficient $\sigma < 1$) by active uptake, passive drag, and diffusion, and he derives from Eq. (7)

$$J_v = L_p\left(\Delta P - \frac{2\sigma^2\pi_0}{1+\sigma}\right) + \frac{2\sigma L_p RTJ_s^*}{J_v(1+\sigma)} \tag{13}$$

where π_0 is the external osmotic potential.

Solving for ΔP and calculating the apparent resistance gives the equation

$$R^a = \frac{d\,\Delta P}{dJ_v} = \frac{1}{L_p} + \frac{2\sigma RTJ_s^*}{(1+\sigma)J_v^2} \tag{14}$$

Generating flow curves at various values of reflection coefficient shows how the relationship between flow rate and total inwardly directed driving force is strongly dependent on this parameter (Fig. 10). As membrane selectivity decreases, nonlinearity increases and the force–flux relationship eventually exhibits a negative slope at low flow rates. This phenomenon is in agreement with the observations of Mees and Weatherley (201), which, as Fiscus suggests, may be explained on the basis of the effect of two differentially effective driving forces rather than by any change in root properties with increasing flux.

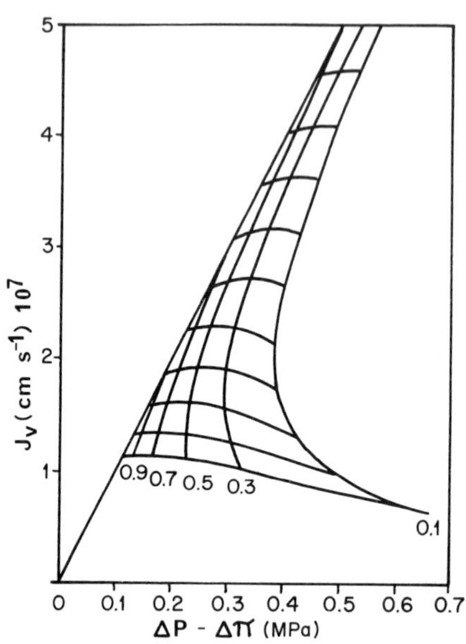

FIG. 10. Effect of the reflection coefficient on the total force–flux curve illustrating the negative resistance region. The straight line is $\sigma = 1$. $J_s^* = 0.1\,(10^{-10})$ mol cm^{-2} s^{-1}; $L_p = 0.1$ (10^{-6}) cm s^{-1} MPa^{-1}; $\pi^0 = 1$ MPa. The horizontal lines connect points of equal hydrostatic pressure and the numbers on the curves denote the reflection coefficients. After Fiscus (89).

Fiscus and Kramer (93) and Fiscus (90, 91) have extended the work from a flat membrane of unit surface area and uniform properties and have developed a line model analogous to a terminal root segment with a non-uniform distribution of properties. Using this model they were able to confirm their earlier work.

According to the Fiscus model, when the osmotic efficiency of the membranes, seen in the values of the reflection coefficient, is taken into account, flux should be a linear function of the effective potential difference (Fig. 10). Newman (232) points out that this does not appear to be true in Mees and Weatherley's (201) experimental data, a fact which he interprets as indicating a changing hydraulic conductivity with increasing ΔP. Fiscus (90) feels that Mees and Weatherley's results may be explained by his model if xylem loading from adjacent tissues within the root membrane is taken into account. At low flow rates the errors associated with xylem loading and unloading will be greatest and the discrepancies between the experiments and the theory will appear to be enhanced.

As the rate of flow of water through the root increases, the influence of solute flux decreases and the right-hand term of the flow rate equation (13) tends to 0. Therefore, the curve is asymptotic to a straight line given by

$$J_v = L_p\{\Delta P - [2\sigma\pi_0/(1 + \sigma)]\} \quad (15)$$

This line (Fig. 9, line A) cuts the abscissa at

$$\Delta P = 2\sigma^2\pi_0/(1 + \sigma) \quad (16)$$

The reflection coefficient of the membrane cannot be greater than 1, therefore $2\sigma^2/1 + \sigma$ cannot be greater than unity. This means that if a tangent drawn to any part of the curve J_v against ΔP (line A, Fig. 9) cuts the abscissa at a higher ΔP than π_0, then the curvature cannot be entirely explained by Fiscus's model. According to Newman (232), this is the case if the model is applied to the data of Lopushinsky (183). It seems likely, however, that there may be some difficulty in establishing a value for the effective external osmotic potential (90). If the endodermis of the root constitutes a significant barrier to solute transport, then it is possible that solute will build up on the outer side of this barrier. A boundary layer will result, which owing to its position cannot be disrupted by stirring the bathing solution. In addition, with inadequate stirring, a significant boundary layer may persist at the outer surface of the root, again raising the effective value of the osmotic concentration outside the root membrane.

A model comprising the factors influencing the flux of water through roots is useful since we know comparatively little about the pathway of water movement (see Section III,A). The nature of the functional root membrane is not defined and it may or may not have properties similar to the typical plant plasmalemma. In addition, many root properties, particularly σ and L_p, would be expected to vary along the root axis as suberization and secondary growth proceed. Values for the various root parameters in the models discussed earlier have been chosen to cover wide ranges of conditions. Nevertheless, the relationships obtained from the models depend upon the root membrane being somewhat leaky, and this fact has not been established beyond doubt. The important effects of active salt uptake on water flux are seen in the model. It is interesting to note that apparent resistance to water uptake by roots may vary diurnally (246). This variation may be a function of differential salt loading in light and dark.

The overall conclusion from this work is that the coupled solute and water flow models (61, 89, 93) may adequately explain a variety of experimentally observed phenomena. Nevertheless, without a change in hydraulic conductivity these models cannot explain a slope of zero in the response of leaf water potential to flow (relationship C, Fig. 7) (15, 40, 177, 312, 325, 326). In considering this problem, Bunce (38) has examined the response of leaf water potential of young soybean and cotton plants to a change in transpiration rate. To vary the rates of water loss from plants, ambient humidity levels were changed. Leaf water potential was measured 1 h after transpiration became constant following a change in humidity. At this time, a wide range of transpiration rates resulted in the same minimum leaf water potential (Fig. 11). Two hours later the relationship between transpiration and leaf water potential had altered, indicating that a considerable time lapse was necessary before water potential came into equilibrium with flow. At equilibrium, which was only reached 3 h after a change in transpiration rate, an increase in transpiration always resulted in a decrease in leaf water potential (Fig. 11). This work suggests that resistance calculations made before leaf water potential has equilibrated with flow will indicate a variation in apparent resistance to water movement. At equilibrium, the data indicate that the resistance is constant over a wide range of flow rates.

The existence of an apparently variable resistance in the root is supported by the demonstration that when the root system is killed by immersion in hot water, the resistance provided by the whole plant becomes

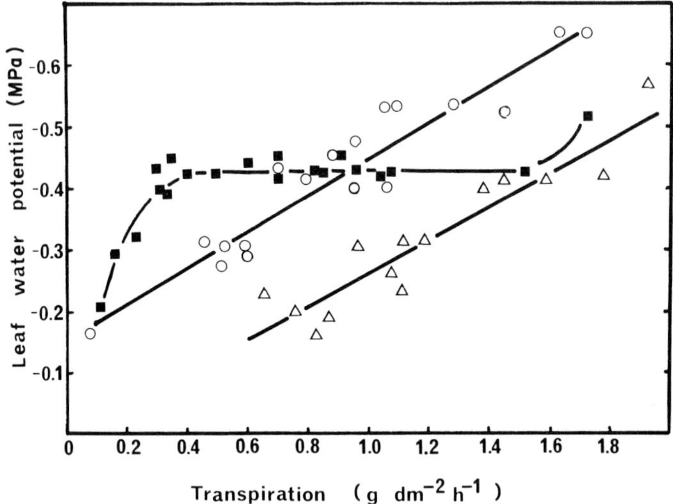

FIG. 11. Dependence of leaf water potential on transpiration in soybean. ■, Leaf water potential measured 1 h after transpiration became constant, 500 μmol m^{-2} s^{-1}; ○, leaf water potential 3 or more hours after transpiration became constant, 500 μmol m^{-2} s^{-1}. △, 3 or more hours after transpiration became constant, 1500 μmol m^{-2} s^{-1}. Transpiration is per unit root surface area; average leaf (one side) to root (total) surface area was 0.8 : 1.0. (Note that 1.0 g dm^{-2} h^{-1} = 27.8 mg m^{-2} s^{-1}.) After Bunce (38).

constant (312). Recent work has indicated, however, that the resistancc to flow in the leaf lamina is not negligible and may decline with increasing transpiration (27, 29, 141, 170, 256). Indeed, Boyer (27) suggests that while root resistance in *Helianthus* varies by a factor of only 2.5 as flow varies, total plant resistance to transport at low flow rates was about 30 times the resistance at high flow rates. He suggests that much of the change may take place in the leaves since similar changes can be demonstrated in detached leaves (Fig. 12).

Boyer's calculations suggest that in *Helianthus* leaves the protoplast resistance is almost 30 times larger than the resistance of the path leading from the soil to the protoplasts. He attributes the decreasing resistance with increasing flux to a decrease in water movement in and out of the cells and an increase in water movement along the low-resistance pathway around the protoplast, i.e., to a change in pathway. Resistances were apparently high at low rates of flow, where only cell growth was occurring (no transpiration). Low resistances apparently occurred at high rates of flow when no growth was occurring and flow consisted solely of transpiration. Boyer concluded that because of the high resistance of the protoplast pathway, leaf water potentials over a considerable range of rates of water absorption were determined more by protoplast water movement than by transpiration. Boyer's analysis highlights the importance of specifying what type of water potential measurement has been made and whether this measurement has been made in growing or nongrowing tissue.

A further complication in this type of analysis arises from the observation of Cosgrove and Cleland (49, 49a) that there may be a significant concentration of solute in the free space (apoplast) of cells in the growing regions of plants. Accurate calculation of resistances to transport of water from the general transport equation (1) will, as described earlier, depend on an accurate assessment of the water potential gradient which is driving flow ($\Delta\psi$). This is usually the difference between the ψ of the plant tissue and the ψ of the water supply. Solutes in the apoplast will lower tissue water potential ψ_t and this will stay depressed as long as the solutes remain in the cell wall space (the wall space, of course, cannot support a positive hydrostatic pressure). As the transpirational flux declines to zero and $\Delta\psi$ remains of a significant magnitude (because of free space solutes), one would calculate an ever-increasing hydraulic resistance (49, 49a). Such an apparent change in resistance has no physical reality but is due to a lack of a cell wall solute term π^{cw} in the general transport equation, which may be more correctly written (49a) as

$$J_v = (\Delta\psi - \pi^{cw})/R \qquad (16a)$$

b. Real Changes in Plant Resistance. Pospisilova (254–256) has pro-

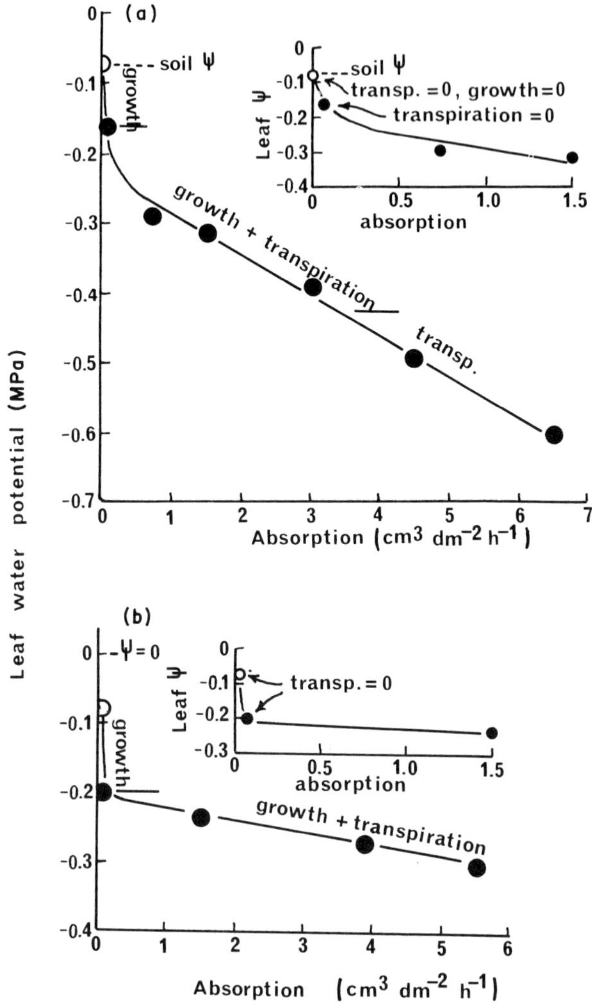

FIG. 12. (a) Leaf water potentials at various steady-state rates of water absorption by sunflower plants in soil having a water potential of −0.07 MPa throughout the experiment. The inset shows an enlarged view of the data at low rates of absorption. ○, Older leaves at the base of the plant; ●, younger leaves toward the top of the plant. Transpiration was varied by altering the evaporative demand. After Boyer (27). (b) Leaf water potentials at various steady-state rates of water absorption by excised sunflower leaves having their petioles in water ($\psi = 0$). The inset shows an enlarged view of the data at low rates of absorption. ○, Older leaves from the base of the plant; ●, younger leaves from the upper part of the plant. Transpiration was varied by altering the evaporative demand. After Boyer (27).

posed that the resistance to water movement in the apoplast will change with changes in leaf water content or flow. Results suggested that at full hydration and a water saturation deficit of 5 to 10% the hydraulic conductance of leaf tissue is high. At greater water deficits, conductances decrease. In addition, it seems likely that membrane permeability might change with plant water status. Zimmermann and Steudle (368) found that the hydraulic conductivity of membranes of internodal cells of *Nitella flexilis, Chara intermedia,* and *Chara fragilis* increased sharply as turgor was reduced (see also Dale and Sutcliffe, Chapter 1 in this volume). Powell (259) noted that if these characteristics were found with the membranes of higher plants, they would provide an explanation for apparent changes in resistances with flow rate. He suggests that flow of water through endodermal cells (and other cells if a large proportion of water is moving via the symplast) might cause their solute content to fall, reducing solute potential and, therefore, turgor. The consequent increase in hydraulic conductivity will act to produce a nonlinear force–flux relationship. This result will depend upon the sensitivity of membrane permeability to cell turgor. In some situations, membranes may be insensitive to turgor changes (244), but recent measurements of the water transport properties of single leaf cells of *Tradescantia virginiana* using the pressure probe (139) suggest that hydraulic conductivities of cell membranes may be reduced at low turgors (369) (see also Dale and Sutcliffe, Chapter 1 in this volume).

Johnson (147) has investigated the effect of possible changes in pathway geometry on flow of water. In this study, vascular bundles of petioles were frozen intact and freeze-fractured for electron microscopy. Cell walls below wilted leaves appeared to be drawn in against the helical thickenings of xylem conduits. Below turgid leaves, the walls of the vascular conduits were straight or bulged outward slightly (Fig. 13). Johnson proposes that the transpiring leaf cells around the vessels in the leaf may tend to contract before the vessels do. The vessel walls may then bulge outward, decreasing the resistance to flow at high rates of flux.

Bunce (38) has shown that low irradiance and decreased leaf area can cause increases in root resistance and that these changes can be correlated with lower rates of root elongation. The data indicate that shoot–root interactions are occurring which influence apparent root resistance to water flow. Reductions in photosynthesis leading to a reduction in assimilate transport can rapidly affect ion uptake by roots (25) as well as plant turgor and growth (293), and therefore such effects on root resistance should not be surprising. These results emphasize the importance of performing experiments with whole plants as well as with isolated root systems.

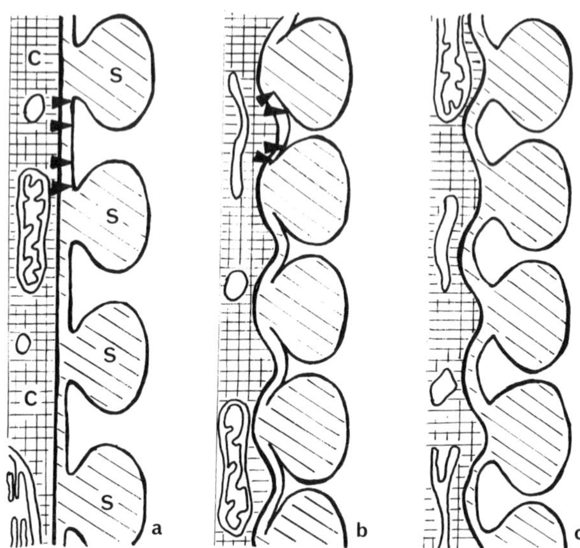

FIG. 13. Diagram showing the shape of the cell wall between a vessel with helical thickenings (S) and a parenchyma cell with cytoplasm (C). (a) With pressure equal on each side of the vessel; (b) with pressure higher in the parenchyma cell than in the vessel; (c) with pressure higher in the vessel than in the parenchyma cell. After Johnson (147).

2. *Soil Resistances*

One of the most perplexing problems in plant and soil water relations has been the definition of the magnitude of the resistance to the flow of water in the soil immediately adjacent to the roots (the perirhizal resistance). If this resistance is large, for example, as in unsaturated soil, the changes in plant resistance discussed earlier may have little influence on leaf water potential (325). Gardner (99) and Cowan (50) have calculated gradients of water potential and hydraulic conductivity close to the root. Gardner writes the equation for unsaturated flow for a cylindrical system with water moving in toward the root from the surrounding soil as

$$\left(\frac{\partial \theta}{\partial t}\right)_r = \frac{1}{r}\frac{\partial}{\partial r}\left(rD_w \frac{\partial \theta}{\partial r}\right) \quad (17)$$

where θ is the volumetric water content, r is the radial distance from the axis of the root, and D_w is the water diffusivity. The solution of this equation is given in Eq. (18).

For boundary conditions given by (1) a constant rate (I_w) of uptake of water per unit length of root, where $I_w = 2a\pi K\, d\psi/dr$ at $r = a$, the radius of the root, and (2) an infinite soil volume

$$\psi_s - \psi_a = \frac{I_w}{4\pi K}\left(\ln \frac{4D_w t}{a^2} - 0.577\right) \tag{18}$$

when ψ_a and ψ_s are, respectively, the water potentials at the root surface and in the bulk soil and K is the capillary conductivity.

Data from this model for three soils indicate that if the bulk soil is moderately dry (−0.2 to −0.3 MPa), a large potential drop between the soil and the root can occur, indicating a large resistance to the movement of water somewhere in the pathway. If the soil is moist, the decrease in potential at the root surface is always trivial. There is, however, uncertainty over some of the assumptions built into Gardner's model. Newman (227, 229) has suggested that the values of water uptake per unit length of root (I_w) used by Gardner and by Cowan were too large and that therefore the root–soil water potential gradient was overestimated by their model. Tinker (324) suggests that the problem may be further complicated by the fact that water uptake per unit length of root (I_w) will vary significantly over the root system and therefore the mean value of I_w may not be a real guide to the local situation. This may be a serious problem when water is not always available at all points in the root system. The local situation may be further complicated by the buildup of solutes near the root surface (198).

Newman (227, 228) argues, on theoretical grounds, that unless the water potential of the soil drops to around −1.5 MPa, the perirhizal resistance will not be great enough to cause a significant drop in water potential in the soil adjacent to the root. He suggests that even in quite moist soil, the rooting zone as a whole may be depleted of water because of the so-called "pararhizal resistance" to water transport from the water table or adjacent masses of soil. Weatherley (350), however, has shown that large differences in water potential can exist between the surface of the root and moderately dry soil in the rooting zone. In addition, experiments by several workers (100, 101, 118, 119, 176, 211, 232, 357) have shown that the total resistance to flow of water increases markedly when soil water potential drops by only a few hundredths of megapascal near field capacity. Gardner and Ehlig (100, 101) concluded that this increase is associated with a similar change in soil moisture conductivity, while Newman (232), Lawlor (176), Hansen (118, 119), and Williams (357) attribute large changes in total resistance to changes in plant resistance. Therefore, while there is some agreement that resistance to the flow of water increases as soils dry, there is little agreement over what causes this increase.

It has been suggested that the high resistance to water movement occurs at the root–soil interface because of incomplete contact between the root and the soil particles. In practice much of the root surface is in contact with air, either because the root grows into a void or because the root and the soil shrink on drying (135). It appears that comparatively large gaps of 10 μm or more may open up between the root and adjacent soil particles.

Weatherley (350) reports experiments which suggest that this gap may have a significant effect on leaf water deficit. Sunflower plants were grown in polyethylene bags and water deficit in the plant was monitored by a β-gauge. When the plant was losing water rapidly, leaf water deficit increased steadily. When the bag was squeezed gently, leaf water deficit immediately decreased. The occurrence of mucigel and the development of root hairs may significantly reduce the effect of gap development under water stress (324), but work by Herkelrath *et al.* (123, 124) suggests that the effect of such a gap on root water uptake may be significant.

In order to assess the effect of poor contact between the root and the soil on root water uptake, Herkelrath *et al.* (123, 124) grew plants in soil columns divided into sections by wax layers. These layers could be penetrated by roots, but they prevented the vertical movement of water (138, 353). The rate of extraction of water from each soil section could be obtained simply from the rate of change of soil water content. In addition, the dependence of the rate of extraction upon soil dryness could be measured in two small "test layers" without dehydrating the plants because the larger supply sections were kept well irrigated. The test layers also provided a measurement of the average root water potential (the soil water potential in the test layer) at which water extraction ceased (211). Herkelrath *et al.* (123) showed that the rate of extraction of water from the test layers decreased with decreasing soil moisture levels below a soil water content of 0.10 cm^3 water cm^{-3} soil and a water potential of -0.01 MPa (Fig. 14). These results are not in agreement with those obtained from Gardner's (99) root uptake model, which predicts that uptake should not decrease at these high water potentials. Applying this model for realistic rooting densities leads to an overestimate of the rate of extraction of water from the soil (Fig. 14).

Herkelrath *et al.* (124) assumed the effective conductivity of a root segment to be proportional to the wetted fraction of the surface area of that segment such that

$$q = f\rho(\psi_1 - \psi_r) \tag{19}$$

where q is the rate of flow through the root membrane, f is the wetted fraction of the root surface (Fig. 15), ρ is the membrane permeability per unit length of root, ψ_1 is the soil water potential at the root surface, and ψ_r is the water potential inside the root.

The factor f may be assumed to be the average volume saturation of the

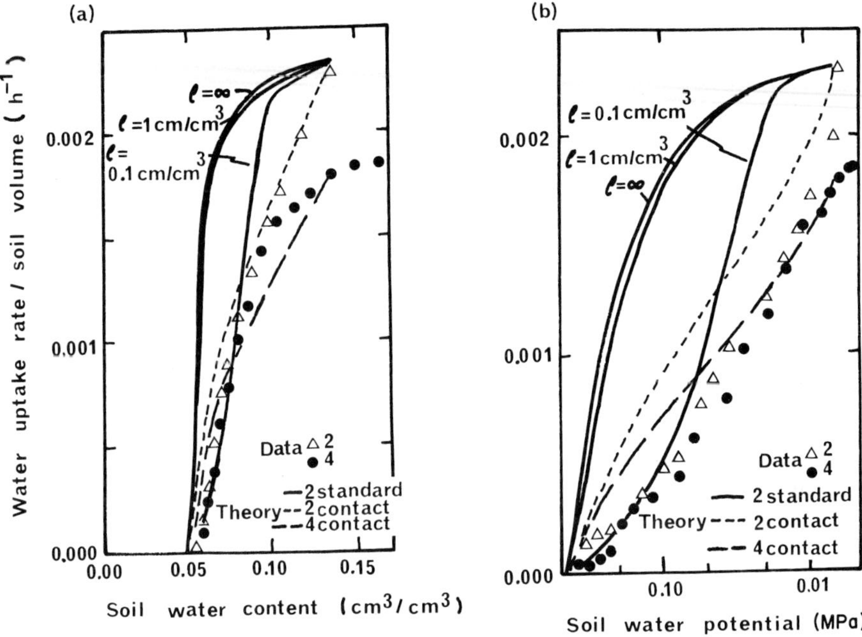

FIG. 14. Root water uptake as a function of soil water status. A comparison of theory and experiment for several theoretical rooting densities (l). (a) Uptake versus soil water content; (b) uptake versus soil water potential. Results from both the standard (Gardner) model and the root contact model are shown. After Herkelrath *et al.* (124).

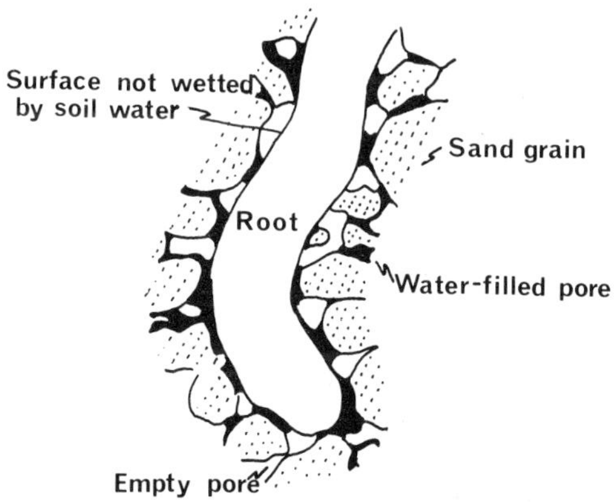

FIG. 15. A diagrammatic representation of the poor contact between the root and the soil. After Herkelrath *et al.* (124). Adapted from *Soil Science Society of America Journal,* Nov.–Dec., 1977, #6, Volume 41, pp. 1040–1041, 1977.

soil pore space (θ/θ_{sat}) such that

$$q = (\theta/\theta_{sat})\rho(\psi_1 - \psi_r) \tag{20}$$

The unknowns ρ and ψ_r are found by forcing the theory to fit the experiments at water contents of 0.048 and 0.135 cm^3 water cm^{-3} and assuming that the root density is high enough that ψ_1 can be taken to be $\overline{\psi}_{soil}$, the average soil water potential. The predicted extraction rate can then be expressed as

$$-\frac{\partial\theta}{\partial t} = ql = \frac{\theta}{\theta_{sat}}\rho l(\overline{\psi}_{soil} - \psi_e) \tag{21}$$

where ρl is the permeability of all roots in a section and ψ_e is the equilibrium test layer potential. As Fig. 14a,b shows, the agreement between the data and this form of the root contact model is much improved, suggesting that poor contact between roots and soil may have a significant effect on root water uptake. Faiz and Weatherley (82, 83) have reached a similar conclusion.

D. Capacitance

The capacitance of a plant or a soil is an indication of how its water content changes with potential, defined as $C = dW/d\psi$, where W is tissue water content. In most soils the water potential is largely a matric potential associated with surface forces. Generally there are some large pores present in soils, which makes little difference to the potential as they fill or drain, so that when the soil is wet, capacitance is very large. In dry soils (water potential -0.5 to 1.5 MPa) the capacitance is considerably smaller since the absorbed water films are very thin and their matric potential changes rapidly with water content.

In plants it is often convenient to define

$$C = \frac{dW}{d\psi} = W_{max}\frac{d\theta_p}{d\psi} = W_{max}C_r \tag{22}$$

where W_{max} is the maximum (turgid) tissue water content and θ_p is the relative water content. The term $d\theta/d\psi$ is called the relative capacitance C_r and is useful for comparing water relations properties of plants of different species, size, and shape.

The relative capacitance of vacuolated plant tissue is very low compared to that of soil. Nevertheless, it is important to know how capacitance varies from plant to plant or from plant part to plant part. There are at least three components which comprise the capacitance of plant tissue. These are changes in turgor within the cell vacuole due to the effect of the restraining cell wall, changes in osmotic potential due to concentration and changes in matric potential of water associated with cell walls and other surfaces. For mesophyll tissue in the normal water content range, the relative capacitance of matric water is small and can be ignored (277, 355).

A modified Höfler diagram (see also Chapter 1) (Fig. 16) shows how the cell solutes and cell wall characteristics combine to influence the water content–potential relationship of the plant leaf. The changes in osmotic potential π with water volume are described by a simple concentration effect on the appropriate volume

$$(\partial\pi/\partial V)V = \pi \tag{23}$$

where V is the volume of osmotic water (total water volume minus the matric fraction). The change in turgor P is defined as

$$(\partial P/\partial V)V = \epsilon_m \tag{24}$$

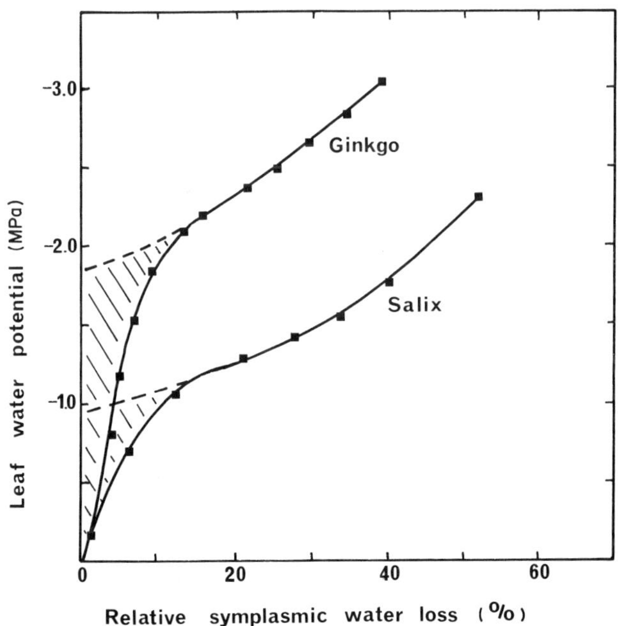

FIG. 16. A modified Höfler diagram showing how leaf water potential, solute potential, and pressure potential depend upon the water content of the plant. ψ is represented by the solid line, π by the broken line, and the amount of leaf turgor by the crosshatched area. After Tyree (333).

where ϵ_m is the volumetric modulus of elasticity for plant tissue (42). The total change of water potential is given from Eqs. (21) and (22) by

$$(\partial\psi/\partial V)V = \epsilon_m - \pi \quad (25)$$

so that plant capacitance is represented by (355)

$$C' = 1/(\epsilon_m - \pi) \quad (26)$$

The bulk volumetric modulus for elasticity for fully turgid leaf tissue is typically 5.0 to 30.0 MPa for a wide range of species (333, 360). In relation to the osmotic properties of the cell the elastic modulus dominates cell water relations (see also Chapter 1). Cells have rigid walls so that large changes in turgor are associated with small changes in volume. The parallel changes in solution concentration may be relatively insignificant, although the turgor and solute relations of some cells, e.g., guard cells, are exceptional (see Mansfield, Chapter 3 in this volume). Some calculations by Zimmermann and co-workers (369) have underlined just how unusual the water relations of guard cells may be. Several workers have shown that the elastic modulus of leaf cells generally increases with increasing turgor. Elastic moduli of guard cells may, however, decrease as cell turgor increases (369). It is now possible, using the miniaturized pressure probe to measure the water transport properties of individual cells (see Chapter 1). Zimmermann *et al.* (369) have reported that the elastic moduli of epidermal, subsidiary, and mesophyll cells from *Tradescantia* leaves are in the range of 4.0 to 24.0 MPa, 3.0 to 20.0 MPa, 0.6 to 1.4 MPa, respectively.

Potential–volume relationships differ markedly among species and also depend on the water relations history of the plant (64). These relationships can give some indication of the drought tolerance of different species. Figure 16 shows a comparison of the potential–volume characteristics of leaves from two trees growing in very different habitats (333). *Ginkgo biloba* has a much lower osmotic potential than does *Salix lasiandra* (−2.2 MPa versus 1.2 MPa). *Gingko* leaves can therefore maintain turgor and probably growth, at low leaf water potentials. In both species, turgor regulation accounts for most of the decline of leaf water potential as water content decreases. The modulus of elasticity of cell walls of *Gingko* leaves is twice that of *Salix* leaves, and this means that when both species have a leaf water potential of, for example, −1.0 MPa, *Gingko* has lost only 4.8% of its osmotic water, whereas *Salix* has lost 10.8% of its osmotic water. Clearly a low osmotic potential and a high modulus of elasticity might be of some advantage in a dry habitat.

Although Fig. 16 describes the dependence of vacuolated tissue water potential on volume, there are also parallel changes in potential not related to water volume per se. Changes in osmotic potential due to solute accumulation may be much larger than the shift due to increased concentration as volume decreases. There is now good evidence that plants can exhibit low water potentials while maintaining turgor in their leaves by accumulating solutes in the expanding cells (2, 3, 58, 59, 132, 133, 222, 238, 290, 328). Such adaptations are necessary if growth is to continue in plants or plant parts which must generate low water potentials to obtain water, e.g., plants in dry or saline soils or the leaves at the tops of tall trees (see Section V,C,2).

The total change in water volume can be described by using capacitance only if we subtract the change in water potential due to osmotic adjustment. Thus, the effect of increases in solute concentration with drying is to decrease the "apparent capacitance" defined by the observed volume and potential changes. It was concluded earlier that changes in osmotic potential with water volume have a small effect on capacitance as long as the modulus of elasticity is large. This may not necessarily be the case if the plant has the capacity to osmoregulate.

The contribution from storage in the plant to the transpiration stream is represented diagrammatically in Fig. 17 (349). The solid line represents the situation in a rigid "plant" with no storage capacity. Following an increase in transpiration in the plant represented in Fig. 6, i.e., with storage capacity, absorption also starts immediately but at a slower rate (dashed line). The stippled area represents the accumulating water deficit as a result of flow out of storage. Uptake of water continues after transpiration has ceased until the water deficit has been relieved. Diurnal changes in plant water content can cause significant shrinking and swelling of plant parts (Fig. 18).

Jarvis (142) has estimated the water storage capacity of different plant

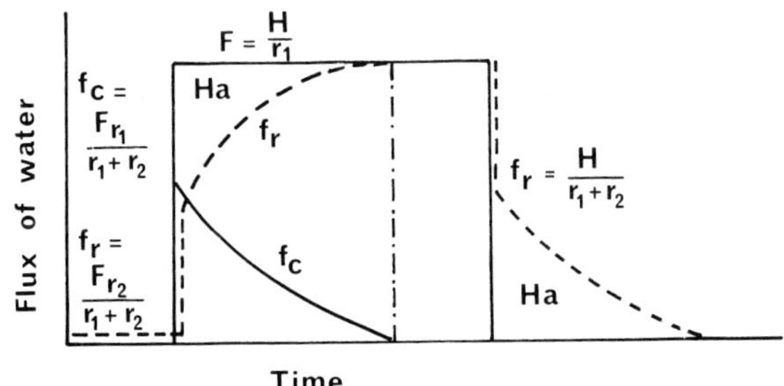

FIG. 17. Changes in water flux, predicted by the model shown in Fig. 6, on sudden starting and stopping of transpiration (F). Flux through roots f_r and movement of water out of the cells f_o are also shown. After Weatherley (349).

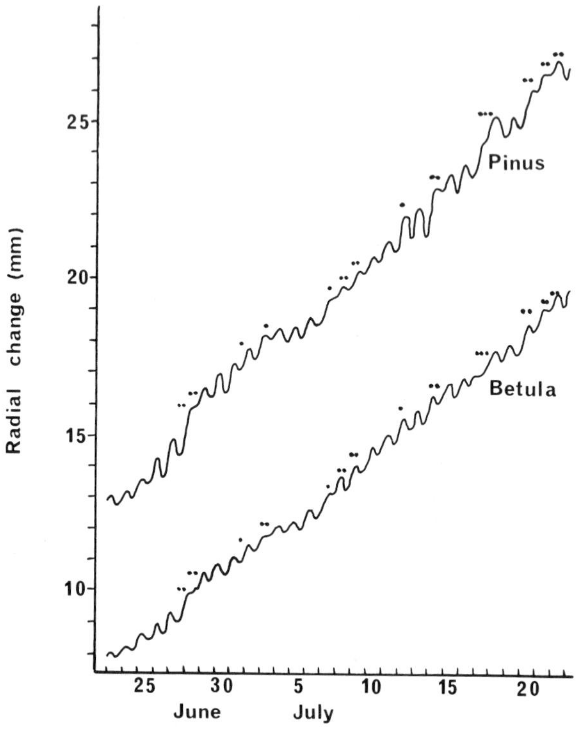

FIG. 18. Shrinking and swelling of stems of *Pinus resinosa* and *Betula papyrifera* trees during June and July of 1972 at stem height of 1.3 m. Dots indicate the incidence and intensity of precipitation. After Braekke and Kozlowski (33).

parts. The results of his calculations indicate that the roots of both herbaceous and woody plants could be important sources of stored water. Huck *et al.* (135) have reported that on a sunny day the roots of cotton shrank to as little as 60% of their turgid diameter. If this change is repeated in other plants with large root systems, it could represent an enormous quantity of tissue water supplied to the transpiration stream. In stems, both living and woody tissues change water content (122, 218). The water content of heartwood in trees does not change significantly throughout the year (102), but the water content of the sapwood can change dramatically. Waring and Running (344) have shown that in large Douglas fir trees, withdrawal of water from the stemwood can reach 1.7 mm day^{-1} on clear days after cloudy and rainy weather. Recharge could be almost as fast after rain, suggesting that water stored in the sapwood can make an important contribution to transpiration by this plant. Roberts (276) claims that the amount of water stored in trunks or Scotch pine is quite small. The importance of the stem as a reservoir in herbaceous plants is unknown.

IV. Plant Water Loss

A. PHYSICAL AND PHYSIOLOGICAL CONTROL

Transpiration from leaves and canopies is clearly under both physical and physiological control, and while rates of water loss from single small groups of plants can be measured, a means of estimating transpiration rate and analyzing its variations now exists. Equation (27), modified by Monteith (220) from the original work of Penman (250), describes the dependence of transpiration rate E on available energy A; vapor pressure deficit D; boundary layer conductance g_b, a function of wind speed, s (de_s/dT, where e_s is the saturation vapor pressure and T is temperature); and canopy conductance g_c:

$$E = \frac{sA + c_p \rho D g_b}{\lambda[s + \gamma(1 + g_a/g_c)]} \tag{27}$$

where c_p is the specific heat of air at constant pressure, ρ is the density of air, γ is the psychrometric constant, and λ is the heat of vaporization of water. A simple derivation of the equation is shown by Jarvis (145).

Since the mid-1960s, Eq. (27) has been used by ecologists and crop physiologists to determine transpiration rates of crops and individual species in mixed communities, and Jarvis (145) and Jones (149) describe the various important uses to which it may be put.

Equation (27) can be applied successfully over periods of minutes, hours, days, and weeks, provided that adequate meteorological data are available. Boundary layer conductances are estimated, while leaf conductance can be measured with a diffusion porometer (see e.g., 149). The conductance of a leaf varies with the leaf's age, the position on the plant, and the position in the canopy as well as with recent and current weather and time of day or year. Canopy conductance g_c is the arithmetical sum of the conductances of all the individual leaves in the canopy and estimates may be made by stratifying with respect to age, position, and branching hierarchy (145) and by approximating the leaf area index of each subsample L_i. Then

$$g_c = \sum g_{s,i} L_i \tag{27a}$$

where g_{si} is the average leaf conductance of a sample of leaves of area index L_i.

In this chapter we are concerned with how different variables, particularly plant factors which we can control, influence transpiration. We will confine ourselves therefore to a consideration of water loss from individual leaves, but it is important to bear in mind that canopy structure also has an important influence on water loss, most crucially through an effect of structure on boundary layer resistance/conductance. Jarvis (145) has pointed out that in shortish vegetation in which the boundary layer resistance is high, the influence of a change in stomatal aperture on transpiration may be small, at least until the stomata are nearly closed (Fig. 19). Under these conditions, transpiration tends to proceed at an equilibrium rate and is determined almost wholly by the available radiant energy rather than by the stomata. If the stomata close a little, the air in the vicinity of the leaves becomes less humid and transpiration is driven harder, coming back to the same rate determined by the available energy. Conversely, if the stomata open a little, the air in the vicinity of the leaves becomes more humid and transpiration is damped down. Evaporation is much more sensitive to changes in stomatal aperture in aerodynamically rough vegetation, e.g., forests, in which boundary layer resistances are low, and thus attempts to control transpiration of short crops via manipulation of stomatal behavior may be less successful.

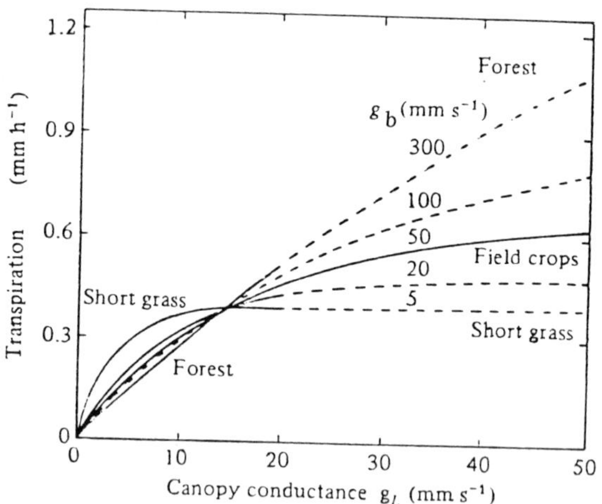

FIG. 19. Calculated relationships between transpiration rate and canopy conductance at different boundary layer conductances for 400 W m^{-2} available energy, 1 kPa vapor pressure deficit, and 15°C. The solid lines represent the probable range of values of canopy conductance for different crops, being up to 50 mm s^{-1} for some field crops. Boundary layer conductance g_b is included. After Jarvis (145).

B. RESISTANCE TO PLANT WATER LOSS

The arrangement of resistances in the path of water vapor diffusing from one surface of a single leaf is shown in Fig. 20. The resistances to diffusion provided by the substomatal cavity and by the cell walls are small compared to the other resistances in the pathway and are ignored for most purposes (54). In this section, the discussion will be confined to the effect on plant water balance of variation in the magnitude of stomatal, leaf boundary layer, and cuticular resistance.

In the complex series and parallel arrangement of components shown in Fig. 20, it is expedient to use resistances to describe the characteristics of each pathway. Since the stomatal resistance–aperture relationship is hy-

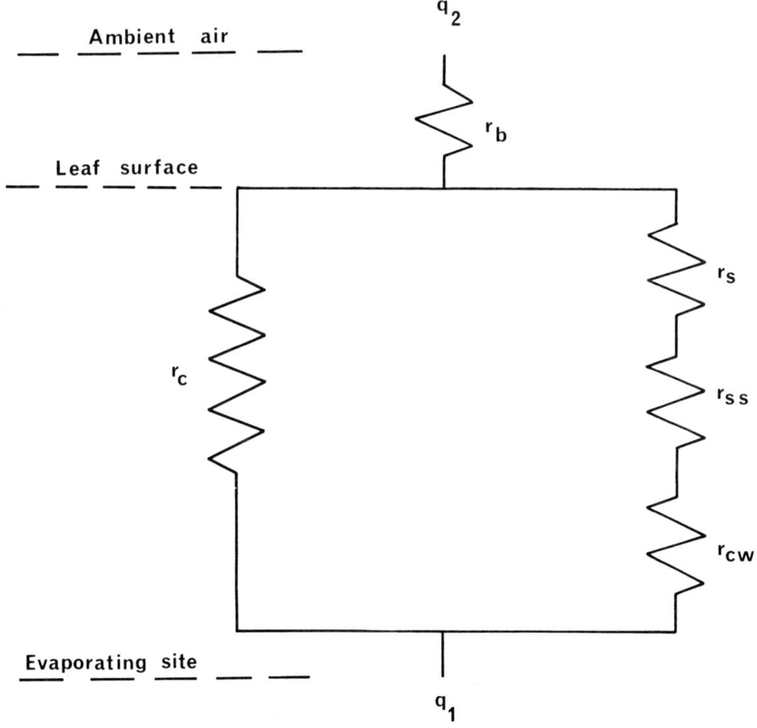

FIG. 20. Arrangement of resistances in the path of water vapor diffusing from one surface of a leaf. q_1 and q_2 are the vapor concentrations at the evaporating surface and in the ambient air, and r_b, r_c, r_s, r_{ss}, and r_{cw} are the resistances of the boundary layer, cuticle, stomata, substomatal cavity, and cell walls, respectively.

perbolic, small apparent changes in resistance may represent appreciable responses by the guard cells. The relationship between conductance (the reciprocal of resistance) and aperture may be nearly linear (73), and therefore, where the emphasis is placed on functional relationships, conductance may be a more appropriate expression of the proportionality between flux and concentration difference. Dimensions for resistance and conductance are shown in Table I.

1. *Cuticular Resistance*

Cuticular resistances are often 30 – 50 times the minimum stomatal resistance and usually are about 100 times the external resistance. Since the stomatal and cuticular resistances are arranged in parallel, cuticular water loss is an insignificant proportion of the whole when the stomata are open. Because of the relatively low cuticular water loss, full or partial closure of the stomata will lead to a very substantial reduction in transpiration rate. Since it is impossible to ensure that the stomata are ever completely closed, it is difficult to measure the cuticular water loss of a surface that also possesses stomata. Cowan and Milthorpe (54) have suggested that a satisfactory estimate may be given by determining the transpiration rate per unit vapor pressure difference over a range of measured stomatal conductances and extrapolating the curve to zero stomatal conductance. Measured values of cuticular resistance are between 30 and 50 s cm^{-1} for most mesophytes, but may be much higher for some xerophytes, e.g., 700 s cm^{-1} for *Opuntia basilaris* (317). Meidner (203) has suggested that cuticular resistance to vapor loss may increase in dry air. He proposes that dehydration leads to a decrease in cuticle thickness and an increase in diffusion resistance. In moist air, the cuticle remains hydrated, has a lower resistance to vapor loss, and can sustain a relatively high rate of water loss. Cuticular thickness and wax development will vary with the conditions of growth, but once the plant is fully developed, the only effective control over water loss that the plant has is via the closure of stomata.

2. *Boundary Layer Resistance*

Exchange of heat, water vapor, and carbon dioxide between the leaf and the surrounding air take place through the boundary layer. Boundary layers may be laminar, when the exchange process is molecular diffusion, or turbulent, when the transport process is eddy-assisted diffusion. In this situation, the chaotic motion of parcels of air serves to sweep away the air near to the leaf surface and mix it with the ambient air. Even in fully developed turbulent boundary layers a thin laminar sublayer will occur; adjacent to the leaf surface and across this layer, the mode of transport must be molecular diffusion.

Grace (107) has described the factors influencing the rates of diffusion in boundary layers. Decreases in wind speed and increases in the leaf size result in increased boundary layer resistance r_b, and the relationships may be described by Eqs. (28) and (29):

$$r_b = x/D_j \tag{28}$$

where x is the equivalent boundary layer thickness (in centimeters) and D_j is the diffusion coefficient (in square centimeters per second). Based on hydrodynamic theory for laminar flow adjacent to a flat surface as modified by observations with leaves, there is an approximate expression for equivalent boundary layer thickness x (233):

$$x \simeq 0.4(l^{leaf}/v^{wind}) \tag{29}$$

where l^{leaf} is the linear dimension of the leaf in centimeters in the downwind direction and v^{wind} is the ambient wind velocity. The factor 0.4 is determined from wind tunnel measurements and has dimensions on cm $s^{-1/2}$. Various considerations including irregular shape, leaf curl, and leaf dimensions perpendicular to the wind may require other factors to be used (233).

Under all conditions boundary layer resistance is a significant component of total diffusion resistance and must be taken into account in gas exchange studies of canopies and single leaves. Its neglect has lead to erroneous conclusions about stomatal control of water loss (54, 145) and sensitivity of a range of plants to atmospheric pollutants (8).

By using an energy balance approach (220), it can be shown that changes in boundary layer resistance will have different effects on transpiration rate depending upon the available energy and the stomatal resistance. Decreases in boundary layer resistance will increase transpiration rate only when the stomatal resistance is small and the available energy is low (Fig. 21) (67, 107). In some situations, a decrease in boundary layer resistance can lead to a reduction in transpiration rate, and in intermediate conditions a change in boundary layer resistance will have no effect on leaf water loss. A low degree of stomatal control of transpiration when boundary layer resistance is high (12) can be seen in Fig. 21.

3. *Stomatal Resistance*

Compared to the water content of the leaf, the rate of water loss through open stomata is extremely high. Leaves often lose water at a greater rate than it can be supplied via the xylem, and unless the rate of water loss is reduced, the leaf may quickly suffer the severe consequence of desiccation. Although certain structural features, for example, cuticular development, will act to reduce the rate of transpiration, a fully grown leaf cannot

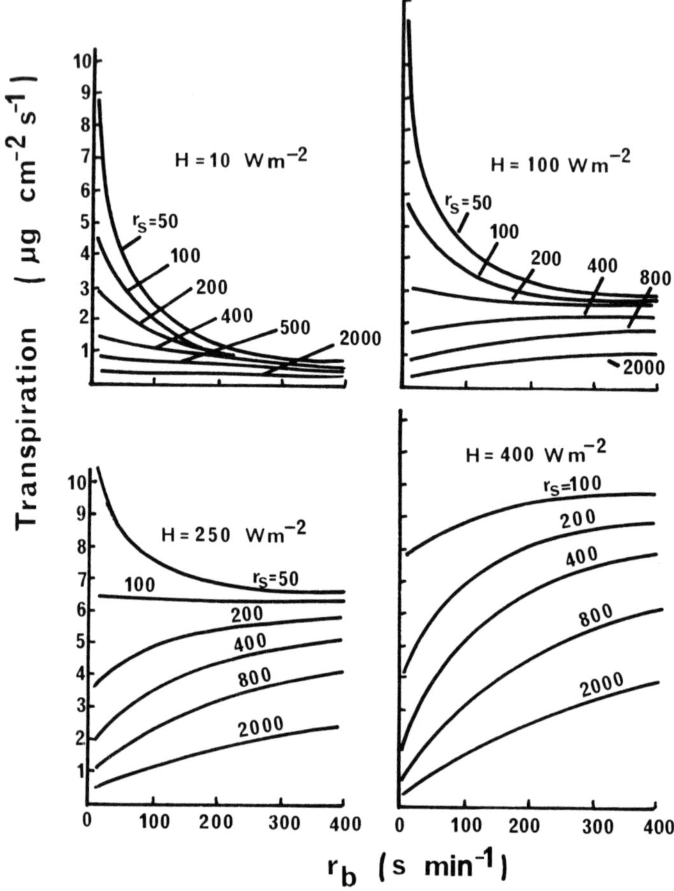

FIG. 21. Calculated rates of transpiration at various levels of available energy H and at different stomatal resistance r_s (s min^{-1}) as affected by changes in boundary layer resistance r_b (s min^{-1}). Atmospheric saturation vapor deficit 8 mb; air temperature 15°C. After Grace (107).

change its structure in order to conserve water and thus stomatal control of water loss can assume enormous importance.

The limited reserves of water held in a leaf and the consequent precarious water balance dependent upon precise stomatal control are shown by a simple experiment described by Mansfield and Davies (195) (Fig. 22).

An illuminated plant of *Commelina communis* is sealed in an inverted beaker such that gas exchange is restricted and a low carbon dioxide concentration and high humidity result. Under these conditions the sto-

FIG. 22. *Commelina communis* plants potted in moist soil. (a) One plant sealed in an inverted beaker for 1 to 2 h to restrict gas exchange with the ambient air. (b) The same plants 5 min after the beaker was removed. The leaves of both plants were exposed to a light wind. After Mansfield and Davies (195).

mata open widely (see Mansfield, Chapter 3 in this volume). Upon removal of the beaker, the leaves are exposed to a light wind. Within a few minutes the leaves are severely wilted and permanent damage to the lamina can result. As a result of the experimental treatment, the stomata were unable to close rapidly enough to maintain a favorable water balance in the leaves under conditions of high evaporative demand. In a control plant, a somewhat reduced stomatal aperture when the plants were subjected to wind resulted in the maintenance of leaf turgor. The importance of stomatal control of water loss is therefore illustrated by subjecting the plant to an "unnatural" change in ambient conditions.

Slatyer (303) has suggested that increases in resistance to water movement anywhere within the plant will ultimately reduce the flux through the whole system by influencing the water balance of the leaf and therefore influencing stomatal behavior. It is well know that stomata generally close when leaf water potentials are reduced. Many authors have reported a threshold level of water deficit above which stomata conductance increases markedly (327), although other authors have reported a continuous, gradual decline in leaf conductance as leaf water potentials decrease (79). It is possible that this second type of response is brought about by solute accumulation and therefore by turgor maintenance in the leaves as water potential falls. With this type of response it seems likely that some stomatal closure in response to water deficit will occur as a result of the normal daily fluctuations in water potential (18). Certainly, threshold responses seem more common in plants grown in less extreme environments, e.g., growth chambers (Fig. 23) (64, 153). It seems likely that differences among threshold responses of stomata in different species, in plants of different ages, and in plants grown under different conditions may be smaller if leaf conductances are used rather than leaf resistances. Moreover, if conductances are plotted as a function of leaf turgor rather than as a function of leaf water potential, differences are again minimized. The latter result suggests that stomata may act to maintain a positive leaf turgor rather than a particular level of leaf water potential (72).

Stomatal closure in response to an increase in leaf water deficit will generally reduce the rate of water loss, although as noted earlier this will depend upon canopy structure. If enough water is available in the soil to restore leaf turgor, closure may be followed by the reopening of stomata. An oscillation in stomatal conductance and in transpiration may result if the reopening of stomata in dry, hot conditions leads to a further increase in leaf water deficit (see Mansfield, Chapter 3 in this volume and 51, 52).

Cowan (52) has noted that the threshold responses of stomata, as well as the capacity to accumulate solutes in leaves as water potentials decline (see Section V,C,2), and the decreases in resistance to water movement in roots

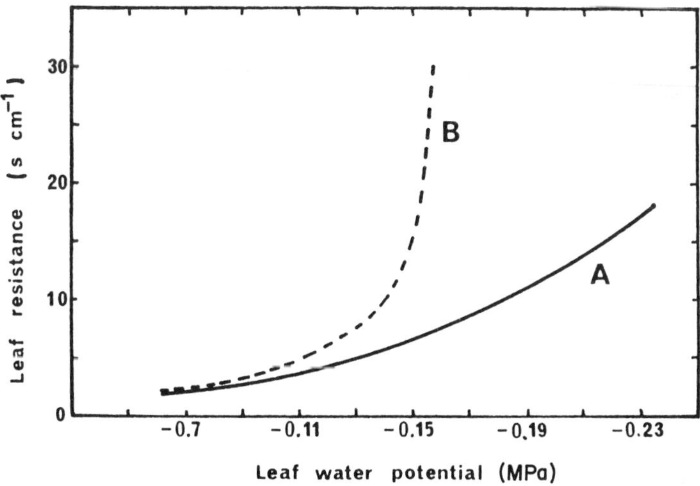

FIG. 23. Stomatal responses to plant water stress in cotton plants either grown in dry soil at a high vapor pressure deficit (i.e., subjected to water stress during their growing period) (A) or grown in moist soil at a low vapor pressure deficit (i.e., not subjected to water stress) (B).

and leaves of stressed plants (see Section III,C) are characteristics which tend to maintain the flux of water through the plant and thus preserve turgor rather than conserve water. Such characteristics are present in many crop plants and have probably been accentuated by the plant breeder (68) in that turgor maintenance can be combined with continued carbon dioxide exchange. Nevertheless, stomatal responses to other environmental factors, e.g., changes in irradiance and carbon dioxide concentration (see Mansfield, Chapter 3 in this volume), might be interpreted as turgor-preserving *and* water-conserving responses. Certainly the stomatal response to changes in humidity of the air exhibited by several plants (116, 282–286) can lead to an extremely effective conservation of water, but only at some expense to carbon dioxide exchange (268).

An increase in evaporation from the moist cell surfaces of the leaf following an increase in leaf–air vapor pressure deficit (VPD) will often result in leaf water deficit and stomatal closure in consequence. Schülze and coworkers (282–286) observed that the effect of an increase in the VDP between the leaf of *Prunus armeniaca* (apricot) and the air was a *decrease* in leaf conductance but an *increase* in the relative water content of the leaf (Fig. 24). From these results one must conclude that in this plant the water relations of the epidermis are to some extent independent of those of the mesophyll cells. The stomata of a large number of plants have now been

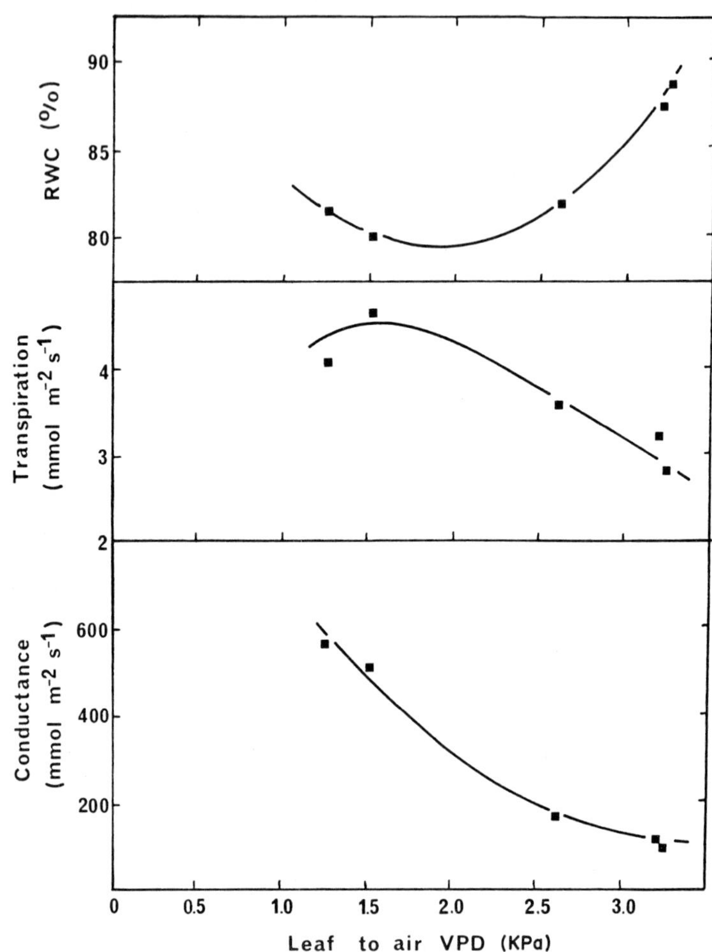

Fig. 24. Steady–state responses by irrigated *Prunus armeniacea* to changes in vapor pressure difference with a constant leaf temperature of 30°C and natural light conditions. After Hall *et al.* (116); data from Schülze *et al.* (282).

shown to be sensitive to a greater or lesser degree to changes in humidity (184) and the mechanism of this response is discussed in detail elsewhere (see 145, 146, and Mansfield, Chapter 3 in this volume). To explain the observed responses it is necessary to postulate a relatively large resistance to water movement between the mesophyll cells and the epidermis and the loss of some water from the outside of the guard cells or the cells immediately adjacent to the guard cells. This latter characteristic is necessary since the stomata of humidity-sensitive plants *stay* closed at high VPD despite the resulting decrease in transpiration and increase in relative water content (Fig. 24) (6, 144). Maier Maercker (186–191) and others (146) have demonstrated the presence of evaporation sites on the outer surfaces of leaves of a number of species, while Appleby and Davies (6, 7) have proposed a mechanism whereby an evaporating site in the guard cell walls of comparatively turgid *Ulmus* and *Quercus* leaves may be exposed to the desiccating influence of the air even though the stomata are closed (Fig. 25; see also Fig. 5). When water deficit develops, the guard cells may change orientation in the epidermis and the evaporating sites may be masked within the stomatal pore, greatly restricting total water loss.

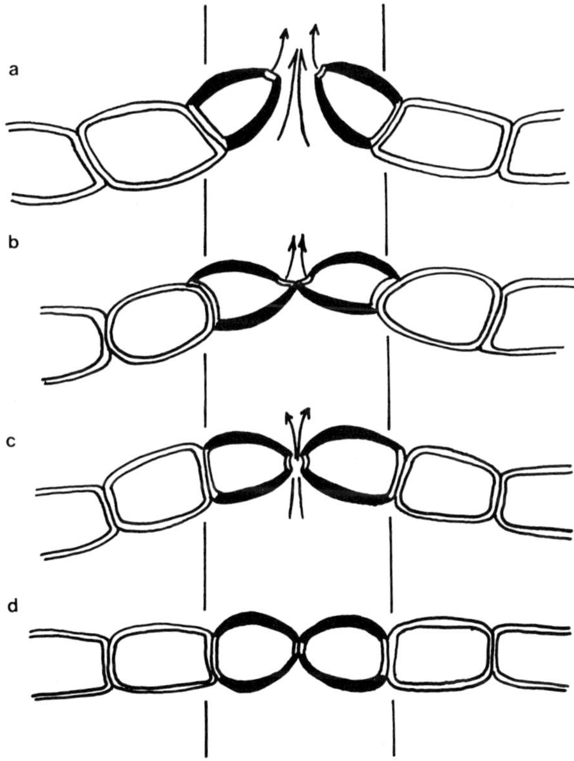

FIG. 25. Diagrammatic representation of orientation of guard cells and water loss from leaves of *Ulmus* seedlings which are (a) well watered in moist air, (b) well watered in dry air, (c) mildly stressed in dry air, and (d) severely stressed in moist or dry air. After Appleby and Davies (6).

Mansfield and Davies (195) and Mansfield *et al.* (196) have proposed that the humidity response may be part of the plant's first line of defense against developing water deficit, i.e., a response to an increase in evaporative demand. A stomatal response to humidity before excessive water loss occurs and *before* water deficit develops in the leaf is a highly efficient

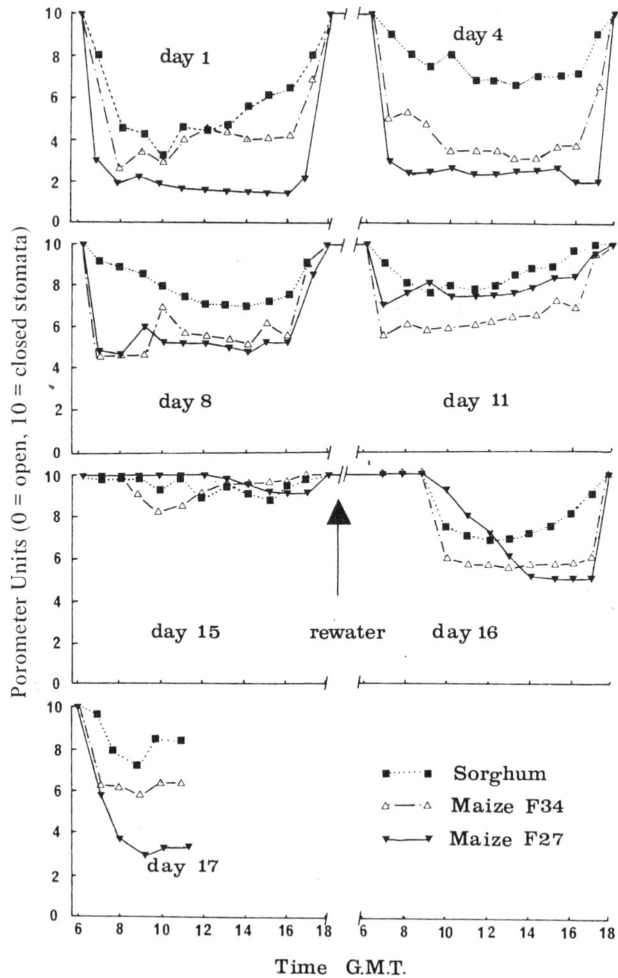

FIG. 26. The effect of increasing plant water stress on the stomatal behavior of three maize and sorghum cultivars. Plants were not watered after day 1 and were rewatered at the end of day 16. After Davies *et al.* (72).

strategy for water conservation and the avoidance of water deficit. If the first line of defense is inadequate, a second line of defense may be necessary, i.e., the direct response of the stomata to decreased ψ or P. This may involve a distress signal (possibly abscisic acid; see Mansfield, Chapter 3 in this volume) moving from the deficit-sensitive areas (the mesophyll chloroplasts?) to the guard cells to promote stomatal closure (195). Such a distress signal may provide a ceiling of stomatal conductance below which other stomatal responses operate (Fig. 26) or enhance the effect of the first line of defence (195, 239). If the stomata respond to both the conditions inside the leaf and the external conditions, a balanced regulation of production and water loss may be possible. Such a strategy may enable the plant to maintain a high level of metabolic activity in the mesophyll such that a considerable carbon dioxide gain may occur whenever the stomata are open.

The humidity response is an example of partial stomatal closure despite the occurrence of a favorable leaf water balance. Another well-documented example of such an occurrence is the delayed reopening of stomata following rewatering after deficit. After high leaf turgor has been reestablished, stomata may not open fully for hours or even days (72, 121) (Fig. 26). The mechanism of this response is discussed by Mansfield elsewhere in this volume (Chapter 3).

The delay in stomatal opening after low turgor has been experienced has been interpreted as a safety mechanism allowing the plant to regain full turgor more rapidly. This explanation is not entirely convincing since turgor is often restored several days before stomata exhibit full opening (358). Mansfield *et al.* (196) have proposed that partial stomatal closure in this situation may result in a highly beneficial increase in the efficiency of water use. If the first rain after a drought provides only a limited, temporary supply of water, an efficient, reduced use of water will be of some advantage to the plant.

C. STOMATAL BEHAVIOR AND WATER USE EFFICIENCY

Cowan (52) has proposed a simple extension of the general transport equation (1) to describe the factors which influence the rate of water loss from a leaf,

$$E = \frac{W_o - W}{r_l + \epsilon r_b^\# + r_b} \tag{30}$$

where W_o is an estimate of the humidity which would result in the leaf if leaf conductance were zero and ϵ were constant, ϵ is the rate of increase in the latent heat content of saturated air with increase in sensible heat content, $r_b^{\#}$ is the boundary layer resistance to heat transfer, W is the ambient humidity, and r_l and r_b are the leaf and boundary layer resistances to water loss.

The rate of assimilation A is described by

$$A = \frac{c - \Gamma}{r_b' + r_l' + 1/K} \tag{31}$$

where c is the ambient carbon dioxide concentration, Γ is the carbon dioxide compensation point, r_b' and r_l' are the boundary layer and leaf resistances to carbon dioxide exchange, and $1/K$ is the internal resistance to carbon dioxide exchange (the resistance to diffusion of carbon dioxide through the liquid phase of the mesophyll cells).

Allowing for the relationships $r_l' = 1.6r_l$ and $r_b' = 1.37r_b$, it follows that

$$\frac{c - \Gamma}{A} - 1.6(W_o - W) = \frac{1}{K^*} \tag{32}$$

with

$$\frac{1}{K^*} = \frac{1}{K} - 1.6\epsilon r_b^{\#} - 0.2r_b$$

$$\simeq \frac{1}{K} - 1.6\epsilon r_b^{\#} \tag{33}$$

Here $1/K^*$ is an expression of the "supraresistance," i.e., the additional resistance to transfer of carbon dioxide. It is not exactly analogous to the internal resistance since it includes what is equivalent to a resistance to transfer of water vapor that has no counterpart in carbon dioxide transfer, i.e., that associated with the term ϵ in Eq. (33).

Cowan writes $R = E/A$ for the transpiration ratio and rearranges Eq. (32) as

$$R = R_o + \frac{E}{K^*(c - \Gamma)} = \frac{R_o}{1 - A/K^*(c - \Gamma)} \tag{34}$$

where

$$R_o = \frac{1.6(W_o - W)}{c - \Gamma} \tag{35}$$

and the limit of R as leaf conductance approaches zero. Equation (34) shows how the transpiration ratio will vary with leaf conductance. If $1/K^*$ is positive, the transpiration ratio will increase with an increase in leaf conductance.

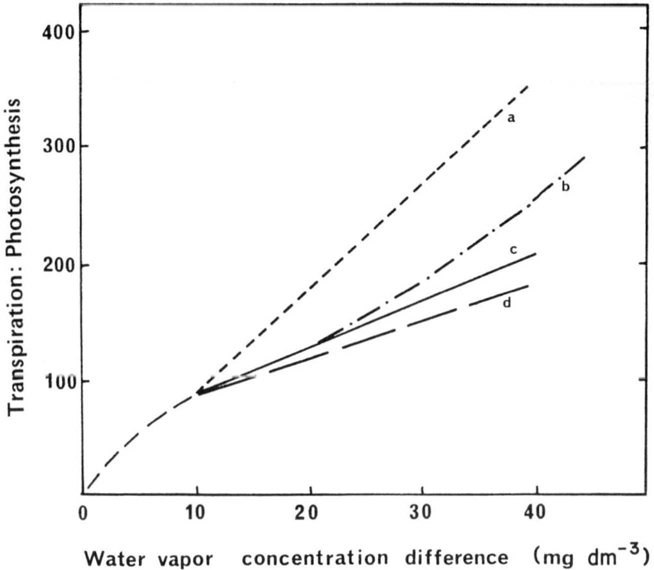

FIG. 27. The transpiration : photosynthesis (T : P) ratio as related to the change in water vapor concentration difference. (Note the concurrent increase in leaf temperature from 25°C at 10 mg H_2O dm^{-3} to 40°C at 40 mg H_2O dm^{-3}.) (a) Simulated T/P response at a constant diffusion resistance for water vapor and at constant photosynthetic activity; (b) observed values of the T : P ratio; (c) simulated T/P response calculated under the assumption that the stomata respond to both changing temperature and humidity at a constant photosynthetic activity; (d) simulated T/P response calculated with the assumption that stomata only respond to changing humidity at a constant photosynthetic activity. After Schülze *et al.* (286).

For most plants having C_3 metabolism, the internal resistance may be taken as having a minimum magnitude of about 6 m^2 s mol^{-1} (at sea level and at 25°C, a resistance expressed in terms of molar flux (m^2 s mol^{-1}) = 2.5 × a resistance expressed in s cm^{-1} and = 0.025 × a resistance expressed in s m^{-1}). In C_4 plants the internal resistance is often around 1 m^2 s mol^{-1}. Table IV shows the influence of wind speed, leaf size, and temperature on the difference between internal resistance and supraresistance and shows that in most C_3 species transpiration ratio will increase with increasing leaf conductance unless the boundary layer conductance is very small or the temperature very high (52). In C_4 plants $1/K^*$ will be negative and the transpiration ratio will decrease with stomatal opening unless the boundary layer conductance is large and the temperature relatively low. Sinclair *et al.* (300) have shown that the water use efficiency of field-grown maize decreases as stomatal resistance increases. Partial stomatal closure caused by ABA application to C_3 plants can lead to a decrease in the transpiration ratio [Table V (194)]. In addition, humidity-induced closure of stomata appears to decrease the transpiration ratio of some plants (286) (Fig. 27). Jones (148) has concluded that particularly in plants in which cuticular conductance is high, the most efficient use of water may occur at an intermediate stomatal aperture. Conclusions of this type are complicated when water loss from canopies rather than single plants is considered (see above).

Equations (31) and (32) show clearly that the instantaneous efficiency of water use by the plant (E/A) is a function of both the environment and the magnitude of the resistances to diffusion and that under some circumstances, the environment will have a dominating influence on E/A. For example, as a result of diurnal variation in humidity and temperature, the transpiration ratio of individual plants will often reach a maximum around midday. Incident irradiance also has an important effect on transpiration ratio (82). Because of rising leaf temperature and falling r_s, transpiration of individual plants increases in a linear or curvilinear fashion with increasing irradiance. Net photosynthesis, especially of C_3 species, shows downward curvilinearity with decreased irradiance and is negative at zero irradiance. Thus, there is an optimum irradiance for maximum efficiency of water use which is usually less than the irradiance incident upon a leaf which is normal to the sun's rays (148). Leaf movements, e.g., rolling and flagging, are common in arid zone plants. These movements can be considered adaptations to reduce the effective incident irradiance and increase the efficiency of water use (82). Increased reflectivity of leaves would have the same effect.

We have shown above how a change in stomatal resistance will influence the efficiency of water use and how the magnitude of this effect will depend upon the magnitude of the boundary layer and internal resistances. Under most conditions, boundary layer resistance is small when compared to the stomatal or internal resistances. It is unlikely, therefore, that changes in r_b will influence markedly the water use efficiency of exposed leaves. Although the changes may only be small, calculations suggest that transpiration ratio will decrease with decreases in r_b unless radiation levels are low or VDP and air temperatures are high (148). Parkhurst and Loucks (245) have suggested that since boundary layer resistance is dependent upon leaf size, a major factor explaining leaf size variation across plant communities may be the need to maximize the efficiency of water use.

The calculations and conclusions with regard to the effect of stomatal closure on water use efficiency depend upon the assumption that the internal resistance remains relatively constant as the stomatal resistance

TABLE IV

Estimates of 1.6 ϵr_b in $m^2\ s\ mol^{-1}$ for Various Combinations of Wind Speed u, Leaf Breadth b, and Temperature[a]

Temperature (°C)	ϵ	u/b (s^{-1}) 2	5	10	20	50	100	200	500	1000
16	1.75	3.5	2.7	2.1	1.7	1.1	0.8	0.6	0.4	0.3
32	4.06	7.6	5.9	4.7	3.7	2.5	1.9	1.4	0.9	0.7
48	8.49	14.0	11.5	9.4	7.4	5.2	3.9	2.9	1.9	1.4

[a] After Cowan (52).

TABLE V

Effects of Chemically Induced Closure of Stomata on Transpiration and Growth[a]

Author(s)/plant species/ treatments	Observations
Zelitch and Waggoner (366a)/ *Nicotiana tabacum*/ PMA: 3.3×10^{-4} M, 1×10^{-4} M, and 3.3×10^{-5} M	$(P_t/P_o)/(T_t/T_o)$ ranged from 1.40 to 0.96 and was significantly above unity; ($P < 0.05$); P_t, P_o = photosynthesis in treated and control plants; T_t, T_o = transpiration in treated and control plants
Shimshi (299a)/ *Nicotiana tabacum*/ PMA: 9×10^{-5} M	Ratio of growth over transpiration increased from 0.78 to 1.00 in well-watered plants and from 0.83 to 0.93 in water-stressed plants; growth was indicated by leaf area
Slatyer and Bierhuizen (305)/ *Gossypium* sp./ PMA: 10^{-4} M	Grams transpiration/grams dry tissue produced was 245 for controls and 180 for treated plants
Davenport (63)/ *Festuca rubra*/ PMA: $10^{-3.5}$ M	Grams transpiration/grams dry tissue produced was 2052 for controls and 1734 for treated plants
Jones and Mansfield (152a)/ *Hordeum vulgare*/ ABA and its methyl and phenyl esters, all at 10^{-4} M	Grams transpiration/grams dry tissue produced was 560 in controls, 410, 390, and 390, respectively, after treatment with ABA, methyl-ABA, and phenyl-ABA
Mizrahi *et al.* (217a)/ *Hordeum vulgare*/ ABA: 3.8×10^{-4} M	Cumulative transpiration over 20 days was approximately 13 cm^3 in controls and 5 cm^3 in treated seedlings; total dry mass was 24% greater in treated plants (over unspecified period), these figures applied to nonwatered plants
Raschke (267)/ *Xanthium strumarium*/ ABA: 10^{-5} M	Grams transpiration/grams dry tissue was reduced by one-half, while net photosynthesis was reduced only by one-seventh

[a] After Mansfield (194).

changes. There is evidence (28) that as leaf water potential falls, an increase in internal resistance will occur in concert with an increase in stomatal resistance. If such increases occur, this means that the efficiency of water use will be unchanged or reduced when stomata close due to increasing water deficit. Boyer (28) has suggested that following rewatering after drought, internal resistance will decrease more rapidly than stomatal resistance, leading to an increase in the efficiency of water use.

Although low internal resistance to carbon dioxide diffusion in C_4 plants means that the effect of partial stomatal closure on transpiration ratio is not as favorable as in C_3 plants, the efficiency of water use in C_4 plants when the stomata are open can be extremely high. This is because the ratio of resistances to water loss compared to the resistances to carbon dioxide exchange is 0.7 or 0.8 in C_4 plants but only 0.2 or 0.4 in C_3 plants (323). Under the same ambient conditions the efficiency of water use of C_4 plants may be two to three times higher than that of C_3 plants (72, 304). Fischer and Turner (88) have pointed out that C_4 plants generally occupy warmer, drier habitats than C_3 plants and therefore the difference between the two groups in efficiency of water use may not be as great when they are compared under their respective field conditions.

The majority of the estimates of water use efficiency reported here are the result of measurements made at a individual leaves (e.g., 78). It is important to note that differences are often reflected largely unchanged at the canopy level and in terms of dry matter gain per unit of water lost (88). The exact relationship between water use efficiency in terms of carbon dioxide exchange and dry matter increment will depend upon the carbon content of dry matter produced and nighttime respiration and transpirational losses. Water use efficiency based on dry matter accumulation will be lower as a result.

Cowan (52) and Cowan and Farquhar (55) have considered the possibility that stomata control gas exchange so that the total *diurnal* loss of water by the plant is the minimum possible for whatever total amount of carbon is taken up. This corresponds to minimizing

$$\int_0^t E\,dt \tag{36}$$

subject to the constraint of

$$\int_0^t A\,dt = \text{const} \tag{37}$$

where E and A are the rates of transpiration and carbon dioxide assimilation, respectively, and t is a period of time. A certain pattern of stomatal behavior is optimal if the ratio

$$\frac{\partial A/\partial g_1}{\partial E/\partial g_1} \tag{38}$$

of the sensitivities of rates of transpiration and assimilation to changes in conductance g_1 remains constant during time t. This ratio may also be written as $\partial E/\partial A$ and called the gain ratio (84) so that stomatal behavior is considered to be optimal if the gain ratio is constant. By applying this

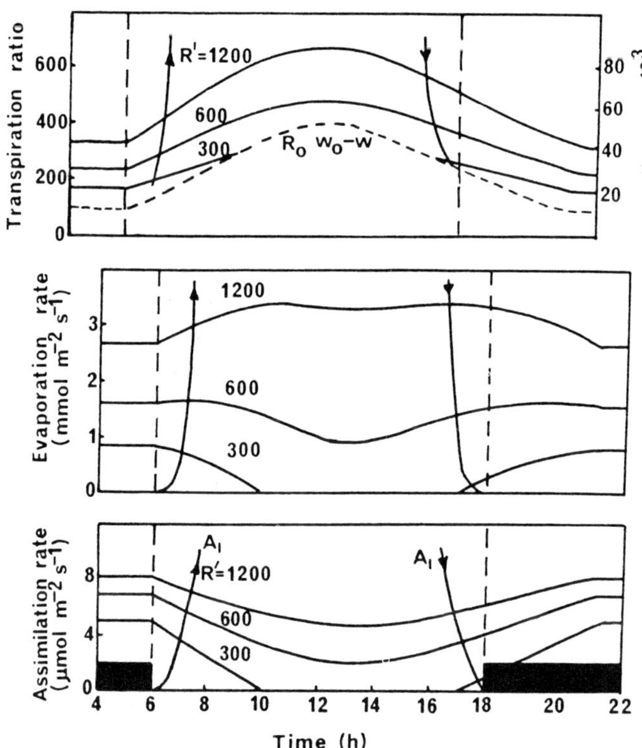

FIG. 28. Diurnal variation of transpiration ratio $R(E/A)$, rate of evaporation E, and rate of assimilation A, based on a model described by Cowan (52), for the three magnitudes of R', the gain ratio ($\partial E/\partial A$) shown (R' is an indication of the plant's potential for growth or its need to conserve water). The line R_0 represents the limit of R as leaf conductance approaches zero; it also represents the difference in humidity ($W_o - W$) between leaf and air that would obtain if leaf conductance were zero. The lines marked A_1 and their counterparts in relation E and R represent limitations associated with the influence of light intensity on rate of assimilation and leaf conductance.

criterion to equations describing the rate of evaporation and the rate of assimilation in a leaf [Eqs. (30) and (31)], Cowan (52) has determined an optimal time course for leaf conductance and the corresponding rates of gas exchange in terms of the metabolic characteristics of the leaf and the environment to which it is exposed (Fig. 28). Despite some anomalies, the solutions obtained are in qualitative agreement with certain aspects of stomatal behavior observed in the field. For C_3 plants, optimization requires that midday stomatal closure should occur. Such behavior has been observed on many occasions (Fig. 29) and is probably attributable to the effect of water deficit perhaps combined with the effect of carbon dioxide on stomata (206) and/or the direct stomatal response to the difference in VPD between leaf and air (284).

When the conclusions reached previously are borne in mind, Cowan's analysis requires that for optimal efficiency of water use in C_4 plants, some stomata should be completely open and others should be completely closed. There is little evidence that such behavior occurs, although observation of rapidly transpiring leaves often reveals a percentage of stomata that are completely closed (243). These stomata have generally been assumed to be nonfunctional.

Farquhar and co-workers (85) have monitored the gas exchange of

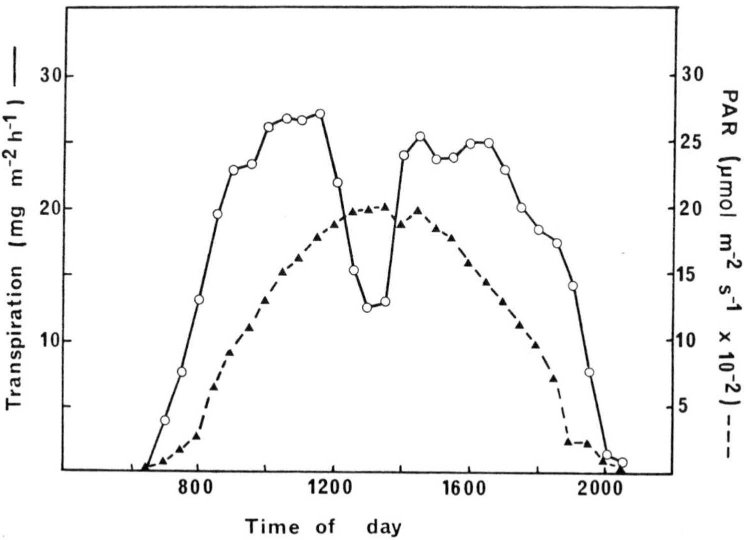

FIG. 29. Diurnal variation in transpiration by a single leaf from a cotton plant growing in the field. Diurnal variation in photosynthetically active radiation (PAR) is also shown.

intact leaves of plants of *Nicotiana glauca* and *Corylus avellana* exposed to a range of VPDs. The gain ratio $\partial E/\partial A$ of these plants remained substantially constant as VPD was increased from 0.8 to 3.2 kPa. These results therefore exhibit the hypothetical requirements of Cowan and Farquhar (55) for optimal regulation of gas exchange. It is clear from Fig. 30 that the stomatal behavior of *Nicotiana* over the range of experimental humidities conformed most closely with a hypothesis that $\partial E/\partial A$ is kept constant, and the measured data show strong deviation from the hypothetical responses required if assimilation A or transpiration E were kept constant. Interestingly, when VPD was increased, constant $\partial E/\partial A$ demanded increasing E/A. While some plants apparently exhibit some degree of "optimal" regulation of gas exchange, it is now becoming clear that this is not always the case (e.g., 357a).

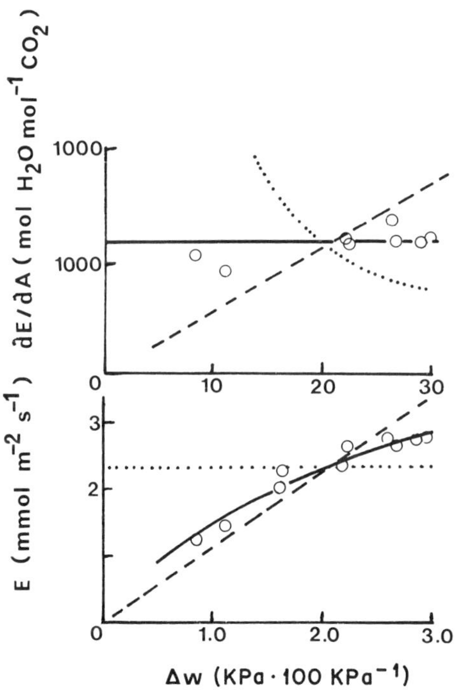

FIG. 30. $\partial E/\partial A$ and rate of transpiration E as related to humidity difference between leaf and air Δw for *N. glauca*. The circles represent the measured data. The relationships represented by the various lines were derived by assuming constant $\partial E/\partial A$ (1200 mol H_2O mol^{-1} CO_2), constant E (2.36 mmol m^{-2} s^{-1}), and constant A (7.7 μmol m^{-2} s^{-1}). Constant E is represented by the dotted line, constant $\partial E/\partial A$ by the solid line, and constant A by the broken line. After Farquhar *et al.* (85).

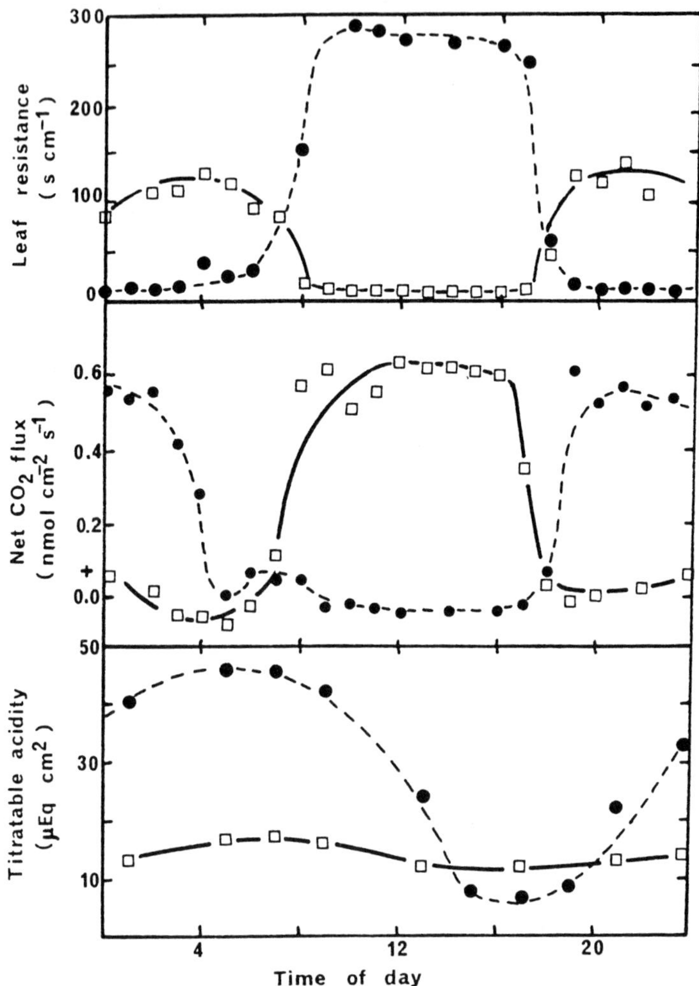

FIG. 31. Conversion from crassulacean acid metabolism (CAM, ●) to daytime CO_2 uptake (□) after watering *Agave deserti.* Data were obtained in identical conditions using attached leaves, 2 weeks (CAM), and 12 weeks after raising soil water potential from less than −9.0 MPa to −0.01 MPa. After Hartsock and Nobel (120).

Crassulacean acid metabolism (CAM) plants provide an extreme example of how stomatal behavior may maximize daily water use efficiency (237). It is now well known that under certain conditions CAM plants can exhibit complete closure of stomata during the daylight hours but open their stomata and fix carbon dioxide during the dark period (Fig. 31). During the dark period, VPD and temperatures are low, and this com-

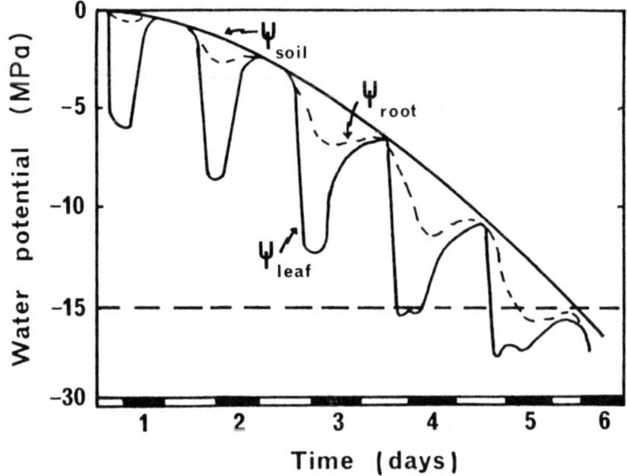

FIG. 32. Diagram showing likely changes in leaf water potential (ψ_{leaf}) of root water potential (ψ_{root}) of a transpiring plant rooted in soil allowed to dry from a soil water potential (ψ_{soil}) near zero (wet soil) to a water potential at which wilting occurs. The dark bars indicate periods of darkness. After Slatyer (303).

bined with a resistance ratio similar to that exhibited by C_4 plants means that the efficiency of water use by CAM plants can be extremely high (226). When there is plenty of water in the soil, CAM plants have the capacity to switch to "daytime fixation" thereby increasing total carbon dioxide uptake but decreasing the efficiency of water use (120).

V. Development and Effects of Plant Water Deficit

A. PROGRESSIVE DEVELOPMENT OF PLANT WATER DEFICIT

Figure 32 (50, 303) shows the daily variation in water potential of a plant growing in drying soil. In the absence of added soil water, both plant and soil water potential decrease over a number of days. The daily variation in plant water potential occurs because of the daily variation in transpiration and because absorption lags behind transpiration as water is removed from the cells of the plant. Leaf water potential is given by Eq. (9a). When the soil water potential is relatively high, plant water potential returns during the night to a value equal to soil water potential. As the soil dries, the hydraulic conductivity of the soil decreases and the rate of water movement toward the roots is too slow to completely replace daytime losses

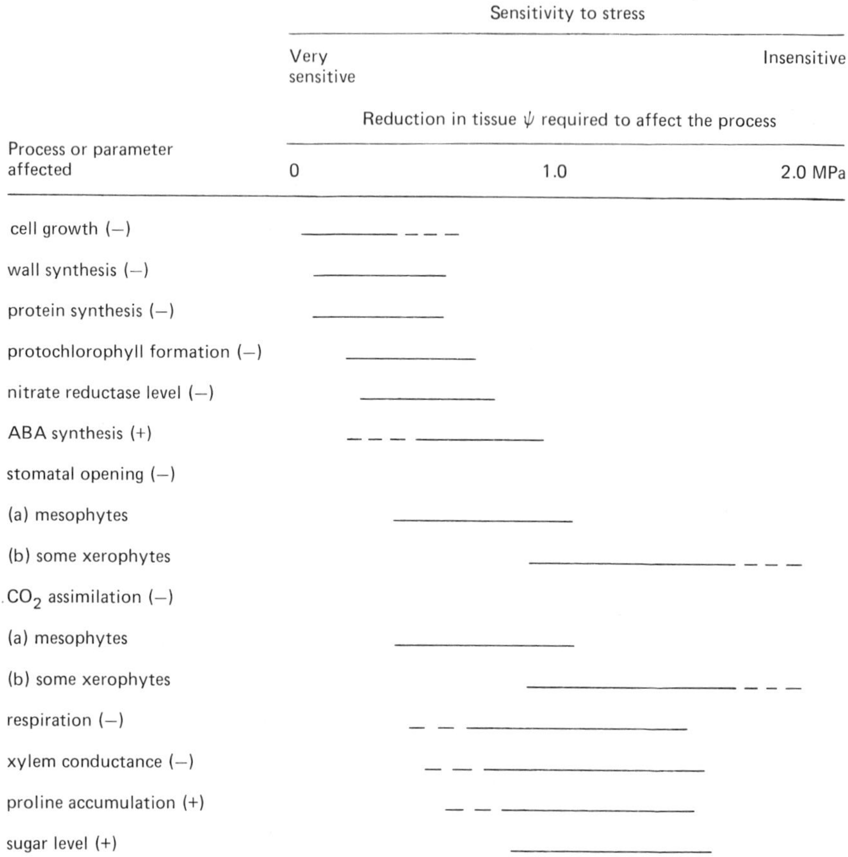

Process or parameter affected	Sensitivity to stress: Very sensitive — Insensitive Reduction in tissue ψ required to affect the process 0 — 1.0 — 2.0 MPa
cell growth (–)	
wall synthesis (–)	
protein synthesis (–)	
protochlorophyll formation (–)	
nitrate reductase level (–)	
ABA synthesis (+)	
stomatal opening (–)	
(a) mesophytes	
(b) some xerophytes	
CO_2 assimilation (–)	
(a) mesophytes	
(b) some xerophytes	
respiration (–)	
xylem conductance (–)	
proline accumulation (+)	
sugar level (+)	

FIG. 33. The sensitivity to water stress of plant processes or parameters. Length of horizontal lines represents the range of stress levels within which a process first becomes affected. Dashed lines signify deductions based on more tenuous data. Plus signs indicate that water stress causes an increase in the process or parameter and minus signs signify a decrease. Reference to individual observations found in Hsiao (129).

(303). When leaf water potential is sufficiently low to cause stomatal closure (and on an extremely hot, dry day this may occur even when the soil is wet), transpiration will be reduced and water potential may increase. Unless water is added to the soil, however, the plant will eventually reach zero turgor (plant water potential = plant osmotic potential = soil water potential). This point, often referred to as the permanent wilting point, is a function of the osmotic potential of the plant (301). Although an oversimplification, Fig. 32 summarizes many of the interrelationships described in the previous sections. Clearly, spatial distribution of leaves and roots and the water relations of different cells and tissues should be considered in any detailed treatment.

B. PHYSICAL AND CHEMICAL LINKS BETWEEN PLANT WATER STATUS AND PLANT PROCESSES

When water is withheld from a plant, the effects on plant water balance are commonly expressed in terms of water potential. The loss of water from plant tissue has several effects which may influence plant metabolism, the most important being (1) reduction in the chemical potential or activity of water, (2) concentration of macromolecules and solutes, (3) changes in the spatial relationship of organelles and membranes, and (4) reduction in hydrostatic pressure inside the cell. Hsiao (129) has concluded that effects (1) and (3) are unlikely to be of importance when the plant is subjected to mild water deficit. The chemical activity of water will be reduced by less than 1% when the water potential falls to 1.0 MPa, and we know that halophytes can grow and develop rapidly at very low water potentials. Volume changes of cells, membranes, and organelles due to water deficit may be important. Kramer (168) has suggested that future research will prove that decreased hydration and changes in conformation of macromolecules are important with respect to the effects of water deficit on enzyme activity. Effect (2) is probably only important if the biochemical reactions are catalyzed by allosteric enzymes which are sensitive to small changes in concentration of specific effectors. In contrast, effect (4) can be shown to directly influence crucial physiological processes (see later) (129). Significantly, changes in turgor are most pronounced when a fully turgid plant loses water (Fig. 16). Although water potential values may not by themselves be crucial in determining plant behavior, tissue water potential is still used as a convenient indicator of plant water deficit. In view of the important effects of reductions in plant turgor, it is logical that turgor measurements should be made in studies of the effects of water deficit, particularly where investigators are concerned with drought effects on growth. We shall see later how it may be important to measure the water relations of growing cells.

C. EFFECTS OF WATER DEFICIT ON PLANT PROCESSES

1. *Reduction in Leaf Expansion, Photosynthesis, and Yield*

The indications are that almost every plant constituent and process can be altered if plant water deficit is severe enough and lasts long enough. If the deficit develops slowly relative to the speed of molecular events, responses occurring first can be considered to be most sensitive. Hsiao (129) has provided an indication of the generalized sensitivity to water deficit of a number of plant processes shown in Fig. 33, which assumes an absence of osmotic adjustment.

The exceptional sensitivity of leaf enlargement to water deficit can be seen in the data of Boyer (24, 25) and Acevedo *et al.* (1) (Fig. 34). Most of this response is an effect of water deficit on cell enlargement with cell division being generally somewhat less sensitive (209). From the viewpoint of dry matter production, expansion growth deserves special attention since it is the means of developing leaf area for intercepting radiation and carrying out photosynthesis. It is well known that for an irreversible increase in the dimensions of a cell to occur, a minimum cell turgor is a prerequisite (46). This is shown clearly in the approach used by Lockhart (see Chapter 1) (180, 181) to describe the growth process in simple terms.

In its simplest form the theory states that both water uptake and irreversible expansion of the cell wall are required for cell enlargement. In principle, either one of these processes may limit growth. Lockhart (180, 181) has argued that during steady-state growth, the rate of water flux into

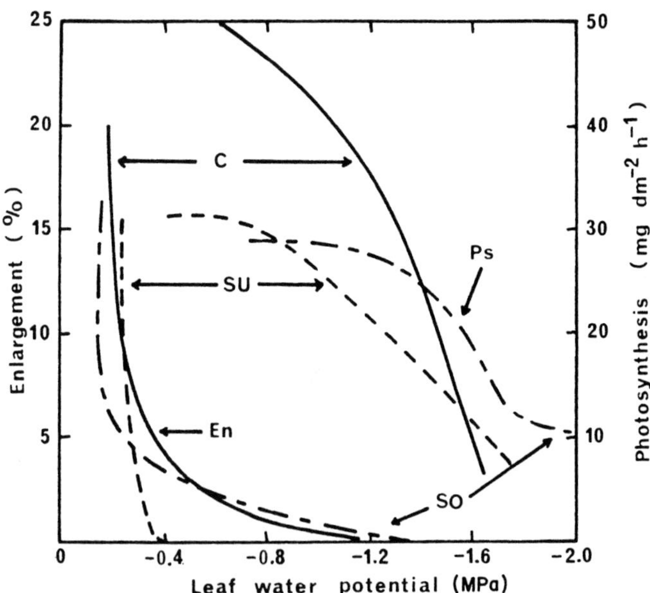

FIG. 34. The relation of leaf enlargement (En) and net photosynthesis (Ps) to leaf water potential for three crops as measured in a growth chamber. C, corn; SO, soybean; SU, sunflower. (Note: 10 mg dm^{-2} h^{-1} ≡ 0.28 mg m^{-2} s^{-1}.) After Boyer (25).

a cell must equal the rate of irreversible volumetric expansion of the cell wall chamber.

Combining an equation describing water influx with one describing wall expansion, the general equation for growth rate in the steady state V_s is

$$V_s = \frac{L\phi}{L+\phi}(\sigma\,\Delta\pi - Y) \tag{39}$$

where L is hydraulic conductivity, ϕ is wall extensibility, Y is wall yield threshold, $\Delta\pi$ is the osmotic concentration gradient across the plasmalemma, and σ is the reflection coefficient.

Following some perturbation of the properties described in Eq. (39), water influx may not equal cell wall expansion. Equation (39) is therefore valid only during steady-state growth. Cosgrove (48) has provided an example of such a situation in which an instantaneous doubling of wall extensibility will cause the rate of wall expansion to double instantaneously but will have no immediate effect on the rate of water influx. The higher rate of wall expansion will necessarily cause turgor to decrease, thereby increasing the driving force for water influx and hence the rate of water influx. Turgor will continue to decrease until the rate of water influx and wall expansion again equal each other. At this point, a new steady-state growth rate is restored and Eq. (39) is again valid. The difference between the volume generated by water influx and the volume generated by irreversible wall expansion is accounted for by reversible changes in the cell walls during the adjustment in turgor pressure.

In Eq. (39) the coupling of turgor and growth is implicit, but in an extension of this approach by Cosgrove (48) the coupling becomes explicit. In addition, his analysis describes the time-dependent behavior of growing cells, and although the Lockhart equations strictly apply only to single cells, Cosgrove's treatment includes the more complicated case of growing tissue in which the resistance to water flow is distributed throughout the tissue. The conclusions drawn from the two treatments are, however, rather similar.

In the steady state, turgor is closely linked to growth rate and the concept of turgor-driven cell expansion has been supported by numerous experimental results (Fig. 35) (24, 112, 113, 318). Withholding water from the plant can clearly directly influence cell turgor, but in addition to this, other components of the growth equation may also be influenced. For changes in water permeability to influence growth, water uptake must at least in part be limiting for cell expansion. Several studies have suggested that this may be the case (112, 113), while others suggest that growing cells are essentially in osmotic equilibrium with their water source (24, 219). Water potentials induced by growth (290, 356) may be an indication of

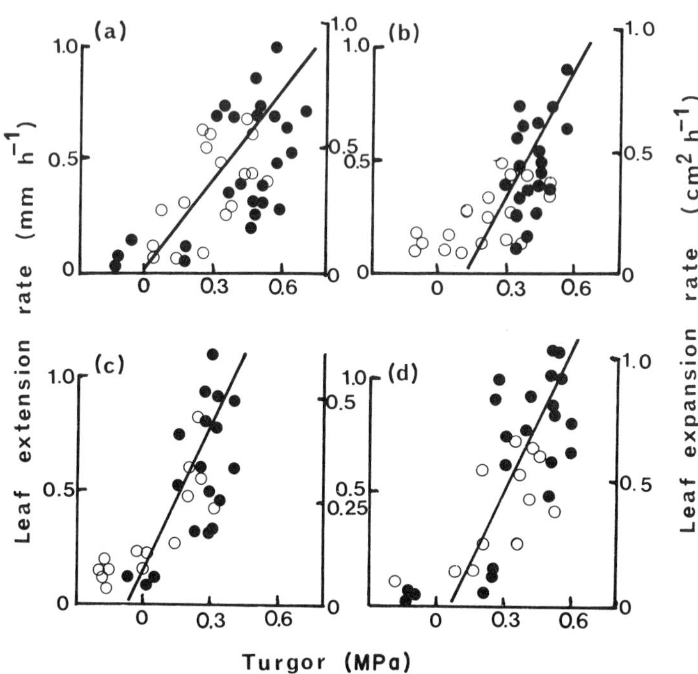

FIG. 35. Relationship between leaf extension rate and turgor pressure for four sunflower cultivars: (a) Hysun 31, (b) Havasupai, (c) Hopi, and (d) Seneca, during the light (○) and dark (●) periods. The leaf extension rates were averaged over 1 h prior to measurement of turgor pressure. Comparable leaf expansion rates are also presented; differences among cultivars in the relationships between leaf extension and leaf expansion arise from differences in leaf shape. The coefficients of determination r^2 for the fitted linear regressions are 0.50, 0.40, 0.65, and 0.53 for Hysun 31, Havasupai, Hopi, and Seneca, respectively. After Takami *et al.* (318).

limiting resistance to water flow in plant tissues, but Cosgrove and Cleland (49, 49a) have suggested that these potentials may develop due to the existence of a significant quantity of solute in the apoplast of the cell. Concentrations of auxin may be reduced in tissue which develops some water deficit (363), and there is some evidence that auxins can influence the water permeability of growing tissue (31), although this point is controversial (49a). In addition, auxins may influence leaf growth by altering the extensibility of cell wall materials (45, 47), and ABA may also influence this property (271, 340). Wall stiffness can be increased in plants experiencing plant water deficit with the result that plants previously exposed to water deficit may grow more slowly when turgor is restored following rewatering than do control plants which have remained well watered throughout the experiment.

Several studies have shown that water deficit can act to increase the threshold turgor for leaf growth *Y* (200, 334), while others note no apparent change in *Y* with increasing water deficit (60) or even a decrease in *Y* with deficit (37). The threshold turgor for leaf growth is not easy to measure, but it is important to attempt to do so because metabolism can change the value of *Y*. Since the driving force for growth is $P - Y$ (where *P* is leaf turgor), this means that turgor can be a very insensitive indicator of long-term growth. It seems likely that as leaves grow toward full expansion, *Y* will increase with respect to *P*.

In addition to a marked effect of plant water deficit, the growth rate of leaves can be greatly influenced by several other environmental factors. Watts (346, 347) has shown that growth rates of maize leaves are greatly influenced by temperature, and Van Volkenburgh and Cleland (338, 339) and Takami *et al.* (318) have described a stimulation of leaf growth by light in *Phaseolus* and *Helianthus,* respectively. Interestingly, light stimulation of leaf growth in *Helianthus* was particularly marked in plants subjected to drought. Leaf growth of maize and wheat is apparently not stimulated by light (69, 291); this may be because the growing regions of monocotyledons are surrounded by the sheath of other leaves and may even be below the surface of the soil and therefore not subjected to larger changes in irradiance. Van Volkenburgh and Cleland (339) have shown that increased growth of *Phaseolus* leaf segments in light was not associated with an increase in cell turgor. Rather, these authors found that exposure to light doubled the rate of cell enlargment by increasing cell wall extensibility. This was due to light-stimulated acidification of the cells walls (339) resulting in acid-induced wall loosening (270). Cosgrove (48) has used Lockhart's analysis (180, 181) to show how changes in cell wall extensibility and cell permeability can influence growth rate. It is clear from this treatment that increased wall extensibility can increase cell growth while cell turgor is decreasing, underlining the fact that we might expect to find correlations between growth rate and turgor only when these measurements are made in the steady state.

Acevedo *et al.* (1) have shown that reduced growth during a period of very mild water deficit can be offset completely by a rapid but transitory phase of growth following release from stress. The result of such a burst of growth is no net reduction in total growth for the period. This suggests that during the period when water availability is reduced growth is limited only by the physical force required for expansion and that metabolic processes necessary for growth are unaffected. Most evidence indicates that cell division is generally less sensitive to increasing water stress than is cell expansion (156) so that small cells produced during the stress period may expand when water is supplied although it is reasonable to assume that a prolonged period of mild water deficit or a short period of more severe deficit will result in a growth reduction that is only partially reversible. Undoubtedly, protein synthesis, cell wall synthesis, and membrane proliferation will be reduced if the deficit is more severe, although the mechanism of this limitation is not well defined. If cell expansion is stopped for an extended period, the accumulation of metabolites resulting from reduced growth may have a feedback effect on carbohydrate and nitrogen metabolism. In addition, the rate of leaf initiation may be reduced if water deficit is severe or prolonged.

Figures 33 and 34 show that in many plants cell growth may be much more susceptible to water deficit than is carbon dioxide assimilation. As Hsiao (129) points out, it cannot be assumed that dry matter production is not affected if plant water status does not fall to a level that reduces stomatal opening and photosynthesis. Whether a mild water deficit will affect dry matter yield depends on whether leaf area is limiting the crop's assimilation of carbon dioxide (87). In a crop in which leaf area index (LAI) is high and photosynthesis is not limited by light interception or by an accumulation of assimilates (225), a mild deficit should have little effect on dry matter yield. In a crop in which LAI is low, the effect of deficit on final yield may be much more severe. Similar reasoning suggests that yield should be less sensitive to a shortage of water during the filling of grain or storage organ (130).

Leaf water deficit of a more severe nature will eventually have many direct effects at the cellular level (178). Boyer (28) has concluded that photosynthesis is reduced by the lowered availability of water both as a result of stomatal closure and as a result of effects at the chloroplast level. In general, stomatal effects on photosynthesis are likely to be most important at high irradiance, since under these conditions there is a large flux of carbon dioxide to the chloroplasts. Carboxylation activities, which may be influenced by water deficit (130, 136), are unlikely to affect photosynthesis to a large degree at high irradiance. As radiation decreases, chloroplast effects should become increasingly predominant; for example, the photochemical activity of chloroplasts may be reduced as water deficit increases (30, 155, 258). Both dark respiration and photorespiration are influenced by changes in plant water status (26, 28).

When yield consists of reproductive structures, the partition of assimilates among various plant parts becomes extremely important (217). Hsiao *et al.* (132) have pointed out the complexity of the situation viewed in terms of size and strength of sinks for assimilates. Influorescence development, fertilization, abortion of fertilized ovaries, fruit filling and abscission, and leaf senescence are all important processes that may be influenced by water deficit, but there is comparatively little comprehensive information available on their sensitivity to deficit. Clearly, the timing of any reduction in water availability is likely to be important, along with a consideration of whether the harvest can be delayed so that compensatory growth can take place.

2. *Adjustment of Plant Properties to Sustain Growth and the Efficient Use of Soil Water as Plant Water Potentials Decline*

a. Solute Adjustment and the Growth of Roots and Shoots. Figures 33 and 34 suggest that leaf expansion may be substantially reduced when leaf water potential falls by less than 1 MPa. In certain situations, however, e.g., at the top of tall trees, leaves *never* attain high water potentials. The fact that these leaves grow at all suggests that they are capable of adjusting their properties in some way. We have already seen that as water potential falls, the threshold turgor for cell expansion may adjust to maintain cell growth. For turgor to be maintained, solutes must be accumulated at a substantial rate by growing cells. A further increase in the rate of accumulation must occur if turgor of growing cells is to be maintained as water potential falls. Some reports suggest that this may be the case.

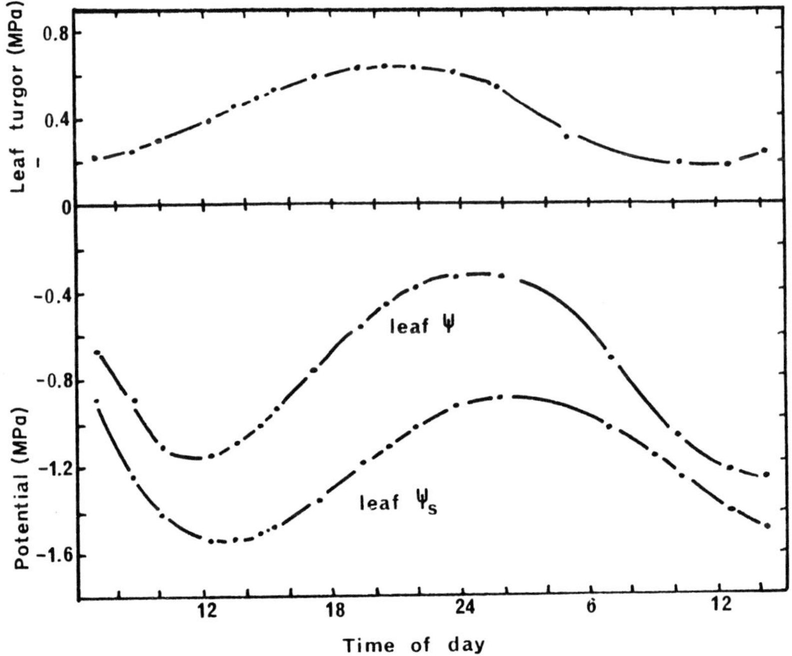

FIG. 36. Diurnal patterns of water potential, solute potential, and pressure potential in expanding leaves of unirrigated maize. After Fereres *et al.* (86).

A lowering of osmotic potential in response to increasing water deficit can arise simply as a result of the concentration of solutes in the vacuole as water is withdrawn from the cell or as a result of the active *accumulation* of solutes in the cell. In higher plants this latter process is referred to as solute or osmotic adjustment. Although this mechanism of turgor maintenance has long been recognized in halophytes (95, 311) and evidence of osmotic adjustment in response to water deficits has been available for many years, it is only comparatively recently that its importance to droughted plants has been recognized. Considerable evidence has accumulated that osmotic potentials of leaves, hypocotyls, roots, and reproductive organs of several plants are lowered in response to developing water deficit (106, 209, 223, 290, 328). It is likely that in several of these many reports, solutes are merely concentrated rather than accumulated actively, but in those plants in which turgor remains constant as water potential falls (328) there is little doubt that active accumulation has taken place. Reports of only limited solute accumulation are common (328), but in many of these the water relations of fully expanded leaves have been monitored. Jones and Turner (151) and Michelena and Boyer (212) have noted that there may be signifi-

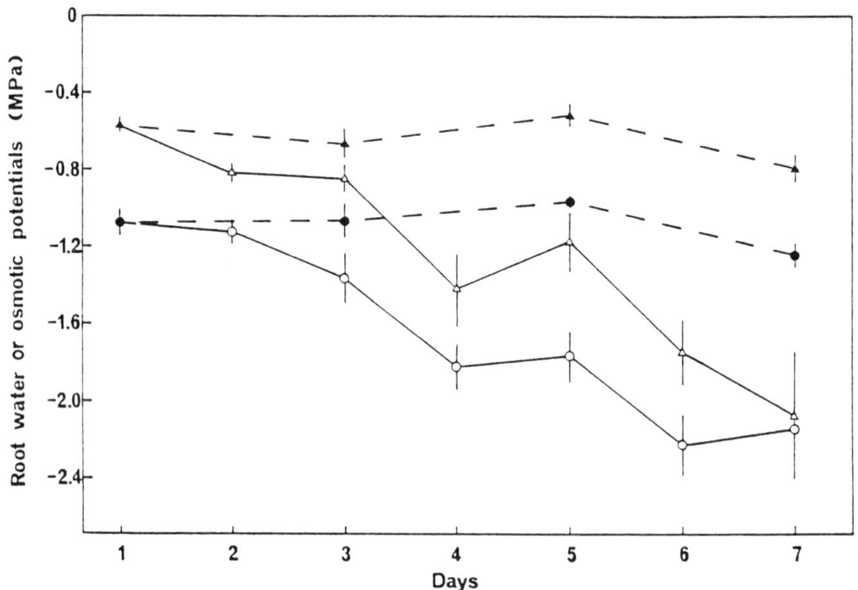

FIG. 37. Root water (▲, △) and solute (●, ○) potentials of maize plants watered every day (——) and plants not watered after day 0 (——). Points are means ± standard error. After Sharp and Davies (290).

cant differences in the capacity for solute accumulation in expanding and fully expanded cells. Matsuda and Riazi (199) have shown the importance of measuring the water relations of expanding cells if water relations measurements are to be related to measurements of growth, but problems with growing cells such as relaxation of cell walls in the psychrometer make this difficult (49b).

Fereres *et al.* (86) have reported a diurnal variation in osmotic potential in leaves of maize plants such that a minimum level of leaf turgor is maintained throughout the day when water potentials are low (Fig. 36). As a result of this turgor maintenance, the reduction in plant water potential had little effect on leaf growth. This type of adaptation may be particularly important for plants in the field, where leaf growth during the night period may be limited by low temperatures (346, 347). Interestingly, Michelena and Boyer (212) report complete turgor maintenance at low water potentials in the elongating regions of maize leaves. Despite this, leaf

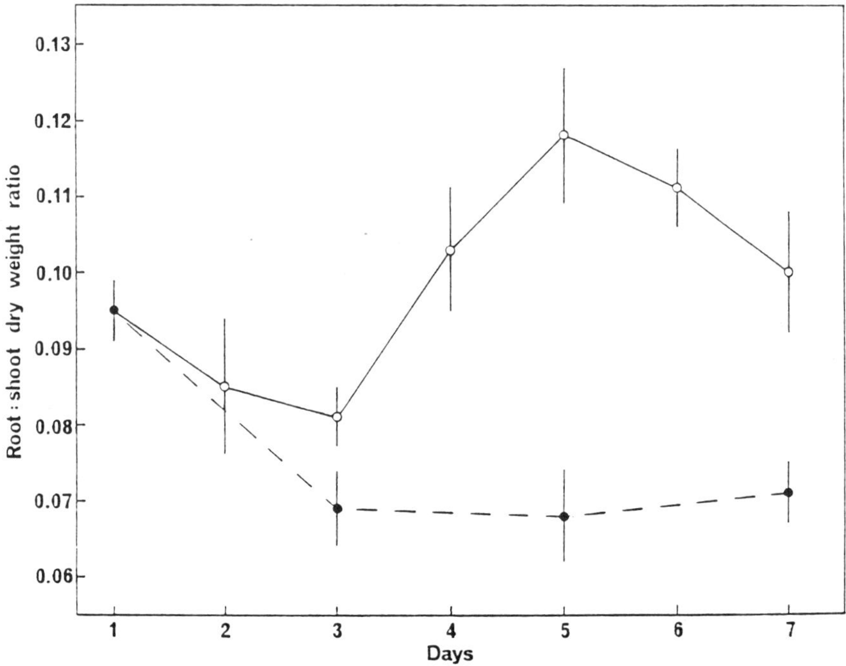

FIG. 38. The root : shoot dry weight ratio of maize plants watered every day (●------●) and plants not watered after day 0 (○——○). Points are means ± standard error. After Sharp and Davies (290).

growth rates are reduced, indicating that some factor other than turgor also affected leaf growth and caused most of the reduction in growth under dry conditions. Solute adjustment in maize laminae was incomplete. Sharp and Davies (290) also noted that solute adjustment in maize laminae is incomplete and is accompanied by reduced rates of daytime leaf expansion. High conductances indicate that photosynthesis is taking place even though rates of leaf expansion are limited (69). One result of the continued availability of carbohydrate coupled with the reduction in sink size is increased solute concentrations in roots, particularly in root tips (Fig. 37) (290). Maintenance of root tip turgor means that root growth may continue even in drying soil (110) (Fig. 38). In this study, there was a net increase in the growth of roots in drying soil. Such a response has been observed by several workers (19, 80, 130, 193). Mild water deficit stimulated the growth of nodal roots to maize and also caused seminal roots to penetrate deeper into the soil (289), presumably resulting in the exploration of a greater volume of soil with a consequent increase in the availability of soil water (Fig. 39). Certainly soil water depletion rates deep in the profile were greatly increased when plants were not watered for several weeks (Fig. 40).

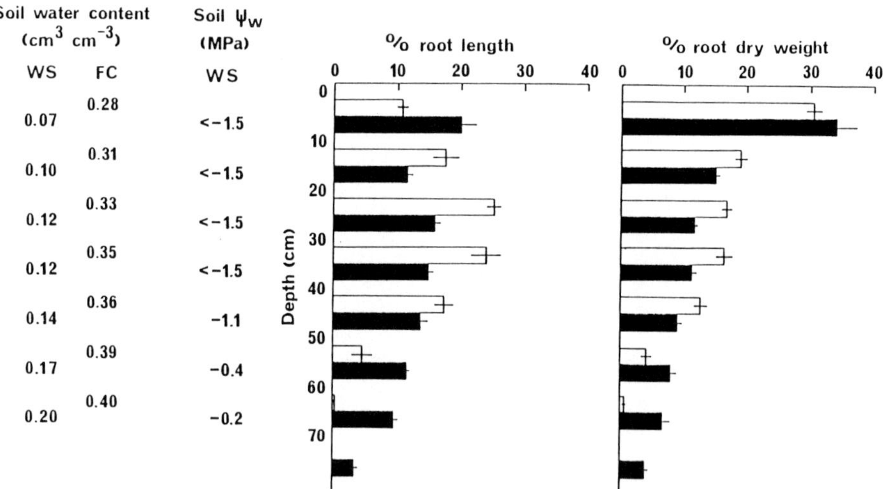

Fig. 39. The percentages of total root length and of total root dry weight in consecutive 10-cm soil layers after an 18-day period during which plants were either watered daily (open bars) or not watered after day 0 (shaded bars). Values are means of 5 ± standard error. Mean soil water content and approximate bulk soil water potential of each layer are indicated for the water-stressed (WS) plants and those kept at field capacity (FC). After Sharp (289).

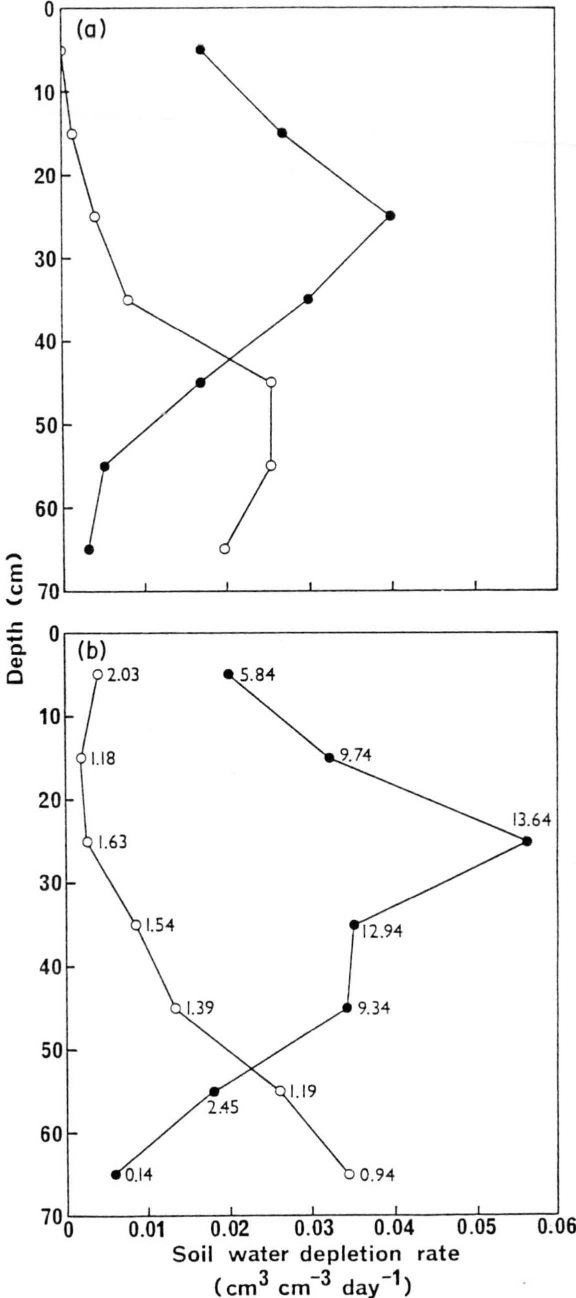

FIG. 40. The mean soil water depletion rate from consecutive 10-cm soil layers of tubes containing maize plants watered daily (●——●) and of tubes not watered after day 0 (○——○) during (a) day 15 (well watered) and days 13–15 (unwatered) and (b) day 18 (well watered) and days 16–18 (unwatered) of an experimental treatment. The numbers against the points in (b) are the mean root length densities in each soil layer on day 18 (cm root cm^{-3} soil). After Sharp (289).

It is now clear (328) that the capacity for a solute adjustment varies markedly among cultivars of individual species (222, 330), is less marked as water deficits become more severe (150), and is most marked when water deficits develop slowly. Adjustment of solute concentrations may be somewhat transient, however, since studies by Wilson *et al.* (360) and Turner *et al.* (329) showed that 0.4 to 0.7 MPa of accumulated solute in buffel grass *(Cenchrus ciliaris)*, siratro *(Macroptilium atropurpureum)*, sunflower, and *Sorghum* had disappeared within 10 days of the relief of plant water deficit. These results suggest that under many circumstances a prestressing treatment to promote solute accumulation may be of no benefit in a subsequent drying cycle. In contrast to these results, Ackerson (2) has reported some carryover effect of soil drying on solute amounts in cotton leaves, such that adapted plants maintained turgor at lower leaf water potentials than did nonadapted plants. Several workers have reported a decrease in the sensitivity of stomata of adapted plants to decreasing water potential, suggesting some carryover effect of water deficit on solute accumulation.

Although solute adjustment may apparently maintain the turgor of expanding leaves as water potential falls, decreases in growth rate may not be prevented by such an adjustment (209, 219). Although one explanation for such a response may be a problem with psychrometric determination of turgor (49b), other explanations may be a decrease in cell wall extensibility (340) and/or an increase in yield turgor as water deficits increase. Clearly, these responses may differ among species. Breeding for solute adjustment and continued growth at low water potentials may be successful in some plants, but in others additional limitations on leaf growth must be overcome before any benefits will be seen. While osmotic adjustment may not sustain growth at low water potentials or contents it may lower the water potential at which deficit-induced death occurs.

While complete turgor maintenance may not maintain growth in water-stressed leaves, we have seen that growth of roots may actually increase as their water potential falls (130, 290). This is not always the case, however. For example, Molyneux and Davies (217b) showed that although two species of pasture grass both showed complete solute adjustment in roots, only one was able to maintain root growth when compared with well-watered plants.

In halophytes, solute regulation as an adaptation to saline conditions is associated with metabolic adjustments leading to the accumulation of specific organic solutes. (See Chapter 5 in this volume.) The most important of these are imino acids (311), betaines such as β-homobetaine, oxo-6-trimethylammonio-2-heptanoic acid (175), and glycine betaine (314), methylated sulfonium compounds (174), the amino acid Δ'-acetylornithine (5), and the sugar alcohol sorbitol (4). We have already seen that as plants rooted in drying soil develop an increasing water deficit, photosynthesis may continue as turgor falls and leaf growth is at least partially restricted. Hsiao and Acevedo (130) have suggested that such a combination of events may lead to the accumulation of carbohydrates in roots. These carbohydrates can act both as the substrate and, by lowering osmotic potential and maintaining turgor, as the driving force for growth. Sharp (289) (Table VI) has shown that carbohydrate is a substantial proportion of the solute accumulating in the roots of mildly stressed maize plants. Decreased osmotic potentials in the leaves of sorghum (152) were fully accounted for by sugars, potassium, and chloride, and the contributions of total inorganic

TABLE VI

CONTENTS OF SEVERAL MAJOR SOLUTES IN THE YOUNGEST EXPANDING LEAVES AND NODAL ROOT TIPS OF MAIZE PLANTS WHICH WERE WATERED DAILY OR UNWATERED AFTER DAY 0.

	Well watered			Unwatered		
Day:	1	4	8	3	5	7
Youngest expanding leaf						
Sugars						
Sucrose	0.11	0.06	0.10	0.07	0.14	0.13
Reducing sugars	0.09	0.17	0.19	0.12	0.41	0.28
Total	0.20	0.23	0.28	0.19	0.55	0.41
Total free α-amino nitrogen	0.19	0.17	0.14	0.18	0.20	0.40
Total free amino acids	0.12	0.09	0.07	0.08	0.12	0.22
Inorganic ions						
K^+	1.77	1.77	1.05	1.93	1.64	1.69
Na^+	NA	NA	NA	NA	NA	NA
Ca^{2+}	0.03	0.04	0.01	0.04	0.01	0.02
Mg^{2+}	0.07	0.06	0.05	0.07	0.05	0.06
Malic acid	0.09	0.04	0.03	0.06	0.06	0.02
Nodal root tips						
Sugars						
Sucrose	0.60	0.36	0.44	0.55	0.68	0.79
Reducing sugars	1.25	1.36	1.76	1.26	2.13	2.19
Total	1.85	1.72	2.20	1.81	2.80	2.98
Total free α-amino nitrogen	1.64	0.70	0.77	0.95	0.92	2.05
Total free amino acids	0.82	0.56	0.48	0.58	0.65	1.63
Inorganic ions						
K^+	2.56	1.91	1.80	2.10	1.72	1.57
Na^+	NA	NA	NA	NA	NA	NA
Ca^{2+}	<0.01	<0.01	<0.01	<0.01	<0.01	<0.01
Mg^{2+}	0.07	0.05	0.05	0.03	0.05	0.05
Malic acid	0.21	0.36	0.21	0.16	0.09	0.11

[a] After Sharp (289).

[b] Contents in millimoles per gram ethanol-insoluble dry matter. Values are means. NA, Not available.

ions and sugars were approximately equal. Sugars did not contribute to the decrease in leaf osmotic potential in sunflower leaves when water was withheld from roots, but inorganic ions and free amino acids did increase in concentration. Munns *et al.* (223) and Ackerson (2, 3) have reported that in drying soil carbohydrates made major contributions to increasing solute concentration in wheat and cotton, respectively.

b. Rooting Patterns in Drying Soil. When a plant needs to take up water quickly, the best means of doing so is to have a large rooting density which will ensure that even though the root system as a whole may be taking up water rapidly, the uptake per unit length of root is small.

The result of this will be a small drop in water potential across the roots and in the perirhizal zone (Section III,C,2). Larcher (173) has noted that some desert plants may have root : shoot ratios as high as 10 : 1, and from the preceding discussion it is clear that one effect of drought is to increase root : shoot ratio. Osonubi and Davies (240) and Sharp (289) have shown how deeper rooting patterns can result in effective use of water in the soil profile (Fig. 40), but Passioura (248) has suggested that such deep rooting patterns may only be an advantage to a wild plant which is in competition with plants of other species for a limited supply of water. Competition for water may lose its pertinence in a community where it is the performance of the community rather than the performance of the individual that is important.

Sauerbeck and Johnen (see 248) note that when roots are grown in soil, the weight of the root system at harvest may be less than 25% of the assimilates that have been transported to it. This shows that the cost of growing and maintaining roots may be rather more than the distribution of dry matter within the plant suggests. One might expect therefore that in a given environment, a crop species selected for high yield whose growth is limited by water would have an optimum ratio of root to shoot substantially less than that of similar but individual plants that are competing with each other.

Passioura (248) has noted that plants seem to have evolved two contrasting strategies in their use of water stored in the soil. Plants of most wild species use water very rapidly. If it does not rain, the plant is faced with the problem of survival rather than of maximizing dry matter production. Such plants can compete for water most effectively with extensive, high-density root systems with highly permeable roots. The alternative strategy, which is appropriate for weed-free crops, is the conservative use of water, which lessens the risk of developing very severe water deficits and makes it likely that water will be available during flowering and seed filling later in the season. Crops that have water for use after flowering are likely to have better yields than those that do not (88). Water will be conserved if a plant has a sparse and poorly permeable root system or a high rooting density with a slow rate of extension (248). Passioura has noted that cereals with seminal roots offering high resistance to water flow have such systems but may be able to respond to good conditions by producing nodal roots in the top soil. As already discussed, solute regulation in nodal roots of maize allows continued extension of nodal roots through very dry soil (289). Presumably, if the top soil is recharged with water after a light rain, these roots are able to utilize this water effectively.

D. An Integrated Reaction to a Reduction in Plant Turgor

Mansfield (Chapter 3 in this volume) has described how withholding water from plants can result in a rapid buildup in the amount of abscisic acid in leaves (361, 362). There is growing evidence that this ABA, perhaps in conjunction with other plant growth regulators, may act to regulate water use and turgor in plants subjected to intermittent periods of drought (77, 242). Mansfield *et al.* (196) have pointed out that ABA can close stomata (194), increase the flux of water through roots (104, 105), decrease cell size (263), and decrease the rate of transport of sugars from leaves (192), thereby promoting some degree of solute regulation. All of these effects can act to maintain cell turgor. In addition, ABA may promote the growth of roots (43, 278, 345, 366), which can increase the availability of water in drying soil. Markhart *et al.* (197) and Davies *et al.* (70) have noted that ABA may decrease the hydraulic conductance of *Phaseolus* and wheat roots (Fig. 41) but ABA-stimulated increase in flux through roots can still occur despite this decrease in conductance if ABA promotes the accumulation of solutes inside the "root membrane" and thereby increases the driving force for water uptake (70, 92). However, it is unlikely that this accumulation could occur where it is needed in a rapidly transpiring plant (70, 92). In addition, since the resistances in the flow pathway of water into roots from drying soil will be dominated by the root interface resistance (see Section III,C,2), any change in the root segment of the pathway will therefore have only a minimal effect on the total water flux. An ABA-stimulated growth of roots away from drying soil into moist soil will act to slow the development of a large root–soil interface resistance and thereby act to maintain a comparatively high rate of water uptake; this response in a plant with a developing water deficit may be more important than any effect of ABA on the permeability of roots.

It is now apparent that it is a reduction in cell turgor rather than water

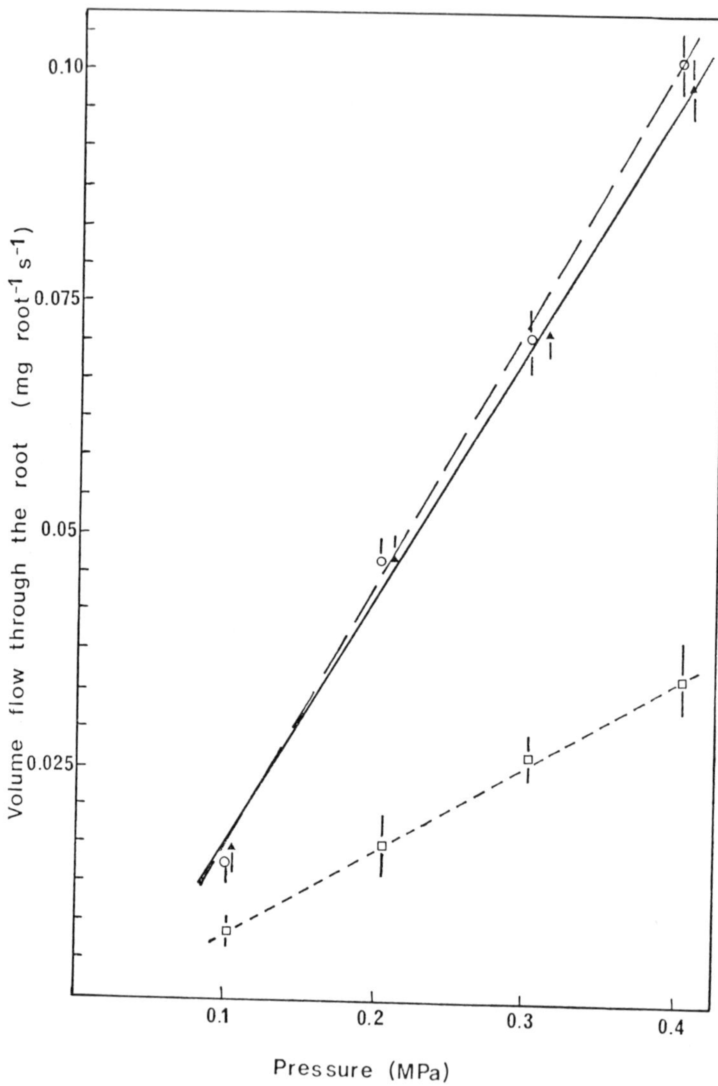

FIG. 41. The rate of volume flow through excised root systems of wheat plants subjected to different hydrostatic pressures. ▲, Roots bathed in nutrient solution only; ○, roots bathed in nutrient solution (20 dm^3) plus 1 cm^3 ethyl alcohol (95%); □, roots bathed in nutrient solution, ethyl alcohol, and ABA to a concentration of 10^{-3} mol m^{-3}. Points are means ± standard errors. After Davies *et al.* (70).

potential that stimulates the redistribution and enhanced production of ABA in the leaf (72, 251). Pierce and Raschke (251) have reported that enhanced ABA concentrations are not recorded until the mesophyll cells reach zero turgor. It is generally held (see Mansfield, Chapter 3 this volume) that most ABA is produced in the mesophyll cells of the leaf. Clearly, many of the plant's turgor maintenance mechanisms are initiated long before zero turgor is generally reached (69), and therefore if ABA does have a coordinating role, it must be redistributed before zero turgor is reached in more than a few cells (72) or perhaps produced elsewhere in the plant closer to several different sites of action. There is some suggestion that ABA may be produced in the roots (343), although this is controversial (125), and a recent report shows that enhanced amounts of ABA are found in guard cell protoplasts subjected to a dehydration treatment (354). Because of the very large turgor changes experienced by guard cells, it has been considered unlikely that ABA would be produced in these cells. For example, humidity responses of stomata involve large changes in guard cell turgor which are rapidly reversible; this would not necessarily be the case if the guard cells produced ABA. Clearly, confirmation is required of this important possibility.

Our initial understanding of the role of ABA in the plant developing water deficit was of a compound increasing in concentration over a short period of time and acting to restrict almost totally the loss of water from the plant. More recently it has become apparent that many responses to water deficit depend on the plant's "stress prehistory" and that certainly stomatal behavior and probably other processes within the plant can exhibit varying sensitivity to ABA. Potassium concentrations in and around the guard cell apparatus (359), concentrations of other growth regulators (307, 308), temperature (279), and water stress prehistory (71) are several factors which can influence the sensitivity of stomata to ABA. These observations, coupled with the possibility that ABA may be produced at different sites within the plant, suggest that the plant has sensitive means of detecting and responding to environmental perturbations which might result in an integrated reaction to plant water deficit.

A central role for ABA in the plant's deficit responses is shown by work by Quarrie and co-workers (264), who compared the growth and water relationships of homozygous F_5 generation high and low ABA lines under drought conditions. The observation that high ABA lines outyield low ABA lines both when water was freely available and when water availability was low (139a) argues for the involvement of ABA in processes contributing to drought resistance.

Several growth regulators other than ABA have been shown to influence stomatal behavior (20–23, 307, 308) and other aspects of the plant's

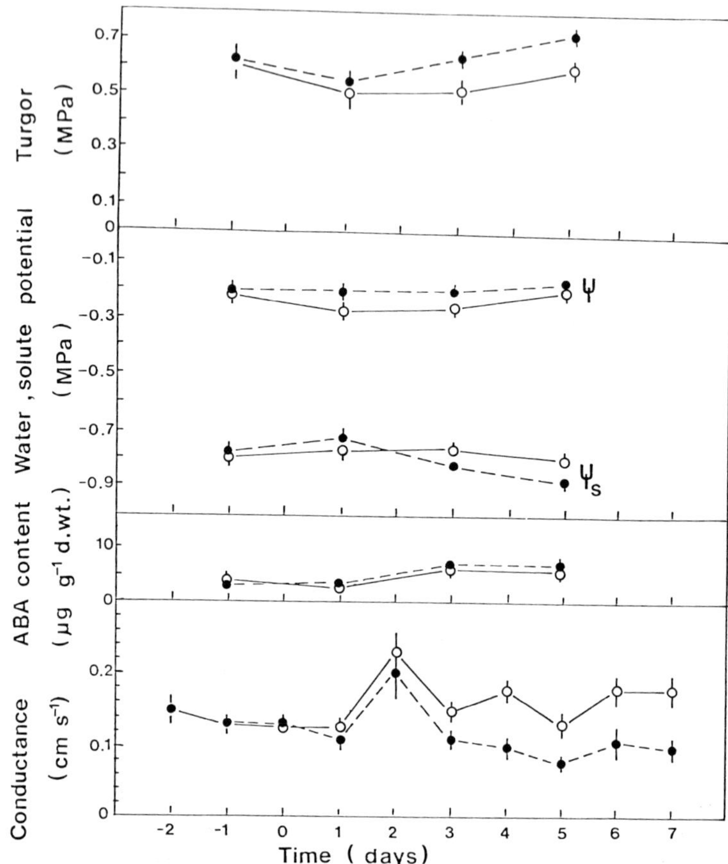

Fig. 42. Water relations, ABA content, and stomatal behavior of maize leaves in plants for which water was withheld from part of the root system, which was divided and grown in two separate rooting containers. Control plants have both parts of the root system well watered (○——○). Plants for which part of root system was dried (●------●) were last watered on day 0. After Blackman and Davies (23).

water balance. Synthesis and transport of these compounds may be influenced by water deficit (62, 336). Nevertheless, such data are rather few and far between and unlike the situation with ABA, which is effective on stomata of most plants, there are many reports that, for example, auxins and cytokinins have no effects on the stomata of certain species. Davies and Mansfield (66) have argued that this may be because inappropriate conditions have been used in experiments to detect activity of different compounds. For example, Blackman and Davies (20–22) report that although cytokinins have no effect on stomata on young leaves of well-watered maize plants, these compounds will open stomata as leaves age, if leaves are suffering a slight water deficit or if leaves are treated with ABA. In an experiment in which soil was dried around only part of a maize root system, stomata showed restricted opening despite water potentials and turgors which were favorable for opening (Fig. 42). In fact, even when leaves were incubated on water under ideal conditions for stomatal opening (high irradiance, low carbon dioxide concentration), apertures remained low and were not increased until cytokinin was added to the incubation medium (Fig. 43) (23). Blackman and Davies have interpreted this result as indicating that soil drying restricted the synthesis and/or transport of cytokinins from roots so that as a result stomatal opening was restricted. This observation suggests the possibility (16, 65, 210) that under certain circumstances roots may have the capacity to signal perturbations in the root environment to the shoots. Such signals may enable the plant to optimize the use of water as the soil dries over the long term (53).

Clearly, for cytokinin to act as a signal in a situation such as this, it is necessary for cytokinin synthesis to be largely restricted to the root. This is

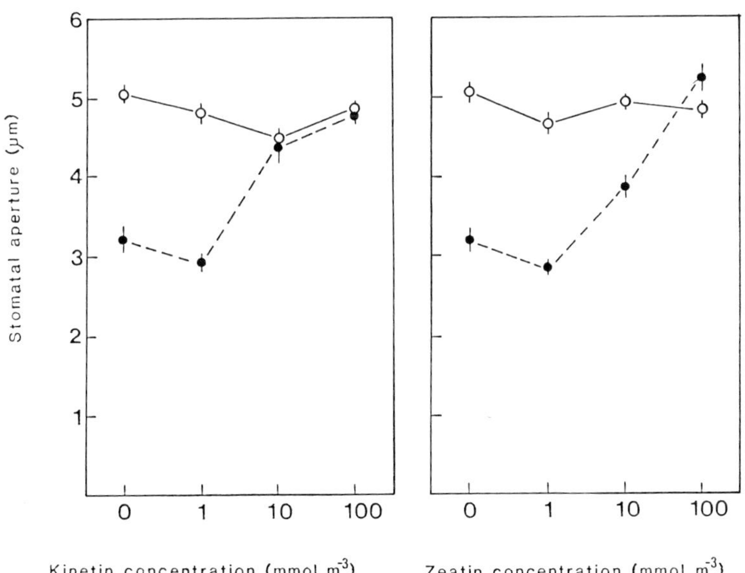

FIG. 43. Apertures of stomata of maize leaves incubated in the light in CO_2-free air on distilled water, kinetin, or zeatin solutions. Leaves which were taken from well-watered (○——○) or partly dried plants (●------●) were sampled 8 h into the light period on day 5 after water was first withheld from part of the root system of half of the plants. Points are means of 60 observations. After Blackman and Davies (23).

a controversial point (280), but in many plants this may be the case. Certainly there is circumstantial evidence that cytokinin transport from roots is reduced as a result of water deficit (140), but this evidence resulting from bioassays must be treated with caution until new evidence obtained from physicochemical methods is available. Nonetheless, the idea that stomatal behavior is the result of an interaction between the influence of external and endogenous factors is now gaining acceptance along with the view that these endogenous factors can be promoters and inhibitors. Research of this type raises new possibilities for the artificial control of stomatal behavior, both to maintain plant turgor of crop species during drought (66) and to reduce turgor with a view to killing weed species through severe water deficit (71).

VI. Concluding Remarks

In the 25 years since the publication of Kramer's chapter in Volume II of this treatise (166), we have gained an improved understanding of soil–plant–water relations. To some extent this has followed the development of convenient, practical methods of measuring gas exchange in plants and the water status of plants and soil. In addition, there is now a better understanding of how the rate of transpiration is controlled, although important questions still remain to be resolved in this area (55). Despite some signal advances, there are still some significant and often surprising gaps in our knowledge. For example, recent publications have raised questions about the relative importance of different pathways of water movement through the root and the leaf (207, 297, 298, 335). Despite an increasing interest in root growth and physiology, our understanding of root function is limited. The development of models of water movement in plants is restricted by uncertainty about driving forces and fluxes and therefore resistances in different components of the system. In this regard, the relationships between salt and water uptake by roots (see Chapter 4 in this volume) are still ill defined, as is the occurrence of real or apparent variation in resistance with flux. Many of these problems may be resolved if driving forces and fluxes of water in different systems can be accurately determined.

It is now clear that plant growth can be limited by very mild water deficits. Unfortunately, it is not always clear why this is so and how a water deficit of less than 1 MPa can influence an enzyme-mediated process. While the most rapidly expanding area of research in plant water relations during the past 20 years has been a consideration of the effects of water deficit on the physiological processes of plants, the various effects of water deficit are so interrelated that it is often difficult to determine how a reduction in growth is brought about. In this regard, the involvement of growth regulators in the mediation of the effects of water stress is becoming increasingly evident (e.g., 66), and it will be interesting to see what future research reveals, particularly in the general area of root-to-shoot communications of perturbations in the soil.

An elucidation of the primary and secondary effects of water deficit may result from studies of the comparative drought tolerance of a number of species. As Kramer (168) has suggested, future studies should quantify those aspects of water deficit which are likely to be limiting. In the past, a degree of deficit has often been inferred from observations of wilting or from measurements of soil moisture status, whereas measurements of turgor as well as plant water potential are required. Considerations of the effects of deficits on several plant processes at various stages of growth, rather than a consideration of a single effect at a single growth stage, may well be rewarding. Investigations of the water relations of growing leaf cells rather than fully expanded cells in the lamina are more likely to provide some explanations of the limitation of plant growth by water deficit.

References

1. Acevedo, E., Hsiao, T. C., and Henderson, D. W. (1971). Immediate and subsequent growth responses of maize leaves to changes in water status. *Plant Physiol.* **48,** 631–636.
2. Ackerson, R. C. (1981). Osmoregulation in cotton in response to water stress. II. Leaf carbohydrate status in relation to osmotic adjustment. *Plant Physiol.* **67,** 489–493.
3. Ackerson, R. C., and Herbert, R. R. (1981). Osmoregulation in cotton in response to water stress. I. Alterations in photosynthesis, leaf conductance, translocation and ultrastructure. *Plant Physiol.* **67,** 484–488.
4. Ahmad, I., Larher, F., and Stewart, G. R. (1979). Sorbitol, a compatible osmotic solute in *Plantago maritima. New Phytol.* **82,** 671–678.
5. Ahmad, I., Larher, R., and Stewart, G. R. (1981). The accumulation of Δ′-acetylornithine and other solutes in the salt marsh grass *Puccinellia maritima. Phytochemistry* **20,** 1501–1504.
6. Appleby, R. F., and Davies, W. J. (1982). A possible evaporation site in the guard cell wall and the influence of leaf structure on the humidity response by stomata of woody plants. *Oecologia* **56,** 30–40.
7. Appleby, R. F., and Davies, W. J. (1983). The structure and orientation of guard cells in plants showing stomatal responses to changing VPD. *Ann. Bot. (London)* [N.S.] **52,** 459–468.
8. Ashenden, T. W., and Mansfield, T. A. (1976). Influence of wind speed on the sensitivity of ryegrass to SO_2. *J. Exp. Bot.* **28,** 729–735.
9. Askenasy, E. (1895). Ueber das Satsteigen. *Bot. Centralbl.* **62,** 237–238.
10. Ayres, P. G. (1978). Water relations of diseased plants. *In* "Water Deficits and Plant Growth" (T. T. Kozlowski, ed.), Vol. 5, pp. 1–60. Academic Press, New York.
11. Baker, D. A., and Moorby, J. (1969). The transport of sugar, water and ions into developing potato tubers. *Ann. Bot. (London)* [N.S.] **33,** 729–741.

12. Bange, G. G. J. (1953). On the quantitative explanation of stomatal transpiration. *Acta Bot. Neerl.* **2**, 255–296.
13. Barrs, H. D. (1968). Determination of water deficits in plant tissue. *In* "Water Deficits and Plant Growth" (T. T. Kozlowski, ed.), Vol. 1, pp. 235–368. Academic Press, New York.
14. Barrs, H. D. (1971). Cyclic variations in stomatal aperture, transpiration and leaf water potential under constant environmental conditions. *Annu. Rev. Plant Physiol.* **22**, 223–236.
15. Barrs, H. D. (1973). Controlled environmental studies of the effects of variable atmospheric water stress on photosynthesis, transpiration and water status of *Zea mays* L. and other species. *Ecol. Conserv.* **5**, 249–258.
16. Bates, L. M., and Hall, A. E. (1981). Stomatal closure with soil water depletion not associated with changes in bulk leaf water status. *Oecologia* **50**, 62–65.
17. Baughn, J. W., and Tanner, C. B. (1976). Leaf water potential: comparison of pressure chamber and *in situ* hygrometer on five herbaceous species. *Crop Sci.* **16**, 181–184.
18. Beadle, C. L., Jarvis, P. G., and Neilson, R. E. (1979). Leaf conductance as related to xylem water potential and carbon dioxide concentration in Sitka spruce. *Physiol. Plant.* **45**, 158–166.
19. Bennett, O. L., and Doss, B. D. (1960). Effect of soil moisture level on root distribution of cool-season forage species. *Agron. J.* **52**, 204–207.
20. Blackman, P. G., and Davies, W. J. (1983). The effects of cytokinins and ABA on stomatal behaviour of maize and *Commelina. J. Exp. Bot.* **34**, 1619–1626.
21. Blackman, P. G., and Davies, W. J. (1984). Modification of the CO_2 responses of maize stomata by abscisic acid and by naturally occurring and synthetic cytokinins. *J. Exp. Bot.* **35**, 174–179.
22. Blackman, P. G., and Davies, W. J. (1984). Age-related changes in stomatal response to cytokinins and ABA. *Ann. Bot. (London)* [N.S.] **54**, 121–125.
23. Blackman, P. G., and Davies, W. J. (1985). Root to shoot communication in maize plants of the effects of soil drying. *J. Exp. Bot.* **36**, 39–48.
24. Boyer, J. S. (1968). Relationships of water potential to growth of leaves. *Plant Physiol.* **42**, 213–217.
25. Boyer, J. S. (1970). Leaf enlargement and metabolic rates in corn, soybean and sunflower at various leaf water potentials. *Plant Physiol.* **46**, 233–235.
26. Boyer, J. S. (1971). Recovery of photosynthesis in sunflower after a period of low leaf water potential. *Plant Physiol.* **47**, 816–820.
27. Boyer, J. S. (1974). Water transport in plants: Mechanism of apparent changes in resistance during absorption. *Planta* **117**, 187–207.
28. Boyer, J. S. (1976). Photosynthesis at low water potentials. *Philos. Trans. R. Soc. London, Ser. B* **273**, 501–512.
29. Boyer, J. S. (1977). Regulation of water movement in whole plants. *Symp. Soc. Exp. Biol.* **31**, 455–470.
30. Boyer, J. S., and Bowen, B. L. (1970). Inhibition of oxygen evolution in chloroplasts isolated from leaves with low water potentials. *Plant Physiol.* **45**, 612–615.
31. Boyer, J. S., and Wu, G. (1978). Auxin increase the hydraulic conductivity of auxin sensitive hypocotyl tissue. *Planta* **139**, 227–237.
32. Bradford, K. J., and Hsiao, T. C. (1982). Physiological responses to moderate water stress. *In* "Encyclopedia of Plant Physiology" (O. L. Lange, P. S. Nobel, C. B. Osmond, and H. Ziegler, eds.), Vol. 12B, pp. 325–378. Springer-Verlag, Berlin and New York.
33. Braekke, F. H., and Kozlowski, T. T. (1975). Shrinkage and swelling of stems of *Pinus resinosa* and *Betula papyrifera* in northern Wisconsin. *Plant Soil* **43**, 387–410.

34. Briggs, G. W. (1967). "Movement of Water in Plants." Blackwell, Oxford.
35. Brouwer, R. (1953). Water absorption by the roots of *Vicia faba* at various transpiration strengths. II. Causal relation between suction tension, resistance and uptake. *Proc. K. Ned. Akad. Wet., Ser. C.* **56,** 129–136.
36. Brouwer, R. (1965). Water movement across the root. *Symp. Soc. Exp. Biol.* **19,** 131–149.
37. Bunce, J. A. (1977). Leaf elongation in relation to leaf water potential in soybean. *J. Exp. Bot.* **28,** 156–161.
38. Bunce, J. A. (1978). Effects of shoot environment on apparent root resistance to water flow in whole soybean and cotton plants. *J. Exp. Bot.* **29,** 595–601.
39. Byrne, G. F., Begg, J. E., and Hansen, G. K. (1977). Cavitation and resistance to water flow in plant roots. *Agric. Meteorol.* **18,** 21–25.
40. Camacho-B, S. E., Hall, A. E., and Kaufmann, M. R. (1974). Efficiency and regulation of water transport in some woody and herbaceous species. *Plant Physiol.* **54,** 169–172.
41. Campbell, G. S., and Campbell, M. D. (1974). Evaluation of a thermocouple hygrometer for measuring leaf water potential *in situ. Agron. J.* **66,** 24–27.
42. Cheung, Y. N. S., Tyree, M. T., and Dainty, J. (1975). Water relations parameters on single leaves obtained in a pressure bomb and some ecological interpretations. *Can. J. Bot.* **53,** 1342–1346.
43. Chin, T. Y., Meyer, M. M., and Beevers, L. (1969). Abscisic acid stimulated rooting of stem cuttings. *Planta* **88,** 192–196.
44. Clarkson, D. T., Robards, A. W., and Sanderson, J. (1971). The tertiary endodermis in barley roots: Fine structure in relation to radial transport of ions and water. *Planta* **96,** 296–305.
45. Cleland, R. E. (1959). Effect of osmotic concentration on auxin-action and on irreversible expansion of the *Avena* coleoptile. *Physiol. Plant.* **12,** 809–825.
46. Cleland, R. E. (1967). A dual role of turgor pressure in auxin-induced cell elongation in *Avena* coleoptile. *Planta* **77,** 182–191.
47. Cleland, R. E. (1971). Cell wall extension. *Annu. Rev. Plant Physiol.* **22,** 197–222.
48. Cosgrove, D. J. (1981). Analysis of the dynamic and steady-state responses of growth rate and turgor pressure to changes in cell parameters. *Plant Physiol.* **68,** 1439–1446.
49. Cosgrove, D. J., and Cleland, R. E. (1983). Solutes in the free space of growing stem tissues. *Plant Physiol.* **72,** 326–331.
49a. Cosgrove, D. J., and Cleland, R. E. (1983). Osmotic properties of cell internodes in relation to growth and auxin action. *Plant Physiol.* **72,** 332–338.
49b. Cosgrove, D. J., Van Volkenburgh, E., and Cleland, R. E. (1984). Stress relaxation of cell walls and the yield threshold for growth. *Planta* **162,** 46–54.
50. Cowan, I. R. (1965). Transport of water in the soil-plant-atmosphere system. *J. Appl. Ecol.* **2,** 221–239.
51. Cowan, I. R. (1972). Oscillations in stomatal conductance and plant functioning associated with stomatal conductance: Observations and a model. *Planta* **106,** 185–219.
52. Cowan, I. R. (1977). Stomatal behaviour and environment. *Adv. Bot. Res.* **4,** 117–228.
53. Cowan, I. R. (1982). Regulation of water use in relation to carbon gain in higher plants. *In* "Encyclopedia of Plant Physiology" (O. L. Lange, P. S. Nobel, C. B. Osmond, and H. Ziegler, eds.), Vol. 12B, pp. 589–614. Springer-Verlag, Berlin and New York.
54. Cowan, I. R., and Milthorpe, F. L. (1968). Plant factors influencing the water status of plant tissues. *In* "Water Deficits and Plant Growth" (T. T. Kozlowski, ed.), Vol. 1, pp. 137–193. Academic Press, New York.
55. Cowan, I. R., and Farquhar, G. D. (1977). Stomatal function in relation to leaf metabolism and environment. *Symp. Soc. Exp. Biol.* **31,** 471–505.

56. Cox, E. F. (1968). Cyclic changes in transpiration of sunflower leaves in a steady environment. *J. Exp. Bot.* **19,** 167–175.
57. Crowdy, S. H., and Tanton, J. W. (1970). Water pathways in higher plants. 1. Free space in wheat leaves. *J. Exp. Bot.* **21,** 102–111.
58. Cutler, J. M., Shahan, K. W., and Steponkus, P. L. (1980). Dynamics of osmotic adjustment in rice. *Crop Sci.* **20,** 310–314.
59. Cutler, J. M., Shahan, K. W., and Steponkus, P. L. (1980). Influence of water deficits and osmotic adjustment on leaf elongation in rice. *Crop Sci.* **20,** 314–318.
60. Cutler, J. M., Steponkus, P. L., Wach, M. J., and Shahan, K. W. (1980). Dynamic aspects of enhancement of leaf elongation in rice. *Plant Physiol.* **66,** 147–152.
61. Dalton, F. N., Raats, P. A. C., and Gardner, W. R. (1975). Simultaneous uptake of water and solutes by plants. *Agron. J.* **67,** 334–339.
62. Darbyshire, B. (1971). Changes in indoleacetic acid oxidase activity associated with plant water potential. *Physiol. Plant.* **25,** 80–84.
63. Davenport, D. C. (1967). Effects of chemical antitranspirants on transpiration and growth of grass. *J. Exp. Bot.* **18,** 332–347.
64. Davies, W. J. (1977). Stomatal responses to water stress and light in in plants grown in controlled environments and in the field. *Crop Sci.* **17,** 735–740.
65. Davies, W. J., and Sharp, R. E. (1981). The root: A sensitive detector of a reduction in water availability? *In* "Mechanisms of assimilate distribution and Plant Growth Regulators" (J. Kralovic, ed.), pp. 53–67. Slovak Acad. Agric. Piestany.
66. Davies, W. J., and Mansfield, T. A. (1986). Auxins and stomata. *In* "Stomatal Function" (E. Zeiger *et al.,* eds.). Stanford Univ. Press, Stanford, California (in press).
67. Davies, W. J., Gill, K., and Halliday, G. (1978a). The influence of wind on the behaviour of stomata of photosynthetic stems of *Cytisus scoparius* (L.) Link. *Ann. Bot. (London)* [N.S.] **42,** 1149–1154.
68. Davies, W. J., Mansfield, T. A., and Orton, P. J. (1978b). Strategies employed by plants to converse water: can we improve on them? *In* "Opportunities for Chemical Plant Growth Regulation" (E. F. George, ed.), pp. 45–54. B.C.P.C., London.
69. Davies, W. J., Mansfield, T. A., and Wellburn, A. R. (1979). A role for abscisic acid in drought endurance and drought avoidance. *In* "Plant Growth Substances 1979" (F. Skoog, ed.), pp. 242–253. Springer-Verlag, Berlin and New York.
70. Davies, W. J., Rodriguez, J. L., and Fiscus, E. L. (1982). Stomatal behaviour and water movement through roots of wheat plants treated with abscisic acid. *Plant, Cell Environ.* **5,** 485–493.
71. Davies, W. J., Blackman, P. G., and Mansfield, T. A. (1983). Manipulation and plant water status to increased herbicide effects. *Aspects Appl. Biol.* **4,** 197–205.
72. Davies, W. J., Wilson, J. A., Sharp, R. E., and Osonubi, O. (1981). Control of stomatal behaviour in water stressed plants. *In* "Stomatal Physiology" (P. G. Jarvis and T. A. Mansfield, eds.), pp. 247–279. Cambridge Univ. Press, London and New York.
73. De Michele, D. W., and Sharpe, P. J. H. (1973). An analysis of the mechanics of guard cell motion. *J. Theor. Biol.* **41,** 77–96.
74. Dimond, A. E. (1966). Pressure and flow relations in vascular bundles of the tomato plant. *Plant Physiol.* **41,** 119–131.
75. Dixon, H. H. (1914). "Transpiration and the Ascent of Sap in Plants." Macmillan, New York.
76. Dixon, H. H., and Joly, J. (1895). The path of the transpiration current. *Ann. Bot. (London)* [N.S.] **9,** 416–419.
77. Dörffling, K., Streich, J., Kruse, W., and Muxfeldt, B. (1977). Abscisic acid and the after-effect of water stress on stomatal opening potential. *Z. Pflanzenphysiol.* **81,** 43–56.

78. Downes, R. W. (1970). Effect of light intensity and leaf temperature on photosynthesis and transpiration in wheat and sorghum. *Aust. J. Biol. Sci.* **23,** 775–782.
79. Driessche, R. van den, Connor, D. J., and Tunstall, B. R. (1971). Photosynthetic response of brigalow to irradiance, temperature and water potential. *Photosynthetica* **5,** 210–217.
80. El Nadi, A. H., Brouwer, R., and Locher, J. Th. (1969). Some responses of the root and the shoot of *Vicia faba* plants to water stress. *Neth. J. Agric. Sci.* **17,** 133–142.
81. Esau, K. (1977). "Anatomy of Seed Plants." Wiley, New York.
82. Faiz, S. M. A., and Weatherley, P. E. (1977). The location of the resistance to water movement in the soil supplying the roots of transpiring plants. *New Phytol.* **78,** 337–347.
83. Faiz, S. M. A., and Weatherley, P. E. (1978). Further investigations into the location and magnitude of the hydraulic resistances in the soil: plant system. *New Phytol.* **81,** 19–28.
84. Farquhar, G. D. (1979). Carbon assimilation in relation to transpiration and fluxes of ammonia. *In* "Photosynthesis and Plant Development" (R. Marcelle *et al.,* eds.), pp. 321–328. Junk, The Hague.
85. Farquhar, G. D., Schülze, E.-D., and Küppers, M. (1980). Responses to humidity by stomata of *Nicotiana glauca* L. and *Corylus avellana* L. are consistent with the optimization of carbon dioxide uptake with respect to water loss. *Aust. J. Plant Physiol.* **7,** 315–327.
86. Fereres, E., Acevedo, E., Henderson, D. W., and Hsiao, T. C. (1978). Seasonal changes in water potential and turgor maintenance in sorghum and maize under water stress. *Physiol. Plant.* **44,** 261–267.
87. Fischer, R. A., and Hagan, R. M. (1965). Plant water relations, irrigation management and crop yield. *Exp. Agric.* **1,** 233–241.
88. Fischer, R. A., and Turner, N. C. (1978). Plant productivity in the arid and semi arid zones. *Annu. Rev. Plant. Physiol.* **29,** 277–337.
89. Fiscus, E. L. (1975). The interaction between osmotic- and pressure-induced water flow in plant roots. *Plant Physiol.* **55,** 917–922.
90. Fiscus, E. L. (1977). Determination of hydraulic and osmotic properties of soybean root systems. *Plant Physiol.* **59,** 1013–1020.
91. Fiscus, E. L. (1977). Effects of coupled solute and water flow in plant roots with special reference to Brouwer's experiment. *J. Exp. Bot.* **28,** 71–77.
92. Fiscus, E. L. (1981). Effects of abscisic acid on the hydraulic conductance of and the total ion transport through *Phaseolus* root systems. *Plant Physiol.* **68,** 169–174.
93. Fiscus, E. L., and Kramer, P. J. (1975). General model for osmotic- and pressure-induced flow in plant roots. *Proc. Natl. Acad. Sci. U.S.A.* **72,** 3114–3118.
94. Fiscus, E. L., Parsons, L. R., and Alberte, R. S. (1973). Phyllotaxy and water relations in tobacco. *Planta* **112,** 285–292.
95. Flowers, T. J., Troke, P. F., and Yeo, A. R. (1977). The mechanism of salt tolerance in halophytes. *Annu. Rev. Plant Physiol.* **28,** 89–121.
96. Frey-Wyssling, A., and Mühlethaler, K. (1965). "Ultrastructural Plant Cytology." Elsevier, Amsterdam.
97. Gaastra, P. (1959). Photosynthesis of crop plants as influenced by light, carbon dioxide, temperature, and stomatal diffusion resistances. *Meded. Landbouwhogesch. Wageningen* **59,** 1–68.
98. Gaff, D. F., Chambers, T. C., and Markus, K. (1964). Studies of extrafascicular movement of water in the leaf. *Aust. J. Biol. Sci.* **17,** 581–586.
99. Gardner, W. R. (1960). Dynamic aspects of water availability to plants. *Soil Sci.* **89,** 63–73.

100. Gardner, W. R., and Ehlig, C. F. (1962). Impedance to water movement in soil and plant. *Science* **138,** 522–523.
101. Gardner, W. R., and Ehlig, C. F. (1963). The influence of soil water on transpiration by plants. *J. Geophys. Res.* **68,** 5719–5724.
102. Gibbs, R. D. (1958). Patterns in the seasonal water content of trees. *In* "The Physiology of Forest Trees" (K. V. Thimann, ed.), pp. 43–69. Ronald Press, New York.
103. Ginsburg, H., and Ginzburg, B. Z. (1970). Radial water and solute flows in roots of *Zea mays.* I. Water flow. *J. Exp. Bot.* **21,** 580–592.
104. Glinka, Z., and Reinhold, L. (1971). Abscisic acid raises the permeability of plant cells to water. *Plant Physiol.* **48,** 103–105.
105. Glinka, Z., and Reinhold, L. (1972). Induced changes in permeability of plant cell membranes to water. *Plant Physiol.* **49,** 602–606.
106. Goode, J. E., and Higgs, K. H. (1973). Water, osmotic and pressure potential relationships in apple leaves. *J. Hortic. Sci.* **48,** 203–215.
107. Grace, J. (1978). "Plant Response to Wind." Academic Press, London.
108. Gradmann, H. (1928). Untersuchungen über die Wasserverhältnisse des Bodens als Grundlage des Pflanzenwachstums. *Jahrb. Wiss. Bot.* **69,** 1–100.
109. Graham, J., Clarkson, D. T., and Sanderson, J. (1974). Water uptake by the roots of marrow and barley plants. *Annu. Rep.—Agric. Res. Counc., Letcombe Lab.* pp. 9–11.
110. Greacen, E. L., and Oh, J. S. (1972). Physics of root growth. *Nature (London), New Biol.* **235,** 24–25.
111. Green, P. B. (1968). Growth physics in *Nitella* a method for continuous *in vivo* analysis of extensibility based on a micro-manometer technique for turgor pressure. *Plant Physiol.* **43,** 1169–1184.
112. Green, P. B., and Cummins, W. R. (1974). Growth rate and turgor pressures. Auxin effect studies with an automated apparatus for single coleoptiles. *Plant Physiol.* **54,** 863–869.
113. Green, P. B., Erickson, R. O., and Buggy, J. (1971). Metabolic and physical control of cell elongation rate—*in vivo* studies in *Nitella. Plant Physiol.* **47,** 423–430.
114. Grier, C. C., and Waring, R. H. (1974). Conifer foliage mass related to sapwood area. *For. Sci.* **20,** 205–206.
115. Haines, F. M. (1935). Observations on the occurrence of air in conducting tracts. *Ann. Bot. (London)* **49,** 367–379.
116. Hall, A. E., Schülze, E.-D., and Lange, O. L. (1976). Current perspectives of steady-stage stomatal responses to environment. *Ecol. Stud.* **19,** 169–188.
117. Handley, W. R. C. (1936). Some observations on the problem of vessel length determination in woody dicotyledons. *New Phytol.* **35,** 456–471.
118. Hansen, G. K. (1974). Resistance to water flow in soils and plants, plant water status, stomatal resistance and transpiration of Italian ryegrass, as influenced by transpiration demand and soil water depletion. *Acta Agric. Scand.* **24,** 83–93.
119. Hansen, G. K. (1974). Resistance to water transport in soil and young wheat plants. *Acta Agric. Scand.* **24,** 37–48.
120. Hartsock, T. L., and Nobel, P. S. (1976). Watering converts a CAM plant to daytime CO_2 uptake. *Nature (London)* **262,** 574–576.
121. Heath, O. V. S., and Mansfield, T. A. (1962). A recording porometer with detachable cups operating on four separate leaves. *Proc. R. Soc. London, Ser. B* **156,** 1–13.
122. Hellkvist, J., Richards, G. P., and Jarvis, P. G. (1974). Vertical gradients of water potential and tissue water relations in Sitka spruce trees measured with the pressure chamber. *J. Appl. Ecol.* **11,** 637–668.
123. Herkelrath, W. N., Miller, E. E., and Gardner, W. R. (1977a). Water uptake by plants. 1. Divided root experiments. *Soil Sci. Soc. Am. J.* **41,** 1033–1038.

124. Herkelrath, W. N., Miller, E. E., and Gardner, W. R. (1977b). Water uptake by plants. II. The root contact model. *Soil Sci. Soc. Am. J.* **41**, 1039–1043.
125. Hoad, G. V. (1965). Effect of osmotic stress on abscisic acid levels in xylem sap of sunflower (*Helianthus annuus* L.). *Planta* **124**, 25–29.
126. Honert, T. H. van den (1948). Water transport as a catenary process. *Discuss. Faraday Soc.* **3**, 146–153.
127. Hopmans, P. A. M. (1969). Types of stomatal cycling and their water relations in bean leaves. *Z. Pflanzenphysiol.* **60**, 242–253.
128. House, C. R., and Findlay, N. (1966). Water transport in isolated maize roots. *J. Exp. Bot.* **17**, 344–354.
129. Hsiao, T. C. (1973). Plant responses to water stress. *Annu. Rev. Plant Physiol.* **24**, 519–570.
130. Hsiao, T. C., and Acevedo, E. (1974). Plant responses to water deficits, water-use efficiency and drought resistance. *Agric. Meteorol.* **14**, 59–84.
131. Hsiao, T. C., Acevedo, E., and Henderson, D. W. (1970). Maize leaf elongation: Continuous measurements and close dependence on plant water status. *Science* **168**, 590–591.
132. Hsiao, T. C., Acevedo, E., Fereres, E., and Henderson, D. W. (1976). Stress metabolism, water stress, growth and osmotic adjustment. *Philos. Trans. R. Soc. London, Ser. B* **273**, 479–500.
133. Hsiao, T. C., Fereres, E., Acevedo, E., and Henderson, D. W. (1976). Water stress and dynamics of growth and yield of crop plants. *Ecol. Stud.* **19**, 281–305.
134. Huber, B. (1937). Wasserumsatz und Stoffbewegungen. *Fortschr. Bot.* **7**, 197–207.
135. Huck, M. G., Klepper, B., and Taylor, H. M. (1970). Diurnal variations in root diameter. *Plant Physiol.* **45**, 529–530.
136. Huffaker, R. C., Radin, T., Kleinkopf, G. E., and Cok, E. L. (1970). Effects of mild water stress on enzymes of nitrate assimilation and of the carboxylative phase of photosynthesis in barley. *Ann. Bot. (London)* [N.S.] **34**, 393–408.
137. Huisinga, B., and Knijff, A. M. W. (1974). On the function of the casparian strip in roots. *Acta Bot. Neerl.* **23**, 171–175.
138. Hunter, A. S., and Kelley, O. J. (1946). A new technique for studying the absorption of moisture and nutrients from soil by plant roots. *Soil. Sci.* **62**, 441–450.
139. Hüsken, D., Steudle, E., and Zimmermann, U. (1978). Pressure probe technique for measuring water relations of cells in higher plants. *Plant Physiol.* **61**, 158–163.
139a. Innes, P., Blackwell, R. D., and Quarrie, S. A. (1984). Some effects of genetic variation in drought-induced abscisic acid accumulation on the yield and water use of spring wheat. *J. Agric. Sci.* **102**, 341–351.
140. Itai, C., and Vaadia, Y. (1965). Kinetin-like activity in root exudate of water-stressed sunflower plants. *Physiol. Plant.* **18**, 941–944.
141. Janes, B. E. (1970). Effect of carbon dioxide, osmotic potential of nutrient solution and light intensity on transpiration and resistance to flow of water in pepper plants. *Plant Physiol.* **45**, 95–103.
142. Jarvis, P. G. (1975). Water transfer in plants. *In* "Heat and Mass Transfer in the Environment of Vegetation" (D. A. de Vries, ed.), pp. 369–394. Scripta, Washington, D.C.
143. Jarvis, P. G. (1976). The interpretation of the variations in leaf water potential and stomatal conductance found in canopies in the field. *Philos. Trans. R. Soc. London, Ser. B* **273**, 593–610.
144. Jarvis, P. G. (1980). Stomatal response to water stress in conifers. *In* "Adaptation of Plants to Water and High Temperature Stress" (N. C. Turner and P. J. Kramer, eds.), pp. 105–122. Wiley, New York.

145. Jarvis, P. G. (1981). Stomatal conductance, gaseous exchange and transpiration. *In* "Plants and their Atmospheric Environment" (J. Grace *et al.*, eds.), pp. 171–240. Blackwell, Oxford.
146. Jarvis, P. G., and Morison, J. I. L. (1981). The control of transpiration and photosynthesis by the stomata. *In* "Stomatal Physiology" (P. G. Jarvis and T. A. Mansfield, eds.), pp. 247–279. Cambridge Univ. Press, London and New York.
147. Johnson, R. P. C. (1977). Can cell walls bending round xylem vessles control water flow? *Planta* **137,** 187–194.
148. Jones, H. G. (1976). Crop characteristics and the ratio between assimilation and transpiration. *J. Appl. Ecol.* **13,** 605–622.
149. Jones, H. G. (1983). "Plants and Microclimate." Cambridge Univ. Press, London and New York.
150. Jones, M. M., and Turner, N. C. (1978). Osmotic adjustment in leaves of sorghum in response to water deficits. *Plant Physiol.* **61,** 122–126.
151. Jones, M. M., and Turner, N. C. (1980). Osmotic adjustment in expanding and fully expanded leaves of sunflower in response to water deficits. *Aust. J. Plant Physiol.* **7,** 181–192.
152. Jones, M. M., Osmond, C. B., and Turner, N. C. (1980). Accumulation of solutes in leaves of *Sorghum* and sunflower in response to water deficits. *Aust. J. Plant Physiol.* **7,** 193–205.
152a. Jones, R. J., and Mansfield, T. A. (1972). Effects of abscisic acid and its esters on stomatal aperture and the transpiration ratio. *Physiol. Plant* **26,** 321–327.
153. Jordan, W. R., and Ritchie, J. T. (1971). Influence of soil water stress on evaporation, root absorption, and internal water status of cotton. *Plant Physiol.* **48,** 783–788.
154. Kamiya, N., Tazawa, M., and Takata, T. (1962). Water permeability of the cell wall in *Nitella. Plant Cell Physiol.* **3,** 285–292.
155. Keck, R. W., and Boyer, J. S. (1974). Chloroplast response to low leaf water potentials. III. Differing inhibition of electron transport and photophosphorylation. *Plant Physiol.* **53,** 474–479.
156. Kirkham, M. B., Gardner, W. R., and Gerloff, G. C. (1972). Regulation of cell division and cell enlargement by turgor pressure. *Plant Physiol.* **49,** 961–962.
157. Kozlowski, T. T., ed. (1968). "Water Deficits and Plant Growth," Vol. 1. Academic Press, New York.
158. Kozlowski, T. T., ed. (1968) "Water Deficits and Plant Growth," Vol. 2. Academic Press, New York.
159. Kozlowski, T. T., ed. (1972). "Water Deficits and Plant Growth," Vol. 3. Academic Press, New York.
160. Kozlowski, T. T., ed. (1976). "Water Deficits and Plant Growth," Vol. 4. Academic Press, New York.
161. Kozlowski, T. T., ed. (1978). "Water Deficits and Plant Growth," Vol. 5. Academic Press, New York.
162. Kozlowski, T. T., ed. (1981). "Water Deficits and Plant Growth," Vol. 6 Academic Press, New York.
163. Kozlowski, T. T., ed. (1983). "Water Deficits and Plant Growth," Vol. 7. Academic Press, New York.
164. Kozlowski, T. T., ed. (1984). "Water Deficits and Plant Growth," Vol. 8. Academic Press, New York.
165. Kozlowski, T. T., Hughes, J. F., and Leyton, L. (1966). Patterns of water movement in dormant gymnosperm seedlings. *Biorheology* **3,** 77–85.
166. Kramer, P. J. (1959). Transpiration and the water economy of plants. In "Plant Physiology" (F. C. Steward, ed.), Vol. 2, pp. 607–730. Academic Press, New York.

167. Kramer, P. J. (1983). "Water Relations of Plants." Academic Press, New York.
168. Kramer, P. J. (1974). Fifty years progress in water relations research. *Plant Physiol.* **54,** 463–471.
169. Kramer, P. J., and Bullock, H. C. (1966). Seasonal variations in the proportions of suberized and unsuberized roots of trees in relation to the absorption of water. *Ann. J. Bot.* **53,** 200–204.
170. Landsberg, J. J., Blanchard, T. W., and Warrit, B. (1976). Studies in the movement of water through apple trees. *J. Exp. Bot.* **27,** 579–596.
171. Lang, A. R. G., Klepper, B., and Cumming, M. J. (1969). Leaf water balance during oscillation of stomatal aperture. *Plant Physiol.* **44,** 826–830.
172. Lange, O. L., Kappen, L., and Schülze, E.-D., eds. (1976). "Water and Plant Life," Ecol. Stud., Vol. 19. Springer-Verlag, Berlin and New York.
173. Larcher, W. (1975). "Physiological Plant Ecology." Springer-Verlag, Berlin and New York.
174. Larher, F., Hamelin, J., and Stewart, G. R. (1977). L'acide dimethylsulfonium-3 propanvique de *Spartina anglica. Phytochemistry* **16,** 2019–2020.
175. Larher, F., and Hamelin, J. (1975). L'acide trimethylamino-propionique des rameaux de *Limonium vulgare. Phytochemistry* **14,** 1798–1800.
176. Lawlor, D. W. (1974). Growth and water use of *Lolium perenne.* 1. Water transport. *J. Appl. Ecol.* **9,** 79–98.
177. Lawlor, D. W., and Milford, G. F. J. (1975). The control of water and carbon dioxide flux in water-stressed sugar beet. *J. Exp. Bot.* **26,** 657–665.
178. Levitt, J. (1980). "Responses of Plants to Environmental Stresses," 2nd ed., Vol. 2. Academic Press, New York.
179. Lewis, F. J. (1945). Physical condition of the surface of the mesophyll cells walls of the leaf. *Nature (London)* **156,** 445–453.
180. Lockhart, J. A. (1965a). An analysis of irreversible plant cell elongation. *J. Theor. Biol.* **8,** 264–276.
181. Lockhart, J. A. (1965b). Cell extension. *In* "Plant Biochemistry" (J. Bonner and J. E. Varner, eds.), pp. 827–849. Academic Press, New York.
182. Lopushinsky, W. (1961). Effect of water movement on salt movement through tomato roots. *Nature (London)* **192,** 994–995.
183. Lopushinsky, W. (1964). Effect of water movement on ion movement into the xylem of tomato roots. *Plant Physiol.* **39,** 494–501.
184. Lösch, R., and Tenhunen, J. D. (1981). Stomatal responses to humidity-phenomenon and mechanism. *In* "Stomatal Physiology" (P. G. Jarvis and T. A. Mansfield, eds.), pp. 137–162. Cambridge Univ. Press, London and New York.
185. Maercker, U. (1965). Water movement in plant cell walls. *Protoplasma* **60,** 61–78.
186. Maier-Maercker, U. (1979). "Peristomatal transpiration" and stomatal movement: A controversial view. I. Additional proof of peristomatal transpiration by hydrophotography and a comprehensive discussion in the light of recent results. *Z. Pflanzenphysiol.* **91,** 25–43.
187. Maier-Maercker, U. (1979). "Peristomatal transpiration" and stomatal movement: A controversial view. II. Observation of stomatal movements under different conditions of water supply and demand. *Z. Pflanzephysiol.* **91,** 157–172.
188. Maier-Maercker, U. (1979). "Peristomatal transpiration" and stomatal movement: A controversial view. III. Visible effects of peristomatal transpiration on the epidermis. *Z. Pflanzenphysiol.* **91,** 225–238.
189. Maier-Maercker, U. (1979). "Peristomatal transpiration" and stomatal movement: A controversial view. IV. Ion accumulation by peristomatal transpiration. *Z. Pflanzenphysiol.* **91,** 239–254.

190. Maier-Maercker, U. (1980). "Peristomatal transpiration" and stomatal movement: A controversial view. VI. Lanthanum deposits in the epidermal apoplast. *Z. Pflanzenphysiol.* **100,** 121–130.
191. Maier-Maercker, U. (1981). "Peristomatal transpiration" and stomatal movement: A controversial view. V. Rubidium-86 in the epidermal transpiration stream. *Z. Pflanzenphysiol.* **101,** 447–459.
192. Malek, F., and Baker, D. A. (1978). Effect of fusicoccin on proton co-transport of sugars in phloem loading of *Ricinus communis* L. *Plant. Sci. Lett.* **11,** 233–239.
193. Malik, R. S., Dhankar, J. S., and Turner, N. C. (1979). Influence of soil water deficits on root growth of cotton seedlings. *Plant Soil* **53,** 109–115.
194. Mansfield, T. A. (1976). Chemical control of stomatal movements. *Philos. Trans. R. Soc. London, Ser. B* **273,** 541–550.
195. Mansfield, T. A., and Davies, W. J. (1982). Stomata and stomatal mechanisms. *In* "The Physiology and Biochemistry of Drought Resistance in Plants" (L. G. Paleg and D. Aspinall, eds.), pp. 315–346. Academic Press, Sydney.
196. Mansfield, T. A., Wellburn, A. R., and Moreira, T. J. S. (1978). The role of abscisic acid and farnesol in the alleviation of water stress. *Philos. Trans. R. Soc. London, Ser. B* **284,** 471–482.
197. Markhart, A. H., Fiscus, E. L., Naylor, A. W., and Kramer, P. J. (1981). Effect of abscisic acid on root hydraulic conductivity. *Plant Physiol.* **64,** 611–614.
198. Marriott, F. H. C., and Nye, P. H. (1969). The importance of mass flow in the uptake of ions by roots from soil. *Trans. Int. Congr. Soil Sci., 9th, 1968,* Vol. 1, pp. 127–134.
199. Matsuda, K., and Riazi, A. (1981). Stress induced osmotic adjustment in growing regions of barley leaves. *Plant Physiol.* **68,** 571–576.
200. Matthews, M. A., Van Volkenburgh, E., and Boyer, J. S. (1984). Acclimation of leaf growth to low water potentials in sunflower. *Plant, Cell Environ.* **7,** 199–206.
201. Mees, G. C., and Weatherley, P. E. (1957). The mechanism of water absorption by roots. I. Preliminary studies on the effects of hydrostatic pressure gradients. *Proc. R. Soc. London, Ser. B* **147,** 367–380.
202. Meidner, H. (1975). Water supply, evaporation, and vapour diffusion in leaves. *J. Exp. Bot.* **26,** 666–673.
203. Meidner, H. (1976). Vapour loss through stomatal pores with the mesophyll tissue excluded. *J. Exp. Bot.* **27,** 172–174.
204. Meidner, H. (1976). Water vapour loss from a physical model of a substomatal cavity. *J. Exp. Bot.* **27,** 691–694.
205. Meidner, H. (1977). Sap exudation via the epidermis of leaves. *J. Exp. Bot.* **28,** 1408–1416.
206. Meidner, H., and Mansfield, T. A. (1968). "Physiology of Stomata." McGraw-Hill, London.
207. Meidner, H., and Sheriff, D. W. (1976). "Water and Plants." Blackie, Glasgow.
208. Mer, C. L. (1940). The factors determining the resistance to the movement of water in the leaf. *Ann. Bot. (London)* [N.S.] **4,** 397–401.
209. Meyer, R. F., and Boyer, J. S. (1972). Sensitivity of cell division and cell elongation to low water potentials in soybean hypocotyls. *Planta* **108,** 77–87.
210. Meyer, R. E., and Gingrich, J. H. (1964). Osmotic stress: Effects of its application on a portion of wheat root systems. *Science* **144,** 1463–1464.
211. Michel, B. E., and El Sharkawi, H. M. (1970). Investigation of plant water relations with divided root systems of soybean. *Plant Physiol.* **46,** 728–731.
212. Michelena, V. A., and Boyer, J. S. (1982). Complete turgor maintenance at low water potentials in the elongating region of maize leaves. *Plant Physiol.* **69,** 1145–1149.

213. Milburn, J. (1966). The conduction of sap. I. Water conduction and cavitation in water stressed leaves. *Planta* **69**, 34–42.
214. Milburn, J. A. (1973). Cavitation in *Ricinus* by acoustic detection: Induction in exised leaves by various factors. *Planta* **110**, 253–265.
215. Milburn, J. A., and Johnson, R. P. C. (1966). The conduction of sap. II. Detection of vibrations produced by sap cavitation in *Ricinus* xylem. *Planta* **69**, 43–52.
216. Milburn, J. A. (1979). "Water Flow in Plants." Longman, London.
217. Milthorpe, F. L., and Moorby, J. (1974). "An Introduction to Crop Physiology." Cambridge Univ. Press, London and New York.
217a. Mizrahi, Y., Scherings, S. G., Malis Arad, S., and Richmond, A. E. (1974). Aspects of the effect of A.B.A. on the water status of barley and wheat seedlings. *Physiol. Plant.* **31**, 44–50.
217b. Molyneaux, D. E., and Davies, W. J. (1983). Rooting pattern and water relations of three pasture grasses growing in drying soil. *Oecologia* **58**, 220–224.
218. Molz, F. J., and Klepper, B. (1973). On the mechanism of water-stress-induced stem deformation. *Agron. J.* **65**, 304–306.
219. Molz, F. J., and Boyer, J. S. (1978). Growth-induced water potentials in plant cells and tissues. *Plant Physiol.* **62**, 423–429.
220. Monteith, J. L. (1965). Evaporation and environment. *Symp. Soc. Exp. Biol.* **29**, 205–234.
221. Monteith, J. L., and Owen, P. C. (1958). A thermocouple method for measuring relative humidity in the range of 95–100%. *J. Sci. Instrum.* **35**, 443–446.
222. Morgan, J. M. (1977). Difference in osmoregulation between wheat genotypes. *Nature (London)* **270**, 234–235.
223. Munns, R., Brady, C. J., and Barlow, E. W. R. (1979). Solute accumulation in the apex and leaves of wheat during water stress. *Aust. J. Plant Physiol.* **6**, 379–389.
224. Nagahashi, G., Thomson, W. W., and Leonard, R. T. (1974). The Casparian strip as a barrier to the movement of lanthanum in corn roots. *Science* **183**, 670–671.
225. Neales, T. F., and Incoll, L. D. (1968). The control of leaf photosynthesis rate by the level of assimilate concentration in the leaf: A review of a hypothesis. *Bot. Rev.* **34**, 107–125.
226. Neales, T. F., Hartney, V. J., and Patterson, A. A. (1968). Physiological adaptation to drought in the carbon assimilation and water loss of xerophytes. *Nature (London)* **219**, 469–472.
227. Newman, E. I. (1969). Resistance to water flow in soil and plant. I. Soil resistance in relation to amounts of root. Theoretical estimates. *J. Appl. Ecol.* **6**, 1–12.
228. Newman, E. I. (1969). Resistance to water flow in soil and plant. II. A review of experimental evidence on rhizosphere resistance. *J. Appl. Ecol.* **6**, 261–272.
229. Newman, E. I. (1973). Permeability to water of the roots of five herbaceous species. *New Phytol.* **72**, 547–555.
230. Newman, E. I. (1974). Root and soil water relations. *In* "The Plant Root and its Environment" (E. W. Carson, ed.), pp. 363–440. Univ. Press of Virginia, Charlottesville.
231. Newman, E. I. (1976). Water movement through root systems. *Philos. Trans. R. Soc. London, Ser. B* **273**, 463–478.
232. Newman, E. I. (1976). Interaction between osmotic- and pressure-induced water flow in plant roots. *Plant Physiol.* **57**, 738–739.
233. Nobel, P. S. (1983). "Biophysical Plant Physiology and Ecology." Freeman, San Fransisco, California.
234. Norris, R. I., and Bukovac, M. J. (1968). Structure of the pear leaf cuticle with special reference to cuticular penetration. *Am. J. Bot.* **55**, 975–983.

235. O'Brien, T. P., and McCully, M. E. (1969). "Plant Structure and Development." Macmillan, London.

236. Oertli, J. J. (1969). Terminology of plant-water energy relations. *Z. Pflanzenphysiol.* **61,** 264–265.

237. Osmond, C. B. (1978). Crassulacean acid metabolism. A curiosity in context. *Annu. Rev. Plant Physiol.* **29,** 379–414.

238. Osonubi, O., and Davies, W. J. (1978). Solute accumulation in leaves and roots of woody plants subjected to water stress. *Oecologia* **32,** 323–332.

239. Osonubi, O., and Davies, W. J. (1980). The influence of plant water stress on stomatal control of gas exchange at different levels of atmospheric humidity. *Oecologia* **46,** 1–6.

240. Osonubi, O., and Davies, W. J. (1981). Root growth and water relations of oak and birch seedlings. *Oecologia* **51,** 343–350.

241. Paleg, L. G., and Aspinall, D., eds. (1982). "The Physiology and Biochemistry of Drought Resistance in Plants." Academic Press, Sydney.

242. Pallas, J. E., and Kays, S. J. (1982). Inhibition of photosynthesis by ethylene—a stomatal effect. *Plant Physiol.* **70,** 598–601.

243. Pallas, J. E., Michel, B. E., and Harris, D. G. (1967). Photosynthesis, transpiration, leaf temperature and structural activity of cotton plants under varying water potentials. *Plant Physiol.* **42,** 76–88.

244. Palta, J. P., and Stadelmann, E. J. (1977). Effect of turgor pressure on water permeability of *Allium cepa* epidermis cell membranes. *J. Membr. Biol.* **33,** 231–247.

245. Parkhurst, D. F., and Loucks, O. L. (1972). Optimal leaf size in relation to environment. *J. Ecol.* **60,** 505–537.

246. Parsons, L. R., and Kramer, P. J. (1974). Diurnal cycling in root resistance to water movement. *Physiol. Plant.* **30,** 19–23.

247. Passioura, J. B. (1980). The meaning of matric potential. *J. Exp. Bot.* **31,** 1161–1169.

248. Passioura, J. B. (1982). Water collection by roots. *In* "The Physiology and Biochemistry of Drought Resistance in Plants" (L. G. Paleg and D. Aspinall, eds.), pp. 39–53. Academic Press, Sydney.

249. Passioura, J. B. (1982). Water in the soil-plant-atmosphere continuum. *In* "Encyclopedia of Plant Physiology" (O. L. Lange, P. S. Nobel, C. B. Osmond, and H. Ziegler, eds.), Vol. 12B, pp. 5–33. Springer-Verlag, Berlin and New York.

250. Penman, H. L. (1948). Natural evaporation from open water, bare soil and grass. *Proc. R. Soc. London, Ser. A* **193,** 120–145.

251. Pierce, M., and Raschke, K. (1980). Correlation between loss of turgor and accumulation of abscisic acid in detached leaves. *Planta* **148,** 174–182.

252. Pizzolato, T. D., Burbano, J. L., Berlin, J. D., Morey, P. R., and Pease, R. W. (1976). An electron microscope study of the path of water movement in transpiring leaves of cotton *(Gossypium hirsutum* L.). *J. Exp. Bot.* **27,** 145–161.

253. Plymale, E. L., and Wylie, R. B. (1944). The major veins of mesomorphic leaves. *Am. J. Bot.* **31,** 99–106.

254. Pospisilova, J. (1969). Role of water transport in the origin of water stress. *Biol. Plant.* **11,** 130–138.

255. Pospisilova, J. (1970). The relationship between the resistance to water transport and the water saturation deficit in leaf tissue of kale and tobacco. *Biol. Plant.* **12,** 78–80.

256. Pospisilova, J. (1972). Variable resistance to water transport in leaf tissue of kale. *Biol. Plant.* **14,** 293–296.

257. Postlethwaite, S. N., and Rogers, B. (1958). Tracing the path of the transpiration stream in trees by the use of radioactive isotopes. *Am. J. Bot.* **45,** 753–757.

258. Potter, J. R., and Boyer, J. S. (1973). Chloroplast response to low leaf water potentials. II. Role of osmotic potential. *Plant Physiol.* **51,** 993–997.

259. Powell, D. B. B. (1978). Regulation of plant water potential by membranes of the endodermis in young roots. *Plant Cell Environ.* **1,** 69–76.
260. Powell, D. B. B., and Thorpe, M. R. (1977). Dynamic aspects of plant-water relationships. *In* "Environmental Effects on Crop Physiology" (J. J. Landsberg and C. V. Cutting, eds.), pp. 259–279. Academic Press, London.
261. Preston, R. D. (1938). The contents of the vessels of *Fraxinus americana* L. with respect to the ascent of sap. *Ann. Bot. (London)* [N.S.] **2,** 1–22.
262. Puritch, G. S., and Petty, J. A. (1971). Effect of Balsam woolly aphid, *Adelges picae* (Ratz), infestation on the xylem of *Abies grandis* (Derry.) Linde. *J. Exp. Bot.* **22,** 946–952.
263. Quarrie, S. A., and Jones, H. G. (1977). Effects of abscisic acid and water stress on development and morphology of wheat. *J. Exp. Bot.* **28,** 192–203.
264. Quarrie, S. A., and Lister, P. G. (1983). Characterization of spring wheat genotypes differing in drought-induced abscisic acid accumulation. 1. Drought-stressed ABA production. *J. Exp. Bot.* **34,** 1260–1270.
265. Raschke, K. (1970). Leaf hydraulic system: Rapid epidermal and stomatal responses to changes in water supply. *Science* **167,** 189–191.
266. Raschke, K. (1970). Stomatal responses to pressure changes and interruptions in the water supply of detached leaves of *Zea mays* L. *Plant Physiol.* **45,** 415–423.
267. Raschke, K. (1974). Simultaneous requirement of ABA and CO_2 for the modulation of stomatal conductance in *Xanthium strumarium. Plant Physiol.* **53,** 55.
268. Raschke, K. (1976). How stomata resolve the dilemma of opposing priorities. *Philos. Trans. R. Soc. London, Ser. B.* **273,** 551–560.
269. Rawlins, S. L. (1963). Resistance to water flow in the transpiration stream. *Bull.—Conn., Agric. Exp. Stan., New Haven* **664,** 69–85.
270. Rayle, D. L. (1973). Auxin-induced hydrogen-ion secretion in *Avena* coleoptiles and its implications. *Planta* **114,** 63–73.
271. Rehm, M. M., and Cline, M. G. (1973). Inhibition of low pH-induced elongation in *Avena* coleoptiles by abscisic acid. *Plant Physiol.* **51,** 946–948.
272. Renner, O. (1915). Die Wasserversorgung der Pflanzen. *Handwörterbuch Naturwiss.* **10,** 538–557.
273. Richards, L. A., and Ogata, G. (1958). Thermocouple for vapour pressure measurement in biological and soil systems at high humidity. *Science* **128,** 1089–1090.
274. Robards, A. W., Jackson, S. M., Clarkson, D. T., and Sanderson, J. (1973). The structure of barley roots in relation to the transport of ions into the stele. *Protoplasma* **77,** 291–311.
275. Roberts, B. R., and Schreiber, L. R. (1977). Influence of Dutch elm disease on resistance to water flow through roots of American elm. *Phytopathology* **67,** 56–59.
276. Roberts, J. (1976). An examination of the quantity of water stored in mature *Pinus sylvestris* L. trees. *J. Exp. Bot.* **27,** 473–479.
277. Roberts, S. W., and Knoerr, K. R. (1977). Components of water potential estimated from xylem pressure measurements in five tree species. *Oecologia* **28,** 191–202.
278. Rodriguez, J. L. (1982). A role for ABA in the water relations of *Zea mays* L. Ph.D. Thesis, University of Lancaster.
279. Rodriguez, J. L., and Davies, W. J. (1982). The effects of temperature and ABA on stomata of *Zea mays* L. *J. Exp. Bot.* **33,** 977–987.
280. Salama, A. M. EL-O.A., and Wareing, P. F. (1979). Effects of mineral nutrition on endogenous cytokinin in plants of sunflower *(Helianthus annuus* L.) *J. Exp. Bot.* **30,** 971–981.
281. Scholander, P. T., Hammel, H. T., Hemmingsen, E. A., and Bradstreet, E. D. (1964). Hydrostatic pressure and osmotic potential in leaves of mangroves and some other plants. *Proc. Natl. Acad. Sci. U.S.A.* **52,** 119–125.

282. Schülze, E.-D., Lange, O. L., Buschbom, U., Kappen, L., and Evenari, M. (1972). Stomatal responses to changes in humidity in plants growing in the desert. *Planta* **108,** 259–270.
283. Schülze, E.-D., Lange, O. L., Kappen, L., Buschbom, U., and Evenari, M. (1973). Stomatal responses to changes in temperature at increasing water stress. *Planta* **110,** 29–42.
284. Schülze, E.-D., Lange, O. L., Evenari, M., Kappen, L., and Buschbom, U. (1974). The role of air humidity and temperature in controlling stomatal resistance of *Prunus armeniaca* L. under desert conditions. I. A simulation of the daily course of stomatal resistance. *Oecologia* **17,** 159–170.
285. Schülze, E.-D., Lange, O. L., Kappen, L., Evenari, M., and Buschbom, U. (1975). The role of air humidity and leaf temperature in controlling stomatal resistance of *Prunus armeniaca* L. under desert conditions. II. The significance of leaf water status and internal carbon dioxide concentration. *Oecologia* **18,** 219–233.
286. Schülze, E.-D., Lange, O. L., Evenari, M., Kappen, L., and Buschbom, U. (1975). The role of air humidity and temperature in controlling stomatal resistance of *Prunus armeniaca* L. under desert conditions. III. The effect on water use efficiency. *Oecologia* **19,** 303–314.
287. Scott, F. M. (1950). Internal suberization of tissues. *Bot. Gaz. (Chicago)* **111,** 378–394.
288. Scott, F. M. (1964). Lipid deposition in intercellular space. *Nature (London)* **203,** 164–168.
289. Sharp, R. E. (1981). Mechanisms of turgor maintenance in *Zea mays.* Ph.D. Thesis, Lancaster University.
290. Sharp, R. E., and Davies, W. J. (1979). Solute regulation and growth by roots and shoots of water-stressed maize plants. *Planta* **147,** 43–49.
291. Sharp, R. E., Osonubi, O., Wood, W. A., and Davies, W. J. (1979). A simple instrument for measuring leaf extension in grasses, and its application in the study of the effects of water stress on maize and sorghum. *Ann. Bot. (London)* [N.S.] **44,** 35–45.
292. Sheriff, D. W. (1972). A new apparatus for the measurement of sap flux in small shoots with the magnetohydrodynamic technique. *J. Exp. Bot.* **23,** 1086–1095.
293. Sheriff, D. W. (1973). Significance of the occurrence of time lags in the transmission of hydraulic shock waves through plant stems. *J. Exp. Bot.* **24,** 796–803.
294. Sheriff, D. W. (1974). A model of plant hydraulics under non-equilibrium conditions. I. Stems. *J. Exp. Bot.* **25,** 552–561.
295. Sheriff, D. W. (1974). A model of plant hydraulics under non-equilibrium conditions. II. Leaves. *J. Exp. Bot.* **25,** 562–574.
296. Sheriff, D. W. (1977). The effect of humidity on water uptake by, and viscous flow resistance of, excised leaves of a number of species. Physiological and anatomical observations. *J. Exp. Bot.* **28,** 1399–1407.
297. Sheriff, D. W., and Meidner, H. (1974). Water pathways in leaves of *Hedera helix* L. and *Tradescantia virginiana* L. *J. Exp. Bot.* **25,** 1147–1156.
298. Sheriff, D. W., and Meidner, H. (1975). Water movement into and through *Tradescantia virginiana* (L.) leaves. I. Uptake during conditions of dynamic equilibrium. *J. Exp. Bot.* **26,** 897–902.
299. Sheriff, D. W., and Sinclair, R. (1973). Fluctuations in leaf water balance with a period of 1 to 10 minutes. *Planta* **113,** 215–228.
299a. Shimshi, D. (1963). Effect of soil moisture and phenylmercuric acetate on stomatal aperture, transpiration and photosynthesis. *Plant Physiol.* **38,** 713–721.
300. Sinclair, T. R., Bingham, G. E., Lemon, E. R., and Hartwell, A. L. (1975). Water use efficiency of field grown maize during moisture stress. *Plant Physiol.* **56,** 245–248.

301. Slatyer, R. O. (1957). Significance of the permanent wilting percentage in studies of plant and soil relations. *Bot. Rev.* **23,** 585–636.
302. Slatyer, R. O. (1957). Significance of the progressive increases in total soil moisture stress, on transpiration, growth, and internal water relationships of plants. *Aust. J. Biol. Sci.* **10,** 320–336.
303. Slatyer, R. O. (1967). "Plant-Water Relationships." Academic Press, New York.
304. Slatyer, R. O. (1970). Comparative photosynthesis, growth and transpiration of two species of *Atriplex. Planta* **93,** 175–189.
305. Slatyer, R. O., and Bierhuizen, J. F. (1964). The effect of several foliar sprays on transpiration and water use efficiency of cotton plants. *Agric. Meteorol.* **1,** 42–53.
306. Slatyer, R. O., and Taylor, S. A. (1960). Terminology in plant-soil-water relations. *Nature (London)* **187,** 922–924.
307. Snaith, P. J., and Mansfield, T. A. (1982). Control of the CO_2 responses of stomata by indol-3-yl acetic acid and abscisic acid. *J. Exp. Bot.* **33,** 360–365.
308. Snaith, P. J., and Mansfield, T. A. (1982). Stomatal sensitivity to abscisic acid: Can it be defined? *Plant Cell Environ.* **5,** 309–312.
309. Spanswick, R. M. (1975). Symplasmic transport in tissues. *In* "Encyclopedia of Plant Physiology, New Series" (U. Lüttge and M. G. Pitman, eds.), Vol. 2B, pp. 35–49. Springer-Verlag, Berlin and New York.
310. Steudle, E., and Zimmermann, U. (1974). Determination of the hydraulic conductivity and the reflection coefficients in *Nitella flexilis* by means of direct cell-turgor pressure measurements. *Biochim. Biophys. Acta* **332,** 399–412.
311. Stewart, G. R., and Lee, J. A. (1974). The role of proline accumulation in halophytes. *Planta* **120,** 279–289.
312. Stoker, R., and Weatherley, P. E. (1971). The influence of the root system on the relationship between the rate of transpiration and depression of leaf water potential. *New Phytol.* **70,** 547–554.
313. Stone, E. C., Went, F. W., and Young, C. L. (1950). Water absorption from the atmosphere by plants growing in dry soil. *Science* **111,** 546–548.
314. Storey, R., and Wyn Jones, R. G. (1979). Betaine and proline levels in plants and their relationship to NaCl stress. *Plant Sci. Lett.* **4,** 161–168.
315. Strugger, S. (1939). Die lumineszenzmikroscopische Analyse des Transpirationsstromes in Parenchymen. *Flora (Jena)* **133,** 56.
316. Swanson, R. H. (1966). Seasonal course of transpiration of lodgepole pine and Engelmann spruce. *In* "Forest Hydrology" (N. E. Sopper and H. W. Lull, eds.), pp. 419–434. Pergamon, Oxford.
317. Szarek, S. R., and Ting, I. P. (1975). Photosynthetic efficiency of CAM plants in relation to C_3 and C_4 plants. *In* "Environmental and Biological Control of Photosynthesis" (R. Marcelle, ed.), pp. 289–297. Junk, The Hague.
318. Takami, S., Rawson, H. R., and Turner, N. C. (1982). Leaf expansion of four sunflower (*Helianthus annuus* L.) cultivars in relation to water deficits. II. Diurnal patterns during stress and recovery. *Plant Cell Environ.* **5,** 279–286.
319. Talboys, P. W. (1968). Water deficits in vascular disease. *In* "Water Deficits and Plant Growth" (T. T. Kozlowski, ed.), Vol. 2, pp. 255–311. Academic Press, New York.
320. Tanton, T. W., and Crowdy, S. H. (1972). Water pathways in higher plants. II. Water pathways in roots. *J. Exp. Bot.* **23,** 600–618.
321. Taylor, S. A., and Slatyer, R. O. (1962). Proposals for a unified terminology in studies of plant-soil-water relations. *Arid Zone Res.* **16,** 339–349.
322. Teoh, C. T., and Palmer, J. H. (1971). Nonsynchronized oscillations in stomatal resistance among sclerophylls of *Eucalyptus umbra. Plant Physiol.* **47,** 409–411.

323. Ting, I. P. (1976). Crassulacean acid metabolism in natural ecosystems in relation to annual CO_2 uptake patterns and water utilization. *In* "CO_2 Metabolism and Plant Productivity" (R. H. Burns and C. C. Black, eds.), pp. 251–268. University Park Press, Baltimore, Maryland.

324. Tinker, P. B. (1976). Roots and water. Transport of water to plant roots in soil. *Philos. Trans. R. Soc. London, Ser. B* **273,** 445–461.

325. Tinklin, R., and Weatherley, P. E. (1966). On the relationship between transpiration rate and leaf water potential. *New Phytol.* **65,** 509–517.

326. Tinklin, R., and Weatherley, P. E. (1968). The effect of transpiration rate on the leaf water potential of sand and soil rooted plants. *New Phytol.* **67,** 605–615.

327. Turner, N. C. (1974). Stomatal responses to light and water under field conditions. *Bull.—R. Soc. N.Z.* **12,** 423–432.

328. Turner, N. C., and Jones, M. M. (1980). Turgor maintenance by osmotic adjustment: A review and evaluation. *In* "Adaptation of Plants to Water and High Temperature Stress" (N. C. Turner and P. J. Kramer, eds.), pp. 87–103. Wiley, New York.

328a. Turner, N. C., and Kramer, P. J., eds. (1980). "Adaptation of Plants to Water and High Temperature Stress." Wiley, New York.

329. Turner, N. C., Begg, J. E., and Tonnet, M. L. (1978). Osmotic adjustment of sorghum and sunflower crops in response to water deficits and its influence on the water potential at which stomata close. *Aust. J. Plant Physiol.* **5,** 597–608.

330. Turner, N. C., Begg, J. E., Rawson, H. M., English, S. D., and Hearn, A. B. (1978). Agronomic and physiological responses of soybean and sorghum crops to water deficits. III. Components of leaf water potential, leaf conductance, $^{14}CO_2$ photosynthesis and adaptation to water deficits. *Aust. J. Plant Physiol.* **5,** 179–194.

331. Tyree, M. T. (1969). The thermodynamics of short-distance translocation in plants. *J. Exp. Bot.* **20,** 341–349.

332. Tyree, M. T. (1970). The symplast concept: A general theory of symplastic transport according to the thermodynamics of irreversible processes. *J. Theor. Biol.* **26,** 181–214.

333. Tyree, M. T. (1976). Physical parameters of the soil-plant atmosphere system: Breeding for drought resistance characteristics that might improve wood yield. *In* "Tree Physiology and Yield Improvement" (M. G. R. Cannell and F. T. Last, eds.), pp. 329–348. Academic Press, London.

334. Tyree, M. T., and Jarvis, P. G. (1982). Water in tissues and cells. *In* "Encyclopedia of Plant Physiology" (O. L. Lange, P. S. Nobel, C. B. Osmond, and H. Ziegler, eds.), Vol. 12B, pp. 35–78. Springer-Verlag, Berlin and New York.

335. Tyree, M. T., Cruiziat, P., Benis, M., Logullo, M. A., and Salleo, S. (1981). The kinetics of rehydration of detached sunflower leaves from different initial water deficits. *Plant Cell Environ.* **4,** 309–317.

336. Vaadia, Y. (1976). Plant hormones and water stress. *Philos. Trans. R. Soc. London, Ser. B* **273,** 513–522.

337. Van Alfen, N. K., and Turner, N. C. (1975). Influence of a *Ceratocystis ulmi* toxin on water relations of elm *(Ulmus americana). Plant Physiol.* **55,** 312–316.

338. Van Volkenburgh, E., and Cleland, R. E. (1980). Proton excretion and cell expansion in bean leaves. *Planta* **148,** 273–278.

339. Van Volkenburgh, E., and Cleland, R. E. (1981). Control of light induced bean leaf expansion: Role of osmotic potential, wall yield stress and hydraulic conductivity. *Planta* **153,** 572–577.

340. Van Volkenburgh, E., and Davies, W. J. (1983). Inhibition of light-stimulated leaf expansion by abscisic acid. *J. Exp. Bot.* **34,** 835–845.

341. Viets, F. G. (1966). Increasing water use efficiency by soil management. *In* "Plant

Environment and Efficient Water Use" (W. H. Pierre *et al.*, eds.), pp. 259–274. Am. Soc. Agron., Madison, Wisconsin.

342. Waggoner, P. E., and Dimond, A. E. (1954). Reduction in water flow by mycelium in vessels. *Am. J. Bot.* **41,** 637–640.
343. Walton, D. C., Harrison, M. A., and Cote, P. (1976). The effects of water stress on abscisic acid levels and metabolism in roots of *Phaseolus vulgaris* L. and other plants. *Planta* **131,** 141–144.
344. Waring, R. H., and Running, S. W. (1978). Sapwood water storage: Its contribution to transpiration and effect upon water conductance through the stems of old-growth Douglas fir. *Plant, Cell Environ.* **1,** 131–140.
345. Watts, S., Rodriguez, J. L., Evans, S. E., and Davies, W. J. (1981). Root and shoot growth of plants treated with abscisic acid. *Ann. Bot. (London)* [N.S.] **47,** 595–602.
346. Watts, W. R. (1971). Leaf extension in *Zea mays.* I. Leaf extension and water potential in relation to root-zone and air temperatures. *J. Exp. Bot.* **23,** 704–712.
347. Watts, W. R. (1971). Leaf extension in *Zea mays.* II. Leaf extension in response to independent variation of the temperature of the apical meristem of the air around the leaves and the root zone. *J. Exp. Bot.* **23,** 713–721.
348. Weatherley, P. E. (1963). The pathway of water movement across the root cortex and leaf mesophyll of transpiring plants. *In* "The Water Relations of Plants" (A. J. Rutter and F. H. Whitehead, eds.), pp. 85–100. Wiley, New York.
349. Weatherley, P. E. (1970). Some aspects of water relations. *Adv. Bot. Res.* **3,** 171–206.
350. Weatherley, P. E. (1975). Water relations of the root system. *In* "The Development and Function of Roots" (J. G. Torrey and D. T. Clarkson, eds.), pp. 397–413. Academic Press, London.
351. Weatherley, P. E. (1976). Introduction: Water movement through plants. *Philos. Trans. R. Soc. London, Ser. B* **273,** 435–444.
352. Weatherley, P. E. (1982). Water uptake and flow in roots. *In* "Encyclopedia of Plant Physiology" (O. L. Lange, P. S. Nobel, C. B. Osmond, and H. Ziegler, eds.), Vol. 12B, pp. 79–110. Springer-Verlag, Berlin and New York.
353. Weaver, J. E., Jean, F. C., and Crist, J. W. (1922). Development and activities of roots of crop plants. *Carnegie Inst. Washington Publ.* **316.**
354. Weiler, E. W., Schnabl, H., and Hornberg, C. (1982). Stress-related levels of abscisic acid in guard cell protoplasts of *Vicia faba* L. *Planta* **154,** 24–28.
355. Wenkert, W., Lemon, E. R., and Sinclair, T. R. (1978). Water content-potential relationship on soya bean: Changes in component potentials for mature and immature leaves under field conditions. *Ann. Bot. (London)* [N.S.] **42,** 295–307.
356. Westgate, M. E., and Boyer, J. S. (1984). Transpiration and growth-induced water potentials in maize. *Plant Physiol.* **74,** 882–889.
357. Williams, J. (1974). Root density and water potential gradients near the plant root. *J. Exp. Bot.* **25,** 669–674.

357a. Williams, W. E. (1983). Optimal water-use efficiency in a California shrub. *Plant, Cell Environ.* **6,** 145–152.

358. Wilson, J. A., and Davies, W. J. (1979). Farnesol-like antitranspirant activity and stomatal behaviour in maize and sorghum lines of differing drought tolerance. *Plant, Cell Environ.* **2,** 49–57.
359. Wilson, J. A., Ogunkanmi, A. B., and Mansfield, T. A. (1978). Effects of external potassium supply on stomatal closure induced by abscisic acid. *Plant, Cell Environ.* **1,** 199–201.
360. Wilson, J. R., Ludlow, M. M., Fisher, M. J., and Schúlze, E.-D. (1980). Adaptation to water stress of the leaf water relations of four tropical forage species. *Aust. J. Plant Physiol.* **7,** 207–220.

361. Wright, S. T. C. (1969). An increase in the 'inhibitor B' content of detached wheat leaves following a period of wilting. *Planta* **86,** 10–20.
362. Wright, S. T. C. (1977). The relationship between leaf water potential and the levels of abscisic acid and ethylene in excised wheat leaves. *Planta* **134,** 183–189.
363. Wright, S. T. C. (1978). Phytohormones and stress phenomena. *In* "Phytohormones and Related Compounds—A Comprehensive Treatise (D. S. Letham *et al.*, eds.), Vol. 2, pp. 495–536. Elsevier North-Holland Publ., Amsterdam.
364. Wylie, R. B. (1938). Concerning the conductive capacity of the minor veins of foliage leaves. *Am. J. Bot.* **25,** 567–572.
365. Wylie, R. B. (1943). The role of the epidermis in foliar organization and its relations to the minor venation. *Am. J. Bot.* **30,** 273–280.
366. Yamaguchi, T., and Street, H. E. (1977). Stimulation of the growth of excised cultured roots of soya bean by abscisic acid. *Ann. Bot. (London)* [N.S.] **41,** 1129–1133.
366a. Zelitch, I., and Waggoner, P. E. (1962). Effect of chemical control of stomata on transpiration and photosynthesis. *Proc. Natl. Acad. Sci. U.S.A.* **48,** 1101–1108.
366b. Zimmermann, M. H. (1984). "Xylem Structure and Ascend of Sap." Springer-Verlag, Berlin and New York.
367. Zimmermann, M. H., and Brown, C. L. (1974). "Trees: Structure and Function." Springer-Verlag, Berlin and New York.
368. Zimmermann, U., and Steudle, E. (1975). The hydraulic conductivity and volumetric elastic modulus of cells and isolated cell walls of *Nitella* and *Chara* spp. Pressure and volume effects. *Aust. J. Plant Physiol.* **2,** 1–12.
369. Zimmermann, U., Hüsken, D., and Schülze, E.-D. (1980). Direct turgor pressure measurements in individual leaf cells of *Tradescantia virginiana. Planta* **149,** 445–453.

CHAPTER THREE

The Physiology of Stomata: New Insights into Old Problems

T. A. MANSFIELD

I. Introduction

The past 25 years have seen a considerable volume of research into the physiology of stomata. Much of this has stemmed from the study of plant water relations by crop physiologists and ecologists, whose main concern has been the role of stomata in controlling gas exchange. There have also, however, been advances in our understanding of the physiology of guard cells, and it is with these that this chapter is mainly concerned.

In his notable contribution to Volume II of this treatise, O. V. S. Heath (51) discussed at length the responses of stomata to environmental factors. This emphasis reflected the large volume of research done on these responses prior to 1959, but it could also be justified by the fact that reactions

Plant Physiology
A Treatise
Vol. IX: Water and Solutes in Plants

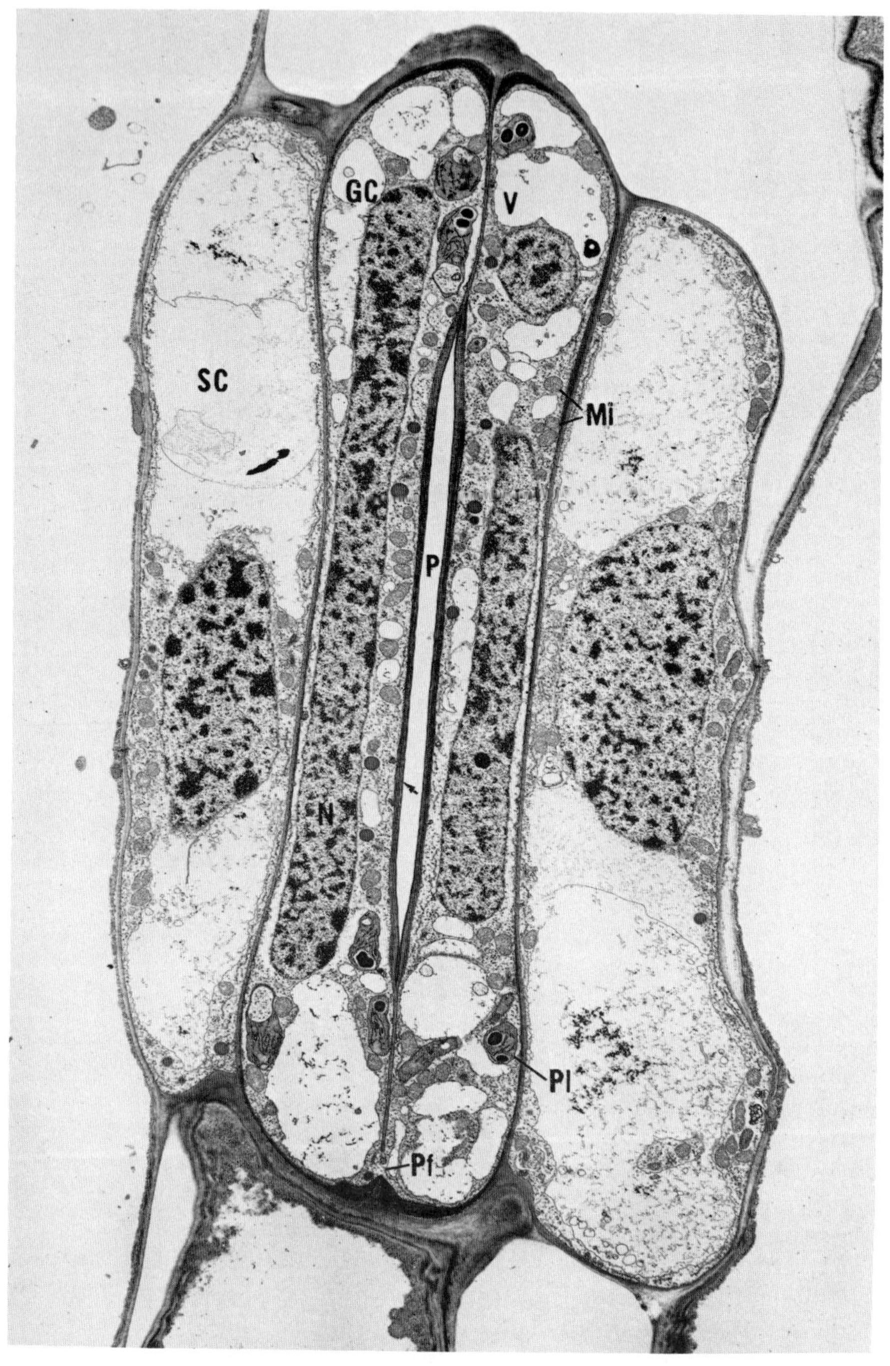
GC
V
SC
Mi
P
N
Pl
Pf

to the environment are of central importance in the physiology of guard cells. It is through these reactions that stomata exert the fine control of gas exchange that is so essential to the survival of the majority of land plants. Furthermore, it is the complexity of these environmental responses that presents the main challenge to the physiologist today, just as in 1959. The emphasis in this chapter remains, therefore, much the same as in Heath's but no attempt has been made to cover all the important literature prior to 1959. The reader is strongly advised to refer to Heath's chapter to gain a full understanding of the many significant earlier researches into the responses of stomata. Heath expressed the hope that his chapter would ". . . serve to indicate the complexity of such an apparently simple matter as a turgor operated cell movement and hearten research workers attracted to this field with the conviction that all is not discovered." The reader of this present chapter may feel that the main achievement of the past 25 years is nothing more than a strengthening of this conviction. If this is the case, then our efforts have not been in vain. We can now claim

---

FIG. 1. An electron micrograph of a stomatal complex of timothy grass, *Phleum pratense* L. The stomatal complexes of graminaceous species such as timothy consist of two guard cells (GC) and two flanking subsidiary cells (SC). The guard cells are bone shaped at maturity, with a narrow, constricted midzone and swollen, bulbous ends. The pore (P) between them is long and slitlike. The guard cell nucleus (N) is elongate and traverses the constricted midzone of the cell. Each guard cell contains many mitochondria (Mi) and plastids (Pl). Although internal membranes are present in the latter, grana are not well developed. However, the plastids do contain prominent, dense starch grains. An elaborate vacuole system is present in each guard cell and appears to be especially concentrated in the bulbous ends. The vacuoles act as osmotically important storage sites for various organic and inorganic solutes. Another characteristic of these cells is a much thickened, lignified wall in the narrow midzone. A thickened ventral wall lining the pore is evident (arrow); although the outer and inner tangential (paradermal) walls are also thickened, they are not visible in this figure. The cellulose microfibrils of the wall are oriented almost axially in the narrow midzone but flare out in a radial manner at the bulbous ends. Both the pattern of thickening and the orientation of wall cellulose appear to be important in the movement and shape changes of guard cells that cause the opening and closing of the pore. Another characteristic of these complexes are large (1.0 μm or larger) perforations (Pf) in the common wall between guard cells at the bulbous ends. Serial sectioning reveals that these wall regions are literally fenestrated by such perforations and that organelles such as plastids and endoplasmic reticulum can span the openings. These perforations may ensure that sister guard cells function as a unit or syncitium. Interestingly, however, plasmodesmata which are present between guard, subsidiary, and neighboring epidermal cells early in development become rare or nonexistent as the cells mature, a modification which may ensure apoplastic transport and membrane regulation of the flow of solutes between guard cells and subsidiary cells and between the stomatal complex and surrounding cells. Note that, in addition to prominent nuclei, the subsidiary cells contain large vacuoles and many mitochondria. Plastids, though present, are relatively undifferentiated. Magnification: $\times 2940$. (Micrograph and legend supplied by Dr. Barry A, Palevitz, State University of New York.)

with confidence that stomata are among the most versatile of sense organs, and if we were describing their responses in an animal, we would say that they represented the senses of sight, smell, touch, and taste. A realization of the complexity of the system under investigation is an essential component of physiological enquiry, and I have regarded a presentation of this complexity as being of more importance than an attempt to produce compact hypotheses.

Space will not allow all the recent advances in stomatal physiology to be covered in detail in this chapter. Several important reviews have appeared recently, and the reader wishing to explore the subject in greater depth will find that most of the areas receiving scant coverage here have been treated more extensively by Allaway and Milthorpe (6), Cowan (17), Raschke (134), Farquhar and Sharkey (29a), and Zeiger (176a).

II. Cytology of Guard Cells

Light microscopy reveals that guard cells have features which readily distinguish them from the rest of the epidermis. They generally possess chloroplasts while epidermal cells often do not, and the behavior of these organelles has always attracted great attention. Their carbohydrate metabolism is peculiar because the amount of starch they contain is usually greater at night than during the day, precisely the opposite of the changes that occur in mesophyll chloroplasts. Guard cell chloroplasts are rarely completely free of starch (see, however, page 179) and they hold it tenaciously even when a leaf is starved for long periods (110). All this suggests that the primary role of the starch is not as a store of carbohydrate. Electron micrographs reveal that a much larger proportion of their volume is occupied by starch than is the case with mesophyll chloroplasts (6), so that structural organization associated with photosynthesis is less dominant, although thylakoids and small grana can be seen (Fig. 1). A feature of guard cell chloroplasts which has attracted considerable attention is the occurrence of many invaginations of the inner of the two boundary membranes, resembling the peripheral reticulum of chloroplasts of C_4 plants (6) (Fig. 2a). Comparable invaginations of the plasmalemma are also seen (Fig. 2b), and it is likely that these are a means of increasing the surface areas of membranes through which massive solute transfer occurs, namely, the exchange of K^+ for H^+ ions through the plasmalemma and the passage of malate in and out of chloroplasts (see page 209). Mitochondria are always well developed in guard cells and may be up to four times as numerous as the chloroplasts, whereas in mesophyll these two organelles are approximately equally abundant (7). Vacuoles which are small and insigificant when the stomata are closed become very prominent when they are open

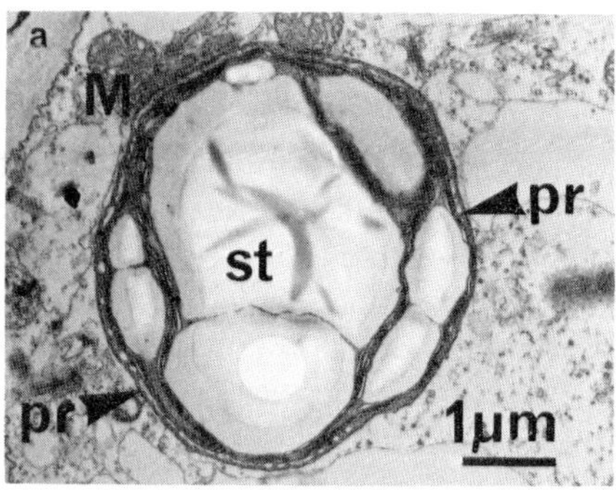

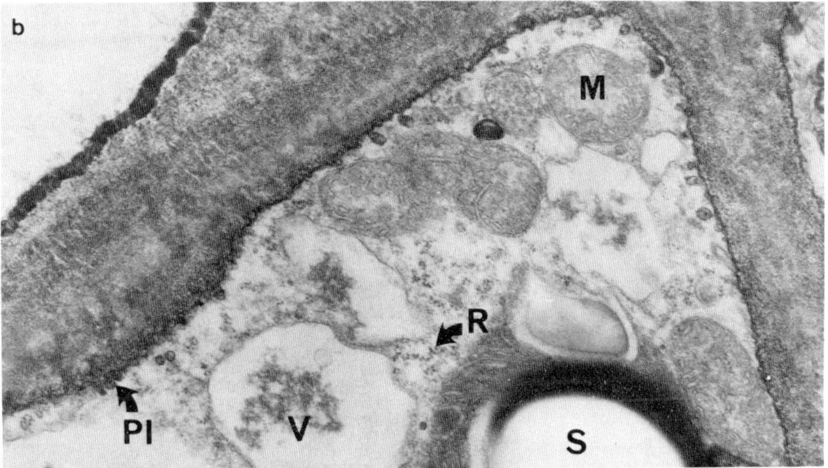

Fig. 2. (a) A guard cell chloroplast from *Vicia faba* showing large starch grains (st) and peripheral reticulum (pr). Bar equals 1 μm. After Allaway and Milthrope (6). (b) Detail of part of a guard cell of *Pelargonium* × *hortorum* to show the numerous invaginations of the plasmalemma (Pl). Also present are vacuoles (V), mitochondria (M), ribosomes (R), and starch (S). Magnification: ×28,000. Reproduced by kind permission of Dr. C. Humbert, Université de Dijon.

(58). Various other cytological changes can be detected in guard cells as stomata open and close, but the significance of these in relation to the turgor changes is not known (55, 59, 60).

An anatomical feature which may be of physiological importance is the apparent absence of plasmodesmata between guard and subsidiary cells. Carr (14) has carefully reviewed the evidence relating to these symplastic connections in guard cells and has concluded that they are probably absent in most mature stomata. The functional significance of plasmodesmata is still uncertain, but one view is that they facilitate metabolic cooperation between cells. A degree of metabolic independence from other leaf cells may be essential for the functioning of guard cells. As already indicated, there is clear evidence of such independence in the starch metabolism of the chloroplasts of guard cells. They are completely out of phase with mesophyll chloroplasts, for usually they form starch at night and lose it during the day (110). We shall see below that guard cells also accumulate certain ions as their neighbors lose them and vice versa. The cell membranes across which ion transport is occurring therefore need to be out of phase in guard and subsidiary cells, which may be possible only if they are structurally independent.

III. Mechanics of Stomatal Functioning

The opening and closing of stomata depend principally on the geometry of the guard cells and the special structure of their cell walls. The micellation (i.e., the orientation of cellulose microfibrils) is distinctive and is thought to play an important part in achieving the opening of the stomatal pore as the turgor of the guard cells increases (179). Since the observations of Schwendener (145) it has been widely accepted that the thickening on the ventral wall of the guard cells (the wall adjacent to the pore) is mainly responsible for the bending of the cells as their turgor rises. However, Aylor *et al.* (9) challenged this view and used a series of physical models to demonstrate that the radial orientation of micellae could play an important part. These micellae limit the extensibility of the guard cells so that the distance between the dorsal and ventral walls cannot change greatly. If the guard cells were free to expand in length, no bending would occur. But they are not free to expand. Aylor *et al.* concluded that the important constraint is the common wall between the two guard cell pairs. This wall remains almost constant in length as the stoma opens, preventing a marked increase in length of either guard cell. However, the thickening on the ventral wall found by Schwendener is a widespread feature of guard cells, and this must also help to restrain the cells' tendency to increase in length. Models made in the present author's laboratory by A. J. Travis have suggested that the thickening on the ventral wall and the rigidity of the walls common to the two guard cells might *both* contribute to the physical mechanism of stomatal opening. One such model is shown in Fig. 3.

Estimates of osmotic potentials in guard cells vary considerably from species to species but may be as low as −4.0 MPa when stomata are open, rising to less negative values by 0.3 to 1.8 MPa as closure occurs (110).

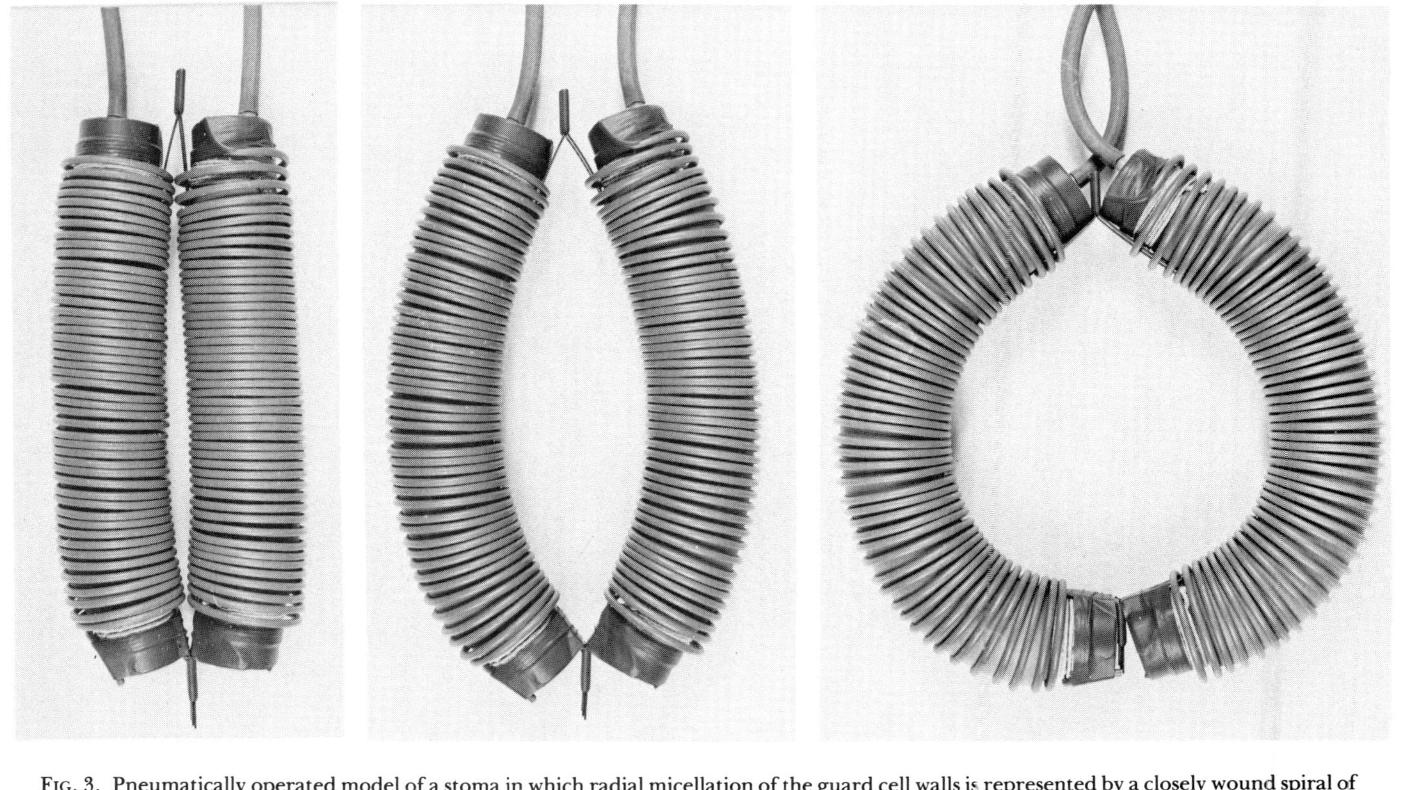

Fig. 3. Pneumatically operated model of a stoma in which radial micellation of the guard cell walls is represented by a closely wound spiral of stiff wire. In this model the guard cells are fixed together at their ends, and there is a strip of stiff plastic inside the wire along the wall adjacent to the pore. This is pressed tightly against the spiral of wire as the rubber tube is inflated, and it prevents lengthwise expansion of that wall. A model in which the guard cell ends are fixed can function without the inclusion of the plastic stiffener, and a single guard cell will bend without being fixed at its ends if the stiffener is present. This suggests that both the anchoring together of the two guard cells at their ends and ventral wall thickening may contribute to the mechanics of stomatal opening. Model constructed by A. J. Travis (unpublished).

Insertion of microneedles into guard and subsidiary cells (29) has enabled estimates to be made of the pressures required to effect stomatal movements (see Fig. 32), and these are found to fall within the range that could be derived from the estimated changes in solute potentials (see also Section VI, B). The guard cells themselves can effect changes in the width of the stomatal pore, as shown by observations of isolated guard cell pairs (114, 153), but in the intact leaf the turgor of the subsidiary and epidermal cells plays an important part (41, 48). The pressure exerted by the turgor of these cells restricts stomatal opening, and it is thought that some stomatal movements are mainly, or even entirely, the result of changes in the water relations of the epidermis. These are often referred to as "transient" movements because they can occur quickly in response to a change in water supply and are essentially the result of epidermal water relations being thrown out of equilibrium. When a rapidly transpiring leaf is excised, the transient movements follow the course shown in Fig. 4. First of all the stomata close because the release of tension in the xylem momentarily increases the water supply to the epidermis, and the cells gain turgor and press upon the guard cells. Thereafter there is a period of opening of several minutes' duration because as transpiration continues the epidermal cells lose water more rapidly than the guard cells do. Eventually closure sets in, probably because the guard cells themselves start to lose turgor by the normal processes involved in stomatal closure. This last phase would not, therefore, be classified as a transient movement.

The occurrence of transient movements may be the outcome of the high

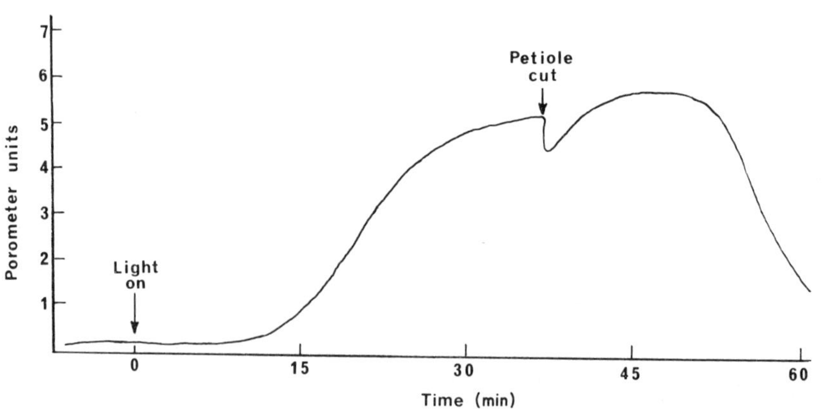

FIG. 4. Changes in stomatal aperture after leaf excision in *Xanthium strumarium*. The first closing movement is clearly seen, followed by a period of wide opening. The author's data (previously unpublished).

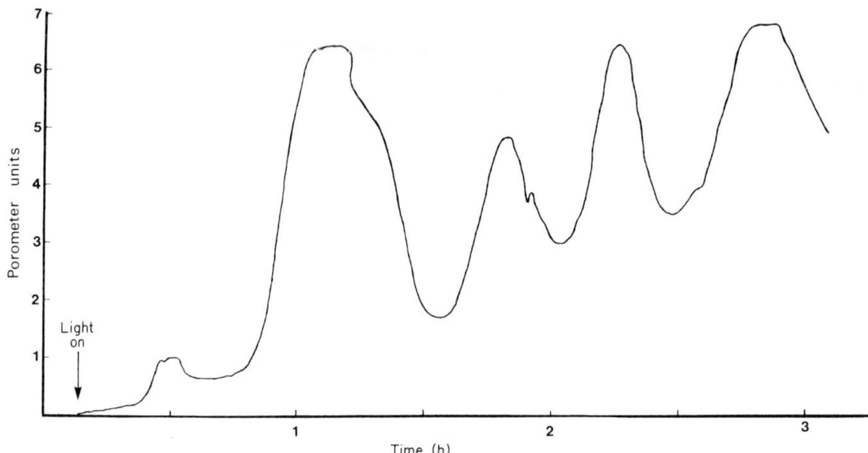

FIG. 5. Short-period oscillations of stomatal aperture in a 6-week-old plant of *Helianthus annuus*. The plant was in a stream of fast-moving air at 23°C and approximately 50% RH. The light of 36 J m^{-2} s^{-1} was switched on at the time shown. Record obtained by J. Miller in the author's laboratory.

hydraulic conductivity of the epidermis and the fact that it is directly connected to the water supply in the xylem (149). If the epidermis were to obtain most of its water via the mesophyll, this tissue would provide a buffer against rapid turgor changes in the epidermis. The direct connection to the xylem conduits means that the epidermis, with small water reserves of its own, will quickly experience the effect of a change in water potential in the xylem. This is why transient movements can be important in a rapidly transpiring plant and can contribute to a rhythmic pattern of stomatal movement of great amplitude (Fig. 5). If a plant loses water quickly by evaporation through wide open stomata so that the supply from the roots cannot meet the demands, the turgor of the guard cells probably falls more rapidly than that of neighboring cells. This is thought to be the case because there is a higher rate of transpiration from the guard cells than from cells further away from the open stomatal pores (see Section VI, B). The consequence is a fall in guard cell turgor relative to that of their neighbors, which produces a closing movement of the stomata. A period of closure allows the water balance of the plant to recover, whereupon the stomata reopen and the whole cycle begins again. This phenomenon has been found to occur on plants in the field (78), which leads us to question the value of spot determinations of stomatal aperture in crops; short-period rhythms could contribute to the enormous variability that is encountered in such measurements.

This chapter is primarily concerned with those stomatal movements that result from the special ability of the guard cells to change their turgor in response to a variety of stimuli. Some readers may feel that transient movements have not been given the space they deserve, but it is the author's opinion that most of the evidence points to metabolic events in the guard cells as the main cause of the cycle of stomatal movements that accompanies the daily alternation of light and darkness and the attendant environmental changes. An important exception may be the responses to atmospheric humidity. These have not yet been properly explained, but it is likely that they are the result of changes in the relative rates of evaporation from guard cells and their neighbors (see Section VI, B).

IV. Mechanisms behind the Turgor Changes of Guard Cells

A. THE PRINCIPAL CLUES: CHANGES IN POTASSIUM AND STARCH

Early in the present century two observations of great significance were made by Macallum (85) and Lloyd (80). Our understanding of the functioning of guard cells has undoubtedly suffered from the neglect of Macallum's discoveries and a preoccupation with Lloyd's which persisted until about 15 years ago. Macallum found that a simple histochemical test for potassium (using sodium cobaltinitrite) showed high levels of K^+ in guard cells of open stomata and low levels in those of closed stomata. Lloyd discovered that the amount of starch in guard cell chloroplasts of plants growing outdoors varied according to time of day and was inversely related to stomatal aperture. The starch content diminished as the stomata opened in the morning and rose again when they closed at night. He suggested that stomatal opening resulted from the stimulation by light of an enzyme which hydrolyzed the starch in guard cell chloroplasts to sugar, which would lower the osmotic potential and lead to increased turgor. Subsequent workers obtained strong circumstantial evidence in favor of Lloyd's hypothesis (e.g., 141) and it soon occupied a dominant place in textbooks. There are many full accounts of it (see 51 and 110) and repetition here is unnecessary.

Surprisingly, Macallum's observation did not lead to an alternative view which seems obvious with hindsight, namely, that accumulation of potassium ions might be responsible for, or contribute to, the turgor increase of guard cells. Important work on this question was done by Imamura (64) and Yamashita (175), but their contribution was not widely recognized until publication of a paper in English by another Japanese worker, Fujino, in 1967 (36).

1. Uptake of Potassium Ions and Sources of Anions[1]

Fujino showed that the potassium content of guard cells was closely correlated with the degree of stomatal opening. This was true whether the opening and closing were induced by environmental factors such as light or suppressed by metabolic inhibitors such as cyanide or azide. On the basis

[1] In assessing the literature on the movement of ions into and out of guard cells, one should be aware of the difficulties of precise quantitative determinations and of the relative merits of the methods adopted by different investigators. None of the methods available gives an unambiguous estimate of ions in critical locations in cells of the stomatal complex. The main techniques that have been employed are as follows:

(a) Histochemistry has been used chiefly for localizing potassium, but other ions can also be detected (36). Macallum's (85) test for potassium depends on the rapid penetration of sodium cobaltinitrite to the intracellular locations of K^+ ions where if the temperature is below 2°C, potassium cobaltinitrite is precipitated. Treatment with ammonium sulfide then leads to a precipitate of black cobalt sulfide, which is easily seen in the light microscope. The intensity of black precipitate should be directly proportional to the original amount of free potassium, and when used carefully the method can provide useful semiquantitative estimates (35). There are problems, however, in knowing whether the sodium cobaltinitrite penetrates quickly to the places where K^+ is located without causing its leakage to other parts of the cells and into the surrounding medium. The technique works best with freshly prepared reagent which is still evolving fumes of nitrogen dioxide, a toxic gas which can damage cell organelles and their membranes (163). It may be that free NO_2 disrupts membranes and speeds up the entry of sodium cobaltinitrite. If this is so, then K^+ ions might be relocated rapidly in the cells, and some could escape altogether.

(b) Estimates of the levels of ions in epidermis have been attempted using simple chemical or enzymatic analyses (e.g., 1, 159). Techniques are available for killing all epidermal cells other than guard cells (3, 153), and an analysis of the epidermis treated in this way should reflect mainly the contents of the guard cells. There is always, however, some uncertainty about whether there is contamination from other epidermal cells or even from adhering mesophyll. Raschke and Dittrich (136) considered that epidermis free of mesophyll could be obtained from *Commelina communis,* a species which has been popular for studies of this kind. Outlaw and Lowry (121) have successfully dissected individual guard cell pairs from freeze-dried leaves of *Vicia faba* and have analyzed the content of K^+ using a new method based on the requirements of pyruvate kinase for this cation. They have shown that it is possible to obtain reproducible results from analyses of such small samples, not only for K^+, but also for organic acids. Zeiger and Hepler (178) have succeeded in obtaining isolated protoplasts from guard cells of onion. These are distinguished from those from other cells by their size, and if they can be separated in sufficient quantity, chemical assays should be rewarding.

(c) X-ray microanalysis is a technique which can be used for estimating most of the important elements (those with atomic numbers greater than fluorine). Individual cells or even organelles can be analyzed. The method depends on the X-ray spectrum emitted by elements when they are bombarded by electrons in the electron microscope. Several good accounts are available of the application of the technique to biological material (15, 46, 140). Despite the claims of some authors (e.g., 63), it is doubtful whether strictly quantitative

of his studies with metabolic inhibitors and his demonstration of a requirement for ATP for rapid stomatal opening, Fujino postulated that active transport of K^+ ions into and out of the guard cells is responsible for stomatal opening and closing and that the driving mechanism for this transport resides in the guard cells themselves.

The elegant work of Fischer and Hsiao (31–33) came quickly to the support of Fujino, firmly establishing his hypothesis as an alternative to the notion of starch–sugar interconversion. It was shown that potassium chloride at concentrations up to 100 m*M* stimulated stomatal opening in isolated epidermis of *Vicia faba,* even in conditions such as darkness and normal air which are not usually conducive to wide opening (Fig. 6). The use of Macallum's histochemical test for potassium (Fig. 7a) shows a clear correlation between density of staining and stomatal aperture (Fig. 7b).

Vicia faba has become a favorite object of study in this connection and there are now numerous quantitative and semiquantitative determinations of the accumulation of potassium and other ions in open and closed stomata. One of the most important was by Humble and Raschke (63), who used the technique of microprobe analysis to determine amounts of potassium, chlorine, and phosphorus in guard cells (Fig. 8). The massive accumulation of potassium in guard cells of open stomata was apparently not balanced by a sufficient intake of inorganic anions, judging from the amounts of Cl, P, and S that were found. Some balancing anions must be present, however; otherwise the pH in the guard cells would rise to an intolerable level. Raschke and Humble (137) presented isolated stomata of *V. faba* (the surrounding epidermal cells had been ruptured) with K^+ in association with nonabsorbable anions and with potassium chloride as a control. Stomatal opening was stimulated by increasing K^+ concentration whether the anions were absorbable or not (Fig. 9). They therefore looked for evidence of the export of protons (H^+) in exchange for K^+ ions absorbed, which would be necessary for the maintenance of electroneutrality. Epidermal strips were floated on solutions of potassium salts and the pH change was observed as the stomata opened. When the mean aperture increased from 1 to 8 μm, the pH of the solution (2.5 cm^3 for each 275 mm^2

---

estimates can yet be achieved. Material for analysis is freeze-dried or rapidly frozen in liquid nitrogen. Freeze-drying can cause relocation of ions in cells (38), but such problems are thought to be minimal in very quickly frozen material.

(d) Ion-sensitive microelectrodes have been used for detecting potassium, chloride, and hydrogen ions in guard cells (126–128). The difficulty with this technique is that of knowing the precise intracellular location of the tip of the electrode. It is usually assumed that it is in contact with vacuolar sap, but this is not necessarily the case. The mechanical disturbance resulting from the insertion of an electrode might also cause a redistribution of ions in the cells.

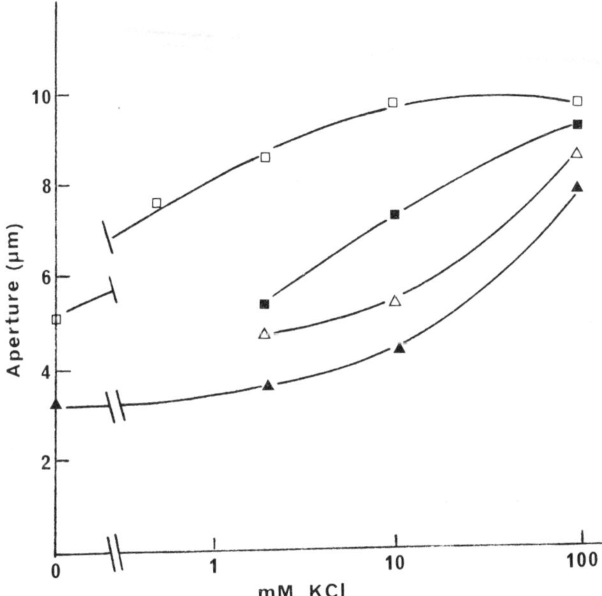

FIG. 6. Responses of stomata on detached epidermis of *Vicia faba* to light and carbon dioxide at various potassium chloride concentrations. □, light and carbon dioxide-free air; ■, darkness and carbon dioxide-free air; △, light and normal air; ▲, darkness and normal air. After Fischer and Hsiao (33).

of epidermis) fell from 7.2 to 6.0. The quantities of proton released during the period of ion exchange were between 0.2 and 1.1 pEq per stoma as the aperture increased by 1 μm, which is of the same order of magnitude as the estimated uptake of K^+ ions.

Extrusion of protons in exchange for K^+ ions is a well-established process in other plant cells (see Chapter 4). Roots have been shown to synthesize organic acid anions, e.g., malate, as a counterion for K^+ (66). It is therefore no surprise to find that malate accumulates in guard cells of *V. faba* during stomatal opening (1) (Fig. 10). The level of malate, like that of K^+, is positively correlated with stomatal aperture, and Allaway (1) found sufficient potassium malate in guard cells of open stomata to decrease the osmotic potential by about 0.7 MPa, which would make an appreciable contribution to the turgor increase required for opening. Further contributions may come from other organic acids, e.g., glyceric and citric, levels of which were also found to increase in epidermal tissue as stomata opened (124). It appears, however, that there can be some contribution from inorganic anions because double-labeling experiments with ^{42}K and ^{36}Cl

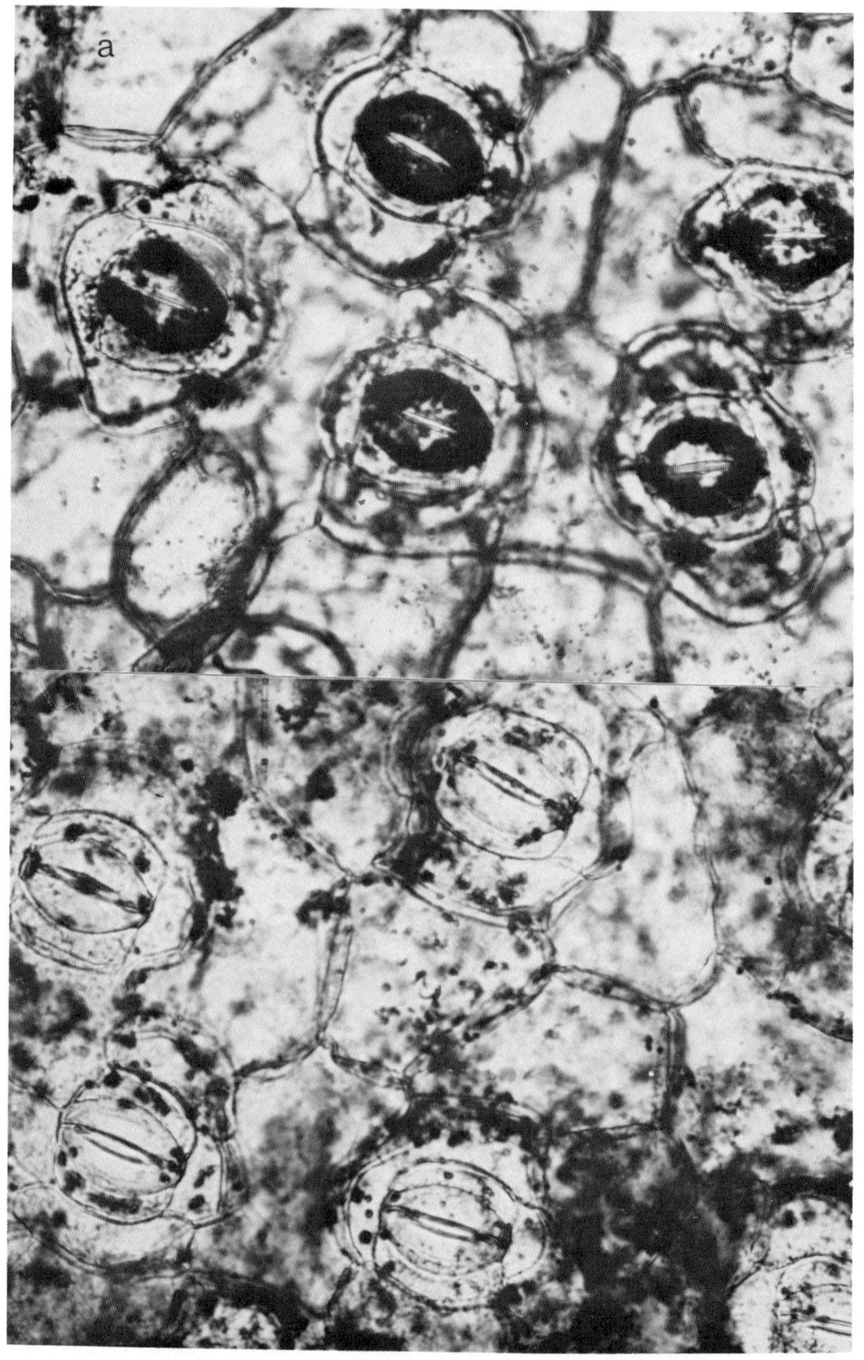

FIG. 7a.

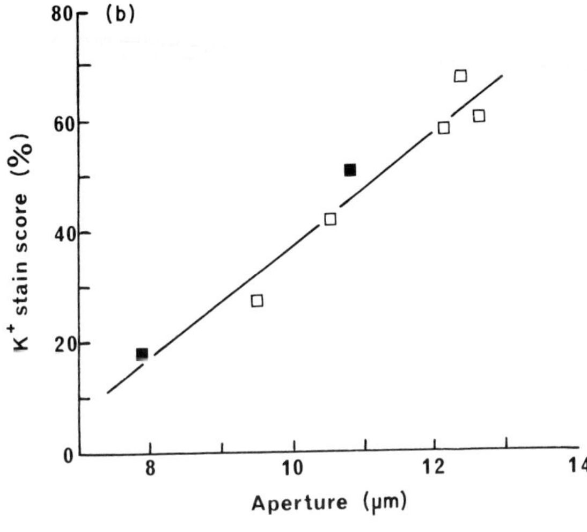

Fig. 7b.

have shown that isolated epidermis floating on a 10 m*M* KCl solution takes in one Cl^- ion for every three K^+ ions as the stomata open (123). There is evidence that the relative roles of internally generated and imported anions can vary according to the availability of the latter (161). It is therefore not possible to deduce, from experiments with epidermal strips, the relative contributions of inorganic and organic anions to turgor increases of guard cells on an intact leaf, but the balance of evidence favors the view that organic anions play the major part at least in *V. faba.* This conclusion depends largely on the work of Humble and Raschke (63), who found that chloride ions contributed only a small amount to the osmotic potential in guard cells of open stomata on the intact leaf of *V. faba.*

Willmer and Pallas (168) used Macallum's histochemical test to survey the distribution of K^+ in the epidermis of species drawn from different sections of the plant kingdom. In some of the ferns and a range of angiosperms spanning many families, there was a clear accumulation of K^+ in the guard cells of open stomata. In most cases when the stomata were closed, there was additional K^+ either in the subsidiary cells or in the surrounding

Fig. 7. (a) Results of Macallum's histochemical test for potassium in open (top) and closed (bottom) stomata of *Commelina communis.* Stomatal closure was induced with 10^{-4} *M* abscisic acid (see p. 202). (b) Relation between stomatal aperture and potassium staining in *Vicia faba* expressed as a percentage of guard cell area. The closed and open squares indicate observations in darkness and light, respectively. From Fischer (35).

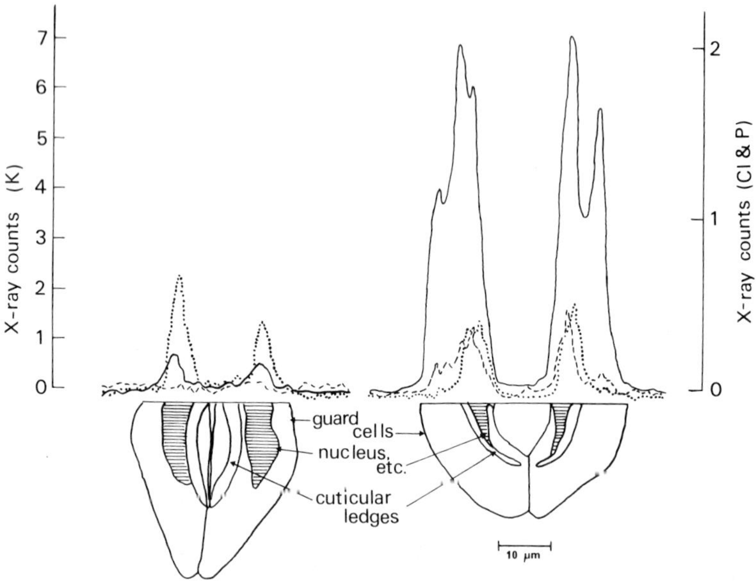

FIG. 8. Profiles obtained by electron problem microanalysis of relative amounts of potassium (——), chlorine (------), and phosphorus (····) across an open and a closed stoma of *Vicia faba*. Scanning was by means of a 0.5-μm-diameter beam crossing the stomata drawn beneath the traces. From Humble and Raschke (63).

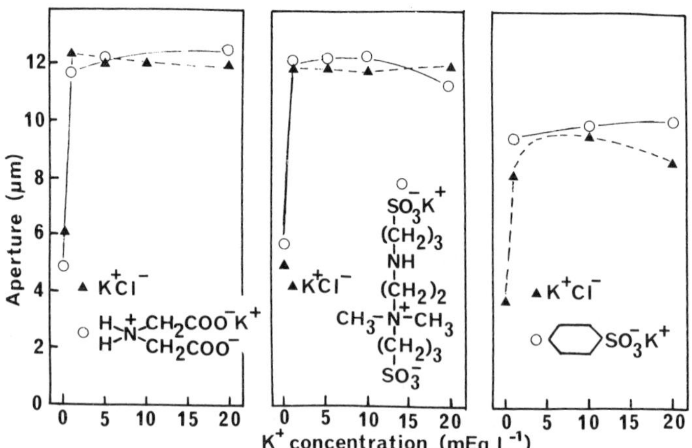

FIG. 9. Responses of stomata of *Vicia faba* to increasing K^+ concentration, when the K^+ ions were in association with the nonabsorbable anions shown. In each case potassium chloride was included for comparison. From Raschke and Humble (137).

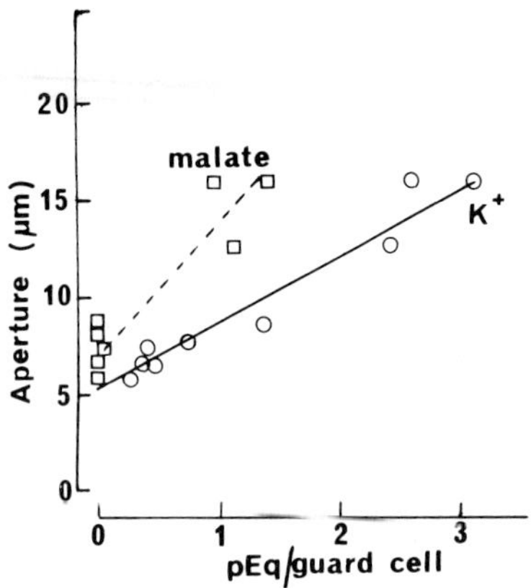

FIG. 10. Correlation between stomatal aperture and amounts of malate (□) and potassium (○) in the epidermis of *Vicia faba*. From Allaway (1).

epidermal cells. In grasses it appears that there is a shuttle of K^+ between the guard and subsidiary cells (Fig. 11) (135, 168), but in other species an intracellular reservoir for the K^+ extruded from the guard cells is not always so clear. A possible solution to this problem has been offered by Stevens and Martin (154, 155) as a result of careful studies of the location of K^+ within the stomatal complexes of a fern, *Polypodium vulgare,* and several members of the monocotyledon family Commelinaceae. Scanning electron micrographs revealed that there were endocuticular sacs beneath the poles of the guard cells at points where K^+ was seen to be deposited (Figs. 12b and 13). It is suggested that these extracellular sites are ion adsorbent and maintain a store of bound potassium which can be released for uptake into the guard cell vacuoles.

In attempting to deduce the functional significance of these "endocuticular sacs," Stevens and Martin noted that the structures are very well developed in a fern which exhibits only small stomatal movements. For this reason they suggested a dual role: in addition to ion storage sites, the sacs could be concerned with water conservation or the trapping of unwanted or toxic ions which arrive in the transpiration stream. There was clear evidence of hollow trabeculae within the cuticle which could be channels for the movement of water and associated ions into the sacs from the surrounding epidermis.

The precise role of these structures in the potassium economy of the stomatal complex is difficult to ascertain. One of the Commelinaceae in which they were observed, *Commelina communis,* has been carefully studied by Penny and Bowling (126). Potassium-sensitive microelectrodes were

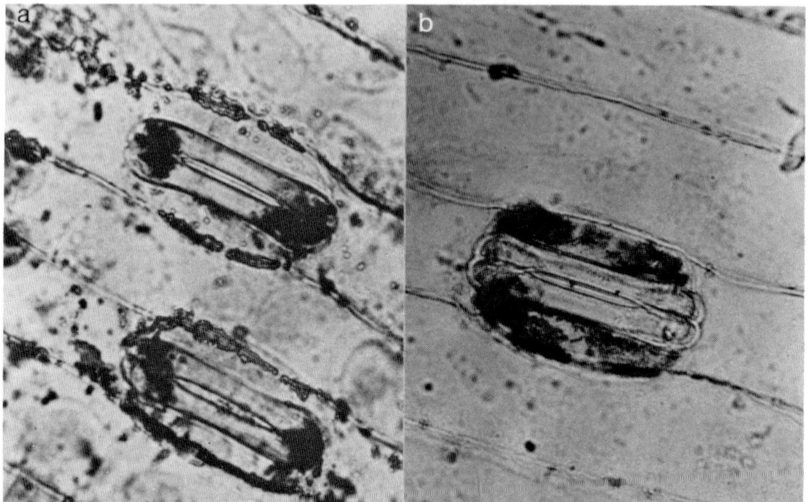

FIG. 11. Distribution of potassium between the guard cells and subsidary cells of *Avena sativa* when stomata were open (a) and closed (b). Staining by means of Macallum's method (p. 165). From Willmer and Pallas (168). Reproduced by permission of the National Research Council of Canada from the *Canadian Journal of Botany*.

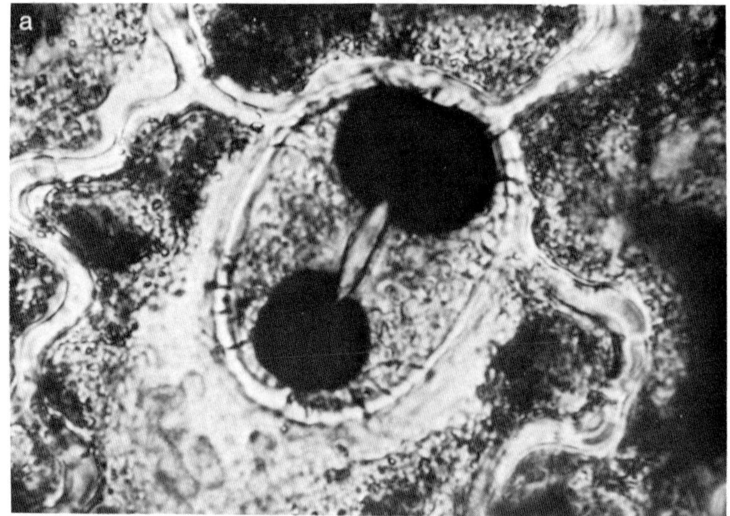

FIG. 12a.

FIG. 12. (a) Light micrograph of a stoma of *Polypodium vulgare* viewed from the underside. The Macallum stain has revealed the polar accumulation of potassium, thought to be located in discrete sacs. The stain has also revealed the potassium-containing trabeculae. (b) Scanning electron micrograph showing the location of the sacs. From Stevens and Martin (154).

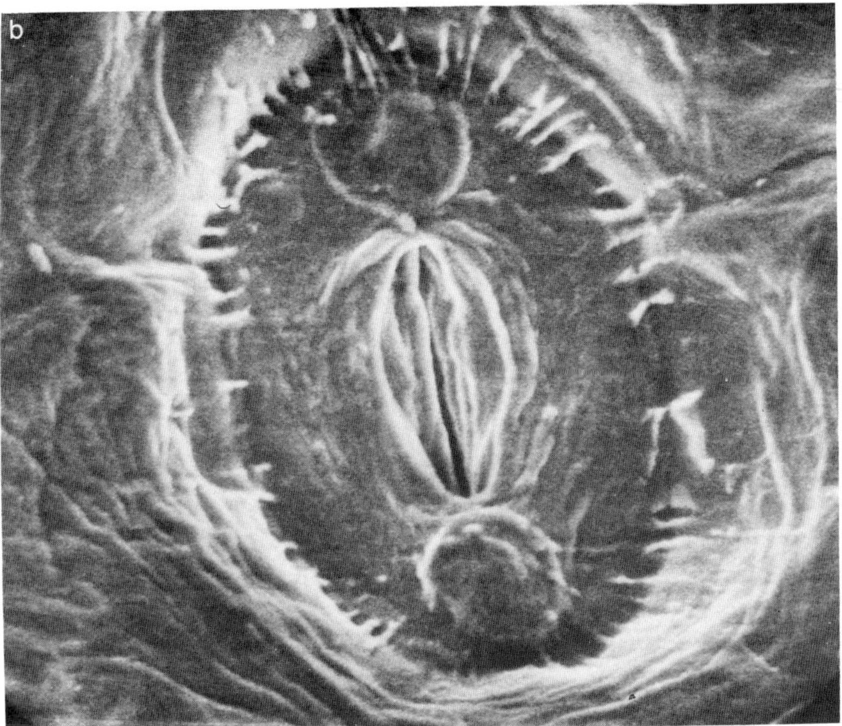

FIG. 12b.

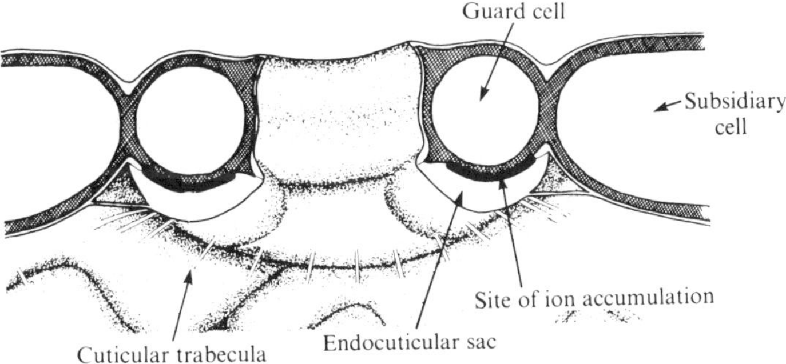

FIG. 13. Drawing of a vertical longitudinal section through a stoma (an attempt to interpret the structures in Fig. 12). The ion-adsorbent sites are indicated by the solid black areas. From Stevens and Martin (154).

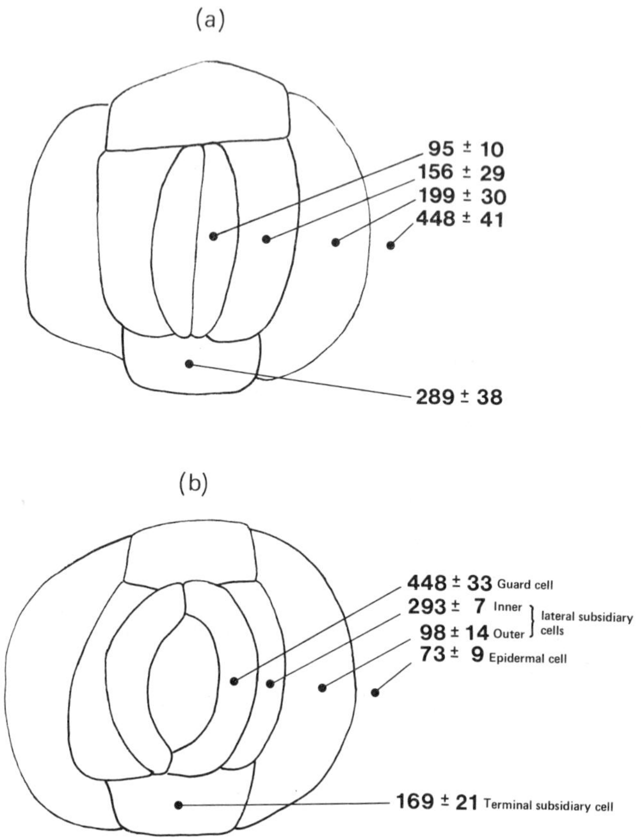

FIG. 14. Vacuolar potassium concentrations (millimolar) in guard cells and their neighbors in *Commelina communis:* (a) stomata closed and (b) stomata open. Means and standard errors are shown. After Penny and Bowling (126).

used to determine the potassium content of individual cells in the epidermis when stomata were open and closed. It was shown that vacuolar concentrations of K^+ rose appreciably in the outer lateral subsidiary cells and the adjacent epidermal cells when the stomata were closed, clearly suggesting that the main export of K^+ ions is from the guard cells to other nearby cells (Fig. 14).

2. *The Role of the Subsidiary Cells*

The work of Penny and co-workers (126–128) focuses attention on the role of the subsidiary cells and adjacent epidermal cells in the stomatal mechanism. Using the data shown in Fig. 14, Penny and Bowling (126) were able to calculate the driving forces on potassium between the cells (i.e., the electrochemical potential differences), and these are shown in Fig. 15. The greatest difference in electrochemical potential when the stomata are open is between the outer and inner lateral subsidiary cells, but there is clearly an active transport of K^+ even from the adjacent epidermal cells, as would be expected judging from the sites of K^+ accumulation when the stomata are closed (cf. Fig. 7a). Penny *et al.* (128) made comparable determinations of levels of chloride in epidermis with open and closed stomata and found electrochemical potential differences indicative of active transport (Fig. 16). There were, however, important differences between the active transport of K^+ and Cl^-. For K^+ the greatest driving force when the stomata were open was between the inner and the outer lateral subsidiary cells, whereas in the case of Cl^- it was between the guard cells and inner lateral subsidiary cells. Furthermore, a comparison of open and closed stomata revealed that there was no change of direction of active Cl^- transport between the outer lateral subsidiary and adjacent epidermal cells, as there was in the case of K^+ transport. It appears, therefore, that active transport operates independently for the two ions. The concentrations of Cl^- in all the cells of the stomatal complex of *C. communis* are well below those of K^+ (Table I), and so Cl^- cannot play a dominant role as a counter ion for K^+. This raises again the question of organic anions contributing to the osmotic potential of guard cells. Bowling (11), stimulated by the foregoing work on K^+ and Cl^- fluxes, put forward an ingenious idea which

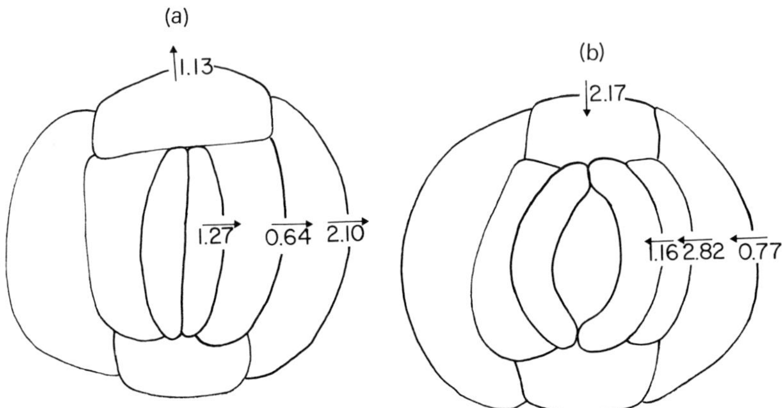

FIG. 15. Electrochemical potential differences for potassium (kilojoules per mole) between cells of the leaf epidermis, calculated from the data in Fig. 14: (a) stomata closed and (b) stomata open. The arrows indicate the direction of active transport. After Penny and Bowling (126).

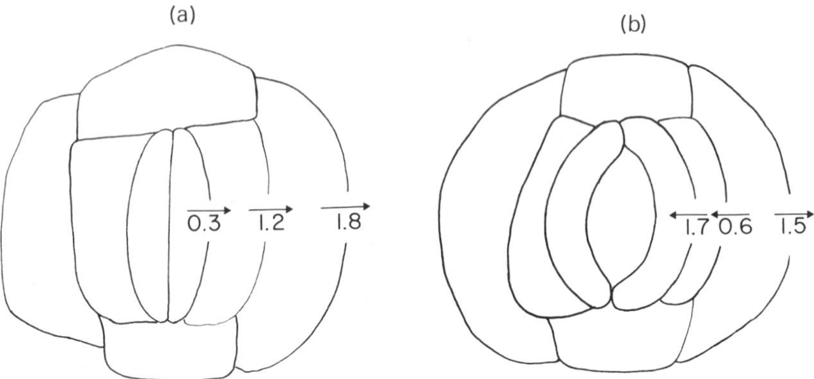

FIG. 16. Electrochemical potential differences for chloride (kilojoules per mole) between guard cells and their neighbors in *Commelina communis:* (a) stomata closed and (b) stomata open. Arrows indicate the direction of active transport. After Penny *et al.* (128).

attributed an all-important role to malate. He presented data showing that levels of malate in the whole epidermis only increased from 61.3 to 64.0 mol m^{-3} when stomata opened, concluding that it is not synthesized or broken down to any great extent but moves with K^+ from cell to cell. He suggested that only malate in the monovalent form is able to move, divalent malate being locked in the cells. The changes in pH occurring when stomata open and close (127) would effect the necessary change in valence. Stomatal closure would occur when the pH of the guard cell changed from around 5.8 to 5.1, causing a conversion of the divalent malate, which was locked in them, to the monovalent form, which could then diffuse out along a concentration gradient until an equilibrium state was reached.

TABLE I

RATIO OF K:Cl IN EPIDERMIS OF LEAVES OF *Commelina communis* WITH OPEN OR CLOSED STOMATA[a]

Type of cell	Stomata closed	Stomata open
Guard	2.9	3.7
Inner lateral subsidiary	4.4	4.7
Outer lateral subsidiary	3.6	2.1
Epidermal	3.8	1.1

[a] Values for potassium concentration are taken from Penny and Bowling (126). After Penny *et al.* (128).

Stomatal opening would be brought about in the reverse manner, i.e., a rise in pH in the guard cells, and conversion of mono- to divalent malate, leading to a concentration gradient for intake of the monovalent form by diffusion, which on entering the cells would be converted to the divalent form and become locked in. The energy required to effect this equilibrium switch would be expended in causing the necessary pH changes.

A critical test of Bowling's hypothesis was attempted by Travis and Mansfield (159), who used intact epidermis and isolated guard cells of *C. communis.* It was found that during stomatal opening induced by fusicoccin, the malate content of guard cells which are not in contact with living subsidiary cells increases to the same extent as in those surrounded by living cells (Fig. 17). It was already known that isolated stomata can open and closed in response to external stimuli such as light and darkness (153).

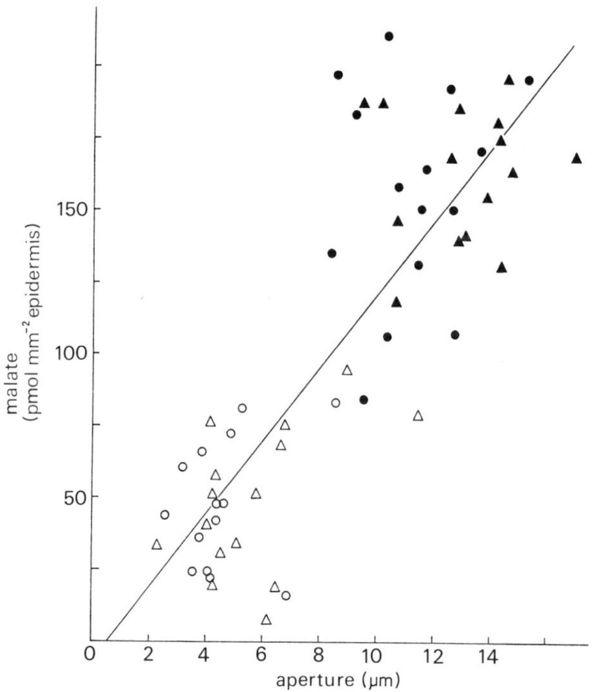

FIG. 17. Correlation between total epidermal malate and stomatal aperture for *Commelina communis.* The regression line is fitted to pooled data from four different treatments, as follows: ○, intact stomata, minus fusicoccin; ●, intact stomata, plus fusicoccin; △, isolated stomata, minus fusicoccin; ▲, isolated stomata, plus fusicoccin. The fungal toxin fusicoccin was used as a means of stimulating wide stomatal opening. After Travis and Mansfield (159).

There thus seems little doubt that guard cells are able if necessary to synthesize all the malate which appears in them as stomata open; transport from surrounding cells is not an absolute requirement, and consequently, movements of malate and its ability to switch its valence seem unlikely to be part of the basic mechanism of ionic movements.

MacRobbie and Lettau (88) used a different kind of potassium-sensitive microelectrode from that employed by Penny and Bowling (126) and reached similar conclusions about the changes in potassium activity in the guard cells of *C. communis* as the stomata opened: the activity rose from about 60 m*M* in the closed state to 300 to 400 m*M* when the aperture was 12 to 14 μm. When, however, they attempted to produce a balance sheet for potassium changes within the epidermis as stomata opened, there were discrepancies, and it was concluded that in the early stages of opening potassium ions could not account for all the osmotic changes in the guard cells. Their findings agree with those of Outlaw and Manchester (122), who found evidence for an increase in the level of sugars in guard cells of *V. faba* during opening. Allaway (2) has pointed out that different solutes may be required to adjust the osmotic strengths of cytoplasm and vacuole as stomata open, and this may account for the discrepancies in the potassium balance sheet of MacRobbie and Lettau.

3. Rates and Pathways of Ion Transport

Penny and Bowling (126) estimated that the flux of potassium across the junction between the guard cells and inner lateral subsidiary cells in *C. communis* is between 150 and 190 pmol cm^{-2} s^{-1} when the guard cell volume increases by between 50 and 100% over a period of 1 h, rates which are 50 to 100 times higher than the normal transmembrane ionic fluxes in plant cells. Their conclusion that the rates are therefore too high for the flux to be transmembrane and that symplastic connections (plasmodesmata) must be present does not, however, find support from ultrastructural studies. Plasmodesmata are usually absent between mature guard cells and their subsidiary cells.

The potassium concentration in the medium required to support stomatal opening in epidermal strips of *C. communis* is high (50–100 m*M*) if the subsidiary cells are intact, but much lower (10 m*M*) if they have been killed and the living guard cells are essentially isolated (153). This suggests that the guard cells are not equipped for rapid uptake of external K^+ except at their junction with the subsidiary cells. The pattern of uptake of the vital stain neutral red lends support to this interpretation. If epidermis of *C. communis* with all its cells intact is placed in neutral red, the epidermal cells become stained almost immediately, but there is no visible uptake into the guard cells (165, 166). Only after a delay of 10 to 15 min do the guard cells become stained. That this is due to transport across the epidermis was shown by removing the external supply of neutral red after the initial uptake into the epidermal cells. The movement which then occurred into the guard cells in 10 to 15 min did not involve any visible accumulation of the dye in the inner lateral subsidiary cells (refer to Fig. 14 for structural details) even though they must have been an essential link in the pathway of transport. Thus the only regions of the guard cells capable of efficient uptake of K^+ and neutral red seem to be those which abut onto the subsidiary cells. Since there is a lack of protoplasmic connections between the mature, functional guard cells, and their neighbors, it appears that the capacity for transmembrane transport at the border between guard and subsidiary cells is indeed remarkably high, perhaps 100 times that in other plant cells.

4. *The Role of Starch*

Outlaw and Manchester (122) have used quantitative histochemical techniques to determine the carbohydrate levels in guard cells from open and closed stomata of *V. faba.* There was a correlation between starch concentration and stomatal aperture, and the observed decrease in starch with opening could account for all the organic anion synthesis. Their study provides strong support for the view that one role for the high starch content of many guard cells is to provide a source of organic anions.

Two species, *V. faba* and *C. communis,* have been intensively studied because their stomata continue to function after the epidermis has been detached, a feature which is not commonly found. The extent to which stomatal functioning in these species is representative of others is not known. There is, however, evidence that the production of an organic anion to provide electroneutrality as K^+ enters the guard cells is not universal. Schnabl and Ziegler (143) found that in *Allium cepa* (onion), potassium ions entering the guard cells are accompanied by enough chloride ions to balance the charges completely. The stomata of onion have been objects of curiosity for many years because they lack starch. Heath (50) discovered that they responded to light in the manner of the stomata of other species, and this provided a strong challenge to Lloyd's starch–sugar hypothesis (though the possibility was recognized that a soluble polysaccharide might substitute for starch). The important findings of Schnabl and Ziegler provide further evidence that the role of the starch, which is so prominent in the guard cells of most other plant species, is to provide a source for the massive amounts of organic anions needed as counterions for potassium. Onion has obviously evolved a different mechanism which would render unnecessary the retention of chloroplasts in guard cells if their sole function were to store starch. Chloroplasts are retained, however, and contain well-developed grana (6). Presumably their pigmentation could be important in the light responses of stomata, and their functions could contribute to the energetics of active ion transport (see Section VII).

B. WHAT IS SPECIAL ABOUT GUARD CELLS?

MacRobbie (86) has put forward a viewpoint which challenges some of the established thought patterns among stomatal physiologists. She points out that the behavior of guard cells as stomata open is not unlike that of other plant cells which accumulate ions—in many other situations cells acquire K^+ and Cl^- from outside or generate organic anions as they take up K^+. The critical question is, she suggests, not what causes stomata to open, but what causes guard cells to lose their ability to accumulate solutes so that their turgor drops as they close. It is in producing stomatal closure that guard cells show behavior that is atypical of plant cells in general.

This is an important suggestion which merits careful consideration, and it draws attention to the dearth of information about events during stomatal closure. Whether or not there is active transport of ions from guard cells during closure, as the data of Figs. 15 and 16 appear to indicate, or whether the loss of solutes is due partly or entirely to an increase in the passive permeability of membranes remains uncertain.

There are several observations which are usually interpreted as suggest-

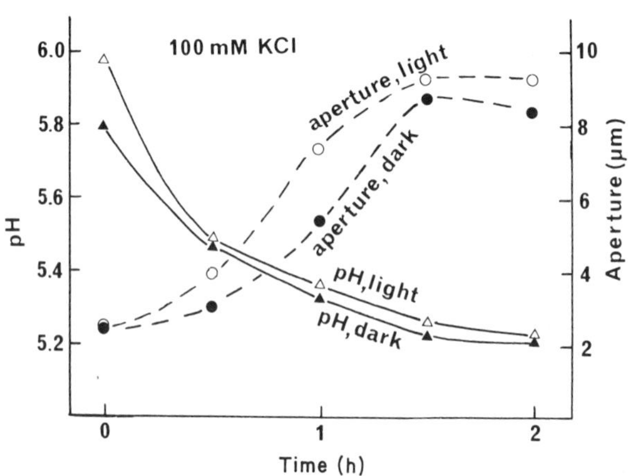

FIG. 18. Changes in stomatal aperture and pH of the surrounding medium with time for epidermal strips of *Vicia faba*. In each treatment 275 mm^2 of epidermis was floated on 2.5 cm^3 of 100 m*M* KCl and 0.1 m*M* $CaCl_2$ in CO_2-free air. After Raschke and Humble (137).

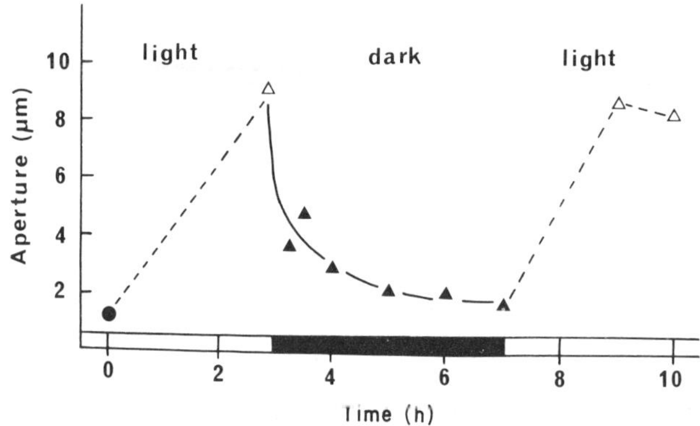

FIG. 19. Response of stomata on isolated epidermis of *Vicia faba* to light and darkness. Epidermis was floated on 10 m*M* KCl in CO_2-free air. After Humble and Hsiao (61).

ing that a distinct opening process is activated by external stimuli. In epidermal strips of *V. faba* stomata open almost as widely in darkness as in light if they are presented with sufficient K^+ ions (61, 137). The time course of opening in 100 m*M* KCl is virtually the same in light and darkness, as is the excretion of H^+ ions shown by pH changes in the bathing medium (Fig. 18). Stomata on epidermal strips from the same species show a marked response to light and darkness in the presence of only 10 m*M* KCl (Fig. 19). An interpretation of these data is that K^+ uptake is usually light activated, but can equally well be activated by K^+ ions. Guard cells do not normally accumulate ions in darkness as they do in light, but can be induced to do so by an abundant supply of K^+. It is arguable that the emphasis here should be on the ability of the cells to respond to light as a signal to take up K^+ from a medium containing the ion in low concentration.

There are, however, other possible interpretations of these data and MacRobbie's views will be worth pursuing in the future. It may be the ability to vary the rate *and direction* of transmembrane transport and/or the passive permeability of membranes in response to external signals that is the special feature of guard cells.

V. Responses to Light

In most mesophytes stomata open during the day and close at night, and it is generally believed that these movements are caused by light and darkness. There is, however, plenty of evidence that diurnal movements can take place without the external stimuli of light and darkness, and this suggests that some reappraisal of the role of light is necessary.

A. IS LIGHT THE ONLY CAUSE OF STOMATAL OPENING?

In many plants stomata begin to open before dawn if the night is long enough and often start to close toward the end of the day even if the light intensity is kept constant. Such movements are the result of endogenous rhythms which are circadian in nature; that is, the period for one complete cycle is around 24 h. Circadian rhythms occur in most organisms and there are very few physiological processes in which they are absent. Their involvement in higher plants has been extensively studied (71, 164), and stomata have been shown to exhibit rhythms whose pattern of behavior is not unlike that found elsewhere in plants. The important characteristics in the present context are the persistence in spite of external conditions and the mechanisms by which phase is controlled.

If a plant is placed in prolonged darkness commencing at the end of a day, the stomata close initially in the normal manner, but then open slightly at, or just before, the time when light would normally be experienced. The aperture achieved during this period is often small compared with that in light (Fig. 20). The endogenous changes that take place in the dark are, however, important in determining the preparedness of the

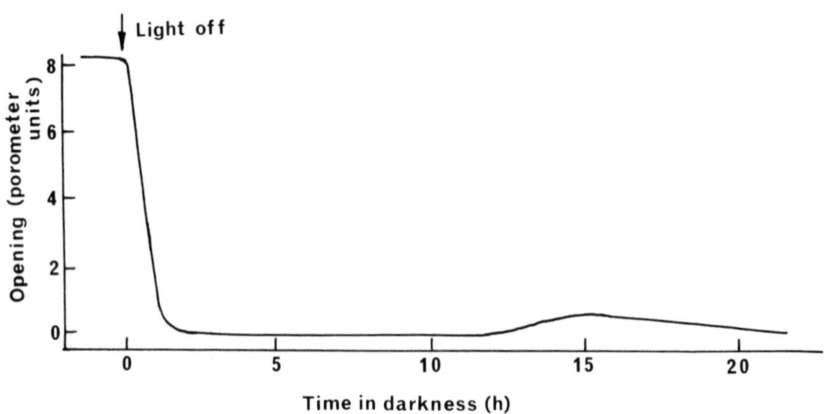

FIG. 20. The small opening that occurs in *Xanthium strumarium* during prolonged darkness at about the time when light would normally have been received. Darkness commenced at the time indicated by the arrow, and prior to this time the irradiance was 60 J m^{-2} s^{-1}. After Mansfield (95).

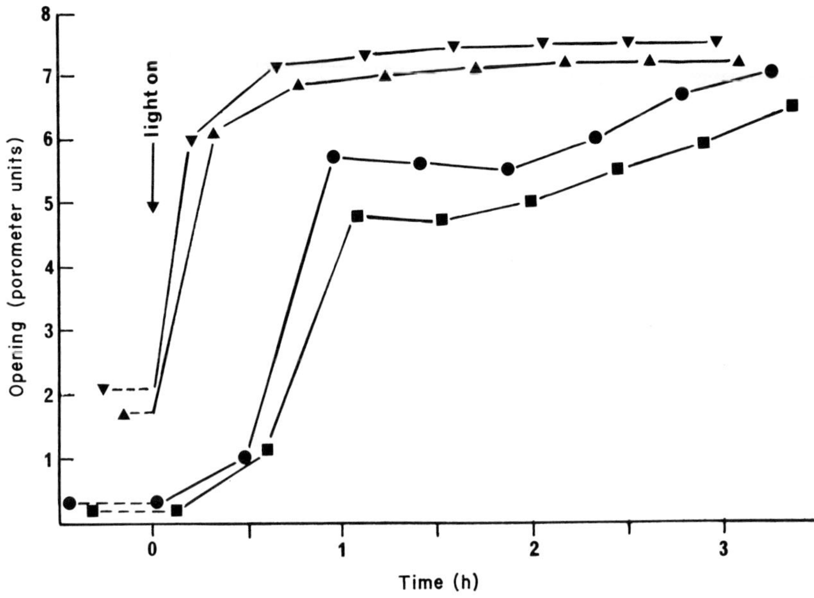

FIG. 21. Stomatal opening in light after 3-h (● and ■) and 16-h (▼ and ▲) nights in *Tradescantia virginiana*. After Martin and Meidner (103).

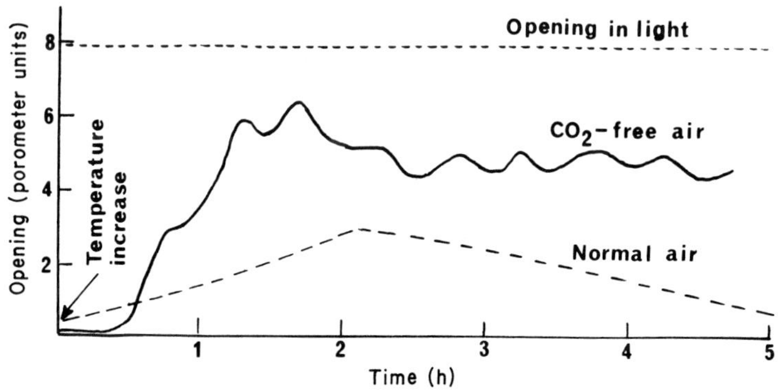

FIG. 22. The small degree of opening exhibited by stomata of *Xanthium strumarium* in darkness (see Fig. 20), which approaches that in light if the temperature is increased from 27 to 36°C and the leaf is flushed with carbon dioxide-free air. The continuous record from an automated porometer shows short-period fluctuations such as are often found in light (see Fig. 5). Also shown are the opening attributable to the temperature increase alone in normal air and the opening in light of 60 J m^{-2} s^{-1}. After Mansfield (95).

stomata to open, as shown by the rate of their response to light given at different times during the dark period (97, 103) (Fig. 21). When the small opening in darkness has occurred, further opening can be induced both by an increase in temperature and by removal of carbon dioxide (Fig. 22). The aperture achieved is not as great as that in light, but it is clear that these two factors which normally accompany the daytime illumination of leaves can make a substantial contribution to opening. Although a distinct response to light is detectable, it must be regarded as only one of a number of contributors to the daily cycle of stomatal movements. Just from these few experimental data we can see four factors which are concerned in producing opening during the day: light, rising temperature, lowered carbon dioxide concentration due to photosynthetic activity, and the circadian rhythm. Each of these factors can thus trigger or otherwise influence the processes that lead to the turgor changes of guard cells, which means, presumably, that they can all regulate active ion transport in some way.

B. Two Distinct Photoreactions

The established nonphotosynthetic effects of light on stomata are of two discrete types which apparently involve different photoreceptors. These are responses to low irradiances, which may be chiefly responsible for determining the phase of circadian rhythms, and to higher irradiances, which have a direct stimulatory effect on opening. These two distinct photoreactions can be compared with the low- and high-irradiance reactions of photomorphogenesis (150) and may involve the same pigment systems.

1. The Low-Irradiance Reaction

The phase of circadian rhythms in plants is often determined by the time of commencement or termination of light of low irradiance. Figure 27 shows in diagrammatic form the kind of treatments that have been used in investigations of this response. For the purposes of illustration a repeating rhythm of opening in darkness is shown. In practice it usually damps out after one cycle (97, 102) although there is evidence of second and third cycles (19). Determinations of the action spectrum for this shift of phase induced by light of low irradiance have shown (for *Xanthium strumarium*) that the red region is most effective, with a peak response in the region of 700 nm. The energy required is low (<0.15 J m^{-2} s^{-1}), but illumination must be continuous or there must be bursts of higher irradiance (>1.0 J m^{-2} s^{-1}) of a few minutes' duration every half-hour or less for phase shift to occur (93).

Phase control of the circadian rhythm shown by *Tradescantia virginiana* stomata has been studied in detail by Martin and Meidner (102–104). The rhythm in continuous light is more persistent than that in darkness and is susceptible to phase shift if a dark perturbation is given at an appropriate time (Fig. 23). The great amplitude of this rhythm is worthy of note; in identical external conditions the stomata vary from wide open at peaks of the rhythm to very small apertures in the troughs. The forces behind the rhythm can thus modify or overrule the expected response of the stomata to conditions normally considered conducive to opening. For this reason experiments into stomatal physiology have to be performed at set times of the day using plant material which has received a known, consistent illumination treatment.

2. *The High-Irradiance Reaction*

Investigations of the stimulation of stomatal opening by light are complicated by the fact that opening is also achieved by a reduction in carbon dioxide concentration. The removal of carbon dioxide within the leaf by photosynthesis thus constitutes a mechanism for stomatal opening, but this response is not truly light dependent because it can be induced in darkness if carbon dioxide is artificially removed. It is important to recognize, however, that part of the opening during the day is the consequence of carbon dioxide depletion.

Evidence that there is a light effect which is totally independent of carbon dioxide concentration has accumulated steadily since the idea was first suggested by Heath and Russell (54). The strongest support comes from a comparison of the action spectra for photosynthesis and stomatal opening. The red and blue regions of the spectrum are both highly effi-

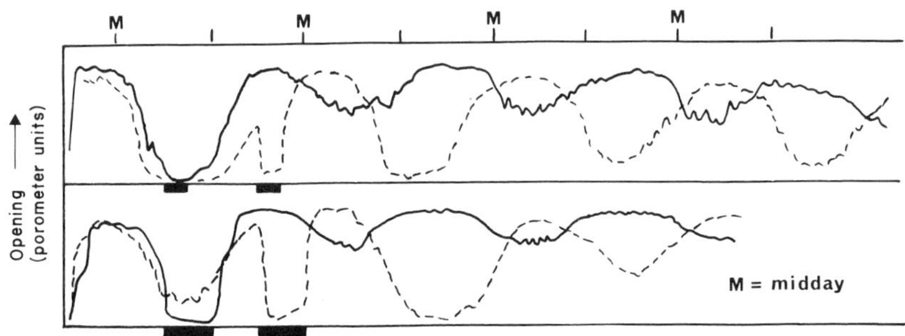

FIG. 23. Phase shift in the stomatal rhythm of *Tradescantia virginiana* in light (5 klx) induced by short dark periods at the times shown. The continuous lines are for leaves not receiving the dark periods. After Martin and Meidner (102).

cient in inducing photosynthesis in the intact leaf (52), but blue has been shown to be considerably more effective than red in stimulating the opening of stomata (67, 76, 99, 131) (Fig. 24). Hsiao *et al.* (57) determined action spectra for stomatal opening and uptake of radioactive rubidium (^{86}Rb) using epidermal strips of *V. faba* in which all cells apart from the guard cells had been destroyed. Uptake of rubidium was assumed to indicate that of potassium. At a photon fluence which was low compared with that required for maximum photosynthesis, both opening and ion uptake were greatly stimulated by the blue region of the spectrum, but not by the red (Fig. 25). These experiments were performed in carbon dioxide-free air so that the response is most unlikely to have involved any appreciable changes in carbon dioxide concentration induced by light. Zeiger and Hepler (178) prepared isolated protoplasts from guard cells of onion by means of enzymatic digestion of cell walls and found that they were caused to swell when illuminated with blue light. This swelling, which amounted to a 35 to 60% increase in volume, was dependent on a supply of K^+ ions. Blue was the only effective spectral region, and the response was saturated by 1.15×10^4 erg cm^{-2} s^{-1} (2.4×10^{15} quanta cm^{-2} s^{-1}). This is a small quantum flux compared with full sunlight and much less than that required for the stomatal response to red light. Light-induced swelling was not found in protoplasts from neighboring epidermal cells. This is further evidence that a peculiar property of guard cells is their ability to accumulate ions in response to external stimuli, of which one is light.

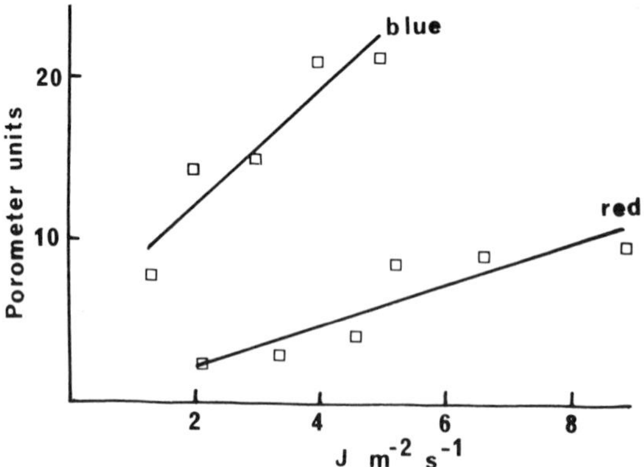

FIG. 24. Stomatal opening in *Xanthium strumarium* in red (650-nm) and blue (450-nm) light of different irradiances. The measurements were in a closed system at the carbon dioxide compensation point established by the leaf. After Mansfield and Meidner (99).

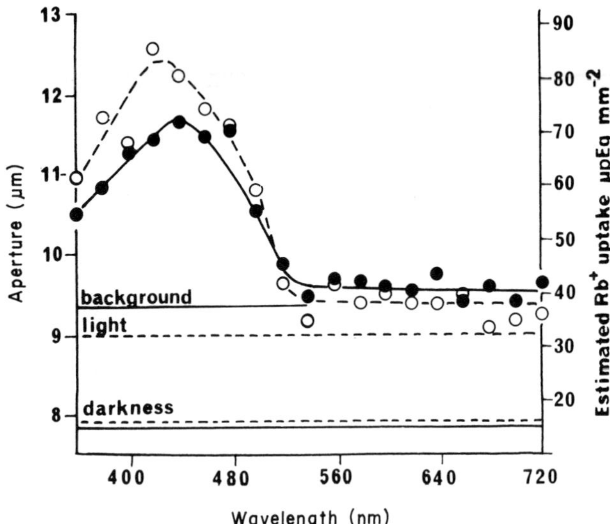

FIG. 25. Action spectra for stomatal opening and uptake of labeled rubidium in epidermal strips of *Vicia faba*. Also shown are uptake and opening in background light (scattered light outside the spectograph beam) and in darkness. ●——●, aperture; ○------○, uptake. After Hsiao *et al.* (57).

C. PHOTORECEPTORS

The movements of stomata are so intimately connected with a leaf's photosynthetic activity that it is not surprising that people have tried to invoke photosynthesis as a major, even perhaps the only cause of stomatal opening. There are, however, many well-established nonphotosynthetic effects of light on plants operating at all stages of development from germination to reproduction, and it is no surprise to discover that similar reactions are involved in stomatal movements.

There are thought to be at least two photoreceptor molecules involved in photomorphogenesis in higher plants. One of these, phytochrome, has been isolated (13) and much is known about the reactions to red and far-red light in which it is involved (150). Although the precise mechanism of its action remains unresolved, there is good evidence that it interacts with membranes to cause rapid photoresponses. It is believed that some phytochrome may be located in cell membranes and that red and far-red light can affect rates of transmembrane transport, for example, the efflux of K^+ ions from the motor cells of *Albizia* pulvini (39) and the release of gibberellins from etioplasts (16).

The first studies of the reaction of stomata to red light of low irradiance (92, 93) did not reveal any photoreversibility as would be expected in a phytochrome-mediated response, but Habermann (43) has since been able to demonstrate red/far-red antagonism in the light responses of sunflower stomata (see page 189). The importance of transmembrane ion transport into and out of guard cells suggests that phytochrome located in the plasma membranes brings about some change in their properties which can alter the rate and even the direction of ion transfer. Further understanding of this phenomenon, and indeed many of the unresolved problems in the wider field of photomorphogenesis, must await elucidation of the basic involvement of phytochrome in the functioning of membranes.

The other photoreceptor in photomorphogenesis is well established in the minds of physiologists but its real nature has not been determined. It has been named "cryptochrome," and it absorbs light in the blue and UV regions of the spectrum between 350 and 450 nm. Light absorption by some carotenoids and flavoproteins nearly matches the action spectra, and it is now strongly believed that cryptochrome belongs in the latter group (12). The action spectrum for the first positive curvature of phototropism shows a main response region between 420 and 480 nm (150), and stomatal opening displays a similar peak (Fig. 25). It seems that cryptochrome may be active in stomatal opening and in a variety of photomorphogenetic phenomena, including phototropism. Preliminary evidence suggests that cryptochrome, like phytochrome, is associated with membranes, and so we can speculate that its mechanism of action in guard cells may be comparable with that suggested above for phytochrome. In this context it is noteworthy that when malate is the counterion for the potassium accumulated by guard cells, solutes will have to be transferred across the outer membranes of the chloroplasts. This, then, is another site at which photocontrol could be exerted in addition to the plasma membrane. The work of Meidner (105) raises the question of whether this is the site of the high-irradiance photoreaction. He found that blue light was less effective in promoting opening of onion stomata, which do not convert starch into malate (see page 179), than in *X. strumarium* guard cells which contain starch.

D. The Roles of the Two Photoreactions

Evaluation of the way in which the two photoreactions control stomatal movements would be difficult enough without the involvement of a response to carbon dioxide. The fact that stomata open as the concentration of carbon dioxide falls adds yet another dimension to the light responses of stomata. The action spectrum for this response must surely follow that for photosynthesis; and it is immaterial whether the guard cells themselves fix carbon dioxide, for it is known that changes in the carbon dioxide concentration in the substomatal cavity alter stomatal aperture (49). Such changes are mainly the result of photosynthesis in the nearby mesophyll.

No single piece of research has succeeded in coming to grips with all the known facets of the responses of stomata to light. A model for the kind of approach required could well be the brilliant work by Heath and Russell (54) of 30 years ago. These authors extracted an amazing amount of information from one factorial experiment involving just six levels of carbon dioxide and four of light, including a clear recognition of a response to light independent of carbon dioxide concentration. The only more recent study with a comparable conceptual framework was done by Morison and Jarvis (113a). They forcibly controlled the intercellular carbon dioxide concentration while observing both the rate of photosynthesis and stomatal conductance. They convincingly demonstrated that there was no unique relationship between stomatal conductance and either irradiance or intercellular carbon dioxide concentration.

Habermann (43) has evaluated the role of the two photoreactions using the nonphotosynthetic "xantha" mutant of *Helianthus annuus* (sunflower), which contains only trace levels of chlorophyll but does exhibit a stomatal response to white light. There was, however, no response to broad-band blue light (peak about 450 nm), although the normal sunflower showed a large response as irradiance increased (Fig. 26). Both types showed no reaction at all to the irradiance of red light used (650 nm), but there was a clear opening, though small in magnitude, to low irradiances of far red (700 – 750 nm). Habermann was able to show that a short

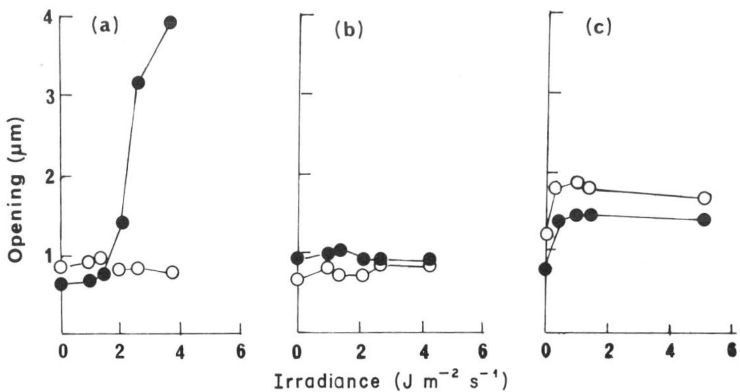

FIG. 26. Responses of stomata to (a) blue, (b) red, and (c) far-red light in mutant and wild-type sunflowers. ○, mutant; ●, wild type. After Habermann (43).

period of red light (5 – 10 min) preceding the far red inhibited this opening reaction, thus providing a strong indication that phytochrome is the photoreceptor involved.

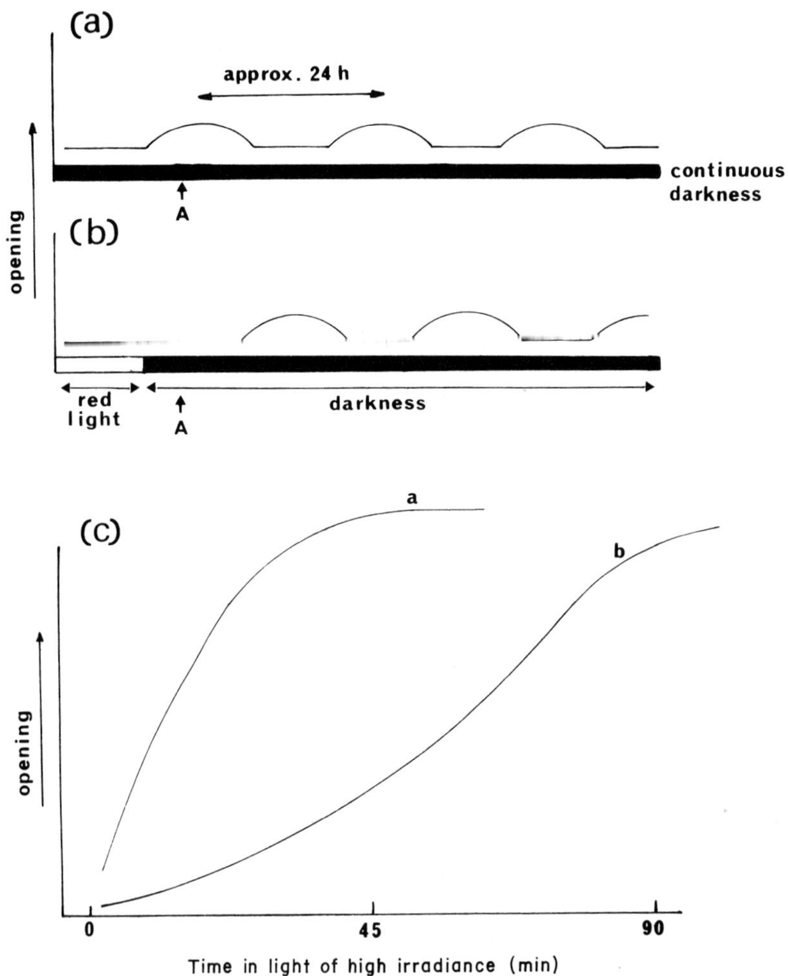

FIG. 27. How a phase shift in the stomatal rhythm as induced by low-irradiance red light can modify the subsequent opening reaction to high-irradiance light. (a) An opening rhythm in continuous darkness, (b) the phase of the rhythm in (a) shifting if the time of commencement of darkness is delayed by low-irradiance red light, and (c) the speed of the reaction to high-irradiance light given at time A [as marked on (a) and (b)] reflecting the physiological state of the stomata as determined by the phase of the rhythm. Diagrammatic but based on data for *Xanthium strumarium* and *Tradescantia virginiana* (cf. Fig. 21 and p. 182).

The phase control of stomatal circadian rhythms by low-irradiance red light can result in a suppression of the high-irradiance photoreaction. The way in which this operates is shown diagrammatically in Fig. 27. A period of low-irradiance red light given to plants of *X. strumarium* to shorten the length of night results in a much slower response of the stomata to high irradiance next morning. This effect is *not,* however, achieved by infrequent short flashes of red light (93). In the photoperiodism of *X. strumarium* it is well established that a short interruption of a long night by red light nullifies its effectiveness in the induction of flowering. It is surprising that the stomatal response to red light in the same leaves is so different, and this fact alone emphasizes just how difficult it is to unravel the photoresponses of plants.

The stomatal reaction to blue light in *X. strumarium* is partly suppressed after short night lengths (99). The slower opening in the morning (cf. Fig. 21) can probably therefore be attributed to this, and we can conclude that the capacity to respond to high-irradiance blue light varies according to the phase of a circadian rhythm.

VI. Agents Causing Stomatal Closure

A. CARBON DIOXIDE

This is the area of stomatal physiology about which most has been written over the past 30 years, but there is still no understanding of the basic mechanism by which carbon dioxide acts.

The quantitative response to carbon dioxide varies from species to species, and in some it is very small in magnitude or absent altogether. Typically, however, stomata close in response to increasing carbon dioxide concentrations over the range they normally encounter due to the action of photosynthesis and respiration in the leaf as a whole. Guard cells are able to sense the carbon dioxide concentration in the substomatal cavity (49, 130, 142). and it has been suggested (134) that by adjusting the diffusion resistance to carbon dioxide entering from the atmosphere, they can effectively regulate the concentration available for photosynthesis from the intercellular air.

The most detailed studies of stomatal responses to carbon dioxide are those of Heath and co-workers, and the important experiment of Heath and Russell (54) still presents the clearest available picture of the interplay between light and carbon dioxide concentration (Fig. 28). Note from this

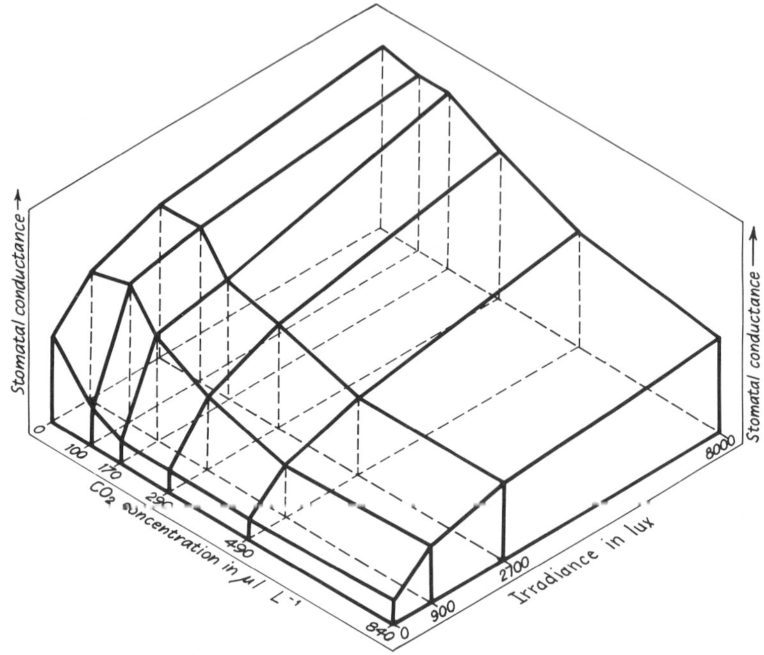

FIG. 28. Recalculated data from an experiment by Heath and Russell (54) showing effects of CO_2 concentration and irradiance on stomatal aperture in wheat.

three-dimensional presentation how the carbon dioxide response curves show evidence of saturation at very low concentrations. The level needed for saturation varies with light intensity, and in darkness it is achieved with a carbon dioxide concentration of only 170 μl L^{-1}. The fact that the response exists in the dark and shows saturation at such a low carbon dioxide concentration has important implications when we attempt to understand the response in biochemical terms. In succulent plants, which can assimilate massive amounts of carbon dioxide in the dark, it has been shown that the opening of stomata which takes place at night is the result of their sensitivity to changing carbon dioxide concentration (115). In this case the level of saturation is only reached at about 1500 μl L^{-1}, i.e., almost an order of magnitude above that in wheat (Fig. 29). Raschke (132) found virtually identical responses to carbon dioxide concentration in light and darkness for *Zea mays* (Fig. 30), with saturation at approximately the same level as that observed by Neales. Another worker found, however, that the stomata of the two leaf surfaces of *Z. mays* showed different sensitivities to carbon dioxide, and there appeared to be differences in the saturation level (25).

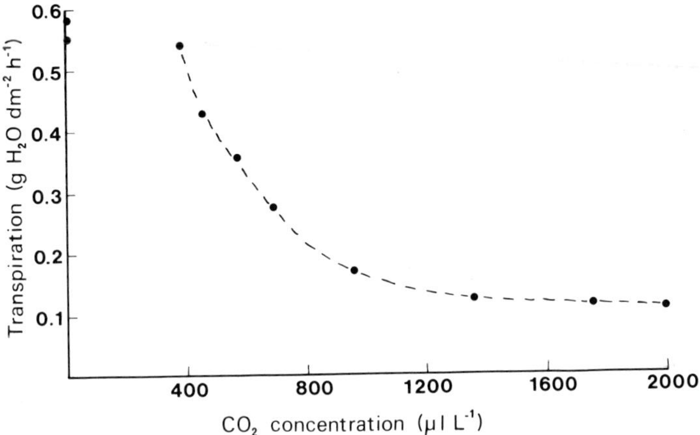

FIG. 29. Effect of ambient carbon dioxide concentration on the dark transpiration rate of *Agave americana*. After Neales (115).

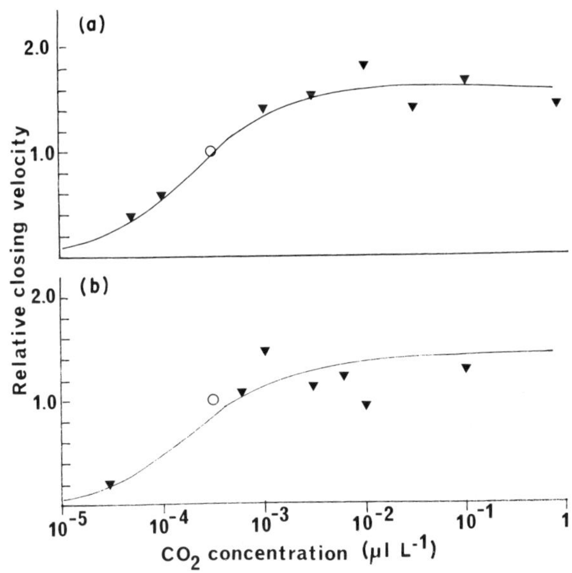

FIG. 30. Relative velocities of stomatal closing in *Zea mays* as a function of carbon dioxide concentration: (a) dark and (b) light. The reaction of the stomata between 2 and 3 min after exposure to carbon dioxide was used as a measure of velocity, and the response to 300 µl L^{-1} was the basis for comparison. O, reference. After Raschke (132).

B. Water Status of the Plant

1. Water Vapor Pressure of the Atmosphere

During the past decade the accumulated evidence leaves little doubt that one of the major environmental factors that can trigger stomatal movements in some plants is ambient humidity. Changes in atmospheric vapor pressure saturation deficit (VPD) do, of course, alter the rate of transpiration from leaves, and so it is possible that any stomatal response could be the result of a change in the bulk water potential. Is there evidence that VPD has a direct effect on the activities of guard cells, thereby inducing stomatal movements?

Lange *et al.* (77) and Lösch (81) performed an important series of experiments using epidermal strips from a fern, *Polypodium vulgare,* and from *Valerianella locusta.* Exposure of the outer side of the epidermis to dry air caused the stomata to close, while moist air caused opening. The responses were readily reversible and the opening and closing of individual stomata in response to alternating high and low VPD could be followed (Fig. 31). They attributed these rapid reactions to local changes in water potential following water loss through the cuticle above the guard cells, known as "peristomatal transpiration" (146, 180), and proposed that guard cells could act as humidity sensors, essentially measuring the water status of the atmosphere outside the leaf.

If guard cells are to act as humidity sensors effective in protecting the leaf from excessive water loss, it is important that they have a means of detecting high VPD before other cells with a vital metabolic role, especially those in the mesophyll, begin to experience severe water stress. A changing rate of evaporation from the guard cells according to the VPD of the atmosphere could be the essential physical means for their operation as sensors. Jarvis and Morison (68) and Appleby and Davies (8) have found unthickened areas in the walls of the guard cells of Scotch pine and Wych elm *(Ulmus glabra),* suggesting that specific sites exist at which evaporation to the atmosphere would occur.

Whether a higher rate of water loss from guard cells must necessarily result from *external* peristomatal transpiration has been critically examined by Meidner (106, 107), who has also challenged traditional views about evaporation sites within leaves. Edwards and Meidner (28, 29, 108) achieved the difficult feat of inserting water-filled microneedles (less than 1 μm in diameter) into guard and subsidiary cells and were able to measure the pressures required to induce varying amounts of stomatal opening (Fig. 32). The range of turgor pressures required to operate the stomatal apparatus was surprisingly small compared with the osmotic potentials of

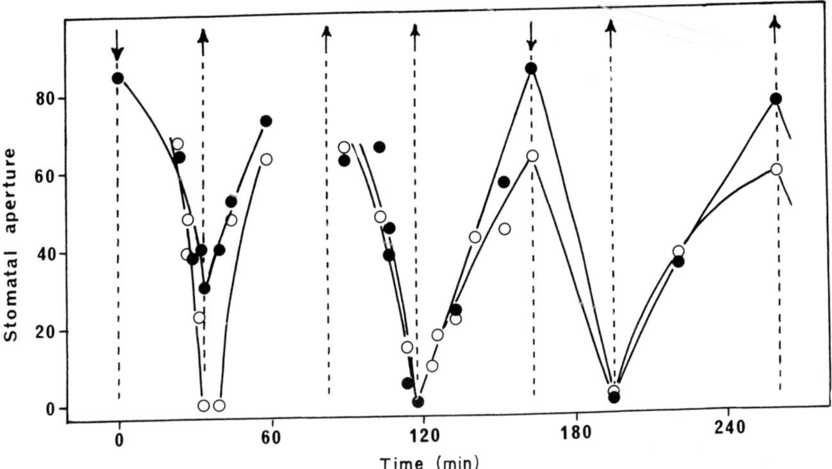

FIG. 31. Effects of moist and dry air applied to the outer side of the epidermis of *Valerianella*. Arrowheads pointing downward indicate the beginning and arrowheads pointing upward the end of the treatment with dry air. After Lange *et al.* (77).

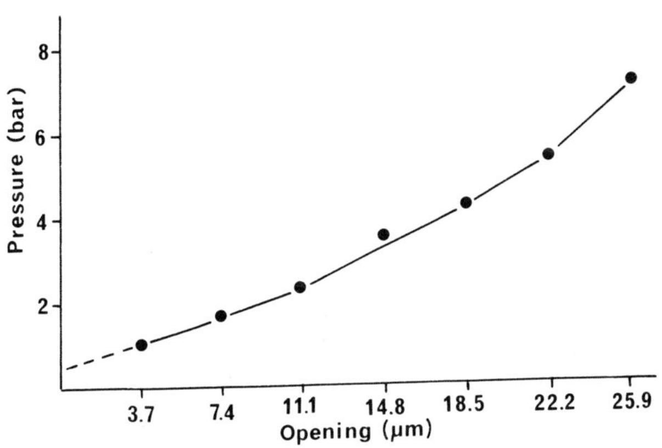

FIG. 32. Pressures applied in the subsidiary cells of *Tradescantia virginiana* required to cause closure of stomata initially open to the extent indicated on the abscissa. Pressures of the same magnitude were required within the guard cells to reopen almost closed stomata. After Edwards and Meidner (28).

guard cell saps estimated plasmolytically, and they suggested that a common assumption could be mistaken, namely, that the guard cells of open stomata are at or near the maximum turgor predicted from their osmotic potentials. The attainment of full turgor could be prevented by a rapid rate of transpiration from the epidermis itself; in other words, the guard cells may lose water more rapidly by evaporation than they can acquire it from the neighboring cells. Meidner (106) visualizes that the main sites of evaporation in a leaf are from the *inner walls* of guard, subsidiary, and epidermal cells which are in contact with the substomatal cavity (Fig. 33). This is because the shortest paths, and therefore those of least diffusion resistance, are from the epidermal cells to the stomatal pore, rather than from mesophyll cells. Meidner (107) performed an ingenious experiment with detached epidermis and showed that the evaporation sites on the inner walls of epidermal cells could sustain a rate of transpiration comparable with that from an intact leaf (Fig. 34). It is suggested that the epidermis obtains its main supply of water directly from the vascular bundles, not from the mesophyll cells, which do not make frequent contact with the epidermis (this is why it is so easily stripped from many leaves). The epi-

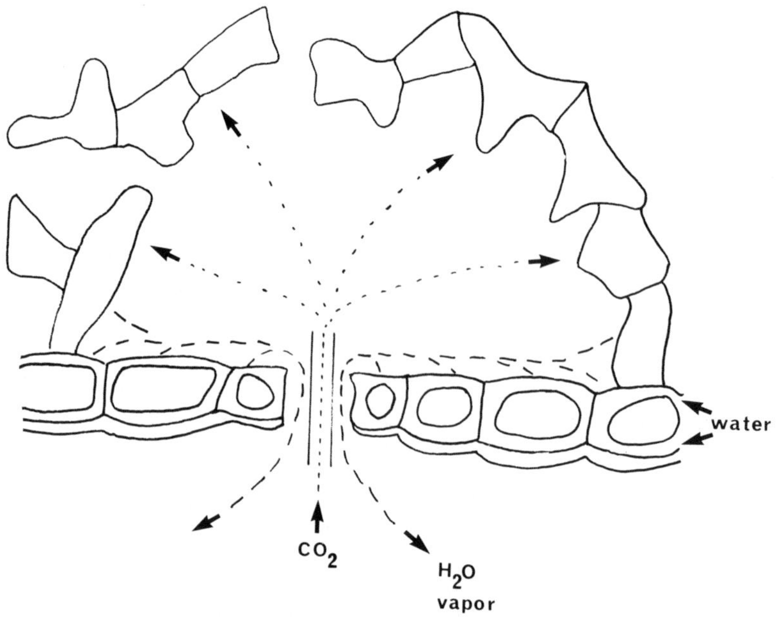

FIG. 33. Diagrammatic representation to show liquid flow paths and gas diffusion paths in leaf tissue in the vicinity of a stomatal pore. After Meidner (106).

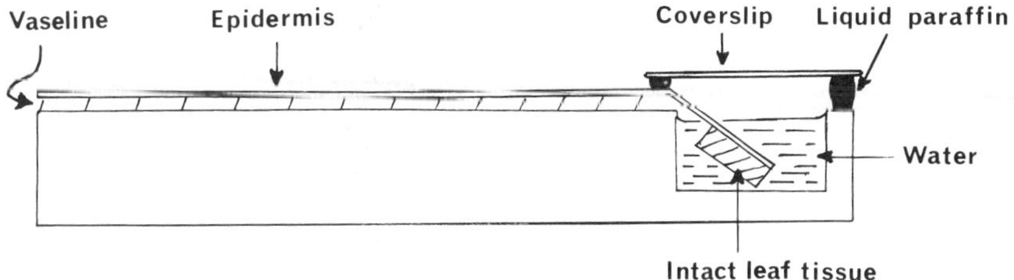

FIG. 34. Slide with water well as used for the measurement of evaporation from epidermal strips of *Tradescantia virginiana*. After Meidner (107).

dermis would thus be a major route for movement of the transpiration stream and should have a high conductivity to water, as is indeed the case (149).

It is now much easier to visualize a mechanism for stomatal responses to VPD by making use of Meidner's concepts. The guard cells are at the very end of the liquid phase part of the transpiration stream, and their situation, nearest to the ambient air, means that they are likely to exhibit a higher rate of evaporation per unit area of cell wall than any other cells in the leaf. Both these facts will operate together to deprive them of water in conditions of high VPD. The extent to which they are water stressed by being in a terminal position will depend on the resistance to water movement from their neighbors. Evidence of the sites of deposition of gold sols and lead chelates and the location of tritiated water, used as labels in the transpiration stream, suggests that there is free evaporation from the guard cells (37, 89, 158) and does not point to the existence of a very high resistance to lateral movement into the guard cells.

The important question of whether the evaporation from guard cells and their neighbors is from their outer or inner surfaces has been examined in an experiment by Sheriff (148). When he prevented evaporation from the inner surfaces of epidermal strips, the magnitude of the stomatal response to a change in humidity was reduced by about half. This suggested that evaporation from both inner and outer walls are similarly effective in producing the response to VPD, though the situation in the intact leaf, where there is also evaporation from the mesophyll, could be different. Sheriff (147) has shown that when a leaf is subjected to a high radiation load, which increases the temperature of the pigmented mesophyll more than that of the nearly colorless epidermis, there can be a substantial condensation of water droplets on the inner walls of the guard cells and their immediate neighbors (Fig. 35). These cells may be cooler than other epidermal cells because they are thinner. The occurrence of

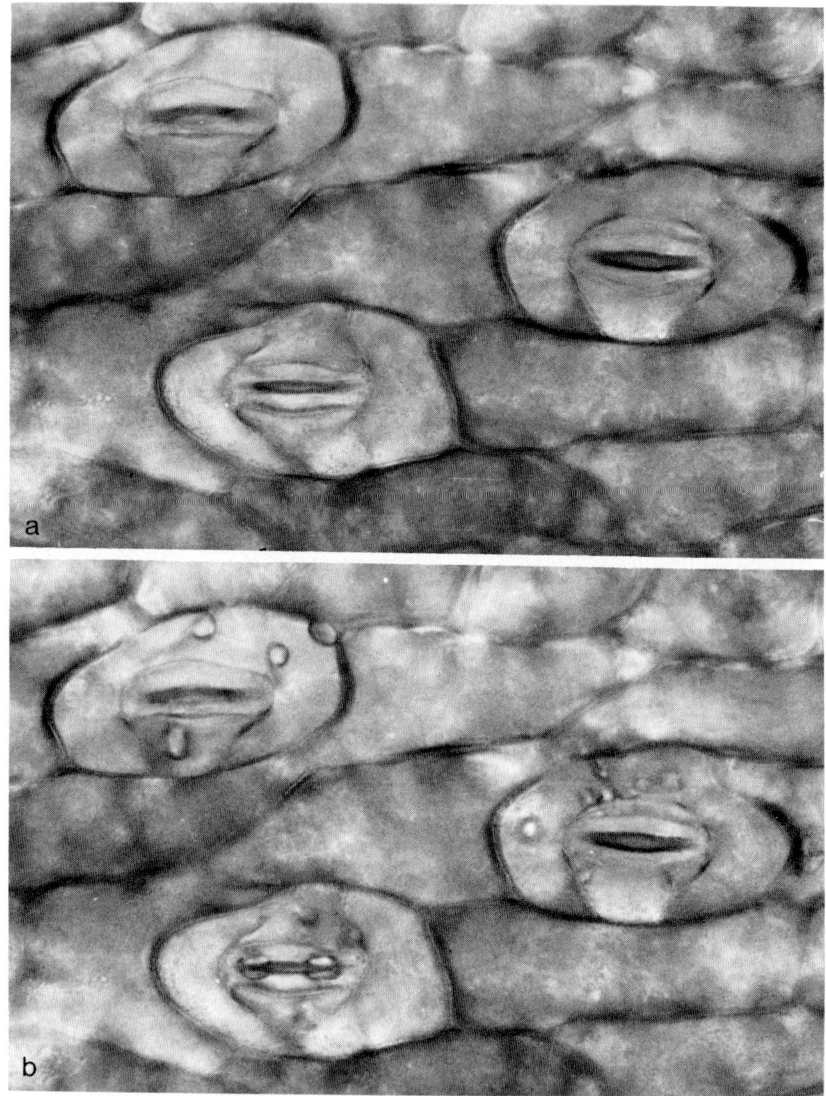

FIG. 35a, b.

FIG. 35. Formation of water droplets, by distillation from the mesophyll, on the inner walls of guard and subsidiary cells of *Tradescantia virginiana*. Air was blown from a fine jet above a leaf illuminated with 819 μE m^{-2} s^{-1}. (a) Immediately before starting the air stream; (b)–(f) 10, 20, 30, 40, and 60 s after the air stream began. After Sheriff (147).

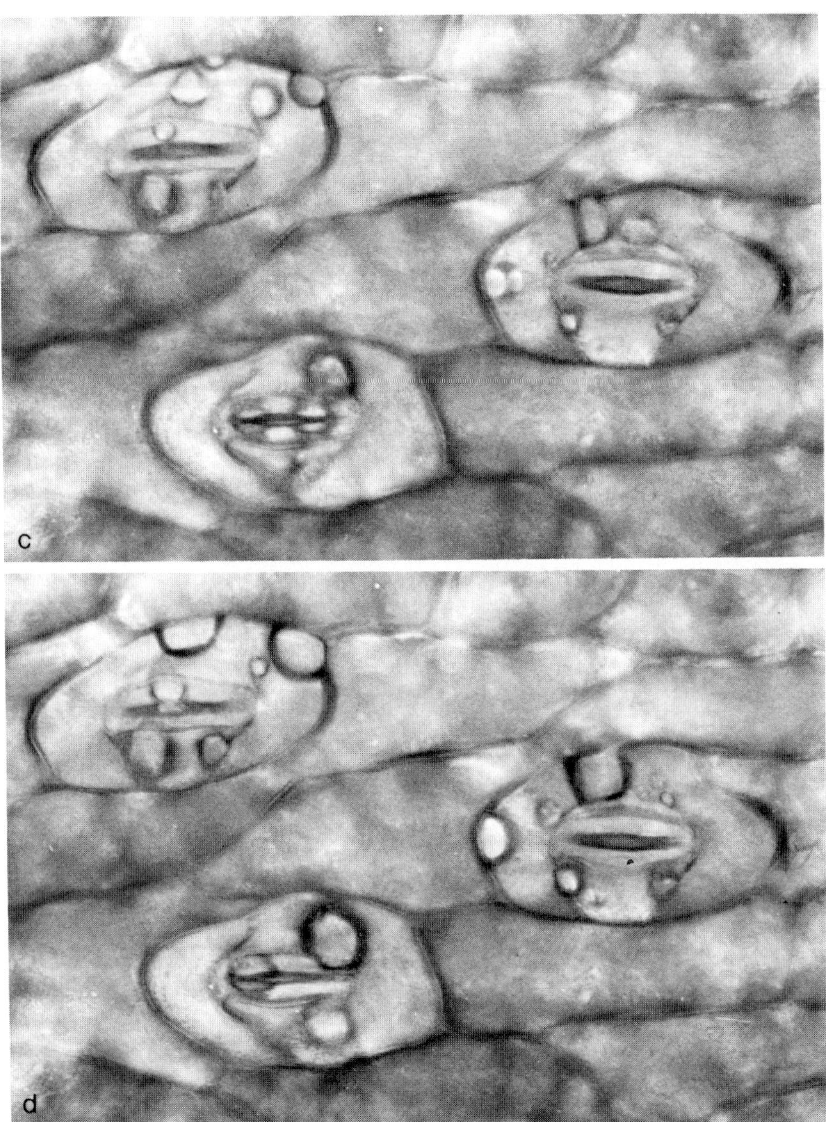

FIG. 35c, d.

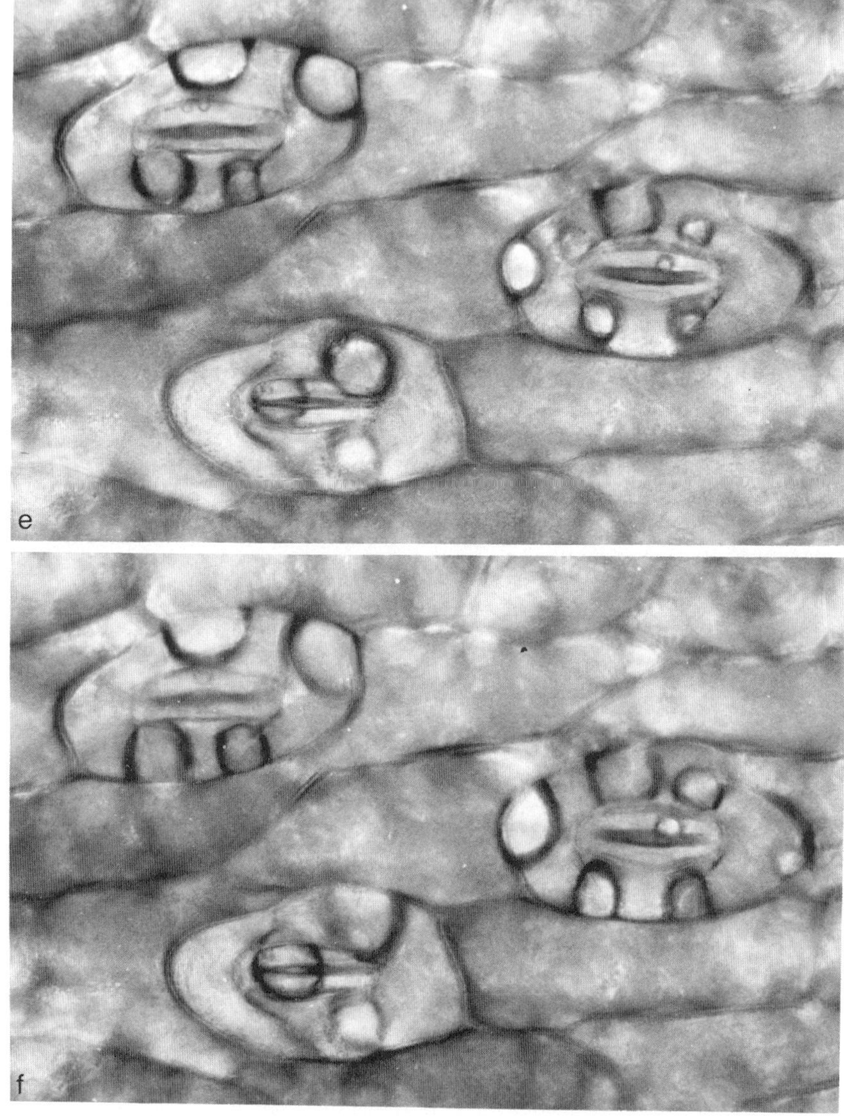

FIG. 35e, f.

this phenomenon of condensation suggests that stomatal responses to VPD might be substantially altered depending on ambient conditions. If the atmosphere has a high VPD when there is a low radiation load, there should be no tendency for condensation to occur, and the evaporation from the guard cells should therefore produce a response to a change in VPD. On the other hand, if the radiation load is high but the VPD is low, condensation will occur and might prevent or reduce the stomatal response to VPD.

If the epidermis acquires most of its water directly from the vascular bundles, then clearly its water relations can have a measure of independence from those of the mesophyll. It may be this independence which enables a stomatal response to VPD to give effective protection to the mesophyll and other vulnerable areas. Schülze *et al.* (144) found that the stomata of apricot *(Prunus armeniaca)* growing in its natural environment closed when VPD was high and opened when it decreased. This opening was in spite of a decreased leaf water content, the result of increased transpiration when the stomata were open. This observation has been interpreted as indicating that stomata respond *directly* to VPD, independently of the bulk water potential of the leaf. It is clear from the work of Meidner, however, that we should distinguish between the water potential of the epidermis and that of the underlying tissues. It is unlikely that the response of the stomata is independent of the former. Such independence would require a high resistance to water movement from subsidiary to guard cells, the existence of which is not supported by abundant evidence of high rates of evaporation from guard cells. Nevertheless, by virtue of their unique situation, the guard cells must experience water potential changes of greater amplitude than the remainder of the epidermis, and in the light of this we might be justified in regarding them as VPD sensors.

Is the evaporative loss of water by guard cells in conditions of high VPD the whole cause of their loss of turgor? Or does a decrease in their water potential trigger a metabolic reaction which subsequently reduces turgor? Ingenious experiments by Lösch and Schenk (82) have shown that the closure of stomata in response to low humidity occurs, initially at least, without changes in the potassium content of the guard cells. This strongly suggests that evaporative loss of water is the primary factor causing the guard cells to lose turgor. If a metabolic process were involved at the outset, we might expect a fall in potassium content to precede the loss of turgor. Maier-Maercker (90) has urged caution in assuming that the effects of peristomatal transpiration can be fully explained in terms of changing turgor pressures. She has drawn attention to the fact that the electrogenic K^+ pump in *Valonia* is pressure dependent (see also Chapter 1 in this volume). Thus changes in guard cell turgor resulting from peristomatal transpiration could influence the balance between K^+ influx and efflux. Hall and Kaufmann (44) found that the conductance of the mesophyll to carbon dioxide did not change appreciably when stomata closed in response to high VPD. This suggests that internal changes in carbon dioxide concentration in the leaf as a whole are not involved.

The functional significance of stomatal closure in response to high VPD appears obvious: in dry air a reduction in water consumption by the plant is required. At first sight some observations do, however, conflict with this simple view. Davies and Kozlowski (22) and Kaufmann (74) found that stomata are less responsive to change in VPD when light intensity is high, and Hall and Kaufmann (45) found a fall in their sensitivity at higher temperatures. This seems to provide for a maximum response to VPD at those times when water stress is least likely to be a problem for plants in natural habitats. It may be that conditions which cause high transpiration rates induce sufficient water stress in the plant as a whole to bring into action hormonal responses which override short-term responses to VPD. Alternatively, it is possible that in some conditions the guard cells are less able to sense the level of VPD because of the condensation of water on their inner walls (see Fig. 35).

Rapid reactions to changes in VPD must clearly be regarded as one of the more important environmental responses of stomata. They have now been observed in many species and in one, Sitka spruce, a response to carbon dioxide appears to be absent (84, 116), but there are large responses to VPD and light, which together may be the main factors that influence stomatal movements in this plant (57).

2. *Hormone-Mediated Stress Responses*

It has been known for many years that after a plant has experienced water stress the stomata do not open fully even when full turgor has been regained. The aftereffect varies in duration from a few days to several weeks depending on the severity of the water stress, and in extreme cases full recovery never occurs (5, 34, 42, 53). This phenomenon remained unexplained until the discovery that wheat leaves which had been allowed to wilt contained greatly increased amounts of abscisic acid (ABA) (171, 174). External applications of ABA were found to induce stomatal closure (79) and the aftereffect of one application of the hormone followed a pattern already familiar from the studies of water stress (72, 73). An appreciable suppression of transpiration in barley seedlings was obtained with an application of ABA of $<0.1\ \mu g\ cm^{-2}$ of leaf surface and the effect remained for 9 days (Fig. 36). The buildup of ABA in stressed plants and the course of stomatal movements have been examined very closely, and while there are good general correlations, there are some unexplained discrepancies which need to be resolved. The data of Beardsell and Cohen

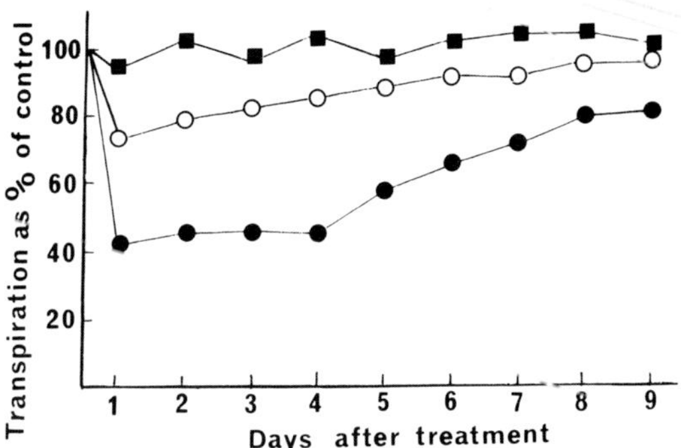

FIG. 36. Effects of externally applied abscisic acid on the transpiration from young barley plants. 1 μM (■), 10 μM (○), and 100 μM (●) ABA applied on day 0. After Jones and Mansfield (73).

(10) for *Z. mays* show that ABA levels rise steeply as water potential declines from -0.7 to -1.0 MPa and fall again rapidly after rewatering. Stomatal closure, however, begins before the level of ABA has risen appreciably, and recovery to full opening lags behind the disappearance of ABA (Fig. 37). Similar discrepancies were found by Dörffling *et al.* (26) for *Pisum sativum,* but there were better correlations for *Helianthus annuus* and *V. faba.*

There are reasonable explanations for the lack of an exact correlation between ABA content and stomatal aperture. First, other compounds may be involved in addition to ABA. It has been shown that externally applied indol-3-ylacetic acid (IAA) or cytokinins can affect the responses of stomata to ABA (102, 125, 151). Although changes in IAA content of plants under water stress have been little studied, there are some indications of a fall in IAA level (173). It thus seems likely that the pattern of stomatal behavior during and after water stress will be determined not only by ABA, but also by other growth substances such as IAA and cytokinins. The involvement of cytokinins in stomatal responses may be restricted to particular taxonomic groups (e.g., the Gramineae) but will nevertheless have to be taken into account alongside effects of ABA in some species (70). Second, the estimates of endogenous amounts of ABA are usually for the leaf tissues as a whole, whereas the critical levels are those in the guard cells. Loveys (83) concluded that detached epidermal tissue of *V. faba* was unable to synthesize ABA even though the amount in the epidermis of intact leaves increased greatly under the influence of water stress. It is possible that some ABA is contained in the chloroplasts of nonstressed mesophyll cells. When they are stressed, both this store of the hormone

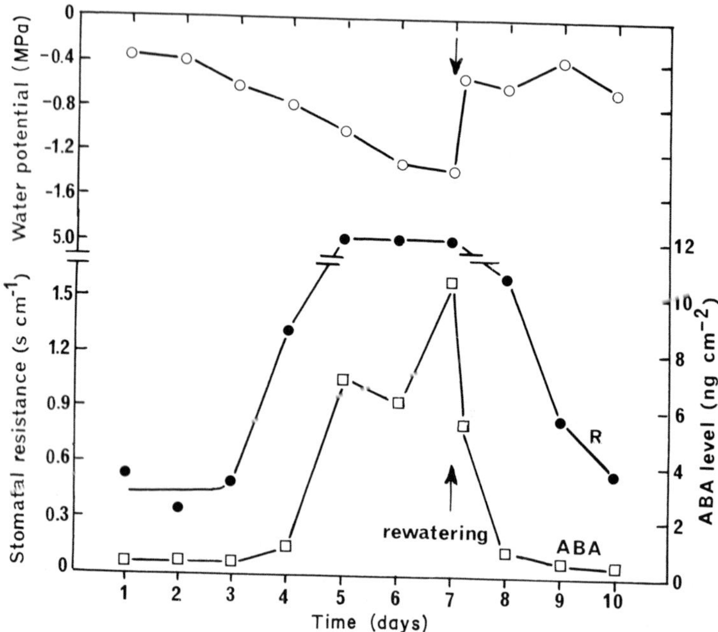

FIG. 37. Changes in leaf water potential, stomatal resistance, and ABA content during and after water stress in *Zea mays*. After Beardsell and Cohen (10).

and newly synthesized ABA might migrate to the epidermis, where selective accumulation could occur in the guard cells. The mechanism of release of ABA from mesophyll chloroplasts could thus play a major regulating role in its distribution to target areas such as the guard cells. Such assumptions about the movement of ABA are, however, based on preconceptions about the role of hormones in plants which have been strongly challenged recently (160a). Speculation is hazardous in the absence of further information on the transport of ABA.

The site of ABA formation in leaves is a controversial topic. Milborrow (112) has interpreted his own data as indicating that ABA is synthesized within the chloroplasts, but Hartung *et al.* (47) have proposed that it is synthesized in the cytoplasm and afterward moves into the chloroplasts, where it is trapped, and they allege that Milborrow's chloroplast preparations were contaminated with cytoplasm. There is, however, agreement between Milborrow and Hartung *et al.* that in unstressed leaves there are relatively high concentrations of ABA inside the chloroplasts. An early event in the plant's response to stress may, therefore, be the release of this stored ABA and its movement to the guard cells. The fact that stomatal closure begins before the total level of ABA in the leaves has increased

(Fig. 37) could be the outcome of the release of ABA already present in the chloroplasts, and the failure of the stomata to reopen when bulk tissue ABA levels have declined might be either the result of residual ABA which is still active in the guard cells or of changes in other growth substances such as IAA or cytokinins.

Farnesol, a compound closely related to ABA (both are sesquiterpenoids), has been detected in leaves of *Sorghum* after water stress (118). This compound causes reversible closure of stomata (30), and it has been suggested that its function in the leaf may be to regulate the permeability of chloroplast membranes to ABA and effect its release during the initial stages of water stress (100). Some support for this suggestion comes from Milborrow's observation (111) that application of farnesol to intact leaves causes an increase in ABA level. It is proposed (100, 111) that when ABA leaks out of the chloroplasts into the cytosol, biosynthesis of ABA within the chloroplasts increases so that a net increase in the ABA content of the leaves results.

The synthesis of ABA in water-stressed leaves has now received a lot of attention, and the relationship with bulk water potential is known for a number of species. Some authors have suggested that there is a "threshold" water potential which triggers off the production of ABA (10, 176),

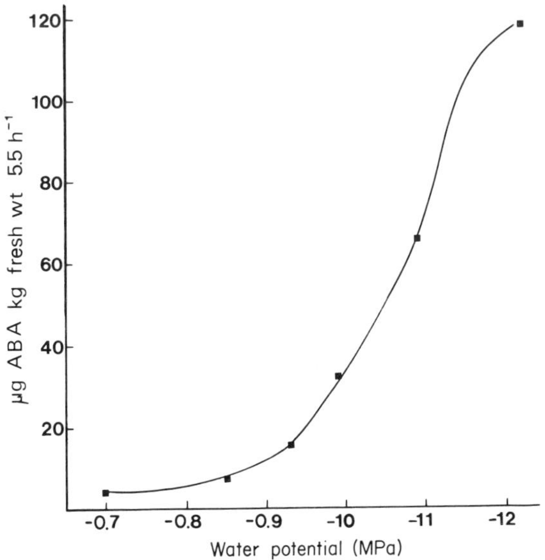

FIG. 38. Effect of leaf water potential on ABA levels on detached wheat leaves. After Wright (172).

estimates of which have varied from −0.8 to −1.2 MPa for different species (cf. Fig. 37). Wright (172) has, however, shown that the relationship between water potential and ABA production shows a smooth sigmoid relationship (Fig. 38) with a very steep rise between −0.95 to −1.1 MPa which could be mistaken for a threshold.

Data such as those in Figs. 37 and 38 show the rise in total ABA content in the leaf tissue as a whole. As we have seen, however, critical processes in rapid stomatal responses to water stress could be the release of stored ABA from mesophyll chloroplasts and its transport to the guard cells. The chloroplasts are probably the most sensitive organelles to water stress, and it is therefore appropriate that they should control the transmission of any hormonal signal to the guard cells. The rapid response of stomata to ABA [Cummins *et al.* (18) detected a response 5 min after injecting ABA into the transpiration stream] means that this hormone may mediate the rapid stomatal adjustments which accompany changing environmental conditions. Further studies of the release of ABA from chloroplasts (cf. 162) should prove rewarding in the future.

The sensitivity of the stomata of a given species to externally applied ABA is not constant but varies according to pretreatment. Kriedemann *et al.* (75) found that there was a smaller stomatal response to ABA in sugar beet which had been previously water stressed. On the other hand, Davies (21) was unable to obtain any appreciable effect of ABA on well-watered plants of *V. faba,* but exposure to water stress rendered them sensitive to the hormone. These two conflicting findings emphasize the difficulty in evaluating how ABA might operate to improve the water use efficiency of plants exposed to periodic stress in the field. Different response patterns would be expected in species adapted to different environments, and in the case of crop plants unusual (even disadvantageous) patterns may have emerged among varieties developed mainly for their potential yield. The vital importance of ABA in a plant's water economy is shown by the work of Tal and colleagues (156, 157) on the *flacca* variety of tomato. This mutant wilts easily because its stomata do not close as water stress increases, and this was found to be due to a hormonal imbalance, in particular a small endogenous supply of ABA. When the plant was given an external supply of ABA, its stomatal behavior reverted to the usual pattern.

C. The Mechanism of Stomatal Responses to Carbon Dioxide and Abscisic Acid

Carbon dioxide and abscisic acid are two naturally occurring compounds which when applied externally to plants induce closure of stomata and which seem to be involved in the natural responses of stomata to environmental variables. Therefore, is there any interdependence in the mode of action of these two chemicals?

Raschke (133) has produced strong evidence that carbon dioxide and ABA do not act independently of one another. He found that the stomata of glasshouse-grown plants of *X. strumarium* were insensitive both to carbon dioxide and to abscisic acid applied alone, but that they closed if both were given simultaneously. He suggested that stomata are "sensitized" to carbon dioxide by the presence of ABA. Thus well-watered plants with a minimum ABA content showed little or no response to carbon dioxide but became sensitive to it if ABA formation were induced by water stress. This would provide an attractive mechanism for water conservation because the control of the stomata by carbon dioxide will be most effective in windy conditions (when the concentration in the substomatal cavity must be greater than in still air), which is when water stress is most likely to be severe.

It is not known whether Raschke's findings are generally applicable. Well-watered plants of *X. strumarium* grown by the present author were sensitive both to carbon dioxide and ABA, and a factorial experiment revealed no interaction between the two in the rate of closure induced (96). There was, however, some interdependence between the two compounds because in carbon dioxide free air there was a delay in the reaction to ABA, which became more marked at higher light intensity. This delay was also evident in some of Raschke's closure curves (Fig. 2 in Raschke, 133) and so the two sets of data are not entirely incompatible. Studies by Snaith and Mansfield (151) have shown that IAA may play an important part in determining the responses of stomata to carbon dioxide. In isolated epidermis of *C. communis,* carbon dioxide and ABA appeared to act independently, but when IAA was incorporated in the incubation medium, a carbon dioxide × IAA × ABA interaction was revealed. The discrepancies between this author's experimental results and Raschke's may have been due to differences in endogenous IAA content of the tissue at the time when the stomatal responses to carbon dioxide and ABA were observed. The situation is further complicated by the discovery that cytokinins may also interact with ABA in governing the response of stomata to carbon dioxide (10a,b). The significance of a *variable* carbon dioxide sensor in the guard cells in relation to wind speed and transpiration raises some important new questions (96a).

Abscisic acid has been shown to inhibit the accumulation of potassium by guard cells and the disappearance of starch which normally accompanies stomatal opening (56, 98). It cannot be assumed, however, that its primary mode of action must be on these processes, both of which are probably inevitably associated with changes in guard cell turgor. Control at the level of starch hydrolysis might require changes in enzyme levels, which it seems unlikely could be effected with the known rapidity of stomatal responses to ABA. Some method of control at the membrane level is more feasible. This might be on the plasmalemma, the site of K^+ accumulation and H^+ extrusion, or on the tonoplast or the chloroplast envelopes, through which malate is presumably transferred. MacRobbie (87) has found a transient but marked stimulation of Rb^+ (analogous to K^+) and Cl^- efflux out of guard cells treated with ABA. There was evidence of effects on fluxes at both plasmalemma and tonoplast.

The mode of action of carbon dioxide has attracted much attention, but little progress has been made toward explaining it. The range of carbon dioxide concentrations involved has led to the belief that a carboxylation reaction could be the basis for the detection of carbon dioxide by the guard cells. The most likely candidate is the incorporation of carbon dioxide into oxaloacetate by the enzyme phosphoenolpyruvate (PEP) carboxylase:

$$\text{Phosphoenolpyruvate} + CO_2 + H_2O \rightarrow \text{oxaloacetate} + H_3PO_4$$

This reaction is familiar as the carboxylation mechanism which is so efficient in trapping carbon dioxide in C_4 plar The oxaloacetate does not accumulate, but it is converted to other C_4 s such as malate and aspartate.

High levels of PEP carboxylase have been detected in leaf epidermis, and its association with stomata is suggested by the fact that its activity is proportional to the stomatal density in upper and lower epidermes (169). It has, therefore, been suggested that PEP carboxylation is a source of malate required for stomatal opening (133). However, some apparently contrary pieces of information need to be accommodated: (1) Malate levels in guard cells increase as stomata open, (2) increases in carbon dioxide concentration over the range 0–1000 μl liter^{-1} cause stomatal *closure*, and (3) increases in carbon dioxide concentration over this same range increase the rate of malate production via PEP carboxylase. Raschke (134) has used great ingenuity to try to circumvent these apparently incompatible findings. Essentially his approach challenges the view that carbon dioxide inhibits stomatal opening over the whole range from 0 to 1000 μl L^{-1}. He suggests that between 0 and 100 μl L^{-1}, carbon dioxide acts to stimulate stomatal opening (27) and that only above 100 μl L^{-1} does it become inhibitory (133).

In essence, Raschke's interpretation is as follows: The incorporation of carbon dioxide into PEP in the cytoplasm of the guard cells leads to malate accumulation and acidification, which then inhibits further action of PEP carboxylase. Malate formation ceases and solutes are lost from the guard cells. He believes that at low carbon dioxide concentrations (<100 μl L^{-1})

the carboxylation of PEP contributes significantly to the malate requirements of guard cells and stimulates opening. Only at higher concentrations is acidification of the cytoplasm sufficient to act in an inhibitory manner.

Dittrich and Raschke (23, 24) have made further studies of malate metabolism in epidermis and have shown that starch is formed in guard cells presented with an exogenous supply of labeled malate. In parallel studies, Willmer and Rutter (170) found that the rate of incorporation of labeled malate into starch in epidermis depended on whether the stomata were opening in light or closing in darkness (Fig. 39). The formation of starch is preceded by conversion of malate to sugars via the process of glucogenesis, which involves the loss of one carbon from malic acid by decarboxylation (23, 24). Dittrich and Raschke also showed that the carbohydrate reserves of guard cells are not derived from their own photosynthesis, but from material imported from the mesophyll. Thus the conversion of starch to malate and vice versa, as stomata open and close, emerges as a central metabolic event in guard cells, but how this process can be influenced by carbon dioxide is by no means clear. Perhaps the greatest difficulties of interpretation arise because malate production in guard cells is unaffected by carbon dioxide concentrations in the range 0–100 μl L^{-1} (159). Even if virtually all the carbon dioxide required for malate production were generated by mitochondrial activity and then efficiently trapped by cytoplasmic PEP carboxylase, some effect of exter-

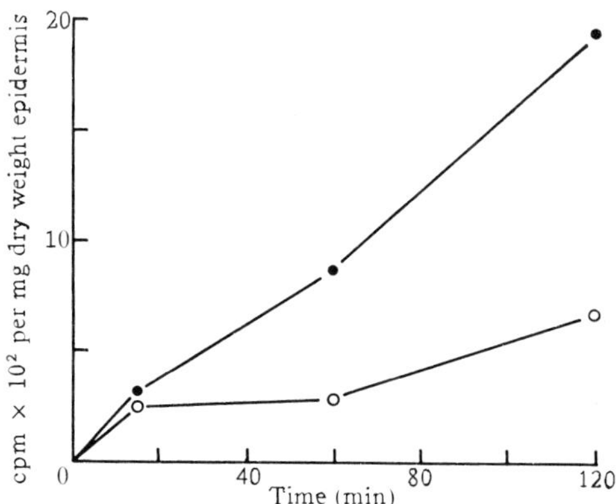

FIG. 39. Time course of incorporation of ^{14}C from L-[U-^{14}C] malic acid into starch in abaxial epidermis of *Commelina communis*. ○, closed stomata opening in light in carbon dioxide-free air; ●, open stomata closing in darkness and normal air. After Willmer and Rutter (170).

nal carbon dioxide on the rate of malate formation would be expected. Exogenous carbon dioxide is known to be incorporated in guard cells in both light and darkness (Fig. 40), and it is therefore surprising that malate formation is not enhanced as ambient carbon dioxide concentration is increased from 0 to 100 μl L^{-1} (159). One seemingly important observation suggests that attempts to link stomatal responses to carbon dioxide with malate metabolism could be misguided. This is the finding that onion guard cells do not accumulate malate as a counterion for K^+ (143); it is known that onion stomata show a normal sensitivity to carbon dioxide (109).

A possible solution of some of these problems may be found in more recent observations of the carbon dioxide responses of stomata treated with fusicoccin (160). Fusicoccin is a fungal toxin which is known to promote proton efflux from cells (101), and it is a powerful stimulant of stomatal opening which can overcome the effects of natural closing agents such as darkness and ABA. In the absence of fusicoccin carbon dioxide causes stomata to close, but in its presence carbon dioxide causes them to open. These effects suggest that carbon dioxide exerts two quite different effects on stomata. The first is the familiar closing response to carbon dioxide, and this is overcome by fusicoccin. Once this response is annulled, a second, opposing one is revealed, namely, a stimulation of opening as increased carbon dioxide concentration permits more production of malate. The fact that fusicoccin overcomes the closing response to carbon dioxide may indicate that carbon dioxide can affect processes of ion transport in guard cells directly.

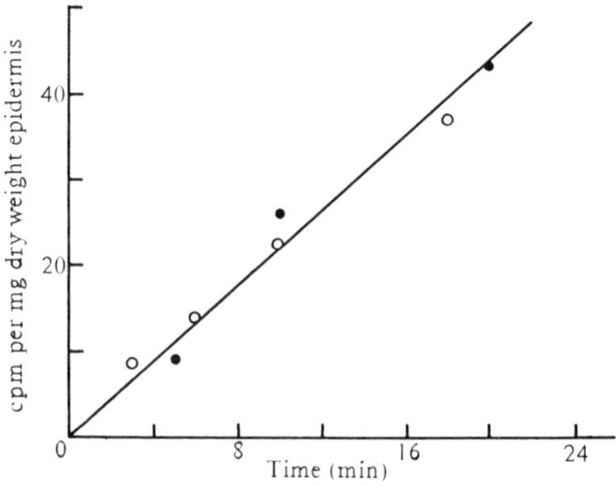

FIG. 40. Kinetics of carbon dioxide fixation into abaxial epidermis of *Commelina communis* in light (○) and darkness (●). After Willmer and Rutter (170).

VII. Energy Requirements for Stomatal Functioning

Direct measurements of ionic concentrations and pH in cells of the stomatal complex of *C. communis* have indicated that there are active driving forces for both anions and cations, not only between the guard cells and their neighbors, but also between subsidiary and epidermal cells (see Section IV). These ionic fluxes will thus require the expenditure of metabolically derived energy. There may also be an energy requirement for the intracellular movement of malate in guard cells and for the interconversion of starch and malate. Raschke (133) has pointed out that the formation of malate from starch via PEP is exergonic.

Guard cells have an abundance of mitochondria and they usually possess chloroplasts, although sometimes these are poorly developed. When chloroplasts are absent altogether, stomata may still function (117), which suggests that energy from photosynthetic sources is not essential for stomatal movements. Another consideration brings us to the same conclusion: the fact that stomata can open widely in the dark under appropriate conditions (see Section V.A) must indicate that the energy demands can be sustained by the mitochondria without a contribution from the chloroplasts. Zeiger (176a) found that the kinetics and saturating levels of stomatal opening which had been stimulated in darkness by fusicoccin were almost identical to those in light.

In spite of this evidence some writers have continued to ascribe a major role to chloroplast functions. Often their conclusions have been based on the results of a carefree use of metabolic inhibitors on whole leaves. It should be obvious (although apparently it is not) that the application of a compound which inhibits any activity of chloroplasts so as to reduce the rate of photosynthesis will be likely to cause stomatal closure by raising the carbon dioxide concentration in the leaf. The reader should beware of the extensive literature on chemicals which affect stomatal movements. Mostly such compounds act by disturbing the carbon dioxide balance in the leaf, and rarely has it been shown that the activities of guard cells are directly influenced.

Metabolic inhibitors of oxygen evolution and the noncyclic electron flow of photosystem 2 in chloroplasts [e.g., 3-(3,4-dichlorophenyl)-1,1-dimethylurea (DCMU)] have little direct effect on stomatal apertures (4, 62, 167). This is also the case in light under anaerobic conditions (which prevent oxidative phosphorylation in mitochondria), suggesting that the energy requirements for opening might be met by cyclic electron flow of photosystem 1 or ATP produced in cyclic photophosphorylation. Several studies with inhibitors of photosystem 1 have suggested that it has a special significance in stomatal opening in light (20, 62, 129, 167). Humble and Hsiao (62) found that far-red light, which can sustain photosystem 1 but not photosystem 2, was as effective as white light in inducing opening. Active cation transport in algae can be supported by ATP synthesized in cyclic photophosphorylation (138), and stomata are not unusual in showing a preference for energy derived from one photosystem over the other. What is surprising, however, is that this preference exists when there is not an absolute requirement for light for stomatal opening. Might the absence of any inhibitory effect of photosystem 2 indicate that it is not present at all in guard cell chloroplasts? This has been considered to be a distinct possibility because guard cells apparently do not reduce carbon dioxide photosynthetically (24). However, recent studies of fluorescence emission spectra have indicated that both photosystems are present in the guard cells (117a, 122a, 176b).

Thus it may be futile to search for a link between specific energy-supplying processes and stomatal movements. The simplest interpretation of the data available may be that there is no real preference for energy from a particular source: the activities of mitochondria and chloroplasts are equally suitable.

The way in which energy is made available to drive specific ion transport processes is fundamental to our understanding of stomatal physiology. Zeiger *et al.* (177) have discussed possible chemiosmotic mechanisms in guard cells, in particular asymmetrical ion transport involving the extrusion of protons and subsequent H^+-K^+ exchange. They consider the possibility that if chemiosmosis is the energy-transducing mechanism behind ion transport in the stomatal complex, the motive force for proton extrusion could be provided by ATP or alternatively by the direct stimulation of electron flow by the absorption of light quanta by a photoreceptor located in the plasmalemma or tonoplast. This would explain the nonphotosynthetic action of blue light in stimulating opening (Section V, B).

Understanding such energy-transducing mechanisms will not necessarily provide an answer to the important question of how stomatal aperture is regulated by various stimuli. If the energy available for ion transport were the sole factor in determining aperture, we would expect stomata always to be open in high-intensity light. They can, however, close under the influence of factors (e.g., ABA) which have no known effect on ATP production in chloroplasts or mitochondria.

There are some indications that factors which induce changes in stomatal aperture might have a direct effect on the transport of K^+ to or from the guard cells. Jarvis and Mansfield (69) found that Na^+ could replace K^+ and support stomatal opening, but that closing responses to darkness, carbon dioxide, and ABA were then much reduced. Thus these agents may inhibit stomatal opening by affecting the intake of K^+ or enhancing its leakage. MacRobbie (87) has obtained evidence that ABA stimulates efflux of Rb^+ (analogous to K^+) but has little effect on influx. The ability of a potassium ionophore (benzo-18-crown-6) to inhibit stomatal opening also suggests that the balance between intake and efflux of K^+ may determine guard cell turgor (139). It is suggested that compounds of this nature can partition into membranes and act as channels for potassium ions to move down their chemical gradient, i.e., leak out of the cell (40). The fact that there is mutual interference between benzo-18-crown-6 and ABA could indicate common features in their modes of action (139).

Thus the control of stomatal opening by environmental and other factors may depend on a complex relationship between energy provision for proton extrusion or other processes at the plasmalemma or tonoplast and changing balances in rates of influx or efflux of K^+ or perhaps other materials. In the light of experimental results it is difficult to visualize a system which involves a single point of control.

VIII. Antitranspirants

A knowledge of energy sources for guard cell turgor changes is important in relation to the search for means of controlling stomatal movements artificially so that gaseous diffusion between leaves and the atmosphere can be regulated. Such regulation could benefit the water relations of crop plants and might also be employed as a means of excluding air pollutants over short time periods.

The theory behind the use of chemical inhibitors of stomatal opening ("antitranspirants") to improve the efficiency with which crops use available water assumes that a given degree of stomatal closure would exert a larger proportional effect on diffusion of water vapor out of the leaf than on carbon dioxide entering because of differences in the lengths of their diffusion pathways (110). The theoretical predictions have been confirmed in small-scale experimental trials of a number of chemicals, there being in some cases a substantially bigger effect on transpiration than on photosynthesis (94). In spite of these successes, however, there are major difficulties in the practical application of compounds of known activity. These difficulties have ruled out realistic trials of chemical "antitranspirants" in the field, where partial stomatal closure could lead to increases in leaf temperature, which would tend to maintain transpiration.

A. METABOLIC INHIBITORS

If stomatal opening were dependent on metabolic events in the guard cells susceptible to selective disruption by inhibitors, these compounds might be employed in the role of antitranspirants. Ideally we would hope to discover an inhibitor affecting guard cell turgor specifically, with no detrimental effect elsewhere in the leaf. Alternatively, a less specific inhibitor with limited permeability might be used, i.e., a compound which remained within the epidermis after application to the leaf surface.

No specific inhibitor of functions unique to the guard cells is known. A better knowledge of the metabolic events behind the turgor changes is an essential prerequisite to an informed search for a compound with the desired properties. In the absence of such an ideal substance, people have turned to more general metabolic and growth inhibitors in the hope that the dose required to affect the guard cells might be insufficient to have serious effects elsewhere in the plant. Many substances which cause stomata to close and reduce transpiration have an equal or greater effect on photosynthesis (94, 120). In carefully controlled doses, however, some can affect the turgor of guard cells without major effects elsewhere. The most interesting such compound is phenylmercuric acetate (PMA), which can affect a useful fall in water consumption. Unfortunately, PMA is unsuitable for application to crops. Apart from the fact that it is toxic to animals, it is an inhibitor of photophosphorylation and also affects membrane permeability. Under laboratory conditions controlled doses can be applied which do not have severe effects on photosynthesis in the mesophyll, but transport into the leaf's interior readily occurs (152), and there is little hope of preventing this after application in the field. Another undesirable effect of PMA is the prevention of stomatal closure in darkness (91), which is probably the result of damage to subsidiary and epidermal cells, the turgor of which is often necessary for complete closure (41). This may happen with any metabolic inhibitor which is not restricted to the guard cells. Experiences with PMA therefore lead to the conclusion that this form of control over the stomata will not be practical until compounds with very specific effects on guard cells have been found.

B. Hormones

The discovery that ABA applied to the surfaces of leaves can affect stomatal behavior for several days (see page 202) has encouraged exploration of the possibility of its use either as the hormone itself or as its derivatives or analogs as antitranspirants. A major physiological role postulated for ABA is that of an "endogenous antitranspirant," and application of ABA analogs to control stomata would be comparable to the use of auxin analogs for a variety of practical ends.

Although ABA has other roles in the plant, the small doses required to cause significant closure of stomata are apparently insufficient to produce side effects of great importance, though there may be some slight inhibition of growth (113). The main drawback from a practical point of view is that its effects are not sufficiently prolonged. Presumably, it is enzymatically broken down over a period of time. Although a number of analogs and derivatives have been tested, none has been found with appreciably more prolonged effects than ABA (73, 119). Nevertheless, the hope remains that a compound of practical interest may emerge, for as yet only a small number of the possible variations on the ABA molecule have been tested.

IX. Concluding Remarks

In spite of the considerable advances that have been made in the past two and a half decades in our understanding of the mechanism behind the turgor changes of guard cells, little progress has been made toward discovering the mechanisms by which stomata respond to environmental stimuli. The further research that has been reported since the publication of Heath's chapter over 25 years ago (51) has done more to establish the complexity of these responses than to elucidate the mechanism behind them. It has been found, for example, that stomata exhibit some of the intricate nonphotosynthetic light reactions that have long exercised the minds of workers in the field of photomorphogenesis (150). It seems improbable that real progress will be made in understanding the way in which light triggers an increase in guard cell turgor until more is known about the blue light photoreceptor from studies of its action in other situations in the plant. Although guard cells show a measure of independence from other leaf cells, their physiological properties must have many features in common with other cells. The task in the future is to elucidate those special mechanisms which cause guard cells to gain and lose turgor in a different way than the cells around them. In some specific way they are different from their neighbors. Despite visible differences in ultrastructure it is not yet possible to identify the mechanism of turgor change in response to stimuli.

The position occupied by stomata in the pathway of water movement between the soil and the atmosphere — they are the major and sometimes the only resistance in this pathway over which the plant has control — gives them a central importance in plant water relations. If the mechanisms by which guard cells operate can be unravelled, this may well lead to some control over the water relations of crops. Thus the further study of stomata has a practical value quite apart from their intrinsic physiological interest. It is to be hoped that more plant physiologists will feel drawn to this area in the future. The present state of the subject is such that studies of stomata in relation to physical ecology have far outpaced our physiological understanding: the task now must be to close the gap.

References

1. Allaway, W. G. (1973). Accumulation of malate in guard cells of *Vicia faba* during stomatal opening. *Planta* **110,** 63–70.
2. Allaway, W. G. (1981). Anions in stomatal operation. *In* "Stomatal Physiology" (P. G. Jarvis and T. A. Mansfield, eds.), pp. 71–85. Cambridge Univ. Press, London and New York.
3. Allaway, W. G., and Hsiao, T. C. (1973). Preparation of rolled epidermis of *Vicia faba* L. so that stomata are the only viable cells. *Aust. J. Biol. Sci.* **26,** 309–318.
4. Allaway, W. G., and Mansfield, T. A. (1967). Stomatal responses to changes in carbon dioxide concentration in leaves treated with 3-(4-chlorophenyl) — 1,1-dimethyl urea. *New Phytol.* **66,** 57–63.
5. Allaway, W. G., and Mansfield, T. A. (1970). Experiments and observations on the after-effect of wilting on stomata of *Rumex sanguineus. Can. J. Bot.* **48,** 513–521.
6. Allaway, W. G., and Milthorpe, F. L. (1976). Structure and functioning of stomata. *In* "Water Deficits and Plant Growth" (T. T. Kozlowski, ed.), Vol. 4, pp. 57–102. Academic Press, New York.
7. Allaway, W. G., and Setterfield, G. (1972). Ultrastructural observations on guard cells of *Vicia faba* and *Allium porrum. Can. J. Bot.* **50,** 1405–1413.
8. Appleby, R. F., and Davies, W. J. (1983). A possible evaporation site in the guard cell wall and the influence of leaf structure on the humidity response by stomata of woody plants. *Oecologia* **56,** 30–40.
9. Aylor, D. E., Parlange, J.-Y., and Krikorian, A. D. (1973). Stomatal mechanics. *Am. J. Bot.* **60,** 163–171.
10. Beardsell, M. F., and Cohen, D. (1975). Relationships between leaf water status, abscisic acid levels and stomatal resistance in maize and sorghum. *Plant Physiol.* **56,** 207–212.

10a. Blackman, P. G., and Davies, W. J. (1983). The effects of cytokinins and ABA on stomatal behaviour of maize and *Commelina. J. Exp. Bot.* **34,** 1619–1626.

10b. Blackman, P. G., and Davies, W. J. (1984). Modification of the CO_2 responses of maize stomata by abscisic acid and by naturally-occurring and synthetic cytokinins. *J. Exp. Bot.* **35,** 174–179.

11. Bowling, D. J. F. (1976). Malate-switch hypothesis to explain the action of stomata. *Nature (London)* **262,** 393–394.
12. Briggs, W. R. (1976). The nature of the blue light photoreceptor in higher plants and fungi. *In* "Light and Plant Development" (H. Smith, ed.), pp. 7–18. Butterworth, London.
13. Butler, W. L., Norris, K. H., Siegelman, H. W., and Hendricks, S. B. (1959). Detection, assay and preliminary purification of the pigment controlling photoresponsive development of plants. *Proc. Natl. Acad. Sci. U.S.A.* **45,** 1703–1708.
14. Carr, D. J. (1976). Plasmodesmata in growth and development. *In* "Intercellular Com-

munication in Plants: Studies on Plasmodesmata" (B. E. S. Gunning and A. W. Robards, eds.), pp. 243–289. Springer-Verlag, Berlin and New York.

15. Chandler, J. A. (1977). "X-ray Microanalysis in the Electron Microscope." North-Holland Publ., Amsterdam.
16. Cooke, R. J., Saunders, P. F., and Kendrick, R. E. (1975). Red light induced production of gibberellin-like substances in homogenates of etiolated wheat leaves and in suspensions of intact etioplasts. *Planta* **124,** 319–328.
17. Cowan, I. R. (1977). Stomatal behaviour and environment. *Adv. Bot. Res.* **4,** 117–228.
18. Cummins, W. R., Kende, H., and Raschke, K. (1971). Specificity and reversibility of the rapid stomatal response to abscisic acid. *Planta* **99,** 347–351.
19. Dale, J. E. (1961). Investigations into the stomatal physiology of Upland cotton. I. *Ann. Bot. (London)* [N.S.] **25,** 39–52.
20. Das, V. S. R., and Raghavendra, A. S. (1974). Role of cyclic photophosphorylation in the control of stomatal opening. *In* "Mechanisms of Regulation of Plant Growth" (R. L. Bieleski, A. R. Ferguson, and M. M. Cresswell, eds.), pp. 455–460. Royal Society of New Zealand, Wellington.
21. Davies, W. J. (1978). Some effects of abscisic acid and water stress on stomata of *Vicia faba* L. *J. Exp. Bot.* **29,** 175–182.
22. Davies, W. J., and Kozlowski, T. T. (1974). Stomatal responses of five woody angiosperms to light intensity and humidity. *Can. J. Bot.* **52,** 1525–1534.
23. Dittrich, P., and Raschke, K. (1977). Malate metabolism in isolated epidermis of *Commelina communis* L. in relation to stomatal functioning. *Planta* **134,** 77–81.
24. Dittrich, P., and Raschke, K. (1977). Uptake and metabolism of carbohydrates by epidermal tissue. *Planta* **134,** 83–90.
25. Domes, W. (1971). Different CO_2-sensitivities of the gas exchange of the two leaf surfaces of *Zea mays. Planta* **89,** 47–55.
26. Dörffling, K., Streich, J., Kruse, W., and Muxfeldt, B. (1977). Abscisic acid and the after-effect of water stress on stomatal opening potential. *Z. Pflanzenphysiol.* **81,** 43–56.
27. Drake, B., and Raschke, K. (1974). Prechilling of *Xanthium strumarium* reduces net photosynthesis and, independently, stomatal conductance, while sensitizing the stomata to CO_2. *Plant Physiol.* **53,** 808–812.
28. Edwards, M., and Meidner, H. (1975). Micromanipulation of stomatal guard cells. *Nature (London)* **253,** 114–115.
29. Edwards, M., Meidner, H., and Sheriff, D. W. (1976). Direct measurements of turgor pressure potentials of guard cells. *J. Exp. Bot.* **27,** 163–171.

29a. Farquhar, G. D., and Sharkey, T. D. (1982). Stomatal conductance and photosynthesis. *Annu. Rev. Plant Physiol.* **33,** 317–345.

30. Fenton, R., Davies, W. J., and Mansfield, T. A. (1977). The role of farnesol as a regulator of stomatal opening in *Sorghum. J. Exp. Bot.* **105,** 1043–1053.
31. Fischer, R. A. (1968). Stomatal opening: Role of potassium uptake by guard cells. *Science* **160,** 784–785.
32. Fischer, R. A. (1968). Stomatal opening in isolated epidermal strips of *Vicia faba.* I. Response to light and CO_2-free air. *Plant Physiol.* **43,** 1947–1952.
33. Fischer, R. A., and Hsiao, T. C. (1968). Stomatal opening in isolated epidermal strips of *Vicia faba.* II. Response to KCl concentration and role of K^+ absorption. *Plant Physiol.* **43,** 1953–1958.
34. Fischer, R. A. (1970). After effects of water stress on stomatal opening potential: II. Possible causes. *J. Exp. Bot.* **21,** 386–404.
35. Fischer, R. A. (1971). Role of potassium in stomatal opening in the leaf of *Vicia faba. Plant Physiol.* **47,** 555–558.

36. Fujino, M. (1967). Role of adenosinetriphosphate and adenosinetriphosphatase in stomatal movement. *Sci. Bull. Fac. Educ., Nagasaki Univ.* **18,** 1–47.
37. Gaff, D. F., Chambers, J. C., and Markus, K. (1964). Studies of extrafascicular movement of water in the leaf. *Aust. J. Biol. Sci.* **17,** 581–586.
38. Galpin, M. F. J., Jennings, D. H., Oates, K., and Hobot, J. A. (1978). Localization by X-ray microanalysis of soluble ions, particularly potassium and sodium, in fungal hyphae. *Exp. Mycol.* **2,** 258–269.
39. Galston, A. W., and Satter, R. L. (1976). Light, clocks and ion flux: An analysis of leaf movement. *In* "Light and Plant Development" (H. Smith, ed.), pp. 159–184. Butterworth, London.
40. Georgiou, P., Richardson, C. H., Simmons, K., Truter, M. R., and Wingfield, J. N. (1982). Inhibition of stomatal opening by cyclic 'crown' polyethers in *Commelina communis,* and a correlation with lipophilicity and bonding. *Inorg. Chim. Acta* **66,** 1–6.
41. Glinka, Z. (1971). The effect of epidermal cell water potential on stomatal responses to illumination of leaf discs of *Vicia faba. Physiol. Plant.* **24,** 476–479.
42. Glover, J. (1959). The apparent behaviour of maize and sorghum stomata during and after drought. *J. Agric. Sci.* **53,** 412–416.
43. Habermann, H. M. (1973). Evidence for two photoreactions and possible involvement of phytochrome in light-dependent stomatal opening. *Plant Physiol.* **51,** 543–548.
44. Hall, A. E., and Kaufmann, M. R. (1975). Stomatal response to environment with *Sesamum indicum* L. *Plant Physiol.* **55,** 455–459.
45. Hall, A. E., and Kaufmann, M. R. (1975). Regulation of water transport in the soil-plant-atmosphere continuum. *Ecol. Stud.* **12,** 187–202.
46. Hall, T. A., Echlin, P., and Kaufmann, R., eds. (1974). "Microprobe Analysis as Applied to Cells and Tissues." Academic Press, New York.
47. Hartung, W., Heilmann, B., and Gimmler, H. (1981). Do chloroplasts play a role in abscisic acid synthesis? *Plant Sci. Lett.* **22,** 235–242.
48. Heath, O. V. S. (1938). An experimental investigation of the mechanism of stomatal movement, with some preliminary observations on the response of the guard cells to shock. *New Phytol.* **37,** 385–395.
49. Heath, O. V. S. (1948). Control of stomatal movement by a reduction in the normal carbon dioxide content of the air. *Nature (London)* **161,** 179–181.
50. Heath, O. V. S. (1952). The role of starch in the light response of stomata. Part II. *New Phytol.* **51,** 30–47.
51. Heath, O. V. S. (1959). The water relations of stomatal cells and the mechanism of stomatal movement. *In* "Plant Physiology: A Treatise" (F. C. Steward, ed.), Vol. 2, pp. 193–250. Academic Press, New York.
52. Heath, O. V. S. (1969). "The Physiological Aspects of Photosynthesis." Heinemann, London.
53. Heath, O. V. S., and Mansfield, T. A. (1962). A recording porometer with detachable cups operating on four separates leaves. *Proc. R. Soc. London, Ser. B* **156,** 1–13.
54. Heath, O. V. S., and Russell, J. (1954). An investigation of the light responses of wheat stomata with the attempted elimination of control by the mesophyll. Part I. Effects of light independent of carbon dioxide. *J. Exp. Bot.* **5,** 1–15.
55. Heller, F. O., and Kausch, W. (1971). Licht und ultraviolet-mikroskopische Beobachtungen an Schliesszellen nach ihrer Reaktion auf verschiedene Umweltbedingungen. *Ber. Dtsch. Bot. Ges.* **84,** 541–549.
56. Horton, R. F., and Moran, L. (1972). Abscisic acid inhibition of potassium influx into stomatal guard cells. *Z. Pflanzenphysiol.* **66,** 193–196.

57. Hsiao, T. C., Allaway, W. G., and Evans, L. T. (1973). Action spectra for guard cell Rb^+ uptake and stomatal opening in *Vicia faba. Plant Physiol.* **51**, 82–88.
58. Humbert, C. (1976). Recherches sur la différenciation et la cytophysiologie des stomates. Doctoral Thesis, University of Dijon.
59. Humbert, C., and Guyot, M. (1972). Modifications ultrastructurales des cellules stomatiques d'*Anemia rotundifolia* Schrad. *C.R. Hebd. Seances Acad. Sci.* **274**, 380–382.
60. Humbert, C., Louguet, P., and Guyot, M. (1975). Etude ultrastructurale comparée des cellules stomatiques de *Pelargonium* × *hortorum* en relation avec un état d'ouverture ou de fermeture des stomates physiologiquement défini. *C.R. Hebd. Seances Acad. Sci.* **280**, 1373–1375.
61. Humble, G. D., and Hsiao, T. C. (1969). Specific requirement of potassium for light-activated opening of stomata in epidermal strips. *Plant Physiol.* **44**, 230–234.
62. Humble, G. D., and Hsiao, T. C. (1970). Light-dependent influx and efflux of potassium of guard cells during stomatal opening and closing. *Plant Physiol.* **46**, 483–487.
63. Humble, G. D., and Raschke, K. (1971). Stomatal opening quantitatively related to potassium transport. *Plant Physiol.* **48**, 447–458.
64. Imamura, S. (1943). Untersuchungen über den mechanismus der turgorschwankung der spaltöffnungesschliesszellen. *Jpn. J. Bot.* **12**, 251–346.
65. Itai, C., and Meidner, H. (1978). Functional epidermal cells are necessary for abscisic acid effects on guard cells. *J. Exp. Bot.* **29**, 765–770.
66. Jacoby, B., and Laties, G. G. (1971). Bicarbonate fixation and malate compartmentation in relation to salt-induced stoichiometric synthesis of organic acid. *Plant Physiol.* **47**, 525–531.
67. Jarvis, P. G., James, G. B., and Landsberg, J. J. (1975). Coniferous forest. *In* "Vegetation and the Atmosphere" (J. L. Monteith, ed.), Vol. 2, pp. 171–240. Academic Press, London.
68. Jarvis, P. G., and Morison, J. I. L. (1981). The control of transpiration and photosynthesis by the stomata. *In* "Stomatal Physiology" (P. G. Jarvis and T. A. Mansfield, eds.), pp. 247–279. Cambridge Univ. Press, London and New York.
69. Jarvis, R. G., and Mansfield, T. A. (1980). Reduced stomatal responses to light, carbon dioxide and abscisic acid in the presence of sodium ions. *Plant, Cell Environ.* **3**, 279–283.
70. Jewer, P. C., and Incoll, L. D. (1980). Promotion of stomatal opening in the grass *Anthephora pubescens* Nees by a range of natural and synthetic cytokinins. *Planta* **150**, 218–221.
71. Jones, M. B., and Mansfield, T. A. (1975). Circadian rhythms in plants. *Sci. Prog. (Oxford)* **62**, 103–125.
72. Jones, R. J., and Mansfield, T. A. (1970). Suppression of stomatal opening in leaves treated with abscisic acid. *J. Exp. Bot.* **21**, 714–719.
73. Jones, R. J., and Mansfield, T. A. (1972). Effects of abscisic acid and its esters on stomatal aperture and the transpiration ratio. *Physiol. Plant.* **26**, 321–327.
74. Kaufmann, M. R. (1976). Stomatal responses of Engelmann spruce to humidity, light and water stress. *Plant Physiol.* **57**, 898–901.
75. Kriedemann, P. E., Loveys, B. R., Fuller, G. L., and Leopold, A. C. (1972). Abscisic acid and stomatal regulation. *Plant Physiol.* **49**, 842–847.
76. Kuiper, P. J. C. (1964). Dependence upon wavelength of stomatal movement in epidermal tissue of Senecio odoris. *Plant Physiol.* **39**, 952–955.
77. Lange, O. L., Lösch, R., Schulze, E.-D., and Kappen, L. (1971). Responses of stomata to changes in humidity. *Planta* **100**, 76–86.

78. Levy, Y., and Kaufmann, M. R. (1976). Cycling of leaf conductance in citrus exposed to natural and controlled environments. *Can. J. Bot.* **54,** 2215–2218.
79. Little, C. H. A., and Eidt, D. C. (1968). Effects of abscisic acid on budbreak and transpiration in woody seedlings. *Nature (London)* **220,** 498–499.
80. Lloyd, F. E. (1908). The behaviour of stomata. *Carnegie Inst. Washington Publ.* **82,** 1–142.
81. Lösch, R. (1977). Responses of stomata to environmental factors—experiments with isolated epidermal strips of *Polypodium vulgare.* I. Temperature and humidity. *Oecologia* **29,** 85–97.
82. Lösch, R., and Schenk, B. (1978). Humidity responses of stomata and the potassium content of guard cells. *J. Exp. Bot.* **29,** 781–787.
83. Loveys, B. R. (1977). The intracellular location of abscisic acid in stressed and non-stressed leaf tissue. *Physiol. Plant.* **40,** 6–10.
84. Ludlow, M. M., and Jarvis, P. G. (1971). Photosynthesis in Sitka spruce. I. General characteristics. *J. Appl. Ecol.* **8,** 925–953.
85. Macallum, A. B. (1905). On the distribution of potassium in animal and vegetable cells. *J. Physiol. (London)* **32,** 95–128.
86. MacRobbie, E. A. C. (1977). Functions of ion transport in plant cells and tissues. *In* "Plant Biochemistry II" (D. H. Northcote, ed.), Vol. 13, pp. 211–247. University Park Press, Baltimore, Maryland.
87. MacRobbie, E. A. C. (1981). Effects of ABA on 'isolated' guard cells of *Commelina communis* L. *J. Exp. Bot.* **32,** 563–572.
88. MacRobbie, E. A. C., and Lettau, J. (1980). Potassium content and aperture in "intact" stomatal and epidermal cells of *Commelina communis* L. *J. Membr. Biol.* **56,** 249–256.
89. Maercker, U. (1965). Zur Kentniss der Transpiration der Schliesszellen. *Protoplasma* **60,** 61–78.
90. Maier-Maercker, U. (1979). Peristomatal transpiration and stomatal movement: A controversial view. I. Additional proof of peristomatal transpiration by hygrophotography and a comprehensive discussion in the light of recent results. *Z. Pflanzenphysiol.* **91,** 25–43.
91. Majernik, O. (1970). Responses of stomata of barley and maize to phenylmercuric acetate. *Biol. Plant.* **12,** 368–372.
92. Mansfield, T. A. (1964). A stomatal light reaction sensitive to wavelengths in the region of 700 mμ. *Nature (London)* **201,** 470–472.
93. Mansfield, T. A. (1965). The low intensity light reaction of stomata-effects of red light on rhythmic stomatal behaviour in *Xanthium pennsylvanicum. Proc. R. Soc. London, Ser. B* **162,** 567–574.
94. Mansfield, T. A. (1976). Chemical control of stomatal movements. *Philos. Trans. R. Soc. London, Ser. B* **273,** 541–550.
95. Mansfield, T. A. (1976). Mechanisms involved in turgor changes of guard cells. *Perspect. Exp. Biol.* **2,** 453–462.
96. Mansfield, T. A. (1976). Delay in the response of stomata to abscisic acid in CO_2-free air. *J. Exp. Bot.* **27,** 559–564.
96a. Mansfield, T. A. (1983). Movements of stomata. *Sci. Prog. (Oxford)* **68,** 519–542.
97. Mansfield, T. A., and Heath, O. V. S. (1963). Studies in stomatal behaviour. IX. Photoperiodic effects on rhythmic phenomena in *Xanthium pennsylvanicum. J. Exp. Bot.* **14,** 334–352.
98. Mansfield, T. A., and Jones, R. J. (1971). Effects of abscisic acid and starch content of stomatal guard cells. *Planta* **101,** 147–158.
99. Mansfield, T. A., and Meidner, H. (1966). Stomatal opening in light of different

wavelengths: Effects of blue light independent of carbon dioxide concentration. *J. Exp. Bot.* **17,** 510–521.

100. Mansfield, T. A., Wellburn, A. R., and Moreira, T. J. S. (1978). The role of abscisic acid and farnesol in the alleviation of water stress. *Philos. Trans. R. Soc. London, Ser. B* **284,** 471–482.
101. Marrè, E. (1979). Fusicoccin: A tool in plant physiology. *Annu. Rev. Plant Physiol.* **30,** 273–288.
102. Martin, E. S., and Meidner, H. (1971). Endogenous stomatal movements in *Tradescantia virginiana. New Phytol.* **70,** 923–928.
103. Martin, E. S., and Meidner, H. (1972). The phase-response of the dark stomatal rhythm in *Tradescantia virginiana* to light and dark treatments. *New Phytol.* **71,** 1045–1054.
104. Martin, E. S., and Meidner, H. (1975). The influence of night length on stomatal behaviour in *Tradescantia virginiana. New Phytol.* **75,** 507–511.
105. Meidner, H. (1968). The comparative effects of blue and red light on the stomata of *Allium cepa* L. and *Xanthium pennsylvanicum. J. Exp. Bot.* **19,** 146–151.
106. Meidner, H. (1975). Water supply, evaporation, and vapour diffusion in leaves. *J. Exp. Bot.* **26,** 666–673.
107. Meidner, H. (1976). Vapour loss through stomatal pores with the mesophyll tissue excluded. *J. Exp. Bot.* **27,** 172–174.
108. Meidner, H., and Edwards, M. (1975). Direct measurements of turgor pressure potentials of guard cells. I. *J. Exp. Bot.* **26,** 319–330.
109. Meidner, H., and Heath, O. V. S. (1959). Stomatal responses to temperature and carbon dioxide concentration in *Allium cepa* L. and their relevance to midday closure. *J. Exp. Bot.* **10,** 206–219.
110. Meidner, H., and Mansfield, T. A. (1968). "Physiology of Stomata." McGraw-Hill, London.
111. Milborrow, B. V. (1979). Antitranspirants and the regulation of abscisic acid content. *Aust. J. Plant Physiol.* **6,** 249–254.
112. Milborrow, B. V. (1981). Abscisic acid and other hormones. *In* "The Physiology and Biochemistry of Drought Resistance in Plants" (L. G. Paleg and D. Aspinall, eds.), pp. 347–388. Academic Press, Sydney.
113. Mizrahi, Y., Scherings, S. G., Malis Arad, S., and Richmond, A. E. (1974). Aspects of the effect of ABA on the water status of barley and wheat seedlings. *Physiol. Plant.* **31,** 44–50.
113a. Morison, J. I. L., and Jarvis, P. G. (1983). Direct and indirect effects of light on stomata. II. In *Comme-lina communis* L. *Plant, Cell Environ.* **6,** 103–109.
114. Mouravieff, I. (1956). Action du gaz carbonique et de la lumière sur l'appareil stomatique isolé. *C.R. Hebd. Seances Acad. Sci.* **242,** 926–927.
115. Neales, T. F. (1970). Effect of ambient carbon dioxide concentration on the rate of transpiration of *Agave americana* in the dark. *Nature (London)* **228,** 880–882.
116. Neilson, R. E., and Jarvis, P. G. (1975). Photosynthesis in Sitka spruce. VI. Response of stomata to temperature. *J. Appl. Ecol.* **12,** 879–891.
117. Nelson, S. D., and Mayo, J. M. (1975). The occurrence of functional nonchlorophyllous guard cells in *Paphiopedilium* spp. *Can. J. Bot.* **53,** 1–7.
117a. Ogawa, T., Grantz, D., Boyer, J., and Govindjee (1982). Effects of cations and abscisic acid on chlorophyll *a* fluorescence in guard cells of *Vicia faba. Plant Physiol.* **69,** 1140–1144.
118. Ogunkanmi, A. B., Wellburn, A. R., and Mansfield, T. A. (1974). Detection and preliminary identification of endogenous antitranspirants in water-stressed *Sorghum* plants. *Planta* **117,** 293–302.

119. Orton, P. J., and Mansfield, T. A. (1974). The activity of abscisic acid analogues as inhibitors of stomatal opening. *Planta* **121,** 263–272.

120. Orton, P. J., and Mansfield, T. A. (1976). Studies of the mechanism by which Daminozide (B9) inhibits stomatal opening. *J. Exp. Bot.* **27,** 125–133.

121. Outlaw, W. H., Jr., and Lowry, O. H. (1977). Organic acid and potassium accumulation in guard cells during stomatal opening. *Proc. Natl. Acad. Sci. U.S.A.* **74,** 4434–4438.

122. Outlaw, W. H., Jr., and Manchester, J. (1979). Guard cell starch concentration quantitatively related to stomatal aperture. *Plant Physiol.* **64,** 79–82.

122a. Outlaw, W. H., Jr., Mayne, B. C., Zenger, V. E., and Manchester, J. (1981). Presence of both photosystems in guard cells of *Vicia faba* L.: Implications for environmental processing. *Plant Physiol.* **67,** 12–16.

123. Pallaghy, C. K., and Fischer, R. A. (1974). Metabolic aspects of stomatal opening and ion accumulation by guard cells in *Vicia faba. Z. Pflanzenphysiol.* **71,** 332–344.

124. Pallas, J. E., and Wright, B. G. (1973). Organic acid changes in the epidermis of *Vicia faba* and their implication in stomatal movement. *Plant Physiol.* **51,** 588–590.

125. Pemadasa, M. A. (1982). Differential abaxial and adaxial stomatal responses to indole-3-acetic acid in *Commelina communis* L. *New Phytol.* **90,** 209–219.

126. Penny, M. G., and Bowling, D. J. F. (1974). A study of potassium gradients in the epidermis of intact leaves of *Commelina communis* L. in relation to stomatal opening. *Plant* **119,** 17–25.

127. Penny, M. G., and Bowling, D. J. F. (1975). Direct determination of pH in the stomatal complex of *Commelina. Planta* **122,** 209–212.

128. Penny, M. G., Kelday, L. S., and Bowling, D. J. F. (1976). Active chloride transport in the leaf epidermis of *Commelina communis* in relation to stomatal activity. *Planta* **130,** 291–294.

129. Raghavendra, A. S., and Das, V. S. R. (1972). Control of stomatal opening by cyclic photophosphorylation. *Curr. Sci.* **41,** 150–151.

130. Raschke, K. (1965). Die Stomata als Glieder eines schwingungsfähigen CO_2-Regelsystems. *Z. Naturforsch., B: Anorg. Chem., Org. Chem., Biochem., Biophys., Biol.* **20B,** 1261–1270.

131. Raschke, K. (1967). Der Einfluss von Rot und Blau Licht auf die Öffnungs und Schliess Geschwindigkeit der Stomata von *Zea mays. Naturwissenschaften* **54,** 73.

132. Raschke, K. (1972). Saturation kinetics of the velocity of stomatal closing in response to CO_2. *Plant Physiol.* **49,** 229–234.

133. Raschke, K. (1975). Simultaneous requirement of carbon dioxide and abscisic acid for stomatal closing in *Xanthium strumarium* L. *Planta* **125,** 243–259.

134. Raschke, K. (1975). Stomatal action. *Annu. Rev. Plant Physiol.* **26,** 309–340.

135. Raschke, K., and Fellows, M. P. (1971). Stomatal movement in *Zea mays:* Shuttle of potassium and chloride between guard and subsidiary cells. *Planta* **101,** 296–316.

136. Raschke, K., and Dittrich, P. (1977). (^{14}C) Carbon dioxide fixation by isolated leaf epidermes with stomata closed or open. *Planta* **134,** 69–75.

137. Raschke, K., and Humble, G. D. (1973). No uptake of anions required by opening stomata of *Vicia faba:* Guard cells release hydrogen ions. *Planta* **115,** 47–57.

138. Raven, J. A. (1969). Effects of inhibitors on photosynthesis and the active influxes of K and Cl in *Hydrodictyon africanum. New Phytol.* **68,** 1089–1113.

139. Richardson, C. H., Truter, M. R., Wingfield, J. N., Travis, A. J., Mansfield, T. A., and Jarvis, R. G. (1979). The effect of benzo-18-crown-6, a synthetic ionophore, on stomatal opening and its interaction with abscisic acid. *Plant, Cell Environ.* **2,** 325–327.

140. Sauberman, A. J., and Echlin, P. (1975). The preparation, examination and analysis of

frozen hydrated sections by scanning transmission electron microscopy and X-ray microanalysis. *J. Microsc. (Oxford)* **105,** 155.

141. Sayre, J. D. (1926). Physiology of stomata of *Rumex patientia. Ohio J. Sci.* **26,** 233–266.
142. Scarth, G. W., and Shaw, M. (1951). Stomatal movement and photosynthesis in *Pelargonium.* I. Effects of light and CO_2. *Plant Physiol.* **26,** 207–225.
143. Schnabl, H., and Ziegler, H. (1977). The mechanism of stomatal movement in *Allium cepa* L. *Planta* **136,** 37–43.
144. Schülze, E.-D., Lange, O. L., Buschbom, U., Kappen, L., and Evanari, M. (1972). Stomatal responses to changes in humidity in plants growing in the desert. *Planta* **108,** 259–270.
145. Schwendener, S. (1881). Über Bau and Mechanik der Spaltöffnungen. *Monatsber. K. Akad. Wiss. Berlin* **46,** 833–867.
146. Seybold, A. (1961). Ergebnisse und Probleme pflanzlicher Transpiration-sanalysen. *Jahresh. Heidelb. Akad. Wiss.* **6,** 5–8.
147. Sheriff, D. W. (1977). Evaporation sites and distillation in leaves. *Ann. Bot. (London)* [N.S.] **41,** 1081–1082.
148. Sheriff, D. W. (1977). Where is humidity sensed when stomata respond to it directly? *Ann. Bot. (London)* [N.S.] **41,** 1083–1084.
149. Sheriff, D. W., and Meidner, H. (1975). Water movement into and through *Tradescantia virginiana* leaves. *J. Exp. Bot.* **26,** 897–902.
150. Smith, H. (1975). "Phytochrome and photomorphogenesis." McGraw-Hill, London.
152. Snaith, P. J., and Mansfield, T. A. (1982). Control of the CO_2 responses of stomata by indol-3ylacetic acid and abscisic acid. *J. Exp. Bot.* **33,** 360–365.
152. Squire, G. R., and Jones, M. B. (1971). Studies on the mechanism of action of the antitranspirant, phenylmercuric acetate and its penetration into the mesophyll. *J. Exp. Bot.* **22,** 980–991.
153. Squire, G. R., and Mansfield, T. A. (1972). A simple method of isolating stomata on detached epidermis by low pH treatment: Observations of the importance of the subsidiary cells. *New Phytol.* **71,** 1033–1043.
154. Stevens, R. A., and Martin, E. S. (1977). New structure associated with stomatal complex of the fern *Polypodium vulgare. Nature (London)* **265,** 331–334.
155. Stevens, R. A., and Martin, E. S. (1977). Ion absorbent substomatal structures in *Tradescantia pallidus. Nature (London)* **268,** 364–365.
156. Tal, M., and Imber, D. (1970). Abnormal stomatal behaviour and hormonal imbalance of *flacca,* a wilty mutant of tomato. II. Auxin and abscisic acid-like activity. *Plant Physiol.* **46,** 373–376.
157. Tal, M., and Nevo, Y. (1973). Abnormal stomatal behaviour and root resistance, and hormonal imbalance in three wilty mutants of tomato. *Biochem. Genet.* **8,** 291–300.
158. Tanton, T. W., and Crowdy, S. H. (1972). Water pathways in higher plants. III. The transpiration stream within leaves. *J. Exp. Bot.* **23,** 619–625.
159. Travis, A. J., and Mansfield, T. A. (1977). Studies of malate formation in isolated guard cells. *New Phytol.* **78,** 541–546.
160. Travis, A. J., and Mansfield, T. A. (1979). Reversal of the CO_2 responses of stomata by fusicoccin. *New Phytol.* **83,** 607–614.

160a. Trewavas, A.-J. (1981). How do plant growth substances work? *Plant, Cell Environ.* **4,** 203–228.

161. Van Kirk, C. A., and Raschke, K. (1978). Presence of chloride reduces malate production in epidermis during stomatal opening. *Plant Physiol.* **61,** 361–364.
162. Wellburn, A. R., and Hampp, R. (1976). Fluxes of gibberellic and abscisic acids,

together with that of adenosine 3′, 5′-cyclic phosphate, across plastid envelopes during development. *Planta* **131**, 95–96.

163. Wellburn, A. R., Majernik, O., and Wellburn, F. A. M. (1972). Effects of SO_2 and NO_2 polluted air on the ultrastructure of chloroplasts. *Environ. Pollut.* **3**, 37–49.
164. Wilkins, M. B. (1969). Circadian rhythms in plants. *In* "Physiology of Plant Growth and Development" (M. B. Wilkins, ed.), pp. 647–671. McGraw-Hill, London.
165. Willmer, C. M. (1976). Some observations and interpretations on the vital staining of leaf epidermal tissue by neutral red. *Protoplasma* **87**, 253–262.
166. Willmer, C. M., and Mansfield, T. A. (1969). A critical examination of the use of detached epidermis in studies of stomatal physiology. *New Phytol.* **68**, 363–375.
167. Willmer, C. M., and Mansfield, T. A. (1970). Effects of some metabolic inhibitors and temperature on ion-stimulated stomatal opening in detached epidermis. *New Phytol.* **69**, 983–992.
168. Willmer, C. M., and Pallas, J. E. (1973). A survey of stomatal movements and associated potassium fluxes in the plant kingdom. *Can. J. Bot.* **51**, 37–42.
169. Willmer, C. M., Kanai, R., Pallas, J. E., and Black, C. C. (1973). Detection of high levels of phosphoenolpyruvate carboxylase in leaf epidermal tissue and its significance in stomatal movements. *Life Sci.* **12**, 151–155.
170. Willmer, C. M., and Rutter, J. C. (1977). Guard cell malic acid metabolism during stomatal movements. *Nature (London)* **269**, 327–328.
171. Wright, S. T. C. (1969). An increase in the 'inhibitor-β' content of detached wheat leaves following a period of wilting. *Planta* **86**, 10–20.
172. Wright, S. T. C. (1977). The relationship between leaf water potential and the levels of abscisic acid and ethylene in excised wheat leaves. *Planta* **134**, 183–189.
173. Wright, S. T. C. (1978). Phytohormones and stress phenomena. *In* "Phytohormones and Related Compounds—A Comprehensive Treatise" (D. S. Letham, P. B. Goodwin, and T. J. V. Higgins, eds.), Vol. 2, pp. 495–536. Elsevier North-Holland, Amsterdam.
174. Wright, S. T. C., and Hiron, R. W. P. (1969). (+) – abscisic acid, the growth inhibitor in detached wheat leaves following a period of wilting. *Nature (London)* **224**, 719–720.
175. Yamashita, T. (1952). Influences of potassium supply upon various properties and movement of the guard cell. *Sieboldia* **1**, 51–70.
176. Zabadal, T. J. (1974). A water potential threshold for the increase of abscisic acid in leaves. *Plant Physiol.* **53**, 125–227.
176a. Zeiger, E. (1983). The biology of stomatal guard cells. *Annu. Rev. Plant Physiol.* **34**, 441–475.
176b. Zeiger, E., Armond, P., and Melis, A. (1981). Fluorescence properties of guard cell chloroplasts. *Plant Physiol.* **67**, 17–20.
177. Zeiger, E., Bloom, A. J., and Hepler, P. K. (1978). Ion transport in stomatal guard cells: A chemiosmotic hypothesis. *What's New Plant Physiol.* **9**, 29–32.
178. Zeiger, E., and Hepler, P. K. (1977). Light and stomatal function: Blue light stimulates swelling of guard cell protoplasts. *Science* **196**, 887–888.
179. Ziegenspeck, A. (1955). Das Vorkommen von Fila in radialer Anordnung in den Schliesszellen. *Protoplasma* **44**, 385–388.
180. Ziegler, H. (1967). Wasserumsatz und Stoffbewegungen. *Fortschr. Bot.* **29**, 68–77.

CHAPTER FOUR

Salt Relations of Cells, Tissues, and Roots

D. H. JENNINGS

Plant Physiology
A Treatise
Vol. IX: Water and Solutes in Plants

I. Introduction

This chapter builds on that by Steward and Sutcliffe in Volume II of this treatise (518), which was an important landmark in the literature of plant salt relations because it was the first overview of the subject since Hoagland's important book "Lectures on the Inorganic Nutrition in Plants," published in 1944 by Chronica Botanica. The time span between the two publications does not accurately indicate the extent of the advance of the subject since Hoagland's book relates to knowledge gathered by plant physiologists before the outbreak of World War II in 1939, after which work in laboratories came to a virtual halt until hostilities ceased. Steward and Sutcliffe summarized the knowledge then available and presented this in a valuable historical perspective. Equally important, they provided guides to future work.

Certain topics which emerged as particularly amenable to future experimental attack included these: (1) With the presence of carriers established, individual molecules might be identified and their mode of action elucidated. (2) The relationship between metabolism and transport ought to be capable of a more exact biochemical description both in nongrowing and growing systems. (3) With the advent of a clearer understanding of the subcellular architecture of cells, there ought to be a clearer idea of the location of specific carriers within cells, particularly in relation to movement of ions into the vacuole. (4) The concept of "free space" could be defined more precisely. (5) Transport of ions through the symplast required further investigation as did the release of ions from the symplast into the stele of a root.

These themes have been taken up in the present chapter. Recent publications by Baker and Hall (24) and Lüttge and Pitman (303) have dealt with the ion relations of particular plants, their cells, tissues, and organs. Reference should be made to these volumes and to that by Lüttge and Higinbotham (302) for further information about topics that are also discussed here. The three volumes, especially that edited by Lüttge and Pitman, are valuable sources of data of a comparative nature.

II. Practical Aspects

A. Systems Used for Experimental Investigations

Steward and Sutcliffe (518) described in considerable detail the various systems whose salt relations had received detailed study at that time. Attention will be restricted here mainly to new developments since 1959.

1. *Sterility of Material*

There has been some concern about the extent to which some studies on ion uptake by higher plant storage tissue discs and excised roots might be affected by microbial contamination.[1] With storage tissue discs, if no special precautions are taken to ensure sterility, as many as 10^7 to 10^9 bacteria per gram of tissue may be present after the preliminary washing of discs before they are placed in the experimental solution (396). This may affect the rate of development of ion absorptive capacity when tissue is aged in a limited volume of solution (305). But it is clear from experiments carried out under asceptic conditions (21, 117, 121) that the increase in respiratory activity, the changes in the activity of invertase, and the rate of protein synthesis which takes place when slices of storage tissue are washed are not due to the activities of microorganisms. Further, Palmer (378) has shown that the development of an increased ability of red beet discs to absorb phosphate (six- to ninefold increase) from 10^{-6} mol L^{-1} phosphate is barely affected by the presence of bacteria. However, the washing period was for only 3 h with hourly changes of distilled water. Results from experiments involving longer washing periods with less frequent changes need always be viewed with some degree of caution.

Virtually all experiments with excised roots have involved the use of nonsterile material. D. A. Barber (27) has reviewed the literature on the influence of microorganisms on ion uptake by roots up to 1967. Two types of experiments have been carried out. Those exemplified by Bowen and Rovira (48) have involved addition of an inoculum of known microorganisms, usually from the rhizosphere, to sterile roots and a comparison is made between the ability of such roots and that of sterile controls to absorb ions. The other type of experiment used, for example, by Barber and Loughman (29), has been a comparison between sterile roots and roots contaminated accidentally with the microflora occurring in a laboratory. The former studies are of great importance with respect to our understanding of how roots behave in the soil, but the latter are of more concern here. They provide valuable information about the extent to which results of experiments with nonsterile material—and these have provided the bulk of the data concerned with ion transport into roots—can be related to the properties of the root cells themselves rather than to the accompanying microorganisms.

Some of the findings with barley roots which are of particular interest

[1] The conditions found to be most conducive to accumulation of inorganic ions by discs of potato tuber were not affected by this problem. These conditions were the use of very thin ($\simeq$0.6 mm) discs in a large volume of vigorously aerated dilute, single-salt solutions that remained clear throughout the experiments. The potato discs activated under these conditions increased in fresh weight, showed a minimum of phenolase activity, and actually resisted direct inoculation with active bacterial *(Erwinia carotovora)* cultures (516).

are the following: (1) At a low concentration of phosphate (3.1×10^{-5} mol L^{-1}) labeled with ^{32}P, there is a fourfold increase in the incorporation of radioactivity into nucleic acids, phospholipid, and phosphoprotein in unsterile compared with sterile roots (29). (2) The uptake of manganese is stimulated by microorganisms irrespective of whether they originated by casual laboratory contamination or from the rhizosphere microorganisms in the soil (28). (3) The breakdown of EDTA chelates occurs at an accelerated rate with nonsterile roots when compared with sterile roots (28).

2. *Leaves*

Submerged leaves of hydrophytes have been favored over aerial leaves for salt uptake studies because of the absence of cuticle and the fact that whole leaves or parts of them can be immersed in experimental solutions under conditions that more nearly correspond to those occurring in nature. Chapter 4 in Volume II of this treatise summarizes the work prior to 1959. In recent years, more attention has been paid to aerial leaves. Research on ion uptake by the cells of such leaves is necessary for a better understanding of their role as sinks for ions coming up the transpiration stream and as sources of ions moving to other parts of the plant via the phloem (see Chapter 6 of this volume). Although ion uptake through the cuticle of leaves may be of some importance in the nutrition of a plant (see 139, 545, 576), the rate is too slow to allow whole leaves or indeed parts of leaves, where cuticle forms a substantial portion of the area exposed to a solution, to be used for experimental purposes. As an alternative, leaf slices have to be used in which ions diffuse into the tissues via the cut surfaces. The rate of uptake is related to the length of the diffusion path. This and the proportion of damaged cells in the slice must be considered when deciding upon the optimal size of the slice. This consideration is pertinent not only to movement of ions into the interior of the slice, but also of carbon dioxide for photosynthesis. Jeschke (233) has pointed out that it is also necessary to consider any limitation brought about by the diffusion of hydroxyl ions out of the leaf slice when it is bathed in bicarbonate buffer.

Jones and Osmond (238) have shown that the rate of photosynthesis of 400- to 500-μm thick slices of leaves of cotton *(Gossypium hirsutum), Sorghum sudanense, Atriplex hastata,* and *Atriplex spongiosa* in solution was comparable to that of whole leaves. Jones and Osmond emphasized that to avoid damaging the tissue excessively care was needed in the preparation of slices, it being necessary to renew the cutting edge in their microtome after every 20 cuts. With respect to ion transport, MacDonald and Macklon (307) found that in the range 1.0 – 10.0 mmol L^{-1} potassium chloride, maximum uptake (μmol 24 h^{-1} 100-mm slice^{-1}) was obtained with 1.5-

mm-long segments of wheat *(Triticum aestivum)* leaves. Above 1.5 mm, ion diffusion through the cut edge was rate limiting, uptake being proportional to the amount of the cut edge; infiltration of the tissue with distilled water before suspension in the salt solution removed this limitation. It should be noted that with both infiltrated and sliced tissue, there was a considerable reduction in the rate of photosynthesis compared with that of large slices of otherwise untreated lamina (306).

MacDonald and Macklon (308) argue that stomata could be a route by which ions in solution penetrate leaf tissue. Jacoby *et al.* (213) from their work on bean *(Phaseolus vulgaris)* leaf slices have argued to the contrary. This needs resolving, if possible, by direct observations on the stomatal aperture and by examination as to whether the substomatal cavity becomes injected with solution. However, a difficulty is to decide whether an effect of some factor, e.g., light, in stimulating ion uptake is due to an influence on uptake per se or is a result of increased penetration of the tissue by the ion due to an effect on the stomata.

The enzymatic separation of leaf cells has also been attempted (240), but the procedure appears to damage the plasmalemma (214). On the other hand, use of leaf discs without an epidermal layer (previously stripped off the intact leaf) (350) needs to be explored further.

3. Filamentous Fungi

Filamentous fungi can often be grown readily on solid media or in stirred liquid culture. They possess the virtue of being susceptible to genetic analysis and an impressive assemblage of transport mutants in filamentous fungi is now available (488; this review also contains information about transport mutants in *Saccharomyces cerevisiae,* bacteria, and other organisms). It has been possible to insert electrodes into some of the large hyphae (Section IV,A,2,a) to measure the potential across the plasmalemma and conductance of the membrane. However, an investigator using filamentous fungi should be aware that they are not as robust as higher plant cells with respect to any dramatic lowering in the concentration (increase in the osmotic potential, see Chapter 1) of the medium. Washing mycelium free of growth medium, particularly with distilled water, prior to experimental treatments, can lead to hyphal bursting (227, 534). Also, it should not be assumed that a fungal mycelium is homogenous with respect to its ion relations. Sodium–potassium selectivity, for instance, can certainly change with the age of the culture (225) and along a hypha (140).

4. Cultured Tissues of Higher Plants

In their chapter, (Chapter 4, Vol. II), Steward and Sutcliffe pointed out the potential usefulness of cultured cells for ion transport studies, especially when examining the relationships between salt uptake and growth. Advantages of this material include the ease with which the cultures can be handled and growth controlled by altering the composition of the bathing medium. By selecting particular clones of cells a high degree of uniformity can be achieved.

Steward and Sutcliffe described some of the early observations made in Steward's laboratory on cesium accumulation in carrot root tissue cultures in which cells were either allowed to grow mainly by cell expansion or stimulated to divide rapidly by provision of coconut milk in the nutrient medium. This work has been extended since, e.g., by Sutcliffe and Counter (526), Smithers and Sutcliffe (498, 499), and Steward and Mott (517). An account of the contribution that this work has made to our understanding of the salt relations of plants is presented in Chapter 7, this volume.

Suspensions of cultured cells clearly provide the most suitable material technically for ion transport studies. Given that the cells are of uniform size, they are amenable to flux measurements (37), although their changing size with growth and cell division could confound these measurements. Problems may also arise when cells are transferred from the growth to the experimental medium. Dorree *et al.* (101) and Thoiron *et al.* (531) report a "shock" effect caused by such a transference of *Acer pseudoplatanus* cells. The shock effect is manifested in marked changes of cell permeability and appears to be associated with the dramatic lowering of carbon dioxide concentration in the medium brought about by opening the culture vessels and separating the cells from the growth medium. The cells can recover inasmuch as the rates of solute uptake become comparable to the rates in nonshocked cells; protein synthesis appears to be necessary for this recovery (532, 533).

One must anticipate callus tissue to be much more intractable than cell suspension for precise measurements of membrane fluxes. The cells are unlikely to be uniform in size and physiological activity due to differentiation and cell turnover.

5. *Membrane Vesicles*

The studies of Kaback (241) and his group on sugar transport into membrane vesicles formed from the boundary membrane of *Escherichia coli* have demonstrated the value of this approach. Christensen and Cirillo (81) and Scarborough (456) have produced plasma membrane vesicles from *Saccharomyces cerevisiae* and *Neurospora crassa,* respectively. The isolation of vesicles from the former organism is of particular interest since they were obtained from normal walled cells; those from *N. crassa* were obtained from the wall-less (s 1) mutant. Once isolated (in a medium of suitable tonicity to prevent bursting) the vesicles can be treated like a suspension of unicells and their ability to absorb solutes investigated (457). Plasma membrane vesicles have been produced from higher plants (203) but have not been used in this way. There is a possibility that the plasmalemma may not be the only membrane system contributing to vesicle production.

The heterogeneity of plasmalemma vesicle preparations could be a barrier to their use in transport studies. Hodges and colleagues (201–203) use the following criteria to support their belief that they have isolated plasma membranes: (1) presence of a glucan synthase (glycosyltransferase, EC 2.4.1.12, cellulose synthase) in the preparation, (2) a high sterol to phospholipid ratio, (3) staining with periodic acid, chromic acid, and phosphotungstic acid (PTAC) that is supposed to be specific for plant plasma membranes.

The actual procedures used are described by Hodges and Leonard (202). They used the following markers for other membrane systems: cytochrome-*c* oxidase for mitochondrial membranes, inosine diphosphatase (IDPase) for Golgi apparatus, NADPH-cytochrome-*c* reductase for endoplasmic reticulum, and NADH-cytochrome-*c* reductase for tonoplasts. Glucose-6-phosphatase and 5′-nucleotidase, which are classic markers for endoplasmic reticulum and plasma membrane, respectively, of animal cells, could not be used because of the low activity of these enzymes in the preparations.

There is doubt about the criteria used by Hodges and colleagues. Thus Hall and Flowers (167) have shown both that IDPase can be associated with the plasmalemma and that the plasmalemma is not always stained by PTAC. Furthermore, the Golgi apparatus itself does not always show IDPase activity (97), and it can show glucan synthase activity of the type used as a marker by Hodges and colleagues (429). While it is true that current evidence indicates that cellulose synthesis does not occur to any significant extent in the Golgi apparatus, there is no doubt that it produces β-1,4-glucan from UDPglucose or glucose diphosphate glucose (428). Therefore, unless it can be proved that cellulose is being produced, the ability of a membrane preparation to synthesize β-1,4-glucan does not seem to be adequate evidence that it consists of plasmalemma.

Though a high sterol-to-phospholipid ratio may be a suitable criterion for the presence of plasmalemma of a eukaryote (389), this in itself is an insufficient "marker" for the membrane. More recent studies (388) suggest that cellulase is a more promising candidate. There are two other possibilities: either a specific inhibitor of a transport enzyme, e.g., ATPase, is found, or the plasmalemma is labeled before cell disruption. The use of ouabain in the isolation of the animal Na^+, K^+-ATPase which is located in the plasmalemma is an example of the former approach (Section III,E). The plasmalemma of an intact cell can be labeled by exposing it either to a radioactive compound which binds to the membrane but does not penetrate the cell, e.g., diazotized sulfanilic acid, or to a visual marker, e.g., concanavalin A (455). For these latter procedures it may be necessary to prepare protoplasts first so that the plasmalemma can be more readily labeled. Techniques are now available for producing protoplasts for ion transport studies (281, 348, 349, 459, 529). Perlin and Spanswick (385) have produced labeled (diazotized [^{125}I]iodosulfanilic acid) plasma membranes from protoplasts from corn *(Zea mays)* leaves.

6. *Isolated Vacuoles*

Little is yet known about the properties of isolated vacuoles, but techniques are now available for their routine isolation. Two procedures have been used: Wagner and Siegelman (558) isolated vacuoles from the petals and leaves of *Tulipa* (cultivar Red Shine) and *Hippeastrum* (cultivar Red Christmas Amaryllis). The procedure was to take strips of tissue and dissolve the cell walls enzymatically in the presence of 0.6 mol L^{-1} mannitol. Vacuoles were produced by suspending protoplasts for a very short period of time (15 s) in 0.2 mol L^{-1} phosphate at pH 8.0 containing dithiothreitol. Some initial studies on the properties of these isolated vacuoles have been reported by Lin *et al.* (281, 282), the most important finding being the presence of a membrane-bound ATPase. A similar procedure has been used to isolate vacuoles from *Kalanchoe daigremontiana* leaf cells (75).

The other procedure was devised by Leigh and Branton (273) for the isolation of vacuoles from red beet hypocotyl tissue, which is tougher than, for example, petals and leaves and therefore more refractory to enzyme treatment. Their technique was based on the observations that small numbers of vacuoles can be prepared by slicing tissue into a suitable osmoticum on a microscope slide (393). Essentially, Leigh and Branton devised an apparatus in which larger pieces of tissue can be cut continuously in the presence of the osmoticum. Gross cytoplasmic contamination can be detected by the use of the vital stain fluorescein diacetate, which is enzymically broken down in the cytoplasmic contamination to yield fluorescein, whose location can be detected fluormicroscopically (5). It has been shown that red beet vacuole preparations contain a membrane-bound ATPase (274).

Ion absorption studies on vacuoles obtained with either procedure have yet to be reported. However, there are some results from studies on the small vacuoles in *Hevea brasiliensis* latex (which are readily isolated by centrifugation) which indicate that citrate uptake into them may be dependent on ATP (20). An ATPase has been demonstrated in the vacuolar membrane (5, 19).

Vacuoles can be observed frequently in fungi (74, 379) and have been isolated from *S. cerevisiae* (338) and *N. crassa* macroconidia (336). Subsequent studies on isolated vacuoles of the former fungus have shown that they are the location of polyphosphate, the negative charges of which are balanced at least in part by the bulk of the arginine present in the cell (114). Subsequent studies have also shown that vacuoles of *S. cerevisiae* contain a number of hydrolytic enzymes, such as an acid proteinase, a serine proteinase, carboxypeptidase, and ribonuclease (337, 461). The presence of such enzymes within vacuoles led to the concept of them as lysosomes (338). There is increasing evidence that vacuoles of higher plants can contain a variety of enzymes (339).

Critical interpretation of data obtained from the study of fluxes into and out of isolated vacuoles is necessary. Racusen *et al.* (403) showed that protoplasts from leaf tissue of a number of plants isolated in 0.57 mol L^{-1} mannitol had slightly (6.4–12.5 mV) positive (inside) membrane potentials, whereas the intact cells had significant (65.0–85.7 mV) negative potentials. Removal of the wall will, of course, reduce the potential, but Racusen *et al.* were able to make the protoplast more electronegative by reducing the concentration of the external medium. This suggests that the membrane potential is being regulated by turgor (Section VI,F). The transport properties of the membranes of isolated vacuoles may also be different from the tonoplast of the intact cell.

B. ELECTRICAL MEASUREMENTS

The importance of electrical measurements in the study of membrane transport is discussed later (Section III,B), but they present certain practical difficulties, e.g., insertion of electrodes, particularly with regard to the measurement of the potential difference across the plasmalemma. Calomel half-cells are used. The potential difference E which is measured is given by

$$E = E_{2-1} + E_{\mathrm{KCl}/2} - E_{\mathrm{KCl}/1} \quad (1)$$

Thus the potential is the algebraic sum of the potentials across the plasmalemma (E_{2-1}) and the two junction potentials—that between potassium chloride in the electrode and the external medium (1) and that between potassium chloride in the other electrode and the cytoplasm (2). If E is to be determined and since we cannot measure $E_{\mathrm{KCl}/2}$, we have to assume that the junction potential $E_{\mathrm{KCl}/2}$ and $E_{\mathrm{KCL}/1}$ are either very small or equal. There may not be equality if the tip of the electrode becomes "plugged" with material such as protein with ion exchange properties. It is therefore a routine practice to compare the potential between the two electrodes in an appropriate salt solution before and after one has been inserted into a cell.

A number of factors need to be considered when interpreting the values for potential differences which are obtained by such measurements: (1) Unless the location of the tip can be seen microscopically, it should not be assumed that it is necessarily in one specified part of the cell rather than in another. This is particularly so for higher plant cells, since it has been assumed that one can measure the potential difference between the medium and the vacuole. There are reasons for thinking that the tip of an electrode inserted in a vacuole may still be covered with cytoplasm (133, 223). There is good visual evidence from the pioneering work of Walker (562) that this can readily happen in giant algal cells. Visual inspection can also sometimes reveal plugging, that is, the presence of material adhering to the microelectrode, particularly on the inside of the tube. The only reliable test is to measure the resistance of the electrode tip (11), and if it is high assume that plugging has occurred. (2) Insertion of the electrode, particularly into a small cell, may significantly change the turgor because it changes the internal volume and the potassium concentration of the vacuolar sap due to diffusion of the ion out of the tip (376). (3) If attempts are made to insert microelectrodes into a tissue to investigate the electrical properties of cells lying there, erroneous readings may be due to the bending of the electrode or the release of ions due to the piercing of cells in the more superficial layers (11).

Findlay and Hope (133) provide a good introduction to the techniques for investigating the electrical properties of plant cells.

The electrical potential across a membrane can be determined indirectly by the use of lipophilic cations as introduced by Liberman and Topali (278). The ions are made lipophilic by surrounding the charge-bearing moiety by hydrophobic groups. The ions can penetrate lipid membranes, and it is assumed that the ions move across biological membranes via the lipid regions and without the aid of a carrier molecule. This means that equilibrium distribution across the membrane should depend on the membrane potential as predicted by the Nernst equation. The charge of the cation must be independent of pH and the cation must not be bound intracellularly. Miller and Budd (342) have used such ions for estimating the membrane potential in the fungus *Neocosmospora vasinfecta*, while Hauer and Höfer (175) have used the ions with the yeast *Rhodotorula gracilis*. The amount of the ion was determined by the use of radioactive label. No attempt has yet been made to compare the membrane potential of a plant cell as determined with lipophilic cations and that determined electrically.

Finally, the role of ion pumps, i.e., those systems directly responsible for the transport of ions across a membrane against their electrochemical potential gradient (see Section V), in the electrical properties of a membrane calls for some comment. If the pump is neutral, i.e., it transports zero net charge across the membrane and only this type of pump is in the membrane, then any potential across that membrane is due to the gradient of the ions which are able to diffuse across it. The potential is then said to be a diffusion potential. On the other hand, if the pump transports net charge, then the pump is said to be electrogenic. If such a pump makes the major contribution to the potential across a membrane, then inhibition of the pump leads to rapid depolarization of the membrane. In the case of a diffusion potential, inhibition of the neutral pump has little immediate effect.

C. Measurement of Fluxes

To understand how an ion moves into or out of a plant cell, it is necessary to determine the fluxes of that ion across, and its concentration (moles per liter) on each side of, the relevant membrane.

Total analysis of the cell to give the amount (moles) of an ion in that cell divided by the amount of water present (in liters) in that cell rarely provides accurate information about the concentration of the ion within any part of the cell. Just occasionally, as in the case of organic acids in succulent cells with large vacuoles, the concentration of acid so determined can be equated with its concentration in the vacuole. Usually the determination of ion concentrations in cytoplasm and vacuole is difficult. In the giant algal cells it is possible because of their size to separate the vacuolar contents from the cytoplasm and the cell wall. The ion content of the cytoplasm and cell wall together can then be determined. Since the ion concentration within the wall can be determined by an independent procedure (free space; see Section VII), it is possible to estimate the ion concentration within the cytoplasm if its volume or its water content are known. That estimate will, of course, represent an average value; there may be different concentrations within individual organelles, e.g., mitochondria, and not all the ions are in free solution.

Various attempts are being made to determine ion concentrations within various cellular compartments by other means, for example, by histochemistry, X-ray microanalysis, ion-specific microelectrodes. As yet, the first two methods have not been sufficiently developed to yield reliable quantitative data. Histochemical procedures will always be suspect, since they depend on treating the cells in aqueous media and it is assumed that the chemicals used precipitate the ion at its original location. It is difficult to see how ion movement over a substantial distance can be avoided before precipitation occurs. X-ray microanalysis is more promising. Essentially it depends on the X-ray spectrum emitted by elements when they are bombarded by electrons in an electron microscope. The sensitivity decreases with atomic number. Successful specimen preparation depends on the very rapid freezing of the tissue at the temperature of liquid nitrogen followed by cutting and analyzing the tissue at similarly low temperatures. Läuchli (268), Hall *et al.* (168), and Saubermann and Echlin (453) furnish details of the method and its applications. The method has been used successfully for determination of the relative amounts of potassium and sodium along fungal hyphae (140).

Spanswick and Miller (507) compared values for the cytoplasmic pH of *Nitella translucens* as obtained using a pH microelectrode and by an alternative method using a weak acid or base which is highly permeable in the associated form. They used the weak acid, 5,5-dimethyloxazolidine-2,4-dione (DMO), previously used for the same purpose by Walker and Smith (569). There was reasonable agreement between the two methods. Microelectrodes which respond only to the presence of a particular ion (ion specific) have been used to determine what is believed to be the vacuolar concentrations of potassium and sodium in stomata (see Chapter 3 in this volume) and in the root of *Zea mays* (111, 112) and of *Helianthus annuus* (50).

If the experimental conditions are carefully chosen, tracer flux measurements can provide an estimate of cytoplasmic and vacuolar ion concentration. In brief, cells have to be in a steady state; i.e., there is no net uptake or loss of the ion. Also the cells are loaded with radioactivity for a time at least five times greater than the rate constant for loss of isotope across the plasmalemma; such loading leads to there being very much more isotope in the vacuole than in the cytoplasm. Given these conditions, the amount of ion in the cytoplasm can be determined from the characteristics of the loss of radioactivity from the cells when they are put into unlabeled medium of the same composition as that used for loading. Cram (89) and Walker and Pitman (568) give very full descriptions of the various relevant equations.

Fluxes are usually given in moles per kilogram second for tissues but can be converted to the customary moles per square centimeter second from a knowledge of the cell water content and surface area (it is assumed that the surface area of the tonoplast is the same as that of the plasmalemma). But while the cytoplasmic ion concentration can be calculated (moles per kilogram), it can only be given as a physiologically meaningful value (moles per liter of cytoplasm) by making estimates of the volume of the cytoplasm, and for many cells and tissues this is difficult.

But there are other points which need to be made with respect to flux analysis: (1) Most models—certainly those which are applied to higher plant cells—assume that the compartments are in series and that the rate of movement across a compartment is limited by the rate of movement across its boundary membrane. It is likely that the latter supposition is correct—with the observed rates of exchange between compartments, the rate of diffusion of an ion would have to be at least 10^{-7} that of its rate of diffusion in water for exchange to be limited by diffusion within the compartment from one membrane to another. On the other hand, the cell system under investigation could deviate quite considerably from the compartments-in-series model. There is evidence from large algal coenocytes that ions may move directly from the medium to the vacuole without equilibrating with the cytoplasm (Section VIII). (2) A system of compartments in parallel also behaves in a manner similar to a system in series. The same will be true for a tissue in which the cells are of unequal size (and therefore unequal area per unit volume). MacRobbie and Dainty (323) believe that such differences in cell size explained their tracer efflux curves obtained with *Rhodymenia palmata* (5). Despite the number of compartmental analyses made on excised roots, there are reservations about the validity of the procedures. These arise because of the different tissues and cell types, but also because there is a large net flux of ions through the cells to the xylem and thence into the medium again via the cut surface of the stele (Section X). To separate and estimate the flux of tracer from the surface of the root and that from the cut stele, excised roots have been placed in chambers with separate compartments so that the bulk of the root is in one compartment (to collect tracer from the outer surface) while the cut end is in another (392). (4) So far the cell wall (free space; see Section VII) has not been considered in relation to compartmental analysis. The cell wall will influence flux measurements both because of its charge and because of the water-filled spaces between the cellulose micelles. Thus tracer which may be absorbed or lost from the wall may influence the estimated amount going to or from the plasmalemma. Walker and Pitman (568) have shown how free space exchange can alter the cytoplasmic specific activity S_c of an ion during elution (Fig. 1). Bivalent ions, especially of calcium, present particular difficulties. Thus in *Nitella translucens* nondiffusible anions are present in cell walls at a concentration of 0.74 Eq L^{-1}, which leads to a high concentration of calcium in the wall and 9% of the calcium exchanges at a rate ($t_{1/2} = \sim 12 \times 10^4$ s) which is sufficient to obscure fluxes from the cytoplasm (509).

One way to test whether a transport process is active or not is to examine whether a maintained electrochemical potential gradient for ions can exist

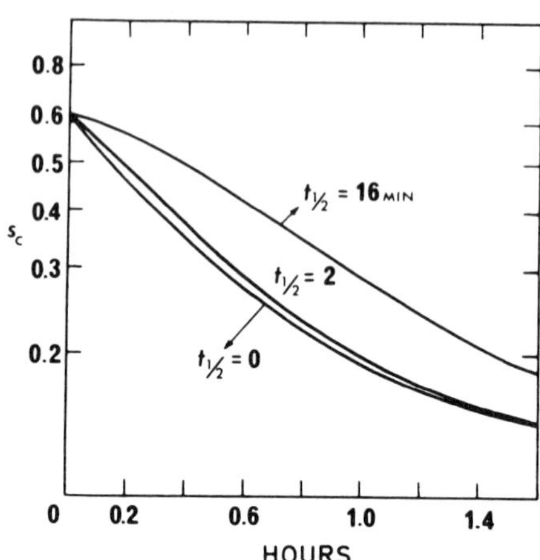

FIG. 1. Effect of free space exchange on the cytoplasmic specific activity of the ion s_c during elution based on computer simulation. The curves were calculated assuming that the flux from medium to cytoplasm = the flux from cytoplasm to vacuole = 2 μmol g fresh weight^{-1} h^{-1}, the cytoplasmic content of the ion = 2 μmol g fresh weight^{-1}, the vacuolar content of the ion = 70 μmol g fresh weight^{-1}. Free space exchange was set to have half-exchange ($t_{1/2}$) in 16 min, in 2 min, or with no free space at all. Note the initial delay in decrease of s_c and change in slope of ln s_c due to diffusion in the free space. From Walker and Pitman (568).

in the absence of any net flux except for the flow through a particular chemical reaction (cf. Section III,D). If the system were initially in equilibrium, inhibition of that specific "energy-yielding" chemical reaction should cause the electrochemical potential gradient to disappear. Such an approach would be possible if one had a specific inhibitor for energy-yielding reactions. However, energy-yielding reactions may depend upon membrane-located processes in mitochondria or chloroplasts, and so it may be difficult to measure those flux changes which might be the direct consequence of the cessation of provision of energy.

The problem of investigating flux changes brought about by an inhibitor is also complicated by the time required to complete a full compartmental analysis. The longer a cell or a tissue is exposed to an inhibitor, the more probable it is that processes other than active fluxes will be affected. Gross changes in cell composition could affect turgor or total ion balance, e.g., rise or fall in the level of organic acids. However, short-term experiments may yield valid data on fluxes across the plasmalemma if effects due to tracer being associated with the free space can be minimized (89).

III. The Hunt for Carriers

A. INTRODUCTION

At its simplest, a carrier is a protein which bring about the movement of an ion across an otherwise impermeable membrane. Steward and Sutcliffe (518) showed how earlier thinking led to current ideas. The concept of a carrier has been associated with that of active transport, namely, that metabolic energy can cause the carrier to move an ion up its electrochemical potential gradient. Therefore, if we find that a solute is actively transported across a membrane, the carrier must be one that permits the intervention of metabolic energy. Thus, such a carrier possesses more complex properties than one which only facilitates movement of an ion from one side of the membrane to another. Therefore the establishment of an active solute movement across a membrane indicates the nature of the carrier to be investigated. It is first necessary to say something about the biophysical basis of ion transport across membranes.

The movement of an ion from one point to another must ultimately be brought about by physical forces, although chemical reactions may be involved in the generation of these forces. The movement of an ion will always be accompanied by the movement in the same direction of an ion of opposite charge or an ion of the same charge in the opposite direction. The difficulty of generating any substantial separation of ions can be demonstrated by a simple calculation. Suppose that there is a 10-L spherical vessel containing an 0.1 mol L^{-1} solution of sodium chloride and there is available some mechanism whereby all the chloride ions could be removed, leaving behind the sodium ions. Since there would be 1 g ion of sodium in the vessel, the charge would be approximately 10^5 C. Taking the radius of the vessel as 13 cm, its capacity would therefore be 14×10^{-12} F. The potential on it would be 7×10^{15} V. This expresses in quantitative terms the fact, familiar in inorganic chemistry, that considerable electrical energy is required to make metallic sodium from sodium chloride!

This can be put another way. If charge separation to the extent of 1 V is generated, the difference in the number of ions of one sign over those of opposite sign will be no more than one part in 10^{15}. This will mean that if a plant cell were to generate a potential of 1 V between the cell interior and the external solution, the charge separation which results would not be detectable by chemical analysis. Thus, for most practical purposes there is an almost exact balance of positive and negative charges at any site within

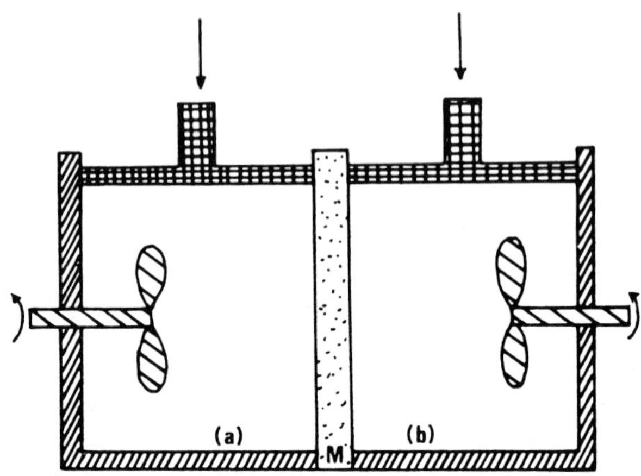

FIG. 2. Diagrammatic representation of the system to which the equations in this section apply. A uniform membrane M separates two aqueous phases a and b. They have the same temperature and solvent (water) but have different hydrostatic pressures, compositions, and electrical potentials. From Walker (565).

the cell. If a plant cell is accumulating salts against a concentration gradient, the question to be asked is on which ion is the work performed?

Walker (565) has described the model that we use for most descriptions of membrane transport (Fig. 2). The membrane forms the only connection between two aqueous phases which are of uniform concentration (hence the stirring arrangements) and these phases may differ with respect to pressure (especially in plants) and electrical potential, but not with respect to temperature. Sets of equations can be derived to describe the model.

The equations can be classified as either mechanism independent or mechanism dependent. Classic or equilibrium thermodynamics and nonequilibrium thermodynamics provide equations in the first category, with nonequilibrium thermodynamics providing kinetic equations of a general kind. Most kinetic descriptions are, however, based on some model which has a mechanistic basis.

B. THERMODYNAMIC DESCRIPTION OF ION TRANSPORT ACROSS A MEMBRANE

Biologists have usually defined active transport as being the process by which an ion is moved against an electrochemical potential gradient (221). Therefore it has been argued that if under equilibrium conditions no electrochemical potential gradient exists for a particular ion across a membrane, then that ion is passively distributed across that membrane. That is, the ion is not actively transported.

The situation in which there is no electrochemical potential gradient across a membrane can be defined thermodynamically as

$$\bar{\mu}_a - \bar{\mu}_b = 0 \tag{2}$$

where $\bar{\mu}$ is the electrochemical potential and the subscripts a and b refer to the two compartments on either side of the membrane. This can be expanded as

$$(\mu_a^\circ - \mu_b^\circ) + \bar{V}(P_a - P_b) + RT\ \ln(Y_a x_a/Y_b x_b) + ZF(E_a - E_b) = 0 \tag{3}$$

where μ° is the electrochemical potential at the standard state, $\bar{V}$ is the partial molar volume, P is the hydrostatic pressure, R is the gas constant, T is the absolute temperature, Y is the activity coefficient of the ion whose mole fraction is x and whose valency is Z, F is Faraday's constant, and E is the electrical potential.

For the system which we are considering, the first term $(\mu_a^\circ - \mu_b^\circ)$ is equal to zero, and we can take the second as being insigificant because there is negligible difference in hydrostatic pressure, so that the equation simplifies to

$$RT\ \ln(Y_a x_a/Y_b x_b) + zF(E_a - E_b) = 0 \tag{4}$$

This can be simplified further by replacing activities by concentrations (C_a, C_b), giving

$$E_a - E_b = E_{ab} = -RT/zF\ \ln C_a/C_b \tag{5}$$

This is the customary form of the Nernst equation often used by plant physiologists to attempt to identify whether the movement of ion between the medium and the interior of the plant cell is brought about by active transport. The neglect of activity coefficients by plant physiologists is not often justified, especially when the internal phase may be the cytoplasm. It applies to the equilibrium for the *isolated* transport of an ion; if this transport is enzymatically coupled to another transport process or chemical reaction, there will be quite different equilibrium conditions. Let us consider the implications of coupling.

If we take X and Y as being the species transported between a and b, the transport reactions can be represented like chemical reactions:

$$X_a \leftrightharpoons X_b$$

and

$$Y_a \leftrightharpoons Y_b$$

These can be summed by using stoichiometric coupling coefficients n_x and n_y so that we get the overall reaction

$$n_x X_a + n_y Y_a \leftrightharpoons n_x X_b + n_y Y_b \qquad (6)$$

The coefficient represents the possible coupling or interaction between the two transport reactions and the two flows between a and b. At equilibrium,

$$n_x RT \ln \frac{C_{xa}}{C_{xb}} + n_y RT \ln \frac{C_{ya}}{C_{yb}} + (n_y z_y + n_y z_y) F E_{ab} = 0 \qquad (7)$$

where C is the concentration of the species and R, T, z, F, and E are as given above.

Thus

$$E_{ab} = \frac{n_x}{n_x z_x + n_y z_y} \frac{RT}{F} \ln \frac{C_{xa}}{C_{xb}} + \frac{n_y}{n_x z_x + n_y z_y} \frac{RT}{F} \ln \frac{C_{ya}}{C_{yb}} \qquad (8)$$

From this one can see that lack of agreement between the calculated Nernst potential $[(RT/zF) \ln (C_{xa}/C_{xb})]$ and the measured potential across the membrane E_{ab} will not necessarily mean active transport. On the other hand, if the flow of x across the membrane does not interact with the flow of any other ion, then

$$E_{ab} = -\frac{RT}{zF} \ln \frac{C_{xa}}{C_{xb}} \qquad (9)$$

The coupling just described is called exchange diffusion or cotransport. The flow of an ion (vectorial reaction) can also be coupled with a chemical (scalar) reaction. Thus if the chemical reaction has a free energy change ΔG_c, the equilibrium will be

$$n_c \Delta G_c + n_x(\bar{\mu}_{xa} - \bar{\mu}_{xb}) = 0 \qquad (10)$$

This is true active (or primary) transport; that is, the free energy of a chemical reaction drives the movement of an ion against its electrochemical potential gradient. If the ion flows which are generated are coupled with the flows of another ion or solute, then the movement of that other ion or solute may also be against, respectively, its electrochemical or chemical potential gradient. Such movement can be termed *secondary* active transport.

Thus the customary approach adopted by many plant physiologists to identify whether a movement is brought about by primary active transport may be of dubious value. This judgment is supported by studies of chloride transport into the internodal cells of Characean algae *Chara corallina, Nitella translucens,* and *Nitellopsis obtusa* (see Section IV,A,3,b). MacRobbie and Dainty (324) were the first to propose, as a result of the application of the Nernst equation, that chloride is actively transported into cells of *N. obtusa*. Later workers, for example, Spanswick and Williams (508), working with *N. translucens* also came to the same conclusion after applying the equation. More recent studies, described in Section IV,A,3,b, on the fluxes of the ion across the plasmalemma of the cells of *C. corallina* and *N. translucens* which have been aimed at trying to elucidate the nature of the transport process for chloride at this membrane still do not allow any firm conclusion as to whether energy is directly involved in the transport of the ion.

To show that active transport occurs it is necessary to establish *either* that the vectorial flow of an ion J_x occurs in the absence of an electrochemical gradient $\Delta\bar{\mu}_x$ and any other flow except the chemical reaction or scalar flow which we can call J_c *or* that a maintained gradient $\Delta\bar{\mu}_x$ can exist in the absence of any net flow (including that of x) except J_c. This is a definition arrived at from nonequilibrium thermodynamics and it should be noted that the flows that have to be considered are not only those of the ion x and other ions, but also other solutes. The flow of electric current and water must also be considered. Walker (565) presents the mathematical formulation for these flows.

Electrokinetic phenomena are good examples of coupled flows. Figure 3 gives a very simplified statement of possible interrelationships. Thus if we consider charge under the influence of an electrical potential gradient, electrophoresis is the movement of charged particles with no net flow of

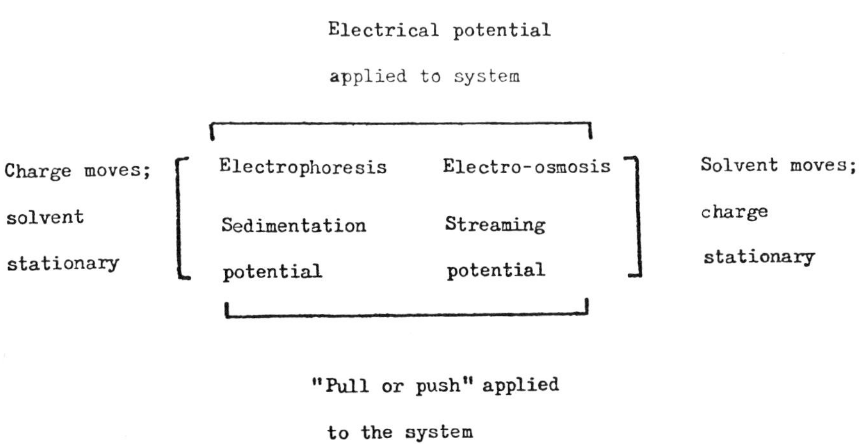

FIG. 3. Diagrammatic representation of the interrelationships of electrokinetic phenomena.

solvent, whereas electro-osmosis occurs when there is stationary charge with consequent bulk flow of solvent. Likewise a streaming potential is generated when solvent moves over a charged surface as the result of a nonelectrical force, e.g., pressure, and a sedimentation potential is produced when a charged particle drops by gravity in a stationary solvent. Such coupled flows can be described by irreversible thermodynamics, and the reader should consult Walker (565) for further details.

Thus if we are to identify active transport systems within a plant cell, we need to be able to measure ion flow or fluxes; to control the other flows, i.e., flows of other solutes, water, and current; and to set the electrochemical potential $\Delta\bar{\mu}_x$ at zero. This has not been attempted for plant cells; some reasons for this were indicated in Section II,C. There the discussion concerned a single membrane, whereas in plant cells there are, of course, two major membranes—the plasmalemma and the tonoplast—across which solutes move into vacuoles, and it is difficult to separate them physiologically.

There is a particular problem in measuring or indeed identifying J_c (317), particularly in green cells in which either respiration or photosynthesis may drive active transport (Section V). It is often the practice to attempt to alter J_c by the use of metabolic inhibitors. But these compounds may do more than inhibit J_c; for example, compounds which inhibit oxidative phosphorylation may act as if they are proton carriers in the membrane, thus altering the flux of hydrogen ions across it (see also Section V).

C. Rate Equations

Rate equations attempt to describe the rate of movement or flow or flux (moles per square centimeter second) of an ion across a membrane in terms of the forces acting upon it. Using them it is possible to find whether the calculated flux equates with that which is measured. Any equations so derived will involve a diffusion and an active transport term and a permeability coefficient which is characteristic of the solute and the membrane (=solvent). Estimation of that coefficient ought to tell us something about the system in which the ion is diffusing. However, an interpretation of the coefficient depends upon the assumptions made in their derivation, e.g., the implied structure of the membrane. Because of these assumptions, such a coefficient cannot necessarily be equated with the permeability coefficient of generalized flux equations such as the Fick equation. In addition, there is uncertainty about what constitutes the best way of representing the active transport term.

It is because of these problems that there were initially many difficulties in the application of rate equations to ion transport across plant cell membranes. Indeed, at one time it was assumed that the active flux neither contributed to the membrane potential nor was under its influence. This is now known not to be so [see Spanswick (502) for an admirable discussion of the problem]. It is only since Spanswick's study (502–505) of the major ion fluxes of *N. translucens*, taking into account the interaction between active pumping and the membrane potential, that we have begun to make progress in relating flux measurements to some coherent, albeit very simple, theory of what is occurring in the membrane.

However, rate equations for ion fluxes can also be derived from a consideration of the rate of flow of electric current across a membrane. In the present state of knowledge of the electrical properties of plant cell membranes, it is not appropriate to say anything here about the electrical conductance of a membrane. Nevertheless, electrical studies do provide a physical description of the plant cell membrane in terms of the ability of ions to move across it and eventually a basis to compare it with artifical systems of known characteristics. We now have extensive knowledge of transport of ions and molecules across artificial systems which are based on ultrathin black lipid films (118, 119, 535) which have many of the characteristics of biological membranes. Recently, isolated proteins from cell membranes have been inserted into such artificial membranes (466, 467). These proteins have been termed "ionophores" because they enhance the movement or incorporation of an ion from the aqueous phase into the hydrophobic phase.

D. KINETIC ANALYSIS OF ION TRANSPORT

Kinetic equations can be considered as a special form of rate equations. They are derived on the premise that for an ion to move across a membrane it must do so via a carrier. That carrier facilitates or catalyzes the movement of the ion across the membrane. Thus movement can be described in terms of the kinetics of ion binding to a carrier and the movement across the membrane of that protein either alone or accompanied by the ion. There is a problem of how to describe this movement within the membrane in kinetic terms. Usually, however, such movement is assumed to be instantaneous.

By 1959, ion transport into plant cells was already being analyzed in simple kinetic terms. Epstein and Hagen (125) had begun the well-known studies on the description of ion uptake by barley roots using Lineweaver–Burk enzyme kinetics. The actual technique then used now seems rather crude since uptake over 3 h was equated with initial rate of uptake and no correction was made for uptake into the free space. The later procedures are less open to this criticism (123). Further work showed that for monovalent cations there appear to be two mechanisms (I and II). Details are given in Table I. Of particular interest is the fact that sodium shows very little competition with potassium for mechanism I, but there is considerable competition for mechanism II (127). In spite of the large number of papers (124) which have been published on this particular topic, the actual kinetic analysis has been rather superficial, with little advance in our understanding of what the data mean in mechanistic terms. The complex Michaelis kinetics found by Epstein and colleagues come from an examination of a plot of rate of uptake v against external concentration S (Fig. 4). The series of rectangular hyperbolae are considered to be related to the presence for the ion of a number of carriers exhibiting different properties. Studies of how the kinetics change under different conditions have led to the kind of conclusions outlined earlier.

The controversy which this work has aroused will not be reviewed here, but there are some important points to be made. First, a fit to a rectangular hyperbola is not unequivocal evidence for carrier mediation and particularly not for active transport. Second, a transport system may fit a rectangular hyperbola in one range of concentration but not in another. Cram (87) has referred to the results of Szabo *et al.* (528), who showed that the potassium influx across an artificial phospholipid bilayer membrane mediated by the ionophore nonactin (at 10^{-7} mol L^{-1} in the external solution) exhibits a rectangular hyperbolic relationship with concentrations up to 0.05 mmol L^{-1} but a linear relationship at higher concentrations. And third, the use of a wide range of concentrations introduces difficulties, especially at the higher concentrations: No account is taken of the fact, and this could be so at the higher concentrations, that the cytoplasmic concen-

TABLE I

PROPERTIES OF MECHANISMS I AND II OF K^+ AND Na^+ ABSORPTION BY BARLEY ROOTS[a]

Property	Mechanism I	Mechanism II	Reference(s)
Kinetics	Michaelis–Menten	Complex	126
Affinity for K^+	High	Low	127, 187, 407
Affinity for Na^+	Low	Moderate to high	127, 405–407, 574
Influence of the anion	Small	Large	127, 187
Response to calcium at high Ca^{2+} concentrations (>1 m*M*)	No inhibition	Inhibition	407

[a] From Epstein (123).

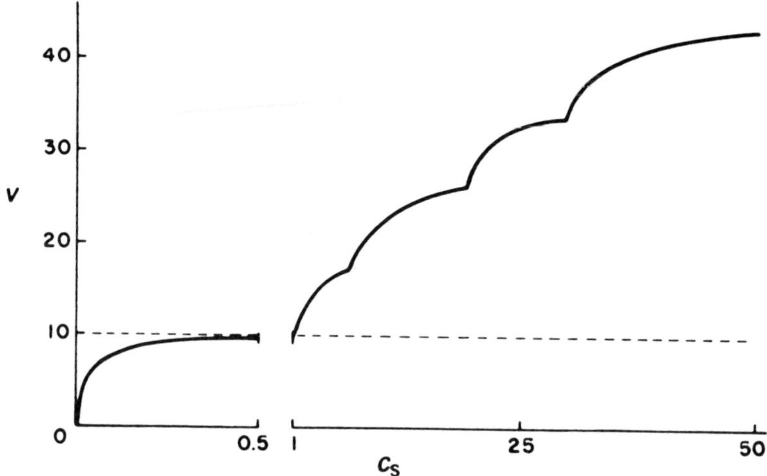

Fig. 4. Generalized diagram showing the complex relationship between the rate of absorption v by a plant tissue of an alkali cation or halide as a function of its concentration c_s in the external solution in millimoles per liter. The units for the rate of absorption are arbitrary. From Epstein (124).

tration of an ion might rise even in the short time over which the uptake is being measured. If this is so, there might be feedback effects on the rate of transport of the ion (Section VI,C). Also, increasing the external concentration of most monovalent cations depolarizes cells. A reduction in the electropotential gradient will result in a decrease in the diffusional influx of cations and an increase of anion influx. The simple kinetics referred to earlier are not then applicable.

Nissen (357), using an analogous kinetic approach, has found that double reciprocal plots of uptake $1/v$ against external concentration $1/S$ for a wide variety of ions and organic solutes and plant tissues are a series of straight lines with sharply defined inflection points (Fig. 5). The series is postulated to represent a series of phases. The molecular basis of what is postulated has not been made completely clear, although Nissen (357, 358) has indicated some possibilities related to conformational change in the carriers. Whatever the mechanistic interpretation resulting from the kinetic treatment, the preceding comments apply almost equally to it as they do to the treatment used by Epstein and colleagues.

Perhaps the most satisfactory kinetic transport studies have been made on the potassium–hydrogen exchange in *Neurospora crassa* and *Saccharomyces cerevisiae* (Sections IV,A,2,c and VI,C). The analyses used are sufficiently sophisticated to yield information which could be used to charac-

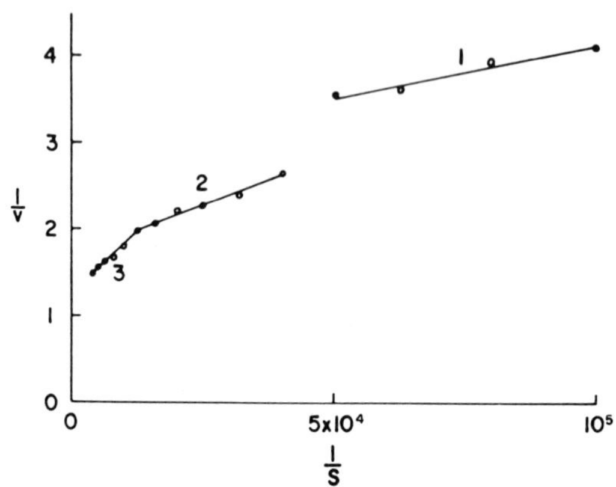

FIG. 5. Double reciprocal plot for uptake of SO_4^{2-} by barley roots in the range 10^{-5}–2.5×10^{-4} mol L^{-1} showing the presence of three phases. From Nissen (357). Reproduced, with permission, from the *Annual Review of Plant Physiology*, Volume 25, © 1974 by Annual Reviews Inc.

terize the transport system if it were to be isolated for study *in vitro;* i.e., affinity constants for the various ion-binding sites have been determined — a point which is dealt with further in the next section — and at the same time provide a possible interpretation of the dual isotherms and multiphasic kinetics.

E. Molecular Basis of Ion Transport

The processes that bring about the transport of a molecule or an ion across a living membrane may be clarified when the molecular basis of that process becomes known. The effectiveness of this approach can be seen by reference to three transport systems — active transport of sodium and potassium in red blood cells (141), sugar transport in facultatively anaerobic bacteria (401), and sucrose transport in the mammalian small intestine (465).

In red blood cells sodium is pumped out and potassium pumped in against their respective electrochemical potential gradients. Powerful support for the physiological evidence that this is so came from the isolation of a membrane-bound ATPase which is believed to be the "pump molecule." The properties of the enzyme have now been studied in considerable detail (154, 479). The enzyme hydrolyzes ATP to ADP and orthophosphate; in the cell it is presumed that the free energy of hydrolysis of ATP brings about the vectorial movement of sodium and potassium across the cell membrane against their electrochemical potential gradients.

The evidence (110, 400) that the enzyme is the ion pump is as follows: (1) Both are located in membranes. (2) Both use ATP. The evidence that the pump requires ATP comes from the study of red cell ghosts. These are cells which have been lysed and have lost most of their hemoglobin and a variety of other compounds by exposure to hypotonic solutions. As a result, they lose their ability to transport actively sodium and potassium. If suitable quantities of ATP are reintroduced into the cells during this period, they regain their active transport ability. (3) Sodium and potassium are required by both for maximum activity. (4) The half-maximal concentration of sodium and potassium required for ATPase activity and ion transport are very similar. (5) The cardiac glycoside ouabain inhibits both the ATPase and the pump, but has no effect on general metabolism. (6) The inhibition of ATPase by very low concentrations of ouabain is prevented by raising the potassium concentration just as the inhibitory effects on potassium influx in intact cells can be reversed by a high external potassium concentration. (7) Activation of the ATPase by potassium is partly inhibited by a high sodium concentration. This effect is comparable to the greater uptake of potassium from choline (sodium-free) media compared to that from normal media.

The facultative anaerobes *Escherichia coli, Salmonella typhimurium,* and *Staphylococcus aureus* contain enzymes which catalyze the following reactions.

The first catalyzed by the so-called enzyme I is

$$\text{Phosphoenolpyruvate} + \text{HPr} \xrightarrow{Mg^{2+}} \text{P–HPr} + \text{pyruvate}$$

where HPr represents a low-molecular-weight (ca. 9500) protein containing two histidine residues and P–HPr the phosphorylated form.

The second reaction catalyzed by the so-called enzyme II is

$$\text{Sugar} + \text{P–HPr} \rightarrow \text{sugar P} + \text{HPr}$$

In most instances the sugar is phosphorylated in the 6 position. This is a slightly simplified description of the steps leading to sugar phosphorylation.

Further work (85) showed that these two enzyme steps are involved in the movement of sugar across the membrane, leading to the production of phosphorylated sugars within the cells. The evidence for this is as follows: (1) Mutants containing a defective enzyme II for a particular sugar cannot transport the sugar into the cell. Other sugars can be transported because the relevant enzymes II are present. (2) Mutants defective in enzyme I and/or HPr are unable to utilize a wide range of sugars. This could be expected on the basis of the fact that the reaction catalyzed by enzyme I is responsible for producing P–HPr, which is the source of ~P for sugar phosphorylation. (3) Cells which are subjected to osmotic shock (cells suspended in 0.5 mol L^{-1} sucrose and EDTA for a period, separated out, and rapidly dispersed in cold dilute magnesium chloride) lose their ability to take up sugars. Most of the cells are still viable, but they have lost most of their HPr. Presentation of HPr to these cells restores their ability to take up sugars.

A sucrose-isomaltase complex can be isolated from the brush border membrane of hamster small intestine. It can be solubilized by papain treatment and obtained as a water-soluble protein which has two subunits — one splitting sucrose and the other splitting isomaltose. This enzyme is believed to provide the molecular basis to explain the movement of glucose across the intestinal wall. In the presence of sucrose, maltose, lactose, or trehalose in the medium, more glucose is transported across the intestinal wall even if the medium contains enough free glucose for complete saturation of the transport system(s) for free monosaccharide. The isolated enzyme can be inserted in an artificial lipid membrane (not otherwise permeable to sucrose, fructose, or glucose), and sucrose will move across the membrane with only glucose and fructose being found on the other side. The permeability of *free* glucose and *free* fructose is not affected. This appears to be one of the first reports that a homogeneous protein solubilized from a biological membrane can produce a reconstitution of the original transport system in an artificial lipid membrane, thereby providing the functional demonstration of a membrane carrier.

These three specific cases show how biochemical studies can provide strong evidence for the way in which a solute penetrates a membrane. Some features of these studies deserve special mention.

1. The availability of an inhibitor specific for a particular transport system provides a means for checking whether a protein isolated from a cell is responsible for the transport process under investigation. Although ouabain is known to inhibit active sodium efflux and active potassium influx in *Hydrodictyon africanum* (410), no other compound directly inhibits specific transport systems in plants. But recent work on phytoalexins and fungal metabolites is suggestive. There is evidence that the phytoalexin phaseolin may exert its effect directly upon the potassium–hydrogen exchange pump of *N. crassa* (482). Also the toxin fusicoccin, produced by the fungus *Fusicoccum amygadali,* which causes blight of almonds and peach, stimulates loss of hydrogen ions by higher plant cells. Its action is discussed later (Section IV,A,2,a). Other phytotoxins appear to act like fusicoccin in that they induce an early hyperpolarization of the plasmalemma and enhance some transport process, e.g., the loss of hydro-

gen ions as just indicated or the uptake of nitrate in the case of the toxin produced by *Helminthosporium carbonum* (329). Another group (such as victorin produced by *H. oryzae* and the *Cercospora beticola* toxin) increase the leakage of electrolytes from cells and where it has been investigated, cause rapid depolarization of the plasmalemma. A third group contains the *Helminthosporium maydis* and *Phyllosticta* toxins which are reported to interact directly with other cell membranes such as those of mitochondria and chloroplasts (542). Strobel (520, 521) and Marrè (329) have provided valuable surveys of toxins produced by microbial plant pathogens and their effects on higher plant membrane processes.

2. Production of transport mutants can provide a very powerful way of analyzing a transport system. This is shown by those genetic studies (85) of the phosphotransferase system described earlier. Pateman and Kinghorn (382) have reviewed the many genetic studies which have helped our understanding of how nitrogen-containing compounds enter fungal hyphae. Genetic studies concerned with transport of solutes into green plant cells are still at a very rudimentary stage. A few but rather ill-defined transport mutants in higher plants have been isolated (Table II). Most of these mutants relate to micronutrient deficiencies which are easier to select than mutants concerned with the transport of macronutrients, but if such mutants were to be found, then an important field of study would open up. As it is, even for micronutrients, the full potential of transport mutants has not yet been realized.

3. There is no doubt that if the study of isolated carrier proteins in artificial membrane systems can be achieved, it will aid our understanding of molecular events *in vivo*.

Much of what has been written in this section has been based on the belief that proteins in a relatively static membrane are responsible for the transport of ions from one side to the other. But an important alternative is through the formation of vesicles, as suggested by Sutcliffe (524) and others. Baker and Hall (23) have argued persuasively for the formation of vesicles in the transport of ions into plant cells. If in fact ions enter or leave plant cells by vesiculation of the membrane, then the procedures for establishing the mechanism involved are likely to be very difficult. Even if it should be possible to isolate a receptor molecule which triggers the formation of a vesicle, it is nevertheless difficult to visualize how the mechanism of the subsequent release of ions into the cytoplasm could be established. However, the vesicle hypothesis is currently *nonproven*. Even in salt glands, in which according to electron microscope evidence vesicles might be involved in the expulsion of salt, there is good reason to believe that carrier transport is the process involved (196) (see Chapter 5 in this volume). Further, Cram (91) has examined in detail the many requirements which

TABLE II

SOME ION TRANSPORT MUTANTS IN HIGHER PLANTS (SINGLE-GENE MUTANTS)[a]

Species	Mutant	Physiological expression of mutation	Inheritance	Transport process possibly involved	Reference(s)
Glycine max (soybean)	PI-54619-5-1 (PI)	Inefficient in Fe utilization	Recessive	Fe uptake after reduction of Fe^{3+} to Fe^{2+}; Fe transport to shoot in PI	61, 64, 66
	Hawkeye (HA)	Efficient in Fe utilization	Dominant		
Lycopersicon esculentum (tomato)	T3238fe	Inefficient in Fe utilization	Recessive	Low uptake and transport in Fe	65, 68, 572, 573
	T3238	Inefficient in B utilization	Recessive	B transport to shoot, controlled in the root	
	Rutgers	Efficient in B utilization	Dominant		
Apium graveolens (celery)	Utah 10B	Mg deficiency chlorosis	Recessive	Mg^{2+} uptake or transport to shoot?	378
Glycine max (soybean)	Jackson	Salt sensitive	Recessive	Cl^- transport to shoot	1, 2, 271
	Lee	Salt tolerant	Dominant	Regulation of salt transport to shoot by root xylem parenchyma	

[a] From Läuchli (269).

would have to be fulfulled if pinocytosis is to be the feasible and principal means of transporting major nutrients across the plasmalemma. (These requirements concern particularly selectivity in uptake, competition among substances, differential effects on influx of changes in internal states, and the associated electrical currents.) The evidence from studies on animal cells also indicate that pinocytosis cannot meet these requirements. Cram also points out that if pinocytosis involves selective binding of the ion prior to uptake, then the required density of binding sites is at least three orders of magnitude greater than is possible. Also the water flow which would be associated with pinocytosis would generate high values of turgor which probably could not be sustained.

IV. Transport Systems

A. SYSTEMS WHICH ARE CLASSIFIABLE

1. Basis of Classification

This section deals with those transport systems which can be categorized with some degree of certainty. It is an attempt to show, by examples, the mechanistic basis for ion uptake by plant cells. Since the list is short, caution must be exercised in extrapolating from these few examples, especially since our knowledge about them is still somewhat meager. This description of transport systems in plants is an attempt to produce a taxonomy for them. It rests heavily on that of Mitchell (345), which in turn is a logical outcome of his chemiosmotic hypothesis (Section V).

Primary translocation[2] *reactions* are reactions which involve the exchange of primary bonds between different pairs of chemical groups or the donation and acceptance of electrons. This class of reaction may be divided into two subclasses: (1) *Group translocation reactions* are those in which chemical groups or electrons pass from one side of an osmotic barrier to the other. The reaction catalyzed by the phosphoenolpyruvate–sugar phosphotransferase system is one such reaction. (2) *Enzyme-linked solute translocation reactions* are those in which the translocation of one or more solutes through the osmotic barrier is coupled to the transference of chemical groups or electrons on only one side of that barrier.

[2] In this context the term *translocation* should not be confused with its longstanding meaning in plant physiology, in which it denotes the massive movements over long distances in vascular tissues. For the movements across membranes here in question the terms *transport* and *transfer* might be preferable even though *translocation* and *translocators* are now widely used.

Secondary translocation reactions are those which do not involve primary bond exchange between different pairs of chemical groups or the donation and acceptance of electrons, which places the catalytic proteins apart from the classic concept of an enzyme. For this reason Mitchell (345) proposed that systems catalyzing secondary translocation systems be referred to as porters.

Secondary translocation reactions can be divided into three subclasses: (1) *Noncoupled solute translocation,* or *facilitated diffusion* in its original sense, or *uniport* reactions are those in which a single solute equilibrates across an osmotic barrier. (2) *Sym-coupled solute translocation,* or *symport* reactions, are those in which two solutes equilibrate across an osmotic barrier and the translocation of one solute is coupled to the translocation of the other in the same direction. (3) *Anticoupled solute translocation,* or *countertransport* or antiport reactions, are those in which two solutes equilibrate across an osmotic barrier and the translocation of one solute is coupled to the translocation of the other in the opposite direction.

The taxonomy can be extended further in certain instances, depending on whether the translocation is via a carrier circulating in the membrane (either by actual movement of the whole molecule or by rotation) or is via a port of channel. It is not appropriate to discuss this further since reference may be made to the extensive article by Mitchell (345). At this stage only the broad concept of the preceding classification will be used to systematize some of the information that is available for the movement of ions across the membrane of plant cells. The fact that a great deal of information cannot be so systematized indicates (1) the difficulties of examining ion transport into plant cells when compared to some other cells and (2) the fact that the information which has been obtained by plant physiologists has not been readily assimilated into the developing views of biologists concerned with membranes.

2. *Primary Translocation*

a. H^+/Monovalent Cation Antiporter-ATPase. The most impressive evidence for the presence of such a translocase comes from studies on *N. crassa,* principally those by C. L. Slayman and C. W. Slayman and colleagues. Potassium uptake by this fungus is essentially a unidirectional process and is accompanied by net extrusion of either hydrogen or sodium ions (486). At pH 5.8, all three net fluxes are exponential with time and obey Michaelis kinetics as a function of external potassium. This indicates that there is a close interconnection between the fluxes, which is confirmed by genetic studies showing that there is a single transport system linking potassium uptake with loss of hydrogen and sodium ions from the cell (491).

The potassium–hydrogen–sodium system is electrogenic (see Section II,B). Metabolic inhibitors [cyanide, carbon monoxide, dinitrophenol (DNP), azide, anoxia, and low temperature] all cause rapid depolarization of the membrane (480) (Fig. 6). The extent of depolarization at any one DNP concentration is similar to the extent to which the three ion fluxes are inhibited (Fig. 7). Slayman *et al.* (485) showed that the decay in potential brought about by a metabolic inhibitor (in this case 1 mmol L^{-1} cyanide) is paralleled by a drastic fall in the mycelial concentration of ATP (from 2.7 to 0.25 mmol kg cell water^{-1} in 30 s) such that the voltage–time curve is superimposable upon the ATP–time curve with the rate constants for both corresponding to a half-time of 3.7 s. Subsequent work (484) confirmed this and showed that the rate of depolarization is severalfold too slow to be directly linked to electron transfer, as judged by the rate of pyridine nucleotide reduction.

These data very strongly suggest that the electrogenic pump in the plasma membrane of *N. crassa* is an ATPase. The reason for classifying it as an H^+/K^+ antiporter lies in the fact that the system translocates in one direction (outside to inside) almost exclusively potassium, whereas hydrogen ions cause the electrogenicity. The potential cannot be changed, for instance, by changing the sodium content of the mycelium (481), suggesting that this ion does not play an integral role in the activity of the ATPase.

The same system may be present in other fungi, particularly *Saccharomyces cerevisiae,* the organism whose potassium uptake has been extensively studied by Conway and Rothstein and their colleagues (226).

In 1968, Kitasato (248) suggested that there is an electrogenic pump in

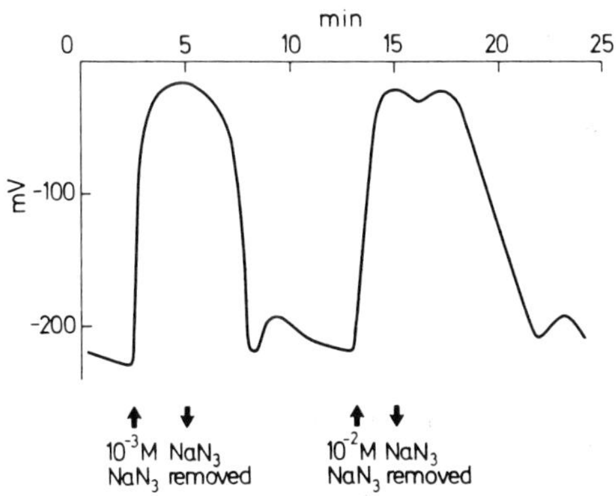

FIG. 6. The response of the membrane potential of *Neurospora crassa* to sodium azide. From Slayman (480).

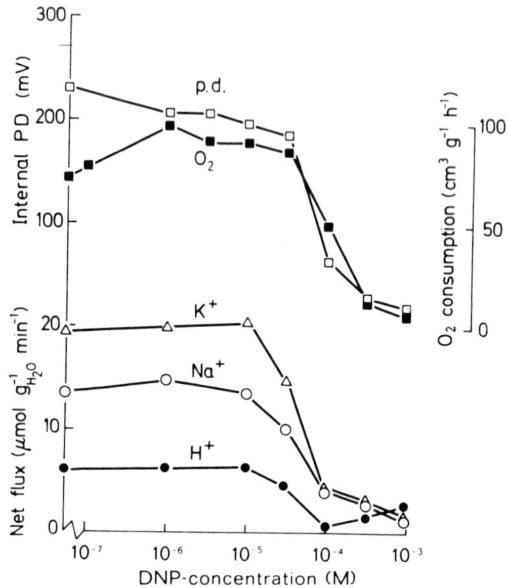

FIG. 7. The effect of 2,4-dinitrophenol on the oxygen consumption and membrane potential (data from 480) and on the initial rates of net K^+ and net Na^+ and H^+ effluxes of low K^+ mycelium (data from 490) of *Neurospora crassa.* From Jennings (226).

Nitella clavata. The suggestion was based on the fact that the internal hydrogen ion concentration measured by various methods was not such as would be predicted from the external hydrogen ion concentration and the membrane potential using the Nernst equation. Further, DNP caused a reduction in the resting potential, suggesting that some energy-consuming mechanism maintains the membrane potential at the resting level. Since the difference between this potential and the hydrogen ion equilibrium potential was always maintained, notwithstanding a possible continuous inward current of hydrogen ions which would result from the potential difference, Kitasato proposed that hydrogen ions are pumped out of the cells.

Spanswick (502–505) by an extensive series of investigations showed that there is an electrogenic proton extrusion pump in *Nitella translucens.* This was established by demonstrating that the membrane potential in the light could be more negative than the theoretical negative maximum for the diffusion potential. An estimate of the flux through the pump was made by measuring the current required to depolarize the membrane to the negative limit of the diffusion potential set by the potassium equilibrium potential. The estimated flux was an order of magnitude too large to be due to any of the major ions found within the cell. The flux must be due to the active pumping out of hydrogen ions.

Studies with metabolic inhibitors indicated that ATP is involved. Thus the following points were observed: (1) Carbonyl cyanide *m*-chlorophenylhydrazone (CCCP) depolarized the cell to the dark level. This inhibitor is known to inhibit cyclic photophosphorylation (250). (2) 3,3-Dichlorophenyl-*N*,*N*-dimethylurea (DCMU), which blocks electron flow from system 2 to system 1 in photosynthesis, had no significant effect on the membrane potential in the light. The negligible effect of this compound at concentrations that inhibit carbon dioxide fixation suggests that electron flow and hence NADPH or an equivalent reductant is not involved in driving the pump. However, since cyclic photophosphorylation may occur in the presence of DCMU, it is likely that ATP is synthesized in the presence of this inhibitor (229).

As always, inferences from results obtained with inhibitors about the source of energy driving a pump should be treated with caution. However, Keifer and Spanswick (245) have now measured ATP levels in the cells in the presence of a range of inhibitors and have confirmed what had been inferred previously. The identification of hydrogen ions as the major ions exported by the pump was made by Spanswick through a process of elimination. Nevertheless, Spear *et al.* (510) have shown in *Nitella clavata* and Lucas and Smith (300) have shown in *Chara corallina* that there is light-stimulated acidification of the external medium. It seems legitimate to assume that hydrogen ions are being extruded. The key question is what other ions are moving at the same time?

Spear *et al.* (510) suggested that since the acidification of the medium and a major portion of chloride influx were both light dependent, the chloride flux might be "mechanistically" linked to proton extrusion. However, Lucas and Smith (301) have shown that there is a lag in chloride uptake which is greater than the lag in net hydrogen ion efflux. Possibly the hydrogen ion efflux is coupled to the influx of a cation such as potassium or sodium as in *N. crassa,* and as has been proposed by Smith (494, 496) for *C. corallina.* The actual evidence is slight for the reason that it is very difficult to quantify the situation in the algae as has been done for *N. crassa.* This is because the acidification of the medium is the net flux of hydrogen ions out of the cell, namely, the sum of effluxes of hydrogen and hydroxyl ions (influx of hydrogen ions). For *N. crassa,* under the experimental conditions used by Slayman and Slayman (486), the acidification of the medium appears to be due to true efflux of hydrogen ions. It is relevant that Kitasato (249) suggested from voltage clamp studies that potassium permeates the plasmalemma of *N. clavata* by channels whose properties are unaffected by the membrane potential, i.e., potassium permeates by a uniporter.

The first convincing evidence for the exchange of potassium or sodium for hydrogen ions in higher plants was reported in the important paper of Jackson and Adams (212). They showed with excised roots of 6-day-old barley *(Hordeum vulgare)* seedlings grown in 2×10^{-4} mol L^{-1} calcium sulfate in the dark that the rates of uptake of potassium or sodium could be independent of identities, concentrations, and rates of absorption of the anions in the external solution, including bicarbonate. They concluded that potassium and sodium are absorbed in exchange for hydrogen ions and anions in exchange for hydroxyl ions. Further, the release of hydrogen ions reflects a specificity of the uptake system for potassium and sodium such that it appears that hydrogen ions are exchanged in the specific rate-limiting reactions of cation absorption. This is the first indication that a specific translocation reaction is involved. The findings of Jackson and Adams were confirmed by Pitman (391).

Subsequent work by Higinbotham *et al.* has shown the presence of an electrogenic pump in cells from dark-grown roots and epicotyls of pea *(Pisum sativum), Zea mays* roots, and oat *(Avena sativa)* coleoptiles (10, 189, 190, 192). The most striking evidence came from the observations that rapid depolarization caused by carbon monoxide is readily reversed by light such that repeated depolarization and repolarization can occur. Since there was no change in membrane resistance as a result of these treatments, the most likely explanation is the presence of a pump transporting ions unidirectionally. The ion causing electrogenicity was not identified, but hydrogen ion was considered a distinct possibility.

There is now fairly conclusive evidence that the electrogenic pump in higher plant tissues is a hydrogen ion extrusion pump. Poole (395) has shown for red beet *(Beta vulgaris)* that potassium uptake and the cell membrane potential are dependent upon the hydrogen ion concentration in the medium, and not upon bicarbonate or chloride concentration. More recently, Mercier and Poole (341) have shown that a plot of membrane potential of red beet cells against tissue ATP concentration approximates a Michaelis relationship with a K_m of around 2 mmol L^{-1}. On the other hand, Petraglia and Poole (386) have shown that, in the presence of various inhibitors, the potassium influx across the plasmalemma shows a linear correlation with ATP concentration with no sign of saturation.

However, some of the strongest evidence for a hydrogen ion extrusion pump comes from the observation that indoleacetic acid (IAA) and fusicoccin (FC) can enhance hydrogen ion extrusion in oat coleoptiles (266, 328, 330, 331). Hydrogen ion extrusion promoted by IAA or FC is dependent upon the nature of the monovalent cations in the medium. Potassium and to a lesser extent sodium stimulate extrusion, while cesium and lithium have little activity or inhibit it. These observations and other data point to a coupling between potassium influx and hydrogen ion efflux (330).

Thus, there is fairly substantial evidence that distributed widely in the plant kingdom a hydrogen extrusion pump is coupled to the influx of monovalent cations. As shown earlier, the best-authenticated case is that of *N. crassa* in which data favor the hypothesis that ion transport is driven by the free energy of hydrolysis of ATP brought about by an ATPase which is activated by the ions which are transported. The presence of an ATPase in the plasma membrane of *N. crassa* has been shown unambiguously by Scarborough (455–457), who used vesicles of plasma membrane referred to earlier (Section I,A,5). Other information has been obtained from studies of *in vitro* preparations of disrupted mycelium (53, 54). The enzyme requires divalent cations for activity, magnesium and cobalt being the most effective. Sodium ions stimulate activity by about 20% and potassium by about 60%. However, these two monovalent cations do not act synergistically as they do with animal ATPases; nor does ouabain have any effect on the enzyme. Thus it is unlike the potassium–sodium–stimulated ATPase of animal cells. This appears to have discouraged studies to relate the activity of the enzyme *in vitro* with the *in vivo* transport system in plants. Thus there are no data on the affinity constant for potassium ion stimulation of ATPase activity which can be compared with an analogous constant for the affinity of the transport system for the same ion. Such comparative information is badly needed (see Section III,E).

There are numerous reports of cation-stimulated ATPases from a variety of plant tissues [see Hodges (201) for a review]. In spite of many studies, the evidence that ATPases bring about the hydrogen ion–monovalent cation exchange in plant cells is not very convincing. The reason is that most of the studies do not meet the criteria outlined in Section III,E. There are two major reasons why progress is handicapped. First, it is difficult to measure the active fluxes associated with the ATPases *in vivo*, and consequently it is difficult to obtain kinetic constants to compare with those from *in vitro* studies. Second, it is difficult to decide how far the ATPase activity of any membrane preparation is due to the specific translocase under investigation. The extensive comparative study by Bowman *et al.* (53) of the effects of inhibitors on the plasma membrane and mitochondrial ATPase of *N. crassa* suggests how the second problem may be overcome, but the first problem seems at present to be intractable.

b. Na^+/K^+-Antiporter-ATPase. This is the classic primary translocation system of animal cells in that it was the first to be identified and to be described mechanistically. The work has already been referred to in this chapter (Section III,E).

There is circumstantial evidence that a similar system is present in a number of algae, including *N. translucens* (see 421, 423). The most detailed characterization is for *Hydrodictyon africanum* (410). Sodium efflux is related to external potassium concentration in a manner describable by Michaelis kinetics (Fig. 8). Active potassium influx (that which occurs only in the light) depends on external potassium concentrations in exactly the same way. The fluxes of sodium and potassium are thus coupled and both are inhibited by ouabain, although at rather high concentrations (5×10^{-4} mol L^{-1} for maximum inhibition). The inhibitor has no effect on the plasmalemma potential differences, which is what would be expected from an electrically neutral pump. Raven (417) has argued from inhibitor studies that the coupled active fluxes of potassium and sodium are driven by ATP, but relevant biochemical studies have not yet been made.

It is important that in *Chlorella pyrenoidosa*, for example, though sodium efflux is sensitive to the presence of potassium in the external medium, there is no effect of ouabain at 5×10^{-4} mol L^{-1} (32). In *C. corallina* neither addition of ouabain nor removal of potassium from the external medium inhibits efflux of sodium (134); this is unlike what has been observed for the related *N. translucens* (310). It seems that (1) the use of ouabain may not necessarily give a clue to the presence of coupled potassium–sodium exchange fluxes and (2) from the work on *C. corallina*, one should not generalize about the presence or absence of such an exchange mechanism even within a group of related algae.

It seems from the studies of higher plant cells by Jeschke (230–235) that

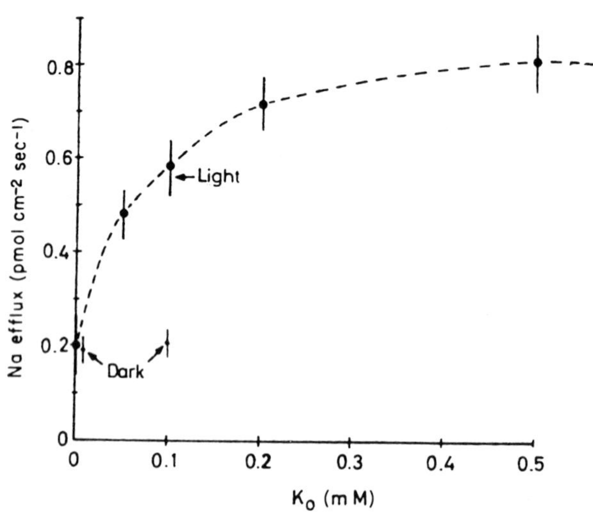

FIG. 8. Effect of the concentration of external K^+ on Na^+ efflux in *Hydrodictyon africanum* in the light and in the dark. From Raven (410).

there is a potassium–sodium exchange system in barley *(H. distichon)* roots. Although not all potassium uptake is linked to sodium efflux, nevertheless part of the influx appeared to be so linked, and compartmental analysis indicated that the mechanism was located in the plasmalemma. The affinity of the external site of the exchange system ($K^+ > Rb^+ > Cs^+ \gg Li^+$) corresponds qualitatively with that of isolated plant (and animal) ATPases. More recent studies (236) suggest that potassium–sodium exchange is not directly mediated by an ATPase activated by the two cations, but is brought about by a proton pump in association with a high-affinity potassium uptake system and a H^+/Na^+-antiporter moving sodium into the external medium. A Na^+, K^+-ATPase has been extracted from sugar beet *(B. vulgaris)* (243, 265).

c. Cl^--Uniporter-ATPase. Gradmann and Bentrup (157) and Saddler (447–449) showed that the membrane potential of the alga *Acetabularia* appears to be controlled by the electrogenic influx of chloride ions, the fluxes of which are very large, much larger than those of other ions (Table III). Gradmann (156) has shown that the membrane potential of *A. mediterranea* consists of an electrogenic pump in parallel with a passive system (giving rise to a diffusion potential). Chloride is believed to be taken in by the electrogenic pump which is driven by ATP. The evidence for the involvement of ATP comes from the qualitatively similar effects of CCCP in reducing the level of ATP and depolarizing the membrane.

In the steady state, charge balance is maintained by an appropriately high chloride efflux (Table III). A drastic transient depolarization of the membrane can be triggered by transfer from light to darkness in the presence of choline chloride, the response taking about 3 min. Gradmann *et al.* (159) have shown that the depolarizing inward current is carried by

TABLE III

Fluxes of Major Ions in *Acetabularia mediterranea*[a,b]

Ion	Influx	Efflux
Na^+	11–49 (20)	15–37 (8)
K^+	11–40 (36)	?
Cl^-	200–790 (35)	290–850 (20)
SO_4^{2-}	0.165 ± 0.003 (4)	?

[a] From Saddler (448).
[b] Fluxes in pmol cm^{-2} s^{-1}.

an efflux of chloride ions. This confirms the idea that there is a close linkage between active chloride influx and passive efflux of the ion.

A similar type of active transport system appears to be present in *Halicystis parvula* (161, 162). There is also strong evidence from the work of Hill and Hill (196, 197, 199; see Chapter 5) that chloride extrusion from the salt glands of *Limonium vulgare* is brought about by an electrogenic chloride pump (ATPase) driven by ATP, which is now becoming well characterized (198). It seems likely that there is present in the glands a transport system similar to that of *Acetabularia* but working in the reverse direction (moving chloride ions out of the cell) and with charge balance maintained by the accompanying movement of sodium ions.

3. Secondary Translocation

a. Glucose/H⁺ Symporter. Powerful support for the presence of such a transport system in plants comes from the work of Slayman and Slayman (487) on *N. crassa*. Using mycelium which had been starved of carbon substrate for 3 to 5 h, they showed that when 3-*O*-methyl-D-glucose was added to the external medium, the plasma membrane became depolarized. The sugar was not metabolized, nor were there marked changes in mycelial ATP levels. Thus the depolarization of the membrane seems to depend upon the transport of 3-*O*-methyl-D-glucose. This sugar also causes the medium to become more alkaline at a rate approximately the same as the uptake of the sugar. Glucose causes a similar response and the peak (maximal) depolarization for varying concentration of this sugar in the medium fits a Michaelis relationship with a K_m similar to that for glucose uptake. All the data support the idea that glucose and hydrogen ions are cotransported under the influence of the membrane potential which is sustained by an ATP-driven hydrogen ion extrusion system (Section IV,A,2,a).

The same system appears to be present for the transport of glucose and other monosaccharides into the yeast *Rhodotorula gracilis* (175, 204). The method used for measuring the potential across the plasmalemma has been described earlier (Section II,B).

There is also very good evidence that the inducible hexose transport system of *Chlorella vulgaris* is also a glucose/H^+ symporter (253, 257, 258, 261, 262). However, for this organism it is not clear whether ATP is involved in the maintenance of the proton gradient across the membrane (259, 260).

It is likely that the same transport system is present in higher plant cells. Hexoses have been shown to depolarize the membrane of *Lemna gibba* cells, the effect and the recovery of the potential being dependent upon external pH (361, 362, 550). Fusicoccin stimulates 3-*O*-methylglucose uptake into maize coleoptiles and roots and pea stems, and the effect has been ascribed to alkalinization of the cytoplasm, which accelerates the release of protons from a hexose/H^+ symporter (83) (see Section IV,A,2,a).

b. Cl^-/OH^- Antiporter. Smith (494) postulated the presence of such a translocase in *C. corallina* as a result of experiments aimed at investigating the effect on the chloride influx of changing either the external or internal pH. Large stimulations were obtained by the additions of imidazole and Tris buffers and ammonium sulfate in both light and dark (Table IV). These compounds have little effect on photosynthesis or respiration, and so the stimulation seemed unlikely to be due to an effect on the energy supply. Smith pointed out that they act either through their ability to enter the cell in exchange for hydrogen ions or as a result of the ready penetration in the uncharged form through the plasmalemma. The former mechanism seemed the more likely for ammonium, since the amounts of free NH_3 in solutions of ammonium sulfate are very small, but the latter is more likely for imidazole and Tris (see, however, Section IV,A,3,c). Whatever the actual process of entry, the net result would be an alkalization of the cytoplasm. With the former mechanism, there would be a decrease in the hydrogen ion concentration of the cytoplasm; with the latter, uncharged base would dissociate, thus increasing the hydroxyl ion concentration in the cytoplasm. Thus the stimulation of chloride influx is believed to be due to an increased availability of hydroxyl ions in the cytoplasm for exchange.

The hypothesis that there is a Cl^-/OH^- antiporter is difficult to test. Lucas and Smith (301) have made parallel studies on the influence of light

TABLE IV

EFFECT OF IMIDAZOLE, TRIS, AND AMMONIUM SULFATE ON Cl^- INFLUX IN *Chara corallina* IN LIGHT AND DARKNESS[a,b]

Experiment	Conditions	Influx in light	Influx in darkness
1	Controls (basic solution)	0.64 ± 0.12	0.19 ± 0.01
	+1 m*M* imidazole (pH 7.1)	2.46 ± 0.46	1.34 ± 0.30
2	Controls	0.47 ± 0.09	0.22 ± 0.03
	+1 m*M* Tris (pH 7.2)	2.04 ± 0.19	0.84 ± 0.16
3	Controls	0.59 ± 0.12	0.27 ± 0.08
	+0.05 m*M* $(NH_4)_2SO_4$	2.31 ± 0.23	0.54 ± 0.06

[a] From Smith (494).

[b] The cells were all pre-treated in the light for 1 hour. Fluxes in pmol $cm^{-2}\ s^{-1}$.

on net hydrogen ion efflux as measured by pH changes at the cell surface (see 288, 300) and ^{36}Cl influx. The rationale behind their experiments was as follows. In the light and according to Smith's (494) original hypothesis, chloride/hydroxyl ion exchange is driven by the primary active extrusion of hydrogen ions (Fig. 9). This process leads to alkalization of the cytoplasm, and the Cl^-/OH^- antiporter would be driven by the pH (and electrical potential) gradient across the plasmalemma. Smith also proposed that chloride influx would be limited by the rate of hydrogen ion extrusion.

The work by Lucas and Smith (301) provided information about the time relationship between chloride and hydrogen ion efflux. Thus, following a dark-to-light transition, there was a lag of 8 to 15 min before a net hydrogen ion efflux was observed, while there was a lag of 40 to 60 min before the chloride transport system attained maximum activity. These facts, together with calculations of the Gibbs free energy involved in the coupled flow of hydroxyl and chloride ions under the conditions of the experiment, suggest that hydrogen ion extrusion per se by the primary active transport process is not the immediate trigger for bringing about

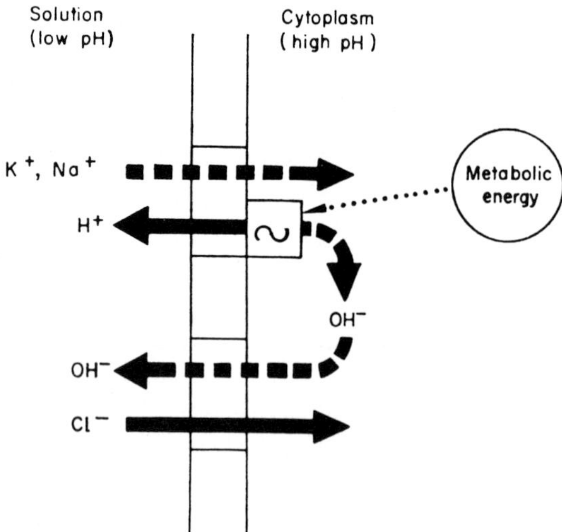

FIG. 9. Hypothetical scheme of Smith (494) for salt uptake across the plasmalemma involving two ion exchange sites. The system is dependent on a supply of metabolic energy from chloroplasts or mitochondria. This is used primarily to separate H^+ and OH^- at a site indicated by ~. Active (thermodynamically uphill) ion fluxes are shown by solid arrows and passive (downhill) fluxes are shown by broken arrows.

exchange of chloride and hydroxyl ions. However, the results are not inconsistent with the presence of the antiporter itself; its activity may be determined by some product of photosynthesis which takes 40 to 60 min to exert its maximal effect.

The real answer could be that this particular transport system is wrongly classified. The ideas of Mitchell (Section V,A), which underly the hypothesis of Smith, may, however, in this context be inappropriate. Mitchell's ideas concern the energy-conserving membranes of mitochondria, chloroplasts, and bacteria and exchange–diffusion systems are a necessary requisite for their proper functioning, i.e., the generation of ATP. Whether this is also true for the plasmalemma is not clear. So there is no a priori need for the presence of an exchange diffusion Cl^-/OH^- system in this membrane. It may well be that chloride is transported into algal cells through the mediation of an ATPase whose activity is under the influence of, amongst other internal factors, cytoplasmic pH; i.e., the relation between chloride and hydroxyl ion exchange is indirect.

More recent work gives some support to this view. Ammonium is now known to enter *C. corallina* and *H. africanum* by a passive electrogenic uniporter (Section IV,A,3,c). Moreover, measurements of cytoplasmic pH in *C. corallina* in light and dark indicates that if the Cl^-/OH^- system exists, two protons (hydroxyl ions in the reverse direction) must accompany each chloride ion (497, 569). Recent work by Sanders (450) with *C. corallina* has shown that chloride influx in *C. corallina* is stimulated by a factor of 2 to 4 by starvation of chloride, suggesting that the transport of the anion across the plasmalemma may be under the control of cytoplasmic chloride concentration (feedback inhibition; see Section VI,C). He goes on to point out that some of the studies by Smith reported earlier can be explained as satisfactorily in terms of changes in cytoplasmic chloride concentration as they can in terms of changing cytoplasmic pH. Since Sanders (451) has shown that chloride influx can be increased by perfusing the interior of the cell (minus its tonoplast and streaming cytoplasm) with ATP and that chloride influx appears to be directly affected by cytoplasmic pH (452), chloride uptake by *C. corallina* may be mediated by an ATPase (see Section IV,A,2,c) whose activity is sensitive to small changes in pH within the range which might be expected in the living cell.

c. Methylammonium/Ammonia Uniporter. This system is responsible for transporting ammonium ions and derivatives such as methylammonium. It has been shown to be present in *C. corallina* (497) and *H. africanum* (495). The studies have relied heavily upon the use of methylammonium, which is a stronger base (pK_a of about 10.65) than ammonium and can be obtained labeled with ^{14}C. The evidence in favor of the system is as follows:

1. Methylammonium influx is not directly proportional to methylammonium concentration in the external solution. This argues most strongly against diffusion of the undissociated molecule into the cell, as does the fact that there is accumulation of methylammonium inside the cell vacuole when the external pH is equal to, or lower than, the vacuolar pH.
2. There are large depolarizing currents and increases in conductance associated with methylammonium uptake which are consistent with the flow of positive charge into the cell, and they are of a magnitude expected for an influx of methylammonium cation.
3. The flux values are such as to rule out passive diffusion of the cation across a lipid membrane. A carrier must be involved.
4. There is no need to invoke an active mechanism since the concentration inside the cell appears to be what would be expected for a Nernst equilibrium.
5. Methylammonium influx is not affected by the presence of chloride, potassium, or sodium in the external medium, but is decreased by the addition of ammonium.

The evidence for a uniporter is convincing. The evidence that it also transports positively charged ammonium comes from an analysis of current flows across the plasmalemma of cells of *C. corallina* when the potential difference across the membrane is kept constant (voltage clamped) (566, 571). The dependence of the current flow on concentration of positively charged ammonium in the external medium follows a Michaelis relationship giving a K_m of 3 μmol L^{-1}. This value is very similar to the K_i (20 μmol L^{-1}) for the inhibition by positively charged ammonium of [^{14}C]methylammonium uptake. Further, such uptake is inhibited by ammonium even when the potential across the plasmalemma is kept constant. These data indicate that ammonium competes for the same carrier as methylammonium and that inhibition of its uptake by ammonium is not due to depolarization of the membrane.

Since a positive charge is being taken across the plasmalemma by this uniporter, charge balance has to be maintained. This is by net efflux of potassium in *C. corallina* and potassium, sodium, and hydrogen ions in *H. africanum*.

More recently, using methods similar to those already described, Felle (131) demonstrated the presence of the uniporter in the plasmalemma of thallus cells of the liverwort *Riccia fluitans*. There is evidence that the same transport system may be present in fungi (165, 383).

4. *Comment*

Essentially six transport systems have been described and classified according to the taxonomy of Mitchell (345). Only for these six systems is the evidence sufficient to indicate with some certainty the major features of the mechanism. These restrictions, together with uncertainty about the operation of some of these mechanisms, demonstrate the paucity of information underlying our understanding of the mechanism of transport of ions in plant cells. On the other hand, there are considerable difficulties in probing specific transport mechanisms in these cells. It is particularly difficult to couple nutrient ion movements to the movement of hydrogen or hydroxyl ions. Measurement of individual fluxes of these latter ions may be impossible because for most plant cells one cannot distinguish between the movement of the two ions and even the kinetics of net movement of hydrogen ions across the membrane (remember that the movement of hydrogen ions can be considered as the movement of hydroxyl ions in the reverse direction) is complicated by the buffering capacity of the cell wall. The work on *N. crassa* indicates that progress may be possible if the organism is in a metabolic state in which a flux of hydrogen ions is predominant. It is also clear from the preceding examples that, if the transport system is electrogenic, the tempo of change of the potential when the pump is inhibited can provide important qualitative information. Likewise identification of the involvement of ATP allows for the search for an ATPase with properties which when found can be compared with the transport process in the intact cell.

Before concluding this discussion of known transport systems, it is appropriate to make some points about other transport systems not described here, but which can be fitted into the above taxonomy.

1. There is almost certainly a glycine/H^+ symport in *Saccharomyces carlsbergensis* (462). Other amino acids and carbohydrates also seem to be cotransported with protons in a number of fungi (115, 116, 226, 227, 463, 577).
2. Convincing evidence is coming forward that the loading of sucrose into the sieve tube occurs via a sucrose/H^+ symporter (22, 142, 143, 209, 210, 254–256, 326, 559). Also amino acids appear to be cotransported with protons in a variety of higher plant cells (25, 36, 128, 239, 280, 360).
3. There are a number of reports of the specific stimulation by sodium of phosphate uptake into plant cells. The relevant organisms are the chlorococcal alga *Ankistrodesmus braunii* (478, 549, 551) and the fungi *Thraustochytrium roseum* (472, 473) and *Saccharomyces cerevisiae* (43, 442). The system may be widespread in algae (419, 552), but has not yet been established in higher plants.
4. Upake of silica [$Si(OH)_4$] by the diatom *Navicula pelliculosa* is stimulated by monovalent cations, sodium being the most effective (522). This suggests that there may be cotransport of a cation with silica as it moves into the cell. Roots are known to transport the essentially nonpolar silica

against a concentration gradient (30); this may be through the involvement of a cotransport system—movement of silica up its concentration gradient being driven by the cation moving down its electrochemical potential gradient.

5. During plant cell growth there is likely to be exchange of ammonium and hydrogen ions and nitrate and hydroxyl ions (427). We need to know for higher plants in particular whether the two exchanges occur via common translocation reactions.

With regard to transport of nitrate into cells, it may be difficult to distinguish between a simple secondary transport system (in terms of nitrate) by the chemical potential gradient into the cell created by the action of nitrate reductase and nitrate reductase itself being involved in the transport process. The latter possibility has been suggested and explained in terms of the known molecular structure of nitrate reductase (76). Pateman and Kinghorn (382) report no indication of a nitrate transport system independent of nitrate reduction in *Aspergillus nidulans,* and they have argued that this is so from work on other fungi. On the other hand, Heimer and Filner (178), using cultured tobacco cells, have shown the presence of an energy-dependent uptake system independent of nitrate reductase. They made the observations that in the presence of tungstate the reductase is nonfunctional, but nitrate transport can continue.

B. Systems Which Are Not Yet Classifiable

There are three transport processes which are not readily classifiable but about which there is a considerable amount of information. They are of interest in a broader context than the straightforward issue of how an ion moves across a membrane; for this reason it is appropriate to deal with these processes separately.

1. HCO_3^-–OH^- Exchange

Most plants use carbon dioxide for photosynthesis; it seems that bicarbonate assimilation is confined to certain algae and to some aquatic angiosperms (416, 513). The principle method used to assess the relative rates of carbon dioxide and bicarbonate assimilation has been to compare the relationship between unhydrated carbon dioxide concentration and rate of photosynthesis at low pH at which 90% or more of the inorganic carbon species is as carbon dioxide and at a much higher pH at which more than 95% of the inorganic carbon is in the form of bicarbonate. When bicarbonate assimilation takes place, the ions are dehydrated as a result of carbon dioxide fixation and hydroxyl ions are produced inside the cell. In contrast, those plants which cannot use bicarbonate produce hydroxyl ions outside the cell.

Knowing this, there are a number of a priori questions which need to be answered: (1) How is bicarbonate uptake coupled to hydroxyl ion loss from within the cell? (2) Are fluxes of other ions involved in bicarbonate uptake, e.g., influx of cations? (3) Are any of the fluxes active?

The last question can be disposed of relatively easily. Both the influx of bicarbonate and the efflux of hydroxyl are likely to be active. The arguments in favor of the active influx of bicarbonate are based on a calculation using the Goldman equation (155), with passive permeability for the ion just sufficient to support its measured rate of use. The potential difference across the membrane has to be known, and it must be assumed that the internal bicarbonate concentration is zero. Raven (411) calculated a permeability value for bicarbonate of 10^{-4} cm s^{-1} for *H. africanum,* while Smith (493) calculated values between 5×10^{-4} and 5×10^{-5} for four species taken from different genera of the Characeae, compared with values for chloride of between 7×10^{-9} and 4×10^{-10} for the same species. It was argued that if the permeability for chloride represents the intrinsic permeability of membranes for an anion, it was legitimate to conclude that bicarbonate ions are pumped into these coenocytes. Facilitated diffusion can be ruled out because the process could not produce an internal concentration of carbon dioxide of sufficient magnitude to match the known K_m values for that molecule of ribulose-bisphosphate carboxylase/oxygenase (416).

The loss of hydroxyl ions is also unlikely to be passive. Algae such as *H. africanum,* which is known to be capable of active photosynthesis in a medium of pH 10.5, would need a cytoplasmic pH of 9.0 for hydroxyl ions to diffuse passively out of the cell (416).

The coupling of bicarbonate uptake with hydroxyl ions might be through influx of the former and efflux of the latter via a common carrier. If the exchange were one to one, the carrier would be electrically neutral. Walker (563) showed that the addition of bicarbonate to a bathing medium containing chloride as the only other anion caused polarization of the plasma membrane of *C. corallina.* This has been confirmed subsequently by Hope (205), who also showed that in the hyperpolarized state the potential across the plasmalemma was almost unresponsive to potassium and that darkness slowed or inhibited the change in the potential difference (p.d.). A large reversible increase in the resistance of the plasmalemma accompanied the transition from light to dark. From these and other data, Hope proposed that there is an electrogenic bicarbonate pump at the plasmalemma of *C. corallina* which is linked to photosynthetic reactions. However, Spanswick (500) showed that for *N. translucens,* which can also assimilate bicarbonate, the electrical potential changes resulting from the addition of bicarbonate ions can also be produced by the addition of a variety of buffers giving the same pH. Further, bicarbonate ions had no effect when the cell had previously been in a buffered solution at the same pH. The results can be explained in terms of the properties of the electrogenic hydrogen ion extrusion pump described elsewhere (Section IV,A,2,a).

Hence, if bicarbonate influx is coupled to hydroxyl efflux, the coupling must be electrically neutral, a point emphasized by Lucas and Smith (300) when they made their suggestion that the two influxes may be obligately coupled (i.e., a HCO_3^-/OH^- antiporter). However, subsequent experiments by Lucas (291) have shown unequivocally that hydroxyl ion efflux in *C. corallina* can occur in the absence of exogenous bicarbonate. He also found that bicarbonate ions can be transported across the plasmalemma within the dark segment of a partly illuminated cell. Lucas (289, 290) has explained the implication of his findings as follows. The main effect of the proposed system would be on the resistance of the membrane. If the cell is supplied with bicarbonate and it has the capacity to utilize this substrate, the membrane resistance should decrease significantly. In the presence of carbon dioxide no such resistance changes should be observed since there is no operation of the bicarbonate and hydroxyl ion systems. The results of Hope (205) referred to earlier are in keeping with this. Indeed, Lucas (292) in a study of the influence of calcium and potassium on bicarbonate influx in *C. corallina* has obtained data which support the proposal that bicarbonate assimilation contributes toward the electrical properties of the plasmalemma.

The nature of the transport processes has been clarified by subsequent studies (132, 293–299). The most recent findings are as follows:

1. Although previously it has been assumed that hydroxyl ions were lost from the cell, hydrogen influx could achieve the same overall result, in terms of pH changes at the outer suface (Fig. 10). The incoming hydrogen ions could neutralize in the cytoplasm those hydroxyl ions produced by photosynthesis. By changing the pH locally in an alkali-producing area of the cell, it has been possible to show that alkali production is insensitive to pH, showing that it is hydroxyl ion efflux, not hydrogen ion influx, which generates the alkalinity.

2. Simultaneous measurements of the local pH and electrical potential at the surface of the cells of *C. corallina* now confirm that bicarbonate and hydroxyl ion transport occur at different sites (291). Lucas (294) has pointed out that one of the main consequences of coupling of bicarbonate and hydroxyl transport would be the build-up of hydroxyl ions at the outer plasmalemma surface. This would convert bicarbonate to carbonate and

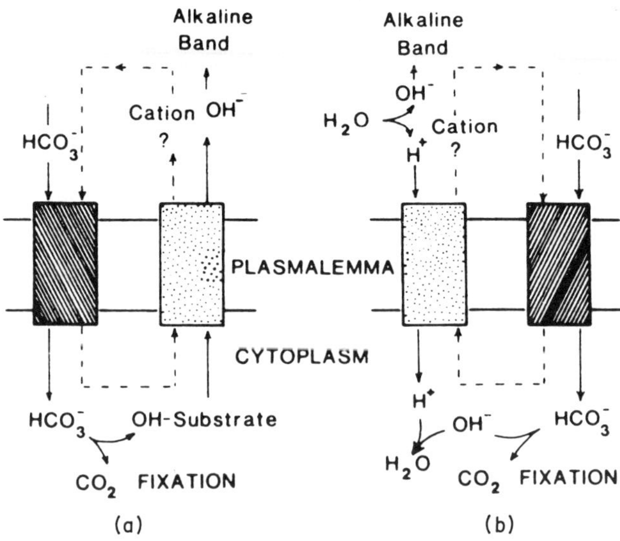

FIG. 10. Schematic representation of two modes by which alkaline bands may be formed on the surface of *Chara corallina* cells. HCO_3^- and OH^- (or H^+) carriers exist at spatially separate sites within the plasmalemma. From Lucas (293).

hence reduce substrate availability. It has been held for some time (14, 512) that the uptake of bicarbonate and the excretion of hydroxyl ions can be spatially separate (See Section IV,B,2,b).

3. Both bicarbonate and hydroxyl ion transport systems require calcium for activity and sulfhydryl groups are involved in the functioning of the system (295).

These intriguing systems require further investigation, particularly of the manner in which metabolic energy is imparted into either one or both of the transport processes and how the two systems act in an integrated manner (294).

2. *Calcification*

There is no evidence that bicarbonate influx is directly coupled to an influx of cations (416). Nevertheless there is evidence for a relationship between bicarbonate/hydroxyl exchange and bivalent cations, particularly calcium, in the process of calcification which is observed in many photosynthetic aquatic plants. The process is generally described as follows: there is a production of hydroxyl ions as a result of photosynthesis (either with carbon dioxide or bicarbonate), which raises the carbonate concentration in the solution to such a level that calcium carbonate is precipitated. This is an oversimplification. Work on the subject is fragmentary and except for the study with the seaweed *Halimeda* dealt with later, no clear picture has emerged to show how calcification occurs. There are two general points to remember in any consideration of the phenomenon: (1) Early work in this field should be accepted with reservations when rates of calcification were determined by using ^{45}C. The procedure overestimates the rate owing to binding of the isotope to cell walls and retention of it in any intercellular space of thalloid organisms (38). (2) Many natural waters are supersaturated with calcium carbonate, and so precipitation requires nucleation rather than an increase in carbonate concentration. Precipitation may then occur by the removal of compounds which inhibit crystal formation such as phosphate (475).

Three areas of research in which there is significant information, merit attention. Pentecost (384) should be consulted for a general overview of calcification in the various plant phyla.

a. Coccolith Formation (96, 375, 474). Coccoliths are scales consisting of calcium carbonate (generally aragonite) deposited on a polysaccharide matrix which are produced inside the Golgi apparatus and transferred to the cell surface. The physiology of the process is not clear. In particular, we need to know how calcium reaches the Golgi apparatus. Coccolith formation is one aspect of intracellular biomineralization which occurs in many living organisms and which Simkiss (476) has argued may be a method for the removal of calcium or protons from the cell, namely,

$$Ca^{2+} + HA^- \rightarrow CaA + H^+$$

b. Precipitation of Calcium Carbonate on the Upper (Adaxial) Surface of the Leaves of Submerged Angiosperms (285–287, 512). These plants, which include several species of *Potamogeton* and *Elodea,* assimilate bicarbonate from both adaxial (upper) and abaxial (lower) leaf surfaces but release hydroxyl ions only on the adaxial surface. Further, carbonate, bicarbonate, and cations as well as hydroxyl ions appear in an initially ion-free solution above the latter surface. The various mechanisms that have been postulated to drive these movements are (1) a hydroxyl ion pump at the adaxial surface (512), (2) a cation (calcium) pump at the same surface (285–287), (3) higher hydroxyl ion permeability at the adaxial compared with the abaxial surface (60).

More work is needed on this very interesting phenomenon, which brings to mind the proposed separation of bicarbonate and hydroxyl transport systems in the Characeae (see Section IV,B,1). A valuable start has been made by Helder and colleagues (179–184, 402). They have characterized in some detail the pH changes over the two leaf surfaces under a variety of conditions. They have also shown that when the leaf becomes polarized with respect to bicarbonate uptake and hydroxyl ion release, the upper surface becomes electrically negative with respect to the lower, and there is net cation movement from the former to the latter surface. However, polarity is not absolute; bicarbonate can be absorbed by the upper surface when it is absent from the lower. As yet firm conclusions are not possible; there is a need for studies of ion movements at the cellular level as well as at the organ level.

c. Calcification in the Macroalgae. There are many studies on calcification in the macroalgae (96, 275), but the most comprehensive study is that by Borowitzka and Larkum (38–41) on the process in *Halimeda tuna.* They showed that calcification is stimulated by light. Not only is photosynthesis necessary, but the site where calcification occurs must be separate from the external medium. A schematic representation of the proposed fluxes in calcification is given in Fig. 11. In considering this figure, it needs to be realized that the algae are made up of loosely packed parallel running coenocytic filaments with large intercellular spaces. The filaments produce branches which turn at right angles and themselves produce short swollen branches (the utricles) which compose the cortex. The branching is controlled in such a way that an apparently fleshy structure (a segment) is produced, the peripheral utricles of which adhere by protrusions of the cell wall to produce a closed outer surface.

Thus an important feature of the alga is the isolation from the external medium of the intercellular space from which carbon dioxide is used during photosynthesis and returned during respiration. This gas exchange is facilitated by the fact that 70% of the total cell surface area faces the intercellular space. Essentially there is a net removal of carbon dioxide and with the limited volume there is a rise in pH and of CO_3^{2-} which because seawater is supersaturated with $CaCO_3$, leads to precipitation of aragonite. Bicarbonate ion uptake by itself will have little effect on the intercellular pH and the concentration of CO_3^{2-}, but if hydroxyl ion loss from the cytoplasm accompanies bicarbonate uptake, this will have the same effect as uptake of carbon dioxide. One of the major uncertainties in the model is the extent of hydrogen ion loss from the cells since this can clearly inhibit calcification. Borowitzka and Larkum (41) attempted to quantify this loss indirectly, but no clear picture has emerged. Further work is needed, perhaps involving the insertion of microelectrodes into the intercellular spaces to obtain a direct measure of pH changes.

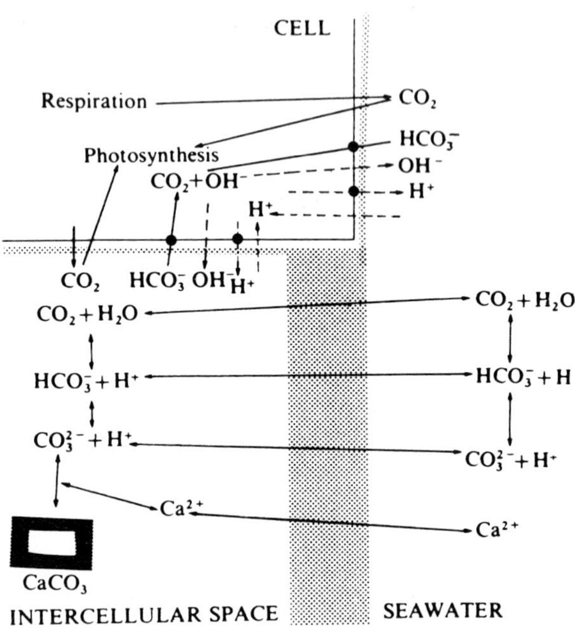

FIG. 11. Schematic representation of the postulated ion fluxes which affect $CaCO_3$ precipitation in *Halimeda tuna*. Fluxes whose existence has been demonstrated are shown by solid arrows, and fluxes postulated but not confirmed are shown by broken arrows. A dot at the plasmalemma indicates that this flux is postulated to be active. Passage of ions from the seawater to the intercellular space is by diffusion via 20 μm or more of the cell wall of the appressed utricles. CO_2 for photosynthesis enters the cell by diffusion from both the external medium and the intercellular space, and CO_2 produced during respiration diffuses out of the cell. HCO_3^- probably enters the cell by an energy-requiring process. After dissociation of the CO_2 and HCO_3^-, OH^- may leave the cell (possibly in very localized regions). A light-stimulated H^+ efflux is also postulated. From Borowitzka and Larkum (40).

3. Iron Uptake

Iron is a micronutrient for plants and when there is a deficiency of the element, higher plants exhibit chlorosis. Such symptoms can occur in plants grown in solution culture in which iron is added in inorganic form. This is because of hydrolysis and oxidation of the iron followed by precipitation. It is now customary to avoid this by presenting the iron in chelated form, particularly with citrate or ethylenediaminetetraacetic acid (EDTA). Investigators have been concerned about establishing how the plant obtains iron presented to it in chelated form. Does the plant separate the ion from the chelate prior to transport across the plasmalemma, or is the iron absorbed in chelated form?

An important point to establish is whether ionic iron can be absorbed. There seems little doubt that ionic iron as Fe^{2+} can be absorbed by excised roots of rice *(Oryza sativa)* (242). Christ (79, 80) tackled the problem by separating the nutrient solution into two components, one containing the nutrient elements except the iron salts and the other containing only the iron salts. The separation can be carried out spatially (split root) or temporally (change of medium). Iron uptake was not measured; the utilization of the element from the solution was assessed by the degree of chlorosis induced. Differences in the degree of chlorosis between species were obtained when they were presented with Fe^{3+}, but these differences disappeared when Fe^{2+} was used. Christ concluded that iron entered plants as Fe^{2+} and that Fe^{3+} could only be utilized if the plant were able to reduce the Fe^{3+} to Fe^{2+}.

Jeffreys *et al.* (220) found that there was equivalent uptake of iron and chelator, suggesting that the two were absorbed as one entity. On the other hand, Tiffin *et al.* (536, 537) showed that this need not be so. In particular, in experiments with sunflower *(Helianthus annuus)* in culture medium presented with iron chelated with ethylenediaminedi(*O*-hydroxyphenylacetic acid) (EDDHA), they found a progressive loss of iron from the medium and an accumulation in the root exudate, reaching eight times that in the medium. With loss of iron from the medium there was a sevenfold increase in the iron chelating capacity of the medium due to an increase of iron-free EDDHA. The difference between the results obtained by Tiffin and colleagues and those of Jeffreys *et al.* (220) appear to be due to different growth conditions for the plants (200). Thus iron can be absorbed separately from the chelator, but further work is needed to see whether the equivalence observed by Jeffries *et al.* was fortuitous.

If iron and the chelator are separated after entry, this may be as a result of metabolism of the organic part of the chelate complex. This can certainly take place, as demonstrated by the presence of breakdown products of ^{14}C-labeled chelator on radiochromatograms of plant extracts (200).

Brown *et al.* (66, 67, 69) suggested that the separation of iron from the chelator before the entry of iron involves the reduction of Fe^{3+} chelate to Fe^{2+} chelate. Chaney *et al.* (78) showed that the ability of a plant to recover from iron deficiency is correlated with the reductive capacity of the roots. This demonstration involved the use of the Fe^{2+} color reagent 4,7-di(4-phenylsulfonate)-1,10-phenanthroline (BPDS), which forms a soluble nonautoxidizable Fe^{2+} chelate of high stability and low exchangeability. It only binds Fe^{3+} weakly; EDDHA binds Fe^{3+} strongly but Fe^{2+} weakly.

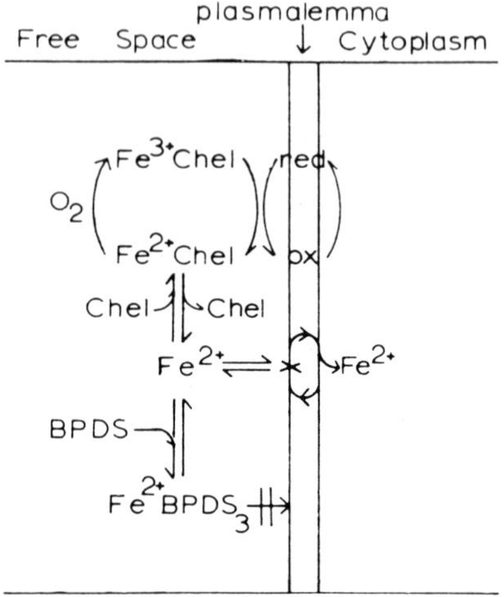

FIG. 12. A scheme for the uptake of iron by higher plants by obligatory reduction of ferric chelate (Fe^{3+} Chel) and its prevention by BPDS. Proposed by Chaney *et al.* (78).

Addition of BPDS to nutrient solutions containing Fe-EDDHA inhibited iron uptake. Figure 12 shows schematically what happens.

Confirmation of this comes from the studies of Christ (79, 80). In the studies referred to earlier, he also showed that the ability of plants to absorb iron from Fe^{3+} chelates was likewise a function of the ability to reduce Fe^{3+} to Fe^{2+}.

The source of the electrons for the reduction of Fe^{3+} chelates is uncertain. Although reducing compounds [probably with oxidizable phenolic groups (368)] are released into the medium in which iron-deficient soybeans are growing, these compounds reduce Fe^{3+} chelates very slowly (63, 78). Chaney *et al.* (78) suggest that a membrane-bound system may be involved, e.g., a cytochrome or flavin.

Plant species and varieties within species may differ in their susceptibility to iron-deficiency chlorosis (62, 67, 368–370). The differences appear to be due to differing abilities (1) to reduce Fe^{3+} to Fe^{2+} at the root surface, (2) to release reducing compounds into the medium, and (3) to release hydrogen ions, thus increasing the solubility of Fe^{3+} in the medium. Interestingly, Kramer *et al.* (263) have found that under conditions of iron deficiency there is an induction of transfer cells in the epidermis of sunflower *(H. annuus)*.

V. Ion Transport and Metabolism

A. DEVELOPMENT OF PRESENT CONCEPTS

For an account of the way our ideas have evolved about how metabolism might drive ion transport, the chapter by Steward and Sutcliffe (Chapter 4, Vol. II) is required reading. There is an excellent summary of the experimental data which led Lundegårdh, Conway, and Robertson to each develop their own ideas about the possibility that ion transport may be driven by a membrane-located redox system. Since the date of the chapter, it is Robertson who continued with this theme and convinced those who were originally sceptical about the validity of the ideas.

The views put forward by Robertson in a review in 1960 (437) were seminal to the present outlook on the relationship between membrane transport and metabolism. In that review, he re-evaluated his own previous views and those of Lundegårdh and Conway that redox systems are directly involved in the transport of ions into plant cells. At the same time, he questioned the newer idea that active transport might be driven by the free energy of hydrolysis of ATP.

Robertson pointed out that evidence for the involvement of ATP rested largely upon the observation that 2,4-dinitrophenol (DNP) allows electron transport to oxygen to proceed but abolishes the simultaneous phosphorylation and active transport. The alternative explanation of Robertson had as its basis the well-established idea that there can be separation of positive and negative charge across a membrane as the result of a hydrogen ion becoming separated from an electron which moves through an electron carrier system leading to the formation of a hydroxyl ion on the other side of the membrane. He postulated that the separation of positive and negative charge drives active transport following the original idea of Lundegårdh and also brings about the formation of ATP and both processes are abolished by the action of DNP. Thus, he believed the active transport mechanism to be intimately connected with one of the steps in oxidative phosphorylation and separation of charge could be part of one of those steps. Prophetically, Robertson said that an understanding of the mechanism of oxidative phosphorylation might lead to an understanding of active transport, but knowledge about active transport may also contribute to understanding oxidative phosphorylation.

It was Mitchell who achieved the intellectual synthesis with his chemiostatic hypothesis for oxidative phosphorylation (cf. Volume VII, Chapter 2). The original ideas were put forward in 1961 (343) and stated in comprehensive form in 1966 (344) [but see Robertson (438) for an elegant summary of how Mitchell's ideas follow from what Lundegårdh, Robertson, and others had proposed previously]. The important new suggestion made by Mitchell was that ATP synthesis is achieved by a membrane-bound ATPase which reverses the reaction

$$ATP^{4-} + H_2O \rightarrow ADP^{3-} + H_2PO_4^{2-} + H^+$$

The reversal takes place because the enzyme is so oriented in the membrane that the oxidation of two protons to form water occurs on the opposite side to that on which ATP is being produced (Fig. 13). It can be described in another way, namely, that proton flow down its electrochemical potential gradient (from right to left in Fig. 13) brings about the formation of ATP. In other words, the free energy inherent in that gradient will drive the synthesis of ATP. Such a proton flow will dissipate the gradient unless it is conserved by some other mechanism. Mitchell proposed that conversation could be achieved by an electron carrier bringing about charge separation such that hydrogen ions are produced on the side

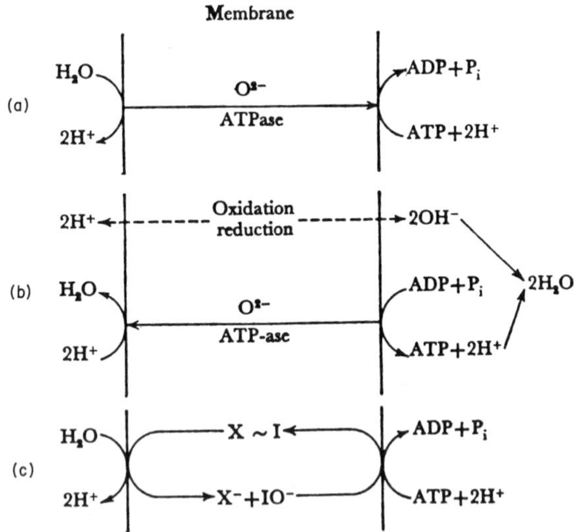

FIG. 13. Principles of Mitchell's hypothesis of the reversible ATPase in the membrane of the mitochondrion. (a) Normal ATPase with two protons extruded for one ATP hydrolyzed; (b) synthesis of ATP due to protons and hydroxyl ions in simultaneous oxidation–reduction reaction; H^+ concentration increases on the left and due to water formation decreases on the right, reversing the ATPase reaction; (c) normal ATPase, showing the possibility that specific substances X and I are translocating O^{2-} to the right and returning as the corresponding anhydride to the left. From Robertson (438).

of the membrane on which depletion is taking place as a result of ATP synthesis and hydroxyl ions on the other (Fig. 13). Thus there is ATP synthesis at the expense of an oxidation reaction mediated via the redox carrier.

The proton electrochemical potential gradient is the algebraic sum of the hydrogen ion and potential gradients across the membrane. This means that in order for the ATPase to synthesize ATP, the hydrogen ion (pH) gradient across the membrane can be relatively small, given that the membrane potential forms a major portion of the total electrochemical potential gradient. A membrane potential of 210 mV will poise the system at an ATP : ADP ratio of unity. However, the presence of a significant membrane potential will allow ions of the opposite sign to leak across the membrane. In order to prevent the swelling and lysis of mitochondria and chloroplasts during the formation of ATP, leakage of ions must be balanced by extrusion of ions against the electrical gradient. It was therefore necessary to postulate that the membrane also contains exchange diffusion systems that couple the exchange of anions against OH^- and of cations against H^+.

The complete system is dependent upon the so-called "coupling membrane" being impermeable to protons and other ions since any leakage of these across the membrane will change the proton electrochemical potential gradient. Indeed, it was also postulated that uncoupling agents such as DNP and azide acted upon the coupling membrane as proton-conducting agents.

This is the bare outline of Mitchell's *chemiosmotic* hypothesis, which has four fundamental features (Fig. 14): (1) the proton-translocating reversible ATPases, (2) the proton-translocating oxidoreduction chain, (3) the exchange diffusion systems coupling proton translocation to that of anions and cations, and (4) the ion-impermeable coupling membrane in which (1), (2), and (3) reside.

The hypothesis injected two concepts into the consideration of ion transport in plants. First, that the extrusion of protons may be a fundamental process in a plant cell. This concept reappears in two other sections—one concerned with the presence of systems by which such extrusion might occur (Section IV,A,2,a) and the other with the role of such extrusion in the control of cytoplasmic pH (Section VI,G). From these concepts developed the realization that, like the membranes of mitochondria and chloroplasts, other membranes, and particularly the plasmalemma and tonoplast, may also contain exchange–diffusion transport systems. Specific evidence is presented in Section IV,A,3.

However, while the Mitchell hypothesis has helped to advance the study of ion transport in plants, there still remains the central dilemma: what is

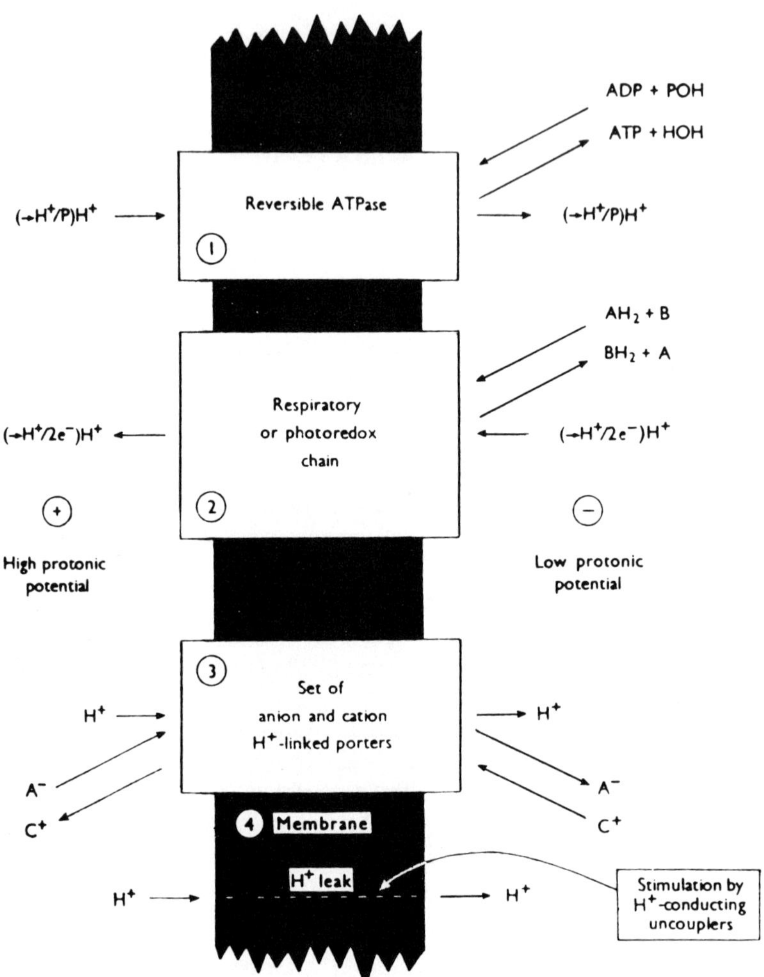

FIG. 14. This diagram summarizes the main features described by the four fundamental postulates of the chemiosmotic hypothesis. For further explanation see the text. From Mitchell (346).

the energy source for ion transport into the cell? Is it provided by ATP acting via an ATPase or a substrate capable of being oxidized by a membrane-treated redox system? In other words, how are ion transport and metabolism interconnected? In particular, how does photosynthesis drive ion transport in green plant cells since it has long been known that light can enhance salt uptake? (See Chapter 4 in Volume II of this treatise.) This question is dealt with further in the next section. The relationship between ion transport and respiration is considered here from the standpoint of the information concerning salt respiration which has accrued since 1959.

The relationship of ion transport to metabolism also reappears in other sections. This is because metabolism exerts its effect on the salt status of a cell or tissue in ways other than the supply of energy for transport. Although important information can be obtained from determining how the rate of ion uptake is affected when the rate of metabolism is changed, this is by no means the only way in which relevant data can be obtained. Important clues come from other studies, e.g., those concerned with isolated transport systems, examples of which have been given in Section IV.

B. Relationship between Ion Transport and Photosynthesis

For the background on the topic of the relationship of ion transport and photosynthesis, the reader should consult Chapter 4 in Volume II of this treatise, in which there is an excellent summary of work done prior to 1959 and reviews by Hope and Walker (206), Jeschke (233), and Raven (421, 423), for information of work done since that date. Much of the work has depended upon a judicious use of inhibitors and reference may be made to Raven (415) for a brief but informative summary of the presumed target sites for the most commonly used inhibitors (Fig. 15).

Interpretation of data obtained with inhibitors is always difficult, but it is particularly so in those studies in which inhibitors have been used to elucidate how light might influence ion transport. This is because respiration may provide an alternative energy supply to photosynthesis. A corollory to this is that one should conclude that if light does not stimulate uptake of an ion, then photosynthesis cannot drive its transport (439, 440). There is also another complication. We now know that neither ATP nor NADPH per se is exported from chloroplasts at a significant rate and the need for these compounds for participation in external metabolic reactions is met by their resynthesis from other compounds that do come from the chloroplast (177, 561). Thus ATP appears to be generated from dihydroxyacetone phosphate while NADH is formed from dihydroxyacetone phosphate and malate. There is a similar situation in mitochondria, although in these organelles ATP can exchange for ADP and orthophosphate (169, 185). Walker (561) has stressed the importance of orthophosphate concentrations in the cytoplasm in controlling the rate of production of ATP. The orthophosphate is transported into the chloroplast, where it is incorporated via ATP into triosephosphate. But the supply of orthophosphate to

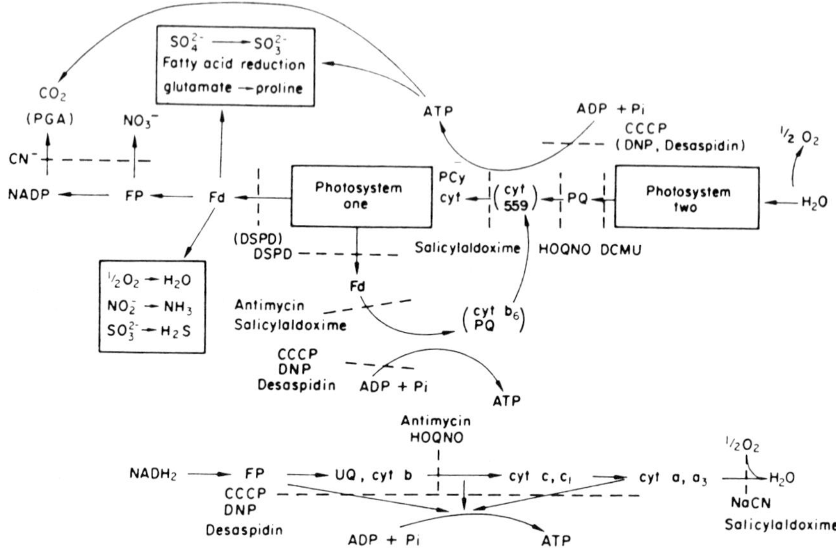

FIG. 15. Sites of action of inhibitors in photosynthesis and respiration. CCCP, carbonyl cyanide *m*-chlorophenyl-hydrazone; DCMU, 3′ (3,4-dichlorophenyl) 1′,1′ dimethylurea; DPN, 2,4 dinitrophenol; DSPD, disalicylidenepropanediamine; Fd, ferrodoxin; FP, flavoprotein; HOQNO, 4 (*n*-heptyl)hydroxyquinoline-*N*-oxide; NAD, nicotinamide adenine dinucleotide; NADP, nicotinamide adenine dinucleotide phosphate; PCy, plastocyanin; P_i, inorganic orthophosphate; PGA, 3-phosphoglyceric acid; PQ, plastoquinone; UQ, ubiquinone. From Raven (415).

the chloroplast is also an important determinant of the rate of export of elaborated carbon from the chloroplast. High external concentrations of orthophosphate encourage export, while low internal concentrations aid retention. Although there is a seemingly coherent view, as expressed in the reviews referred earlier, of what is happening, it should be emphasized that the hypothesis is based on data from studies on isolated chloroplasts from a restricted number of higher plants, e.g., spinach *(Spinacia oleracea)* and *Elodea densa.* Studies on glucose uptake and metabolism by *Hydrodictyon africanum* (422) indicate that conversion to triose phosphate may not be essential for transport of hexose across the outer chloroplast membranes.

Despite the complexity of the situation, progress is being made toward identifying the energy coupling between photosynthesis and respiration and ion transport. Perhaps the most detailed and informative studies have been those of Raven (410, 412–415, 418) on potassium transport in *H. africanum.* The conclusion from these studies is that potassium influx (which is linked to sodium efflux; Section IV,A,2,b) is ATP driven. The evidence is as follows:

1. Potassium influx is stimulated by white light of which far red between 700 and 740 nm is more effective than red light at 680 nm. At the former wavelengths, carbon dioxide fixation is much less stimulated than potassium influx. This indicates that ATP is generated by cyclic photophosphorylation in far-red light but NADPH for carbon dioxide fixation is not.
2. Light stimulation of potassium influx is less sensitive to cyanide than is carbon dioxide fixation and also less sensitive than is the influx in the dark. Cyanide inhibits carbon dioxide reduction and oxidative phosphorylation but not photophosphorylation. On the other hand, when photophosphorylation uncouplers (CCCP, DNP, desaspidin) are used, light-stimulated influx is inhibited equally with carbon dioxide fixation.
3. Influx in the dark can be inhibited by substances which uncouple oxidative phosphorylation (CCCP, HOQNO, antimycin A, salicylaldoxine).
4. At low oxygen concentrations and at low light intensity, light stimulation is less in the presence of carbon dioxide. This observation has been interpreted in terms of competition for ATP between the mechanism of potassium influx and carbon dioxide fixation.

Thus there is considerable evidence in favor of potassium influx being driven by ATP. However, certain inhibitors (DCCD and phloridzin) have a greater effect on potassium influx and respiration, while photosynthesis and chloride influx are much less inhibited (412, 413, 417, 420). A number of explanations have been put forward to explain this finding, but they involve assumptions about the regulation of light-stimulated ATP-requiring processes for which the direct experimental verification is meager.

Despite some doubts, the general conclusion is consistent with that from other information (coupling with sodium efflux and ouabain sensitivity) which indicates that the transport protein is an ATPase (Section IV,A,2,b). It seems that phosphate influx in *H. africanum* could be driven also by ATP, since the flux has an inhibitor sensitivity very similar to that of active potassium influx (419, 420).

In the case of chloride it is difficult to speculate about a possible mechanism from the flux data alone. MacRobbie (312, 313) in a series of elegant experiments showed that in *Nitella translucens* the light stimulation of chloride influx appeared to be linked with the electron-transfer reaction associated with photosystem II. The same seems to be the case with other algae, e.g., *Chara australis, Enteromorpha intestinalis, Griffithsia monila, H. africanum, H. reticulatum* (see 423). In *H. africanum* the action spectrum for chloride influx is similar to that for carbon dioxide fixation, showing the red drop for wavelengths beyond 680 nm; this is unlike the situation for potassium influx. In another section (IV,A,3,b) evidence was presented that chloride influx in these algae may be in exchange for hydroxyl ions. The crucial question is whether a gradient of hydroxyl ions across the cell membrane drives chloride influx or whether the flux is driven directly by some energy source generated by photosynthesis. There is no doubt that in *N. translucens* (312, 313) and *H. africanum* (410) conditions in which only cyclic photophosphorylation is possible, e.g., in far-red light or in the presence of DCMU, do not support chloride influx and therefore ATP is not apparently involved. But that is all that can be said with confidence about the energetic coupling. Although the data obtained by Smith and colleagues (Section IV,A,3,b) give support to the view that pH gradients across the cell membrane may be important in controlling chloride influx, this itself does not dispel uncertainty about the mechanism whereby the flux is linked to metabolism. This is particularly so if one accepts that the system is driven under normal conditions by a hydrogen ion efflux pump which we would expect to be fueled by ATP. Also, MacRobbie (320) has pointed to the need for careful interpretations of experiments which involve changing the external pH. Although such changes may lead to a change in the pH gradient across the membrane, there may also be an effect due to an absolute pH change at the surface of the membrane.

For ion transport in leaves, the problem of energy supply becomes even more intangible. There is the difficulty of actually measuring the relevant fluxes (Section II,A,2), and the extent to which the influx is or is not affected by light can depend upon the physiological state of the leaf. Thus, the following points can be made.

1. For hydrophytes, there is good evidence that the presence of a light-stimulated flux of an ion can be dependent upon its concentration in the pretreatment medium, with only low concentrations leading to a light-stimulated flux (233). An analogous situation seems to hold for leaves of terrestrial plants. Thus Schöch and Lüttge (460) found no stimulation by light ^{86}Rb uptake into slices of *Zea mays* leaves taken from plants grown in soil. On the other hand, Rains (404) grew plants on one-fifth strength Johnson's nutrient solution with the potassium concentration at one-twentieth that normally added. Potassium absorption by leaf slices from such plants was enhanced by light at relatively low intensities.
2. The response of a leaf seems to be age dependent; this is certainly so for potassium–sodium selectivity in leaves of *Phaseolus vulgaris* (213).
3. As with other tissues excised from the parent organ, e.g., storage tissue discs (518), leaf slices exhibit an aging response when kept in aerated

solutions after cutting. This is particularly so for the response of ion transport to light (233). The mechanism underlying the changed response is not known.

In view of these problems, it is not surprising that there is no clear idea about the energy supply for transport into leaves. For example, although Johansen and Lüttge (237) argue that chloride transport in *Tradescantia albiflora* leaf slices under aerobic conditions is powered by a cytoplasmic pool of ATP, this conclusion is based on assumptions about the mode of action of the inhibitors which they use; these assumptions may not be justified and the same may be true for other studies (see 233).

A more fundamental assumption which has been made not only for studies on leaves and leaf slices, but also for studies on algae, needs to be questioned. This assumption is that transport is coupled to respiration and/or photosynthesis either by ATP or NADH and it is the concentration of these compounds within the cytoplasm which controls the rate of transport. Thus the results of experiments using inhibitors are interpreted on the basis of their presumed effect on the concentration of ATP or NADH at the transport site. But overall control of transport may not be mediated in this way, and there are now indications that other mechanisms may override any controls that may be exerted by the concentration of ATP or NADH (cf. Section VI,D).

C. SALT RESPIRATION

Chapter 4 in Volume II of this treatise contains a very full consideration of so-called anion respiration (now more usually called *salt* respiration) (221), since it is known that both cations as well as anions can bring about the respiratory response. The most satisfactory hypothesis to explain this phenomenon postulates that respiration is regulated by the ratio of ATP to ADP in the cells. Prior to the presentation of salt to the tissue, there is a high ATP/ADP, whereas uptake of ion leads to breakdown of ATP and the ratio falls. The ratio is important because negative feedback regulates glycolysis via ATP (264, 354), and there is also a requirement for ADP in two of the steps in the glycolytic pathway. The early evidence for this hypothesis has been reviewed by Jennings (221).

Since then Adams and Rowan (4) have provided further supporting evidence. They measured the amounts of glycolytic intermediates and ADP and ATP in carrot slices before and after addition of potassium and sodium chloride, both of which stimulate respiratory activity. Though assay of phosphorylated compounds in plant tissues can present considerable difficulties (430), the methods seem to be reliable. Adams and Rowan (4) found that within 1 min the ADP level rose by about 30%. The data for the changes at various times within a 10-min period in the amount of the various glycolytic intermediates indicate that this release of ADP stimulates a reaction requiring the coenzyme that is catalyzed by the enzyme phosphoglycerate kinase. Thereafter the point of control alternates between this reaction and the phosphofructokinase reaction. These observations are supported by the studies of Adams (3), which showed that ADP markedly stimulates the respiration of well-washed carrot slices.

On the other hand, Faiz-ur-Rahman *et al.* (130) with the same tissue showed that the control of glycolysis is by a negative feedback system involving phosphoenolpyruvate and phosphoglycerate. Although this may be so, it does not necessarily rule out the importance of ADP in salt respiration. It is necessary to distinguish between the signal for the increased respiratory activity and the control of the process once it is under way. The data of Adams and Rowan (4) are compatible with a rise in ADP concentration initiating an increase in respiration.

It would be easy to conclude from this that the transport of salt into the cells consumes ATP and releases ADP. This is certainly a possibility, but as Poole (396) has pointed out, the increased rate of respiration may continue unabated for a time when salt is removed. This does not necessarily contradict the hypothesis just presented. The continuation of respiratory metabolism at an elevated rate may reflect the new ionic environment in the cytoplasm consequent upon the absorption of salt. One possibility which we must keep in mind is the increased rate of production of hydrogen and/or hydroxyl ions consequent upon the increased rate of metabolism (Section VI,G).

VI. Regulation of Transport

A. Introduction

The regulation of the ion content of plant cells and tissue is a theme running through Chapter 4 in Volume II of this treatise. What follows relates to developments which have taken place since 1959 with respect to control of transport at the cellular level. The control of ion transport in the whole plant is dealt with in Chapter 5 of this volume.

B. PUMP AND LEAK

If there were a constant energy source available to an ion pump and ions were being pumped into the cell against a free-energy gradient and if there were no "channels" through which ions could move passively, then there would be no net flow whenever there was equality between the free energy available from the energy source and the free energy stored in the electrochemical gradient of the ion across the membrane. With the presence of passive permeability channels, the equilibrium concentration of the ion will be lower because the free-energy gradient will drive the ion out of the cell. Raven (424) has discussed the thermodynamic aspects of such a situation. Such regulation may occur in nongrowing storage tissue; Cram (86) has shown that during the accumulation of chloride by carrot slices efflux of the ion across the plasmalemma increases. However, it is likely that there is also a kinetic component, i.e., the internal concentration of the ion, controlling either the rate of pumping or the rate of leakage (but more likely the former).

C. KINETIC CONTROL

By kinetic control is meant the control of transport by the concentration of ions or intermediary metabolites acting directly on the transport system. The ion which is itself being transported may regulate transport as its concentration within the cell increases. This is a case of negative feedback and the possibility of it occurring in plant cells has been considered by Cram (90).

The strongest evidence for the occurrence of negative feedback comes from the work of Glass (151–153) on potassium uptake by barley roots. He demonstrated that the plasmalemma influx of the ion is sigmoidally related to the internal potassium concentration, varied either by growing plants in media of different potassium concentration or by exposing roots in a pretreatment solution for different intervals of time. Glass (152) analyzed the kinetics of the relationship between potassium influx and the external and internal concentrations of the ion, assuming that the carrier might function like an allosteric protein. From the analysis he concluded that the carrier could be envisaged as possessing a single external (transport) site for potassium as well as four internal sites for allosteric control of influx. Glass suggested that when these latter sites are saturated, the conformation of the external binding site is modified, resulting in a lower affinity for potassium (Fig. 16). Jensen and Pettersson (229) have pro-

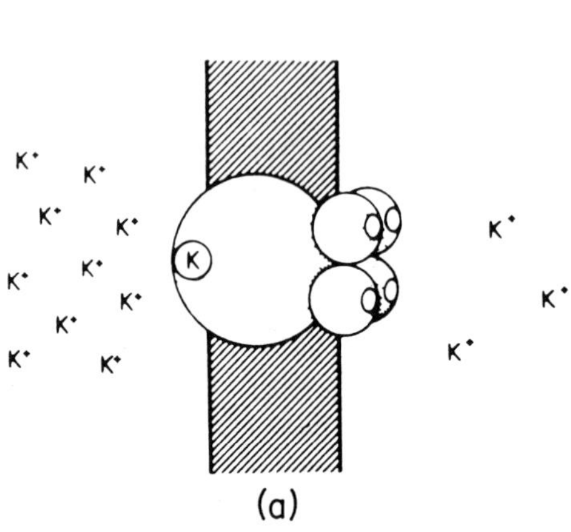

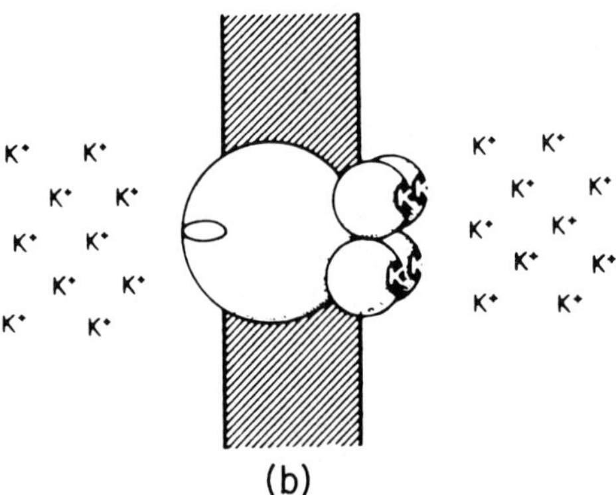

FIG. 16. Model of allosteric control system for K^+ influx into barley root cells: (a) Low-salt roots and (b) high-salt roots. The large centrally located structure represents the K^+ carrier. On the outer surface a single K^+-binding site is represented by the small circle. On the inner surface, four allosteric binding sites located on four subunits are represented by four circles. In low-salt roots, the allosteric sites are vacant, and K^+ can be bound to the external site, giving high initial rates of influx. When the internal concentration of K^+ is high (high-salt roots), the allosteric sites are saturated, and the conformation of the external binding site is modified, resulting in lowered affinity for K^+. From Glass (152).

duced similar evidence for the allosteric regulation of potassium uptake by roots of a number of other higher plants, including *Pinus sylvestris* and *Picea abies.*

Further work (153) showed that the relationship between influx of potassium and internal concentration was independent of DNA, RNA, and protein synthesis during the pretreatment period, indicating that the changes in influx are not brought about by the synthesis of new carrier molecules.

There is good evidence that allosteric control of potassium transport occurs in other plants. Thus the potassium–hydrogen exchange pump of *N. crassa* has been shown by Slayman and Slayman (490) to exhibit complex kinetics at high pH (above 6.0). The data have been fitted satisfactorily by two different two-site models. In one, the transport system is thought to contain both a carrier site responsible for potassium uptake and a modifier site for which there is competition between hydrogen and potassium ions. The other model postulates a transport protein consisting of multiple subunits, each with an active site for potassium ions, hydrogen ions acting as an allosteric activator. How this relates to the regulation of potassium uptake by the fungus is still unknown. The ease with which the fungus can be grown should make it possible to investigate how potassium fluxes in the growing cell are regulated. Although information for *N. crassa* is lacking, there is evidence from the fungus *Dendryphiella salina* that growth of the mycelium occurs with a relatively low (approx. 0.5 pmol cm^{-1} s^{-1}) net influx of potassium, and this has been taken to mean that uptake of the ion is under internal control (227). Extensive work on the kinetics of potassium transport into cells of bakers' yeast *(Saccharomyces cerevisiae)* (16, 17, 42, 44) has indicated that the transport system possesses not only a modifier site, but also an activation site. The former, like that for *N. crassa,* can bind hydrogen ions for which it has a high affinity, while the latter binds monovalent cations, potassium rubidium, and sodium, which stimulate rubidium (potassium) entry.

Negative feedback may be brought about by mechanisms other than the presence of an allosteric transport protein. One possibility for the control of ion transport into plant cells by negative feedback is exemplified by the mechanism of amino acid uptake in a fungus, e.g., *N. crassa* (377). In *N. crassa* there are five transport systems for amino acids (see 226): I, L-neutral; II, D- or L-basic, neutral and acidic; III, L-basic; IV, D- or L-acidic; V, specific for methionine and ethionine. The uptake of amino acids by any one system can be inhibited by the presence of free amino acids within the hyphae; the ability of a particular amino acid to inhibit a transport system is highly correlated with its affinity for that system. Amino acids with high affinity are effective in producing inhibition; those with lower affinity are less effective.

Pall (377) has suggested two possible ways in which inhibition, or what he calls transinhibition, might occur. In the first, transinhibition is a consequence of the protein involved in transport having an allosteric binding site for the amino acids concerned. Binding of the amino acid would inhibit the activity of the transport protein and therefore transport. By this mechanism, the protein would have two binding sites, one binding the amino acids prior to transport into the hyphae and another binding transinhibiting amino acids inside the hyphae.

On the other hand, the other, simpler hypothesis is that transport and transinhibition may be determined by a single site. In this case, transinhibition will be caused by the binding of the amino acid to the active site. Figure 17 shows schematically the model proposed by Pall (377).

The slowness of reaction 4, compared with 3, is equivalent to the carrier-amino acid being in a lower energy state when oriented to the inside of the hypha than when oriented toward the outside. When an amino acid which brings about inhibition accumulates inside the hypha, the equilibrium between the free carrier and carrier–amino acid complex is shifted toward the latter. Thus the concentration of free carrier is reduced. This leads to an inhibition of the energy-dependent recycling process. Recycling cannot occur via reaction 4 because of its slowness. Of course, if amino acid transport is driven by ion gradients (226, 227; Section IV,A,4), an energy-dependent recycling process will not be involved. However, the preceding model still can be reconciled with a secondary active transport

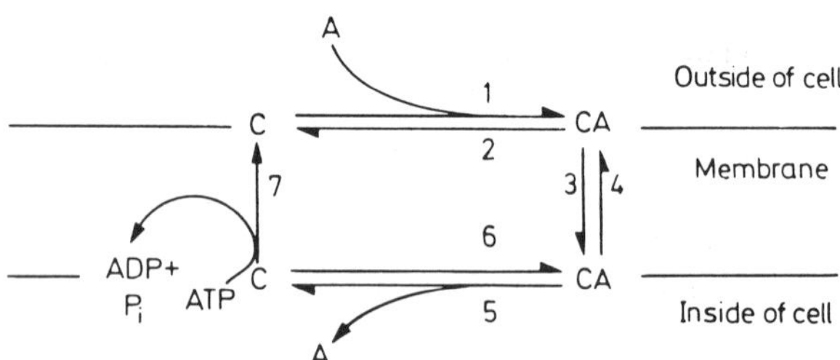

FIG. 17. A model for system-specific transinhibition. 1 and 2 allow reversible association of the amino acid with the carrier. 3 brings about orientation of the carrier from the outside to the inside of the membrane. 4 is the reverse of 3 but very much slower. 5 and 6 allow reversible dissociation of the amino acid from the carrier. 7 brings about recycling of the carrier by a process involving ATP or some other energy-rich compound. A, amino acid; C, protein carrier. From Pall (377).

process. All that need be done is to have the carrier react with an ion (hydrogen ion being the most likely; Section IV,A,4) instead of ATP.

Regulation of influx may be brought about not only by the concentration of an ion at the inner surface of the membrane, but if the ion is metabolized (e.g., nitrate, phosphate, sulfate) by some metabolic product (e.g., in the case of nitrate and sulfate, an amino acid). In fungi, a recurring observation is the repression of sulfate transport by methionine (333, 334, 340, 464, 546).

D. Energy Supply

One might presume from the preceding discussion that since primary active transport can be dependent upon ATP, the concentration within the cytoplasm of this high-energy compound might control the rate of transport. However, recent evidence from *N. crassa* obtained by Slayman *et al.* (158, 483) suggests that extending this assumption to all plant cells may not be justified.

The first clues came from inhibitor studies on the membrane potential and intrahyphal concentration of ATP. The effects were complex, but the salient point to emerge was that in *N. crassa* there are one or more mechanisms for restoring the membrane potential, after initial inhibition, to the level approaching that in the absence of inhibitor. This occurs whether or not the ATP level can be kept high. The mechanism seems to operate in part by suppressing those fluxes which would otherwise depolarize the membrane. These data do not indicate whether pump activity per se is related to the intracellular ATP concentration, but they do show that it is likely that there is a process with overriding control over many events occurring at the plasmalemma.

A clue to the nature of this control comes from studies on the respiratory mutant *poky* (489). This mutant is defective in cytochromes *b* and aa_3 complex, and it was shown that there is a control system which conserves ATP by regulating consumption in concert with its generation. This was revealed by short-term inhibitor and growth experiments which showed that ATP levels in the mycelium are kept high by a feedback process which depresses breakdown (and growth) very quickly after synthesis is inhibited. Concurrent work on the membrane potential of *poky* (158, 483) showed that in the presence of cyanide there are oscillations which become damped. Oscillations indicate that the membrane potential and therefore the transport system are nevertheless under regulation. However, it was clear from measurements of intrahyphal ATP concentration that the regulation is not exerted through this compound. Other experiments involving inhibitors such as theophylline point to control through the level of cyclic (3′,5′-) AMP.

These observations emphasize the note of caution introduced earlier (Section V) that the interpretation of data obtained from studies of the effects of inhibitors of oxidative phosphorylation and photophosphorylation on ion fluxes may not be a simple one. Inhibition of ATP generation may not necessarily lead to a reduction of the cytoplasmic concentration of the compound and the rate of ion transport may be subject to a variety of controls other than through the concentration of ATP or NADH.

E. Number of Transport Sites

Steward and Sutcliffe (518) reviewed the idea that the "aging" of storage tissue disks increases the number of carriers in the cell membranes. Since 1959 further evidence has been obtained that higher plant cells may possess transport systems whose synthesis can be induced or repressed by environmental factors. The systems which are apparently subject to induction and repression are chloride uptake in beet disks (515) and sodium uptake in the same tissue (394, 523), chloride extrusion in *Limonium vulgare* salt glands (196, 197, 199), and nitrate uptake in tobacco tissue culture (178). Of the transport processes that have been investigated, it is only the chloride extrusion system of the salt gland which we can confidently describe as being an inducible system. Insertion of such a transport system into a membrane is a specific case of what must be occurring with constitutive transport systems during the growth of cells. The rate of entry of an ion per unit area of cell surface may change during growth, depending upon the rate at which newly synthesized transport proteins become active after their insertion in the membrane. Regulation may also occur through control of the rate of turnover of these proteins (525).

F. Turgor

The presence of osmotically active solutes within the cell and the presence of a rigid cell wall cause the generation of a hydrostatic pressure (see Chapter 1). This gives rise to cell turgor, and we know that the degree of turgor can regulate membrane ion transport processes (579). It was Jacques (215) who first showed that turgor might regulate ion movements into cells. He impaled mature *Valonia* coenocytes with glass capillaries and found that the rate at which fluid was absorbed (measured by movement in the capillary) was 10 – 15 times greater than control plants, which enlarged at a rate of only 0.8 to 1.0% per day. The only important difference was the turgor pressure, which was 0.5 to 2.0 MPa in intact vesicles and virtually zero in the impaled ones. In other experiments Jacques abolished the turgor pressure by immersing *Valonia* in hypertonic seawater, and the vesicles responded by rapidly absorbing salts and water until the normal turgor pressure was restored. Although removing the turgor pressure creates an inwardly directed water potential gradient, this gradient is nearly abolished by the inward osmotic flow which occurs in the first few hours after impalement. Jacques postulated that the effect of turgidity in reducing uptake of ions was due to dehydration of the protoplasm, which increases its resistance to the movement of ions. This now seems very unlikely since we know that plants possess the capacity to produce solutes which reduce dehydration problems within the cytoplasm to a minimum (see Chapter 6).

Gutknecht (163) reinvestigated the problem and confirmed what Jacques had found and pointed out that pressures of less than 2.0 MPa should not significantly affect either the passive or the metabolic driving forces for fluid transport for the following physical chemical reasons: (1) The electrochemical potential of an ion will only change by 0.2 mV per 1.0 MPa; (2) pressures of more than 1.0 MPa are required to markedly affect the rate of a chemical or metabolic reaction involving solid or liquid phases.

In Gutknecht's experiments the vacuolar pressure was adjusted by means of a mercury manometer attached to an internal perfusion system. This system can only be used for pressure gradients of about 2.0 MPa; with larger gradients, rupture of the cytoplasmic seals around the perfusion pipettes occurs (579). It is preferable if higher pressure gradients can be obtained. This can now be done with the pressure probe developed by Steudle and Zimmermann (514, 580), which allows both the measurement of turgor pressure and changes in it independent of the osmotic pressure (see Chapter 1).

Studies on *V. utricularis* have shown that potassium influx from the external medium into the vacuole decreased markedly (by 80%) with increasing turgor pressure up to a value of around 2.0 MPa, becoming independent above that value. The efflux from vacuole to outside medium increased over the entire pressure range investigated (up to 0.4 MPa). Studies on cells of different sizes have shown this latter flux to be volume dependent, whereas influx is not. The volume dependence of efflux appears to be due to the elastic properties of the wall which are defined by the volumetric elastic modulus ϵ. Zimmermann and Steudle (581, 582) have shown that the absolute value of ϵ and its pressure dependence are functions of the cell volume. Thus the magnitude of ϵ can be supposed to determine the magnitude of the efflux.

Zimmermann (579), on the basis of these and many other observations, provides an electromechanical model of the plant cell membrane which is able to sense (1) absolute pressure, (2) turgor pressure, i.e., the pressure gradient across the membrane wall, (3) stretching of the cell wall, which is a function of the elastic modulus and which operates as a consequence of the wall and membrane being coupled (either physically or chemically), and (4) electric compressive pressure. The forces which may be operating are shown in Fig. 18. The stresses induced in the membrane by these forces lead to membrane compression. It is thought that there are special regions in the membrane which become compressed and that these regions are predominantly protein, which is believed to be more compressible than the lipid portion of the membrane.

Some of the implications of the experimental findings and the model which is based on them are as follows: (1) The possibility that fluxes may be related to cell volume or the elastic properties of the wall must mean that caution is required in the interpretation of compartmental analysis in complex tissues (Section II,C). (2) It would seem from the effects of pressure on efflux from the vacuole that turgor changes can be sensed by a membrane which is not mechanistically associated with a wall, e.g., the tonoplast. (3) Electrical compressive pressure and turgor pressure are equivalent and so the pressure sensing areas of the membrane will respond not only to changes in turgor, but also to changes in membrane potential. This is important because small changes in the concentration of specific

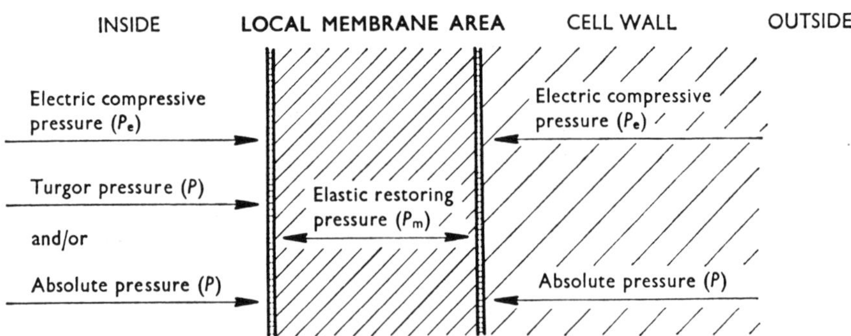

FIG. 18. A schematic representation of the membrane of a plant cell. It is assumed that the membrane can be considered as a capacitor filled with an elastic dielectricum. The potential difference across the membrane creates an electric compressive force P_e which tends to compress the membrane. Similarly, the turgor pressure (or absolute pressure) P leads to a compression of the thickness of the membrane. These compressive forces are counterbalanced at equilibrium by the elastic restoring force P_m generated in the membrane. From Zimmermann (579).

ions through their effects on membrane potential, which may not necessarily contribute to the osmolarity of cells, could be important in regulating turgor.

The whole field of ion transport and turgor regulation in giant algal cells has been admirably reviewed by Gutknecht *et al.* (164).

The studies on turgor regulation have been made almost exclusively with coenocytes. However, too much should not be ascribed to a regulatory mechanism based on sensing the turgor pressure of the cell. When there are difficulties in producing a kinetic explanation of flux changes for an ion brought about by increased concentration inside the cell, it may be attractive to conclude that regulation is via turgor pressure; however, direct evidence that the change of flux is attributable to a change in pressure should be required.

G. CHARGE BALANCE

Steward and Sutcliffe (518) give a wealth of information about how charge balance in plant cells is achieved. Some more recent information is noteworthy. In fungi, it now seems that phosphate groups can often make a significant contribution to the negative charges within mycelia. In *N. crassa,* phosphorus levels of 300 mmol L^{-1} have been detected (486). Some of the most interesting information about how charge balance is achieved comes from studies on halophytes (6, 7, 56, 57). Thus Albert and Popp (7) showed from their studies of plants growing in the Neuseidler Lake region of Austria (where SO_4^{2-}, Cl^-, HCO_3^-, and CO_3^{2-} are the main water-soluble anions balanced by sodium) that, for example, (1) members of the Chenopodiaceae contain very high levels of sodium balanced mainly by chloride and oxalate, with some sulfate; (2) *Plantago maritima* accumulates high levels of sulfate as well as chloride, with malate and citrate being the organic acids, although they play a lesser role in balancing cations; sodium again is the major cation; (3) in the Poaceae, Cyperaceae, and Juncaceae chloride is often the major anion, with much lower, but relatively similar, concentrations of sulfate, organic acids, and phosphate. In some cases, potassium is the major cation, by a substantial amount.

Thus, plants growing in a similar environment may display considerable differences in ion content, a fact which has been known for many years (518). In part, of course, this will be due to different edaphic factors in the area, but the differing ion contents are also a consequence of different plants having different strategies with respect to the generation of turgor. The mechanistic basis of the various strategies is one fundamental problem of plant nutrition. What the preceding results seem to show is that a particular plant might have fewer transport systems for a certain ion than for another one or that if there are the same number of transport systems, they are more tightly regulated by, for instance, the internal concentration of the ion in question.

By 1959 the importance of the production in a cell or tissue of organic anions to balance the excess uptake of cations was well established. Since that date, the corollary has been demonstrated, namely, that when anions are absorbed in excess of cations, the level of organic acids within the cells is reduced (187, 212).

The questions which are still being asked are as follows: How does ion absorption interact with organic acid metabolism? What triggers the synthesis and breakdown of organic acids when an ion enters a cell?

Although there is evidence that a number of organic acids may contribute to charge balance in different tissues, namely, transaconitic acid in *Zea mays* (356, 543), it is only for malic and oxalic acid that we have any detailed information about how the levels of these acids are related to the concentrations of ions inside the cell.

There is general agreement that malate is formed by a pathway involving phosphoenolpyruvate (PEP) carboxylase coupled with malate dehydrogenase (251, 317, 539, 560), although the evidence for the pathway operating *in vivo* in tissues absorbing ions is rather meager (371). The breakdown of malate is believed to be by malic enzyme, which appears to work in the cell only in the decarboxylation mode (138). The probable reaction sequence of carboxylation and decarboxylation is shown in Fig. 19.

1. *Effect of pH*

Hiatt (187) made a theoretical analysis of the equilibrium between glyceraldehyde 3-phosphate and malate via phosphoenolpyruvate carboxykinase, i.e.,

$$\text{Glyceraldehyde-3-P} + 2\text{ADP} + \text{P}_i + \text{CO}_2 \rightarrow \text{malate} + 2\text{ATP}$$

which can be expressed in terms of the equilibrium constant

$$K = \frac{(\text{glyceraldehyde-3-P})[\text{ADP}]^2(\text{P}_i)(\text{CO}_2)}{(\text{malate})[\text{ATP}]^2} \qquad (11)$$

The equilibrium constant is 1.58×10^{10} at pH 5.0 and 2.99×10^6 at pH 6.0. If all other factors are constant, a pH shift from 5.0 to 6.0 results in a 36-fold increase in the malate to glyceraldehyde-3-P ratio at equilibrium. Hiatt ruled out the involvement of PEP carboxylase in malate regulation because the sequence of reactions via this enzyme from glyceraldehyde-3-P to malate is irreversible. Since the metabolic evidence points unambigu-

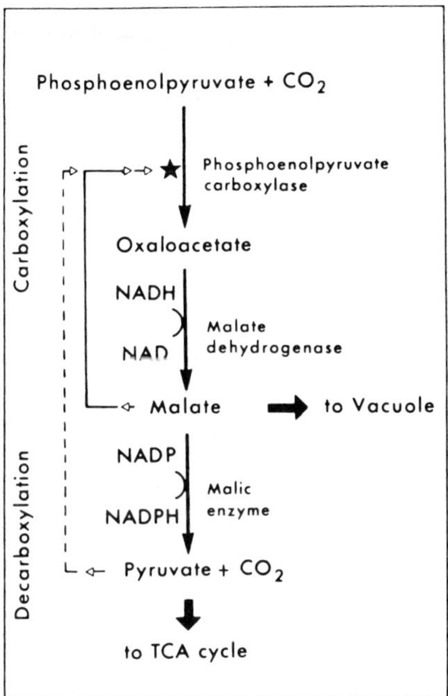

FIG. 19. Probable reaction sequence for malate synthesis and degradation during nonautotrophic CO_2 fixation. Phosphoenolpyruvate carboxylase activity is under feedback control (*) due to the concentration of malate in the cytoplasm. From Osmond (371).

ously to the involvement of PEP carboxylase, we must look elsewhere for the involvement of hydrogen ions in the control of carboxylation and decarboxylation. Nevertheless, as Hiatt and Hendricks (188) have suggested, the following sequence of events in the accumulation of potassium from a solution of potassium sulfate could occur: (1) initial potassium exchange for hydrogen ions is associated with organic acids, and (2) the resulting change in pH controls the rate of synthesis of malic acid, a process which must occur for the continued uptake of potassium.

Another hypothesis for the mechanisms whereby acid production on breakdown is controlled by pH was made by Davies (98). He was concerned initially with control of pH in plant cells, and he pointed out that it could be controlled by regulating the number of carboxyl groups in the cytoplasm, thus:

$$\underset{\text{Decarboxylation}}{RH \xleftarrow{\;CO_2\;} RCOOH} \rightleftharpoons \underset{\text{Carboxylation}}{R'COOH \xleftarrow{\;CO_2\;} R'H}$$

Davies proposed that the processes of carboxylation and decarboxylation would then be directly controlled by pH via the effect of hydrogen ions on the enzymes responsible for the two processes. He postulated that if the cytoplasmic pH were normally 7.0, then carboxylating enzyme should have an alkaline pH optimum. A survey of the literature showed this to be true. Davies showed that the converse was also essentially true, namely, that decarboxylating enzymes have an acidic pH optimum. This being so, the cell can operate as a pH-stat; if the cell pH rises, the decarboxylation system will be reduced in activity and the carboxylation system will be enhanced, producing a net gain of carboxyl groups with a consequential reduction in pH. A fall in pH will lead to a drop in carboxyl groups, which leads to a change of pH toward its original value.

Raven and Smith (425, 426), who were concerned with cytoplasmic pH regulation through the pumping out of hydrogen and hydroxyl ions, integrated their ideas with those of Davies to produce a model very similar to that originally proposed by Hiatt and Hendricks (188). This model combines pumps at the plasmalemma for hydrogen and hydroxyl ions. These are responsible for short-term control, while the pH-stat operates over the long term. In their 1974 paper, Raven and Smith (426) pointed out that many key metabolic processes generate hydrogen or hydroxyl ions which are frequently made manifest as pH changes in the external medium (Table V). When a plant becomes highly differentiated with aerial parts such that cells are located some distance from the nutrient medium, there are other aspects of pH regulation which have to be considered. This matter is dealt with in Chapter 6 of this volume.

Raven and Smith do not believe that the biochemical pH-stat plays a major role in pH regulation. The evidence in support of this view arises from a comparison of the ATP requirements of malate synthesis and hydroxyl ion extrusion. Thus, (1) for malate synthesis, each H^+-producing carboxyl group is the equivalent of 10 ATP molecules. (2) The free energy of one ATP molecule could bring about the extrusion of one hydroxyl ion against a pH gradient of 7.0 (cytoplasm) to 11.0 (medium) when the membrane potential is -120 mV (inside negative), given the energy in the electrochemical potential gradient for the ion across the membrane.

Clearly, energetically, a hydroxyl ion extrusion pump is more effective than neutralization of the ion by the biochemical pH-stat. Further, Osmond (371) has pointed out that measurements of pH optima of the enzyme responsible for carboxylation and decarboxylation are very dependent upon which conditions *in vitro* are chosen for the assay. However, it seems likely that pH shifts of 1 to 2 units between pH 5.5 and 8.5 may favor the synthetic over the degradative system, or vice versa, if both are functioning *in vivo* at substrate concentrations below the K_m.

TABLE V

PROCESSES WHICH ALTER THE pH OF SOLUTIONS SURROUNDING PLANT CELLS, BY PRODUCTION OR CONSUMPTION OF HYDROGEN IONS INSIDE (i) OR OUTSIDE (o) THE CELL[a]

Process	Direction of pH change	Reference(s)
Photosynthetic CO_2 fixation, CO_2 entering cell (o)	Increase	416
Photosynthetic CO_2 fixation, HCO_3^- entering cell (i)	Increase	416
Organic acid assimilation, undissociated form entering (o)	Increase	108
Nitrogen assimilation, NO_3^- as N source (i)	Increase	100
Nitrogen assimilation, NH_4^+ entering the cell (i)	Decrease	304, 519
Nitrogen assimilation, NH_4OH entering the cell (o)	Decrease	72, 73
Excretion of organic acids made in fermentation (e.g., lactate), photosynthesis (glycolate), or cell wall synthesis (polyuronides) in undissociated form (o)	Decrease	84, 137, 359
Excess influx of cations over anions [K^+ organate accumulation (i)]	Decrease	212
Excess influx of anions over cations (i)	Increase	212
Rapid responses to IAA, $P_r \rightarrow P_{fr}$, mechanical stimulation (o)	Decrease	166, 217, 218

[a] From Raven and Smith (426). Reproduced by permission of the National Research Council of Canada from the *Canadian Journal of Botany*, Volume 52, pp. 1035–1048, 1974.

2. *Effect of Inorganic Anions*

There are two possible ways in which inorganic anions may interact with malate production. Either the anions may prevent carboxylation, or if the organic acid is eventually sequestered in some compartment inside the cell such as the vacuole, the anion may inhibit transport into it.

With respect to the former possibility, Osmond and Greenway (372) showed that PEP carboxylase may be very sensitive to inorganic salts. For instance, potassium or sodium chloride at 100 mmol L^{-1} gave almost 100% inhibition of activity of the enzyme from *Atriplex spongiosa* leaves. Although the enzyme from *Zea mays* is much less sensitive, the possibility of control of carboxylation by inorganic salts should not be discounted, particularly since Cram (88) has produced good evidence that chloride inhibits carbon dioxide fixation in carrot root cells.

The other possible way in which carboxylation might be inhibited by inorganic anions could be as a result of an effect on the transport of carbon dioxide or bicarbonate into the cell. There is no evidence from studies on carrot (*Daucus carota* var. *sativus*) root tissue (88), *Vallisneria spiralis* leaves (284), and *Hydrodictyon africanum* (411) that chloride ions have very much effect on bicarbonate uptake.

3. Calcium

Kirkby and DeKock (247) showed that there was a highly significant relationship between calcium and malate in the leaves of Brussels sprouts (*Brassica oleracea* var. *gemmifera*). Phillips and Jennings (387) have shown a significant relationship between calcium and organic acid levels in leaves of *Kalanchoe daigremontiana*. Splittstoesser and Beavers (511) showed that malate synthesis from a medium containing 30 mmol L^{-1} potassium bicarbonate through which was bubbled air and 5% carbon dioxide was dramatically enhanced by the presence of 0.1 mmol L^{-1} calcium sulfate.

Calcium may exert its effect by inhibiting pyruvate kinase. A consequence of such inhibition would be an increase in the concentration of phosphoenolpyruvate (PEP) such that an increased rate of carboxylation would be possible. To investigate this further we need to know (1) the concentration of calcium and magnesium and potassium ions in the cytoplasm, remembering that in animal (244) and fungal (544) cells the latter two ions are activators of pyruvate kinase while the former is an inhibitor, and (2) the extent to which calcium is chelated by the end products of carboxylation—malic and also citric and isocitric acid.

It is clear that a complex interaction exists between inorganic ions and malate synthesis in plant tissues. Thus, potassium–hydrogen exchange may lead to a change in cytoplasmic pH, in which case the biochemical pH-stat will come into action. But if the views of Raven and Smith are accepted, production of hydrogen ions in the cytoplasm may not necessarily bring about a fall in pH, since they are pumped out into the medium in exchange for potassium ions.

Two other important points need to be considered: (1) There is feedback inhibition of PEP carboxylase by malate (252, 352, 538). (2) Malate is compartmented within the cell. The evidence for this comes from kinetic analysis of $^{14}CO_2$ incorporation into, or evolution from, malate (373, 540); comparison of specific activity of carbon dioxide evolved under steady-state conditions with that of malate in the tissue (309); and kinetic analysis of malate exchange between the medium and the tissue (92, 373). All the data show that for the various tissues investigated malate is not distributed uniformly but is in at least two compartments—one almost certainly the vacuole, the other probably the cytoplasm.

Inorganic ions could conceivably affect fluxes of malate between compartments, although Cram (88) has shown that chloride has little effect on these fluxes in carrot tissue. On the other hand, there may be an interaction between bivalent cations and the movement of malate between cytoplasm and vacuole. This could happen as the consequence of the formation of uncharged molecules through chelation. Thus:

$$Ca^{2+} + \text{malate}^{2-} \rightleftharpoons \text{Ca malate}$$

$$Ca^{2+} + \text{citrate}^{3-} \rightleftharpoons \text{Ca H citrate}$$

A similar situation could hold for magnesium. Due to their lipid solubility, the uncharged species would more readily diffuse through the tonoplast than do ions. In this way, the ionic balance of the cytoplasm and thus the ionic milieu of the enzymes involved in carboxylation or decarboxylation and thus the rates of the two processes might be changed. This is a development of the ideas of Wyatt (578).

VII. Free Space

Steward and Sutcliffe in Chapter 4 of Volume II in this treatise review succinctly what is meant by free space. There is no longer any doubt about its location, and it is accepted as being equivalent to the apoplast (468). For a short period subsequent to 1959, that part of the free space under the influence of nonmobile anions (predominantly carboxylic acids) was treated simply as a Donnan system (60). However, Dainty and Hope (94) showed that for *Chara australis* (= *C. corallina*) the exchange characteristic is better described in terms of an electric double layer at a charged surface. Treating the wall as a Donnan system provides a simplified but somewhat qualitative view since it is assumed that the charge and therefore the potential are uniform throughout the wall. Also, the treatment requires that the water free space (WFS) and Donnan free space (DFS) be separate entities.

If the free space is considered as an electrical double layer, these assumptions need not be made. The free space may then be described in terms of charge density. This diverts attention from the free space as an independent entity to a consideration of it as a layer through which solutes must pass to reach the plasmalemma.

When the free space is treated as a double layer, it should also be considered a water-filled porous structure (94). Figure 20 shows schematically a negatively charged pore which is in equilibrium with various ion concentrations in the external medium. When bivalent cations are present, there is a greater concentration of anions than when the cations are monovalent. If one were to put the tissues into pure water, more electrolytes would come out from that which was in equilibrium with a bivalent cationic salt than with a monovalent one. Thus the bivalent cations can be considered as being more effective than monovalent cations in neutralizing negative charges within the pore. If we take the more simplistic view, what has just been said equates with the greater effectiveness of bivalent cations over monovalent cations in changing the potential differences of the DFS as shown by calculations by Walker and Pitman (568) (Table VI). Bivalent cations thus make the free space and therefore the plasmalemma more

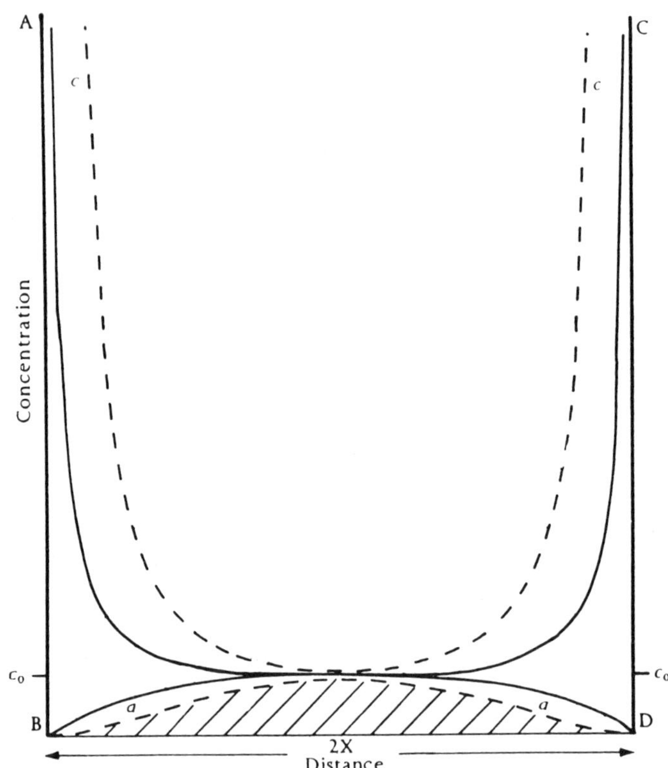

FIG. 20. Variation with distance from surface of concentration of cations (c) and anions (a) within a negatively charged pore in equilibrium with an electrolyte at concentration c_0. Dotted lines: salt of monovalent cation; solid lines: salt of bivalent cation. From Shone and Barber (471).

accessible to anions. Pitman (390) invoked this explanation for the so-called Viet's effect, namely, the observation that salt (or more strictly speaking anion) uptake is stimulated by polyvalent cations (556). Pitman proposed that uptake (by red beet disks) of salt (chloride) is rate limited by uptake of the anion. Calcium by allowing more chloride to diffuse through the negatively charged surface allows a greater concentration to reach the transport sites in the plasmalemma and thus uptake is increased.

The most comprehensive study of the free space of a higher plant has been for barley roots (31, 470, 471). It was shown that uptake of sulfate and iodide ions behaved in accordance to the qualitative requirements of a simplified theoretical treatment for a double layer. Under the specific conditions of low temperature and high pH, application of the theory

TABLE VI

Potential Difference (E, Inside Negative) between the Medium (o) Containing Various Concentrations of K^+, Ca^{2+}, and Cl^- and the Donnan Free Space (w) Containing Fixed Anions at a Concentration of 100 mmol liter^{-1}[a,b]

$[K^+]_o$ (mmol liter^{-1})	$[Ca^{2+}]_o$ (mmol liter^{-1})	$[Cl^-]_o$ (mmol liter^{-1})	E (mV)	$[K^+]_w$ (mmol liter^{-1})	$[Ca^{2+}]_w$ (mmol liter^{-1})	$[Cl^-]_w$ (mmol liter^{-1})
0.1	0.0	0.1	−173.4	100.0	0.0	0.0
1.0	0.0	1.0	−115.6	100.0	0.0	0.0
0.1	0.1	0.3	− 77.7	2.2	48.9	0.0
1.0	0.1	1.2	− 75.2	20.0	40.0	0.1
1.0	1.0	3.0	− 48.3	6.8	46.8	0.4
4.0	1.0	6.0	− 45.7	24.7	38.1	1.0
1.0	4.0	9.0	− 31.6	3.5	49.5	2.6

[a] From Walker and Pitman (568).
[b] Concentrations of K^+, Ca^{2+}, and Cl^- in w are also given.

demonstrated that ions appeared initially to diffuse into a free space consisting of water-filled pores about 10 nm in diameter. The behavior of strontium also accorded qualitatively with the theory, provided that allowance was made for effects due to the hydrolysis of negative sites and for the presence of positive charges in the roots. For rubidium the situation was much more complex. At 0.2°C, it seemed as if the number of exchange sites was small by comparison with those for strontium. Interestingly, there was evidence that there are metabolically produced exchange sites for rubidium; that is, there is a fraction of rubidium held in the free space at 25°C which will exchange with rubidium in the external solution but cannot be washed out with water. This is not the case at 0.2°C.

Although the free space contains nonmobile anions, nothing as yet has been said about their chemical nature. In most green plants it is almost certain that they are mainly carboxylic acid anions belonging to the uronic acids of the pectic material within the wall.[3] Keller and Deuel (246), for instance, conducted a fairly exhaustive investigation into the ion exchange properties of roots of several species. They showed that 70–90% of the cation exchange capacity could be attributed to the free carboxyl groups of pectic substances in the root. Crooke *et al.* (93) showed that there was a strong quantitative agreement between cation exchange capacity and pectic content of disks of several storage organs. A similar situation holds for

[3] Compare the brief "nonmetabolic" binding of cations (e.g., Rb.) that occurs in potato disks which was attributed to sites that contain acidic (uronic acid) groups. Cf. Vol. I, Chapter 4, Section C,2 and references there cited.

oat coleoptiles (219). Dainty *et al.* (95) showed that the fixed charge in walls of *Chara australis* were due to carboxyl groups with a p*K* of 2.2.

The number of fixed negative charges can differ widely from species to species, as shown in Table VII. The exchange abilities of a variety of plants have been measured under constant external conditions. Obviously, the properties of the exchange system will be dependent on the environment; pH and the concentration of bivalent cations will be particularly important. An increase in the former will cause an increase in the number of exchange sites, while an increase in the latter will cause a decrease.

The exchange characteristics of the free space can be changed in a more dramatic way, as has been shown by the work of Rorison (443). He found that aluminum absorbed by sainfoin (*Onobrychis sativa*) roots goes predominantly into the free space. Once there, aluminum seems to be bound in the same way as it is to clay particles such that there is peripheral positive charge capable of binding phosphate. The binding of phosphate in this way appears to lead to a reduction in the amount of the ion reaching the

TABLE VII

EXCHANGE ABILITY OF ROOTS OF VASCULAR PLANTS AND OF WHOLE PLANTS OF BRYOPHYTES

Roots of vascular plant	Exchange	Whole plants of bryophyte	Exchange
Calluna vulgaris	0.079	*Pellia epiphylla*	0.110
Fagus sylvatica	0.131	*Plectocolea crenulata*	0.112
Acer pseudoplatanus	0.162	*Nardia scalaris*	0.171
Eriophorum angustifolium	0.030	*Polytrichum commune*	0.069
Juncus bulbosus	0.051	*Fontinalis antipyretica*	0.121
Juncus squarrosus	0.061	*Leucobryum glaucum*	0.123
Carex lepidocarpa	0.063	*Rhytidiadelphus triquetrus*	0.127
Dryopteris filix-mas	0.140	*Thuidium tamariscinum*	0.131
		Pleurozium schreberi	0.147
		Hypnum cupressiforme	0.154
		Aulacomnium palustre	0.184
		Acrocladium cuspidatum	0.188
		Mnium hornum	0.190
		Sphagnum spp.	0.213 0.242
		Lobaria pulmonaria	0.039
		Cetraria islandica	0.068
		Cladonia sp.	0.069
		Usnea barbata	0.172

[a] From Clymo (82).
[b] In milliequivalents per gram dry weight at equilibrium with a solution of 30 mEq liter^{-1} Ca^{2+} at pH 3.5.

plasmalemma. Thus there is less phosphate taken up by the plant, which goes some way toward explaining why calcifuge plants, those plants incapable of growing in alkaline soils, often show symptoms of phosphorus deficiency when they are grown in soils below pH 5.0 when aluminum is present as the cation (325).

Although carboxylic acid groups contribute most of the fixed negative charges in the free space of green plants, this is not necessarily true for all plants. In yeast *(Saccharomyces cerevisiae)* polyphosphate groups are an important component of the cell surface (≡part of free space) (444). Theuvenet and Borst-Pauwels (530) have shown that polyvalent cations such as UO_2^{2+} and La^{3+} reduce the surface potential generated by these polyphosphate groups and in doing so can inhibit rubidium uptake. In keeping with what has been learned from studies on higher plants, anion (phosphate) uptake in yeast is stimulated by polyvalent cations (in this case, Mn^{2+}).

VIII. Movement across the Cytoplasm to the Vacuole

Little is yet known about how ions are transported across the cytoplasm into the vacuole; yet this process is of key importance to the plant cell. The vacuole may occupy 90% of the cell volume, and if the cell is to remain turgid and grow, much of the salt absorbed must move into the vacuole, a point continually emphasized by Steward and Sutcliffe (518). What are the reasons for our ignorance about the processes involved? It has already been shown that the measurement of ion fluxes for plant cells is difficult, particularly so for the tonoplast. The tracer flux across this membrane is often initially obscured by that across the plasmalemma. There is also a problem, as we shall see, of determining the cytoplasmic concentration of the ion under study.

The information that we do have about ion transport across the cytoplasm comes almost exclusively from tracer studies on *Nitella translucens* by MacRobbie (311, 314–316, 318–320). The earliest observations (311), particularly those concerning chloride, directed attention in a more precise and quantitative way to a point stressed by Steward and Sutcliffe (518), namely, that a distinction can be made between entry of the ion into the cytoplasm and its transfer into the vacuole. A key finding of MacRobbie (311) was the correlation between the rate of active uptake of chloride into the cytoplasm and the rate of transfer of both potassium and chloride from the cytoplasm to the vacuole. A cell with a high influx retains less radioactivity in the cytoplasm (i.e., it transfers a relatively greater proportion into the vacuole), while the converse holds for a cell with a low influx.

The preceding conclusions were obtained from experiments in which the cell was treated as a three-compartment system with wall, cytoplasm, and vacuole in series with respect to movement of tracer. Subsequently (315), it was shown that in uptake experiments of very short duration (a very few minutes) very much more activity reaches the vacuole than would be expected and that the series model is inadequate. A key point is that ions entering the cell do not mix freely with the whole of the cytoplasm and its specific activity cannot be taken as uniform.

MacRobbie showed that there are two components of chloride transfer across the cytoplasm to the vacuole, a fast component which increases linearly with time and a second component with a time course similar to that for the series model, in which the cytoplasmic phase increases in activity compatible with that model. The fast phase also has the following characteristics: (1) The phase is associated with a detectable lag in the appearance of radioactivity in the vacuole. (2) The amount of chloride in this phase is much less than would be expected even if this phase consisted of the cytoplasm minus the chloroplasts. (3) In *N. translucens* the rapid transfer involves sodium as well as chloride but not potassium. But in *Tolypella intricata* it is potassium which accompanies the rapid transfer of chloride. These are the cations which Smith (492, 493) has shown to be linked to chloride influx across the plasmalemma in the two algae. (4) The phase represents a constant fraction of the total uptake into the cell.

Subsequent work (316, 318, 319, 321) demonstrated other properties of the fast phase. The first is that it is independent of the slow component; the second is that it is extremely variable both within and between experiments. The values obtained are believed to be nonrandomly distributed or quantized. The statistics of the analysis has been disputed (135, 206, 564). Third, the contribution of the fast phase can be reduced by bathing the cells in 10 mmol L^{-1} manganese sulfate when the cells become electrically unexcitable.

Although the phase is not likely to be an artifact due to contamination of the vacuolar sap by cytoplasm (315), it may be due to the generation of action potentials during cutting of the cell. Walker (564) has suggested that if action potentials are generated, they cannot produce fluxes of sufficient magnitude to account for the fast phase. MacRobbie (321) appears to believe that what she has described is not a normal physiological phenomenon. Nevertheless, if the fast component is brought about by action potentials, it is of great interest that the tracer only enters a very small compartment of the cytoplast en route to the vacuole.

It has been pointed out that there is a link between the slower movement of an ion across the cytoplasm to the vacuole and the influx of that ion across the plasmalemma. Subsequent work by MacRobbie (321) has confirmed this and from flux analysis has shown that the bulk (water) phase of the cytoplasm has to be included in the compartment from which ion is transferred to the vacuole. She has also shown (318) that there is no discrimination between chloride and bromide in the process of transfer across the cytoplasm and their accumulation in the vacuole in spite of the marked preference for chloride over bromide by the plasmalemma. At this point, the question is whether one should think solely in terms of the movement of individual ions to the vacuole. The coupling of chloride and potassium movement in *N. translucens* indicates that we should be concerned with the movement of salt. MacRobbie (322) describes a process of salt transfer to the vacuole that involves the creation of new vacuoles in the cytoplasm and their discharge into the central vacuole. If this process occurs, the manner of its control by cytoplasmic salt content is a matter of pure speculation. Furthermore such a process is incompatible with the finding that various algae (206, 421, 423) seem to have ion pumps located at the tonoplast.

IX. Movement of Ions in the Symplasm

Since 1959 work on the movement of ions in the symplasm has been done mainly with members of the Characeae. The importance of these studies lies in the attempt to determine the rates at which the movement occurs. From this information and the extent and geometry of the protoplasmic connections, it is possible to calculate the resistance of these connections to such movements. Although caution should be exercised in applying the results to higher plants, the figures obtained have been used to calculate rates of flow through the symplast of tissues.

The relevant observations have been made not only on *Chara* by Walker and Bostrom (567), Tyree *et al.* (548), and Bostrom and Walker (45, 46), but also on *N. translucens* by Williams and Fensom (575). Essentially the experiments have involved feeding radioactive tracer to one internodal (source) cell and measuring the accumulation of tracer in the next internodal (sink) cell, which is separated from the radioactive loading solution by a barrier. Tyree *et al.* measured the accumulation by direct counting of the sink cell; the other authors destructively analyzed it. Williams and Fensom plunged the cell into liquid nitrogen and cut it into measured lengths before analysis.

The interpretation of the experimental results is not simple because of protoplasmic streaming and the difficulty of obtaining an accurate estimate of the specific activity of the ion in the streams of cytoplasm approaching one side and leaving the other side of the node. Walker and Bostrom in their studies adopted a kinetic model which is shown diagrammatically in Fig. 21. The small fast compartment is assumed to be in the cytoplasm, probably the groundplasm, although this is not consistent with the studies made by MacRobbie, which are described in Section VIII. Strong evidence for the model came from the constancy with time of uptake of the tracer into the source cell, of transport into the vacuole of that cell, and of transport into the sink cell (45). The general conclusion from the studies of Walker and Bostrom (567) and Tyree *et al.* (548) is that the rate of transport of ions to the plasmodesmata is at a sufficiently high rate to allow subsequent movement through the plasmodesmata itself to occur by diffusion. Thus it is the rate of transport in the cytoplasm, perhaps involving cyclosis, which limits the rate of movement from the site of uptake into the source cell and not movement through the plasmodesmata, which also does not need to be metabolically mediated. On the other hand, Williams and Fensom (575) believe that active processes are involved, although their conclusions are based on calculations that required several assumptions, notably the number of plasmodesmata in the wall and build-up of tracer at the nodes. Nevertheless, their results are of interest in that there may be a faster-than-streaming translocation of small amounts of tracer in the form of "pulses" which may pass through the nodes.

The conclusion that the rate of movement of material to the node limits the rate of movement across it into the next cell has been supported by the observations of Bostrom and Walker (46), who altered the speed of streaming in the source cell with cytochalasin B and then studied the effect on intercellular transport. While chloride influx into the source cell did not depend upon the streaming, the relative rate of intercellular transport depended markedly on the rate of streaming at all speeds up to those found in untreated cells (Fig. 22).

These experiments with Characean plants are important because they may bear upon the situation in the tissue of higher plants which are more intractable experimentally. The most detailed theoretical treatment of symplastic transport in such tissues has been made by Tyree (547). In his analysis, he made the following assumptions: (1) that the rate-limiting step for symplastic transport is located at the plasmodesmata, (2) that cyclosis and self-diffusion keep a solute at a uniform concentration in the cytoplasm (near-perfect mixing), and (3) that diffusion will be the predominant mechanism for movement of transport through the plasmodesmata for all small nonelectrolytes and for small electrolytes, provided that there are no substantial electrostatic forces within the pore. Tyree's analysis, unlike some of the experiments described earlier, considers not only solute, but also solvent movement and the degree of coupling between the two flows.

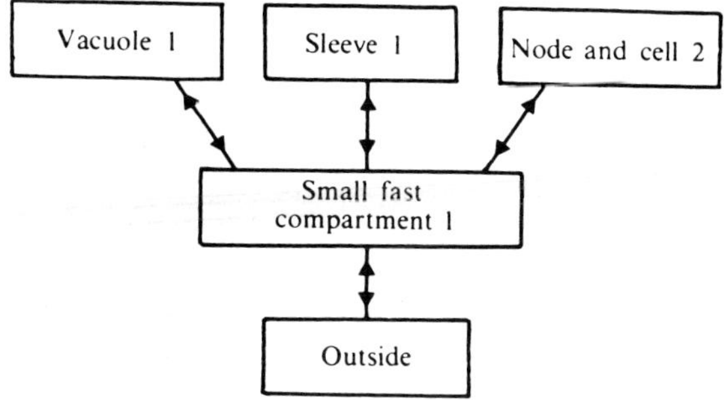

FIG. 21. The kinetic model used by Bostrom and Walker (45) to explain their experimental findings for the movement of $^{36}Cl^-$ across a node of *Chara corallina.*

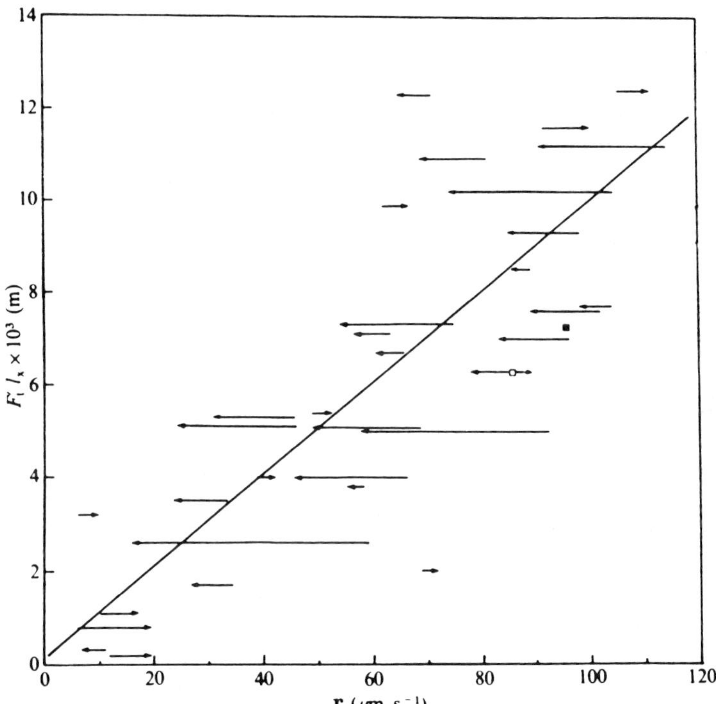

FIG. 22. Scatter diagram of the relative rate of intercellular transport of Cl^- (r) from cell 1 of a pair against speed of streaming in that cell in *Chara corallina.* Vertical axis: product of fraction transported F'_t and exposed cell length l_x, adjusted for transit time, i.e., the time taken for cytoplasm to traverse that part of the cell not exposed to radioactive solution. Solid line: least-squares fit to data. Each data point is represented by a horizontal line giving the range of speeds observed during the uptake time (1800 s). In least-squares fit a mean speed was used for each data point. From Bostrom and Walker (46).

This is, of course, highly relevant to an organ such as a root (Section X).

The analysis shows that coupled flows can be ignored without causing too much distortion in the comparison of calculated and experimental data. The comparison itself is between the solute flux which can be measured (J_s) and one which can be calculated (J_s') from the solute concentration profile C_s multiplied by the resistance to flow through the plasmodesmata. This resistance is the product of the solute permeability of a plasmodesmatal pore (w_s') and the fractional area of the cross-sectional area of the tissue taken up by plasmodesmatal pores (α). Therefore if symplastic transport is occurring,

$$J_s' = \alpha' w_s'(RT\ \Delta C_s) \tag{12}$$

This equation is similar to that for solute flow across a membrane (see Chapter 2). Here w_s' is calculated from the known dimensions of a plasmodesmatal pore on the assumption that solutes diffuse through the pore.

Tyree concludes for onion roots that the salt transported from the outer cortex to the stele is carried by the symplastic pathway. This is assuming that the plasmodesmata perform the rate-limiting step in transport, an assumption which may not be justified (see preceding discussion). Tyree *et al.* (548) report calculations for potassium transport across maize roots in which perfect mixing is not assumed and the rate of cyclosis is taken into account. These calculations still indicate that for roots the plasmodesmata are likely to be the rate-limiting step for symplastic transport. All the evidence available indicates that whether the plasmodesmata are or are not rate limiting, their resistance to solute flow is low enough to accommodate the bulk of the solute traversing a tissue.

It is now well known that plasmodesmata have a complicated internal structure (431–433, and Chapter 1 of Volume XII). The question is whether the physiological information obtained in these studies and other work can throw further light on the properties of the plasmodesmatal matrix. The most useful information comes from electrophysiological studies. Essentially the procedure is to cause the flow of electric current into a single cell. The flow will cause a change in membrane potential of that cell but will not change the potential of an adjoining cell unless there are connections between the cytoplasm of the two cells. Experiments of this sort have been carried out on a number of symplastic systems (501, 506) (Fig. 23). In all cases, which include Characean and higher plant systems, the specific resistance (ohm meter squared) of the intercellular junctions is at least an order of magnitude lower than that of the plasmalemma but is much higher than would be expected if there were unrestricted diffusion of ions through the plasmodesmatal pores, i.e., if they contained a completely aqueous system. Tyree *et al.* (548) came to similar conclusions from their tracer studies, although assumptions had to be

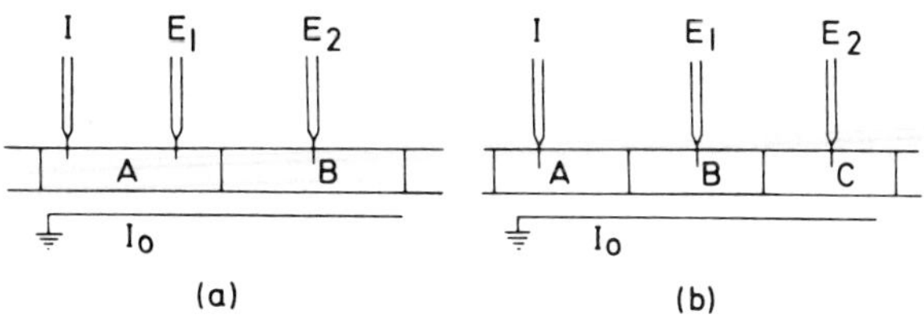

FIG. 23. Microelectrode arrangements used in the demonstration of electrical coupling between cells of *Elodea*. I is the electrode used to pass current between the vacuole of cell A and the electrode I_0 in the external solution. E_1 and E_2 are the electrodes used to measure the electrical potential of the vacuoles relative to a reference electrode in the external solution (not shown). (a) Arrangement used initially to find the coupling ratio for cells A and B. (b) Arrangement used to find the coupling ratio for cells B and C. From Spanswick (506).

made in deriving the final calculated value, which is somewhat lower than that obtained by direct electric measurements (45).

X. Movement of Ions through the Root into the Xylem

There is little doubt that the exudation of fluid from the cut end of a root is dependent upon its metabolic activity. The question has been whether there is active movement of water or whether the root behaves as a purely osmotic system at equilibrium (see Chapter 2).

The following information must be obtained for a proper evaluation of the process of exudation: (1) A simple comparison of the osmotic pressure of the exudate with that of the external medium under steady-state conditions, as originally carried out by Sabinin (446) and which led to the concept of the root as an osmometer, is not sufficient. The studies of van Overbeek (374) show that we need to be able to describe what is happening under non-steady-state conditions; i.e., we need to ask how the osmotic potential of the exudate changes with relatively rapid changes in the osmotic potential of the medium. (2) No adequate description of root exudation can be made without detailed and precise knowledge of root anatomy, e.g., the state of development of the endodermis or the xylem elements at any point on the root axis. (3) In parallel with an anatomical description there must be a physical description of the root, e.g., the resistance of the root to diffusion of oxygen or the resistance of possible pathways across the root to the movement of ions.

The first detailed analysis of root exudation under non-steady-state conditions was that of Arisz *et al.* (15) (see Fig. 24a,b). They assumed that exudation is brought about by two processes: an active secretion of salt into the xylem and water transport effected by the water potential gradient so generated.

Their mathematical analysis (expressed in modern terminology) of the situation was as follows. Under constant conditions, the rate of exudation is found to be relatively constant over a short period. This means that the rate of secretion into the xylem is also constant. So if ψ_{sap} is the osmotic potential of the sap, the rate of salt secretion J_s is given by

$$J_s = -r\psi_{sap} \tag{13}$$

where r is the rate of exudation. If that part of the xylem which is responsible for water transport has a volume V, then the change of ψ_{sap} with time is given by

$$\frac{d\psi_{sap}}{dt} = \frac{J_s - r\psi_{sap}}{V} \tag{14}$$

Thus the extent to which the ψ_{sap} changes will be dependent upon the rate of salt and water secretion. Also we can state that in the steady state

$$r = k(\psi_{sap} - \psi_{medium}) \tag{15}$$

where ψ_{medium} is the osmotic potential of the external medium. This is the relationship found by Sabinin (446); this was also shown to hold by Arisz *et al.* (15). Given this relationship,

$$\frac{d\psi_{sap}}{dt} = \frac{J_s - k(\psi_{sap} - \psi_{medium})}{V}\psi_{sap} \tag{16}$$

This can be integrated for either the preceding situation, where $d\psi_{sap}/dt > 0$ or $d\psi_{sap}/dt < 0$. The equations produced can for the most part describe root exudation of tomato under a variety of conditions. The exceptions to this were the following: (1) The computed osmotic potential of the exudation was sometimes higher (i.e., less negative) than the measured value. (2) With equal concentration of ions in the exudation sap and medium, there could still be exudation. (3) The osmotic potential of the external medium which instantaneously prevents exudation is lower than the osmotic potential of the sap, which is essentially what van Overbeek (374) found but the differences were much smaller ($\simeq$0.05 MPa) because of the sensitivity of analytic procedures which were then available.

Although active water movement could be invoked to explain these discrepancies, the following should be noted:

1. Arisz *et al.* doubted that the composition of the exudation sap which appears at the cut surface is the same as that of the sap present in the "active volume." Thus during upward transport of the sap, exchange of salt and water may occur between a vessel and the surrounding cells.

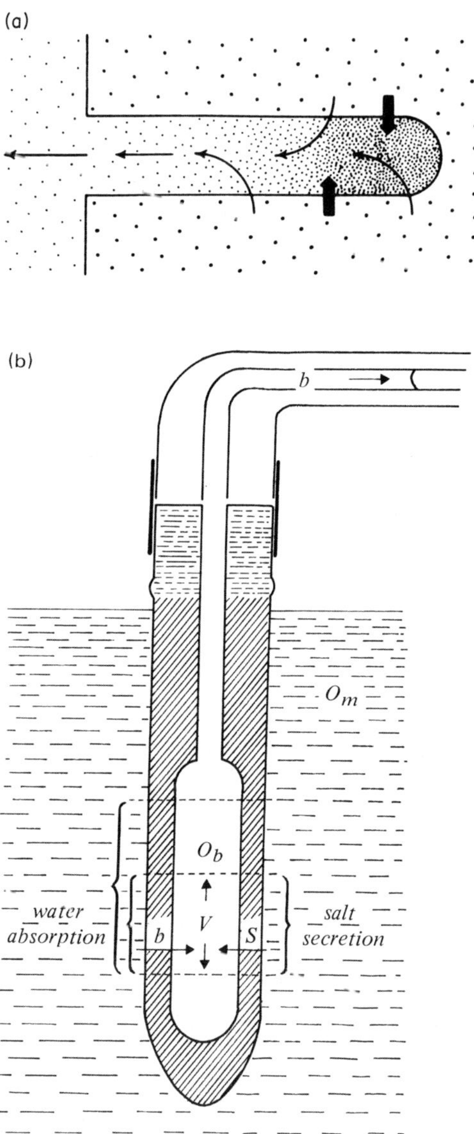

FIG. 24. Models of osmotic flow. (a) Standing gradient osmotic flow model. Ions pumped into the channel (broad arrows) create osmotic gradients drawing water from the cell (thin arrows). Fluid flows out of the open end of the channel. (b) The model of Arisz *et al.* (15) for an exuding root. O_m, osmotic value of the outer solution; O_b, osmotic value of the xylem sap; *S*, salt secretion; *b*, exudation, rate of water absorption; *V*, active volume of the xylem in which salt secretion and water transport take place.

2. In their mathematical formulation, Arisz *et al.* assumed that the salt-secretion zone coincides with the water-absorbing zone. Thus there is a need to look at the individual fluxes (amount moving per unit area) of water and salt across the root. This was first done by House and Findlay (207, 208), who investigated water and salt movement across isolated maize roots. They showed the following: (a) Water movement in isolated roots can be explained by a simple osmotic model with the additional (now considered dubious) possibility that a relatively small nonosmotic water flow occurs. The osmotic flow is driven by the active transport of salt (in their experiments potassium chloride) into some compartment in the root. (b) Analysis of the experimental data indicated that the volume of the compartment was of a similar magnitude to the estimated total volume of the xylem vessels in the root. It should be noted that Arisz *et al.* calculated a compartment volume for tomato roots, but they did not speculate on the physical location of the compartment. (c) There was evidence that the major portion of salt and water absorption occurs within the terminal 5 cm of the maize root. The mathematical formulations of both Arisz *et al.* and House and Findlay assume that the osmotic permeability is constant along the length of the root, although they did admit that this may not be so.

Anderson *et al.* (10) have considered the situation in which the osmotic permeability varies along the length of the root. They used the same equations as those used by Diamond and Bossert (99) to describe what Diamond and Bossert called a standing-gradient osmotic flow in animal epithelia which perform solute-linked water transport, e.g., the avian salt gland or the large intestine. These epithelia have long, narrow, fluid-filled spaces bounded by cell membranes and are generally open at the end facing the solution toward which the fluid is being transported but closed at the end facing the solution from which transported material originates. Diamond and Bossert supposed that active transport moved solute into the channel, lowering the osmotic potential and thus also causing water to flow. The solute moves toward the open end by diffusion, but the solution will be diluted as water enters osmotically along the length of the channel. Solution will emerge from the open end either at the same or lower osmotic potential, depending upon the characteristics of the channel (Fig. 24).

The general expression to describe the osmolarity of the emergent solution is given by

$$\frac{2\pi r \displaystyle\int_{x=0}^{x=L} N(x)\, dx}{2\pi r^2 V(L)} \tag{17}$$

where L is the length of the channel (in centimeters), r is the radius of the channel (in centimeters), V is the linear velocity of flow of the solution (in centimeters per second), $N(x)$ is the rate of active transport of solute across the membrane of the channel at a distance x from the closed end (in osmols per square centimeter second). Passive solute movement across the membrane is assumed to be negligible.

It is not appropriate to discuss the mathematical expansion of the equation, but there is a need to incorporate in the expanded expression the hydraulic conductivity L_p of the membrane and the diffusion coefficient of the solute in water. Further the expanded expression has to be evaluated numerically. However, there are two assumptions implict in the treatment which need to be mentioned: (1) The membrane is completely semipermeable in the matrix surrounding the active transport sites; i.e., the reflection coefficient σ is equal to unity and there is no back diffusion of solute. (2) The hydrostatic pressure within the channel is much less than the osmotic potential gradient across the membranes; i.e., there is no pressure driven backflow of water.

It should be noted that the standing gradient osmotic flow hypothesis has been criticized by Hill (194, 195), who believes that electro-osmotic and not simple osmotic flows are involved.

The application of the theory to root exudation by Anderson *et al.* required little modification, the major one being, as already mentioned, that the osmotic permeability σL_p was allowed to vary along the channel. When this was done, a good fit with experimental data for water flux, salt flux, and salt concentration was obtained (Fig. 25). As far as our consideration of root exudation is concerned, two important results comes out of the analysis. First, the supposed presence of an active water flux is a consequence of too simple a model, and second, the magnitude of $N(x)$ which was calculated on the assumption that the channel is either the stele as a whole or the xylem vessels.

It was calculated that when the root is bathed in a solution of 10 mmol L^{-1} potassium chloride, the flux of salt is 4.36×10^{-10} mol cm^{-2} s^{-1} for the stele (surface area of stele per centimeter of root, 0.105 cm^2) and 3.32×10^{-10} mol cm^{-2} s^{-1} (total surface area of the xylem vessels per centimeter of mature root, 0.137 cm^2). These values are 15 times higher than the largest salt fluxes measured in plant tissue—those of the salt glands of *Limonium vulgare* (193).

The location of the membrane across which ions are transported is still in doubt. There are two possibilities: the plasmalemma of the parenchyma cells of the stele or the same membrane of the developing xylem vessel elements. There are two variations of the latter possibility. The first postulates that salts are first absorbed across the plasmalemma of the element

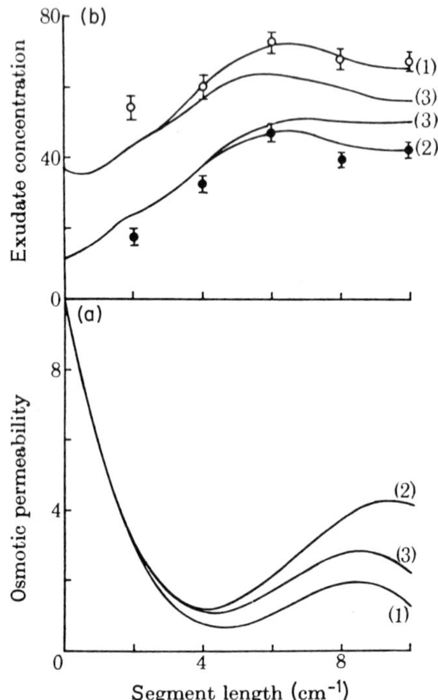

FIG. 25. The exudate ion concentrations C (mmol L^{-1}) and the "osmotic permeability" $L(x)$ ($cm^2\ s^{-1}\ Osm^{-1}$) shown as functions of distance from the root tip x (in centimeters) for *Zea mays* roots bathed in 10 mmol L^{-1} KCl (○) and 1 mmol L^{-1} KCl (●). In (a) the best fits, curves (1) and (2) result in different values of osmotic permeability given by curves (1) and (2) in (b). On the other hand, the mean value of osmotic permeability, curve (3) in (b), results in curve (3) in (a). The points are the experimental values. See the text for further details. From Anderson *et al.* (9).

when the membrane is entire and then released into the vessel column by breakdown of the membrane when the transverse walls are ruptured. The second postulates that even after rupture of the transverse walls the plasma membrane continues to function and the nearly mature vessel element continues to absorb salt—the so-called test-tube hypothesis of Hylmö (211). However, it is hard to understand how the membrane can function without a continuing source of energy. The only evidence in favor of the active role of developing vessel elements in salt accumulation in the xylem is the occurrence of protoplasm with intact membranes and nuclei in metaxylem in the 9 to 10-cm zone of *Zea mays* roots, a level at which vigorous root pressure exudation occurs (191).

The latter observation would suggest that if the plasmalemma of the developing xylem vessel elements is responsible for accumulation of salt in the xylem, root exudation is brought about by the developing metaxylem elements. But we have no evidence that these elements are solely involved and that developing protoxylem elements have no role to play. Also a study by Laüchli *et al.* (272) raises other doubts about the hypothesis involving developing xylem elements. The compound *p*-fluorophenylalanine (FPA) was used to inhibit ion transport to the stele to the xylem vessels at selected regions at varying distances from the apex of barley roots. In the presence of the inhibitor, protein synthesis does not appear to be affected, but proteins are presumed to be produced which are ineffective in their functioning with respect to bringing about ion movement to the xylem (458). Using FPA, Laüchli *et al.* were able to inhibit chloride exudation from the root in parts of the root in which *all* the xylem vessels were mature, suggesting that developing vessels do not play a significant role in exudation.

From these studies, the exudation of roots may be seen as being a process akin to the fluid transfer across animal epithelia. Salt moves into channels which appear to be located in the stele. These channels are most likely to be the walls of the xylem parenchyma plus the xylem vessels, the volume of which would be of a magnitude similar to that determined by experiment by House and Findlay (207) and Anderson and House (13). The questions which now need to be asked are these: (1) How is a flow of ions to the channel membranes of the magnitude indicated maintained through the various pathways which are known to be open to ion movement from the external medium? (2) Is the calculated flux through the channel membrane brought about by active pumping of the ions into the channel? In other words, is the basic assumption for the standing gradient osmotic flow hypothesis of a unidirectional pump located in an otherwise relatively impermeable membrane correct?

An answer to question (1) requires some idea of the pathways of ion movement from the external surface through to the xylem vessels. Unless we can identify the physical pathways, we cannot calculate the resistance to movement or the force necessary to drive ions across the root. A knowledge of the magnitude of the force would allow us to check whether our identification of the pathway is correct and may allow us to suggest a possible mechanism bringing about movement.

The root is divided anatomically into cortex and stele, which are separated by the endodermis. At an early stage in root ontogeny—within a few millimeters from the tip—this layer of cells develops Casparian bands. These are suberized strips which are laid down in the radial walls of every cell within the layer (see Fig. 26). Electron microscopy shows that the plasmalemma is particularly prominent in the vicinity of the band, since

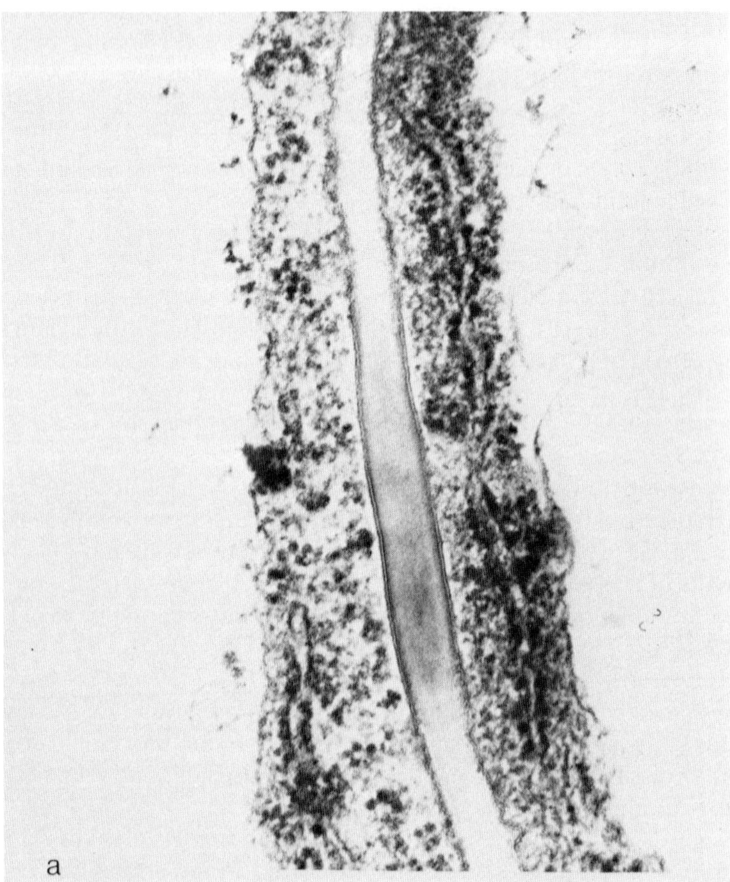

FIG. 26. (a) Transverse section of a young Casparian band in a barley root. The plasmalemma appears well defined and closely appressed to the wall in the region of the band as opposed to the situation elsewhere. Magnification: ×80,000. From Robards *et al.* (435). (b) Casparian band of an endodermal cell in a barley root that has been plasmolyzed with 20% aqueous glycerol. The plasmalemma remains tightly appressed to the Casparian band but is retracted from the remainder of the wall. Magnification: ×37,000. From Robards *et al.* (435). (c) Transverse section of cortical and endodermal cells of a barley root treated with 10^{-5} mol L^{-1} lanthanum nitrate for 18 h. The apoplasm up to the Casparian band is clearly heavily labeled with the opaque tracer but not beyond. Magnification: ×12,000. From Robards and Robb (436).

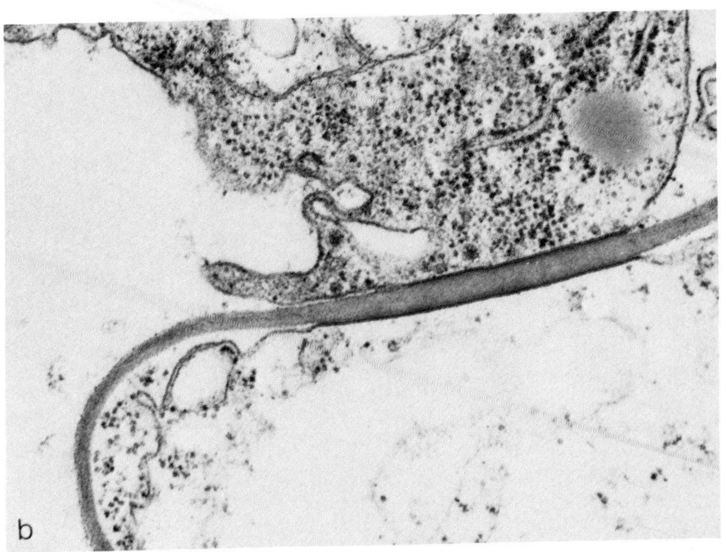

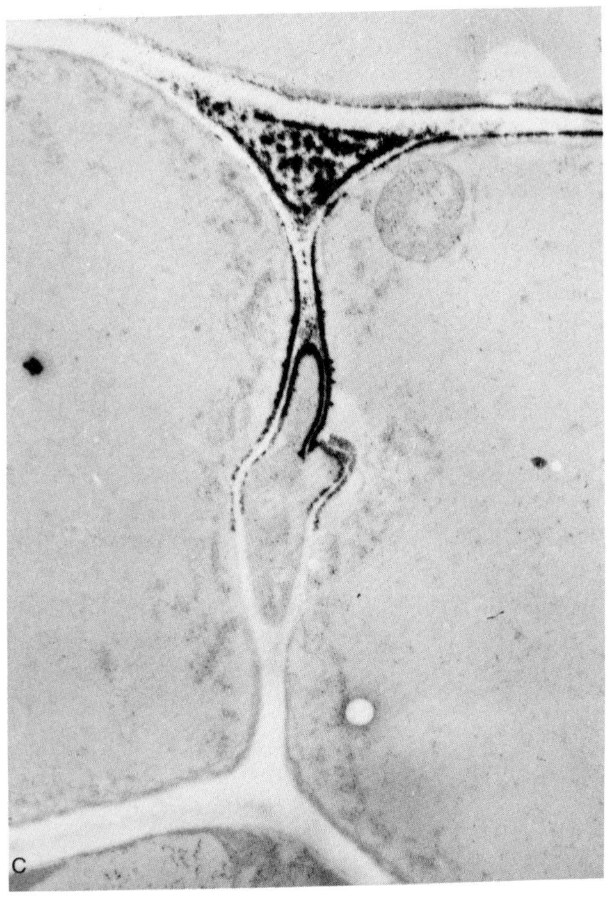

FIG. 26b, c.

the membrane stains there more darkly and is pressed very closely to the band, the surface of which is quite smooth in contrast to the rougher surface of the wall on either side (Fig. 26a). Consistent with much earlier observations, Bryant (71) noted that when endodermal cells with Casparian bands are plasmolyzed, their cytoplasm remains attached to those parts of the wall at which the band is located. This has now been confirmed by electron microscopy (Fig. 26b). As the root develops, further differentiation of the endodermal cells takes place. These changes have been well documented for barley by Robards *et al.* (435), who recognized three successive stages: (1) Casparian bands in the radial walls, (2) suberin lamella laid down around the whole of the wall toward the inside of the cell, and (3) a secondary cellulosic layer laid down inside the suberin lamella toward the inside of the cell, except where deep pits remain which appear to link with others in the pericycle (these pits are joined by plasmodesmata).

It has long been axiomatic (cf. Steward and Sutcliffe, Vol. II, pp 271–274) that the Casparian band is impermeable to water and that in order to reach the stele water and dissolved ions must cross the plasmalemma of at least the endodermal cells once the Casparian band has been formed. This is hard to demonstrate directly, but there is no doubt that material in colloidal suspension capable of penetrating across the root through the wall cannot move past the Casparian band (Fig. 26c) (436).

Since water and ions cannot penetrate the Casparian band and are forced to cross the plasmalemma of the endodermis or reach the endodermal cytoplasm via the symplasm, it would be useful to have information about the number of plasmodesmata per unit area of wall surface at each wall traversed, the fluxes of ions and water across the root, particularly across the endodermis, and the rate of cyclosis in the cells of the pathway and the rate of movement of ions and water through the plasmodesmata.

When movement across the symplasm of the root is being considered, it is important to recognize the flux of sucrose in the opposite direction. This flux across the endodermis may need to be of the order 1×10^{-8} mol m^{-2} s^{-1}. This rate is calculated on the basis of the respiratory rate of isolated fungal sheaths of beech mycorrhizas (173), which must have received their carbohydrate in the form of sucrose from the host root (277). This compares with a calculated flux of 14×10^{-8} mol m^{-2} s^{-1} of potassium across the inner tangential wall of the endodermis of marrow *(Cucurbita pepo)* roots (434). The value given for the flux of sucrose is a minimum estimate because sucrose may be required within the fungal sheath for processes other than respiration. Furthermore there may be hyphae in the soil which, extending from the sheath, would be broken off when the roots were collected for experimental work. Therefore the amount of fungal material requiring sucrose could be underestimated.

This flux of sucrose is moving from the phloem toward the root surface; i.e., it is moving *against* the flow of ions and water across the root. Within the cytoplasm cyclosis could counteract the bulk flow of solutes brought about by the flow of water. But it is difficult to see how this could occur in the plasmodesmata. Therefore, if bulk flow of solution were occurring in the pore in one direction, it is difficult to understand how sucrose might move in the other if the sugar does so via the plasmodesmata. On the other hand, if there is a gellike matrix which is resistant to the bulk flow of water but not to diffusion of solutes, then the bidirectional flow should be possible. Measurement of the electrical resistance of plasmodesmata supports this idea (Section IX). Thus the presence of structures in the plasmodesmata becomes less surprising if they play a rôle in preventing bulk flow of fluid. On the other hand, the ability of viruses to move through plasmodesmata (144) indicates that bulk flow is still possible.

It has been thought that the pathway of movement of ions might be identified by studying the transroot potential. This is the potential which is measured between the exudate from xylem vessels and the external solution. In the initial studies by Bowling and Spanswick (52), Bowling (49), and Bowling *et al.* (51), the measured values of the potential difference across the root were compared with calculated Nernst potentials for individual ions by using known values of their concentrations in the medium and exudate. Conclusions were drawn from such a comparison about whether an ion was actively transported at some stage in its journey across the root or moved from the external solution to the exudate solely by diffusion.

This approach has been criticized by Jennings (222) and Shone (469) on a number of counts.

1. Apart from the caution which is needed in the interpretation of Nernst potential (Section III,B), the Nernst equation ought not to be applied to the exuding root because it is not in an equilibrium state (222).
2. The Nernst equation applies to ions which are moving independently of each other. For a root this is unlikely (469).
3. Effects due to solvent drag ought to be considered; the Nernst equation does not allow for this (469).
4. When the transroot potential is being considered, at least two membranes are involved—the plasmalemmas, respectively, of the outer cortical cells and the parenchyma cells abutting on to xylem vessels. These two membrane systems may have pumps which act in opposing directions; this will make the demonstration of the reality of active mechanisms difficult (222).
5. Ion pumps at tonoplasts could also have a bearing on the interpretation of the transroot potential. Such pumps could lower the cytoplasmic

concentration of an ion such that the concentration appearing in the exudate could seem to have been brought about by diffusion (222). Shone (469) has interpreted data for calcium movement across the root in the presence or absence of 2,4-dinitrophenol (DNP) in terms of a pump at the tonoplast driving the ion into the vacuole. The Nernst potential indicates that calcium diffuses across the root in the absence of DNP but is actively transported when the inhibitor is present.

It should be clear from these criticisms that the transroot potential represents a complicated electrical situation. Shone (470) has listed some of the components which will contribute to the potential. They are (1) a diffusion potential resulting from the existence of a higher concentration of ions in the xylem sap than in the medium (this potential would arise from a tendency of the accumulated ions to diffuse back into the medium at different rates), (2) a Donnan potential resulting from the presence of nondiffusible ions and charged groups in the cell wall, (3) an electro-osmotic potential due to the movement of salts and water through tissues and vessels carrying a fixed charge, (4) a carrier potential which we would now equate with the potential produced by an electrogenic pump, and (5) a transport potential caused by metabolic processes (cyclosis) tending to transport anions and cations across the root at different rates.

A correct analysis of the transroot potential is therefore likely to be complex. The most detailed so far is that of Ginsburg (145). The radial element of the model proposed was treated in terms of two parallel pathways, the symplasm and the apoplasm. The p.d. across the latter pathway is at its simplest a Donnan potential (Section VII). In his model Ginsberg treated the pathway as a cation exchange membrane. The symplasm pathway was treated as a system of three resistances in series: the outer plasmalemma (at the cortex), the symplasm itself, and the inner plasmalemma (its location is not defined, but it could be the xylem parenchyma). In the analysis, the symplasm was assumed to have minimum resistance to the movement of ions (cf. 547) and the potential across the two membranes was assumed to be a diffusion potential.

Figure 27 shows the transroot potential calculated from different values for the various electrical parameters just mentioned. The root is in a solution of potassium chloride. The following information derived from these calculated results is of interest. The linear-log relationship (curves 1 and 4) reflects the p.d. across the extracellular pathway E_e, since $R_e < R_c$, where R_e is the resistance of the extracellular pathway and R_c is that of the symplasmic pathway. The downward concave pattern (curves 3 and 5) is the result of R_e being larger than R_c and therefore the p.d. tends to reflect E_c. The lower values for curve 5 are due to higher values for α, which is the

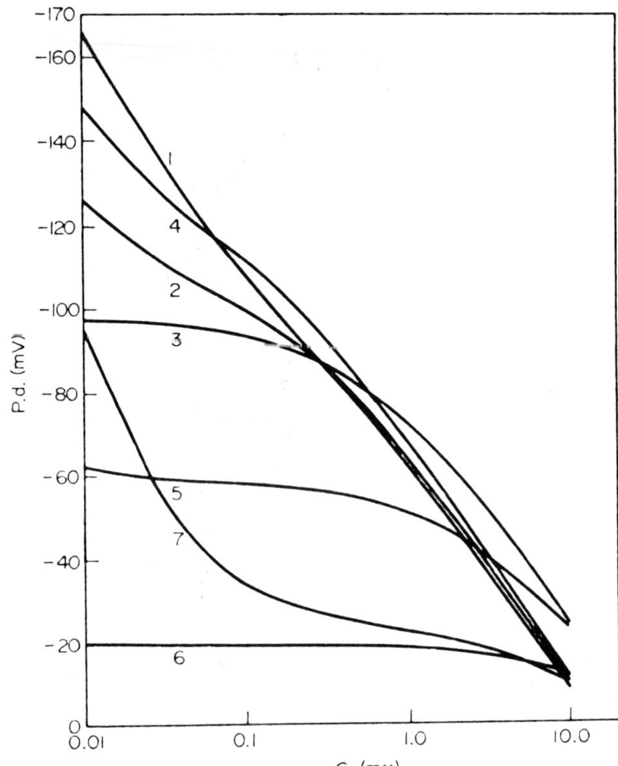

FIG. 27. The overall radial p.d. as a function of concentration of potassium chloride in the external medium for different sets of parameters. See the text for further details. The concentration of salt in the xylem is 10 mmol L^{-1}. From Ginsburg (145).

ratio of P_{Cl} to P_K, the membrane passive permeabilities of chloride and potassium, respectively. The upward-concave pattern of curve 7 is a result of $R_c \simeq R_e$ at the lower concentration range and a large value for α. Curve 6 results from large α and $R_e \ll R_c$.

Ginsberg was unable to provide information for the situation where $R_e \gg R_c$ because the absolute values of P_K and cation mobilities in the extracellular pathway can be varied quite widely without affecting the p.d., provided that their relative magnitudes are unchanged.

This analysis ignores any p.d. along the longitudinal axis of the root. Figure 28 shows the representation of the whole root in terms of a three-dimensional electrical analog, and Fig. 29 a much simpler two-dimensional

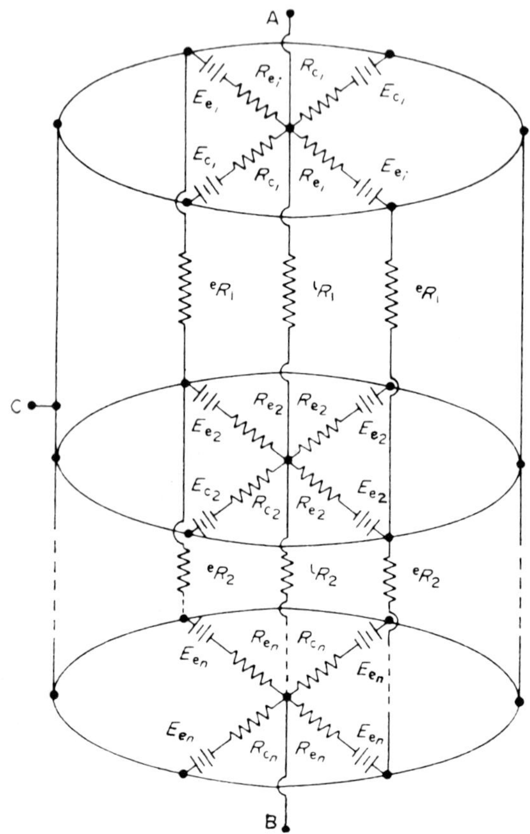

FIG. 28. Three-dimensional electrical analog of a plant root. E_c, E_e = the radial cellular and extracellular electromotive forces of the ith length element of that root. R_c, R_e = radial cellular and extracellular resistances of the ith length element. ${}^{l}R_i$ = the xylem resistance of the ith length element. ${}^{e}R_i$ = the longitudinal extracellular resistance. From Ginsburg (145).

one. The following $(n-1)$ equations for the circuit apply:

$$E_1 - E_2 + I_1(R'_1 + R'_2 + {}^{l}R_1) - I_2R'_2 = 0 \quad (18)$$

$$E_2 - E_3 + I_2(R'_2 + R'_3 + {}^{l}R_2) - I_1R'_2 - I_3R'_3 = 0 \quad (19)$$

$$E_{n-1} - E_n + I_{n-1}(R'_{n-1} + R'_n + {}^{l}R_{n-1}) - I_{n-2}R'_{n-1} = 0 \quad (20)$$

If every R' and every R are equal, then when $R' \ll {}^{l}R$, the longitudinal p.d. is equal to $E_1 - E_n$. Therefore this p.d. should not vary with root length. As R' approaches the value of ${}^{l}R_1$, the relative contribution of the first and the $(n-1)$ elements decreases and that of the intermediary ele-

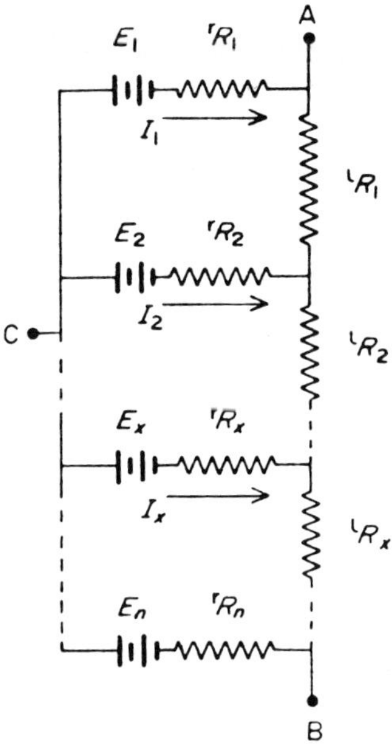

FIG. 29. Merged two-dimensional electrical analog of a plant root. E_i = the overall radial emf of the ith length element. R_i = the parallel combination of R_c and R_e of the ith length element. From Ginsburg (145).

ments increases. The longitudinal p.d. will therefore be a linear function of root length.

These are important points with regard to our understanding of the transroot potential. It has been shown (191) that an electrical potential difference is still maintained between the xylem exudate of an excised maize root and the bathing solution even when successive segments are removed from the root tip. While one might have expected the observed potential to be short-circuited through the open xylem vessels, the preceding analysis indicates that as long as the longitudinal resistance of the root is greater than the radial resistance, then the transroot potential can be maintained. If current is passed down the xylem vessel, it is therefore acting as a very leaky cable.

Ginsburg and Laties (146) and Anderson and Higinbotham (12) have measured the longitudinal resistance of root segments as a function of root length and in the former case also as function of potassium concentration in the external medium. The analysis of the data in both cases used a resistance analog of the root in which there are three parallel resistances, namely, the extracellular space, the intracellular pathway, and the xylem vessels. There is no doubt from both studies that the observed resistance of a root segment is principally a reflection of the resistance of the symplast. Because of the relatively high resistance of the xylem vessels, it is not possible to deduce anything about their internal contents.

Ginsburg (145) has pointed out that the high resistance of the xylem vessels will mean that the p.d. measured between the bathing medium and the exudate of an excised root reflects only the properties of the uppermost segment of the system. Indeed the observations by Ginsburg and Laties and Anderson and Higinbotham raise the question as to whether the so-called transroot potential is in fact truly the p.d. between the external solution and the xylem vessel. It is quite likely that the p.d. is that between the external solution and the fluid exuding from the cut surface of the root, the current being carried through the symplasm, the stele being short-circuited. If this is so, it is not surprising that the transroot potential reflects the transmembrane potential of the outer cortical cells (111–113).

Clearly, studies on the transroot potential tell us nothing about how ions actually get into the xylem vessels. There is still no evidence therefore that salts move out of the living cells of the stele as a result of the activity of a unidirectional pump, i.e., by a process of secretion. It is conceivable that salt diffuses across the membrane by a cation–anion symporter. This would mean that the pumping activity responsible for xylem exudation would reside in the plasmalemma of the parenchyma cells of the stele in which accumulation first occurs (*see* 222). Salt would then move to the xylem vessels down its chemical potential gradient. We are no closer than we were in 1959 to deciding which of these two possibilities is correct.

XI. Ion Uptake by Roots from the Soil

A. Theoretical Aspects

Ion uptake by roots from the soil is a subject which has developed very considerably since 1959. Present ideas owe much to Bray (55), who was initially concerned with a reappraisal with what were then basic ideas of soil fertility. Thus, Liebig's (279) law of the minimum stated that the nutrient present in the least relative amount is the limiting nutrient. Implicit in this law was the belief that all the other nutrients were present in excess until the deficient or limiting one was made adequate, whereupon the nutrient previously next least in relative amount became that which was deficient. The other basic idea was formulated by Mitscherlich (347), who arrived at the conclusion that plant growth follows a diminishing increment type of curve now known as the "yield" curve. To fit this curve he produced an equation which can be expressed as

$$\log(A - y) = \log A - cb \tag{21}$$

where A is the total yield of all portions of the crop when the nutrient is not deficient, y is the yield when b amounts of a given nutrient are present, and c is the proportionality constant. From this equation, if the amount of a nutrient present in the soil and c and y are known, then the yield can be calculated. However, it assumes (and this was asserted when the equation was formulated) that for a given nutrient form, c is constant.

Careful consideration of these two concepts will show that they are in conflict. Thus, given a constant value of c, it follows that one amount of a deficient nutrient always produces the same proportion of that yield which is obtainable when the nutrient is adequate. As Bray (55) put it, "a given amount of nutrient is 'good for' a certain percentage of the obtainable yield as that yield varies in size with other growth factors." Baule (35) concluded that when more than one nutrient was deficient, the final percentage sufficiency is the product of the individual sufficiencies. This is clearly contrary to Liebig's ideas that only one nutrient can limit growth and that other nutrients are in excess.

Bray then pointed out that the problem revolves around the amount of a given nutrient within a soil and this in turn is dependent upon the mobility of the nutrient within the soil. The mobility is dependent upon the soil solution and exchange of the nutrient with the soil as well as actual movement to the root. It also follows that growth of the root is important in exploiting nutrients. He postulated that the extremes of mobility—highly mobile and essentially immobile—created two root sorption zones. One zone, the *root system sorption zone,* included the whole volume of soil within the major part of the root system. From this zone, the root system obtains relatively mobile nutrients. The other zone contains those small volumes of soil adjacent to the root surface and from which the roots obtained relatively immobile nutrients. This zone he called the *root surface sorption zone.* The zones are shown diagrammatically in Fig. 30.

If two plants are growing close together, their sorption zones may overlap; this will be particularly so for their root system sorption zones. There will thus be competition for nutrients. Depending on the planting pattern and rate of planting, plants can grow without competing with each other for any nutrients, can effectively compete for the relatively mobile nutrients only, and can compete for both the relatively mobile and relatively immobile nutrients.

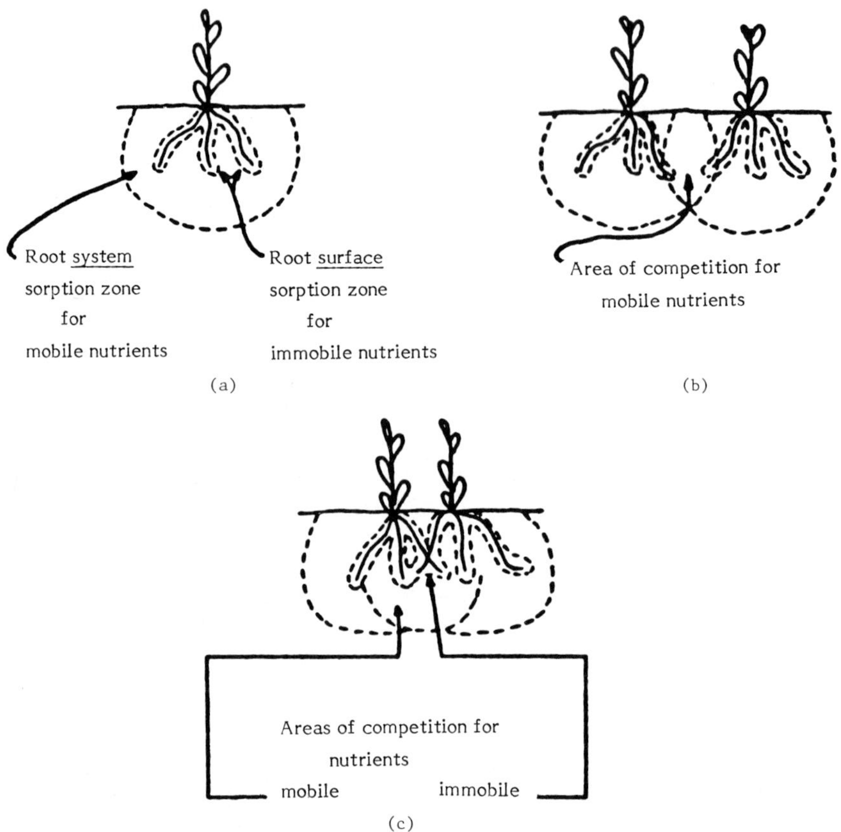

FIG. 30. Diagrams illustrating the root sorption zones proposed by Bray (55). (a) An individual plant; (b) two plants with root system sorption zones overlapping; (c) two plants with both root surface sorption and root system sorption zones overlapping.

Bray then developed the following points which conflict with the assertion that the proportionality constant *c* can be universally constant for a given form of nutrient:

1. Not all nutrient forms will follow a percentage sufficiency concept; the relatively mobile nutrients should follow the law of limiting nutrients.
2. Even a restricted demonstration of the constancy of *c* for a relatively immobile nutrient should be impossible if a relatively mobile nutrient is actually (not potentially) deficient.
3. Since plants vary in the nature of their root systems, the value of *c* will vary with the kind of plant.

4. Variations in the rate of planting and the planting pattern will vary competition for the nutrient and hence vary the value of c.

5. Mitscherlich recognized that variations in the form of the nutrients varied the value of c. But variations in their distribution within the soil, i.e., in the fertility pattern, should also vary the value of c for the relatively immobile nutrients.

Bray's analysis shows that there are limits to the applicability of both the Liebig and Mitscherlich concepts. At one end of the mobility scale, the highly mobile nutrient can be removed almost completely by a plant from the soil. This results in their "net" requirement being almost equal to the crop content at maturity. That is, one amount is sufficient for only one yield of a given percentage composition. Such nutrients can act as "limiting nutrients' in the sense that Liebig's law postulates. On the other hand, when the nutrient is immobile, the roots will absorb only a small portion of what is present in the soil. The amounts necessary to give maximum yield will be many times larger than the crop content. For such a nutrient, the Mitscherlich percentage sufficiency concept can only hold for a particular plant, provided that the planting pattern and rate remain constant and that one is dealing with a similar type of soil. Thus, if competition occurs between root surface sorption zones in one situation but not in the other or in one soil the immobile nutrient is initially distributed homogeneously but not in another soil, then the percentage sufficiency concept will not hold.

Movement of a nutrient through the soil to the root can occur either by diffusion or as a result of bulk flow of solution. Taking the relatively simple case of diffusion first, Nye (364) described mathematically the diffusion of nutrients to a root surface, treating it as a planar sink, i.e., one that is perpendicular to the direction of flow, as

$$Q_t \simeq \frac{2}{2\pi} \Delta C_1 \sqrt{\frac{\Delta C}{\Delta C_1}(v_1 f_1)(D_1 t)} \tag{22}$$

where Q_t is quantity moving into sink; ΔC_1 the difference in concentration between the ion in the soil solution and sink solution, ΔC the corresponding difference in the total concentration of diffusible ion in the soil, $\Delta C/\Delta C_1$ the buffer power of the soil over the concentration ΔC_1, D_1 the diffusion coefficient of the ion in free solution, v_1 the volume fraction of soil moisture (percentage of total volume), f_1 the impedance factor, and t time.

The buffer power of the soil concerns that quantity of nutrients that will be released when the solution concentration is lowered over a given range. The moisture term is involved because the effective cross section for diffusion is reduced when the moisture content falls. The impedence factor is a measure of the "tortuosity" of the path, which is also a function of the moisture content of the soil.

Buffer capacity will be considered when dealing with the release of ions into the soil solution. But if we assume a constant rate of release, then the rate of diffusion will depend upon the water content of the soil, which in turn will be affected by the plants themselves as a result of transpiration. While transpiration is occurring, the soil solution will also move, and this bulk movement will bring ions to the plant. This is usually what is meant by mass flow, but there can be general movement of soil water up or down a profile, depending upon the environmental conditions. However, for this discussion the focus will be on mass flow brought about by the loss of water from the plant as a result of transpiration. The extent to which there will be a lower concentration of ions at the root surface than at some point away will depend (all other factors being discounted) upon the rate of removal by the root itself relative to the rate of movement of water into the plant. Thus it is conceivable that the concentration of a nutrient could increase at the root surface. The extent to which this occurs will be dependent also upon the back diffusion of the nutrient into the solution in the bulk of the soil.

A detailed theoretical analysis has been given by Nye and Marriott (366). They considered 1 cm of a root cylinder in the soil and the balance of solute in a small annular element, thickness δr (in centimeters), of that soil which is being influenced by the activity of the root.

Thus over the time δt (seconds)

$$\frac{\delta\, 2\pi r F}{\delta r} = 2\pi r \frac{\delta C}{\delta t} \tag{23}$$

where F is the inward radial flux of diffusible solute (in grams per square centimeter second) and C the concentration of diffusible solute (in grams per cubic centimeter of soil).

The change in the balance of solute will be the sum of the rate of diffusion of the solute and the rate of movement due to mass flow. Therefore,

$$F = D \frac{\delta C}{\delta r} + v C_1 \tag{24}$$

where D is the diffusion coefficient of solute in the soil (in square centimeters per second), C_1 is the concentration in the soil solution (in grams per cubic centimeter of solution) and v is the inward flux of water (in cubic centimeters per square centimeter second).

If we consider the balance of water,

$$2\pi r v = 2\pi r_0 v_0 \tag{25}$$

r being the radial distance (in centimeters) from the root axis and r_0 the radius of the root (in centimeters).

Hence from these last three equations and putting the differential buffer power $dC/dC_1 = b$, we have

$$\frac{1}{r}\frac{\partial}{\partial r}\frac{rD\,\partial C_1}{r} + \frac{v_0 r_0 C_1}{b} = \frac{\partial C_1}{dt} \tag{26}$$

We need to decide the boundary conditions. The first is given by $t = 0$, $r > r_0$ and $C_1 = C_{1_i}$, where C_{1_i} is the initial uniform concentration in the soil solution (in grams per cubic centimeter) and defines that situation, where the root is present but exerting no influence on the soil.

The second boundary condition relates to root activity. Nye and Marriott assume that the overall rate of movement of the nutrient into the root will be governed by an enzymic process having Michaelis-type kinetics. The equation used by Nye and Marriott is

$$F = \frac{kC_{1_0}}{(1 + kC_{1_0})/F_{\max}} \tag{27}$$

where $F_{\max}$ is the maximal flux. This differs from the usual Michaelis formulation in that there is a rate constant and not a ratio of rate constants.

Accordingly, the second boundary condition is $t > 0$, $r = r_0$ and

$$Db\frac{\delta C_i}{\delta r} + v_0 1_0 = \frac{kC_{1_0}}{(1 + kC_1)/F_{\max}} \tag{28}$$

Since this is after a finite time, we need to consider the concentration at the root surface and that situation in which the rate of supply, taking into account the buffer capacity of the soil, is equal to the rate of entry of nutrient into the root.

In producing these equations it is important to recognize that the following assumptions have been made.

1. It is assumed that $F_{\max}$ and k are independent of v_0, namely, that there is no interaction between nutrient and water flows. This need not be the case for movement across certain membranes, but it seems to be a reasonable assumption with respect to the effect of transpiration (see Chapter 5 in this volume).

2. It is assumed that D is independent of v. This may not be so because we are concerned with a situation in which there is flow through a porous material and thus the flow may be irregular because of changing molecular interactions.

3. No account is taken of root hairs. It has been argued by Passioura (380) that if they are closely spaced, the radius under consideration is not that of the bulk of the root tissue but that of the cylinder joining them at the tips of the root hairs (Fig. 31).

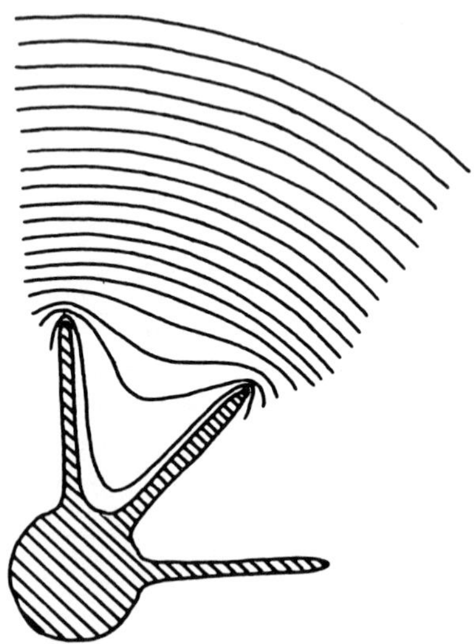

FIG. 31. Diagrammatic transverse section of a root in soil. The lines represent equal concentration of a solute in the surroundings. From Passioura (380).

4. It is assumed that b and D are independent of concentration.

5. There are some other more obvious assumptions, namely, that the soil is a homogenous, unstirred porous medium and that the root absorbs nutrients and water uniformly over its whole surface.

Figures 32a–c show some of the values which were calculated by using the preceding equations. The graphs show how the ratio C_1/C_{1_i} relates to the distance from the root, essentially the extent to which the concentration of nutrient has changed from the initial concentration. One can see that when there is diffusion alone (Fig. 32), the concentration at the root surface continues to fall and the zone of depletion spreads outward with time. The rate at which this depletion occurs depends upon the rate flow v_0 (Fig. 32b,c). Indeed, v_0 can be such that, as indicated already, a zone of accumulation can be produced. On the other hand, if the root absorbing power k is increased (Fig. 33b,c) depletion can again occur.

There is need to clarify the meaning of the diffusion coefficient D. We are referring to diffusion throughout the whole soil mass. So far, all nutrients have been treated in the same way, but, of course, their diffusion will depend upon their physicochemical properties. For an anion such as nitrate, which is not absorbed by the soil, diffusion will be via the bulk phase of the liquid which is out of contact with the fixed negative charges at the soil surface. These negative charges will bind cations, but the sites of

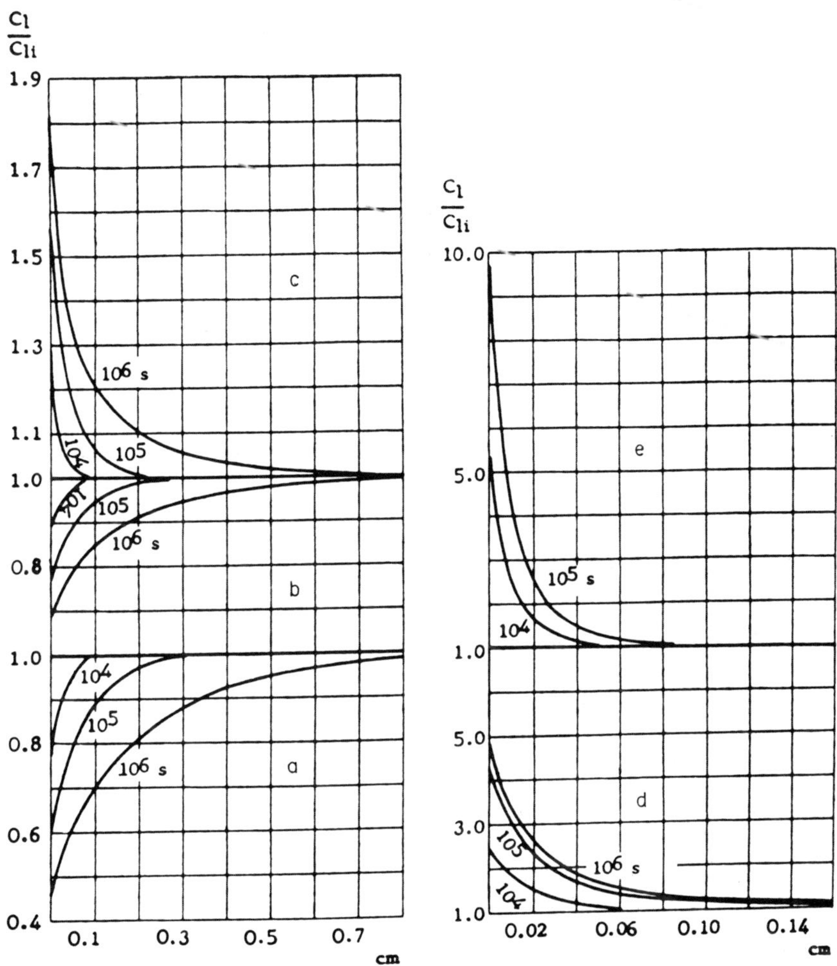

FIG. 32. The effect of the rate of soil solution flow on the relative concentration of ions near a root surface. $k = 2 \times 10^{-7}$ cm s^{-1}; $D = 10^{-7}$ cm^2 s^{-1}; $b = 0.2$; F_{max} high; $r_0 = 0.05$ cm. (a) $v_0 = 0$ cm s^{-1}, i.e., diffusion alone. C_1/C_{1_i} continues to drop at the root surface and the zone of depletion to spread outwards at 10^6 s. (b) $v_0 = 10^{-7}$ cm s^{-1}. C_1/C_{1_i} has not reached the limiting value $v_0/k = 0.5$ after 10^6 s. (c) $v_0 = 4 \times 10^{-7}$ cm s^{-1}. C_{1_o}/C_{1_i} has nearly reached the limiting value $v_0/k = 2$ after 10^6 s, but the zone of accumulation c continues to spread outward since $r_0v_0/Db = 1$. For a steady state r_0v_0/Db must exceed 2. (d) $v_0 = 10^{-6}$ cm s^{-1}. $r_0v_0/Db = 2.5$. After 10^6 s C_{1_o}/C_{1_i} has almost reached the steady-state value $v_0/k = 5$, and the zone of accumulation has ceased to spread outward. (e) $v_0 = 2 \times 10^{-6}$ cm s^{-1}. $r_0v_0/Db = 5$. Both C_{1_o}/C_{1_i} and the zone of accumulation have nearly attained the steady state after only 10^5 s. C_{1_o}/C_{1_i} is increased to $v_0/k = 10$, but the zone of accumulation is compressed. From Nye and Marriott (366).

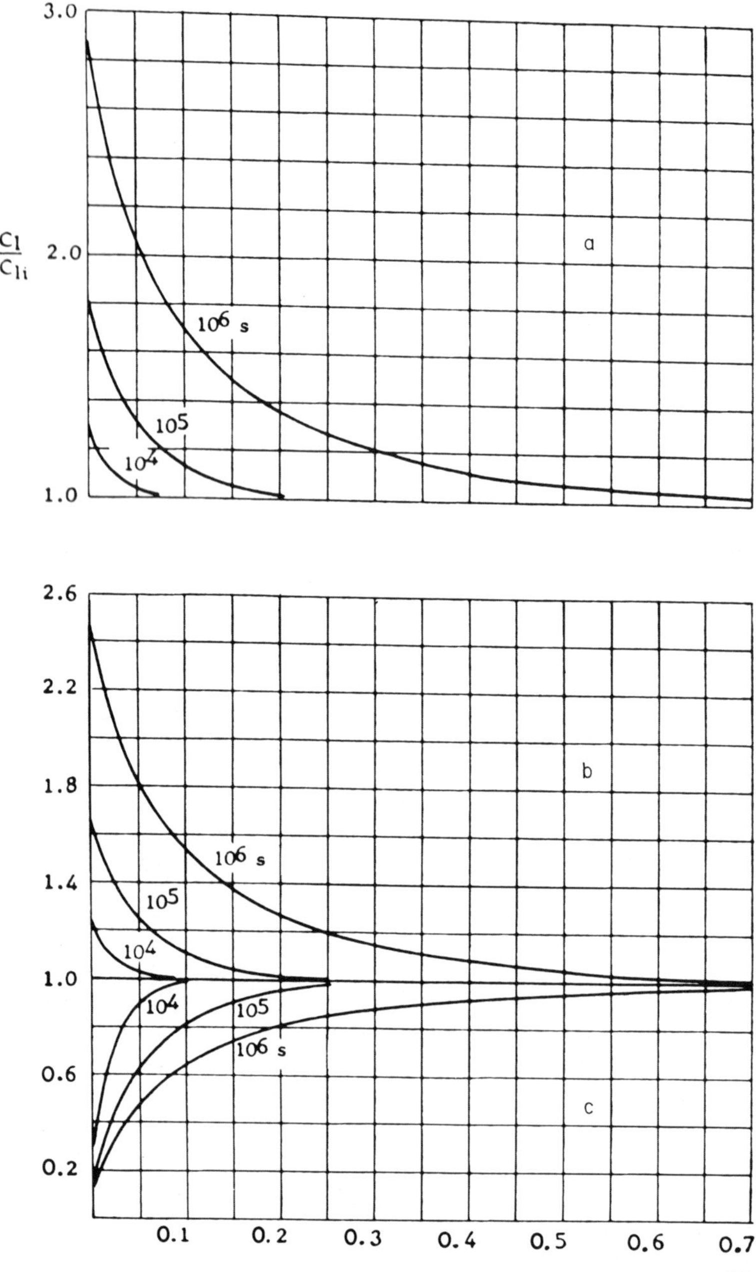

FIG. 33. The effect of the root absorbing power on the relative concentration near the root surface. $v_0 = 2 \times 10^{-6}$ cm s^{-1}; $D = 10^{-7}$ cm^2 s^{-1}; $b = 2$; F_{max} high; $r_0 = 0.05$ cm). (a) $k = 0$, i.e., no absorption by the root. C_{1_o}/C_{1_i} continues to increase and the zone of accumulation continues to spread outward. (b) $k = 2 \times 10^{-7}$ cm s^{-1}. The limiting value of $C_{1_o}/C_{1_i} = 10$ is approached very slowly. (c) $k = 2 \times 10^{-5}$ cm s^{-1}. The limiting value of C_{1_o}/C_{1_i} ($v_0/k = 0.1$) is rapidly approached at this high value of k, but the rate at which the zone of disturbance spreads outward it little affected. From Nye and Marriott (366).

the charges are often sufficiently close together that weakly held cations can change position, and this can lead to overall cation interdiffusion through the bulk phase of the liquid. However, there are indications that the process may not be important (445). Phosphate is strongly adsorbed and the sites are relatively far apart, and so interdiffusion does not occur. Even from this simple consideration, it is not difficult to see that the more the layer of water over the surface of the soil particles is reduced, the less will diffusion of a nutrient in that layer have characteristics of diffusion of the same nutrient in simple solution.

Porter *et al.* (399) have shown that the following relationship exists between the diffusion coefficient D for the movement of nutrient through a soil and for movement in simple solution:

$$D = D_1 v_1 f_1 (dC_1/dC) \tag{29}$$

where f_1 is the impedance factor and depends on the tortuosity of the pathway through the liquid phase; $dC_1/dC = 1/b$ and is referred to as the buffer capacity of the soil. Here v_1 and f_1 are obviously related, since when the water content of the soil decreases, f_1 will also decrease (but more rapidly). The buffer capacity, which is the ratio of the amount of the nutrient in free solution to the total quantity in the soil, is a measure of those interactions briefly referred to earlier. The lower the ratio, the more the soil is holding on to the nutrient and therefore restraining its movement in free solution.

Nye and Marriott have investigated the interaction between D and moisture content (Figs. 34a–c). As the water content of the soil is reduced with a consequent fall in v_1 and f_1, the less the effect of diffusion is apparent; i.e., the tendency for nutrients to accumulate at the root surface increases because they diffuse back less readily into the bulk portion of the soil. The same situation holds for a change in the buffer capacity ($1/b$), as can be seen from inspection of Figs. 32e and 33b. With all other factors being equal (and in this particular situation when there is a relatively high volume flow v), a higher buffer capacity expresses itself in the rapid achievement of a steady-state concentration of nutrient close to the root. When the buffer capacity is 10 times lower, the zone of accumulation continues to spread out from the root as time goes on. What we are seeing is the effect of back diffusion of nutrient away from the root.

In summary, the concentration profile of a nutrient in the soil around a plant root system will be essentially the interaction between the ability of the root to absorb the nutrient, the volume flow to the root, and the rate of diffusion of the nutrient in the soil. What happens in actual plant–soil situations involving specific ions will be discussed later.

Before these situations can be described, something needs to be said

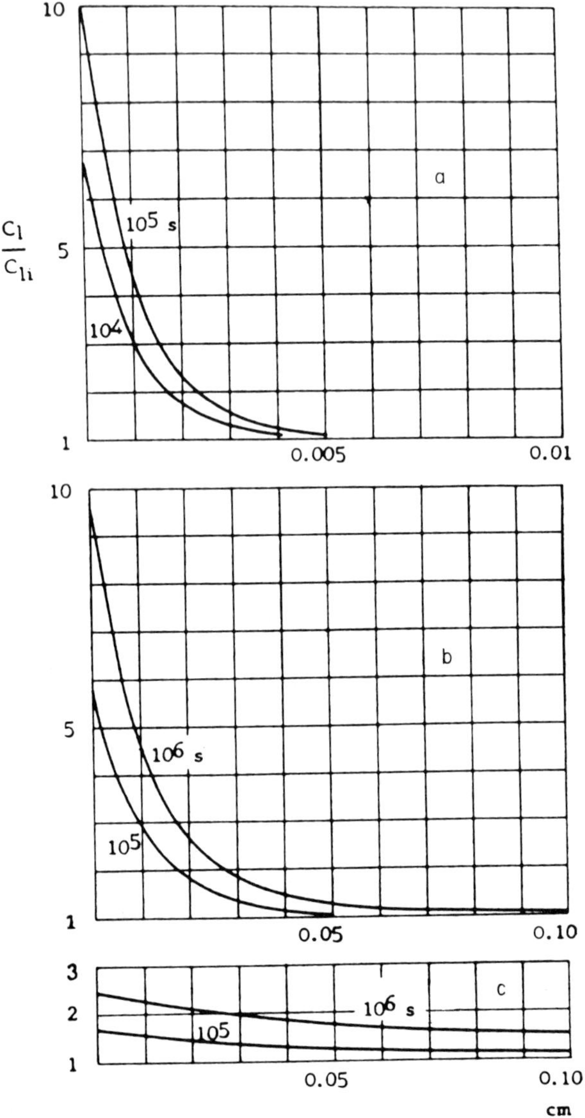

FIG. 34. The effect of Db on the relative concentration near the root surface. Db increases as the moisture level in the soil increases. $v_0 = 2 \times 10^{-6}$ cm s^{-1}; $k = 2 \times 10^{-7}$ cm s^{-1}; $b = 2$; F_{max} high; $r_0 = 0.05$ cm. (a) $Db = 2 \times 10^{-9}$, corresponding to a very dry soil. C_{1_o}/C_{1_i} at the root surface rapidly attains its limiting value, and the zone of disturbance is very compressed. (b) $Db = 2 \times 10^{-8}$ corresponding to a moderately dry soil. C_{1_o}/C_{1_i} has attained its limiting value after 10^{-6} s, and the zone of accumulation is less compressed. (c) $Db = 2 \times 10^{-7}$, corresponding to a moist soil. C_{1_o}/C_{1_i} increases very slowly, and the zone of disturbance spreads outward more rapidly than in (b). No steady state will be reached because $r_0v_0/Db = 0.5$. From Nye and Marriott (366).

about the application of the preceding equations and other theoretical treatments (e.g., 26, 380, 381) to given experimental conditions. It is not too difficult to devise the latter: plants are grown in a homogenous soil under conditions in which the rate of transpiration is known and the nutrient (ion) concentration can be determined.

The values obtained experimentally, e.g., v_0, C_1/C_{1_i}, are substituted in the theoretical equation(s) together with those values describing the properties of the soil, e.g., D and b. Although determination of these values is not without difficulty, the degree of reliability which can be attached to them is relatively unambiguous. This is because the soil can be treated as a physicochemical (albeit a rather complicated) system. On the other hand, there are problems in determining the ability of the root system under consideration to absorb ions.

This ability can be defined in terms of a coefficient α, the so-called root-absorbing power (47, 363, 380), which is defined as

$$F = \alpha C_1 \tag{30}$$

or uptake per unit length

$$2\pi(\alpha r)C_1 \tag{31}$$

It should be noted that α is a coefficient and not a constant and that is it may vary for a given root depending upon the external and internal conditions. Some of these conditions might be the degree of maturity of the tissue, the internal solute content of the root cells, the degree of microbial infection, for example. On the other hand, for the population of roots produced by a plant one might anticipate that αr may show much less variation. Therefore it is more appropriate to consider the mean value of αr, namely, $\overline{\alpha r}$, which can be determined indirectly (367). There is a need here to digress and to indicate how this determination can be made.

Let us consider an absorbing root system under conditions in which there is a constant concentration in the medium of the nutrient C_{1_i}. If we are also considering the mean absorption capacity of the system, the rate of uptake U is given by

$$dU/dt = 2\overline{\alpha r}LC_{1_i} \tag{32}$$

where L is the total length of the root system (in centimeters).

The value of $\overline{\alpha r}$ may be further expressed in terms of more readily measured characteristics of plant growth.

Thus

$$\frac{dU}{dt} = \frac{d(CW)}{dt} = \frac{C\,dW}{dt} + \frac{W\,dC}{dt} \tag{33}$$

where C is the mean concentration of nutrients in the whole plant (in grams per gram) and W is the total dry weight (in grams).

Substituting for dU/dt we get

$$2\pi\overline{\alpha r} = \frac{1}{LC_{1_i}}\left(\frac{C\,dW}{dt} + \frac{W\,dC}{dt}\right) \tag{34}$$

$$= \frac{W}{L}\frac{C}{C_{1_i}}\left(\frac{dW}{dt}\frac{1}{W} + \frac{dC}{dt}\frac{1}{C}\right) \tag{35}$$

where $[(dW/dt)(1/W)]$ is the relative growth rate (RGR) and $[(dC/dt)(1/C)]$ is the relative concentration change rate.

If a constant solution concentration is maintained $[(dC/dt)(1/C)]$ can be neglected; that is, the mean concentration of ion in the plant does not change. This, of course, assumes that under constant conditions of supply a plant uses all its nutrients directly for growth and there is no storage component (but see Chapter 5). Nevertheless, if this assumption is made,

$$\overline{\alpha r} = \frac{W}{2L}\frac{C}{C_{1_i}} \tag{36}$$

Nye and Tinker (367) used the data of Asher and Loneragan (18) and Loneragan and Asher (283), who measured the uptake of phosphate by eight species over 4 weeks in which culture solutions were kept constant, to calculate $\overline{\alpha r}$ (Table VIII). It is noteworthy that $\overline{\alpha r}$ remains fairly steady over the range 0.04–1.0 mol L^{-1}. Only above that value is there a decrease. However, these values are for the situation in which the plant is in the steady state, namely, that its nutrient uptake is "keeping pace" with growth.

The preceding analysis can be expanded into the situation in which root growth is also considered (e.g., 58). However, although root growth is obviously important for a plant as it "forages" for nutrients (55), there is no need here to consider an expanded analysis which involves root growth. The more simple analysis has shown that if we have information about the length of a root system—and this can be determined relatively easily (353)—one can obtain a value of α from work with solution culture and thus determine $\overline{\alpha r}$. It is not much more difficult to determine $\overline{\alpha r}$ for a plant growing in soil (59).

The term α has been treated in broad terms, but as yet without specifying its physiological basis. However, α has been defined in terms of root length. If α is independent of radius, then for a given external concentration of nutrient, α will depend on the surface area. On the other hand, if α is also dependent on the radius of the root, α will depend upon the volume. Nye (365) has considered this situation theoretically, examining the movement of ions through the root of a hypothetical intact plant when there is a

TABLE VIII

PHOSPHORUS UPTAKE BY PLANTS OF EIGHT SPECIES FROM NUTRIENT CULTURE SOLUTION[a,b]

Term	Phosphorus concentration (μmol liter^{-1})				
	0.04	0.2	1.0	5.0	25.0
$\frac{C}{C_{li}}\left(\frac{g\ g^{-1}}{g\ cm^{-3}}\right)$	82×10^4	42×10^4	22×10^4	5.6×10^4	1.3×10^4
$\frac{1}{W}\frac{dW}{dt}$ (g g^{-1} day^{-1})	0.07	0.10	0.12	0.13	0.13
$\frac{W_R}{W}$ (g g^{-1})	0.44	0.28	0.20	0.19	0.19
$\frac{W_R}{L}$ (g cm^{-1})[b]	1.82×10^{-4}	1.36×10^{-4}	1.22×10^{-4}	1.20×10^{-4}	1.20×10^{-4}
$\therefore \frac{W}{L}$ (g cm^{-1})	4.1×10^{-4}	4.8×10^{-4}	6.1×10^{-4}	6.3×10^{-4}	6.3×10^{-4}
$\therefore \overline{\alpha r}$ (cm^2 s^{-1})	43×10^{-6}	37×10^{-6}	30×10^{-6}	8×10^{-6}	2×10^{-6}

[a] Average values of the terms in equation 36 (see text) relating to data derived from Loneragan and Asher (283). From Nye and Tinker (367).
[b] Assuming $r = 0.025$ cm.

quasi-steady-state rate of uptake over an extended period of, for instance, several days. This is the situation when accumulation in root tissues is small compared to throughput to the xylem and hence to the shoot. Nye refers to a model root in which the rate of uptake could be controlled by (1) diffusion through the free space, (2) uptake by the cortical cells, (3) transfer through the symplast, (4) transfer across the endodermis, or (5) transfer into the xylem.

Nye explored the consequences if uptake is controlled by a combination of processes 1 and 2, assuming that 3, 4, and 5 are rapid. He pointed out that the extreme cases are fairly clear. If all the cortical cells are in equilibrium with the soil solution, the rate of uptake per unit length will depend on root volume. On the other hand, if diffusion through the free space is relatively slow, only the cortical cells near the surface will be effective. The rate of uptake per unit length of root will then depend on the surface area of the root; α will be independent of the radius. In an analysis of the situation relating to what may be occurring, Nye has shown by a theoretical analysis that for nutrients in the deficiency range of concentrations (at which the effect of mass flow is likely to be low) root surface area should be more significant than root volume. At high concentrations when α is lower, volume may be more significant.

B. INFORMATION FROM EXPERIMENTS

This section deals with actual situations. Although values for some of the properties of the soil–plant system are needed, simple measurements and calculations can provide useful information. Barber (33), for instance, used data obtained by a number of workers to calculate the ratio of plant content to soil content for a number of elements (Table IX). Taking the quantity of water transpired as 200 g per gram of plant material, Barber pointed out that any ratio less than 200 will represent a situation in which more ions move to the plant root than the plant will absorb. Where the concentration of the element in the soil solutions was high, phosphorus was the only element which would not accumulate at the root surface. At the lowest concentration found, only sulfur would accumulate. Thus it appears from this that phosphorus moves to a root by diffusion, and this may also be so for potassium.

The results for phosphorus have been confirmed in a qualitative but more direct manner by Lewis and Quirk (276), who examined the changing concentration of the element in the soil labeled with ^{32}P around growing wheat roots. Successive autoradiographs showed depletion of phosphate round the roots. A more quantitative approach has been made by Brewster and Tinker (59). They grew leeks *(Allium porrum)* in a sandy loam under conditions of adequate water supply. Their data show (Table X) that the apparent mass flow (water inflow × soil solution concentration,

TABLE IX

RELATION BETWEEN CONCENTRATION OF IONS IN THE EXPRESSED SOIL SOLUTION AND CONCENTRATION WITHIN THE *Zea mays* PLANT[a]

	Concentration (ppm)			Ratio of plant content to lowest and highest soil solution contents	
	Soil solution		Plant		
Element	Low	High	Mean	Low	High
Ca	S	450	2,200	275	4.9
K	3	156	20,000	6,666	128.0
Mg	3	204	1,800	600	8.8
N	6	1700	15,000	2,500	8.8
P	0.3	7.2	2,000	6,000	278.0
S	118	655	1,700	155	2.6

[a] From Barber (33).

i.e., $cm^3\ cm^{-1}\ s^{-1} \times mol\ cm^{-3}$, which is $mol\ cm^{-1}\ s^{-1}$) was small compared with the total inflow for potassium but increased with time; it was much larger than inflow for calcium and slightly larger for magnesium and sodium. They used the equation developed by Passioura (380) to calculate the mean solution concentration at the root surface. The equation is based on a model similar to that of Nye and Marriott (366) and requires the use of suitable values for plant and soil characteristics. Table X shows that the potassium concentration decreased to between one-seventh and one-half of the original concentration but that of calcium, magnesium, and sodium was slightly increased. It would seem therefore that potassium moves to the root for the most part by diffusion, whereas calcium and magnesium move by mass flow.

The results just described show that diffusion is very important in the supply of phosphorus and potassium to plant roots. Drew *et al.* (105) carrying out experiments under fairly precise conditions have shown that uptake of potassium by cylindrical roots of onion and leek is reduced by 25 to 50% of that which would be obtained if the initial soil solution concentration were maintained at the root surface. In a subsequent study, Drew and Nye (103) examined uptake of potassium by roots of Italian rye grass *(Lolium multiflorum).* Calculations showed that the concentration of labile potassium within the soil may be reduced to between 99 and 53% of the initial, depending upon the diffusion characteristics of the soil and nutrient demand by the root. Of the total potassium absorbed by a root in 4 days, the proportion which is supplied from within the root hair cylinder is small (0.8–6.3%). When root demand is high (increased α), diffusion limits uptake, but this uptake is enhanced by root hairs because they increase the effective root diameter. When root demand is low, diffusion does not appear to be a limiting factor and under these conditions hairs add little to the efficiency of the root system; at least theoretical calculations indicate that diffusion to the central root could still account for uptake.

In the preceding experiments, the diffusion coefficient for potassium is readily determined by using ion exchange (in the H^+ form) resin paper adpressed to the soil as a sink maintaining zero concentrations of the diffusate at the soil surface (554). However, it should be noted that D will depend on the buffer capacity (exchangeable potassium level) and water content of the soil (see Section XI,A and 553). The impedance factor f_1 can be determined by measuring the rate of self-diffusion of ^{36}Cl (i.e., the ion exchanges with its own isotope in an otherwise uniform environment; in particular there are no changes in concentration at specific points in the soil) through soil whose water content is known (445). It is assumed that there is virtually no interaction between the chloride ion and the soil.

For phosphate, there are problems in determining the effective diffu-

TABLE X

Mean Inflow of Water and Various Cations, the Mean Fraction of Total Uptake Supplied by Apparent Mass Flow and the Calculated Mean Relative Concentration Changes at the Onion Root Surface during the Three Periods between Harvests[a,b]

Interval	Water inflow ($cm^3\ cm^{-1}\ s^{-1} \times 10^{-6}$)	K^+ Inflow ($mol\ cm^{-1}\ s^{-1} \times 10^{-13}$)	K^+ Apparent mass flow/inflow	K^+ C_{1r}/C_{1i}	Ca^{2+} Inflow ($mol\ cm^{-1}\ s^{-1} \times 10^{-13}$)	Ca^{2+} Apparent mass flow/inflow	Ca^{2+} C_{1r}/C_{1i}	Mg^{2+} Inflow ($mol\ cm^{-1}\ s^{-1} \times 10^{-14}$)	Mg^{2+} Apparent mass flow/inflow	Mg^{2+} C_{1r}/C_{1i}	Na^+ Inflow ($mol\ cm^{-1}\ s^{-1} \times 10^{-14}$)	Na^+ Apparent mass flow/inflow	Na^+ C_{1r}/C_{1i}
June 16–29	0.24	14.8	0.07	0.14	4.4	8.2	1.20	10.1	1.5	1.03	21.5	1.8	1.05
June 29–July 7	0.23	11.8	0.08	0.33	4.8	7.4	1.19	8.5	1.7	1.03	18.4	2.3	1.07
July 7–13	0.24	10.0	0.10	0.40	4.4	8.2	1.20	6.8	2.2	1.06	4.8	8.3	1.11

[a] From Brewster and Tinker (59). Reproduced from *Proceedings of the Soil Science Society of America*, Volume 34, p. 424, 1970.

[b] C_{1i} = soil solution concentration, C_{1r} = soil solution concentrations root surface.

sion coefficients. This is because there is uncertainty about the measurement of labile phosphate and doubt about whether all the phosphate in the labile pool is depleted at the boundary between the resin paper and the soil. This means that one cannot be certain about the total diffusible concentrations C_0 which should be inserted into any equation required to determine the diffusion coefficient, nor can one assume that diffusion occurs toward zero concentration at the boundary between the resin paper and the soil.

However, a relationship exists between the diffusion coefficient for phosphate and the concentration-dependent process which involves the release or *desorption* of the ion from the soil surface, [cf. Larsen (267) and Sutton and Gunary (527) for a discussion of phosphate equilibria in soil]. Vaidyanathan and Nye (555) showed that diffusion coefficients calculated by using the boundary concentration appropriate to the phosphate concentration in the soil under investigation were related to the slopes of the corresponding desorption relationships. Essentially these are between the equilibrium phosphate concentration in the soil solution (μmoles per cubic centimeter) and the exchangeable ^{32}P brought off the soil (previously equilibrated with the isotope) by calcium chloride at the same ionic strength as the solution in the soil pores.

Drew and Nye (104) used such values to calculate the phosphate concentration around the roots of onion, leek, and Italian ryegrass. In all but one of the soil treatments, the average concentration of phosphorus in soil solution was reduced to between 9 and 17% of the initial concentration, which indicates that diffusion is an important determinant of uptake. The limitation of diffusion on uptake was also demonstrated by the fact that onion and leek showed the same uptake of phosphate, although in other experiments (105) potassium uptake was greater for leek than for onion. The results obtained with ryegrass showed that, as with potassium, root hairs can increase uptake efficiency with respect to phosphorus, but for this element enhancement also occurs at lower rates of uptake.

In general, therefore, when a plant grows in a temperate agricultural soil, nitrate is likely to move to the roots by mass flow and phosphate by diffusion, whereas the extent to which potassium will move by one or the other of these processes will depend upon the exchangeability of the ion. Hence, in soils of low phosphorus availability, the more competitive plants would be those which can extend the *length* of their root system, since this is the most effective way of increasing uptake. The increased uptake would lead to increased shoot growth.

The marked ability of most soils to sequester phosphorus has the consequence that in many soils the amount of available phosphorus is very low. Under such conditions, the presence of vesicular–arbuscular mycorrhizae increase uptake of phosphorus (351). Tinker (541) has analyzed the situation in some detail. He pointed out that increased phosphate uptake per plant could most probably arise from (1) morphological changes in the plant, (2) increased ability of the plant's root surface to absorb phosphorus, (3) provision of additional or more efficient absorbing surface in the hyphae with subsequent transfer to the host, (4) ability of the mycorrhizal root or hyphae to utilize sources of phosphorus not available to the uninfected root, and (5) longer duration of mycorrhiza than uninfected root as absorbing organ.

The ability of the mycorrhizal root or hyphae to use sources of phosphorus that are not available to the uninfected root can be tested directly by using an isotopic dilution method. Plants have been grown in soil labeled with ^{32}P and the specific activity of plants with and without mycorrhizae compared (176, 454). The results indicate that both mycorrhizal and nonmycorrhizal plants draw their phosphate from soil solution or the absorbed phosphate in equilibrium with it. Thus it would appear that the fungi do not release more firmly bound phosphorus, as may be the case with ectomycorrhiza (34).

Saunders and Tinker (454) found a fourfold increase in the uptake per unit length (Table XI) for phosphorus when onion roots become mycorrhizal. Nonmycorrhizal roots absorb phosphorus by diffusion (mass flow can account for only about 1% of the movement), and the uptake of phosphorus per unit length of these roots was found to be at the maximum rate possible; i.e., the concentration of phosphorus at the root surface was reduced to near zero. This means that possibilities (1) and (2) pointed out by Tinker (541) can be discounted. The additional phosphorus inflow must have entered via the external hyphae of the mycorrhizal associations; that is, the phosphorus is absorbed by the hyphae and translocated into the host [see Tinker (541) for the evidence concerning the movement by phosphorus into the host]. Although this does not eliminate possibility (3), it does nevertheless mean that there is no need to invoke it to account for the increased uptake of phosphorus when the roots become mycorrhizal.

These investigations indicate that in a soil of uniformly low phosphorus status, it is more appropriate to increase the length of the absorbing system, i.e., root and hyphae, than to increase the efficiency of the adsorption system, i.e., to increase α. However, for ectomycorrhizal roots of beech there is no doubt that α is increased as a result of infection (171, 172). But for this association the situation may be different. The roots are found mostly in the litter layer of the soil, where microbes and probably the mycorrhizal roots themselves are releasing phosphorus from breakdown of organic matter (170). It may be that the local concentration of phosphate rises to quite high levels, in which case the efficiency of uptake will

TABLE XI

INFLOWS AND FLUXES OF PHOSPHORUS DURING UPTAKE BY ONION PLANTS INFECTED WITH YELLOW VACUOLATE ENDOPHYTE[a]

Conditions	Root infected (%)	Inflows: Mean for mycorrhizal onions ($mol\ cm^{-2}\ s^{-1} \times 10^{-14}$)	Inflows: Mean for nonmycorrhizal onions ($mol\ cm^{-2}\ s^{-1} \times 10^{-14}$)	Inflows: Via hyphae[b] ($mol\ cm^{-2}\ s^{-1} \times 10^{-14}$)	Calculated flux in hyphae[c] ($mol\ cm^{-2}\ s^{-1} \times 10^{-3}$)
Expt. A: 44–45 days	50	13	4.2	17.6	3.8 (Expts. A and B)
Expt. B: 39–54 days	45	11.5	3.2	18.5	

[a] From Tinker (541).
[b] Calculated from difference between mycorrhizal and nonmycorrhizal inflows, divided by fraction of root infected.
[c] Based on six main entry hyphae (not actual entry points) per centimeter of infected root.

depend upon the effectiveness of the absorption system, which may well be an active uniport.

The treatment in this section of competition for nutrients by roots has necessarily been a simplistic one. Any such competition has been regarded as a function of properties of the soil and its solution that determine whether specific nutrients are to be regarded as mobile or immobile. But competition may take other forms. One such is the phenomenon of allelopathy, namely, the production of compound(s) by one plant which suppress the growth of another. The phenomenon is difficult to demonstrate, but Glass (148–150) has produced evidence for its mechanistic basis. Considering the views expressed in the literature that hydroxybenzoic acids may be introduced by plants into their surrounding environment, he studied the effects of these compounds on ion uptake by barley roots. He showed that inhibition constants of the benzoic acids were strongly correlated with their octanol–water partition coefficients and their pK_a values, suggesting that the compounds inhibit normal membrane functions as a result of a generalized increase in membrane permeability. More information about the ecological significance of these observations needed.

Competition may also occur through mycorrhizal interactions. Fitter (136) grew *Lolium perenne* and *Holcus lanatus* in a soil of very poor phosphorus-supplying capacity in a factorial experiment combining the two factors mycorrhizal infection and root competition. Both of them gave a slight advantage to *H. lanatus,* but in combination they produced considerable suppression of *L. perenne.* The results could be explained solely in terms of increased mycorrhizal infection reducing uptake ability. The increased infection could be due to *H. lanatus* providing more mycelium or stimulating fungal spore germination.

Finally, as yet, all theoretical treatments of ion movement through the soil to the plant root system require that the soil be considered as a homogenous system. Clearly this may not be so. We need to remember also not only that inhomogeneity will influence the rate of supply from different directions, but also that root morphology and physiology may also be changed (102, 106, 107) (Fig. 35).

FIG. 35. The effect of NO_3^- supply on barley root growth. Nutrient solutions containing high (H, 1.0 mmol L^{-1}) or low (L, 0.01 mmol L^{-1}) concentrations of NO_3^- were supplied to three zones along the treated seminal axis (limits indicated by bars). Capital letters refer to the sequence of these concentrations along the axis (uppermost zone on the left). Thus in treatment (b), LHL, 1.0 mmol L^{-1} NO_3^- was supplied to the middle zone; the remainder of the root system, including the other seminal axes and nodals, received the 0.01 mmol L^{-1} concentration. Plants grew for 26 days in treatments (a) (HHH) and (b) (LHL) and 34 days in a separate experiment in treatments (c) (LLL) and (d) (HLH). From Drew *et al.* (107).

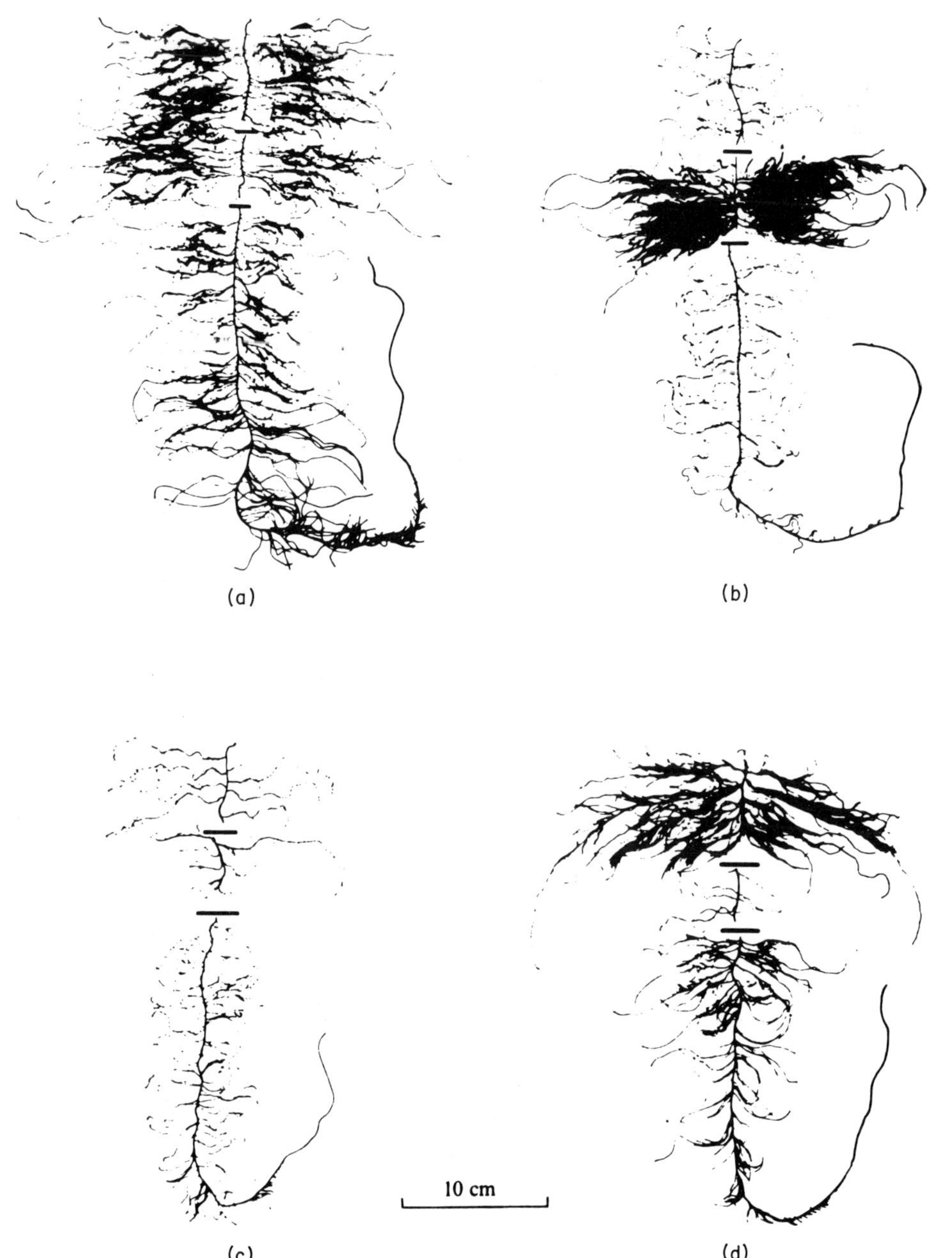
(a)
(b)
(c)
10 cm
(d)

XII. Epilogue

This chapter has continued along lines developed in the earlier chapter by Steward and Sutcliffe in Volume II of this treatise (518). The text attempts to show what is now known and where knowledge is still lacking. The emphasis is upon systems and situations which can be conceived in terms of their underlying mechanisms and described in precise formulations and schemata.

The large body of data and experimental observations on plants describe the contents of cells and organisms in terms of their salts, ions, and total solutes. They also relate these to their ambient media and to the environments and developmental events by which they are initiated and controlled *in vivo.*

Modern knowledge of fine structure in cells describes the resources of organization in terms of their membranes and organelles that create and maintain the distinctive compositions of cells and organisms. Biochemistry and bioenergetics now reveal, at the molecular level, the properties of membranes, proteins, and enzymes that may be harnessed to make cells work as physical systems. Therefore, the preferred approach in this chapter has been to dwell upon formulations and schemata which describe and classify operative systems and mechanisms precisely and even where this is not yet possible to focus attention upon areas of knowledge that still require investigation and interpretation.

Some topics have been purposely avoided, partly because they seem at present to be only at a descriptive stage. Although the phenomena here in question may clearly have an ionic basis, it is either not clear what particular ions may be involved or if this uncertainty has been removed, the nature and quantitative aspects of the ion fluxes have not yet been determined. There are, however, very good reviews of these topics.

Still undeveloped areas of plant cell biology which relate to membranes will remain of interest to plant physiologists because they are so relevant to such areas of current concern as the timing of processes *in vivo* and the sequential events of developmental biology. Such topics are the regulation of biological rhythms (557), the involvement of phytochrome in membrane processes (70, 327), bioelectric phenomena and the growth of plant cells (174, 216, 228), and the involvement of ions in plant movements, e.g., stomatal opening and closure (Mansfield, Chapter 3 this volume; 322, 441).

Although the intention has been to stress mechanisms wherever possible, it is clear that many aspects of the salt relationships of plant cells and tissues are still incompletely known. For example, even though the movement of ions across the plasmalemma is readily accessible to investigation, the molecular basis of the movement of many ions is still obscure.

It is worth redirecting attention to the system for which there is a coherent body of information, namely, the potassium–hydrogen–sodium exchange system of *Neurospora crassa* (Section IV,A,2,a), to seek the underlying causes for the greater progress in establishing the molecular basis of this system, when compared with the progress on other systems. There are, of course, advantages in the use of fungus which can be grown under a variety of conditions to produce different physiological states and with readily selected mutants for particularly biochemical processes. These characteristics have been used to good effect in probing fungal transport systems, but in the systems which have been considered here, namely, the salt relations of higher plant cells and tissues, a wide range of biophysical and biochemical techniques have been employed, albeit with more limited progress. In contrast, studies on transport systems of large algal cells have been largely biophysical. While there can be problems of obtaining sufficient algal material and there are certainly difficulties in growing these cells under defined conditions, it is even more regrettable that their biochemistry has not been studied in more detail. The undue emphasis on the use of inhibitors as indirect probes of metabolic processes has not been as productive as first envisaged, as is evidenced by the continuing uncertainty about how transport processes in these cells are powered.

If a transport protein could be isolated, then there would be no doubt that the study of its properties would be extremely rewarding. The difficulties of unambiguously identifying such a transport protein, once it has been isolated, have been emphasized. Nevertheless, if these difficulties can be overcome, the biochemical properties of a transport protein should show how the transport process is related to the energy-generating systems of cells (cf. Chapter 2, Vol. VII) and indicate how the rate of transport can be controlled.

Thus "carriers" should be isolated if at all possible, but also more basic biochemical information is needed for all those systems whose salt relations are to be studied. This applies to tissues and organs as well as to cells. Although there have been many studies on the ion movement across maize roots, their other biochemical and physiological features are insufficiently known.

Much effort has gone into the elucidation of the transport of potassium, sodium, and chloride across the plasmalemma of plant cells. Less effort has gone into the transport of other ions, particularly phosphate, nitrate, and the bivalent cations calcium and magnesium. The two anions present difficulties because they are metabolized. On the other hand, there can be problems in determining the plasmalemma fluxes of calcium and magnesium because of fluxes associated with binding of the ions to the cell wall. Spanswick and Williams (509) met this problem when studying the calcium fluxes in *Nitella translucens;* the work nevertheless suggested that calcium might be pumped into the vacuole across the tonoplast. However, even had it been possible to determine more accurately the tracer fluxes across the plasmalemma, there would still have been the problem of determining the concentration of free calcium ions in the cytoplasm. The concentration of calcium in the cytoplasm of *N. translucens* was reported to be about 8 mmol L^{-1}, but the concentration of free calcium ions in the ground cytoplasm, i.e., outside the mitochondria, could be many times lower than this. It has been suggested (Section VI,G) that calcium may have an important regulatory function in plant cells, possibly akin to that in animal cells (109). The recent isolation of calmodulin from plant tissue (8, 120, 160) tends to support this suggestion. If this is so, it points to a very low free calcium ion concentration in the cytoplasm. Very low concentrations of free calcium ions would be compatible with the frequently high concentrations in plant cells of organic acids which readily complex bivalent cations, e.g., malic and citric acids.

It has been postulated frequently that calcium has an important role in membrane integrity. Simon (477) has invoked such a role to explain symptoms of calcium deficiency in plants (Fig. 36). There are many recorded instances of the loss of solutes from plant cells, often brought about by monovalent cations in the medium, being reduced by calcium ions (224, 355, 397). It has long been assumed that calcium ions cross-link structures in the membrane so that it becomes less permeable, namely, Campbell and Pitman (77), but other explanations are possible. Thus Nieman and Willis (355) found that pretreatment of carrot disks with 0.3 mol L^{-1} sodium chloride reduces their ability to absorb glucose. The treatment caused loss of protein and here there is a comparison with loss of protein from the periplasmic space of bacteria (186), that protein being involved in transport of solutes into the cell. The protein lost from carrot tissue may have a similar role. Calcium prevents sodium chloride from having its effect, particularly when added in time to prevent release of protein. Here then is a suggestion that calcium ions may be important to keep a carrier molecule functioning within a membrane.

Epstein (122), Rains *et al.* (408), and Läuchli and Epstein (270) believe that calcium is necessary for potassium–sodium selectivity of plant cells, acting specifically at the pump. The so-called unspecific loss of solutes from cells in the absence of calcium ions could be a consequence of these ions being necessary for primary active transport systems. If they become nonfunctional, the electrochemical potential gradient of certain ions, e.g., hydrogen or sodium, across the membrane will be dissipated, such that secondary transport systems driven by the gradient may reverse. There is no doubt that these various phenomena require more precise investigation, e.g., flux analysis.

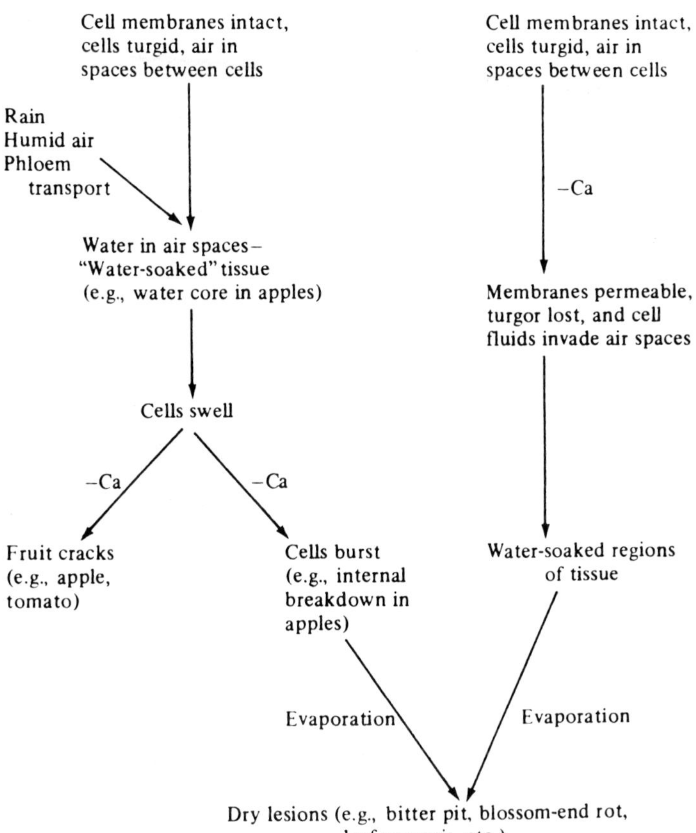

FIG. 36. Pathways of development of calcium deficiency symptoms. The left-hand pathway applies to soft, succulent fruits, the cells of which are liable to burst under hypotonic conditions *in vitro*. It is proposed that *in vivo* the cells swell as they absorb water taken up from the atmosphere or supplied by the phloem. If the fruit is calcium deficient, it may crack or individual cells may burst. The right-hand pathway describes the situation in which calcium deficiency renders membranes permeable so that cell fluids invade the intercellular air spaces of the tissue. The final stage in either pathway may be one of desiccation. From Simon (477).

Finally, the section (IV,B,3) concerned with iron uptake focuses on the need to recognize that the plasmalemma has enzymatic functions (oxidoreductive) other than catalyzing the movement of solutes into the cytoplasm and that these other functions may be important in the ionic relations of cells. A particularly intriguing possible example of such an oxidoreductive function of the plasmalemma is iodine accumulation by larger brown seaweeds, such as *Fucus ceranoides* and *Laminaria digitata*, through the uptake of HIO (221). There is evidence that the weeds can themselves bring about the oxidation of the iodide anion to iodine which is hydrolyzed to HIO. The same section shows how very little is known about trace metal transport into plant cells. For example, although there have been studies on the uptake of zinc into higher plants (147), the studies made with microorganisms (129) are somewhat more illuminating about the characteristics of its transport into the cells. Further studies may well be complicated by the presence in cells of small molecular weight polypeptides (metallothionins) which bind these trace metals (129, 409) and by their capacity to be chelated by organic acids (335).

The preceding paragraph has indicated some of the still-elusive problems of ion transport across the plasmalemma. There are even more intractable problems facing those who are concerned with movement of ions across the cytoplasm into the vacuole (Section VIII) and across the symplasm into the xylem vessels of the root (Sections IX and X). These difficulties have been emphasized in the relevant sections.

References

1. Abel, G. H. (1969). Inheritance of the capacity for chloride inclusion and chloride exclusion by soybeans. *Crop Sci.* **9,** 697–698.
2. Abel, G. H., and Mackenzie, A. J. (1964). Salt tolerance of soybean varieties (*Glycine max* L. Merrill) during germination and later growth. *Crop Sci.* **4,** 157–161.
3. Adams, P. B. (1970). Effect of adenine nucleotides on the respiration of carrot root slices. *Plant Physiol.* **45,** 495–499.
4. Adams, P. B., and Rowan, K. S. (1972). Regulation of salt respiration in carrot root slices. *Plant Physiol.* **50,** 682–686.
5. Admon, A., Jacoby, B., and Goldschmidt, E. E. (1980). Assessment of cytoplasmic contamination in isolated vacuole preparations. *Plant Physiol.* **65,** 85–87.
6. Albert, R., and Kinzel, H. (1973). Unterscheidung von Physiotypen bei Halophyten des Neusiedlerseegebietas (Osterreich). *Z. Pflanzenphysiol.* **70,** 138–158.
7. Albert, R., and Popp, M. (1977). Chemical composition of halophytes from the Neusiedler lake region of Austria. *Oecologia* **27,** 157–170.
8. Anderson, J. M., Charbonneau, H., Jones, H. P., McCann, R. O., and Cormier, M. J. (1980). Characterisation of the plant nicotinamide adenine dinucleotide kinase activator protein and its identification as calmodulin. *Biochemistry* **19,** 3113–3120.
9. Anderson, W. P., Aikman, D. P., and Meiri, A. (1970). Excised root exudation: A standing gradient osmotic flow. *Proc. R. Soc. London, Ser. B* **174,** 445–458.
10. Anderson, W. P., Hendrix, D. L., and Higinbotham, N. (1974). The effect of cyanide and carbon monoxide on the electrical potential and resistance of cell membranes. *Plant Physiol.* **54,** 712–716.
11. Anderson, W. P., and Higinbotham, N. (1975). A cautionary note on plant root electrophysiology. *J. Exp. Bot.* **26,** 533–535.
12. Anderson, W. P., and Higinbotham, N. (1976). Electrical resistance of corn root segments. *Plant Physiol.* **57,** 137–141.

13. Anderson, W. P., and House, C. R. (1967). A correlation between structure and function in the roots of *Zea mays*. *J. Exp. Bot.* **18,** 544–555.
14. Arens, K. (1939). Physiologische Multipolaritate der Zelle von *Nitella* wahrend der Photosynthese. *Protoplasma* **33,** 295–300.
15. Arisz, W. H., Helder, R. J., and Van Nie, R. (1951). Analysis of the exudation process in tomato plants. *J. Exp. Bot.* **2,** 257–297.
16. Armstrong, W. McD., and Rothstein, A. (1964). Discrimination between alkali metal cations by yeast. I. Effect of pH on uptake. *J. Gen. Physiol.* **48,** 61–71.
17. Armstrong, W. McD., and Rothstein, A. (1967). Discrimination between alkali metal cations by yeast. II. Cation interactions in transport. *J. Gen. Physiol.* **50,** 967–988.
18. Asher, C. J., and Loneragan, J. F. (1967). Response of plants to phosphate concentration in solution culture. 1. Growth and phosphorus content. *Soil Sci.* **103,** 225–233.
19. Auzac, J. D'. (1977). ATPase membranaire de vacuoles lysomomales: Les Intoides du latex *Hevea brasiliensis*. *Phytochemistry* **16,** 1881–1885.
20. Auzac, J. D', and Lioret, C. (1974). Misc en évidence d'un mécanisme d'accumulation du citrate dans les Intoides du latex *Hevea brasiliensis*. *Physiol. Vég.* **12,** 617–635.
21. Bacon, J. S. D., MacDonald, I. R., and Knight, A. H. (1965). The development of invertase activity in slices of the root of *Beta vulgaris* L. washed under asceptic conditions. *Biochem. J.* **94,** 175–182.
22. Baker, D. A. (1978). Proton co-transport of organic solutes by plant cells. *New Phytol.* **81,** 485–497.
23. Baker, D. A., and Hall, J. L. (1973). Pinocytosis, ATPase and ion uptake by plant cells. *New Phytol.* **72,** 1281–1291.
24. Baker, D. A., and Hall, J. L. (1975). "Ion Transport in Plant Cells and Tissue." North-Holland Publ., Amsterdam.
25. Baker, D. A., Malek, F., and Dehvar, F. D. (1980). Phloem loading of amino acids from the petioles of *Ricinus* leaves. *Ber. Dtsch. Bot. Ges.* **93,** 203–209.
26. Baldwin, J. P. (1975). A quantitative analysis of the factors affecting plant uptake from some soils. *J. Soil Sci.* **26,** 195–206.
27. Barber, D. A. (1968). Microorganisms and the inorganic nutrition of higher plants. *Annu. Rev. Plant Physiol.* **19,** 71–88.
28. Barber, D. A., and Lee, R. B. (1974). The effect of microorganisms on the absorption of manganese by plants. *New Phytol.* **73,** 97–106.
29. Barber, D. A., and Loughman, B. C. (1967). The effect of microorganisms on the absorption of inorganic nutrients by intact plants. II. Uptake and utilisation of phosphate by barley plants grown in sterile and non-sterile conditions. *J. Exp. Bot.* **18,** 170–176.
30. Barber, D. A., and Shone, M. G. T. (1966). The absorption of silica from aqueous solutions by plants. *J. Exp. Bot.* **17,** 569–578.
31. Barber, D. A., and Shone, M. G. T. (1967). The initial uptake of ions by barley roots. III. The uptake of cations. *J. Exp. Bot.* **18,** 631–643.
32. Barber, J. (1968). Sodium efflux from *Chlorella pyrenoidosa*. *Biochim. Biophys. Acta* **150,** 730–733.
33. Barber, S. A. (1962). A diffusion and mass flow concept of soil nutrient availability. *Soil Sci.* **93,** 39–49.
34. Bartlett, E. M., and Lewis, D. H. (1973). Surface phosphatase activity of mycorrhizal roots of beech. *Soil Biol. Biochem.* **5,** 249–257.
35. Baule, B. (1918). Zu Mitscherlichs Gesetz der physiologischen Bezichungen. *Landwirtsch, Jahrb.* **51,** 363–385.

36. Bell, A. J. E. van, and Erven, A. van (1976). Stimulation of proton influx by amino acid uptake in tomato internode disks. *Z. Pflanzenphysiol.* **80,** 74–76.
37. Bentrup, F. W., Pfrüner, H., and Wagner, G. (1973). Evidence for differential action of indole acetic acid upon ion fluxes in single cells of *Petroselinum sativum. Planta* **110,** 369–372.
38. Borowitzka, M. A. (1977). Algal calcification. *Oceanogr. Mar. Biol.* **15,** 189–223.
39. Borowitzka, M. A., and Larkum, A. W. D. (1976). Calcification in the green alga *Halimeda.* II. The exchange of Ca^{2+} and the occurrence of age gradients in calcification and photosynthesis. *J. Exp. Bot.* **27,** 864–878.
40. Borowitzka, M. A., and Larkum, A. W. D. (1976). Calcification in the green alga *Halimeda.* III. The sources of inorganic carbon for photosynthesis and calcification and a model of the mechanism of calcification. *J. Exp. Bot.* **27,** 879–893.
41. Borowitzka, M. A., and Larkum, A. W. D. (1976). Calcification in the green alga *Halimeda.* IV. The action of metabolic inhibitors on photosynthesis and calcification. *J. Exp. Bot.* **27,** 894–907.
42. Borst-Pauwels, G. W. F. H., Schnetkamp, P., and Well, P. van (1973). Activation of Rb^+ and Na^+ uptake into yeast by monovalent cations. *Biochim. Biophys. Acta* **291,** 274–279.
43. Borst-Pauwels, G. W. F. H., Theuvenet, A. P. R., and Peters, P. H. J. (1975). Uptake by yeast: Interaction of Rb^+ and Na^+ and phosphate. *Physiol. Plant.* **33,** 8–12.
44. Borst-Pauwels, G. W. F. H., Wolters, G. H. J., and Henricks, J. J. G. (1971). The interaction of 2,4-dinitrophenol with anaerobic Rb^+ transport across the yeast cell membrane. *Biochim. Biophys. Acta* **225,** 269–276.
45. Bostrom, T. E., and Walker, N. A. (1975). Intercellular transport in plants. I. The rate of transport of chloride and the electric resistance. *J. Exp. Bot.* **26,** 767–782.
46. Bostrom, T. E., and Walker, N. A. (1976). Intracellular transport in plants. II. Cyclosis and the rate of intercellular transport of chloride in *Chara. J. Exp. Bot.* **27,** 347–357.
47. Bouldin, D. R. (1961). Mathematical description of diffusion processes in the soil-plant system. *Soil Sci. Soc. Am. Proc.* **25,** 476–480.
48. Bowen, G. D., and Rovira, A. D. (1966). Microbial factor in short-term phosphate uptake studies with plant roots. *Nature (London)* **211,** 665–616.
49. Bowling, D. J. F. (1966). Active transport of ions across sunflower roots. *Planta* **69,** 377–382.
50. Bowling, D. J. F. (1972). Measurement of profiles of potassium activity and electrical potential in the intact root. *Planta* **108,** 147–151.
51. Bowling, D. J. F., Macklon, A. E. S., and Spanswick, R. M. (1966). Active and passive transport of the major nutrient ions across the root of *Ricinus communis. J. Exp. Bot.* **17,** 410–416.
52. Bowling, D. J. F., and Spanswick, R. M. (1964). Active transport of ions across the root of *Ricinus communis. J. Exp. Bot.* **15,** 422–427.
53. Bowman, B. J., Mainzer, S. E., Allen, K. E., and Slayman, C. W. (1978). Effects of inhibitors on the plasma membrane and mitochondrial adenosine triphosphatases of *Neurospora crassa. Biochim. Biphys. Acta* **512,** 13–28.
54. Bowman, B. J., and Slayman, C. W. (1977). Characteristics of plasma membrane adenosine triphosphatase of *Neurospora crassa. J. Biol. Chem.* **252,** 3357–3363.
55. Bray, R. H. (1954). A nutrient mobility concept of soil-plant relationships. *Soil Sci.* **78,** 9–22.
56. Breckle, S. W. (1974). Wasser-under Salzperhältnisse bei Halophytin der Saltzsteppe in Utah, USA *Ber. Dtsch. Bot. Ges.* **87,** 589–600.

57. Breckle, S. W. (1975). Ionengehalte halophiler Pflanze Spaniens. *Decheniana* **127,** 221–228.
58. Brewster, J. L., Bhat, K. K. S., and Nye, P. H. (1975). The possibility of predicting solute uptake and plant growth response from independently measured soil and plant characteristics. *Plant Soil* **42,** 171–195.
59. Brewster, J. L., and Tinker, P. B. (1970). Nutrient cation flows in soil around plant roots. *Soil Sci. Soc. Am. Proc.* **34,** 421–426.
60. Briggs, G. E., Hope, A. B., and Robertson, R. N. (1961). "Electrolytes and Plant Cells." Blackwell, Oxford.
61. Brown, J. C. (1963). Iron chlorosis in soybeans as related to the genotype of root stock. *Soil Sci.* **96,** 387–394.
62. Brown, J. C. (1978). Mechanism of iron uptake by plants. *Plant, Cell Environ.* **1,** 249–258.
63. Brown, J. C., and Ambler, J. E. (1973). Reductants released by roots of Fe-deficient soybeans. *Agron. J.* **65,** 311–314.
64. Brown, J. C., Ambler, J. E., Chaney, R. L., and Foy, C. D. (1972). Differential responses of plant genotypes to micronutrients. *In* "Micro-nutrients in Agriculture" (J. J. Mortvedt, P. M. Giordana, and W. L. Lindsay, eds.), pp. 389–413. Soil Sci. Soc. Am., Madison, Wisconsin.
65. Brown, J. C., Chaney, R. L., and Ambler, J. E. (1971). A new tomato mutant inefficient in the transport of iron. *Physiol. Plant.* **25,** 48–53.
66. Brown, J. C., Holmes, R. S., and Tiffin, L. O. (1958). Iron chlorosis in soybeans as related to the genotype of root stalk. *Soil Sci.* **86,** 75–82.
67. Brown, J. C., Holmes, R. S., and Tiffin, L. O. (1961). Iron chlorosis in soybeans as related to the genotype of root stalk. 3. Chlorosis susceptibility and reductive capacity of the root. *Soil Sci.* **91,** 127–132.
68. Brown, J. C., and Jones, W. E. (1971). Differential transport of boron in tomato (*Lycopersicon esculentum* Mill.). *Physiol. Plant.* **25,** 279–282.
69. Brown, J. C., Tiffin, L. O., Specht, A. W., and Resnicky, J. W. (1961). Iron absorption by roots as affected by plant species and concentration of chelating agent. *Agron. J.* **53,** 85–90.
70. Brownlee, C., Roth-Bejerano, N., and Kendrick, R. E. (1979). The molecular mode of phytochrome action. *Sci. Prog. (Oxford)* **66,** 217–229.
71. Bryant, A. E. (1934). A demonstration of the connection of the protoplasts of the endodermal cells with Casparian strips in the roots of barley. *New Phytol.* **33,** 231.
72. Budd, K., and Harley, J. L. (1962). The uptake and assimilation of ammonia by *Neocosmospora vasinfecta*. *New Phytol.* **61,** 244–255.
73. Budd, K., and Harley, J. L. (1962). The uptake and assimilation of ammonia by *Neocosmospora vasinfecta*. II. Increases in the ammonia level in the mycelium during the uptake of ammonia. *New Phytol.* **61,** 244–255.
74. Buller, A. H. R. (1933). "Research on Fungi 5." Longmans, Green, London.
75. Buser, C., and Matile, P. (1977). Malic acid in vacuoles isolated from *Bryophyllum* leaf cells. *Z. Pflanzenphysiol.* **82,** 362–466.
76. Butz, R. G., and Jackson, W. A. (1977). A mechanism for nitrate transport and reduction. *Phytochemistry* **16,** 409–417.
77. Campbell, L. C., and Pitman, M. G. (1971). Salinity and plant cells. *In* "Salinity and Water Use" (T. Talsma and J. R. Philip, eds.), pp. 207–224. Macmillan, London.
78. Chaney, R. L., Brown, J. C., and Tiffin, L. O. (1972). Obligatory reduction of ferric chelates in iron uptake by plants. *Plant Physiol.* **50,** 208–213.

79. Christ, R. A. (1974). A method to compare the effect of ionic iron and iron chelates in nutrient solution cultures. *Plant Physiol.* **54,** 579–581.
80. Christ, R. A. (1964). Iron requirement and iron uptake from various iron compounds by different plant species. *Plant Physiol.* **54,** 582–584.
81. Christensen, M. S., and Cirillo, V. P. (1972). Yeast membrane vesicles: Isolation and general characteristics. *J. Bacteriol.* **110,** 1190–1205.
82. Clymo, R. S. (1963). Ion exchange in *Sphagnum* and its relation to bog ecology. *Ann. Bot. (London)* [N.S.] **27,** 309–324.
83. Columbo, R., DeMichelis, M. I., and Lado, P. (1978). 3-*O*-methyl glucose uptake stimulation by auxin and by fusicoccin in plant materials and its relationship with proton extrusion. *Planta* **138,** 249–256.
84. Conway, E. J., and Brady, E. J. (1950). Biological production of acid and alkali. Quantitative relations of succinic and carbonic acids to the potassium and hydrogen exchanges in fermenting yeast. *Biochem. J.* **47,** 360–369.
85. Cordaro, C. (1976). Genetics of the bacterial phosphoenolpyruvate: glucose phosphotransferase system. *Annu. Rev. Genet.* **10,** 341–359.
86. Cram, W. J. (1968). The control of cytoplasmic and vacuolar ion contents in higher plant cells. *In* "Stofftransport und Stoff verteilung in Zellen hoherer Pflanzen" (K. Mothes, E. Müller, A. Nelles, and D. Neumann, eds.), pp. 117–143. Akademie-Verlag, Berlin.
87. Cram, W. J. (1974). Influx isotherms—their interpretation and use. *In* "Membrane Transport in Plants" (U. Zimmermann and J. Dainty, eds.), pp. 334–337. Springer-Verlag, Berlin and New York.
88. Cram, W. J. (1974). Effects of Cl^- on HCO_3^- and malate fluxes and CO_2 fixation in carrot and barley root cells. *J. Exp. Bot.* **25,** 253–268.
89. Cram, W. J. (1976). Storage tissues. *In* "Ion Transport in Plant Cells and Tissues" (D. A. Baker and J. L. Hall, eds.), pp. 161–191. North-Holland Publ., Amsterdam.
90. Cram, W. J. (1976). Negative feedback regulation of transport in cells. *In* "Encyclopedia of Plant Physiology, New Series" (U. Lüttge and M. G. Pitman, eds.), Vol. 2, Part A, pp. 284–316. Springer-Verlag, Berlin and New York.
91. Cram, W. J. (1980). Pinocytosis in plants. *New Phytol.* **84,** 1–17.
92. Cram, W. J., and Laties, G. G. (1974). The kinetics of bicarbonate and malate exchange in carrot and barley root cells. *J. Exp. Bot.* **25,** 11–27.
93. Crooke, W. M., Knight, A. H., and MacDonald, I. R. (1960). Cation exchange properties and pectin content of storage tissue disks. *Plant Soil* **13,** 55–67.
94. Dainty, J., and Hope, A. B. (1961). The electric double layer and the Donnan equilibrium in relation to plant cell walls. *Aust. J. Biol. Sci.* **14,** 541–551.
95. Dainty, J., Hope, A. B., and Denby, C. (1960). Ionic relations of cells of *Chara australis.* II. The indiffusible anions of the cell wall. *Aust. J. Biol. Sci.* **13,** 267–276.
96. Darley, W. M. (1974). Silicification and calcification. *In* "Algal Physiology and Biochemistry" (W. P. D. Stewart, ed.), pp. 655–675. Blackwell, Oxford.
97. Dauwalder, M., Whaley, W. C., and Kephart, J. C. (1969). Phosphatases and differentiation of the Golgi apparatus. *J. Cell Sci.* **4,** 455–497.
98. Davies, D. D. (1973). Control of and by pH. *Symp. Soc. Exp. Biol.* **27,** 513–529.
99. Diamond, J. M., and Bossert, W. H. (1967). Standing gradient osmotic flow: A mechanism for coupling of water and solute transport in epithelia. *J. Gen. Physiol.* **50,** 2061–2083.
100. Dijkshoorn, W. (1962). Metabolic regulation of the alkaline effect of nitrate utilisation in plants. *Nature (London)* **194,** 165–167.
101. Doree, M., Leguay, J.-J., and Terrine, C. (1972). Flux de CO_2 et modulations de

perméabilité cellulaire chez les cellules d'*Acer pseudoplatanus* L. *Physiol. Vég.* **10,** 115–131.

102. Drew, M. C. (1975). Comparison of the effects of a localized supply of phosphate, nitrate, ammonium and potassium on the growth of the seminal root system and the shoot in barley. *New Phytol.* **75,** 479–490.
103. Drew, M. C., and Nye, P. H. (1969). The supply of nutrient ions by diffusion to plant roots in soil. II. The effect of root hairs on the uptake of potassium by roots of rye grass *(Lolium multiflorum). Plant Soil* **31,** 407–424.
104. Drew, M. C., and Nye, P. H. (1970). The supply of nutrient ions by diffusion to plant roots in soil. III. Uptake of phosphate by roots of onion, leek and rye-grass. *Plant Soil* **33,** 545–563.
105. Drew, M. C., Nye, P. H., and Vaidyanthan, L. V. (1969). The supply of nutrient ions by diffusion to plant roots in soil. I. Absorption of potassium by cylindrical roots of onion and leek. *Plant Soil* **30,** 252–270.
106. Drew, M. C., and Saker, L. R. (1975). Nutrient supply and the growth of the seminal root system of barley. II. Localised compensatory increases in lateral root growth and rates of nitrate uptake when nitrate supply is restricted to only part of the root system. *J. Exp. Bot.* **26,** 79–90.
107. Drew, M. C., Saker, L. R., and Ashley, T. W. (1973). Nutrient supply and the growth of the seminal root system in barley. I. The effect of nitrate concentrations on the growth of axes and laterals. *J. Exp. Bot.* **24,** 1189–1202.
108. Droop, M. R. (1973). Heterotrophy *In* "Algal Physiology and Biochemistry" (W. D. P. Stewart, ed.), pp. 530–559. Blackwell, Oxford.
109. Duncan, C. J., ed. (1976). Calcium in biological systems. *Symp. Soc. Exp. Biol.* (C. J. Duncan, ed.), Vol. 30. Cambridge Univ. Press, London and New York.
110. Dunham, E. T., and Glynn, I. M. (1961). Adenosine triphosphatase activity and the active movements of alkali metal ions. *J. Physiol. (London)* **156,** 274–293.
111. Dunlop, J., and Bowling, D. J. F. (1971). The movement of ions to the xylem exudate of maize roots. I. Profiles of membrane potential and vacuolar potassium activity across the root. *J. Exp. Bot.* **22,** 434–444.
112. Dunlop, J., and Bowling, D. J. F. (1971). The movement of ions to the xylem exudate of maize roots. II. A comparsion of the electrical potential and electrochemical potentials of ions in the exudate and in the root cells. *J. Exp. Bot.* **22,** 445–452.
113. Dunlop, J., and Bowling, D. J. F. (1971). The movement of ions to the xylem exudate of maize roots. III. The location of the electrical and electrochemical potential differences between the exudate and the medium. *J. Exp. Bot.* **22,** 453–464.
114. Durr, M., Urech, K., Boller, T., Wiemken, A., Schwenke, J., and Nagy, M. (1979). Sequestration of arginine by polyphosphate in vacuoles of yeast *Saccharomyces cerevisiae. Arch. Microbiol.* **121,** 169–175.
115. Eddy, A. A. (1978). Proton-dependent solute transport in microorganism. *Curr. Topic. Membr. Transp.* **10,** 279–360.
116. Eddy, A. A. (1980). Some aspects of amino acid transport in yeast. *In* "Microorganisms and Nitrogen Sources" (J. W. Payne, ed.), pp. 35–62. Wiley, Chichester.
117. Edelman, J., and Hall, M. A. (1965). Development of invertase and ascorbate oxidase activities in native storage tissue of *Helianthus tuberosus. Biochem. J.* **95,** 403–410.
118. Eisenman, G. (1973). "Membranes," Vol. 2. Dekker, New York.
119. Eisenman, G. (1975). "Membranes," Vol. 3. Dekker, New York.
120. Eldik, L. J. Van, Grossman, A. T., Iverson, D. B., and Watterson, D. M. (1980). Isolation and characterisation of calmodulin from spinach leaves and in vitro translation mixtures. *Proc. Natl. Acad. Sci. U.S.A.* **77,** 1912–1916.

121. Ellis, R. J., and MacDonald, I. R. (1967). Activation of protein synthesis by microsomes from ageing beet discs. *Plant Physiol.* **42,** 1297–1302.

122. Epstein, E. (1961). The essential role of calcium in selective cation transport by plant cells. *Plant Physiol.* **36,** 437–444.

123. Epstein, E. (1973). Mechanisms of ion transport through plant cell membranes. *Int. Rev. Cytol.* **34,** 457–474.

124. Epstein, E. (1976). Kinetics of ion transport and the carrier concept *In* "Encyclopedia of Plant Physiology, New Series" (U. Lüttge and M. G. Pitman, eds.), Vol. 2, Part B, pp. 70–94. Springer-Verlag, Berlin and New York.

125. Epstein, E., and Hagen, C. E. (1952). A kinetic study of the absorption of alkaline earth cations by barley roots: Kinetics and mechanism. *Am. J. Bot.* **41,** 783–792.

126. Epstein, E., and Rains, D. W. (1965). Carrier-mediated cation transport in barley roots: Kinetic evidence for a spectrum of active sites. *Proc. Natl. Acad. Sci. U.S.A.* **49,** 1320–1324.

127. Epstein, E., Rains, D. W., and Elzam, C. E. (1963). Resolution of dual mechanisms of potassium absorption by barley roots. *Proc. Natl. Acad. Sci. U.S.A.* **49,** 684–692.

128. Etherton, B., and Rubinstein, B. (1978). Evidence for amino acid—H^+ co-transport in oat coleoptiles. *Plant Physiol.* **61,** 933–937.

129. Failla, M. L., and Weinberg, E. D. (1980). Zinc transport and metabolism in microorganisms. *In* "Zinc in the Environment" (J. O. Nriagu ed.), Part II, pp. 439–465. Wiley, New York.

130. Faiz-ur-Rahman, A. T. M., Davies, D. D., and Trewavas, A. J. (1974). The control of glycolysis in aged slices of carrot root tissue. *Planta* **118,** 211–224.

131. Felle, H. (1980). Amino transport at the plasmalemma of *Riccia fluitans. Biochim. Biophys, Acta* **602,** 181–195.

132. Ferrier, J. M., and Lucas, W. J. (1979). Plasmalemma transport of OH^- in *Chara corallina.* II. Further analysis of the diffusion system associated with OH^- efflux. *J. Exp. Bot.* **30,** 705–718.

133. Findlay, G. P., and Hope, A. B. (1976). Electrical properties of plant cells: Methods and findings. *In* "Encyclopedia of Plant Physiology, New Series" (U. Lüttge and M. G. Pitman, eds.), Vol. 2, Part A, pp. 53–92. Springer-Verlag, Berlin and New York.

134. Findlay, G. P., Hope, A. B., Pitman, M. G., Smith, P. A., and Walker, N. A. (1969). Ion fluxes in *Chara corallina. Biochim. Biophys. Acta* **183,** 565–576.

135. Findlay, G. P., Hope, A. B., and Walker, N. A. (1971). Quantization of a flux ratio in Charophytes? *Biochim. Biophys. Acta* **233,** 155–162.

136. Fitter, A. H. (1977). Influence of mycorrhizal infection on competition for phosphorus and potassium by two grasses. *New Phytol.* **79,** 119–125.

137. Fogg, G. E. (1965). "Algal Cultures and Phytoplankton Ecology." Univ. of Wisconsin Press, Madison.

138. Fowler, N. W. (1974). Role of the malic enzyme reaction in plant roots. Utilisation of [2,3-^{14}C] malate, [4-^{14}C] malate and [1-^{14}C] pyruvate by pea root apices and measurements of enzyme activity. *Biochim. Biophys. Acta* **372,** 245–254.

139. Franke, W. (1967). Mechanism of foliar penetration of solutes. *Annu. Rev. Plant Physiol.* **18,** 281–300.

140. Galpin, M. F. J., Jennings, D. H., Oates, K., and Hobot, J. (1978). Localisation by x-ray microanalysis of soluble ions, particularly potassium and sodium, in fungal hyphae. *Exp. Mycol.* **2,** 258–269.

141. Garrahan, P. J. (1970). Ion movements in red blood cells. *In* "Membranes and Ion Transport" (E. E. Bittar, ed.), pp. 185–215. Wiley (Interscience), New York.

142. Giaquinta, R. T. (1979). Phloem loading of sucrose. Involvement of membrane ATP-ase and proton transport. *Plant Physiol.* **63,** 744–748.

143. Giaquinta, R. T. (1980). Mechanism and control of phloem loading of sucrose. *Ber. Dtsch. Bot. Ges.* **93**, 187–201.
144. Gibbs, A. J. (1976). Viruses and plasmodemata. *In* "Intercellular Transport in Plants" (B. E. S. Gunning and A. W. Robards, eds.), pp. 149–164. Springer-Verlag, Berlin and New York.
145. Ginsburg, H. (1972). Analysis of plant root electropotentials. *J. Theor. Biol.* **37**, 389–412.
146. Ginsburg, H., and Laties, G. G. (1973). Longitudinal electrical resistance of maize roots. *J. Exp. Bot.* **24**, 1035–1040.
147. Giordano, P. M., and Mortveldt, J. J. (1980). Zinc uptake and accumulation by agricultural crops. *In* "Zinc in the Environment" (J. O. Briagu, ed.), Part II, pp. 401–413. Wiley, New York.
148. Glass, A. D. M. (1973). Influence of phenolic acids on ion uptake 1. Inhibition of phosphate uptake. *Plant Physiol.* **51**, 1037–1041.
149. Glass, A. D. M. (1974). Influence of phenolic acids upon ion uptake. III. Inhibition of potassium absorption. *J. Exp. Bot.* **25**, 1104–1113.
150. Glass, A. D. M. (1975). Inhibition of phosphate uptake in barley roots by hydroxy-benzoic acid. *Phytochemistry* **14**, 2127–2130.
151. Glass, A. D. M. (1975). The regulation of potassium absorption in barley roots. *Plant Physiol.* **56**, 377–380.
152. Glass, A. D. M. (1976). Regulation of potassium absorption in barley roots. An allosteric model. *Plant Physiol.* **58**, 33–37.
153. Glass, A. D. M. (1977). Regulation of K^+ influx in barley roots: Evidence for direct control by internal K^+. *Aust. J. Plant Physiol.* **4**, 313–318.
154. Glynn, I. M., Hoffman, J. F., and Lew, V. L. (1971). Some 'partial' reactions of the sodium pump. *Proc. R. Soc. London, Ser. B.* **262**, 91–102.
155. Goldman, D. E. (1943). Potential, impedance and rectification in membranes. *J. Gen. Physiol.* **27**, 37–60.
156. Gradmann, D. (1975). Analog circuit of the *Acetabularia* membrane. *J. Membr. Biol.* **25**, 183–208.
157. Gradmann, D., and Bentrup, F. W. (1970). Light-induced membrane potential changes and rectification in *Acetabularia. Naturwissenschaften* **57**, 46–47.
158. Gradmann, D., and Slayman, C. L. (1975). Oscillations of an electrogenic pump in the plasma membrane of *Neurospora. J. Membr. Biol.* **23**, 181–212.
159. Gradmann, D., Wagner, G., and Glassel, R. M. (1973). Chloride efflux during light-triggered action potentials in *Acetabularia mediterranea. Biochim. Biophys. Acta* **323**, 151–155.
160. Grand, R. J. A., Nairn, A. C., and Perry, S. V. (1980). The preparation of calmodulin from barley (*Hordeum* sp.) and basidiomycete fungi. *Biochem. J.* **185**, 755–760.
161. Graves, J. S., and Gutknecht, J. (1977). Chloride transport and the membrane potential in the marine alga. *Halicystis parvula. J. Membr. Biol.* **36**, 65–81.
162. Graves, J. S., and Gutknecht, J. (1977). Current voltage relationships and voltage sensitivity of the Cl^- pump in *Halicystis parvula. J. Membr. Biol.* **36**, 83–95.
163. Gutknecht, J. (1968). Membranes of *Valonia ventricosa:* Apparent absence of water-filled pores. *Science* **158**, 787.
164. Gutknecht, J., Hastings, D. F., and Bisson, M. A. (1978). Ion transport and turgor pressure regulation in giant algal cells. *In* "Membrane Transport in Biology" (G. Giesbisch, D. C. Tosteson, and H. H. Ussing eds.), Vol. 3, pp. 125–174. Springer-Verlag, Berlin and New York.
165. Hackette, S. L., Skye, G. E., Burton, C., and Segal, I. H. (1970). Characterization of an

ammonium transport system in filamentous fungi with methylammonium-^{14}C as the substrate. *J. Biol. Chem.* **245,** 4241–4250.

166. Hager, A., Menzel, H., and Krauss, A. (1971). Versuche und Hypothese zur Primawirkung des Auxins beim Streckungswachstum. *Planta* **100,** 47–75.
167. Hall, J. L., and Flowers, T. J. (1976). Properties of membranes from the halophyte *Suaeda maritima.* 1. Cytochemical staining of membranes in relation to the validity of membrane markers. *J. Exp. Bot.* **27,** 658–672.
168. Hall, T., Echlin P., and Kaufmann, R., eds. (1974). "Microprobe Analysis as Applied to Cells and Tissues." Academic Press, London.
169. Hanson, J. B., and Koeppe, D. E. (1975). Mitochondria. *In* "Ion Transport in Plant Cells and Tissues" (D. A. Baker and J. L. Hall, eds.), pp. 231–266. North-Holland Publ., Amsterdam.
170. Harley, J. L. (1969). "The Biology of Mycorrhiza." Leonard Hill, London.
171. Harley, J. L., and McCready, C. C. (1950). The uptake of phosphate by excised mycorrhizal roots of the beech. *New Phytol.* **49,** 388–397.
172. Harley, J. L., and McCready, C. C. (1952). The uptake of phosphate by excised mycorrhizal roots of the beech. II. Distribution of phosphorus between host and fungus. *New Phytol.* **51,** 56–64.
173. Harley, J. L., McCready, C. C., Brierley, J. K., and Jennings, D. H. (1956). The salt respiration of excised beech mycorrhizas. II. The relationship between oxygen consumption and phosphate absorption. *New Phytol.* **55,** 1–28.
174. Harold, F. M. (1977). Ion currents and physiological functions in microorganisms. *Annu. Rev. Microbiol.* **31,** 181–203.
175. Hauer, R., and Höfer, M. (1978). Evidence for interactions between the energy-dependent transport of sugars and the membrane potential in the yeast *Rhodotorula gracilis (Rhodosporidium toruloides). J. Membr. Biol.* **43,** 335–349.
176. Hayman, D. S., and Mosse, B. (1972). Plant growth responses to vesicular-arbuscular mycorrhiza. III. Increased uptake of labile P from the soil. *New Phytol.* **71,** 41–47.
177. Heber, U. (1974). Metabolite exchange between chloroplasts and cytoplasm. *Annu. Rev. Plant Physiol.* **25,** 393–421.
178. Heimer, Y. M., and Filner, D. (1971). Regulation of the nitrate uptake system. *Biochim. Biophys. Acta* **230,** 363–372.
179. Helder, R. J. (1975). Polar potassium transport and electrical potential difference across the leaf of *Potamogeton lucens.* L. *Proc. K. Ned. Akad. Wet., Ser. C* **78,** 189–197.
180. Helder, R. J. (1975). Flux-ratios and concentration-ratios in relation to electrical potential differences and transport of rubidium ions across the leaf of *Potamogeton lucens. Proc. K. Ned. Akad. Wet., Ser. C* **78,** 376–388.
181. Helder, R. J., and Boerma, J. (1972). Polar transport of labelled rubidium ions across the leaf of *Potamogeton lucens. Acta Bot. Neerl.* **21,** 211–218.
182. Helder, R. J., and Boerma, J. (1973). Exchange and polar transport of rubidium ions across the leaves of *Potamogeton. Acta Bot. Neerl.* **22,** 686–693.
183. Helder, R. J., Boerma, J., and Zanstra, P. E. (1980). Uptake pattern of carbon dioxide and bicarbonate by leaves of *Potamogeton lucens* L. *Proc. K. Ned. Akad. Wet., Ser. C* **83,** 151–166.
184. Helder, R. J., and Zanstra, P. E. (1977). Changes of the pH at the upper and lower surface of bicarbonate assimulating leaves of *Potamogeton lucens* L. *Proc. K. Ned. Akad. Wet., Ser. C* **80,** 421–436.
185. Heldt, H. W. (1976). Transport of metabolites between cytoplasm and mitochondrial matrix. *In* "Encyclopedia of Plant Physiology, New Series" (C. R. Stocking and U. Heber, eds.), Vol. 3, pp. 235–254. Springer-Verlag, Berlin and New York.

186. Heppel, L. E. (1969). The effect of osmotic shock on release of bacterial proteins and on active transport. *J. Gen. Physiol.* **54,** 95S–100S.
187. Hiatt, A. J. (1967). Relationship of cell sap pH to organic acid change during ion uptake. *Plant Physiol.* **42,** 294–298.
188. Hiatt, A. J., and Hendricks, S. B. (1967). The role of CO_2 fixation in accumulation of ions by barley roots. *Z. Pflanzenphysiol.* **56,** 220–232.
189. Higinbotham, N. (1970). Movement of ions and electrogenesis in higher plant cells. *Am. Zool.* **10,** 393–403.
190. Higinbotham, N., and Anderson, W. P. (1974). Electrogenic pumps in higher plant cells. *Can. J. Bot.* **52,** 1101–1022.
191. Higinbotham, N., Davies, R. F., Mertz, S. M., and Shumway, K. K. (1973). Some evidence that radial transport in maize roots is into living vessels. *In* "Ion Transport in Plants" (W. P. Anderson, ed.), pp. 493–706. Academic Press, New York.
192. Higinbotham, N., Graves, J. S., and Davis, R. F. (1970). Evidence for an electrogenic ion transport pump in cells of higher plants. *J. Membr. Biol.* **3,** 210–222.
193. Hill, A. E. (1967). Ion and water transport in *Limonium.* II. Short-circuit analysis. *Biochim. Biophys. Acta* **135,** 461–465.
194. Hill, A. E. (1975). Solute-solvent coupling in epithelia: A critical examination of the standing-gradient osmotic flow theory. *Proc. R. Soc. London, Ser. B* **190,** 115–134.
195. Hill, A. E. (1975). Solute-solvent coupling in epithelia: An electro-osmotic theory of fluid transfer. *Proc. R. Soc. London, Ser. B* **190,** 135–148.
196. Hill, A. E., and Hill, B. S. (1973). The *Limonium* salt gland: A biophysical and structural study. *Int. Rev. Cytol.* **35,** 299–319.
197. Hill, A. E., and Hill, B. S. (1973). The electrogenic chloride pump of the *Limonium* salt gland. *J. Membr. Biol.* **12,** 129–144.
198. Hill, B. S., and Hanke, D. E. (1979). Properties of the chloride-ATPase from *Limonium* salt glands. Activation by, and binding to, specific sugars. *J. Membr. Biol.* **51,** 185–194.
199. Hill, B. S., and Hill, A. E. (1973). ATP-driven chloride pumping and ATPase activity in the *Limonium* salt gland. *J. Membr. Biol.* **12,** 145–158.
200. Hill-Cottingham, D. G., and Lloyd-Jones, C. P. (1965). The behaviour of iron chelating agents with plants. *J. Exp. Bot.* **16,** 233–242.
201. Hodges, T. K. (1976). ATPases associated with membranes of plant cells. *In* "Encyclopedia of Plant Physiology, New Series" (U. Lüttge and M. G. Pitman, eds.), Vol. 2, Part A, pp. 260–283. Springer-Verlag, Berlin and New York.
202. Hodges, T. K., and Leonard, R. T. (1974). Purification of a plasma membrane-bound adenosine triphosphatase from plant roots. *In* "Methods in Enzymology" (S. Fleischer and L. Packer, eds.), Vol. 32, pp. 392–406. Academic Press, New York.
203. Hodges, T. K., Leonard, R. T., Bracker, C. E., and Keenan, T. W. (1972). Purification of aplasma membrane-bound adenosine triphosphatase from plant roots: Association with plasma membranes. *Proc. Natl. Acad. Sci. U.S.A.* **69,** 3307–3311.
204. Höfer, M., and Misra, P. C. (1978). Evidence for a proton/sugar symport in the yeast *Rhodotorula gracilis (glutinis). Biochem. J.* **172,** 15–22.
205. Hope, A. B. (1965). Ionic relations of cells of *Chara australis.* X. Effects of bicarbonate ions on electrical properties. *Aust. J. Biol. Res.* **18,** 789–801.
206. Hope, A. B., and Walker, N. A. (1975). "The Physiology of Giant Algal Cells". Cambridge Univ. Press, London and New York.
207. House, C. R., and Findlay, N. (1966). Water transport in isolated maize roots. *J. Exp. Bot.* **17,** 344–354.
208. House, C. R., and Findlay, N. (1966). Analysis of transient changes in fluid exudation from isolated maize roots. *J. Exp. Bot.* **17,** 627–640.

209. Hutchings, V. M. (1978). Sucrose and proton cotransport in *Ricinus* cotyledons. I. H^+ influx associated with sucrose uptake. *Planta* **138,** 229–235.
210. Hutchings, V. M. (1976). Sucrose and proton cotransport in *Ricinus* cotyledons. II. H^+ efflux and associated K^+ uptake. *Planta* **138,** 237–241.
211. Hylmo, B. (1953). Transpiration and ion absorption. *Physiol. Plant.* **6,** 333–405.
212. Jackson, P. C., and Adams, H. R. (1963). Cation-anion balance during potassium and sodium absorption by barley roots. *J. Gen. Physiol.* **46,** 369–386.
213. Jacoby, B., Abas, S., and Steinitz, B. (1973). Rubidium and potassium absorption by bean-leaf slices compared to sodium absorption. *Physiol. Plant.* **28,** 209–214.
214. Jacoby, B., and Dagan, J. (1967). A comparison of two methods of investigating sodium uptake by bean-leaf cells and the vitality of isolated leaf cells. *Protoplasma* **64,** 325–329.
215. Jacques, A. G. (1938). Kinetics of penetration. XV. The restriction of the cellulose wall. *J. Gen. Physiol.* **22,** 147–163.
216. Jaffe, L. F., and Nuccitelli, R. (1977). Electrical controls of development. *Annu. Rev. Biophys. Bioeng.* **6,** 445–476.
217. Jaffe, M. J. (1970). Evidence for the regulation of phytochrome-mediated processes in bean roots by the neurohumer, acetylcholine. *Plant Physiol.* **46,** 768–777.
218. Jaffe, M. J., and Galston, A. W. (1968). Physiological studies on pea tendrils. V. Membrane changes and water movements associated with contact coiling. *Plant Physiol.* **43,** 537–542.
219. Jansen, E. F., Jang, R., Albersheim, P., and Bonner, J. (1960). Pectic metabolism of growing cell walls. *Plant Physiol.* **35,** 87–97.
220. Jeffreys, R. A., Hale, V. Q., and Wallace, A. (1961). Uptake and translocation in plants of labelled ion and labelled chelating agents. *Soil Sci.* **92,** 268–273.
221. Jennings, D. H. (1963). "The Absorption of Solutes by Plant Cells." Oliver & Boyd, Edinburgh and London.
222. Jennings, D. H. (1967). Electrical potential measurement, ion pumps and root exudation—a comment and a model explaining cation selectivity by the root. *New Phytol.* **66,** 357–369.
223. Jennings, D. H. (1968). Microelectrode experiments with potato cells: A reinterpretation of the experimental findings. *J. Exp. Bot.* **58,** 13–18.
224. Jennings, D. H. (1969). The physiology of the uptake of ions by the growing plant cell. *In* "Ecological Aspects of the Mineral Nutrition of Plants" (I. H. Rorison, ed.), pp. 261–279. Blackwell, Oxford.
225. Jennings, D. H. (1973). Cations and filamentous fungi: Invasion of the sea and hyphal functioning. *In* "Ion Transport in Plants" (W. P. Anderson, ed.), pp. 323–325. Academic Press, New York.
226. Jennings, D. H. (1976). Transport in fungal cells. *In* "Encyclopedia of Plant Physiology, New Series" (U. Lüttge and M. G. Pitman, eds.), Vol. 2, Part A, pp. 189–338. Springer-Verlag, Berlin and New York.
227. Jennings, D. H. (1976). Transport and translocation in filamentous fungi. *In* "The Filamentous Fungi" (J. E. Smith and D. R. Berry, eds.), Vol. 2, pp. 32–64. Edward Arnold, London.
228. Jennings, D. H. (1979). Membrane transport and hyphal growth. *In* "Fungal Walls and Hyphal Growth" (J. H. Burnett and A. P. J. Trunci, eds.), Vol. 2, pp. 279–294. Cambridge Univ. Press, London and New York.
229. Jensen, P., and Pettersson, S. (1978). Allosteric regulation of potassium uptake in plant roots. *Physiol. Plant.* **42,** 207–213.
230. Jeschke, W. D. (1970). Evidence for K^+—stimulated Na^+ efflux at the plasmalemma of barley root cells. *Planta* **94,** 240–245.
231. Jeschke, W. D. (1972). Wirkung von K^+ auf die Fluxe den Transport von Na^+ in

Gerstenwurzeln K^+—stimulierter Na^+—efflux in der Wirzelrinde. *Planta* **106,** 73–90.

232. Jeschke, W. D. (1973). K^+—stimulated Na^+ efflux and selective transport in barley roots. *In* "Ion Transport in Plants" (W. D. Anderson, ed.), pp. 285–296. Academic Press, New York.

233. Jeschke, W. D. (1976). Ionic relations of leaf cells. *In* "Encyclopedia of Plant Physiology, New Series" (U. Lüttge and M. G. Pitman, eds.), Vol. 2, Part B, pp. 160–194. Springer-Verlag, Berlin and New York.

234. Jeschke, W. D. (1977). K^+-Na^+ exchange and selectivity in barley root cells. Effects of K^+, Rb^+, Cs^+, and Li^+ on the Na^+ fluxes. *Z. Pflanzenphysiol.* **84,** 247–264.

235. Jeschke, W. D. (1977). K^+-Na^+ exchange and selectivity in barley root cells: Effect of Na^+ on the Na^+ fluxes. *J. Exp. Bot.* **28,** 1289–1306.

236. Jeschke, W. D. (1980). Involvement of proton fluxes in the K^+-Na^+ selectivity at the plasmalemma: K^+ dependent net extrusion of sodium in barley roots and the effect of anions and pH on sodium fluxes. *Z. Pflanzenphysiol.* **98,** 155–175.

237. Johansen, C., and Lüttge, U. (1974). Respiration and photosynthesis as alternative energy sources for chloride uptake by *Tradescantia albiflora* leaf cells. *Z. Pflanzenphysiol.* **71**, 189–199.

238. Jones, H. G., and Osmond, C. B. (1973). Photosynthesis by thin leaf slices in solution. I. Properties of leaf slices and comparison with whole leaves. *Aust. J. Biol. Sci.* **26,** 15–24.

239. Jung, K. D., and Lüttge, U. (1980). Amino acid uptake by *Lemma gibba* by a mechanism with affinity to neutral L- and D-amino acids. *Planta* **150,** 230–235.

240. Jyung, W. H., Wittwer, S. H., and Bukovac, M. J. (1965). Ion uptake by cells enzymically isolated from green tobacco leaves. *Plant Physiol.* **40,** 410–14.

241. Kaback, H. R. (1972). Transport across isolated bacterial cytoplasmic membranes. *Biochim. Biophys. Acta* **265,** 367–416.

242. Kannan, S. (1971). Kinetics of iron absorption by excised rice roots. *Planta* **96,** 262–270.

243. Karlsson, J., and Kylin, A. (1974). Properties of Mg^{2+}—stimulated and (Na^+ and K^+)—activated adenosine—5′-triphosphatase from sugar beet cotyledons. *Physiol. Plant.* **32,** 136–142.

244. Kayne, F. J. (1973). Pyruvate kinase. *In* "The Enzymes" (P. D. Boyer, ed.), 3rd ed., Vol. 8, Part A, pp. 353–382. Academic Press, New York.

245. Keifer, D. W., and Spanswick, R. M. (1979). Correlation of adenosine triphosphate levels in *Chara corallina* with the activity of the electrogenic pump. *Plant Physiol.* **64,** 165–168.

246. Keller, P., and Deuel, H. (1957). Katimenaustauchkapazitat und Pektingehalt von Pflangenwurzeln. *Z. Pflanzenernaehr. Dueng.* **79,** 119–131.

247. Kirkby, E. A., and DeKock, P. C. (1965). The influence of age on the canion-anion balance in the leaves of Brussels sprout (*Brassica oleracea* var. *gemmifera*). *Z. Pflanzenernaerh. Dueng.* **111,** 197–203.

248. Kitasato, H. (1968). The influence of H^+ on the membrane potential and ion fluxes of *Nitella. J. Gen. Physiol.* **52,** 60–87.

249. Kitasato, H. (1973). K permeability of *Nitella clavata* in the depolarised state. *J. Gen. Physiol.* **62,** 535–549.

250. Klob, W., Kandler, O., and Tanner, W. (1973). The role of cyclic photophosphorylation *in vivo. Plant Physiol.* **51,** 825–827.

251. Kluge, M. (1977). Regulation of carbon dioxide fixation in plants. *Symp. Soc. Exp. Biol.* **31,** 155–175.

252. Kluge, M., and Osmond, C. B. (1972). Studies on phospheonolpyruvate carboxylase

and other enzymes of Crassulacean acid metabolism in *Bryophyllum tubiflorum* and *Sedum praealtum. Z. Pflanzenphysiol.* **66,** 97–105.

253. Komer, E. (1973). Proton-coupled hexose transport in *Chlorella vulgaris. FEBS Lett.* **38,** 16–18.
254. Komor, E. (1977). Sucrose uptake by cotyledons of *Ricinus communis* L.: Characteristics, mechanism and regulation. *Planta* **137,** 119–131.
255. Komor, E., Rotter, M., and Tanner, W. (1977). A proton-cotransport system in a higher plant: Sucrose transport in *Ricinus communis. Plant Sci. Lett.* **9,** 153–162.
256. Komor, E., Rotter, M., Waldhauser, J., Martin, E., and Cho, B. H. (1980). Sucrose proton symport for phloem loading in the *Ricinus* seedling. *Ber. Dtsch. Bot. Ges.* **93,** 211–219.
257. Komor, E., Schwab, W. G. W., and Tanner, W. (1979). The effect of intracellular pH on the rate of hexose uptake in *Chlorella. Biochim. Biophys. Acta* **555,** 524–530.
258. Komor, E., and Tanner, W. (1974). The hexose-proton cotransport system of *Chlorella.* pH-dependent changes in K_m values and translocation constants of the uptake system. *J. Gen. Physiol.* **64,** 567–581.
259. Komor, E., and Tanner, W. (1974). The nature of the energy metabolite responsible for sugar accumulation *Chlorella vulgaris. Z. Pflanzenphysiol.* **71,** 115–128.
260. Komor, E., and Tanner, W. (1974). The hexose-proton symport system of *Chlorella vulgaris.* Specificity, stoichiometry and energetics of sugar-induced proton uptake. *Eur. J. Biochem.* **44,** 219–223.
261. Komer, E., and Tanner, W. (1976). The determination of the membrane potential of *Chlorella vulgaris.* Evidence for electrogenic sugar transport. *Eur. J. Biochem.* **70,** 197–204.
262. Komor, E., Weber, H., and Tanner, W. (1978). Essential sulfhydryl group in the transport-catalysing protein of the hexose-proton cotransport system of *Chlorella. Plant Physiol.* **61,** 785–786.
263. Kramer, D., Romheld, V., Landsberg, E., and Marschner, H. (1980). Induction of transfer-cell-formation by iron deficiency in the root epidermis of *Helianthus annuus* L. *Planta* **147,** 335–339.
264. Krebs, H. A., and Kornberg, H. L. (1957). A survey of the energy transformations in living matter. *Ergeb. Physiol., Biol. Chem. Exp. Pharmakol.* **49,** 212–298.
265. Kylin, A. (1973). Adenosine triphosphatases stimulated by sodium and potassium. Biochemistry and possible significance for salt resistance. *In* "Ion Transport in Plants" (W. D. Anderson, ed.), pp. 369–377. Academic Press, New York.
266. Lado, P., Rasi-Caldogno, F., Colombo, R., De Michelis, M. I., and Marrè, E. (1976). Effects of monovalent cations on 1AA- and FC-stimulated proton-cation exchange in pea stem segments. *Plant Sci. Lett.* **7,** 199–209.
267. Larsen, S. (1967). Soil phosphorus. *Adv. Agron.* **19,** 151–210.
268. Läuchli, A. (1972). Electron probe analysis. *In* "Microautoradiography and Electron Probe Analysis" (U. Lüttge, ed.), pp. 191–236. Springer-Verlag, Berlin and New York.
269. Läuchli, A. (1976). Genotypic variation in transport. *In* "Encyclopedia of Plant Physiology, New Series" (U. Lüttge and M. G. Pitman, eds.), Vol. 2, Part B, pp. 372–393. Springer-Verlag, Berlin and New York.
270. Läuchli, A., and Epstein, E. (1970). Transport of potassium and rubidium in plant roots. *Plant Physiol.* **45,** 639–641.
271. Läuchli, A., Kramer, D., and Stelzer, R. (1974). Ultrastructure of xylem parenchyma cells of roots. *In* "Membrane Transport in Plants" (U. Zimmerman and J. Dainty, eds.), pp. 363–371. Springer-Verlag, Berlin and New York.

272. Läuchli, A., Pitman, M. G., Lüttge, U., Kramer, D., and Ball, E. (1978). Are developing xylem vessels the site of ion exudation from root to shoot? *Plant, Cell Environ.* **1,** 217–224.
273. Leigh, R. A., and Branton, D. (1976). Isolation of vacuoles from root storage tissue of *Beta vulgaris* L. *Plant Physiol.* **58,** 650–662.
274. Leigh, R. A., and Walker, R. R. (1980). ATPase and acid phosphatase associated with vacuoles isolated from storage roots of red beet (*Beta vulgaris* L.). *Planta* **150,** 222–229.
275. Lewin, J. C. (1962). Calcification. *In* "Physiology and Biochemistry of the Algae" (R. A. Lewin, ed.), pp. 457–465. Academic Press, New York.
276. Lewis, D. G., and Quirk, J. P. (1967). Phosphate diffusion in soil and uptake by plants. III. P^{31} movement and uptake by plants as indicated by P^{32} autoradiography. *Plant Soil* **26,** 445–453.
277. Lewis, D. H., and Harley, J. L. (1965). Carbohydrate physiology of mycorrhizal roots of beech. III. Movement of sugars between host and fungus. *New Phytol.* **64,** 256–269.
278. Liberman, E. A., and Topali, V. P. (1969). Selective transport of ions through bromolecular phospholipid vesicles. *Biochim. Biophys. Acta* **163,** 125–136.
279. Liebig, J. von (1855). "Die Grandsatze der Agricultur-chemie." Vieweg, Braunschweig.
280. Lien, R., and Rognes, S. E. (1977). Uptake of amino acids by barley leaf slices: Kinetics, specificity and energetics. *Physiol. Plant.* **41,** 173–183.
281. Lin, W. (1980). Corn root protoplasts. Isolation and general characterisation of ion transport. *Plant Physiol.* **66,** 550–554.
282. Lin, W., Wagner, G. J., Siegelman, H. W., and Hind, G. (1977). Membrane-bound ATPase of intact vacuoles and tonoplasts isolated from mature plant tissue. *Biochim. Biophys. Acta* **465,** 110–117.
283. Loneragan, J. F., and Asher, C. J. (1967). Response of plants to phosphate concentration in solution culture. II. Rate of phosphate absorption and its relation to growth. *Soil Sci.* **193,** 311–318.
284. Lookeren Campagne, R. N. van (1957). Light-dependent chloride absorption in *Vallisneria* leaves. *Acta Bot. Neerl.* **6,** 543–582.
285. Lowenhaupt, B. (1956). The transport of calcium and other cations in submerged aquatic plants. *Biol. Rev. Cambridge Philos. Soc.* **31,** 371–395.
286. Lowenhaupt, B. (1958). Active cation transport in submerged aquatic plants. I. Effect of light upon the absorption and excretion of calcium by *Potamogeton crispus* (L) leaves. *J. Cell. Comp. Physiol.* **51,** 199–208.
287. Lowenhaupt, B. (1958). Active cation transport in submerged aquatic plants. II. Effects of aeration on the equilibrium content of calcium in *Potamogeton crispus* (L) leaves. *J. Cell. Comp. Physiol.* **51,** 209–219.
288. Lucas, W. J. (L975). Analysis of the diffusion symmetry developed by the alkaline and acid bands which form at the surface of *Chara corallina* cells. *J. Exp. Bot.* **26,** 271–286.
289. Lucas, W. J. (1975). Photosynthetic fixation of ^{14}carbon by internodal cells of *Chara corallina. J. Exp. Bot.* **26,** 331–346.
290. Lucas, W. J. (1975). The influence of light intensity on the activation and operation of the hydroxyl efflux system of *Chara corallina. J. Exp. Bot.* **26,** 347–360.
291. Lucas, W. J. (1976). Plasmalemma transport of HCO_3^- and OH^- in *Chara corallina:* Non-antiporter systems. *J. Exp. Bot.* **27,** 19–31.
292. Lucas, W. J. (1976). The influence of Ca^{2+} and K^+ on $H^{14}CO_3^-$ influx in internodal cells of *Chara corallina. J. Exp. Bot.* **27,** 32–42.

293. Lucas, W. J. (1979). Alkaline band formation in *Chara corallina:* Due to OH^- efflux or H^+ influx? *Plant Physiol.* **63,** 248–254.
294. Lucas, W. J. (1980). Control and synchronization of HCO_3^- and OH^- transport during photosynthetic assimilation of exogenous HCO_3^-. *In* "Plant Membrane Transport: Current Conceptual Issues" (R. M. Lucas, W. J. Lucas, and J. Dainty, eds.), pp. 317–327. Elsevier/North-Holland Biomedical Press, Amsterdam.
295. Lucas, W. J., and Alexander, J. M. (1980). Sulfhydryl group involvement in plasmalemma transport of HCO_3^- and OH^- in *Chara corallina. Plant Physiol.* **65,** 274–280.
296. Lucas, W. J., and Dainty, J. (1977). Spatial distribution of functional OH^- carriers along a Characean internodal cell: Determined by the effect of cytochalasin B on $H^{14}CO_3^-$ assimilation. *J. Membr. Biol.* **32,** 75–92.
297. Lucas, W. J., and Ferrier, J. M. (1980). Plasmalemma transport of OH^- in *Chara corallina.* III. Further studies on transport substrate and directionality. *Plant Physiol.* **66,** 46–50.
298. Lucas, W. J., Ferrier, J. M., and Dainty, J. (1977). Plasmalemma transport of OH^- in *Chara corallina:* Dynamics of activation and deactivation. *J. Membr. Biol.* **32,** 49–73.
299. Lucas, W. J., and Nuccitelli (1980). HCO_3^- and OH^- transport across the plasmalemma of *Chara.* Spatial resolution obtaining using extracellular vibrating electrodes. *Planta* **150,** 120–131.
300. Lucas, W. J., and Smith, F. A. (1973). The formation of alkaline and acid regions at the surface of *Chara corallina* cells. *J. Exp. Bot.* **24,** 1–14.
301. Lucas, W. J., and Smith, F. A. (1976). Influence of irradiance on H^+ efflux and Cl^- influx in *Chara corallina:* An investigation aimed at testing two Cl^- transport models. *Aust. J. Plant Physiol.* **3,** 443–456.
302. Lüttge, U., and Higinbotham, N., eds. (1979). "Transport in Plants." Springer-Verlag, Berlin and New York.
303. Lüttge, U., and Pitman, M. G., eds. (1976). "Encyclopedia of Plant Physiology, New Series," Vol. 2, Parts A and B. Springer-Verlag, Berlin and New York.
304. Lycklama, J. C. (1963). The absorption of ammonium and nitrate by perennial ryegrass. *Acta Bot. Neerl.* **12,** 361–423.
305. MacDonald, I. R. (1967). Bacterial infection and ion absorption capacity in beet disks. *Ann. Bot. (London)* [N.S.] **33,** 163–172.
306. MacDonald, I. R. (1975). Effect of vacuum infiltration on photosynthetic gas exchange in leaf tissue. *Plant Physiol.* **56,** 109–112.
307. MacDonald, I. R., and Macklon, A. E. S. (1972). Anion absorption by etiolated wheat leaves after vacuum infiltration. *Plant Physiol.* **49,** 303–306.
308. MacDonald, I. R., and Macklon, A. E. S. (1975). Light-enhanced chloride uptake by wheat laminae. A comparison of chopped and vacuum-infiltrated tissue. *Plant Physiol.* **56,** 105–108.
309. MacLennan, D. H., Beevers, H., and Harley, J. L. (1963). "Compartmentation" of acids in plant tissues. *Biochem. J.* **89,** 316–327.
310. MacRobbie, E. A. C. (1962). Ionic relations of *Nitella translucens. J. Gen. Physiol.* **45,** 861–878.
311. MacRobbie, E. A. C. (1964). Factors affecting the fluxes of potassium and chloride ions in *Nitella translucens. J. Gen. Physiol.* **47,** 859–877.
312. MacRobbie, E. A. C. (1965). The nature of the coupling between light energy and active ion transport in *Nitella translucens. Biochim. Biophys. Acta* **94,** 64–73.
313. MacRobbie, E. A. C. (1966). Metabolic effects on ion fluxes in *Nitella translucens.* I. Active influxes. *Aust. J. Biol. Sci.* **19,** 363–370.

314. MacRobbie, E. A. C. (1966). Metabolic effects on ion fluxes in *Nitella translucens*. II. Tonoplast fluxes. *Aust. J. Biol. Sci.* **19,** 371–383.
315. MacRobbie, E. A. C. (1969). Ion fluxes to the vacuole of *Nitella translucens*. *J. Exp. Bot.* **20,** 236–256.
316. MacRobbie, E. A. C. (1970). Quantized fluxes of chloride in the vacuole of *Nitella translucens*. *J. Exp. Bot.* **21,** 335–344.
317. MacRobbie, E. A. C. (1970). The active transport of ions in plant cells. *Q. Rev. Biophys.* **3,** 251–295.
318. MacRobbie, E. A. C. (1971). Vacuolar fluxes of chloride and bromide in *Nitella translucens*. *J. Exp. Bot.* **22,** 487–502.
319. MacRobbie, E. A. C. (1973). Vacuolar ion transport in *Nitella*. *In* "Ion Transport in Plants" (W. P. Anderson, ed.), pp. 431–446. Academic Press, New York.
320. MacRobbie, E. A. C. (1974). Ion uptake. *In* "Algal Physiology and Biochemistry" (W. D. P. Stewart, ed.), pp. 676–713. Blackwell, Oxford.
321. MacRobbie, E. A. C. (1975). Intracellular kinetics of tracer chloride and bromide in *Nitella translucens*. *J. Exp. Bot.* **26,** 489–507.
322. MacRobbie, E. A. C. (1977). Functions of ion transport in plant cells and tissues. *Int. Rev. Biochem.* **13,** 211–247.
323. MacRobbie, E. A. C., and Dainty, J. (1958). Sodium and potassium distribution and transport in the seaweed *Rhodymenia palmata* (L) Grov. *Physiol. Plant.* **11,** 782–801.
324. MacRobbie, E. A. C., and Dainty, J. (1958). Ion transport in *Nitellopsis obtusa*. *J. Gen. Physiol.* **42,** 335–353.
325. Magistad, O. C. (1925). The aluminium content of the soil solution and its relation to soil reaction and plant growth. *Soil Sci.* **20,** 181–226.
326. Malek, F., and Baker, D. A. (1977). Proton co-transport of sugars in phloem loading. *Planta* **135,** 297–289.
327. Marmé, D. (1977). Phytochrome: Membranes as possible sites of action. *Annu. Rev. Plant Physiol.* **28,** 173–198.
328. Marrè, E. (1979). Fusicoccin: A tool in plant physiology. *Annu. Rev. Plant Physiol.* **30,** 273–288.
329. Marrè, E. (1980). Mechanism of action of phytotoxins affecting plasmalemma functions. *Prog. Phytochem.* **6,** 253–284.
330. Marrè, E., Lado, P., Rasi-Caldogno, F., and Colombo, R. (1974). Fusicoccin-activated proton extrusion coupled with K^+ uptake and its role in the regulation of growth, germination, opening of stomata and mineral nutrition. *Atti Accad. Naz. Lincei, Cl. Sci. Fis., Mat. Nat., Rend.* **55,** 690–700.
331. Marrè, E., Lado, P., Rasi-Caldogno, F., Colombo, R., Cocucci, M., and De Michelis, M. I. (1975). Regulation of proton extrusion by plant hormones and cell elongation. *Physiol. Vég.* **13,** 799–811.
332. Marrè, E., Lado, P., Rasi-Caldogno, F., Colombo, R., and De Michelis, M. I. (1974). Evidence for the coupling of proton extrusion to K^+ uptake in pea internode segments treated with fusicoccin or auxin. *Plant Sci. Lett.* **3,** 365–379.
333. Marzluf, G. A. (1970). Genetical and biochemical studies of distinct sulphate permease species in different developmental stages of *Neurospora crassa*. *Arch. Biochem. Biophys.* **138,** 254–263.
334. Marzluf, G. A. (1972). Control of the synthesis, activity and turnover of enzymes of sulphur metabolism in *Neurospora crassa*. *Arch. Biochem. Biophys.* **150,** 714–724.
335. Mathys, W. (1980). Zinc tolerance by plants. *In* "Zinc in the Environment," (J. O. Nriagu ed.), Part II, pp. 415–438. Wiley, New York.
336. Matile, P. (1971). Vacuoles, lysosomes of *Neurospora*. *Cytobiologie* **3,** 324–330.

337. Matile, P. (1978). Biochemistry and function of vacuoles. *Annu. Rev. Plant Physiol.* **29,** 193–213.
338. Matile, P., and Wiemken, A. (1967). The vacuole as the lysosome of the yeast cell. *Arch. Mikrobiol.* **56,** 148–155.
339. Matile, P., and Wiemken, A. (1976). Interactions between cytoplasm and vacuole. *In* "Encyclopedia of Plant Physiology, New Series" (C. R. Stocking and U. Heber eds.), Vol. 3, pp. 255–287. Springer-Verlag, Berlin and New York.
340. Maw, G. A. (1963). The uptake of inorganic sulphate by a brewer's yeast. *Folia Microbiol. (Prague)* **8,** 325–332.
341. Mercier, A. J., and Poole, R. J. (1980). Electrogenic pump activity in red beet: Its relation to ATP levels and cation influx. *J. Membr. Biol.* **55,** 165–174.
342. Miller, A. G., and Budd, K. (1976). Evidence for a negative membrane potential and for movement of Cl^- against its electrochemical gradient in the ascomycete *Neocosmospora vasinfecta. J. Bacteriol.* **128,** 741–748.
343. Mitchell, P. (1961). Coupling of phosphorylation to electron and hydrogen transfer by a chemiosmotic type of mechanism. *Nature (London)* **191,** 144–148.
344. Mitchell, P. (1966). Chemiosmotic coupling in oxidation and photosynthetic phosphorylation. *Biol. Rev. Cambridge Philos. Soc.* **44,** 445–602.
345. Mitchell, P. (1967). Translocation through natural membranes. *Adv. Enzymol.* **29,** 33–87.
346. Mitchell, P. (1976). Vectorial chemistry and the molecular mechanics of chemiosmotic coupling: Power transmission by proticity. *Biochem. Soc. Trans.* **4,** 399–430.
347. Mitscherlich, E. A. (1909). Das Gesetz des minimums und das Gestez des abnenmenden Bodenertrages. *Landwirtsch. Jahrb.* **38,** 537–552.
348. Morris, P., and Thain, J. F. (1980). Comparative studies of leaf tissue and isolated mesophyll protoplasts. I. O_2 exchange and CO_2 fixation. *J. Exp. Bot.* **31,** 83–96.
349. Morris, P., and Thain, J. F. (1980). Comparative studies of leaf tissue and isolated mesophyll protoplasts. II. Ion relations. *J. Exp. Bot.* **31,** 97–104.
350. Morrod, R. S. (1974). A new method for measuring the permeability of plant cell membranes using epidermis-free leaf discs. *J. Exp. Bot.* **25,** 521–533.
351. Mosse, B. (1973). Advances in the study of vesicular-abuscular mycorrhiza. *Annu. Rev. Phytopathol.* **11,** 171–196.
352. Mukerji, S. K., and Ting, I. P. (1971). Phosphoenolpyruvate carboxylase isoenzymes: Separation and properties of three forms from cotton leaf tissue. *Arch. Biochem. Biophys.* **143,** 297–317.
353. Newman, E. I. (1966). A method of estimating the total length of root in a sample. *J. Appl. Ecol.* **3,** 139–145.
354. Newsholme, E. A., and Start, C. (1973). "Regulation in Metabolism." Wiley, New York.
355. Nieman, R. H., and Willis, C. (1971). Correlation between the suppression of glucose and phosphate uptake and the release of protein from viable carrot root cells treated with monovalent cations. *Plant Physiol.* **48,** 287–293.
356. Nierhaus, D., and Kinzel, H. (1971). Vergleichende Untersuchungen uber die organischen Sauren in Blattern hoherer Pflanzen. *Z. Pflanzenphysiol.* **64,** 107–123.
357. Nissen, P. (1974). Uptake mechanisms: Inorganic and organic. *Annu. Rev. Plant Physiol.* **25,** 53–79.
358. Nissen, P. (1980). Multiphasic uptake of potassium by barley roots of low and high potassium content. Separate sites for uptake and transitions. *Physiol. Plant.* **48,** 193–200.

359. Northcote, D. H. (1972). Chemistry of the plant cell wall. *Annu. Rev. Plant Physiol.* **23**, 113–132.
360. Novacky, A., Fischer, E., Ullrich-Eberius, C. I., Lüttge, U., and Ullrich, W. R. (1978). Membrane potential changes during transport of glycine as a neutral amino acid and nitrate in *Lemna gibba* G 1. *FEBS Lett.* **88**, 264–267.
361. Novacky, A., Ullrich-Eberius, C. I., and Lüttge, U. (1978). Membrane potential changes during transport of hexoses in *Lemna gibba* G 1. *Planta* **138**, 263–270.
362. Novacky, A., Ullrich-Eberius, C. I., and Lüttge, U. (1980). pH and membrane potential changes during glucose uptake in *Lemna gibba* C 1 and their response to light. *Planta* **149**, 321–326.
363. Nye, P. H. (1966). The effect of the nutrient intensity and buffering power of a soil and the absorbing power, size and root hairs of a root, on nutrient absorption by diffusion. *Plant Soil* **25**, 81–105.
364. Nye, P. H. (1968). Processes in the root environment. *J. Soil Sci.* **19**, 205–215.
365. Nye, P. H. (1973). The relation between the radius of a root and its nutrient-absorbing power [α]. *J. Exp. Bot.* **24**, 783–786.
366. Nye, P. H., and Marriott, F. H. C. (1969). A theoretical study of the distribution of substances around roots resulting from simultaneous diffusion and mass flow. *Plant Soil* **30**, 459–472.
367. Nye, P. H., and Tinker, P. B. (1969). The concept of a root demand coefficient. *J. Appl. Ecol.* **6**, 293–300.
368. Olsen, R. A., and Brown, J. C. (1980). Factors related to iron uptake by dicotyledonous and monocotyledonous plants. 1. pH and reductant. *J. Plant Nutr.* **2**, 629–645.
369. Olsen, R. A., and Brown, J. C. (1980). Factors related to iron uptake by dicotyledonous and monocotyledonous plants. II. The reduction of Fe^{3+} as influenced by roots and inhibitors. *J. Plant Nutr.* **2**, 647–660.
370. Olsen, R. A., and Brown, J. C. (1980). Factors related to iron uptake by dicotyledonous and monocotyledonous plants. III. Competition between root and external factors for Fe. *J. Plant Nutr.* **2**, 661–682.
371. Osmond, C. B. (1976). Ion absorption and carbon metabolism in cells of higher plants. *In* "Encyclopedia of Plant Physiology, New Series" (U. Lüttge and M. G. Pitman, eds.), Vol. 2, Part B, pp. 347–372. Springer-Verlag, Berlin and New York.
372. Osmond, C. B., and Greenway, H. (1972). Salt responses of carboxylation enzymes from species differing in salt tolerance. *Plant Physiol.* **49**, 260–263.
373. Osmond, C. B., and Laties, G. G. (1969). Compartmentalisation of malate in relation to ion absorption in beet. *Plant Physiol.* **44**, 7–14.
374. Overbeek, J. van (1942). Water uptake by excised root systems of the tomato due to non-osmotic forces. *Am. J. Bot.* **29**, 677–683.
375. Paasche, E. (1968). Biology and physiology of the Coccolithophorids. *Annu. Rev. Microbiol.* **22**, 71–86.
376. Page, K. R., Kelday, L. S., and Bowling, D. J. F. (1981). The diffusion of KCl from microelectrodes. *J. Exp. Bot.* **32**, 55–58.
377. Pall, M. L. (1971). Amino acid transport in *Neurospora crassa.* IV. Properties and regulation of a methionine transport system. *Biochim. Biophys. Acta* **233**, 201–214.
378. Palmer, J. M. (1970). The influence of microbial contamination of fresh and washed beetroot disks on their capacity to absorb phosphate. *Planta* **93**, 48–52.
379. Park, D., and Robinson, P. M. (1967). A fungal hormone controlling internal water distribution normally associated with cell ageing in fungi. *Symp. Soc. Exp. Biol.* **21**, 323–336.

380. Passioura, J. B. (1963). A mathematical model for the uptake of ions from the soil solution. *Plant Soil* **18,** 225–238.
381. Passioura, J. B., and Freere, H. H. (1967). A numerical analysis of the convection and diffusion of solutes to roots. *Aust. J. Soil Res.* **5,** 149–161.
382. Pateman, J. A., and Kinghorn, J. R. (1975). Nitrogen metabolism. *In* "The Filamentous Fungi" (J. E. Smith and D. R. Berry, eds.), Vol. 2, pp. 159–237. Edward Arnold, London.
383. Pateman, J. A., Kinghorn, J. R., Dunn, E., and Forbes, E. (1973). Ammonium regulation in *Aspergillus nidulans. J. Bacteriol.* **114,** 943–950.
384. Pentecost, A. (1980). Calcification in plants. *Int. Rev. Cytol.* **62,** 1–27.
385. Perlin, D. S., and Spanswick, R. M. (1980). Labelling and isolation of plasma membranes from corn leaf protoplasts. *Plant Physiol.* **65,** 1053–1057.
386. Petraglia, T., and Poole, R. J. (1980). ATP levels and their effects on plasmalemma influxes of potassium chloride in red beet. *Plant Physiol.* **65,** 969–972.
387. Phillips, R. D., and Jennings, D. H. (1976). Succulence, cations and organic acids in leaves of *Kalanchoe daigremontiana* grown in long and short days in soil and water culture. *New Phytol.* **77,** 599–611.
388. Pierce, W. S., and Hendrix, D. L. (1979). Utilisation of enzyme markers to determine the location of plasma membrane from *Pisum* epicotyls on sucrose gradients. *Planta* **146,** 161–169.
389. Pierre, J. W. De, and Karnovsky, M. L. (1973). Plasma membranes of mammalian cells. A review of methods for their characterization and isolation. *J. Cell Biol.* **56,** 275–303.
390. Pitman, M. G. (1964). The effect of divalent cations on the uptake of salt by beetroot tissue. *J. Exp. Bot.* **15,** 444–456.
391. Pitman, M. G. (1970). Active H^+ efflux from cells of low-salt barley roots during salt accumulation. *Plant Physiol.* **45,** 787–790.
392. Pitman, M. G. (1971). Uptake and transport of ions in barley seedlings. I. Estimation of chloride fluxes in cells of excised roots. *Aust. J. Biol. Sci.* **24,** 407–421.
393. Plowe, J. C. (1931). Membranes of the plant cell. I. Morphological membranes at protoplasmic surfaces. *Protoplasma* **12,** 196–221.
394. Poole, R. J. (1971). Development and characteristics of sodium-selective transport in red beet. *Plant Physiol.* **47,** 735–739.
395. Poole, R. J. (1974). Ion transport and electrogenic pumps in storage tissue cells. *Can. J. Bot.* **52,** 1023–1028.
396. Poole, R. J. (1976). Transport in cells of storage tissues. *In* "Encyclopedia of Plant Physiology, New Series" (U. Lüttge and M. G. Pitman, eds.), Vol. 2, Part A, pp. 229–248. Springer-Verlag, Berlin and New York.
397. Poovaiah, B. W., and Leopold, A. C. (1976). Effect of inorganic salts on tissue permeability. *Plant Physiol.* **58,** 182–185.
398. Pope, D. T., and Munger, H. M. (1953). Heredity and nutrition in relation to magnesium deficiency chlorosis in celery. *Proc. Am. Soc. Hortic. Sci.* **61,** 472–480.
399. Porter, L. K., Kemper, W. D., Jackson, R. D., and Stewart, B. A. (1960). Chloride diffusion in soils as influenced by moisture content. *Soil Sci. Soc. Am. Proc.* **24,** 460–463.
400. Post, R. L., Merritt, C. R., Kinsolving, C. R., and Albright, C. D. (1960). Membrane adenosine triphosphatase as a participant in the active transport of sodium and potassium in the erythocyte. *J. Biol. Chem.* **235,** 1796–1802.
401. Postma, P. W., and Roseman, S. (1976). The bacterial phospheonolpyruvate: sugar phosphotransferase system. *Biochim. Biophys. Acta* **457,** 213–257.
402. Prins, H. B. A., Snel, J. F. H., Helder, R. J., and Zanstra, P. E. (1980). Photosynthetic

HCO_3^- utilization and OH^- excretion on aquatic angiosperms. Light-induced pH changes at the leaf surface. *Plant Physiol.* **66,** 818–822.
403. Racusen, R. H., Kinnersley, A. M., and Galston, A. W. (1977). Osmotically induced changes in electrical properties of plant protoplast membranes. *Science* **198,** 405–406.
404. Rains, D. W. (1968). Kinetics and energetics of light-enhanced potassium absorption by corn leaf tissue. *Plant Physiol.* **43,** 394–408.
405. Rains, D. W., and Epstein, E. (1965). Transport of sodium in plant tissue. *Science* **148,** 1611.
406. Rains, D. W., and Epstein, E. (1967). Sodium absorption by barley roots: Role of the dual mechanisms of alkali cation transport. *Plant Physiol.* **42,** 314–318.
407. Rains, D. W., and Epstein, E. (1967). Sodium absorption by barley roots: Its mediation by mechanisms 2 of alkali cation absorption. *Plant Physiol.* **42,** 319–323.
408. Rains, D. W., Schmid, W. E., and Epstein, E. (1964). Absorption of cations by roots. Effects of hydrogen ions and the essential role of calcium. *Plant Physiol.* **39,** 274–278.
409. Rauser, W. E., and Curvetto, N. R. (1980). Metallothionein occurs in roots of *Agrostis* tolerant to excess copper. *Nature (London)* **287,** 563–564.
410. Raven, J. A. (1967). Ion transport in *Hydrodictyon africanum. J. Gen. Physiol.* **50,** 1607–1625.
411. Raven, J. A. (1968). The mechanism of photosynthetic use of bicarbonate by *Hydrodictyon africanum. J. Exp. Bot.* **19,** 193–206.
412. Raven, J. A. (1968). The linkage of light-stimulated Cl influx to K and Na influxes in *Hydrodictyon africanum. J. Exp. Bot.* **19,** 233–253.
413. Raven, J. A. (1968). The action of phlorizin on photosynthesis and light-stimulated ion transport in *Hydrodictyon africanum. J. Exp. Bot.* **19,** 712–723.
414. Raven, J. A. (1969). Action spectra for photosynthesis and light-stimulated ion transport processes in *Hydrodictyon africanum. New Phytol.* **68,** 45–62.
415. Raven, J. A. (1969). Effects of inhibitors on photosynthesis and the active influxes of K and Cl in *Hydrodictyon africanum. New Phytol.* **68,** 1089–1113.
416. Raven, J. A. (1970). Exogenous inorganic carbon sources in plant photosynthesis. *Biol. Rev. Cambridge Philos. Soc.* **45,** 167–221.
417. Raven, J. A. (1971). Inhibitor effects on photosynthesis respiration and active ion transport in *Hydrodictyon africanum. J. Membr. Biol.* **6,** 89–107.
418. Raven, J. A. (1971). Cyclic and non-cyclic photophosphorylation as energy sources for active K influx in *Hydrodictyon africanum. J. Exp. Bot.* **22,** 420–433.
419. Raven, J. A. (1974). Phosphate transport in *Hydrodictyon africanum. New Phytol.* **73,** 421–432.
420. Raven, J. A. (1974). Energetics of active phosphate influx in *Hydrodictyon africanum. J. Exp. Bot.* **25,** 221–229.
421. Raven, J. A. (1975). Algal cells. *In* "Ion transport in Plant Cells and Tissues" (D. A. Baker and J. L. Hall, eds.), pp. 125–160. North-Holland Publ., Amsterdam.
422. Raven, J. A. (1976). Glucose metabolism in *Hydrodictyon africanum* in relations to cell energetics. *New Phytol.* **76,** 195–204.
423. Raven, J. A. (1976). Transport in algal cells. *In* "Encyclopedia of Plant Physiology, New Series" (U. Lüttge and M. G. Pitman, eds.), Vol. 2, Part A, pp. 129–188. Springer-Verlag, Berlin and New York.
424. Raven, J. A. (1977). Regulation of solute transport at the cell level. *Symp. Soc. Exp. Biol.* **31,** 73–99.
425. Raven, J. A., and Smith, F. A. (1973). The regulation of intracellular pH as a fundamental biological process. *In* "Ion Transport in Plants" (W. P. Anderson, ed.), pp. 271–278. Academic Press, New York.

426. Raven, J. A., and Smith, F. A. (1974). Significance of hydrogen ion transport in plant cells. *Can. J. Bot.* **52,** 1035–1048.
427. Raven, J. A., and Smith, F. A. (1976). Nitrogen assimilation and transport in vascular land plants in relation to intracellular pH regulation. *New Phytol.* **76,** 415–431.
428. Ray, P. M., Eisinger, W. R., and Robinson, D. G. (1976). Organelles involves in cell wall polysaccharide formation and transport in pea cells. *Ber. Dtsch. Bot. Ges.* **89,** 121–146.
429. Ray, P. M., Shininger, T. L., and Ray, M. M. (1969). Isolation of β-glucan synthetase particles from plant cells and identification with Golgi membranes. *Proc. Natl. Acad. Sci. U.S.A.* **64,** 605–612.
430. Rees, T. ap (1974). Pathways of carbohydrate breakdown in higher plants. *MTP Int. Rev. Sci. Biochem.* [1] **11,** 89–127.
431. Robards, A. W. (1971). The ultrastructure of plasmodesmata. *Protoplasma* **72,** 315–323.
432. Robards, A. W. (1975). Plasmodesmata. *Annu. Rev. Plant Physiol.* **26,** 13–29.
433. Robards, A. W. (1976). Plasmodesmata in Higher Plants. *In* "Intercellular Communication in Plants" (B. E. S. Gunning and A. W. Robards, ed.), pp. 15–57. Springer-Verlag, Berlin and New York.
434. Robards, A. W., and Clarkson, D. T. (1976). The role of plasmodesmata in the transport of water and nutrients across roots. *In* "Intercellular Communication in Plants" (B. L. S. Gunning and A. W. Robards, ed.), pp. 181–201. Springer-Verlag, Berlin and New York.
435. Robards, A. W., Jackson, S. M., Clarkson, D. T., and Sanderson, J. (1973). The structure of barley roots in relation to the transport of ions in the stele. *Protoplasma* **77,** 291–331.
436. Robards, A. W., and Robb, M. E. (1974). The entry of ions and molecules into roots: An investigation using electron-opaque tracers. *Planta* **120,** 1–12.
437. Robertson, R. N. (1960). Ion transport and respiration. *Biol. Rev. Cambridge Philos. Soc.* **38,** 231–264.
438. Robertson, R. N. (1968). "Protons, Electrons, Phosphorylation and Active Transport." Cambridge Univ. Press, London and New York.
439. Robinson, J. B. (1969). Sulphate influx in Characean cells. I. General characteristics. *J. Exp. Bot.* **20,** 201–211.
440. Robinson, J. B. (1969). Sulphate influx in Characean cells. II. Links with light and metabolism in *Chara australis. J. Exp. Bot.* **20,** 212–220.
441. Roblin, G. (1979). *Mimosa pudica:* A model for the study of the excitability in plants. *Biol. Rev. Cambridge Philos. Soc.* **54,** 135–154.
442. Roomans, G. M., Blasco, F., and Borst-Pauwels, G. W. F. H. (1977). Cotransport of phosphate and sodium by yeast. *Biochim. Biophys. Acta* **467,** 65–71.
443. Rorison, I. H. (1965). The effect of aluminium on the uptake and incorporation of phosphate by excised sainfoin roots. *New Phytol.* **64,** 23–27.
444. Rothstein, A. (1954). Enzyme systems of the cell surface involved in the uptake of sugars by yeast. *Symp. Soc. Exp. Biol.* **8,** 165–201.
445. Rowell, D. L., Martin, M. W., and Nye, P. H. (1967). The measurement and mechanism of ion diffusion in soils. III. The effect of moisture content and soil-solution concentration on the self-diffusion of ions in soil. *J. Soil Sci.* **18,** 204–222.
446. Sabinin, D. A. (1925). On the root system as an osmotic apparatus. *Bull. Inst. Rech. Biol. Perm.* **4,** Suppl. 2, 129–136.
447. Saddler, H. D. W. (1970). The membrane potential of *Acetabularia mediterranea. J. Gen. Physiol.* **55,** 803–821.

448. Saddler, H. D. W. (1970). The ionic relations of *Acetabularia mediterranea*. *J. Exp. Bot.* **21,** 345–359.
449. Saddler, H. D. W. (1971). Spontaneous and induced changes in the membrane potential and resistance of *Acetabularia mediterranea*. *J. Membr. Biol.* **5,** 250–260.
450. Sanders, D. (1980). Control of Cl^- influx in *Chara* by cytoplasmic concentration. *J. Membr. Biol.* **52,** 51–60.
451. Sanders, D. (1980). The mechanisms of Cl^- transport at the plasma membrane of *Chara corallina*. 1. Cotransport with H^+. *J. Membr. Biol.* **53,** 129–141.
452. Sanders, D. (1980). Control of plasma membrane Cl^- fluxes in *Chara corallina* by external Cl^- and light. *J. Exp. Bot.* **31,** 105–118.
453. Saubermann, A. J., and Echlin, P. (1975). The preparation, examination and analysis of frozen hydrated tissue sections by scanning transmission electron microscopy and autoradiography. *J. Microsc. (Oxford)* **105,** 155–191.
454. Saunders, F. E. T., and Tinker, P. B. H. (1971). Mechanism of absorption of phosphate from soil by *Endogene* mycorrhiza. *Nature (London)* **233,** 278–279.
455. Scarborough, G. A. (1975). Isolation and characterization of *Neurospora crassa* plasma membranes. *J. Biol. Chem.* **250,** 1106–1111.
456. Scarborough, G. A. (1976). *Neurospora* plasma membrane ATPase is an electrogenic pump. *Proc. Natl. Acad. Sci. U.S.A.* **73,** 1486–1488.
457. Scarborough, G. A. (1977). Properties of the *Neurospora crassa* membrane ATPase. *Arch. Biochem. Biophys.* **180,** 384–393.
458. Schaefer, N., Wildes, R. A., and Pitman, M. G. (1975). Inhibition by p-flurophenylalanine of protein synthesis and of ion transport across the roots in barley seedlings. *Aust. J. Plant Physiol.* **2,** 61–73.
459. Schmidt, R., and Poole, R. J. (1980). Isolation of protoplasts and vacuoles from storage tissue of red beet. *Plant Physiol.* **66,** 25–28.
460. Schöch, E. V., and Lüttge, U. (1974). Zur Enstehung einer Lichteabhängigen Komponente der Kationaufhalme bei Blattgewebestreifen mit zunehmenden Zeitabstand von der Präparation. *Biochem. Physiol. Pflanz.* **165,** 345–350.
461. Schwencke, J. (1977). Characteristics and integration of yeast vacuole with cellular functions. *Physiol. Vég.* **15,** 491–517.
462. Seaston, A., Carr, G., and Eddy, A. A. (1976). The concentration of glycine by preparations of the yeast *Saccharomyces carlsbergensis* depleted by adenosine triphosphate. Effects of proton gradients and uncoupling agents. *Biochem. J.* **154,** 669–676.
463. Seaston, A., Inkson, C., and Eddy, A. A. (1973). The absorption of protons with specific amino acids and carbohydrates by years. *Biochem. J.* **134,** 1031–1043.
464. Segel, I. H., and Johnson, M. J. (1961). Accumulation of intracellular inorganic sulfate by *Penicillium chrysogenum*. *J. Bacteriol.* **81,** 91–106.
465. Semenza, G. (1974). The transport system for sugars in the small intestine. Reconstitution of one of them. *In* "Drugs and Transport Processes" (B. A. Callingham, ed.), pp. 317–328. Macmillan, London.
466. Shamoo, A. E., and Goldstein, D. A. (1977). Isolation of ionophores from ion transport systems and their role in energy transduction. *Biochim. Biophys. Acta* **472,** 13–153.
467. Shamoo, A. E., and Tirol, W. F. (1980). Criteria for the reconstitution of ion transport systems. *Curr. Top. Membr. Transp.* **14,** 57–126.
468. Shone, M. G. T. (1966). The initial uptake of ions by barley roots. II. Application of measurements on sorption of anions to elucidate the structure of the free space. *J. Exp. Bot.* **17,** 89–95.
469. Shone, M. G. T. (1968). Electrochemical relations in the transfer of ions to the xylem sap of maize roots. *J. Exp. Bot.* **19,** 468–485.

470. Shone, M. G. T. (1969). Origins of the electrical potential difference between the xylem sap of maize roots and the external solution. *J. Exp. Bot.* **20,** 698–716.
471. Shone, M. G. T., and Barber, D. A. (1966). The initial uptake of ions by barley roots. I. Uptake of anions. *J. Exp. Bot.* **17,** 78–88.
472. Siegenthaler, P. A., Belsky, M. M., and Goldstein, S. (1966). Phosphate uptake in an obligately marine fungus: A specific requirement for sodium. *Science* **155,** 93–94.
473. Siegenthalker, P. A., Belsky, M. M., Goldstein, S., and Menna, M. (1967). Phosphate uptake in an obligately marine fungus. II. Role of culture conditions, energy sources and inhibitors. *J. Bacteriol.* **93,** 1281–1288.
474. Sikes, C. S., Roer, R. D., and Wilbur, K. M. (1980). Photosynthetic and coccolite formation: Inorganic carbon sources and net inorganic reaction of deposition. *Limnol. Oceanogr.* **25,** 248–261.
475. Simkiss, K. (1964). Phosphates as crystal poisons of calcification. *Biol. Rev. Cambridge Philos. Soc.* **39,** 487–505.
476. Simkiss, K. (1976). Intracellular and extracellular solutes in biomineralization. *Symp. Soc. Exp. Biol.* **30,** 423–444.
477. Simon, E. W. (1978). The symptoms of calcium deficiency in plants. *New Phytol.* **80,** 1–15.
478. Simonis, W., and Urbach, W. (1963). Uber eine Wirkung von Natriumionen auf die Phosphataufnahme und die licht-abängige Phosphoryleinung von *Ankistrodesmus braunii. Arch. Mikrobiol.* **46,** 265–286.
479. Skou, J. C. (1975). The ($Na^+ + K^+$) activated enzyme system and its relation to transport of sodium and potassium. *Q. Rev. Biophys.* **7,** 401–434.
480. Slayman, C. L. (1965). Electrical properties of *Neurospora crassa.* Respiration and the intracellular potential. *J. Gen. Physiol.* **49,** 93–116.
481. Slayman, C. L. (1970). Movement of ions and electrogenesis in microorganism. *Am. Zool.* **10,** 377–392.
482. Slayman, C. L. (1974). Contribution to discussion. *In* "Membrane Transport in Plants" (U. Zimmermann and J. Dainty, eds.), p. 348. Springer-Verlag, Berlin and New York.
483. Slayman, C. L. (1977). Energetics and control of transport in *Neurospora. In* "Water Relations in Membrane Transport in Plants and Animals" (A. M. Jungreis, T. K. Hodges, A. Kleinzeller, and S. G. Schultz, eds.), pp. 69–86. Academic Press, New York.
484. Slayman, C. L., Long, W. S., and Lu, C. Y.-H. (1973). The relationship between ATP and an electrogenic pump in the plasma membrane of *Neurospora crassa. J. Membr. Biol.* **14,** 305–338.
485. Slayman, C. L., Lu, C. Y.-H., and Shane, L. (1970). Correlated changes in membrane potential and ATP concentrations in *Neurospora. Nature (London)* **226,** 274–276.
486. Slayman, C. L., and Slayman, C. W. (1968). Net uptake of potassium in *Neurospora.* Exchange for sodium and hydrogen ions. *J. Gen. Physiol.* **52,** 424–443.
487. Slayman, C. L., and Slayman, C. W. (1974). Depolarisation of the plasma membrane of *Neurospora* during active transport of glucose: Evidence for a proton-dependent cotransport system. *Proc. Natl. Acad. Sci. U.S.A.* **71,** 1935–1939.
488. Slayman, C. W. (1973). Genetic control of membrane transport. *Curr. Top. Membr. Transp.* **4,** 1–174.
489. Slayman, C. W., Rees, D. C., Orchard, P. P., and Slayman, C. D. (1974). Generation of adenosine triphosphate in cytochrome-deficient mutants in *Neurospora. J. Biol. Chem.* **250,** 396–408.
490. Slayman, C. W., and Slayman, C. L. (1970). Potassium transport in *Neurospora.* Evidence for a multisite carrier at high pH. *J. Gen. Physiol.* **55,** 758–786.

491. Slayman, C. W., and Tatum, E. L. (1965). Potassium transport in *Neurospora.* III. Isolation of a transport mutant. *Biochim. Biophys. Acta* **109,** 184–193.
492. Smith, F. A. (1967). The control of Na uptake into *Nitella translucens. J. Exp. Bot.* **18,** 716–731.
493. Smith, F. A. (1968). Rates of photosynthesis in Characean cells. II. Photosynthetic $^{14}CO_2$ fixation and ^{14}C-bicarbonate uptake by Characean cells. *J. Exp. Bot.* **19,** 207–217.
494. Smith, F. A. (1970). The mechanisms of chloride transport in Characean cells. *New Phytol.* **69,** 903–917.
495. Smith, F. A., Raven, J. A., and Jayasuriya, H. D. (1978). Uptake of methylammonium ions by *Hydrodictyon africanum. J. Exp. Bot.* **29,** 121–134.
496. Smith, F. A., and Walker, N. A. (1976). Chloride transport in *Chara corallina* and the electrochemical potential difference for hydrogen ions. *J. Exp. Bot.* **27,** 451–459.
497. Smith, F. A., and Walker, N. A. (1978). Entry of methylammonium and ammonium into *Chara* internodal cells. *J. Exp. Bot.* **29,** 107–120.
498. Smithers, A. G., and Sutcliffe, J. F. (1967). The influence of external concentration on sodium ion absorption by carrot root tissue. *J. Exp. Bot.* **18,** 752–757.
499. Smithers, A. G., and Sutcliffe, J. F. (1976). Absorption of potassium and sodium ions by carrot root tissue cultures. *Ann. Bot. (London)* [N.S.] **31,** 713–723.
500. Spanswick, R. M. (1970). The effects of bicarbonate ions and external pH on the membrane potential and resistance of *Nitella translucens. J. Membr. Biol.* **2,** 59–70.
501. Spanswick, R. M. (1972). Electrical coupling between cells of higher plants: A direct demonstration of intercellular communication. *Planta* **102,** 215–227.
502. Spanswick, R. M. (1972). Evidence for an electrogenic ion pump in *Nitella translucens.* I. The effects of pH, K^+, Na^+, light and temperature on membrane potential and resistance. *Biochim. Biophys. Acta* **332,** 387–398.
503. Spanswick, R. M. (1974). Evidence for an electrogenic ion pump in *Nitella translucens.* II. Control of the light-stimulated component of the membrane potential. *Biochim. Biophys. Acta* **332,** 387–398.
504. Spanswick, R. M. (1973). Electrogenesis in photosynthetic tissue. *In* "Ion Transport in Plants" (W. P. Anderson, ed.), pp. 113–128. Academic Press, New York.
505. Spanswick, R. M. (1974). Hydrogen ion transport in giant algal cells. *Can. J. Bot.* **52,** 1029–1034.
506. Spanswick, R. M. (1976). Symplastic transport in tissue. *In* "Encyclopedia of Plant Physiology, New Series" (U. Lüttge and M. G. Pitman, eds.), Vol. 2, Part B, pp. 35–53. Springer-Verlag, Berlin and New York.
507. Spanswick, R. M., and Miller, A. G. (1977). Measurement of the cytoplasmic pH in *Nitella translucens.* Comparison of values obtained by microelectrode and weak acid methods. *Plant Physiol.* **59,** 664–666.
508. Spanswick, R. M., and Williams, E. J. (1964). Electrical potentials and Na, K and Cl concentrations in the vacuole and cytoplasm of *Nitella translucens. J. Exp. Bot.* **15,** 193–200.
509. Spanswick, R. M., and Williams, E. J. (1965). Ca fluxes and membrane potentials in *Nitella translucens. J. Exp. Bot.* **16,** 463–474.
510. Spear, D. G., Barr, J. K., and Barr, C. E. (1969). Localisation of hydrogen ion and chloride ion fluxes in *Nitella. J. Gen. Physiol.* **54,** 397–414.
511. Splittstoesser, W. G., and Beevers, H. (1964). Acids in storage tissues. Effects of salts and aging. *Plant Physiol.* **39,** 163–169.
512. Steeman-Nielsen, E. (1947). Photosynthesis of aquatic plants with special reference to carbon sources. *Dan. Bot. Ark.* **12,** 1–71.

513. Steeman-Nielsen, E. (1960). Uptake of CO_2 by the plant. *In* "Encyclopedia of Plant Physiology" (W. Ruhland, ed.), Vol. 5, Part 1, pp. 70–84. Springer-Verlag, Berlin and New York.
514. Steudle, E., and Zimmermann, U. (1971). Hydraulische leitfahigkeit von *Valonia utricularis. Z. Naturforsch., B: Anorg. Chem., Org. Chem., Biochem., Biophys., Biol.* **26,** 1303–1311.
515. Steveninck, R. F. M. van (1964). A comparison of chloride and potassium fluxes in red beet tissue. *Physiol. Plant.* **17,** 757–770.
516. Steward, F. C. (1968). "Growth and Organization in Plants," Chapter 6, pp. 268, 269. Addison-Wesley, Reading, Massachusetts.
517. Steward, F. C., and Mott, R. L. (1970). Cells, solutes and growth: Salt accumulation in plants re-examined. *Int. Rev. Cytol.* **28,** 276–370.
518. Steward, F. C., and Sutcliffe, J. F. (1959). Plants in relation to inorganic salts. *In* "Plant Physiology: A Treatise" (F. C. Steward, ed.), Vol. 2, pp. 253–478. Academic Press, New York.
519. Street, H. E., and Sheat, D. E. G. (1958). The absorption and availability of nitrate and ammonia. *In* "Encyclopedia of Plant Physiology" (W. Ruhland, ed.), Vol. 8, pp. 150–165. Springer-Verlag, Berlin and New York.
520. Strobel, G. A. (1974). Phytotoxins produced by plant pathogens. *Annu. Rev. Plant Physiol.* **25,** 541–566.
521. Strobel, G. A. (1976). Toxins of plant pathogenic bacteria and fungi. *In* "Biochemical Aspects of Plant Parasite Relationships" (J. Friend and D. R. Threlfall, eds.), pp. 135–139. Academic Press, New York.
522. Sullivan, C. W. (1976). Diatom mineralisation of silicic acid. I. Si $(OH)_4$ transport characteristics in *Navicula pelliculosa. J. Phycol.* **12,** 390–396.
523. Sutcliffe, J. F. (1957). The selective uptake of alkali cations by red beetroot tissue. *J. Exp. Bot.* **8,** 36–49.
524. Sutcliffe, J. F. (1962). "Mineral Salt Absorption in Plants." Pergamon, Oxford.
525. Sutcliffe, J. F. (1973). The role of protein synthesis in ion transport. *In* "Ion Transport in Plants" (W. P. Anderson, ed.), pp. 399–406. Academic Press, New York.
526. Sutcliffe, J. F., and Counter, E. R. (1959). Absorption of alkali cations by plant tissue cultures. *Nature (London)* **183,** 1513–1514.
527. Sutton, C. D., and Gunary, D. (1969). Phosphate equilibrium in soil. *In* "Ecological Aspects of the Mineral Nutrition of Plants" (I. H. Rorison, ed.), pp. 127–134. Blackwell, Oxford.
528. Szabo, G., Eisenman, G., and Ciani, S. (1969). The effects of the macrotetralide actin antiobiotics on the electrical properties of phospholipid bilayer membranes. *J. Membr. Biol.* **1,** 346–382.
529. Theuvenet, A. P. R., and Bindels, R. J. M. (1980). An investigation into the feasibility of using yeast protoplasts to study the ion transport properties of the plasma membrane. *Biochim. Biophys. Acta* **599,** 587–595.
530. Theuvenet, A. P. R., and Borst-Pauwels, G. W. F. H. (1976). Surface charge and the kinetics of two-site mediated ion-translocation. The effects of UO_2^{2+} and La^{2+} upon the Rb^+-uptake into yeast cells. *Bioelectrochem. Bioenerg.* **3,** 230–240.
531. Thoiron, A., Thoiron, B., LeGuiel, J., Guern, J., and Thellier, M. (1974). A shock effect on the permeability to sulphate of *Acer pseudoplatanus* cell-suspension cultures. *In* "Membrane Transport in Plants" (U. Zimmermann and J. Dainty, eds.), pp. 234–238. Springer-Verlag, Berlin and New York.
532. Thoiron, B., Thoiron, A., Espejo, J., LeGuiel, J., Lüttge, U., and Thellier, M. (1980). The effects of temperature and inhibitors of protein synthesis on the recovery from gas-shock of *Acer pseudoplatanus* cell cultures. *Physiol. Plant.* **48,** 161–167.

533. Thoiron, B., Thoiron, A., LeGuiel, J., Lüttge, U., and Thellier, M. (1979). Solute uptake by *Acer pseudoplatanus* cell suspensions during recovery from gas shock. *Physiol. Plant.* **46,** 352–356.
534. Thornton, J. D., Galpin, M. F. D., and Jennings, D. H. (1976). The bursting of the hyphal tips of *Dendryphiella salina* and the relevance of the phenomenon to repression of glucose transport. *J. Gen. Microbiol.* **96,** 145–153.
535. Tien, H. T. (1974). "Bilayer Lipid Membranes (BLM) Theory and Practice." Dekker, New York.
536. Tiffin, L. O., and Brown, J. C. (1961). Selective absorption of iron from ion chelates by soybean plants. *Plant Physiol.* **36,** 710–715.
537. Tiffin, L. O., Brown, J. C., and Krauss, R. W. (1960). Differential absorption of metal chelate components by plant roots. *Plant Physiol.* **35,** 362–367.
538. Ting, I. P. (1968). CO_2 metabolism in corn roots. III. Inhibition of P-enolpyruvate carboxylase by L-malate. *Plant Physiol.* **43,** 1919–1924.
539. Ting, I. P. (1971). Nonautotrophic CO_2 fixation. *In* "Photorespiration and Photosynthesis" (M. D. Hatch, C. B. Osmond, and R. O. Slatyer, eds.), pp. 169–185. Wiley (Interscience), New York.
540. Ting, I. P., and Duggar, W. M., Jr. (1966). CO_2 fixation in *Opuntia* roots. *Plant Physiol.* **41,** 500–505.
541. Tinker, P. B. H. (1975). Effects of vesicular-arbuscular mycorrhizas on higher plants. *Symp. Soc. Exp. Biol.* **29,** 325–349.
542. Tipton, C. L., Mondal, M. H., and Uhlig, J. (1973). Inhibition of the K^+-stimulated ATPase of maize root microsomes by *Helminthosporium maydis* race T pathotoxin. *Biochem. Biophys. Res. Commun.* **51,** 725–728.
543. Torii, J., and Laties, G. G. (1966). Organic acid synthesis in response to excess cation absorption in vacuolate and non-vacuolate sections of corn and barley roots. *Plant Cell Physiol.* **7,** 395–405.
544. Tsao, M. U., and Madley, T. I. (1975). Regulation of glycolysis in *Neurospora crassa.* Kinetic properties of pyruvate kinase. *Microbios* **12,** 125–142.
545. Tukey, H. B., Jr. (1970). The leaching of substances from plants. *Annu. Rev. Plant Physiol.* **21,** 305–324.
546. Tweedie, J. W., and Segel, I. H. (1970). Specificity of transport processes for sulphur, selenium and molybdenum anions by filamentous fungi. *Biochim. Biophys. Acta* **196,** 95–106.
547. Tyree, M. T. (1970). The symplast concept. A general theory of symplastic transport according to the thermodynamics of inversible processes. *J. Theor. Biol.* **26,** 181–214.
548. Tyree, M. T., Fischer, R. A., and Dainty, J. (1974). A quantitative investigation of symplastic transport in *Chara corallina.* II. The symplastic transport of chloride. *Can. J. Bot.* **52,** 1325–1334.
549. Ullrich-Eberius, C. I. (1973). Die pH-Abhangigkeit der Aufnahme von $H_2 PO_4^-$, SO_4^{2-}, Na^+ und K^+ und ihre gegenseitige Beeninflussung bei *Ankistrodesmus braunii. Planta* **109,** 161–176.
550. Ullrich-Eberius, C. I., Novacky, A., and Lüttge, U. (1978). Active transport in *Lemna gibba* G1. *Planta* **139,** 149–153.
551. Ullrich-Eberius, C. I., and Simonis, W. (1971). Der Einfluss von Natrium und Kaliumionen auf die Phosphataufnahme de *Ankistrodesmus braunii. Planta* **93,** 214–226.
552. Ullrich-Eberius, C. I., and Yingchol, Y. (1974). Phosphate uptake and its pH-dependence in halophytic and glycophitic algae and higher plants. *Oceologia* **17,** 17–26.
553. Vaidyanathan, L. V., Drew, M. C., and Nye, P. H. (1968). The measurement and mechanism of ion diffusion in soils. IV. The concentration dependence of diffusion

coefficients of potassium in soils at a range of moisture levels and a method for estimation of the differential diffusion coefficient at any concentration. *J. Soil Sci.* **19,** 94–107.

554. Vaidyanathan, L. V., and Nye, P. H. (1966). The measurement and mechanism of ion diffusion in soils. II. An exchange resin paper method for measurement of the diffusive flux and diffusion coefficient of nutrient ions in soils. *J. Soil Sci.* **17,** 175–183.
555. Vaidyanathan, L. V., and Nye, P. H. (1971). The measurement and mechanism of ion diffusion in soils. VII. Counter diffusion of soil phosphate against chloride in a moisture-saturated soil. *J. Soil Sci.* **22,** 94–100.
556. Viets, F. G. (1944). Calcium and other polyvalent cations as accelerators of ion accumulation by excised barley roots. *Plant Physiol.* **19,** 466–480.
557. Wagner, E. (1977). Molecular basis of physiological rhythms. *Symp. Soc. Exp. Biol.* **31,** 33–72.
558. Wagner, G. J., and Siegelman, H. W. (1975). Large-scale isolation of intact vacuoles and isolation of chloroplasts from protoplasts of mature plant tissue. *Science* **190,** 1298–1299.
559. Waldhauser, J., and Komor, E. (1978). Sucrose transport by seedlings of *Ricinus communis* L.: The export of sucrose from the cotyledons to the hypocotyl as a function of sucrose concentration in the cotyledons. *Plant, Cell Environ.* **1,** 45–50.
560. Walker, D. A. (1962). Pyruvate carboxylation and plant metabolism. *Biol. Rev. Cambridge Philos. Soc.* **37,** 215–256.
561. Walker, D. A. (1976). Plastids and intracellular transport. *In* "Encyclopedia of Plant Physiology, New Series" (C. R. Stocking and U. Heber, eds.), Vol. 3, pp. 83–136. Springer-Verlag, Berlin and New York.
562. Walker, N. A. (1955). Microelectrode experiments on *Nitella. Aust. J. Biol. Sci.* **8,** 476–489.
563. Walker, N. A. (1962). Effect of light on the plasmalemma of *Chara* cells. *Annu. Rep. Div. Plant. Ind. C.S.I.R.O.*, p. 80.
564. Walker, N. A. (1974). Chloride transport to the charophyte vacuole. *In* "Membrane Transport in Plants" (U. Zimmermann and J. Dainty, eds.), pp. 173–179. Springer-Verlag, Berlin and New York.
565. Walker, N. A. (1976). Membrane transport: Theoretical background. *In* "Encyclopedia of Plant Physiology, New Series" (U. Lüttge and M. G. Pitman, eds.), Vol. 2, Part A, pp. 36–52. Springer-Verlag, Berlin and New York.
566. Walker, N. A., Beilby, M. J., and Smith, F. A. (1979). Amine uniport at the plasmalemma of charophyte cells. 1. Current-voltage curves, saturation kinetics and effects of unstirred layers. *J. Membr. Biol.* **49,** 283–296.
567. Walker, N. A., and Bostrom, T. E. (1973). Intercellular movement of chloride in *Chara*—a test of models for chloride influx. *In* "Ion Transport in Plants" (W. P. Anderson, ed.), pp. 447–458. Academic Press, New York.
568. Walker, N. A., and Pitman, M. G. (1976). Measurement of fluxes across membranes. *In* "Encyclopedia of Plant Physiology, New Series" (U. Lüttge and M. G. Pitman, eds.), Vol. 2, Part A, pp. 93–126. Springer-Verlag, Berlin and New York.
569. Walker, N. A., and Smith, F. A. (1975). Intracellular pH in *Chara corallina* measured by DMO distribution. *Plant Sci. Lett.* **4,** 125–132.
570. Walker, N. A., and Smith, F. A. (1977). Circulating electric currents between acid and alkaline zones associated with HCO_3^- assimilation in *Chara. J. Exp. Bot.* **28,** 1190–1207.
571. Walker, N. A., Smith, F. A., and Beilby, M. J. (1979). Amine uniport at the plasmalemma of charophyte cells. II. Ratio of matter to change transported and permeability to free base. *J. Membr. Biol.* **49,** 283.

572. Wall, V. R., and Andrus, C. F. (1962). The inheritance and physiology of boron response in the tomato. *Am. J. Bot.* **49,** 758–762.
573. Wann, E. V., and Hills, W. A. (1973). The genetics of boron and iron transport in the tomato. *J. Hered.* **64,** 370–371.
574. Welch, R. M., and Epstein, E. (1968). The dual mechanisms of alkali cation absorption by plant cells: Their parallel operation across the plasmalemma. *Proc. Natl. Acad. Sci. U.S.A.* **61,** 447–453.
575. Williams, E. J., and Fensom, D. S. (1975). Axial and transnodal movement of ^{14}C, ^{22}Na and ^{36}Cl in *Nitella translucens. J. Exp. Bot.* **26,** 783–807.
576. Wittwer, S. H., and Bukovac, M. J. (1969). The uptake of nutrients through leaf surfaces. *In* "Handbuch der Pflanzenernaehrung und Duengung", (H. Linser, ed.), Vol. I, pp. 235–261. Springer-Verlag, Berlin and New York.
577. Wolfinberger, L. (1980). Transport and utilization of amino acids by fungi. *In* "Micro-organisms and Nitrogen Sources" (J. W. Payne, ed.), pp. 63–87. Wiley, Chichester.
578. Wyatt, H. V. (1964). Cations, enzymes and control of cell metabolism. *J. Theor. Biol.* **6,** 441–470.
579. Zimmermann, H. (1977). Cell turgor pressure regulation and turgor pressure mediated transport processes. *Symp. Soc. Exp. Biol.* **31,** 117–154.
580. Zimmermann, H., Raede, H., and Steudle, E. (1969). Kontinuierliche Druckmessung in Pflanzensellen. *Naturwissenschaften* **56,** 634–635.
581. Zimmermann, U., and Steudle, E. (1974). Hydraulic conductivity and volumetric elastic modulus in giant algal cells: Pressure and volume dependence. *In* "Membrane Transport in Plants" (U. Zimmermann and J. Dainty, eds.), pp. 64–71. Springer-Verlag, Berlin and New York.
582. Zimmermann, U., and Steudle, E. (1975). The hydraulic conductivity and volumetric elastic modulus of cells and isolated cell walls of *Nitella* and *Chara* spp.: Pressure and volume effects. *Aust. J. Plant Physiol.* **2,** 1–12.

CHAPTER FIVE

Salt Relations of Intact Plants

J. F. SUTCLIFFE

I. Introduction

"The study of mineral nutrition now enjoys much less prestige than it did formerly, but the need to understand the whole process and its interaction with growth remains as great as it ever was, particularly in view of the demands of an increasing, hungry, human population" [Clarkson and Hanson (52)]. If plant physiological research is to make a useful contribution to agricultural and horticultural science, it must be applicable to crops grown in the field. Because of the structural complexity of most important food plants, many investigators have turned their attention to less complicated systems, including excised roots, tissue slices, stem sections, cell

Plant Physiology
A Treatise
Vol. IX: Water and Solutes in Plants

cultures, microorganisms, and subcellular preparations. Such studies have yielded much information about ion transport, as the preceding chapters show, but our understanding of the uptake and distribution of mineral elements in intact angiosperms remains limited perhaps because the whole organism is more than just the sum of its parts. As an example, the circulation of such elements as sulfur and phosphorus, described by Biddulph and collaborators (29, 31, 32) when radioactive isotopes became available, could not have been deduced from studies of the salt relations of separated plant parts. In this chapter an attempt is made to assess the advances that have been made over the past 25 years in our understanding of the salt relations of intact plants.

A. THE POSITION IN 1959

At the outset, a summary of the position as it appeared when Chapter 4 in Volume II of this treatise (30) was written sets the scene for what is to follow.

A young seedling depends initially on reserves of mineral elements, previously stored in the seed, which are released during the course of germination and early seedling establishment. As the root system makes contact with the surrounding medium and endogenous supplies become depleted, an increasing proportion of the mineral nutrients required for growth originates from exogenous sources.

When a plant is rooted in soil, uptake occurs mainly in ionic form from the soil solution. The solid phase provides a reservoir of mineral elements from which the soil solution is replenished. There was some evidence (134) that plant roots might absorb a part of their cation supply directly from the solid phase ("contact exchange"). It was realized that bulk flow of water through the soil to supply the transpiration stream might bring solutes to the root surface from an appreciable distance and that this could be important in relation to those elements such as phosphorus and nitrogen and micronutrients which are present at low concentrations in many soil solutions. The role of mycorrhizal fungi in facilitating the uptake of certain ions, notably phosphate, was also recognized (111, 145, 219).

Uptake of ions by root cells occurs from the cortical free space, i.e., the water-filled spaces in the cell walls of the cortex which are continuous with films of water on the surface of the root and adjoining soil particles. In this way, the available surface area for absorption was thought to be increased appreciably, perhaps by 100 times. The cortical free space was assumed to be delimited on the inner side of the root by the endodermis, with its

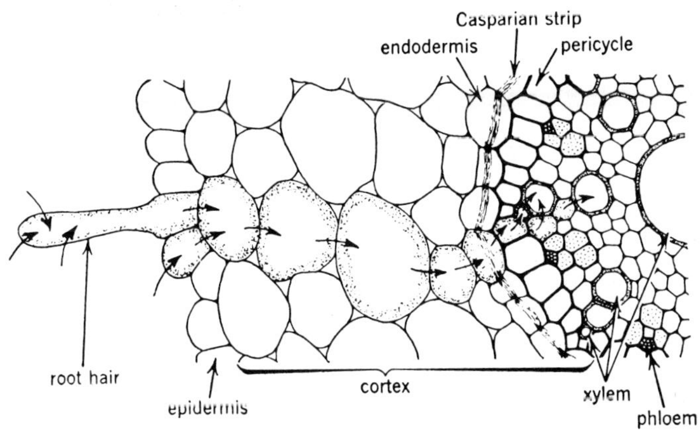

FIG. 1. Diagram of the tissues of a wheat root seen in transverse section illustrating the likely path for movement of water and salts from soil to xylem (×330). From Esau (78a).

lipid-impregnated radial cell walls (Fig. 1). It was realized that some ions can be absorbed passively in the transpiration stream through breaks in the endodermis (passage cells) and transferred directly to the shoot (7). However, the bulk of mineral salts was thought to be taken up actively by root cells as was indicated by the frequent disparity between the absorption of ions and water (42, 43, 125, 225, 254), the selectivity of ion uptake (55), and the influence of environmental factors on the two processes.

The zone of maximum ion absorption in a root was a contentious issue in 1959. Steward and Sutcliffe (232) concluded that root hairs are of significance only insofar as they increase the surface area of the root that is in intimate contact with soil particles and that uptake occurs mainly in the actively growing region distal to the root hair zone. The differential accumulation of radioactively labeled cations (e.g., Cs^{137}) in root apices had been demonstrated (231). Kramer (143, 144) demonstrated that a distinction could be made between a zone of absorption and accumulation.

After ions have been transported across the plasmamembrane into a cortical cell, they may undergo one of several possible fates. Some, particularly anions such as phosphate, sulfate, and nitrate, are converted to soluble organic substances which are either transported as such, metabolized, or incorporated into structural cell constituents. Others, notably potassium, are accumulated mainly in vacuoles in which they help to maintain the osmotic potential of the vacuolar sap and hence control cell turgidity. A third possibility which was recognized is that ions are transported from cell to cell via plasmodesmata inward toward the stele. Thus two alternative routes for ion movement radially across the cortex were identified—one in the cell walls (apoplast) and another in the cytoplasm (symplast). The relative importance of these two pathways was being discussed in the late 1950s (7), but no reliable quantitative data were available upon which definite conclusions could be based. It was suggested that the bulk of ions transported in the cell walls must be diverted into the symplast before they traversed the endodermis because the apoplastic pathway is blocked there by the Casparian bands.

In young plants with an actively growing root system, a relatively high proportion of the ions taken up by the roots are retained there. This is especially true under conditions of low salt supply (121). Later on, as growth of the roots declines, a higher percentage of absorbed salts is transferred to the shoot and, in addition, some of the ions previously stored in root cell vacuoles may be released.

Once ions, or the soluble organic compounds derived from them (for example, amides in the case of nitrogen), have crossed the endodermis, they may be subject to a number of fates within the stele. Some are accumulated or metabolized in living parenchyma cells, some are transferred into the phloem through which they are presumably transported both downward to the root apex and upward toward the leaves, and some are released into the stelar free space from which they find their way into the xylem. Some investigators, e.g., Crafts and Broyer (62) and Lundegårdh (170), believed that release from the symplast was a passive process associated with lower metabolic activity of cells in the stele. An alternative view, for which there was no direct experimental evidence at the time, was that salt is actively secreted from living cells in the stele into the nonliving xylem elements. The observation that transfer of cations from root to shoot in barley plants was more selective than ion accumulation in the root cortex (237, 238) was taken as support for this view.

Solutes that are released into the conducting elements of the xylem are transported upward by bulk flow in the transpiration stream. Although the linear rate of solute transport in the xylem is closely correlated with the rate of transpiration, it was understood that the *quantity* of solutes transported may be largely independent of the rate of water movement because of an increase in concentration of the xylem sap at low rates of transpiration (223, 225). It was also known that the concentration of xylem sap decreases during its passage through the stem, presumably because of the differential absorption of solutes by the xylem parenchyma and other living cells adjoining the conducting elements.

The rapidity with which ions are transferred from the xylem to surrounding tissues as shown by early experiments with radioactive tracers (9,

235) made it difficult to assess the contribution that the phloem makes to upward transport through the stem. Nevertheless, it seemed likely that an appreciable proportion of the mineral salts imported by young unfolding leaves enters through the phloem. Later, the direction of flow of materials in the sieve tubes reverses as photosynthetic activity develops (see Chapter 6, this volume) and eventually a mature leaf becomes a net exporter of

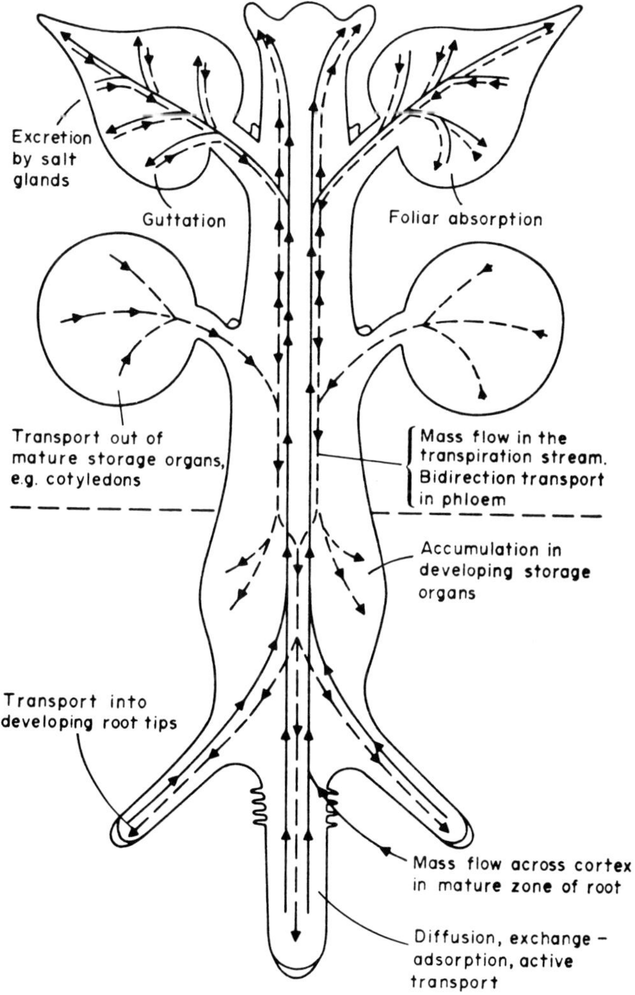

FIG. 2. A general scheme for the uptake and circulation of mineral nutrients in a dicotyledonous plant. After Sutcliffe (240).

some (but not all) mineral elements as well as of assimilates and other organic materials.

As has already been indicated, it was appreciated that there is a circulation of certain elements, notably phosphorus (e.g., 65, 168, 169, 175, 176, 235), between the roots and leaves via the vascular system, at least during the vegetative phase of growth. Elements, such as calcium and silicon, which are relatively immobile in the phloem do not circulate in this way and accumulate in leaves until they fall. Calcium salts are sometimes excreted from leaves passively through modified hydathodes — the so-called chalk glands — whereas active excretion of sodium chloride and other salts occurs through specialized salt glands in halophytes. Some elements, especially potassium, may be leached from leaves by rain.

Secondary distribution of mineral elements within a plant through the phloem was visualized to occur from "sources" to "sinks" (see Chapter 6, this volume). As for assimilates, mature leaves are a major source of inorganic nutrients, and important sinks include shoot and root apices, developing fruits and seeds, and vegetative storage organs such as rhizomes, corms, and bulbs. Individual leaves tend to supply the nearest sinks, but with the onset of the reproductive phase, parts at an increasing distance from a sink, including the roots, may supply previously stored materials to developing fruits in which seeds often become a powerful sink for nutrients during their maturation.

A summary of the major aspects of the salt relations of an intact angiosperm as visualized in 1959 is presented in Fig. 2.

B. Plan of Development

Within the space available, it is impossible to give detailed consideration to any but a few of the many research papers that have been published on salt relations of plants since 1959. Instead, in an attempt to review many of the issues then unresolved, a number of themes are developed; these include the uptake, transport, and redistribution of ions, with an indication of the controlling mechanisms where these are beginning to be understood.

Some topics mentioned in this chapter are also referred to in other chapters in this volume. In fact, Chapters 1, 4, and 7 are especially important in complementing this one. In addition, a large number of books and reviews describe various aspects of the salt relations of intact plants.[1]

[1] Other sources, since 1959, which have dealt with aspects of the salt relations of whole plants may be cited (38, 51, 75, 133, 149, 168, 185, 205, 208, 240).

II. Ion Uptake

A. INFLUENCE OF EXTERNAL CONCENTRATION

The kinetic approach to ion uptake studies, pioneered by Epstein and collaborators (74, 76, 77) mainly with excised barley roots, has aroused renewed interest during the past 20 years in the influence of solute concentration on the rate of uptake of ions by whole plants. This interest has been stimulated further by the development of methods for continuous solution culture, some of which, e.g., the nutrient film technique (NFT), are now finding application in commercial practice (54, 59, 60). Nutrients are supplied at low concentrations, comparable to those occurring in soil solutions, and replenished continuously by maintaining flow rates of sufficient magnitude to ensure that there is no serious depletion of individual elements in the solution. The concentration of some ions can be monitored continuously using ion-specific electrodes, and if the rate of flow is known, uptake can be calculated (54).

Using a continuous culture method, Asher *et al.* (10, 11, 164) studied the effect of nutrient concentration on growth and ion uptake by a number of crop species. A summary of their results from an experiment in which the concentration of potassium ions in the medium was varied from 1 to 1000 μM is given in Table I.[2] They showed that growth expressed as yield

TABLE I

YIELD OF DRY MATTER RELATIVE TO HIGHEST OBTAINED AND RELATIVE POTASSIUM CONTENTS OF 14 SPECIES GROWN IN CONTINUOUS CULTURE WITH DIFFERENT CONCENTRATIONS OF POTASSIUM IN THE EXTERNAL MEDIUM[a]

K^+ Concentration (μM)	Relative yield (%)	K^+ Content (μmol g^{-1} FW)	
		Roots	Shoots
1	7	12	22
8	64	42	110
24	86	66	144
95	93	83	162
1000	97	112	190

[a] Data from Asher and Ozanne (11).

[2] For concentration the recommended SI units are moles per cubic meter; thus 1 to 1000 μM would be 0.001 to 1.0 mol m^{-3}.

TABLE II

YIELD OF DRY MATTER RELATIVE TO HIGHEST OBTAINED AND RELATIVE PHOSPHORUS CONTENTS OF 7 SPECIES GROWN IN CONTINUOUS CULTURE WITH DIFFERENT CONCENTRATIONS OF PHOSPHATE IN THE EXTERNAL MEDIUM[a]

P Concentration (μM)	Relative yield (%)	P Content (μmol g^{-1} FW)	
		Roots	Shoots
0.04	50	3.9	3.8
0.2	76	7.4	10.5
1.0	89	15.0	23.0
5.0	97	20.0	25.0
25.0	98	22.0	27.0

[a] Data from Asher and Loneragan (10).

of dry matter, relative to the maximum obtained for each species, did not increase greatly when the K^+ concentration of the medium was raised above 24 μM, but the potassium content of roots, expressed per gram fresh weight, nearly doubled, and that of the shoots increased by about 30%. At

TABLE III

CONTENT OF CALCIUM IN 16 LEGUMES AND 11 SPECIES OF GRAMINEAE, GROWN IN CONTINUOUS CULTURE WITH DIFFERENT CONCENTRATIONS OF CALCIUM IN THE EXTERNAL MEDIUM[a]

Ca Concentration (μM)	Ca^{2+} Content (μmol g^{-1} FW)			
	Legumes		Graminae	
	Roots	Shoots	Roots	Shoots
0.3	1.1	4.5	0.9	1.7
0.8	1.4	4.4	0.8	1.5
2.5	1.6	4.7	0.7	1.7
10.0	1.7	12.0	0.9	3.3
100.0	2.2	39.0	1.5	10.5
1000.0	3.3	54.0	2.5	18.0

[a] Data from Loneragan and Snowball (165).

every concentration examined, the potassium content of the shoots was about double that of the roots, but the ratio of K^+ content in shoots to roots decreased slightly at the higher concentrations.

A somewhat similar pattern of response was found for phosphorus (Table II); little increase in yield of dry matter occurred when the concentration of phosphate in the solution was increased above 1 μM. There was an increase of about 50% in the phosphorus content per gram fresh weight of the roots over the range 1–25 μM, while the concentration in the shoots over the same range increased by about 20%. There was a smaller difference between the phosphorus content of root and shoot than was observed for potassium. The ability of roots to absorb phosphate efficiently from solutions of very low concentration is well known from earlier studies with both excised roots (121) and intact plants (160, 223). However, the net rate of uptake of phosphorus and potassium at higher concentrations (i.e., above 1 μM for P and 10 μM for K) by the intact plants was largely independent of external supply. In this respect the response differs from that observed in short-term experiments with excised roots when influx continues to increase with concentration up to about 50 μM. High concentrations of mineral salts, especially sodium salts, may cause a reduction in growth of salt-sensitive species; the ionic relations of halophytes are considered in Section VI.

Turning to calcium, Loneragan and Snowball (165, 166) found the content of this element in a range of legumes and grasses to be little affected by calcium concentration over the range 0.3–2.5 μM (Table III). Above this level calcium content increased markedly, especially in the shoots. The legumes accumulated much higher levels of calcium in the shoots than did the grasses, while the concentration in the roots was about the same.

Hoagland and Broyer (121) observed that roots excised from barley plants which had been grown with a minimal supply of mineral nutrients ("low-salt" plant) later absorbed larger amounts of potassium, nitrate, and bromide than did roots of similar plants grown with a generous nutrient supply ("high-salt" plants). Later investigations have confirmed this observation by using a variety of different plant materials and ions. Recently, Lee (160) has studied the influence of limiting the external supply of P, S, Cl, and N on subsequent absorption of these ions by intact barley plants. His results — some of which are presented in Fig. 3 — show that in the case of P, S, and Cl increased absorption was restricted to the nutrient which had been withheld or to a close chemical analog. The uptake of other anions was either unaffected or reduced.

A kinetic analysis of the data (Table IV) indicated that differences in the rates of phosphate, sulfate, and chloride absorption by plants of different

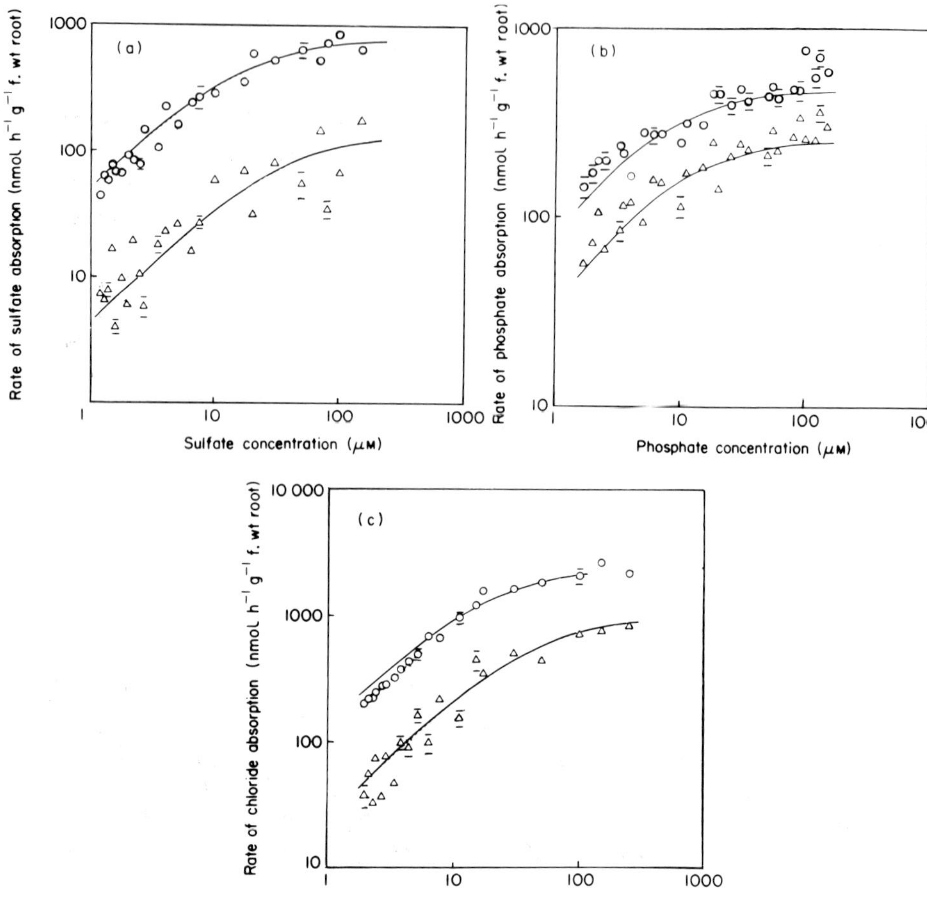

FIG. 3. The rates of absorption of (a) sulfate, (b) phosphate, and (c) chloride (nmol h^{-1} g^{-1} fresh weight of root) by barley plants. Plants were grown in a basal medium containing (△) or without (○) the ion under investigation before being presented with various concentrations of the salts. Redrawn from Lee (160).

nutrient status were principally due to changes in the maximum transport capacity V_{max} for these ions per unit weight of root, although in plants grown without external chloride there was some evidence that the roots also had increased affinity for this ion (i.e., a higher K_m value; see Chapter 4).

TABLE IV

EFFECT OF GROWTH WITHOUT SULFATE, PHOSPHATE, OR CHLORIDE ON THE KINETIC PARAMETERS OF THE SUBSEQUENT ABSORPTION OF LABELED SULFATE, PHOSPHATE, OR CHLORIDE[a]

Nutrition during growth	Labeled ion supplied during measuring period	Parameters for uptake of labeled ion: $K_m \pm$ se (μM)	$V_{max} \pm$ se (nmol g^{-1} FW root h^{-1})
A. With sulfate	Sulfate	13.9 ± 2.5	53.4 ± 6.8
No sulfate	Sulfate	17.6 ± 1.3	758 ± 42
B. With phosphate	Phosphate	6.6 ± 0.6	257 ± 9.2
No phosphate	Phosphate	4.9 ± 0.5	475 ± 17.4
C. With chloride	Chloride	57.4 ± 8.3	1010 ± 119
No chloride	Chloride	23.7 ± 1.6	2600 ± 131

[a] From Lee (160).

B. ZONES OF ABSORPTION AND ACCUMULATION

Ideally a distinction should be made between zones in which absorption may occur and those in which accumulation takes place. Earlier work by Steward and Sutcliffe (232, see refs. 163 and 220 cited therein) had localized in unbranched roots of barley and narcissus the regions of greatest accumulation of anions (Br) and cations (Rb and Cs) and also of their subsequent removal in regions near the root tips.

To determine unequivocally differences in salt uptake by different parts of a root, it is necessary that the whole root system be exposed to the same external concentration and that the zone under examination be supplied, in addition, with radioactive tracer. Apparatus in which it is possible to isolate various parts of the intact root system and supply these separately with salts and water has been used in a number of studies (225, 265).

Results of Russell, Clarkson, and their collaborators (222, 224) have demonstrated that for both dicotyledonous and monocotyledonous species uptake of ions occurs along much of the length of the root (Fig. 4). In the case of *Cucurbita pepo* uptake of potassium and calcium was considerable whatever part of the root was examined. However, in barley *(Hordeum vulgare)*, uptake of calcium decreased with distance from the apex; this may be related to the path of calcium across the cortex to the stele in different parts of the root (see the next section).

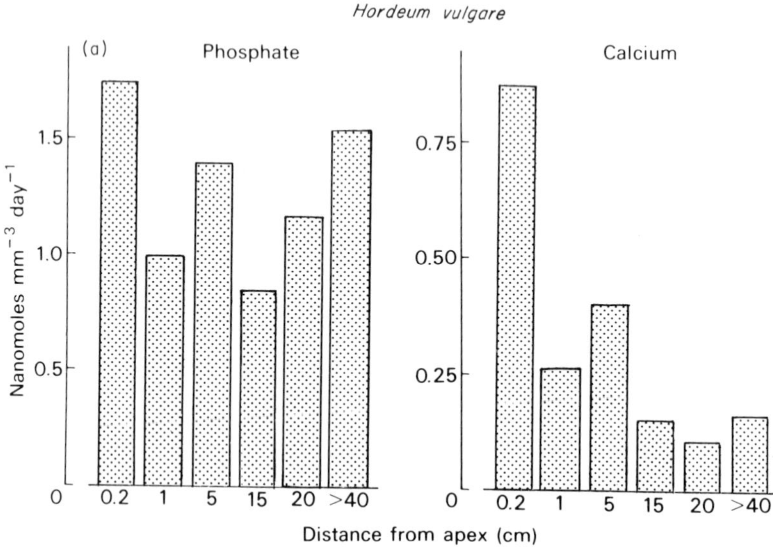

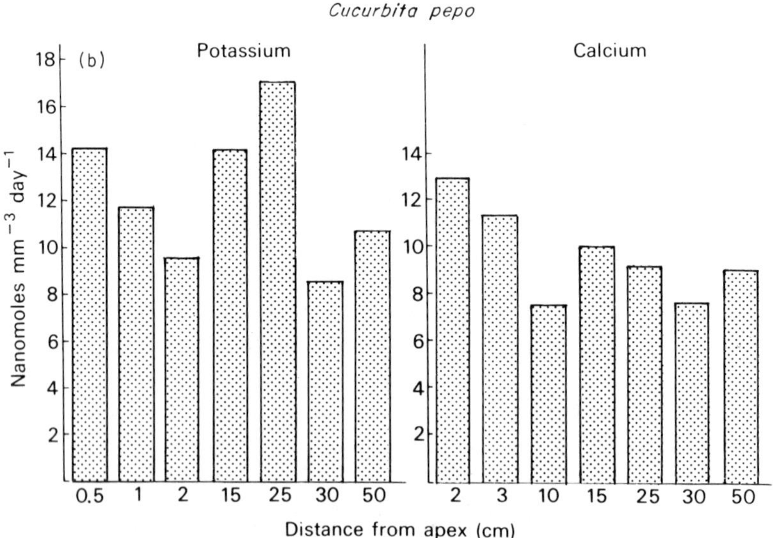

FIG. 4. The rate of absorption of various ions by different parts of (a) seminal roots of barley *(Hordeum vulgare)* and (b) primary roots of squash *(Cucurbita pepo)*. Redrawn from Russell (221).

III. Movement of Ions across the Root

A. THE PATHWAY

The role of the endodermis as a barrier to ion movement in the apoplastic pathway has already been mentioned. In recent years major studies on endodermal development have revealed a number of important points which are now considered.

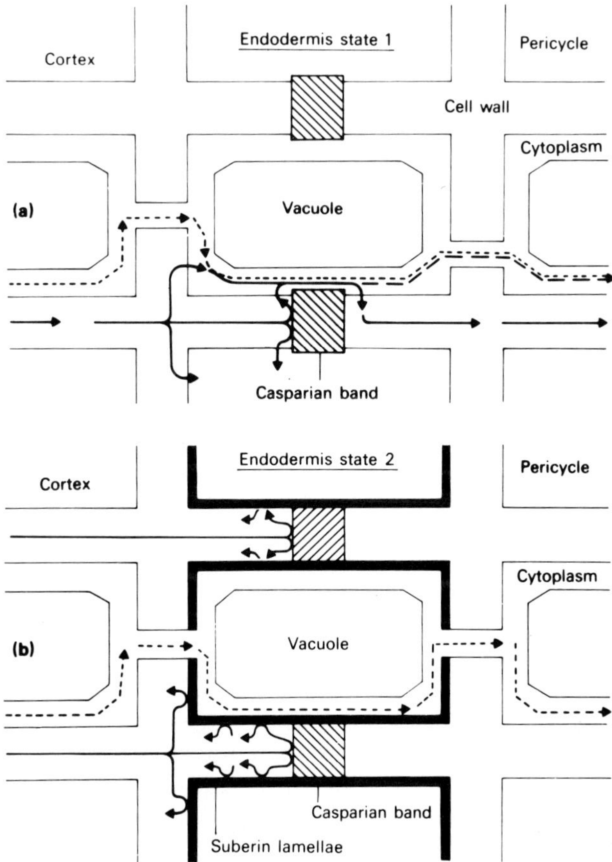

FIG. 5. The pathways for movement of solutes from cortex to stele across an endodermis in state 1 (a) or state 2 condition (b). Note the extensive suberization of the state 2 endodermal cells. Broken arrows indicate symplastic pathways; solid arrows pathways in the apoplast. After Russell and Clarkson (222).

Robards and Jackson (218) identify four stages of endodermal development. Initially, young endodermal cells just cut off from the meristem show no unusual wall features and are referred to as a proendodermis. As development proceeds, cells enter state I in which the Casparian band is formed by a heavy deposition of lignin and suberin material in the radial (anticlinal), longitudinal, and transverse walls. In many species there is little further development of the endodermis, but in grasses suberization of the whole internal surface of the endodermal cell wall occurs (state II), and this may be followed by the deposition of cellulose internal to the suberin (state III).

Since suberin is impermeable to water and ions, this pattern of development is of great significance. Clearly, the proendodermis offers no apparent barrier to ion flow across it in either the apoplastic or symplastic pathway. In state I, however, movement in the apoplast is effectively blocked by the Casparian strip. Symplastic movement through plasmodesmata or across the endodermal plasmalemma is possible. The suberization of state II cells precludes all but plasmodesmatal movement in the symplastic pathway (see Fig. 5). A very high frequency of plasmodesmata has been found for endodermal cells, linking them with each other, and also with cells of the cortex and the pericycle (217, 218). It is reasonable to assume that movement of ions across the symplast will be relatively unaffected by differentiation of the endodermis. Ions moving through the apoplast will, in contrast, be markedly affected, and it has been suggested that the reduced uptake of calcium by proximal portions of barley roots (Fig. 4) is due to a restriction in the apoplastic transport of this element.

B. ROLE OF TRANSPIRATION

The influence of water flux through a plant on the uptake and distribution of mineral elements has been investigated extensively since the 1920s (188). Effects ranging from an exact correlation between water and salt absorption to the absence of any relationship at all have been reported (125–127, 150, 206, 225, 254). Most investigators have observed more rapid ion uptake at higher transpiration rates but have usually failed to establish a quantitative link between the two processes (240). The extent to which water and salt absorption are correlated seems to depend on the species under investigation, the age of the plants, and the conditions under which they are grown, the nature and concentration of the external solutions, and the element in question. The results of experiments published since 1959 support this conclusion.

Jones and Handreck (136) reported that the uptake of silica as undissociated monosilicic acid, H_4SiO_4, by oat plants *(Avena sativa)* is related to the concentration of monosilicic acid in the soil solution and the amount of water transpired. When grown in soils containing 7 and 57 mol^{-1} SiO_2 in solution, the plants at maturity contained 28 and 274 mg SiO_2 per plant, respectively, and each had transpired about 3.9 L of water while producing 7 g of dry matter. The concentration of silica in the xylem sap was found to be similar to that of the external solution, indicating that the roots were exerting little or no influence on the passage of silica in the transpiration stream. It was suggested that measurements of the silicon content of an oat plant can be used to determine the cumulative water loss during its life (183).

On the other hand, the concentration of silica in the xylem sap of the clover, *Trifolium incarnatum,* was invariably lower than in the external solution (110). The silicon content per unit of dry matter in the roots was about eight times that of the corresponding shoots, and it was concluded that there is a mechanism in clover which removes silica from the transpiration stream as it passes through the roots. In contrast, rice *(Oryza sativa)* and bean *(Phaseolus vulgaris)* plants (19) are among those which seem to have the ability to accumulate silica in the xylem sap to higher concentrations than are present in the external solution, as has commonly been observed for other elements. Russell and Shorrocks (225) calculated that the concentration of phosphate in the xylem sap of barley and sunflower plants was many times higher at low than at high rates of transpiration. This reflects the relative rates at which water and dissolved ions are transferred across the root into the xylem and indicates a considerable measure of independence in the two processes.

The uptake of some metabolically important ions, such as K^+, NO_3^-, and $H_2PO_4^-$, appears to be much less affected by the rate of water flow than that of the divalent cations Ca^{2+} and Mg^{2+} (159). This difference may be related to the pathway of transport to the xylem since it has been claimed (53, 112, 218) that whereas K^+ and $H_2PO_4^-$ are transported mainly through the symplast, Ca^{2+} appears to move predominantly in the apoplast, at least as far as the endodermis. For undamaged roots of a number of plant species, it has been calculated that about 1 to 5% of the total flow of solutes bypasses the symplast and moves via the apoplast alone into the xylem (199). Perhaps this fraction is higher in the case of divalent cations and it might be expected to increase linearly with external concentration. This may explain the differing effects of transpiration on uptake of calcium and magnesium by barley plants from solutions containing low and high concentrations of these ions (Fig. 6). In decapitated *Ricinus* plants, the rate of the flow of potassium in the two pathways is dependent on the external concentration (Fig. 7) (18).

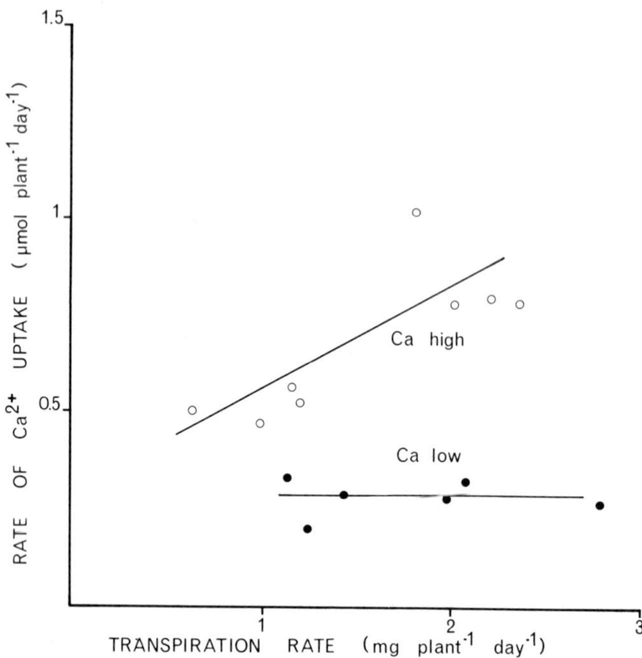

FIG. 6. The relationship between uptake of Ca^{2+} and transpiration rate for barley seedlings exposed to external concentrations of 15 m*M* (○) and 0.5 m*M* (●). Data of Lazaroff and Pitman (159).

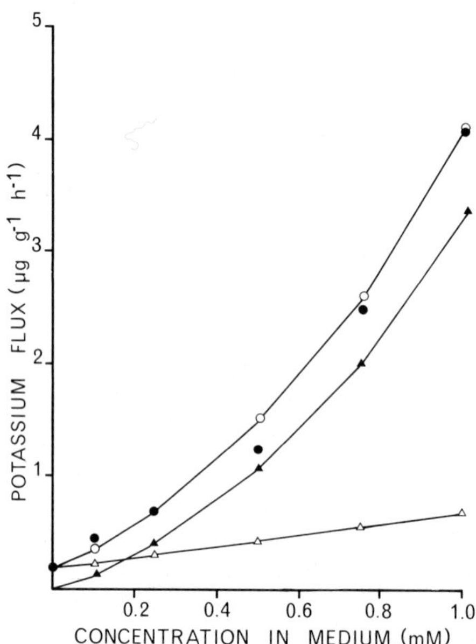

FIG. 7. The relationship between external concentration of potassium and the total flux of potassium (●) in the exudate of a decapitated *Ricinus* plant. The calculated water-dependent (△) and water-independent (▲) fluxes and the sum of the two (○) are also shown. Redrawn from Baker and Weatherley (18).

Christersson and Pettersson (50) found for sunflower plants a correlation between water and calcium uptake at low transpiration rates, and they suggested that calcium ions are transported through the free space of the cortex along an absorption system near the cytoplasmic surfaces. At higher rates of transpiration (and presumably also at high concentrations), this system becomes saturated and the relationship breaks down.

Metabolic inhibitors have often been used in an attempt to distinguish between a passive component of ion uptake which is directly linked to transpiration and an active component which is not so related (e.g., 15, 64, 138, 156). Bengisson (26) studied the relationship between uptake and translocation of ^{45}Ca and water uptake in excised roots and intact cucumber *(Cucumis sativus)* plants. He found that there was no immediate reduction in calcium uptake when water uptake was reduced by excision of the shoot. In the presence of 2,4-dinitrophenol (DNP), calcium uptake was reduced in intact plants, while water uptake remained unchanged. He suggested that regulation of Ca^{2+} uptake occurs primarily in the root and that a DNP-sensitive mechanism is located in the endodermis.

One important effect of transpiration is to regulate the concentration of ions in the vicinity of the inner cortex cells and the endodermis. The greater the rate of solution flow across the cortex, the lower will be the concentration gradient caused through absorption by intervening cells. Bernstein and Nieman (28) pointed out that when a solution is drawn through the cortex and arrives at the endodermis, it will become concentrated if the flux of water through the endodermis exceeds the flux of ions. Weatherley (259–262) has examined the situation mathematically as follows.

The rate of arrival J_s of a solute at the endodermis is given by the equation

$$J_s = J_v C$$

where J_v is the water flux, J_s and J_v are in moles per second, and C is the concentration of solute in the solution in moles per cubic centimeter. If it is assumed that the solute cannot penetrate into the endodermis, then as the concentration there increases, there will be increasing back diffusion of solute into the medium. This is given by $J'_s = Da(dC/L)$, where J'_s is the diffusive flux in moles per second, D is the diffusion coefficient, a is the cross-sectional area of the pathway in square centimeters, L is the length of the pathway between epidermis and endodermis in centimeters, and dC is the concentration difference between the external solution and the endodermis.

At the steady state, back diffusion will be the same as the rate of arrival of solute, i.e., $J_s = J'_s$. Therefore

$$J_v C = Da \frac{dC}{L}$$

or

$$\frac{dC}{C} = \frac{J_v L}{Da}$$

Integrating

$$\int_{C_s}^{C_e} \frac{dC}{C} = \int_{e}^{0} \frac{J_v L}{Da}$$

where C_e and C_s are the concentrations at the endodermis and in the external solution, respectively, gives

$$\ln\left(\frac{C_e}{C_s}\right) = \frac{J_v L}{Da}$$

This formula was applied to data of Brouwer (42, 43) for *Vicia faba* roots. In a particular experiment the rate of absorption of water was 10^{-2} cm^{-3} h^{-1} per centimeter of root length, which gives $J_v = 2.8 \times 10^{-6}$ cm^{-2} s^{-1}. The mean root diameter was 1 mm and the diameter of the stele was about 0.33 mm. It was assumed that the length of the pathway from root surface to endodermis was greater than the diameter of the cortex by a factor of 1.5 because the route taken by water through the cell walls is not straight; L thus becomes 5×10^{-2} cm.

Assuming that the free space of the cortex was 10% of the total volume and that it was filled with water, a would be 0.21 cm^2. Taking these various values and a value for D of 5×10^{-6} cm^2 s^{-1}, the ratio of C_e to C_s is approximately 4. A difference of this magnitude is only likely to be of great significance when roots are in a medium containing very low concentrations of salt.

Transpiration may also have an effect on the rate of transport of ions through the symplast of the root (4). Unfortunately, we do not yet have data on the relative amounts of water (or solutes) moving through the two pathways at different rates of transpiration (see Chapter 2), but there are some indications that the resistance of symplast to solute movement decreases when transpiration increases. Whether this effect is due to acceleration of the movement of ions across membranes, stimulation of bulk flow through the cytoplasm, or reduced resistance of the plasmodesmata, to mention only three possibilities, is not yet clear. The phenomenon may involve the production of streaming potentials and electroosmotic forces. Barry and Hope (22) have discussed the general problem in relation to membrane transport and drawn attention to the complexity of the situation, especially in plant cells in which the cell wall may exert electrokinetic effects (see 41, 253).

A complication in the interpretation of experimental data on the relationship between transpiration and ion uptake is the long delay between the time when transpiration rate is changed and the first effects on absorption, which sometimes are only seen several hours later. Bowling and Weatherley (40) found that the rate of potassium uptake by castor oil *(Ricinus communis)* plants did not change until 7.5 h after transpiration was increased; similarly, when transpiration was reduced, the fall in salt uptake lagged several hours behind the change in water uptake. Graham and Bowling (95) concluded that the delay in response to an alteration in water flux may be due to factors emanating from the shoot. The possible nature of such influences is discussed later.

Some investigators have attempted to simulate the effects of transpiration by applying a pressure gradient to excised root systems. In such experiments it is possible to measure changes in ion flux at different rates of water uptake. It has been demonstrated that under the influence of an applied pressure gradient, the resistance of the root to water movement falls and this is accompanied by an increased flux of solute (see 128, 129, 141, 200). The alteration to the electrochemical potential of the ions because of the pressure is not sufficient to account for the observed changes in flux, and the pressure gradient must be assumed to have an effect on the ion transport mechanism (109).

Fiscus (83) has concluded from a theoretical analysis of the interaction between osmotic and hydrostatic driving forces that transpirational pull should have the same effect as flow induced by a pressure gradient of the same magnitude. However, applied pressure may also have indirect effects on water movement which are not induced by transpiration; e.g., air may be displaced from intercellular spaces, thus affecting root aeration and the volume of the water-filled free space. Meiri and Anderson (177, 178) observed such effects when they applied reduced pressure to the stumps of exuding corn roots.

Another approach to the study of the relationship between water and ion uptake is to alter water flux by changing the water potential of the medium through the addition of solutes. Greenway and Klepper (99) found that there was a marked effect of adding mannitol to the medium on the rate at which previously absorbed phosphate was transferred to the xylem of tomato plants. When the water potential of the medium was reduced from −0.04 to −0.55 MPa, water flux across the root was reduced by a factor of three and phosphate movement to the shoot decreased by a factor of two. There was very little effect on the rate of uptake of phosphate from the medium. In rather similar experiments, Erlandsson (78) studied the effects of changing the water potential of the medium or of the atmosphere on water and ion uptake by young wheat plants. He found that both water and ^{86}Rb uptake were reduced when the water potential of the medium was lowered by addition of polyethylene glycol. On the other hand, when the humidity of the air surrounding the shoots was raised from about 60% RH to 100% RH, water uptake decreased while ion uptake increased temporarily but was not permanently affected. It was concluded that changes in water potential regulate an active ion-transport mechanism in the roots. The way in which this is brought about is still unknown, but given the importance of the turgor term in plant water relations (see Chapter 1), turgor-sensing mechanisms, reviewed by Cram (63) and Zimmermann (278), may be involved.

C. Release of Ions in the Stele

There is abundant evidence that the concentration of ions in the exuding sap from xylem vessels can greatly exceed that in the external medium (75, 225). This accumulation of ions, together with that of organic solutes, is believed to cause the related phenomena of root pressure and guttation.

Although it is evident that energy must be expended to bring about this accumulation, it is not yet certain which ions are actively transported and where the active mechanism is located (see discussion in 4, 173). One view is that ions are accumulated in the symplast of cortical cells and move passively into the stele along a gradient of concentration maintained by leakage into the xylem vessels (62). The measured concentration of potassium ions in root cells (up to 120 mM) is often more than three times that found in xylem exudate (71, 72). The "leakiness" of isolated steles when compared with the cortical tissues of corn roots has been taken as evidence supporting the Crafts–Broyer hypothesis. However, these observations are difficult to reconcile with the fact that living stelar cells contain considerable amounts of ions in intact roots, and damage to the steles or to the endodermis during separation from the cortex may have been responsible for the effect. Electrochemical measurements have also been used to support the leakage hypotheses. Dunlop and Bowling (70–72) measured the electrical potential differences and ion activities in the cortex and stele of corn roots and concluded that movement from the symplast into the xylem vessels was in the same direction as the electrochemical gradient. Anderson and Higinbotham (5) doubted the validity of the technique employed because of effects on injury, but Kelday and Bowling (139) detected an electrogenic component of the potential difference in roots of *Commelina*

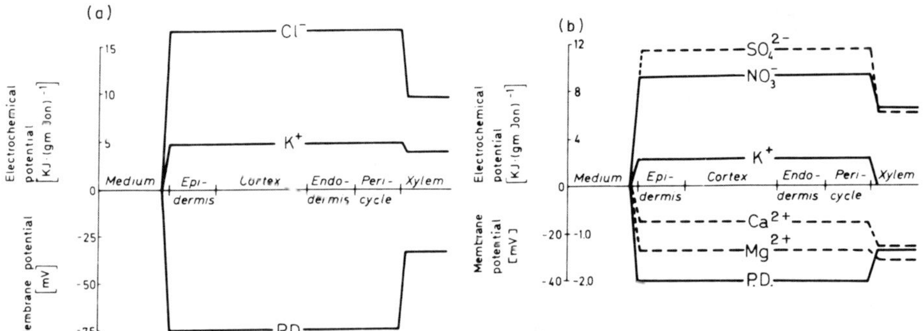

FIG. 8. Variations in membrane potential and electrochemical potential for various ions across the root of (a) *Zea mays* and (b) *Helianthus annuus*. Data of Dunlop and Bowling (72) and Bowling (38).

communis from which they concluded that the cells were not seriously damaged by insertion of the microelectrodes.

Some examples of the results of Bowling and collaborators are shown in Fig. 8. It is of interest that while uptake into the root of all ions, except the divalent cations Ca^{2+} and Mg^{2+}, is against an electrochemical potential gradient, movement into the xylem is downhill with the gradient.

If leakage of ions from living cells into the xylem vessels of the stele does occur, the reason for it is still unexplained. Crafts and Broyer (62) suggested that lowered metabolic activity of stelar cells as a result of oxygen deficiency might be responsible. The conclusion from early investigations (45) that oxygen concentrations in the stele are not much lower than in the cortex has been confirmed by more recent studies (84) and another inducing agent would seem to be involved; a hormonal influence, such as that which instigates release of potassium ions from guard cells under certain conditions (see Chapter 3), is a possibility. Effects of ABA on exudation have been shown, but results are conflicting (56, 57, 210).

An alternative hypothesis to that proposed by Crafts and Broyer is that transfer of ions from the symplast in the stele of roots into the apoplast is an active process akin to secretion.

This view advocated by some of the early investigators has been supported more recently by Läuchli *et al.* (153, 155, 158) and by Pitman (204, 207), who has measured separately ion fluxes into and out of roots and ion transport along the root (see Fig. 9). The "two-pump hypothesis" of Pitman (203) is so-called because it implies the operation of two active transport mechanisms, one inwardly directed into the symplast and located in the plasma membranes of cortical cells and the other outwardly directed

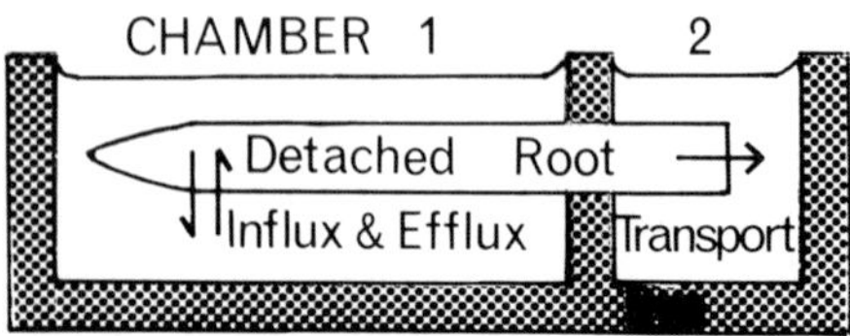

FIG. 9. The double-chamber method used by Weigl and Pitman to study influx, efflux, and transport of salts in detached roots.

across plasma membranes at the symplast–apoplast interfaces in the stele. The situation is analogous to that encountered in phloem transport in which loading and unloading occurs at sources and sinks, respectively. The operation of an active mechanism during loading of the sieve tubes at the source is well authenticated, but the nature of "unloading at a sink" is still uncertain (see Chapter 6). Since phloem transport may be considered to be a highly developed mode of symplastic transport, it is not surprising that the same concept should present itself in both cases. Clarification of one of these situations would help to resolve the other.

The work of Pitman, Läuchli, and their colleagues has shown that release of ions into the xylem can be inhibited by metabolic poisons and inhibitors of protein synthesis under conditions in which uptake by the root is unaffected. Cytokinins and abscisic acid (ABA) have been found to have similar effects (Table V). These compounds will inhibit transport to the xylem in excised roots in short-term experiments, while they have no effect on influx by the roots. Over longer periods uptake may be inhibited, too, but this is thought to be a secondary effect. However, as Bowling (39) has pointed out, the mechanism of action of these substances is uncertain and the observations do not prove that they are directly influencing an active transport process. They might be reducing the passive permeability of the membranes, although since inhibitors generally increase permeability, this possibility is not very probable.

Another approach to the problem of ion release into the xylem vessels is to examine the structural and functional characteristics of the surrounding parenchyma cells. If they have a secretory function, they might be expected to show features of secretory cells, e.g., those of gland cells (171, 172). Läuchli (151, 152), using electron probe microanalysis, showed that there is a high concentration of potassium ions in the xylem parenchyma of corn roots, but Dunlop and Bowling (70–72) could find no evidence of this using K^+-selective microelectrodes (Fig. 10). It would be surprising if living parenchyma cells in the stele did not accumulate ions in their vacuoles, but the ability to accumulate does not prove that they have a secretory

TABLE V

HORMONE EFFECTS ON ION AND WATER TRANSPORT THROUGH EXCISED ROOTS OF MAIZE *(Zea mays)*[a]

Treatment	Water flux ($mm^3\ m^{-2}\ s^{-1}$)	Exudate or transport			Uptake to tissue		
		K^+	Ca^{2+}	Cl^-	K^+	Ca^{2+}	Cl^-
		($mol\ m^{-3}$)[b]			($mol\ m^{-3}$)[b]		
Control (0–24 h)	4.64	22.1	1.8	23.1	69.4	2.9	41.5
Kinetin (1 μM)	1.02	16.1	2.9	15.4	94.2	2.5	69.3
ABA (1 μM)	7.34	20.5	1.1	17.2	91.5	3.7	84.0

[a] Data from Collins and Kerrigan (56).
[b] 1 $mol\ m^{-3} \equiv 1\ mM$.

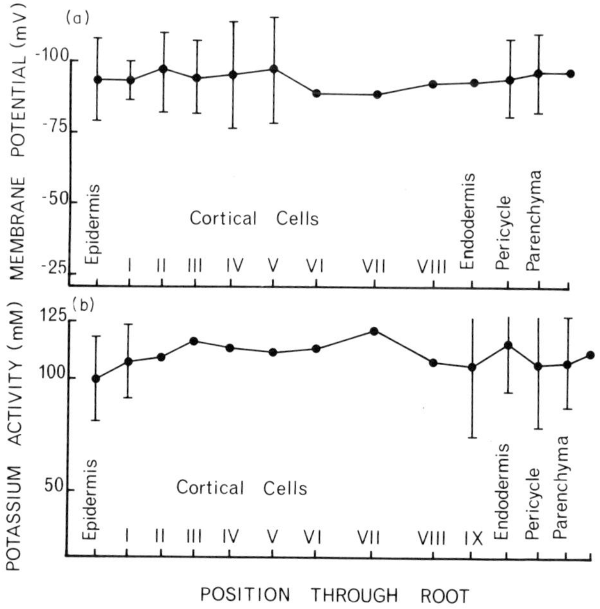

FIG. 10. The concentration of potassium in the vacuoles of cells across the root of corn *(Zea mays)* bathed in 1.0 mM KCl + 0.1 mM $CaCl_2$ (a) and values of membrane potential for similar roots bathed in 1.0 mM KCl (b). There are no significant trends in either potassium concentration or membrane potential. Adapted from Dunlop and Bowling (70, 71).

function. No extensive cell wall development characteristic of transfer cells (see Chapter 6, Fig. 9) has been observed in the stelar parenchyma of corn roots. Läuchli *et al.* (155) found parenchyma cells resembling transfer cells in the proximal region of soybean roots, but they thought that these cells might be accumulating ions from the xylem sap rather than secreting them.

Although attention has been focused mainly on xylem parenchyma adjoining conducting elements, secretion (or leakage) of ions from any living cells into the stelar apoplast may well lead to accumulation in the xylem sap, with which it bears the same physical relationship as the cortical free space does to the external solution. Dunlop (69) observed an abrupt change in electrical p.d. at the endodermis–pericycle interface which he thought might be due to a higher concentration of ions in the walls of the pericycle cells than in the walls of the endodermis outside the Casparian band (Figs. 1, 5). This would be expected if the free space in the stele is bathed in a solution containing ions at a concentration similar to that occurring in xylem sap, while that in the cortex resembles in composition that of the external medium.

There is another way in which ions may accumulate in xylem sap beside the two discussed in the preceding paragraphs. Solutes may be actively absorbed by the differentiating xylem elements themselves during the course of their development while they still have a functional membrane. This idea was suggested by Priestley (213) many years ago and revived by Hylmo (125). It has been termed the "test tube hypothesis" because a developing vessel resembles a test tube in that it is closed at the bottom and open ended at the top (in an excised root). Scott (227) concluded that all the xylem elements are still alive up to a distance of 8 cm from the tip of *Ricinus* roots, and Higinbotham *et al.* (115) showed that metaxylem vessels containing cytoplasm occurred up to 10 cm from the tip of corn roots. So it is conceivable that ion accumulation occurs in living xylem vessels. Anderson and House (6) found that movement of ions into the xylem decreased from the tip to the base of corn roots and that there was a corresponding decrease in the number of metaxylem vessels that contained cytoplasm. Macklon (174), studying uptake and exudation of ions from segments of onion root, observed that calcium ions occurred almost exclusively in the exudate from the proximal end of the segment, whereas sodium ions were exuded mainly from the distal end. He suggested that this polarity could only be explained if there were functional membranes in the xylem vessels transporting ions in different directions. An alternative explanation is that calcium ions were not being transported as readily through the endodermis at the distal end of the root and so were not reaching the xylem vessels, while at the proximal end there may have been reabsorption of sodium ions from the xylem sap by the surrounding parenchyma.

Although it is incontrovertible that ions accumulated during the differentiation of xylem vessels will contribute to the initial solute content of xylem sap, it is doubtful whether this could be quantitatively important, considering the length of time during which vessels remain functional after differentiation has ceased, and the volume of xylem sap involved. Läuchli *et al.* (155) observed that metaxylem vessels were dead 3–4 mm from the tip of barley roots and yet ^{86}Rb was transported into them at this level, even when 1 mm of the root apex was removed. More recently it has been reported (157) that *p*-fluorophenylalanine, an amino acid analog which prevents normal protein synthesis, inhibits chloride transport to the xylem of barley roots in a region in which all the vessels are apparently mature. On balance, it seems probable that ions are released continually into dead xylem elements from the surrounding cells, but the mechanism involved and its control and significance are still uncertain. (See also Chapter 4.)

IV. Transport in the Xylem

Knowledge of the composition of xylem sap comes mainly from analyses of the exudation fluid from decapitated plants or of sap extracted from pieces of stem by suction (36, 37). As already indicated, the concentration of solutes in xylem sap varies greatly, depending particularly on the supply of nutrients and the rate of transpiration. There are marked diurnal and

TABLE VI

COMPARISON OF PHLOEM AND XYLEM SAP COMPOSITION IN TWO SPECIES OF LUPIN[a]

	Lupinus albus		Lupinus angustifolius	
Composition	Xylem sap	Phloem sap	Xylem sap	Phloem sap
Sucrose (mg ml^{-1})	—	154	—	171
Amino acids (mg ml^{-1})	0.70	13	2.6	15
Potassium (μg ml^{-1})	90	1540	180	1820
Sodium (μg ml^{-1})	60	120	50	101
Magnesium (μg ml^{-1})	27	85	8	140
Calcium (μg ml^{-1})	17	21	73	64
Nitrate (μg ml^{-1})	10	—	31	Trace
pH	6.3	7.9	5.9	8.0

[a] Data from Pate (198).

seasonal fluctuations which are related to changes in a variety of environmental factors. Table VI is an informative comparison of the composition of xylem and phloem saps of two species of lupin (195) grown under defined conditions.

Most mineral nutrients are present in xylem sap as free ions, but some are transported in a more complex form. Iron, for example, is often found as iron citrate (242). Immobility of ionic iron in the xylem appears to be related to the level of phosphate, which causes iron deposition as insoluble ferric phosphate (37). The most common organic compound of nitrogen in xylem sap is often glutamine, but a variety of other nitrogen-containing substances have been found. An exceptional case seems to be cocklebur, *Xanthium strumarium (X. pennsylvanicum),* in which about 95% of all the nitrogen in the xylem is nitrate. A full account of the transport of nitrogen compounds in plants is given by Pate in Volume VIII of this treatise.

During its passage through the stem the concentration of xylem sap may decrease considerably as a result of the differential removal of solutes by surrounding cells (142). Divalent ions, notably calcium, appear to be removed from the xylem sap by absorption on the walls of xylem elements (see also p. 395; 25, 31, 32). The pattern of ^{45}Ca movement through bean stems seems to indicate that these ions move to some extent by ion exchange along the wall of xylem cells. Also calcium moves more readily through bean stems when it is supplied together with a chelator, EDTA, which presumably travels in solution (130). Heine (113) studied the movement of labeled phosphate and potassium ions through lengths of poplar *(Populus nigra)* stem in response to suction applied at one end. Arrival of isotope fed from the opposite end was monitored at different distances along the stem and some of the results that were obtained are shown in Fig. 11. When the supply of labeled phosphate was replaced by water, the radioactivity of the xylem sap fell rapidly, indicating that phosphate in the xylem is highly mobile. In contrast, the level of labeled potassium fell much more slowly, suggesting that some potassium becomes bound to negatively charged sites in cell walls.

Similar experiments with apple twigs (80) showed that ^{45}Ca moves much less readily through the xylem than does ^{32}P. There was a marked seasonal effect: retention of calcium in the stem was much greater in the fall (February in New Zealand, where the experiments were carried out) than in the spring. When the labeled calcium chloride solution was replaced by unlabeled salt, the amount of radioactivity released by the stem increased. Addition of EDTA, or citrate, to a labeled calcium chloride solution increased the rate at which ^{45}Ca moved through the stem. It may be that in chelated form calcium does not bind to sites, such as cell walls, as it moves across cells of the stem.

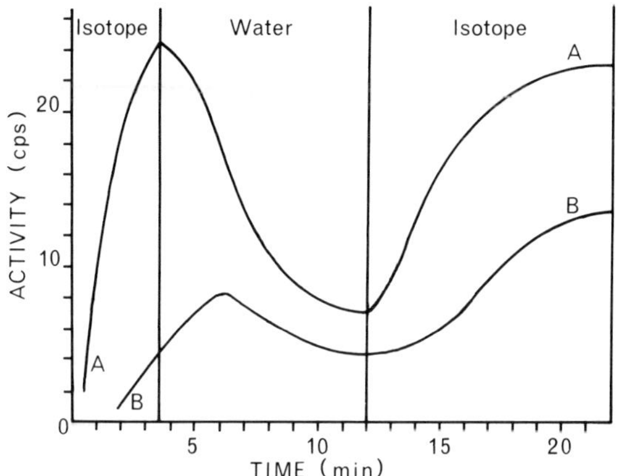

FIG. 11. The flow of radioactive potassium through a 66-cm length of stem of poplar *(Populus nigra)*. A vacuum of about 0.1 MPa was applied to one end of the stem piece. For the first 3.5 min isotope was drawn up, followed by water for 8.5 min, followed again by isotope. Counters were placed 6 cm from the base of the stem (A) and 29 cm from the base (B). Adapted and redrawn from Heine (113).

V. Salt Relations of Specialized Organs

A. STORAGE ORGANS

1. *Fruits and Seeds*

When a plant enters the reproductive phase, the developing inflorescence acts as a powerful sink for assimilates, nitrogen, phosphorus, and other nutrient elements.

In early detailed studies, Williams (268, 269) found that when oat plants were grown with low phosphate supply, the phosphorus accumulated in the ears came mainly from the external solution; in the presence of higher phosphate levels a large proportion was supplied by depletion of leaves, stems, and even roots (Fig. 12). Similar results have been obtained for field-grown wheat plants (184). Neales *et al.* (189) found that there is a greater export of nitrogen from older leaves to the developing grain in wheat plants if the nitrogen supply to the roots is reduced at anthesis. If *Zea mays* plants are deprived of an external supply of nitrogen, phosphorus, potassium, magnesium, and calcium immediately after silking, there is increased movement of all these elements except calcium from the leaves into the developing cobs (140). The leaves surrounding the cob are a

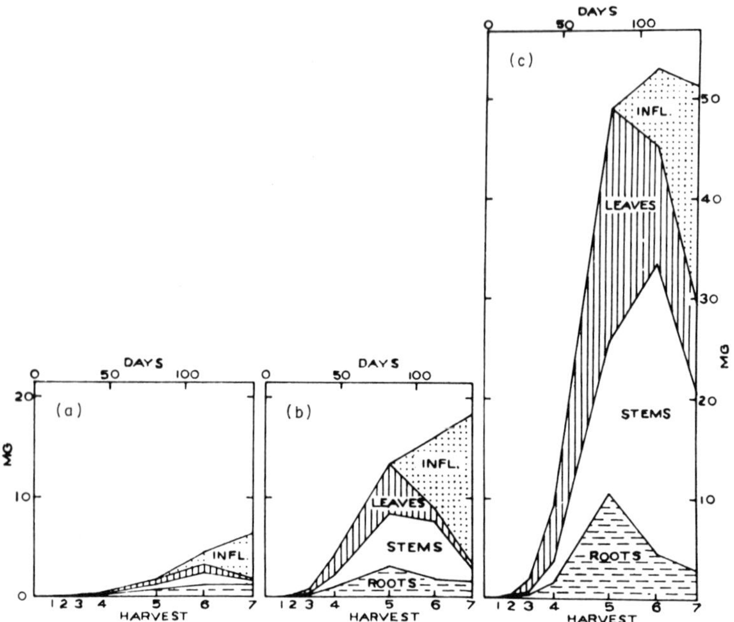

FIG. 12. The effect of phosphate supply on the distribution of phosphorus in oat plants: (a) low phosphate, (b) medium phosphate, and (c) high phosphate. Note the large accumulation of phosphorus in leaves and stems of plants supplied with high phosphate. Redrawn from Williams (268).

major source of assimilates for the developing fruits, as is the flag leaf for grain filling of wheat and barley (see 14, 189), and the extent to which neighboring leaves supply mineral elements to the inflorescence probably depends in part on the nutrient status of the plant as a whole and the available supply. It seems probable that these leaves are important sources of nitrogen, phosphorus, and possibly other elements for developing cereal grains.

Linck (162) made a detailed study of the pattern of distribution of ^{32}P fed to individual leaves of intact pea *(Pisum sativum)* plants, and some of his results are shown in Fig. 13. He showed that movement from lower leaves (L_5 and L_7) on the stem was predominantly toward the roots, while that from upper leaves (L_9 and L_{10}) was mainly upward. A large fraction of the ^{32}P translocated from L_9 moved into the developing pod (P_2) at node 11, which was in the same orthostichy as L_9. On the other hand, ^{32}P applied to leaf 10 was exported mainly into the pod developing in the axis of the same leaf. This pattern of distribution appears to be related to the vascular

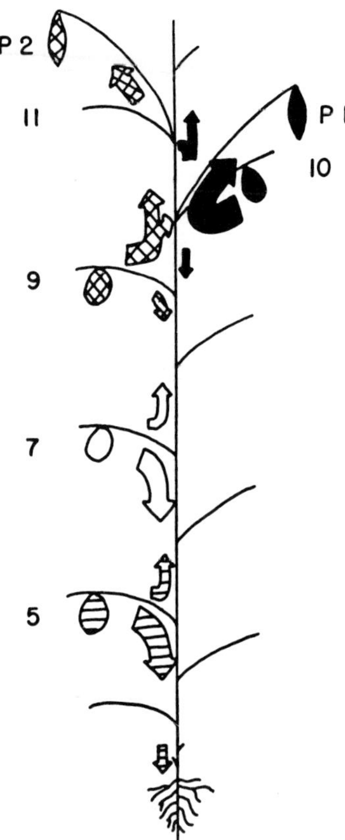

FIG. 13. The distribution of labeled phosphate supplied to particular leaves (L_5-L_{10}) of pea plants *(Pisum sativum).* The arrows and their sizes indicate the amount of transport of phosphate. Redrawn from Linck (162).

connections established in the stem (see also Chapter 6 and that by Pate in Volume VIII of this treatise).

Because transpiration from fruits is relatively low, it would be expected that the phloem is more important than the xylem in supplying mineral elements to them. Pate and collaborators (194–198) have studied in detail the contributions of xylem and phloem to the supply of various mineral elements to the fruits of the field pea *(Pisum arvense)* and the white lupin *(Lupinus albus).* Their technique involved analysis of xylem sap exuded from the stalks of developing pods and estimations of volume flow from which it is possible to calculate the contribution of the xylem to changes in the composition of the pod as a whole. The contribution of the phloem was calculated by difference. It was found that the proportion transported in the phloem varied from 90% or more in the case of phloem-mobile elements such as potassium, nitrogen, and phosphorus to less than 30% for calcium and a number of micronutrients. Data from the balance sheets for nitrogen, carbon, and water are shown in Fig. 14, and further details are given in Chapter 4 of Volume VIII of this treatise.

The dependence of calcium transport into fruits on the transpiration stream can lead to calcium deficiency when plants suffer water stress (94, 180); deficiency diseases (e.g., blossom end rot of tomato) may result in loss of yield and quality and reduction in value of the crop in some cases (79, 93, 94, 181). Methods of dealing with the problem include the use of calcium sprays on leaves and developing fruits and irrigation (20, 21).

The peanut *(Arachis hypogaea)* is a rather special case in that calcium is

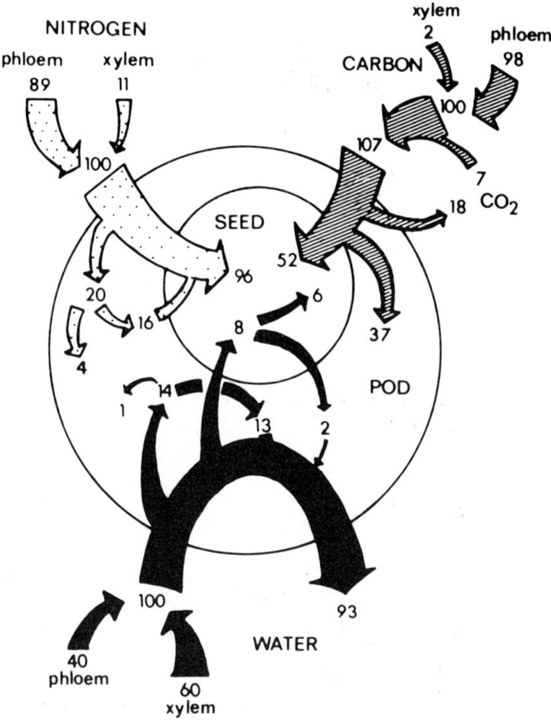

FIG. 14. The movement of carbon, nitrogen, and water into the fruit of white lupin *(Lupinus albus)* during its growth. Values are relative and expressed as percentages. Note that while little of the water reaching the fruit is retained, only a little carbon and no nitrogen is re-exported. Weight ratio: C : N : H_2O – 12 : 1 : 600. Redrawn from Pate *et al.* (196).

taken up directly from the soil by the buried fruit (35). Skelton and Shear (230) demonstrated that there is very little transport of calcium from the parent plant to the fruit.

The concentration of ions in seeds seems to change relatively little over a wide range of nutrient supply (92). This ensures that the embryo in the germinating seed is normally provided with an adequate supply of mineral nutrients. Sometimes these are greatly in excess of actual needs, and the amounts of various trace elements in seeds may be sufficient to supply several generations (236).

2. *Vegetative Storage Organs*

Large quantities of mineral elements as well as carbohydrates are stored in organs such as swollen roots, rhizomes, corms, and bulbs which act as perennating structures. As in the case of fruits, the rate of transpiration, especially of underground organs, is likely to be low and therefore the bulk of the phloem-mobile elements is probably transported through the sieve tubes. It has been shown that when potato plants are actively transpiring under conditions of limited water supply, they may withdraw water from the tubers (17). Under these conditions, ^{89}Sr, which was used as a tracer for calcium, did not move into the tuber from the parent plant. On the other hand, the transport of assimilates and ^{32}P into the tuber, which presumably occurs via the phloem, was unaffected by wide variation in transpiration rate. In view of the evident restriction on calcium transport into potato tubers, it is not surprising that the level of calcium in these organs is low compared with that in other parts of the plant.

Species such as carrot, sugar beet, and red beet in which the tap root becomes a massive storage organ develop fibrous lateral roots through which absorption of water and soluble mineral salts occurs. Some of this material is presumably extracted from the xylem to supply the growing storage tissues, while the remainder is channeled into the shoots. The way in which the distribution of mineral elements in such systems is related to growth of root and shoot requires further investigation.

B. Leaves

The mechanism of absorption of ions by leaf cells has already been discussed in Chapter 4. Here attention will be paid mainly to the regulation of the salt content of leaves—excluding those of halophytes, which will be dealt with later in Section VI.

The salt content of leaves varies considerably among species and with the conditions under which plants are grown, especially of nutrient supply.

It also changes appreciably with time as leaves age and the plant matures.

Yagi (271) made a careful study of changes in the salt content of leaves at different positions on the stem of plants of *Phaseolus vulgaris* grown under controlled conditions with an adequate supply of nutrients. She found that changes in potassium content with time were closely correlated with leaf water content and that the amount did not alter appreciably after a leaf reached its maximum area. On the other hand, calcium content continued to increase throughout the life of the leaf. Sodium content was low relative to that of potassium even when the two ions were present in the medium at comparable concentrations.

A similar exclusion of sodium from leaves has been observed in other plants (133, 201). Pitman found that the $K^+ : Na^+$ ratio in the shoots of barley plants was greatly influenced by the relative concentration of K^+ and Na^+ in the external solution while that in the root was affected much less (Fig. 15). He also found that at higher concentrations, Na^+ uptake by the shoot was increased relative to K^+ when transpiration was increased, although at low concentrations water flux had little effect.

Although the rate of transpiration often has relatively little influence on the quantity of salt transported into leaves, serious curtailment of water flow into a mature leaf, e.g., by its enclosure in a humid atmosphere, reduces movement of ions into it. Such observations indicate that the bulk of the salt entering a mature leaf does so in the transpiration stream.

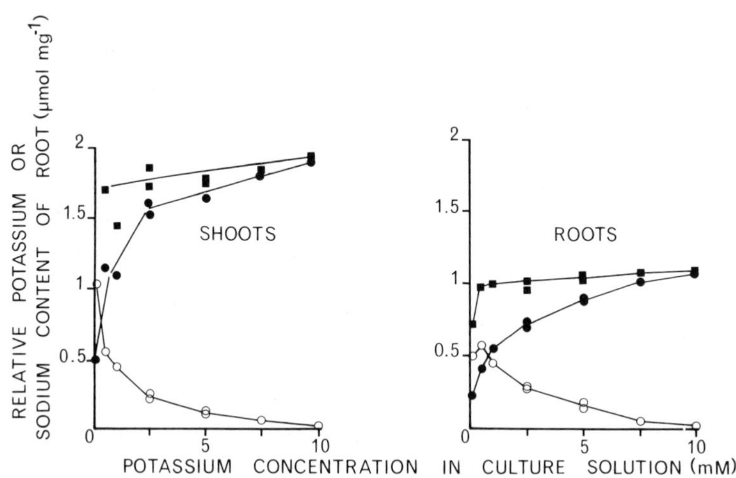

FIG. 15. The effect of varying the ratio of potassium and sodium in the external medium on potassium and sodium content of roots and shoots of barley plants. ○, sodium; ●, potassium; ■, sodium and potassium. Redrawn from Pitman (201).

Evidence that dissolved substances move with water through the apoplast of a leaf comes from the work of Langston (146), who found that ^{60}Co, which is absorbed by leaf cells to only a very limited extent, moved out from the veins of cabbage leaves toward the leaf margins, where it accumulated. Crowdy and Tanton (66) used a lead chelate to trace the pathway of water through wheat leaves and observed that the greatest concentration of the chelate was in the cell walls. Again, there was a tendency for solute to accumulate at the leaf margins. When a leaf is young, before it has developed its full photosynthetic capacity, it is dependent on older leaves or storage organs for a supply of assimilates and to a greater or lesser extent for a variety of mineral elements which are also moved in the phloem (123, 124). As it develops and becomes a net exporter of assimilate, a larger proportion of its mineral supply is acquired via the xylem coincident with increasing transpiration.

Calcium is relatively immobile in leaves. Ringoet *et al.* (216) studied the uptake of ^{45}Ca from calcium salts applied in aqueous solution to the surface of oat leaves. Using microautoradiography they observed that some isotope moved to the distal end of the leaf in the transpiration stream and some was absorbed by mesophyll cells, particularly those close to the bundle sheath. The ^{45}Ca was transferred into sieve tubes only when a large amount of carrier calcium was supplied. This was presumed to saturate binding sites for calcium in the mesophyll and make mobile calcium available for transfer into the phloem (cf. p. 429). A similar result has been shown for potato (181).

Contrasting results have been obtained for the freely mobile element potassium. Pitman *et al.* (209), from a study of potassium uptake by slices of wheat leaves concluded that net flux and potassium accumulation in leaf cells were controlled primarily by the availability and concentration of potassium in the free space of the leaf which in turn is influenced by rates of transpiration and potassium supply.

Widders and Lorenz (263, 264) studied the effect of leaf age on fluxes of potassium (using ^{86}Rb as tracer) in slices of tomato leaf blades. They observed that influx declined rapidly during the phase of leaf expansion and attributed this to a dilution in the concentration of potassium carrier sites with increasing cell volume and surface area. In fully expanded leaf tissue, influx decreased slowly with age while efflux increased slightly. There was no dramatic change in the capacity of cells to accumulate potassium with the onset of leaf senescence. It was suggested that the decline in concentration of potassium in mature tomato leaves is due to a reduction in the availability of ions in the leaf free space which leads to a reduction in influx. Below a certain concentration efflux exceeds influx and potassium content is lowered by translocation in the phloem.

Greenway and colleagues (96–99, 103) have attempted to estimate the relative contributions of the xylem and phloem in the supply of various ions to barley leaves at different stages of development. Changes with time in the total content of each ion in the leaf were used as an estimate of excess export and import in phloem and xylem combined, while measurements of import via the xylem were based on changes in radioactivity of a leaf when a labeled solution was applied to the roots. It was assumed that the specific activity of the labeled salt reaching a leaf was the same as that in the external solution and that no radioactivity entered the leaf via the phloem. Although neither of these assumptions is necessarily valid, the results indicated that barley leaves are net importers of potassium and phosphorus for a prolonged period while the leaves are growing and the supply of these elements via the phloem decreases with leaf age whereas that via the xylem increases (Table VII). Barley leaves become net exporters of potassium and phosphorus at a relatively late stage in their development, but it is likely that some export of these elements via the phloem starts much earlier.

Greenway and Gunn (97) also showed that there was some redistribution of phosphorus within a single leaf lamina. When ^{32}P was applied to the tip of a recently emerged barley leaf, some of the isotope moved into the growing basal region of the same leaf (Table VIII). The existence of a functional meristem at the base of grass leaves may explain why net export from the leaf as a whole is delayed. The transport of phosphorus into a mature leaf in the xylem was markedly reduced when the concentration of phosphate in the external solution was lowered. On the other hand, the rate of export of phosphorus in the phloem was not greatly affected by external concentration and the consequent reduction in the level of phosphorus in the leaf may have contributed to more rapid leaf senescence under conditions of phosphorus deficiency.

TABLE VII

RATES (μmol day^{-1}) OF INTAKE OF POTASSIUM IONS THROUGH THE XYLEM AND PHLOEM IN BARLEY LEAVES OF DIFFERENT AGES[a]

	Age of leaf		
Intake	Young	Intermediate	Old
Intake via xylem	2.0	2.7	1.9
Intake via phloem	1.3	0.7	−1.6
Total intake	3.3	3.4	0.3

[a] Data from Greenway and Pitman (103).

TABLE VIII

EXPORT OF ^{32}P APPLIED TO LEAVES OF BARLEY OF DIFFERENT AGES[a,b]

	Leaf from which export took place		
Individual organs importing ^{32}P	Mature leaf	Growing leaf (about half-expanded)	Rapidly growing leaf (about 4 cm long)
Basal part of treated leaf	30	62	100
Mature leaf	Source	1.5	0
Growing leaf	2.7	Source	0
Rapidly growing leaf	7.1	3.3	Source
Roots	61	32	0
Amount translocated from source as % of amount absorbed by plant	4.5	4.5	4.3

[a] Data from Greenway and Gunn (97).

[b] ^{32}P in each organ is expressed as a percentage of the total amount of ^{32}P in the plant, excluding the section of application.

Observations on a number of dicotyledonous species indicate that a leaf begins to produce assimilates surplus to its requirements by the time it reaches 25–50% of its final area (see Chapter 6) and export attains a maximum at about the time when the leaf reaches its full size. In contrast, net intake of phosphorus appears to continue until the leaf reaches maturity and only during senescence does net export occur (Fig. 16). The phloem remains functional and transports large amounts of various organic substances and inorganic ions after it has stopped transporting significant amounts of assimilates and senescing leaves are probably of greater significance for the rest of the plant as sources of nitrogen and phosphorus than of carbohydrate (140, 269). Bieleski (33, 34) demonstrated a more rapid uptake of phosphorus, potassium, and chloride by sieve tubes from petioles of celery *(Apium graveolens)* and apple than by the surrounding parenchyma. This could be important if ions, previously sequestered in membrane-bound compartments, become available for export as leaves become senescent.

Relatively large amounts of various cations, including calcium, sodium, and magnesium, may be leached by rain from leaves (49, 161, 252). This is important in the case of forest trees before leaf fall, the leachate being an important source of nutrients for the herb layer and especially for underlying bryophytes (73).

In the opposite direction, foliar absorption of mineral nutrients has

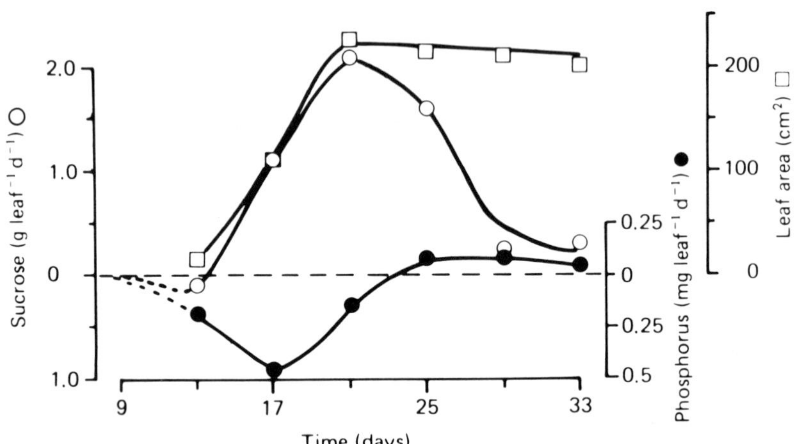

FIG. 16. Export and import of sucrose and phosphate by the second leaf of *Cucumis sativus* during the expansion and postexpansion period. Redrawn from Hopkinson (123).

received considerable investigation because of its possible agricultural importance in the application of fertilizers and particularly for supplying trace elements, particularly where, as in the case of copper, they rapidly become unavailable in certain soils (163). Another potential advantage of foliar application is that response is likely to be more rapid than by addition of fertilizer to the soil.

VI. Salt Relations of Specialized Plants: Halophytes

A. ION ABSORPTION

The salt relations of those plants which tolerate high salinities are considered separately here to highlight the important problems encountered and the solutions adopted by them. Although this treatment concentrates on halophytes, that is, plants which tolerate and are often found in very salty conditions such as estuarine shores and mud and salt flats, reference will also be made to nonhalophytes which may be exposed to high salinity occasionally, as under experimental conditions. Because of increasing salinity in many agricultural soils, the behavior of nonhalophytes is of some importance. The physiology of halophytes has been described by Waisel

(257) and reviewed by Bernstein and Hayward (27) and Flowers *et al.* (87, 89). Greenway and Munns (100) have considered the mechanism of salt tolerance in nonhalophytes (glycophytes), while Yeo (273) has reviewed the problem of salinity resistance in terms of metabolic and energy costs.

Because of the low osmotic potential of the soil solution in which they grow (down to −5 MPa or less), halophytes must have a correspondingly low water potential in order to retain water. Measurements on a number of halophytes give values of sap osmotic potential in the range −2 to −5 MPa, with ionic concentrations of 400 to 700 m*M* (87). Halophytes differ from salt-sensitive plants in their ability to accumulate large amounts of sodium and chloride ions in the cell sap without interference with growth and reproduction. Much of it is accumulated in the leaves in contrast to the situation in glycophytes, which tend to exclude sodium from the shoot. Collander (55) grew a variety of plants in a nutrient medium containing equimolar amounts of potassium and sodium ions and observed that halophytic species absorbed relatively more sodium than potassium in comparison with nonhalophytes. They also transport a higher proportion into the leaves (87).

Obligate halophytes (e.g., *Salicornia* spp.) are species which cannot survive in the absence of large amounts of sodium salts. Little is known about the basis of this requirement, although it has been established that sodium is an essential element for some of these species (46). Others, e.g., *Aster trifolium* and *Suaeda maritima,* can grow in the absence of salt, but are often found in saline habitats because they can tolerate high salinity and so compete successfully with more vigorous but salt-sensitive species (272). Glycophytes differ markedly in their sensitivity to salt, some (e.g., *Zea mays*) being quite tolerant and others very sensitive (249) (Table IX). Even varieties of the same species show differing degrees of salt tolerance, and selection for this characteristic may be necessary when crops are grown in saline soils or need to be irrigated with brackish waters (154).

A possible explanation of salt tolerance is that the enzyme systems of halophytes are less sensitive to high salinity than those of other plants. This possibility has been examined by several investigators (85, 86, 88, 90, 101, 102, 192, 251) and all have indicated that, in contrast to enzymes extracted from halophilic bacteria (147), a range of enzymes extracted from halophytes have similar characteristics to those extracted from glycophytes (see 89). If it was uniformly distributed within the plant, the amount of salt present in halophytes would be sufficient to bring about the almost complete inhibition of salt-sensitive enzymes such as malate dehydrogenase. It has been concluded therefore that low concentrations of salt are maintained in the cytoplasm as a whole, or at least in the vicinity of sensitive enzyme systems, by sequestration of ions in vacuoles.

TABLE IX

CONTENT OF Na^+ IN LEAVES OF 4-WEEK-OLD CONTROL AND SALT-TREATED PLANTS OF CULTIVATED TOMATO, *Lycopersicon esculentum*, *L. peruvianum*, AND HYBRID SALT-TOLERANT PLANTS SELECTED FROM SALT-SENSITIVE *L. esculentum minor*[a]

	Na^+ Content (μmol g^{-1} DW)	
Plant	Control	Treated with NaCl (196 m*M*)
L. esculentum	51	447
L. peruvianum	71	1538
Hybrid plants	33	1034

[a] Data from Tal (249).

Attempts have been made to estimate the amounts of salt in the cytoplasm of halophytes directly by using histochemical methods and by X-ray microanalysis (152, 229), but the technical difficulties are considerable and few reliable estimates of ion concentrations either in the bulk of the cytoplasm or in particular organelles have yet been obtained. Larkum and Hill (148) isolated chloroplasts, in nonaqueous media, from leaf segments of *Limonium vulgare* which had been floated on 100 m*M* NaCl, and found ionic concentrations within them of 500 to 700 m*M* for sodium and chloride. These are higher levels of NaCl than occur in the chloroplasts of other plants but lower than occur in halophyte cells as a whole.

An indirect approach to the measurement of sodium chloride distribution in cells is by the method of flux analysis (see Chapter 4). Jefferies (131) calculated that the sodium concentration in the cytoplasm of root cells of *Triglochin maritima* in equilibrium with 500 m*M* sodium chloride was 148 m*M*. For leaf cells of *Suaeda maritima* plants growing in 240 m*M* sodium chloride, Yeo (272) obtained a cytoplasmic concentration of 165 m*M* and a vacuolar concentration of 600 m*M*.

From the measurements that have been made and the biochemical information available, it seems likely that the concentration of salt in some regions of the cytoplasm of halophytes must be appreciably lower than in the vacuoles. If this is the case, there must be a compensatory mechanism(s) to maintain the water potential of the cytoplasm at the same value as that of the vacuoles. A lower matric potential in the cytoplasm associated with a high concentration of proteins may be one of these factors. Although the concentration of soluble anions, e.g., chloride, in the cytoplasm has not been measured accurately, it is probable that part of the total negative charge (fixed and mobile) is balanced by organic cations. A number of studies have shown that halophytes contain large quantities of either free proline or betaines (132, 234, 250, 270). Stewart and Lee (233) suggested that accumulated proline is important in maintaining the cytoplasmic water potential equal to that of the vacuole in a number of halophytes, and the same role has been postulated for quaternary ammonium compounds such as betaine and choline in *Atriplex spongiosa* and *Suaeda monoica*. Both proline and betaine have been shown to have no significant effect, even at concentrations as high as 50 mmol L^{-1}, on the activity of a number of enzymes (234). Such substances are termed compatible solutes, and the amounts present are often related to the amount of NaCl in the medium and in the cell sap. Current views on the topic of osmoregulation in plants can be found in the papers by Jefferies (132), Wyn Jones (270), and Yeo (273) and in the book edited by Rains *et al.* (214). (See also Chapters 1 and 2.)

At present there is insufficient information to enable one to decide on the location and nature of ion pumps in halophyte cells (cf. Chapter 4). From an electrophysiological study of the root cells of *T. maritima*, Jefferies (131) concluded that at low concentrations of sodium in the external medium (1 – 10 m*M*), sodium ions in the cytoplasm were in electrochemical equilibrium with those in the vacuole. When the external sodium concentration was raised, sodium ions were actively pumped from the cytoplasm into the medium. Jefferies concluded that the accumulation of salt in *Triglochin* root cells is brought about by active uptake of chloride. If the cytoplasmic concentration of monovalent cations is kept low by a pump driving sodium into the medium, then the cytoplasmic concentration of chloride might also be low, with active accumulation occurring at the tonoplast.

B. TRANSFER TO THE SHOOT

Whereas other plants tend to retain sodium ions in the roots (p. 417), halophytes transport a large proportion to the shoots, where they contribute significantly to the osmotic potential and thereby to the uptake of water. Gale *et al.* (91) showed that growth of *Atriplex* in normal air was low when the concentration of sodium chloride in the medium was low and increased to a peak when the concentration was raised to give an osmotic potential in the medium of about −0.5 MPa. In humid air, growth was best in low concentrations of salt and the results were taken to indicate that additional salt was required to lower the water potential of the leaves when the plants were transpiring rapidly. It would have been interesting to compare the salt content of the leaf cells under the various conditions in relation to rates of growth observed.

Dahesh (67) made a detailed study of changes in the salt content of internodal segments of *Salicornia* collected in the field at different times during the growing season. She found that total salt content and the Na : K ratio increased markedly from the stem apex toward the base and also with time. Actively growing regions of the plant were characterized by having a relatively low Na : K ratio, which had earlier been shown to be a feature of the ion relations of carrot root cells grown in culture (248).

The significance of succulence (i.e., a reduction in the surface area : volume ratio) in the physiology of halophytes has been discussed extensively (1, 87, 133). One effect is to reduce the rate of water loss relative to the volume of the shoot and another is to reduce the concentration of salt in the cells through the absorption of water. In some plants it has been observed that despite the continued accumulation of salt in the leaves during growth, salt concentration does not rise appreciably because of increased succulence. In other cases, e.g., *Juncus gerardii,* where the leaves do not become succulent, sodium chloride concentration rises as the leaves age and salt toxicity may eventually be responsible for death of the cells.

Atkinson *et al.* (13) studied the relationships among transpiration, chloride accumulation, and growth in mangroves. In *Rhizophora mucronata,* from measurements of transpiration rates and the chloride concentration in xylem sap over a period of several days, it was calculated that a leaf absorbed about 17 μmol Cl^- day^{-1} and that a dry weight increase of 3% day^{-1} was sufficient to maintain chloride concentration in the leaf at a constant level. On the other hand, the xylem of another mangrove species, *Aegialitis annulata,* delivered some 100 μmol Cl $leaf^{-1}$ day^{-1}, and in this case the growth rate of the leaves was insufficient to prevent chloride concentration from rising. In these plants excess chloride (and sodium) is excreted from the leaf through salt glands. Other plants, e.g., *Atriplex* spp., which appear to take up into their leaves more sodium chloride than they can accommodate, accumulate it in hairs or bladders which discharge their contents at the leaf surface (172, 191, 193). From an ecophysiological study of salt secretion by four halophytes, Rozema *et al.* (220) concluded that species inhabiting the lower parts of salt marshes showed the greatest rates of secretion of sodium and chloride. *Spartina anglica,* which is well adapted to growth in saline habitats, was found to excrete the largest proportion of absorbed salt. Other, less well adapted species excreted a significantly lower proportion.

Presumably, plants which have not developed excretory devices are either more tolerant of high salt levels in their cells or have a more efficient mechanism for the control of salt uptake by the roots. So far we know little of such mechanisms.

C. Salt Glands

Salt glands occur in some, but not all, halophytes. Their morphology varies considerably, from the simple two-celled glands of *Spartina townsendii* to elaborate multicellular structures in *Limonium* (Fig. 17). In recent years, the ultrastructure of salt glands in several species has been examined, revealing that although there are considerable differences in detail, the glands have certain features in common.

1. The walls of all the cells which are exposed to the external air are covered by a cuticle which has pores through which the salt solution is excreted.
2. There is often, but not always, a cuticular layer encircling the gland, with gaps where aggregations of plasmodesmata make symplasmic connections between gland cells and other cells of the leaf. Even where there is

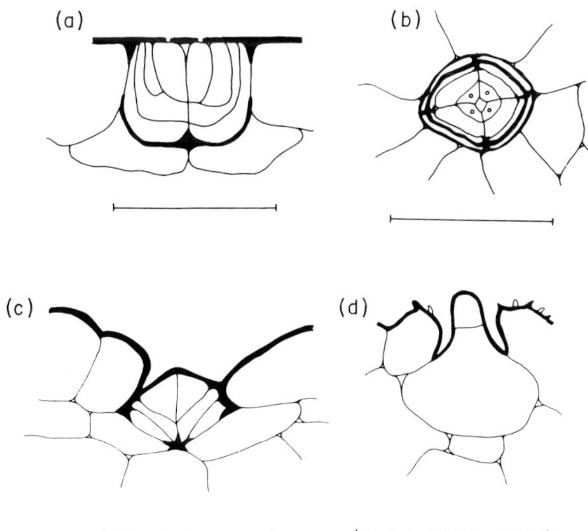

Fig. 17. The structure of the salt glands of *Limonium* in transverse and surface view showing the exit pores through which salt is excreted (a, b) and transverse sections through the salt glands of *Tamarix* and *Spartina* (c, d). The scale line is 50 μm. After Hill and Hill (118).

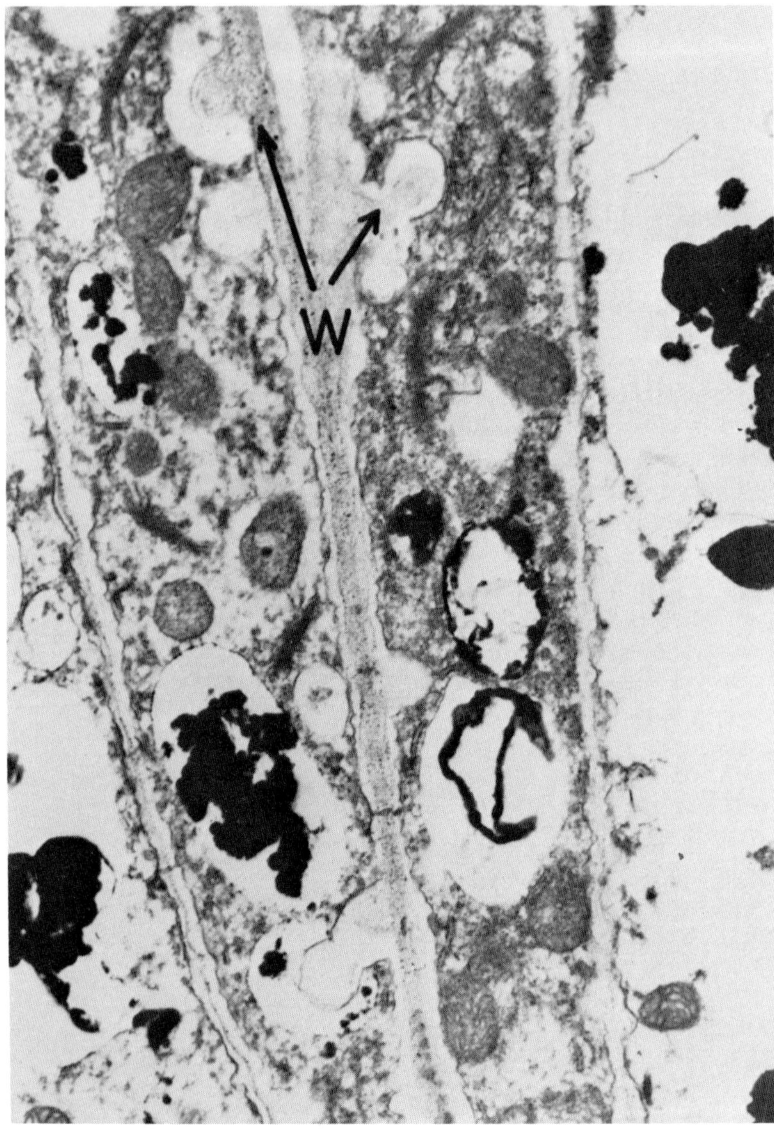

FIG. 18. The fine structure of cells of the salt gland of *Limonium vulgare* showing wall protuberances in adjacent cells (W). From Hill and Hill (117).

no apparent underlying cuticle, there are numerous protoplasmic connections between the gland and surrounding cells through which salt is presumably channeled.

3. The walls of gland cells are sometimes extended in a series of finger-like processes into the cytoplasm (Fig. 18). The effect of this is to increase the surface area of the plasma membranes (cf. transfer cells; see Chapter 6).

4. The gland cells do not usually have a large central vacuole, but there are commonly numerous vesicles in the cytoplasm, many mitochondria, and well-developed endoplasmic reticulum. Leucoplasts are sometimes found, but chloroplasts do not occur.

An early investigation of the functioning of salt glands in leaf disks of *Limonium latifolium* by Arisz *et al.* (8) was described by Steward and Sutcliffe (232). Since 1959, there have been several studies which have clarified some aspects of the problem. It has been confirmed that the relative amounts of sodium and potassium in the excreted fluid reflects the proportions of these ions in the medium in which the plants are grown, but not necessarily that occurring in the leaf (117, 212, 220). On the other hand, the amount of calcium excreted is relatively insensitive to the amount of this ion in the medium.

The structure of glands and their relationship with surrounding cells suggests that ions are transported into them via the symplast. Ziegler and Lüttge (277) investigated the transport of ^{36}Cl into salt glands of *Limonium vulgare* and found indications of movement in both the symplast and apoplast. Evidence that the symplast is the predominant pathway was obtained by Hill (116), who studied the exchange of ^{22}Na and ^{36}Cl in leaf disks of *Limonium vulgare* from which the lower cuticle had been removed. (The salt glands are located on the upper surface of the leaves in this species.)

Hill (116) has confirmed the suspicion that gland cells perform osmotic work by excreting salt against an electrochemical potential gradient by use of the so-called "short-circuit technique" employed by animal physiologists to study ion movements across epithelia. In a further investigation Hill and Hill (119, 120) considered four possible models to explain their observations: (1) an electrogenic pump associated with a similar chloride pump, (2) a neutral salt pump, the negative secretory potential being achieved by shunting the ions through a high-sodium conductance, (3) an electrogenic sodium chloride pump similar to the sodium–potassium exchange pump, and (4) an electrogenic chloride pump, sodium (and other cation) transport being passive.

Measurement of sodium fluxes indicated that the fourth model fit the data best, and it was concluded that salt excretion is driven by active chloride efflux. There is evidence that the chloride pump is an ATPase

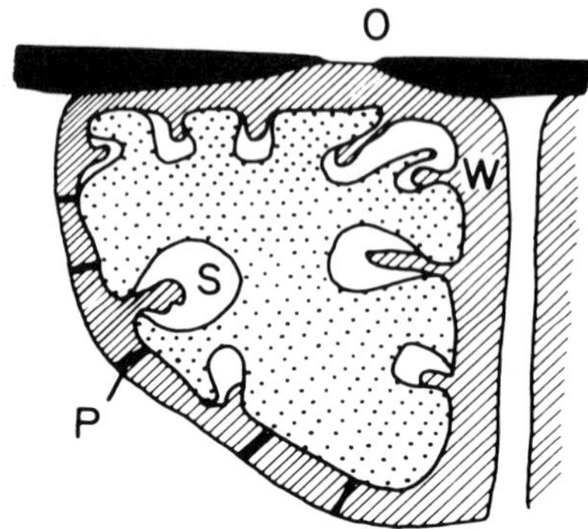

FIG. 19. Diagram for salt pumping in a cell of the salt gland of *Limonium vulgare*. Hill and Hill propose that salt enters the cell through plasmodesmata (P) and is pumped across the plasma membrane into the space(s) adjacent to the wall protuberances. Salts and water then move by bulk flow through the wall (W) and to the exterior through the pore (O). After Hill and Hill (117).

(see Chapter 4), and Hill and Hill (119) suggested that energy may be provided either by respiration in gland cell mitochondria or by photosynthesis in the surrounding mesophyll.

It is proposed that in the *Limonium* gland cells the invaginations of the plasma membranes form channels which act as a standing gradient osmotic system, as shown diagrammatically in Fig. 19. Whether this model will fit other glands, especially those in which wall protuberances do not occur, remains to be established.

VII. Patterns of Redistribution of Ions

A. CEREAL SEEDLINGS

In seeking to establish the principles governing the distribution of materials within such a complicated organism as a typical seed plant, it is useful to examine first the simplest possible system. Such is presented by a young seedling growing under controlled conditions with a minimal external supply of nutrients. Here, a single source (the cotyledons or endosperm) is supplying relatively few sinks in the developing embryo. Surprisingly, there were few studies of this kind before about 1960. Brown (44) had noted that the dry weight of attached embryos of barley seedlings increased less rapidly during the first 48 h after germination than did nitrogen content, suggesting that there was greater export of nitrogen compounds than total dry matter from the endosperm in the early stages of seedling growth. Mer *et al.* (179) confirmed that this was the case in dark-grown oat seedlings, and they supported the suggestion of Albaum *et al.* (2) that growth of these seedlings immediately after germination is controlled by nitrogen supply. However, it is dangerous to reach such a conclusion on the basis of a study of the relationship between transport of nitrogen compounds and dry matter alone. Baset and Sutcliffe (24) studied the export of four elements—potassium, nitrogen, magnesium, and phosphorus—in relation to movement of dry matter from the endosperm during the first seven days of growth of oat seedlings in the dark at 25°C. The plants were grown in various nutrient solutions to determine the influence of an external supply of particular elements on depletion of the endosperm. When a low concentration (10 m*M*) of calcium chloride solu-

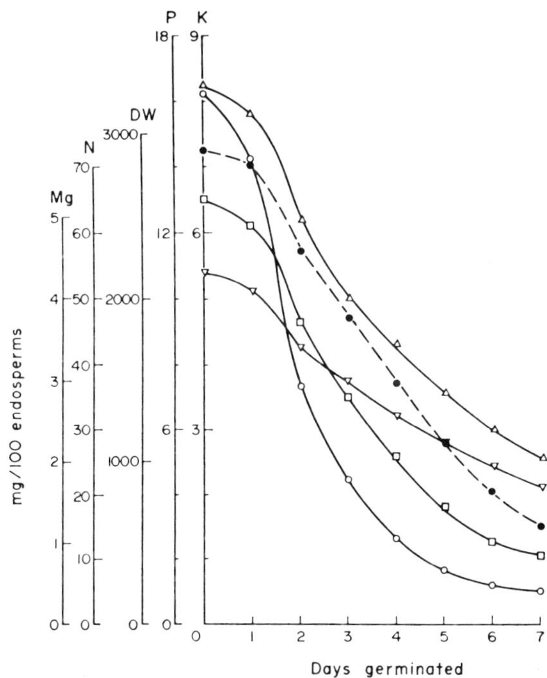

FIG. 20. Changes in mean dry weight (●) and content of nitrogen (□), phosphorus (△), potassium (○), and magnesium (▽) of endosperm of oat seedlings during germination in the dark at 25°C. From Baset (23).

tion was provided, the embryos grew rapidly at the expense of nutrients from the endosperm which was depleted slowly at first, building up to a maximum rate and then declining slowly with time (Fig. 20). Differences in the relative rates of export of individual elements are seen more clearly when the data are plotted as in Fig. 21. It is clear that the rates of export of individual elements relative to dry matter change markedly with time and that the pattern is different for each element.

The simplest curve to analyze is that for potassium. This element was transported from the endosperm relative to dry matter most rapidly during the first day of germination and progressively more slowly thereafter (Fig. 20). This ion appears to be readily available for export soon after hydration of the grain, and it is evident that strong sinks for this element already exist in the growing axis. Because sink strength is likely to increase with time as the axis enters its exponential phase of growth, it is probable that the falling rate of export is related to diminishing supply. After 7 days, the potassium content of the endosperm had been reduced to less than 10% of the original.

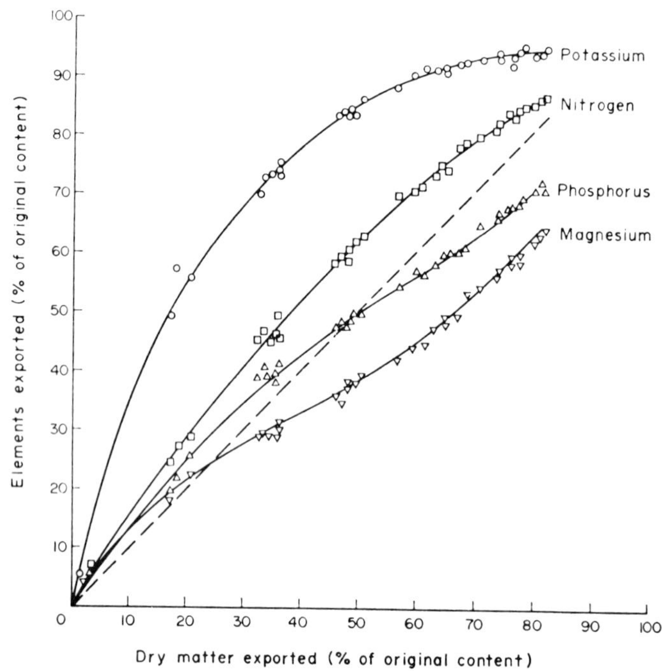

FIG. 21. The relationship between export of dry matter and of nitrogen, phosphorus, potassium, and magnesium from the endosperm of oat seedlings germinated in the dark at 25°C for 7 days. From Baset and Sutcliffe (24).

The initial rate of transport of nitrogen from the endosperm relative to dry matter was lower than for potassium, and it decreased less rapidly with time. This was attributed to a slower rate of mobilization of nitrogen than of potassium because of low proteolytic activity in the newly hydrated endosperm. Protease activity had increased markedly by the second day and did so more rapidly than amylase activity, thus accounting perhaps for the more rapid transport of nitrogen relative to total dry matter in the early phase of seedling growth (246, 247).

The curves for the depletion of magnesium and phosphorus relative to dry matter are more difficult to interpret. The similarities between the two curves (Fig. 21) may be attributed to the fact that both elements are stored in oat endosperm mainly in the form of phytin (magnesium and calcium salts of myoinositol hexaphosphate). Differences may be due to the occurrence of an appreciable amount (20–30%) of phosphorus, but not magnesium, in substances other than phytin, e.g., RNA, which may be mobilized at different rates.

It was concluded from this research that transport of materials from the endosperm of oat seedlings is controlled by growth of the axis through a push–pull system involving both source and sink. Movement of individual elements occurs along a solute potential gradient determined by mobilization of each solute at its source and utilization at the sink. The axis controls sink strength through incorporation of materials into cell constituents or by sequestration in vacuoles. It also regulates the mobilization of reserve materials in response to demand by controlling the rate of synthesis or activation of hydrolytic enzymes in the aleurone layer, presumably through the production of gibberellins (256, 274). Thus both utilization and mobilization of individual elements are under precise control, and transport rate is altered quickly in response to changing needs during early seedling growth.

B. Dicotyledons

The distribution pattern of mineral elements from the cotyledons of dicotyledonous seeds seems to be rather different from that of the endosperm of cereal grains. Bukovac and Riga (48, 215) noted that translocation of cotyledonary phosphorus, calcium, and zinc during germination and early seedling development of *Phaseolus vulgaris* was closely related to seedling growth. Sutcliffe and his co-workers (58, 105–108) found that the rate of transport of a range of ions relative to dry matter from the cotyledons of pea *(Pisum sativum)* seeds remained remarkably constant during the first 4 weeks of growth, each 100 mg of dry matter exported containing 0.8, 0.1, 3.4, 0.01, and 0.5 mg of K, Mg, N, Na, and P, respec-

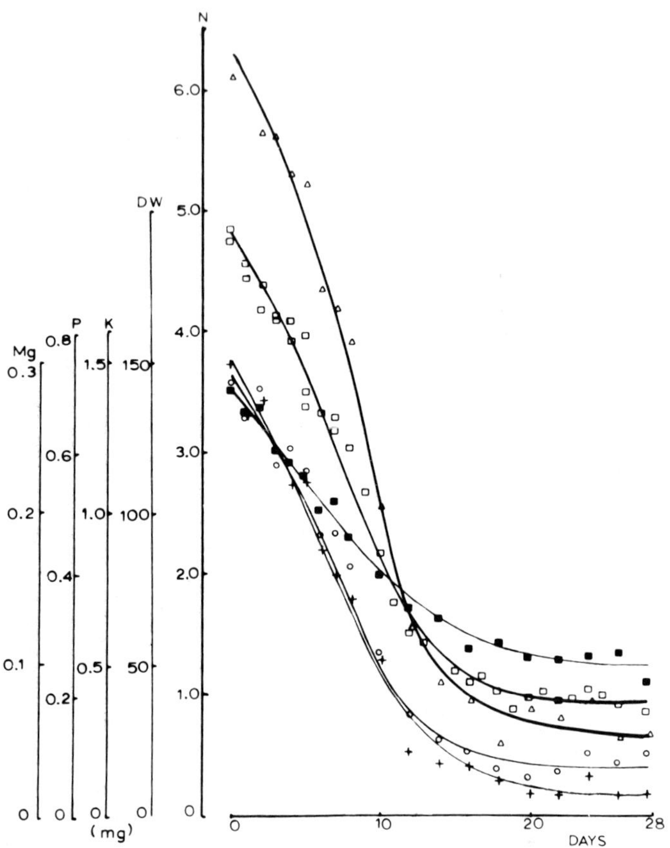

FIG. 22. Changes in mean dry weight (□) and content of nitrogen (△), phosphorus (+), potassium (○), and magnesium (■) of cotyledons of 4-week-old pea seedling. From Collins and Sutcliffe (58).

tively (Figs. 22, 23). Varying the conditions under which seedlings were grown led to a situation in which although the absolute amounts of nutrients exported from the cotyledons altered in response to different rates of growth of the axis, the proportions of the different elements in the total dry matter of the cotyledons remained the same (105).

Calcium differed from the other elements examined in that a relatively high percentage (70–75%) of that found in pea cotyledons was not transported into the axis. This fraction is believed to be associated mainly with cell walls. The mobile fraction appears to occur as calcium phytate, which is broken down by sequential dephosphorylation under the influence of

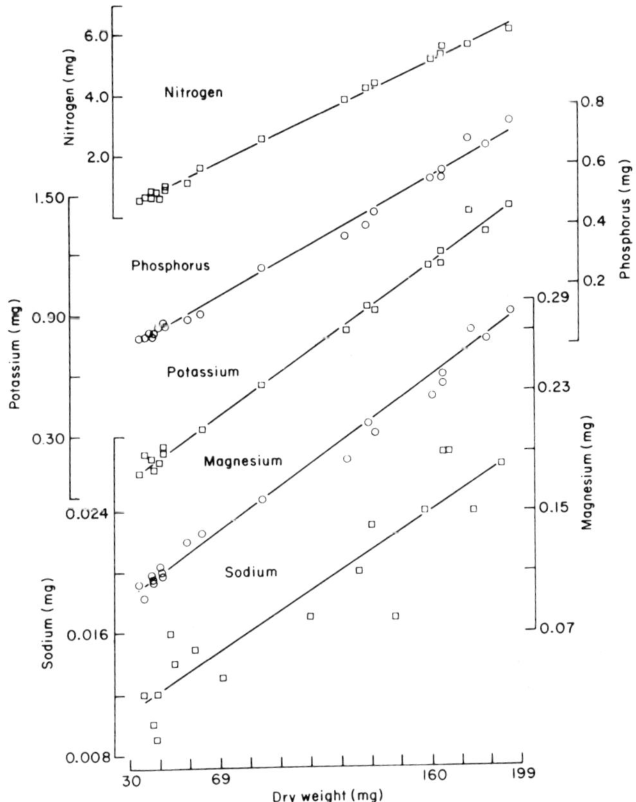

FIG. 23. The relationship between dry matter and content of various elements in the cotyledons of pea seedlings up to 4 weeks old. After Collins and Sutcliffe (58).

phytase (61, 81, 82). Ferguson (80) found that calcium injected as a solution of calcium salts into pea cotyledons was not transported to the axis, presumably because it became bound in cell walls.

Experiments in which the shoots of 9-day-old pea seedlings were steam-girdled (108) showed that movement of calcium and other solutes out of the cotyledons was markedly inhibited and indicated that export occurs in the phloem. When calcium is absorbed by the roots of young pea seedlings, some of it is transported into the cotyledons, presumably through the xylem. Since it is difficult to grow pea seedlings successfully in the complete absence of calcium from the nutrient medium, it is not easy to establish a relationship between calcium export and that of other elements. However, when pea seedlings are grown for a short time with their roots in

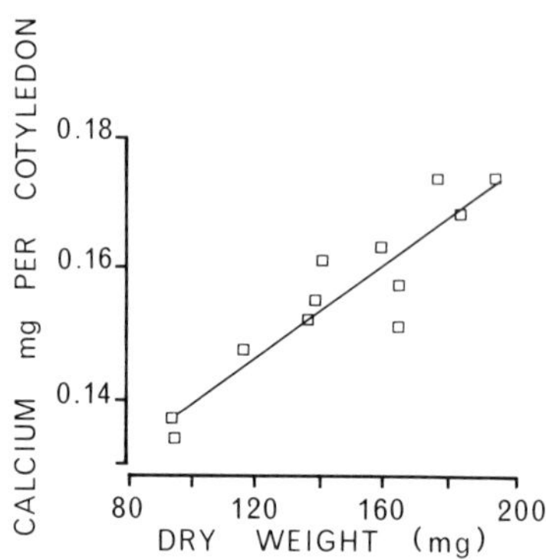

FIG. 24. The amount of calcium in cotyledons of pea seedlings up to 4 weeks old plotted against cotyledon dry weight. Unpublished data of O. D. G. Collins.

distilled water, the indications are that the ratio of mobile calcium to other elements transported remains constant (Fig. 24).

The mechanism by which the ratio of the various elements is maintained constant during cotyledon depletion is not yet clear. There is no qualitative relationship between the activity of proteolytic enzymes in cotyledon extracts and transport of nitrogen (106), or between phytase activity and phosphorus or magnesium transport (107) as was observed in oat seedlings (23). It has been proposed (245) that the rates at which individual elements are transported are determined by the proportions in which they are released from the storage cells and loaded into the sieve tubes at the source.

The view that the rate of transport of individual elements is controlled to some extent by their availability in a mobile form at the point of loading is supported by the finding that injection of potassium chloride into the cotyledons of pea seedlings causes increased movement of potassium relative to other elements (Fig. 25).

Removal of the shoot from a 1-week-old pea seedling did not influence movement of potassium into the roots, suggesting that there is little or no competition between shoot and root for this nutrient in the early stage of seedling growth. Dry matter content of the roots of deshooted plants was lower than in intact plants, but total nitrogen, phosphorus, and sulfur contents were significantly higher. The increases are an indication either that there is competition between shoot and root for these elements in the intact plant or that there is some secondary redistribution from root to shoot which is prevented when the shoot is removed.

The existence of a stoichiometric relationship between the movement of individual elements and total dry matter from cotyledons of a variety of plants has been observed by a number of other investigators (137, 190).

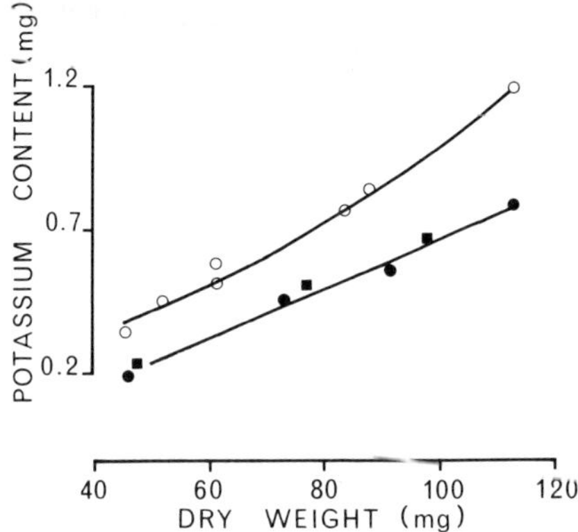

FIG. 25. The effect of injecting various solutions into cotyledons of pea on the relationship between the export of potassium and dry matter to the plumular axis. Unpublished data of O. D. G. Collins.

Hocking (122) studied the transport of dry matter and a range of macro- and micronutrient elements from the cotyledons of two species of *Lupinus* during early seedling growth. Both species showed a linear relationship between the loss of each element and dry matter throughout the growth period and a similar division between root and shoot of nutrients transported from the cotyledons. At the time of cotyledon death, N, K, Mg, and Mn had moved mainly into the shoot, and in the case of Mn it was over 80%, of which 75% was in the leaflets. In contrast, Na accumulated mainly in the hypocotyl and roots (over 92% in *Lupinus angustifolius* and 76% in *Lupinus albus*). The remainder of the elements (including Ca) were distributed more or less evenly between roots and shoots in both species.

The work and data surveyed here indicate the close control that exists over the mobilization of mineral reserves during early seedling development. The mechanisms involved remain largely unknown and are likely to be uncovered only by further detailed studies along the lines of those indicated.

VIII. Salt Relations and Growth

A. REGULATION OF SALT RELATIONS

Early investigators of ion transport in whole plants soon discovered that uptake and distribution were closely related to growth of the plant and its individual parts. Muenscher (188), for example, found that salt uptake by intact barley plants was more exactly correlated with the intensity of photosynthesis and growth than with water absorption. The importance of growth in regulating ion accumulation in the alga *Nitella* can be inferred from the experiments of Hoagland and collaborators in the 1920s. This, and the other work that stemmed from it, is discussed by Steward and Sutcliffe (232), and Chapter 7 in this volume by Steward brings up to date results from cell and tissue culture studies and their application to plants as they grow.

Since growth depends on a spectrum of essential elements, which are provided by the environment in ionic form, it is not surprising that ion uptake increases with increasing demand as growth proceeds. A larger surface area for absorption and multiplication of sites for metabolic utilization are among the contributing factors. A deficiency of any one essential element leads to diminution of growth and a concomitant reduction in the intake of other ions.

However, the relationship between salt uptake and growth cannot be explained simply in terms of supply and demand. This is evident from the fact that nonessential elements such as Br^- and Rb^+ are also absorbed at an enhanced rate when plants are growing rapidly. It has also been demonstrated that plants go on absorbing more ions when external concentration is increased beyond that causing maximum growth. Loneragan (163) termed the minimum concentration at which a plant can maintain its metabolic functions at rates which do not limit growth, the functional nutrient requirement. The lowest concentration in the plant at which the plant can sustain its maximum rate of growth was termed the critical nutrient concentration. Since ion concentrations vary considerably in different parts of a plant, the critical level may be reached in one place and not in another. This is most likely to occur with ions such as calcium which are of limited mobility in the phloem. Even so, concentration is not the only criterion upon which evidence for an adequate salt supply should be judged. Quantity is of considerable importance, and if there is an unlimited supply of a mobile nutrient, a low concentration in the medium can sustain maximum growth. This is shown by the ability of plants to grow well in flowing solution culture at much lower concentrations of ions than are necessary when the solution is static.

It is evident that both mature and actively growing cells have the ability to accumulate ions from the external environment in excess of their immediate needs. This attribute is of fundamental importance because it is the basis of temporary storage within the plant which mitigates the effects of changing supply and demand. Most of the reserve mineral elements are either stored as free ions in the vacuolar sap, where they contribute to the osmotic potential of the cell, or they are deposited as insoluble precipitates in the cytoplasm or cell wall.

In mature cells, the ability to accumulate solutes is limited by the finite size of the vacuolar reservoir, and as the internal concentration rises, net absorption declines. To some extent this can be explained in terms of increased passive or active efflux, but the effect is mainly attributable to feedback control of influx (63). In growing cells, because of increasing

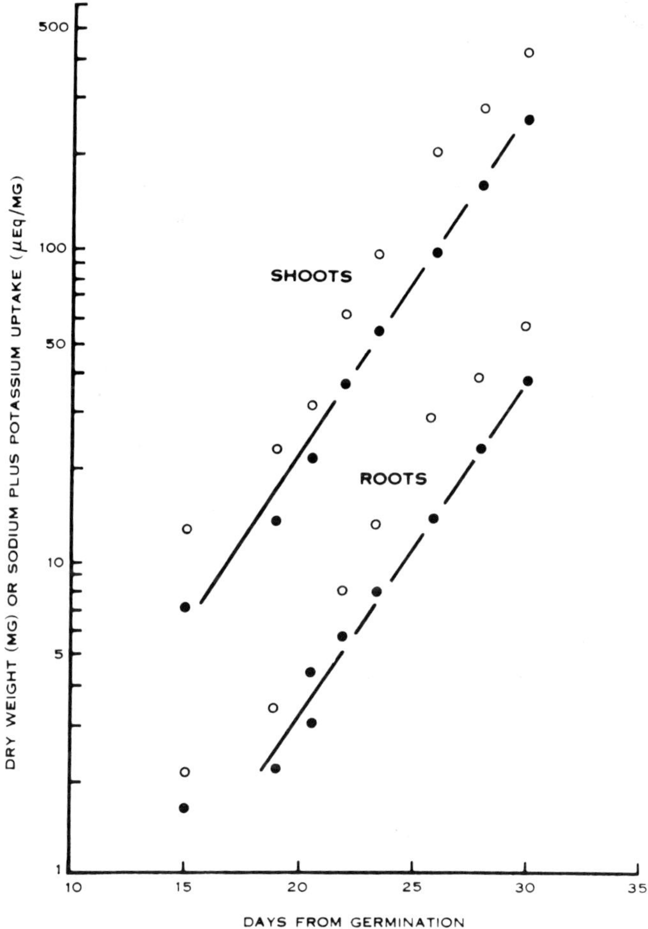

FIG. 26. Changes with time in shoot and root dry weight (●) and content of potassium and sodium (○) of mustard seedlings *[Brassica hirta (Sinapis alba)]*. From Pitman (202).

volume, this restraint is less severe, although again there are indications that ions or their metabolic products may feed back to match the rate of uptake more nearly to that of utilization. The concentration of potassium ions in the vacuolar sap of root cells may indeed decrease during the period of most rapid extension growth.

In whole plants, composed of many different types of cells at various stages of development, there appears to be a close correlation between growth rate, as measured by changes in dry weight, and mineral content. Pitman (202, 207) found that the quantity of potassium and sodium ions relative to dry matter in mustard seedlings *[Brassica hirta (Sinapis alba)]* remained relatively constant over a period of 30 days from germination (Fig. 26). Similar correlations were found between potassium content and growth of barley seedlings with different relative growth rates caused by varying the length of the photoperiod (204). Whereas relative growth rate (RGR) varied from almost zero to 0.22 day^{-1}, potassium concentration varied only from about 190 to 240 $\mu mol\ g^{-1}$ FW. Although there were large differences in rates of transport to the shoot relative to root weight, there was little difference relative to shoot weight for the same RGR. This led to the conclusion that transport of univalent cations from root to shoot is controlled in some way by growth of the shoots. Some of the possible control mechanisms are considered later.

A close relationship has been shown (271) between the accumulation of potassium ions and water content during development of individual leaves in *Phaseolus vulgaris.* When potassium supply is insufficient, leaf potassium content falls progressively with time relative to both fresh and dry weight. Pitman (202) showed that the ratio of $K^+ : Na^+$ is higher in younger than in older leaves of *Brassica hirta,* and this can be attributed to a change in the proportion of ions supplied to the leaf through the phloem and xylem as leaves mature. Phloem sap is much richer in potassium ions relative to sodium ions than is xylem sap, and so young leaves supplied mainly through the phloem accumulate more potassium. In the later stages of leaf growth there may be net export of potassium in the phloem, and this will also tend to reduce the K : Na ratio. Jeschke (135) has used the technique of flux analysis developed for use with excised roots to study the influence of shoot : root ratio on the influx of potassium and sodium into root cells and transport of these ions into the shoot. He showed that when the shoot : root ratio was increased by removal of some of the seminal roots, there was an increase in uptake by the remaining roots. There was also an increase in xylem transport expressed on a root weight basis. He proposed that this stimulation was attributable in part to a relatively higher rate of supply of photosynthates to the root system in plants with fewer roots. The failure of the roots to absorb more ions in plants kept in the dark supports this conclusion. However, Jeschke doubted whether this was an adequate explanation of the whole effect, and he suggested that hormonal control may also be involved.

Abscisic acid (ABA) has been shown to have inhibitory effects on release of potassium and sodium ions into the xylem of barley roots (64, 210, 211) without having any effect on uptake by the root, and it has been suggested that it may be involved in shoot-mediated control of xylem transport. How a high shoot : root ratio could lead to enhanced transport of cations from the root through the mediation of ABA is not at all clear. A high ratio of shoot to root would be expected to increase the amount of ABA supplied to unit weight of root and thus lead to increased inhibition. Jeschke suggested that cytokinins are more likely to be involved, although these hormones have also generally been found to inhibit transport of ions to the shoot (255; but see 204). Possibly, a reduced supply of cytokinins from the roots of plants with low root : shoot ratios triggers a stimulus (increased photosynthesis or reduced ABA synthesis) in the shoot. Further studies are needed to investigate these effects. The influence of root pruning in promoting transport of ions from the remainder of the root system is reminiscent of the effects of source limitation on the movement of assimilates from leaves into developing ears of cereals (258) and may have a similar explanation.

A role for cytokinins in regulating ion accumulation in leaves is more clear-cut. Kinetin and related substances (e.g., benzyladenine) have the capacity to delay senescence of tissues, and when applied to a leaf, they stimulate movement of materials toward the treated region (Fig. 27).

Saeed (226) has shown that application of benzyladenine delayed senescence in a mature leaf of *Xanthium strumarium* and reduced the export of assimilates and of ^{32}P. Growth of the next-higher leaf in the same orthostichy was reduced and smaller amounts of carbohydrates and ^{32}P were transported into it. These effects were thought to be due to an influence of cytokinins on RNA and protein synthesis. Effects of treatment on ion transport may be due to stimulation in the rate of synthesis of carrier protein in membranes which turn over rapidly (243). Inhibitors of protein synthesis have the opposite effect inhibiting ion transport (156, 266, 267).

For many years the varied effects of auxins such as indole acetic acid on plant salt relations were difficult to assess since it was uncertain whether these were due to growth or reflected a direct effect of auxin on ion accumulation or transport. Development of the acid growth hypothesis (see 255, 276) has helped clarify our understanding, and it is now apparent that in a number of systems auxins stimulate the secretion of protons with consequent effects on the transport of other ions, either by exchange or by cotransport (see Fig. 28 and Chapter 6 in this volume). While auxin treat-

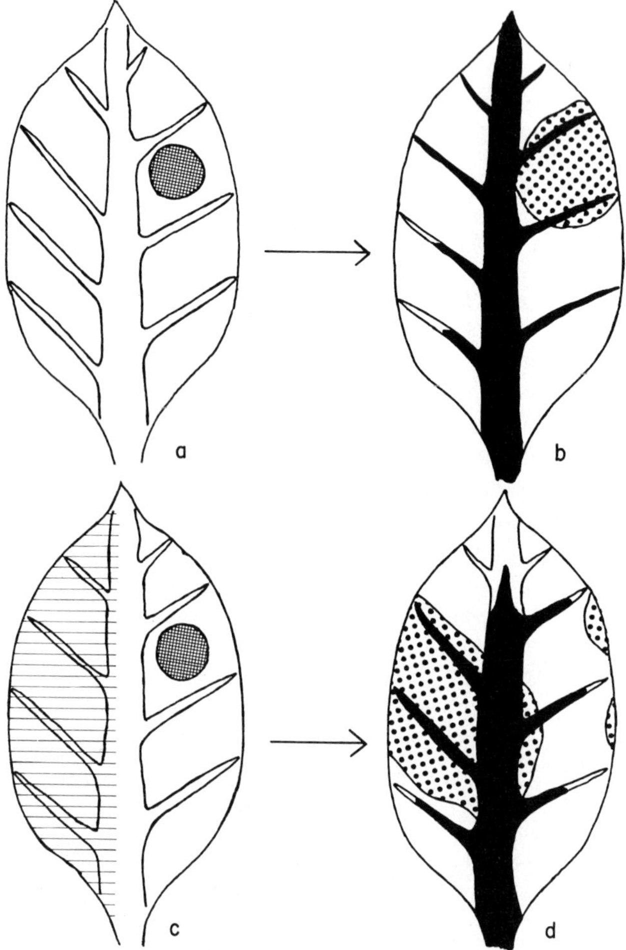

FIG. 27. Diagrammatic summary of the experiment of Mothes (187). In (a) Kinetin and [^{14}C]glycine were applied to the circular area on the right of the leaf blade of *Nicotiana rustica*. 13 days later labeled carbon was found in the major veins (black) and in the treated area and adjacent regions (b). In another treatment (c) kinetin was applied to the left-hand side of the blade (shaded) and [^{14}C]glycine to the circular areas as in (a). After 13 days most labeled carbon was found in the veins and in the kinetin-treated region.

ments may well affect ionic relations of the cell through proton extrusion mechanisms, there may be concomitant and resultant effects on ion transport, for example, into or out of the xylem.

There are also long-term effects of auxin treatment. For example, Mitchell and Martin (182) found a greater accumulation of ions, along

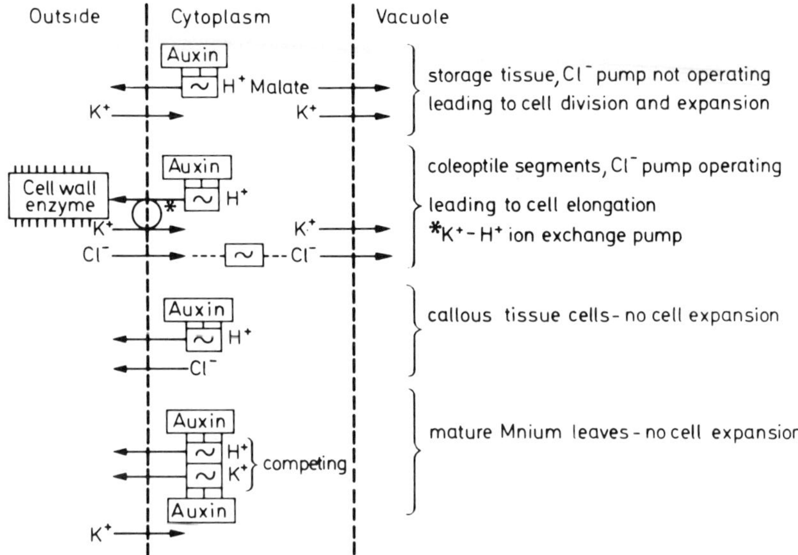

FIG. 28. Possible ways in which IAA may affect ion transport through proton secretion. After van Steveninck (255).

with soluble carbohydrates and nitrogenous compounds in bean seedlings *(Vicia faba)* treated with IAA. In such experiments, conducted over a period of several days, it is possible that growth induced by auxin is the main cause of the greater accumulation of ions. Even in short-term experiments, such as those of Davies and Wareing (68), in which increased transport of ^{32}P toward the top of a decapitated stem of a French bean seedling *(Phaseolus vulgaris)* was detected a few hours after treatment with IAA, the possibility of metabolic activity associated with incipient growth effects cannot be ruled out. However, from experiments involving the use of triiodobenzoic acid, a known inhibitor of auxin transport, Davies and Wareing concluded that basipetal transport of IAA from the cut stump was somehow involved in the stimulation of ^{32}P transport toward the treated region.

Seth and Wareing (228) showed that kinetin and gibberellic acid had no effect on transport of ^{32}P through defruited peduncles of French bean plants, but when these substances were applied together with IAA, the effect of the auxin was markedly enhanced. Whether these effects are due to an influence on transport per se or to an effect on processes, such as RNA or protein synthesis, which lead to a stimulation of growth, is still uncertain. (See also Chapter 6.)

There have been few studies of the effects of gibberellins on uptake and movement of ions in whole plants. Guardiola and Sutcliffe (106, 107) showed that GA_3 stimulated shoot growth in seedlings of *Pisum sativum* cv. "Alaska" without any effect on potassium, phosphorus, nitrogen, or sulfur content on a dry weight basis. Garcia-Luis and Guardiola obtained a much larger effect of GA_3 on shoot growth in another pea variety, Progress, and in this case the nitrogen content in the shoot increased while that of the root was reduced (104). The GA_3 treatment stimulated amylolytic activity in the cotyledons of Progress pea seedlings, and there was increased movement of dry matter and nitrogen from them into the shoot.

B. CONCLUDING REMARKS

The view that emerges from the work presented in this chapter is that a vascular plant functions as an integrated unit in which the distribution of solutes is controlled in such a way that the available supply is utilized most efficiently.

During early growth a seedling is more or less dependent on endogenous reserves of mineral elements as well as of organic solutes. These are mobilized in the storage tissue at a rate determined by growth of the axis and distributed between shoot and root according to the demands of each organ. This movement occurs mainly, if not entirely, in the phloem.

Sooner or later the plant becomes dependent on inorganic solutes absorbed by the roots and carried into the shoot in the transpiration stream. The shoot exerts some control over the absorption of ions by the roots and in the proportion transferred to the shoot. In addition, the shoot may exert an influence by the provision of respiratory substrates upon which absorption into the root and transport to the xylem depends. There is also a possibility that the shoot exerts some control through growth-regulating substances.

In general, these mechanisms result in the supply of ions to the shoot through the xylem being increased when the demand is high and reduced when it is low. However, control of ion transport through the xylem does not seem to be a very effective mechanism for regulating the supply to individual leaves during the course of their development. When a leaf is young, the supply of ions via the xylem is inadequate for the rapidly increasing needs of the expanding cells, and it is supplemented by an additional supply through the phloem. Later, when growth declines and transpiration rate reaches a peak, supply through the xylem is more than sufficient and excess solutes are exported in the phloem. Solutes thus move in the sieve tube along concentration gradients from regions of higher to lower concentration. Solute concentration at the source depends on the supply and in particular on the ability of cells to release stored materials. Sink activity (see Chapter 6, on the other hand, is controlled by the rate at which solutes are removed from the solution to be utilized in growth or accumulated in vacuoles. The amount of an individual solute transported in unit time from a source to an individual sink will be related to the difference in concentration between the two organs and the resistance presented by the connecting sieve tubes.

According to this view, a leaf continues to act as a sink as long as the concentration of the solution bathing its sieve tube terminals is sufficiently low to maintain a gradient in the right direction between it and neighboring leaves. When growth begins to slow down, solutes start to accumulate in the free space and the gradient is reduced until it is finally reversed. The leaf then begins to act as a source, but it does not necessarily do so for all solutes simultaneously because the rates of consumption and the supply of individual substances may be different. In general, the ability of a leaf to function as a sink depends on the intensity of RNA and protein synthesis. This is not surprising because the rate of synthesis of these substances must control the rate of utilization of a variety of organic and inorganic solutes. The degradation of macromolecules, which is a feature of senescing leaves, leads to release of soluble materials, which accumulate in the free space and cause the organ to behave as a source.

One of the most intriguing problems now confronting plant physiologists in the transport field is the distinction between sources and sink. A source cell or tissue is one which releases materials to its surroundings, and a sink is one with the capacity to accumulate them. Most plant cells act successively as sink and source during the course of their life. The difference between a source and a sink might well lie in the direction of operation of proton pumps located in the plasmamembrane. What controls this we do not know, but by analogy with the ion relations of chloroplasts and mitochondria, it seems feasible that a pH gradient is involved. Generation of an excess of hydrogen ions through cellular metabolism may set the mechanism into its extrusion mode, leading to accumulation of other cations by exchange and concomitant uptake of anions or organic molecules by cotransport (see Chapter 6). Thus the cell behaves as a sink. On the other hand, utilization of protons inside the cell may lead to reversal of the pump with associated release of ions and other materials, causing the cell to act as a source.

A second area of research in which progress is vital for further understanding of ion distribution in the whole plant is that of symplastic transport, including that in the sieve tubes. We are aware that ions move through cytoplasm, apparently in a polarized manner, at rates which are in excess of diffusion rates, but we still do not comprehend the mechanisms involved. Various possibilities include the following.

1. Bulk flow of ions in aqueous solution through channels in the cytoplasmic matrix is caused either by a hydrostatic pressure gradient, as seems to be the case in sieve tubes (see Chapter 6), or by some kind of peristaltic or electroosmotic activity generated within the cytoplasm.
2. The ions may be accumulated in small vesicles which are transported through the cytoplasm by bulk flow of the aqueous phase. Such particles need to be small enough to pass through plasmodesmata, i.e., not more than about 30 to 60 nm in diameter, or capable of protean changes of shape which would enable them to squeeze through pores of such dimensions. Anderson (3) has discussed the physicochemical limitations on transport through plasmodesmata.
3. Ions may move by exchange processes along surfaces such as those of the endoplasmic reticulum. If the surfaces are themselves in a state of flux through cytoplasmic streaming, the rate of transport of bound ions might be increased. But since it is improbable that streaming of cytoplasm occurs through plasmodesmata, the main effect of any streaming is likely to be on the rate of equilibration of ions between one part of a cell and another.

Thus despite the great increase in factual knowledge about the salts that cells and organs of plants obtain from their environments and the solutes they retain, there are still some intractable problems and questions that are difficult to comprehend. Some of these are referred to in Chapter 7, especially where they relate to cells not merely as they exist, but as they grow. Although these questions could be deferred for consideration in volumes specifically devoted to growth and development, it is fitting that they should be raised in this volume and in transition to those that are to follow.

References

1. Adriani, M. J. (1958). Halophyten. *In* "Handbuch der Pflanzenphysiologie" (W. Ruhland, ed.), Vol. 4, pp. 709–773. Springer-Verlag, Berlin and New York.
2. Albaum, H. G., Donnely, J., and Korkes, S. (1942). The growth and metabolism of oat seedlings after seed exposure to oxygen. *Am. J. Bot.* **29**, 388–395.
3. Anderson, W. P. (1976). Physico-chemical assessment of plasmodesmatal transport. *In* "Intercellular Communication in Plants. Studies on Plasmodesmata" (B. E. Gunning and A. W. Robards, eds.), pp. 107–120. Springer-Verlag, Berlin and New York.
4. Anderson, W. P. (1976). Transport through roots. *In* "Encyclopedia of Plant Physiology, New Series" (U. Lüttge and M. G. Pitman, eds.), Vol. IIB, pp. 129–156. Springer-Verlag, Berlin and New York.
5. Anderson, W. P., and Higinbotham, N. (1975). A cautionary note on plant electrophysiology. *J. Exp. Bot.* **26**, 523–536.

6. Anderson, W. P., and House, C. R. (1967). A correlation between structure and function in the root of *Zea mays. J. Exp. Bot.* **18,** 544–555.
7. Arisz, W. H. (1956). Significance of the symplasm theory for transport across the root. *Protoplasma* **46,** 5–62.
8. Arisz, W. H., Camphuis, I. J., Heikens, H., and van Tooren, A. J. (1955). The secretion of the salt gland of *Limonium latifolium* Ktze. *Acta Bot. Neerl.* **4,** 322–338.
9. Arnon, D. I., Stout, P. R., and Sipos, F. (1940). Radioactive phosphorus as an indicator of phosphorus absorption at various stages of development. *Am. J. Bot.* **27,** 791–798.
10. Asher, C. J., and Loneragan, J. F. (1967). Response of plants to phosphate concentration in solution culture. 1. Growth and phosphate content. *Soil Sci.* **103,** 225–233.
11. Asher, C. J., and Ozanne, P. G. (1967). Growth and potassium content of plants in solution culture maintained at constant potassium concentrations. *Soil Sci.* **103,** 155–161.
12. Aston, M., and Jones, M. M. (1976). A study of the transpirational surfaces of *Avena sterilis* L. var Algerian leaves using monosilicic acid as a tracer for water movement. *Planta* **130,** 121–129.
13. Atkinson, M. R., Findlay, G. P., Hope, A. B., Pitman, M. G., Saddler, H. D. W., and West, K. R. (1967). Salt regulation in the mangroves *Rhizophora mucronata* Laun and *Aegialitis annulata* R. Br. *Aust. J. Biol. Sci.* **20,** 589–599.
14. Austin, R. B., Edrich, J. A., Ford, M. A., and Blackwell, R. D. (1977). The fate of dry matter, carbohydrates and ^{14}C lost from the leaves and stems of wheat during grain filling. *Ann. Bot. (London)* [N.S.] **41,** 1309–1321.
15. Baker, D. A. (1975). The effect of CCCP on ion fluxes in the stele and centre of maize roots. *Planta* **112,** 293–299.
16. Baker, D. A., and Hall, J. L., eds. (1975). "Ion Transport in Plant Cells and Tissues." North-Holland Publ., Amsterdam.
17. Baker, D. A., and Moorby, J. (1969). The transport of sugar, water and ions into developing potato tubers. *Ann. Bot. (London)* [N.S.] **33,** 729–741.
18. Baker, D. A., and Weatherley, P. E. (1969). Water and solute transport by exuding root systems of *Ricinus communis. J. Exp. Bot.* **20,** 485–495.
19. Barber, D. A., and Shone, M. T. (1966). The absorption of silica from aqueous solutions of plants. *J. Exp. Bot.* **17,** 569–578.
20. Barke, R. E. (1968). Absorption and translocation of calcium foliar sprays in relation to the incidence of blossom end rot in tomatoes. *Ohio J. Agric. Sci.* **25,** 179–197.
21. Barke, R. E., and Menary, R. C. (1971). Calcium nutrition of the tomato as influenced by total salts and ammonium nutrition. *Aust. J. Exp. Agric. Animal Husb.* **11,** 562–569.
22. Barry, P. H., and Hope, A. B. (1969). Electro-osmosis in membranes: Effects of unstirred layers and transport numbers. I. Theory. *Biophys. J.* **9,** 700–728.
23. Baset, Q. A. (1972). Mobilisation and transport of food reserves in etiolated oat seedlings. D. Phil. Thesis, University of Sussex, U.K.
24. Baset, Q. A., and Sutcliffe, J. F. (1975). Regulation of the export of potassium nitrogen, phosphorus, magnesium and dry matter from the endosperm of etiolated oat seedlings (*Avena sativa* cv Victory). *Ann. Bot.* **39,** 31–41.
25. Bell, C. W., and Biddulph, O. (1963). Translocation of calcium. Exchange versus mass flow. *Plant Physiol.* **38,** 610–614.
26. Bengisson, B. (1982). Uptake and translocation of calcium in cucumber. *Physiol. Plant.* **54,** 107–111.
27. Bernstein, L., and Hayward, H. E. (1958). Physiology of salt tolerance. *Annu. Rev. Plant Physiol.* **9,** 925–946.

28. Bernstein, L., and Nieman, R. H. (1960). Apparent free space of plant roots. *Plant Physiol.* **35,** 589–598.
29. Biddulph, O. (1951). The translocation of minerals in plants. *In* "Mineral Nutrition of Plants" (E. Truog, ed.), pp. 261–275. Univ. of Wisconsin Press, Madison.
30. Biddulph, O. (1959). Transport of inorganic solutes. *In* "Plant Physiology: A Treatise" (F. C. Steward, ed.), Vol. 2, pp. 553–603. Academic Press, New York.
31. Biddulph, O., Biddulph, S. F., Cory, R., and Koontz, H. (1958). Circulation patterns for P^{32}, S^{35} and Ca^{45} in the bean plant. *Plant Physiol.* **33,** 293–300.
32. Biddulph, O., Nakayama, F. C., and Cory, R. (1961). Transpiration stream and ascension of calcium. *Plant Physiol.* **36,** 429–436.
33. Bieleski, R. (1966). Accumulation of phosphate, sulphate and sucrose by excised phloem tissues. *Plant Physiol.* **41,** 447–454.
34. Bieleski, R. (1966). Sites of accumulation in excised phloem and vascular tissues. *Plant Physiol.* **41,** 455–456.
35. Bledsoe, R. W., Comar, C. I., and Harris, H. C. (1949). Absorption of radioactive calcium by the peanut fruit. *Science* **109,** 229–230.
36. Bollard, E. G. (1953). The use of tracheal sap in the study of apple tree nutrition. *J. Exp. Bot.* **4,** 363–368.
37. Bollard, E. G. (1960). Transport in the xylem. *Annu. Rev. Plant Physiol.* **11,** 141–166.
38. Bowling, D. J. F. (1976). "Uptake of Ions by Plant Roots." Chapman & Hall, London.
39. Bowling, D. J. F. (1981). Release of ions to the xylem in roots. *Physiol. Plant.* **53,** 392–397.
40. Bowling, D. J. F., and Weatherley, P. E. (1965). The relationship between transportation and potassium uptake in *Ricinus communis. J. Exp. Bot.* **16,** 732–741.
41. Briggs, G. E. (1957). Some aspects of free space in plant tissues. *New Phytol.* **56,** 305–324.
42. Brouwer, R. (1954). The regulating influence of transpiration and suction tension on water and salt uptake by the roots of intact *Vicia faba* plants. *Acta Bot. Neerl.* **3,** 264–312.
43. Brouwer, R. (1956). Investigations into the occurrence of active and passive components in ion uptake by *Vicia faba. Acta Bot. Neerl.* **5,** 287–314.
44. Brown, R. (1946). Studies on germination and seedling growth. III. Early growth in relation to certain aspects of nitrogen metabolism in the seedling of barley. *Ann. Bot. (London)* **10,** 73–76.
45. Brown, R. (1947). The gaseous exchange between the root and the shoot of the seedling of *Cucurbita pepo. Ann. Bot. (London)* **11,** 417–437.
46. Brownell, P. F., and Crossland, C. J. (1972). The requirement of sodium as a micronutrient by species having the C_4 dicarboxylic acid photosynthetic pathway. *Plant Physiol.* **49,** 794–797.
47. Broyer, T. C., and Hoagland, D. R. (1943). Metabolic activity of roots and their bearing on the relation of upward movement of salt and water in plants. *Am. J. Bot.* **30,** 261–273.
48. Bukovac, M. J., and Riga, J. A. (1962). Redistribution of cotyledonary phosphorus, calcium and zinc during germination and early seedling development of *Phaseolus vulgaris* L. *Proc. Int. Hortic. Congr. 16th, 1962,* Vol. 2, pp. 280–285.
49. Carlisle, A., Brown, A. H. F., and White, E. J. (1966). The organic matter and nutrient elements in the precipitation beneath a sessile oak *(Quercus petraea)* canopy. *J. Ecol.* **54,** 87–98.
50. Christersson, L., and Pettersson, S. (1968). Water and sulphate uptake at root reduction in *Ricinus communis. Physiol. Plant.* **21,** 414–422.

51. Clarkson, D. T. (1974). "Ion Transport and Cell Structure in Plants." McGraw-Hill, New York.
52. Clarkson, D. T., and Hanson, J. B. (1981). The mineral nutrition of higher plants. *Annu. Rev. Plant Physiol.* **31,** 239–298.
53. Clarkson, D. T., and Robards, A. W. (1975). The endodermis, its structural development and physiological role. *In* "The Development and Function of Roots" (J. G. Torrey and D. T. Clarkson, eds.), pp. 415–436. Academic Press, London and New York.
54. Clements, C. R., Hopper, M. J., Ganaway, R. J., and Jones, L. H. D. (1974). A system for measuring uptake of ions by plants from flowing solutions of controlled composition. *J. Exp. Bot.* **25,** 81–99.
55. Collander, R. (1941). Selective absorption of cations by higher plants. *Plant Physiol.* **16,** 691–720.
56. Collins, J. C., and Kerrigan, A. P. (1973). Hormonal control of ion movements in plant roots. *In* "Ion Transport in Plants" (W. P. Anderson, ed.), pp. 589–593. Academic Press, London.
57. Collins, J. C., and Kerrigan, A. P. (1974). The effect of kinetin and abscisic acid on water and ion transport in isolated maize roots. *New Phytol.* **73,** 309–314.
58. Collins, O. D. G., and Sutcliffe, J. F. (1977). The relationship between transport of individual elements and dry matter from the cotyledons of *Pisum sativum* L. *Ann. Bot. (London)* [N.S.] **41,** 163–171.
59. Cooper, A. J. (1967). Tiered trough and nutrient film methods of tomato growing. *Annu. Rep. Glasshouse Crops Res. Inst., 1966–1967.*
60. Cooper, A. J. (1979). "The ABC of NFT." Grower Books, London.
61. Cosgrove, D. J. (1966). Chemistry and biochemistry of inositol polyphosphates. *Rev. Pure Appl. Chem.* **16,** 209–224.
62. Crafts, A. S., and Broyer, T. C. (1938). Migration of salts and water into xylem of roots of higher plants. *Am. J. Bot.* **24,** 415–431.
63. Cram, W. J. (1976). Negative feedback regulation of transport in cells. The maintenance of turgor, volume and nutrient supply. *In* "Encyclopedia of Plant Physiology, New Series" (U. Lüttge and M. G. Pitman, eds.), Vol. IIA, pp. 284–316. Springer-Verlag, Berlin and New York.
64. Cram, W. J., and Pitman, M. S. (1972). The action of abscisic acid on ion uptake and water flow in plant roots. *Aust. J. Biol. Sci.* **25,** 1125–1132.
65. Crossett, R. N., and Loughman, B. C. (1966). The absorption and translocation of phosphorus by seedlings of *Hordeum vulgare. New Phytol.* **65,** 459–468.
66. Crowdy, S. H., and Tanton, T. W. (1970). Water pathways in higher plants. 1. Free space in wheat leaves. *J. Exp. Bot.* **21,** 102–111.
67. Dahesh, K. (1977). Aspects of the physiology of *Salicornia* spp. D. Phil. Thesis, University of Sussex, U.K.
68. Davies, C. R., and Wareing, P. F. (1965). Auxin induced transport of radio phosphorus in stems. *Planta* **65,** 135–156.
69. Dunlop, J. (1973). The transport of potassium in the xylem exudate of rye grass. I. Membrane potentials and vacuolar potassium activities in seminal roots. *J. Exp. Bot.* **24,** 995–1002.
70. Dunlop, J., and Bowling, D. J. F. (1971). The movement of ions to the xylem exudate of maize roots. 1. Profiles of membrane potential and vacuoles potassium activities across the root. *J. Exp. Bot.* **22,** 434–444.
71. Dunlop, J., and Bowling, D. J. F. (1971). The movement of ions to the xylem exudate of

maize roots. 2. A comparison of the electrical potential and electrochemical potential of ions in the exudate and in the root cells. *J. Exp. Bot.* **22,** 445–452.
72. Dunlop, J., and Bowling, D. J. F. (1971). The movement of ions to the xylem exudate of maize roots. 3. The location of the electrical and electrochemical potential difference between the exudate and the medium. *J. Exp. Bot.* **22,** 453–464.
73. Duvigneaud, P., and Denaeyer-de Smet, J. (1970). Biological cycling of minerals in temperate deciduous forests. *In* "Analysis of Temperate Forest Ecosystems" (D. E. Reickle, ed.), pp. 199–225. Springer-Verlag, Berlin and New York.
74. Epstein, E. (1966). Dual pattern of ion absorption by plant cells and by plants. *Nature (London)* **212,** 1324–1327.
75. Epstein, E. (1972). "Mineral Nutrition of Plants." Principles and Perspectives." Wiley, New York.
76. Epstein, E., and Hagen, C. E. (1952). A kinetic study of the absorption of alkali cations by barley roots. *Plant Physiol.* **27,** 457–474.
77. Epstein, E., Rains, D. W., and Elzam, O. E. (1963). Resolution of dual mechanisms of potassium absorption by barley roots. *Proc. Natl. Acad. Sci. U.S.A.* **49,** 684–692.
78. Erlandsson, G. (1975). Rapid effects of ion and water uptake induced by changes in water potential in young wheat plants. *Physiol. Plant.* **35,** 256–262.
78a. Esau, K. (1965). Plant Anatomy, 2nd ed. Wiley, New York.
79. Faust, M., and Shear, C. B. (1968). Corking disorders of apples. A physiological and biochemical review. *Bot. Rev.* **34,** 441–469.
80. Ferguson, I. B. (1972). Calcium mobility in plants. Ph.D. Thesis, University of Auckland, New Zealand.
81. Ferguson, I. B., and Bollard, E. G. (1976). The movement of calcium in germinating pea seeds. *Ann. Bot. (London)* [N.S.] **40,** 1047–1055.
82. Ferguson, I. B., and Bollard, E. G. (1976). The movement of calcium in woody stems. *Ann. Bot. (London)* [N.S.] **40,** 1057–1065.
83. Fiscus, E. L. (1977). Effects of coupled solute and water flow in plant roots with special reference to Brouwer's experiment. *J. Exp. Bot.* **28,** 71–77.
84. Fiscus, E. L., and Kramer, P. J. (1970). Radial movement of oxygen in plant roots. *Plant Physiol.* **45,** 667–669.
85. Flowers, T. J. (1972). The effect of sodium chloride on enzyme activities from four halophytic species of Chenopodiaceae. *Phytochemistry* **11,** 1881–1886.
86. Flowers, T. J. (1974). Salt tolerance in *Suaeda maritima* (L) Dum. A comparison of mitochondria isolated from green tissues of *Suaeda* and *Pisum. J. Exp. Bot.* **10,** 101–110.
87. Flowers, T. J. (1975). Halophytes. *In* "Ion Transport in Plant Cells and Tissues" (D. A. Baker and J. L. Hall, eds.), pp. 309–334. North-Holland Publ., Amsterdam.
88. Flowers, T. J., Hall, J. L., and Ward, M. E. (1976). Salt tolerance in the halophyte *Suaeda maritima.* Further properties of the enzyme malate dehydrogenase. *Phytochemistry* **15,** 1231–1234.
89. Flowers, T. J., Troke, P. F., and Yeo, A. R. (1977). The mechanism of salt tolerance in halophytes. *Annu. Rev. Plant Physiol.* **28,** 89–121.
90. Flowers, T. J., Ward, M. E., and Hall, J. L. (1976). Salt tolerance in the halophyte *Suaeda maritima.* Some properties of malate dehydrogenase. *Philos. Trans. R. Soc. London, Ser. B* **273,** 523–540.
91. Gale, J., Naaman, R., and Poljakoff-Mayber, A. (1979). Growth of *Atriplex halimus* in sodium chloride salinated culture solutions as affected by the relative humidity of the air. *Aust. J. Biol. Sci.* **23,** 947–952.

92. Goodall, D. W., and Gregory, F. G. (1947). Chemical composition of plants as an index of their nutritional status. *Tech. Commun.—Imp. Bur. Hortic. Plant Crops.*
93. Goor, B. J. van (1968). The role of calcium and cell permeability in the disease blossom end rot of tomatoes. *Physiol. Plant.* **21,** 1110–1121.
94. Gormley, T. E. (1972). Effect of water composition and feeding method on soil nutrient levels and on tomato fruit yield and composition. *Ir. J. Agric. Res.* **11,** 101–115.
95. Graham, R. D., and Bowling, D. J. F. (1977). The effect of the shoot on the transmembrane potentials of root cortical cells of sunflower. *J. Exp. Bot.* **28,** 886–893.
96. Greenway, H. (1965). Plant responses to saline substrates. VII. Growth and ion uptake throughout plant development in two varieties of *Hordeum vulgare. Aust. J. Biol. Sci.* **18,** 763–781.
97. Greenway, H., and Gunn, A. (1966). Phosphorus retranslocation in *Hordeum vulgare* during early tillering. *Planta* **71,** 43–67.
98. Greenway, H., Gunn, A., Pitman, M. G., and Thomas, D. A. (1965). Plant responses to saline substrates. VI. Chloride, sodium and potassium uptake and distribution within the plant during ontogenesis of *Hordeum vulgare. Aust. J. Biol. Sci.* **18,** 525–540.
99. Greenway, H., and Klepper, B. (1968). Phosphorus transport to the xylem and its regulation by water flow. *Planta* **83,** 119–136.
100. Greenway, H., and Munns, R. A. (1980). Mechanisms of salt tolerance in nonhalophytes. *Annu. Rev. Plant Physiol.* **31,** 149–190.
101. Greenway, H., and Osmond, C. B. (1970). Ion relations, growth and metabolism of *Atriplex* at high external electrolyte concentrations. *In* "The Biology of *Atriplex*" (R. Jones, ed.), pp. 49–56. CSIRO, Canberra, Australia.
102. Greenway, H., and Osmond, C. B. (1972). Salt responses of enzymes from species differing in salt tolerance. *Plant Physiol.* **49,** 256–259.
103. Greenway, H., and Pitman, M. G. (1965). Potassium retranslocation in seedlings of *Hordeum vulgare. Aust. J. Biol. Sci.* **18,** 235–248.
104. Guardiola, J. L. (1973). Growth and accumulation of mineral elements in the axis of young pea (*Pisum sativum* L.) seedlings. *Ann. Bot. Neerl.* **22,** 55–68.
105. Guardiola, J. L. (1975). Effects of gibberellic acid on the transport of nitrogen from the cotyledons of young pea seedlings. *Ann. Bot. (London)* [N.S.] **39,** 325–330.
106. Guardiola, J. L., and Sutcliffe, J. F. (1971). Control of protein hydrolysis in the cotyledons of germinating pea seeds. *Ann. Bot. (London)* [N.S.] **35,** 791–807.
107. Guardiola, J. L., and Sutcliffe, J. F. (1971). Mobilisation of phosphorus in the cotyledons of young seedlings of the garden pea (*Pisum sativum* L.). *Ann. Bot. (London)* [N.S.] **35,** 809–823.
108. Guardiola, J. L., and Sutcliffe, J. F. (1972). Transport of minerals from the cotyledons during germination of seeds of the garden pea (*Pisum sativum* L.). *J. Exp. Bot.* **23,** 322–337.
109. Gutknecht, J. (1968). Salt transport in *Valonia.* Inhibition of potassium uptake by small hydrostatic pressures. *Science* **160,** 68–70.
110. Handreck, K. A., and Jones, L. P. H. (1967). Uptake of monosilicic acid by *Trifolium incarnatum* L. *Aust. J. Biol. Sci.* **20,** 483–485.
111. Harley, J. L., and Brierley, J. K. (1954). Uptake of phosphate by excised mycorrhiza of beech. VI. *New Phytol.* **52,** 240–252.
112. Harrison-Murray, R. S., and Clarkson, D. T. (1973). Relationships between structural development and the absorption of ions by the root system of *Cucurbita pepo. Planta* **114,** 1–16.

113. Heine, R. W. (1970). Absorption of phosphate and potassium ions in Poplar stems. *J. Exp. Bot.* **21,** 497–503.
114. Higinbotham, N. (1973). The mineral absorption process in plants. *Bot. Rev.* **39,** 35–49.
115. Higinbotham, N., Davis, R., Mertz, D., and Shumway, L. (1973). Some evidence that radial transport in maize roots is into living cells. *In* "Ion Transport in Plants" (W. P. Anderson, ed.), pp. 493–506. Academic Press, London.
116. Hill, A. E. (1967). Ion and water transport in *Linonium.* II. Short circuit analysis. *Biochim. Biophys. Acta* **135,** 461–465.
117. Hill, A. E., and Hill, B. S. (1973). The *Limonium* salt gland: A biophysical and structural study. *Int. Rev. Cytol.* **35,** 229–320.
118. Hill, A. E., and Hill, B. S. (1976). Mineral ions. *In* "Encyclopedia of Plant Physiology, New Series" (U. Lüttge and M. G. Pitman, eds.), Vol. IIB, pp. 225–243. Springer-Verlag, Berlin and New York.
119. Hill, B. S., and Hill, A. E. (1973). The electrogenic chloride pump of the *Limonium* gland. *J. Membr. Biol.* **12,** 129–144.
120. Hill, B. S., and Hill, A. E. (1973). ATP-driven chloride pumping and ATP-ase activity in the *Limonium* salt gland. *J. Membr. Biol.* **12,** 145–158.
121. Hoagland, D. R., and Broyer, T. C. (1936). General nature of the process of salt accumulation by roots with description of experimental methods. *Plant Physiol.* **11,** 471–507.
122. Hocking, P. J. (1980). Redistribution of nutrient elements from cotyledons of two species of annual legumes during germination and seedling growth. *Ann. Bot. (London)* [N.S.] **46,** 383–396.
123. Hopkinson, J. M. (1964). Studies on the expansion of the leaf surface. IV. The carbon and phosphorus economy of a leaf. *J. Exp. Bot.* **15,** 125–137.
124. Hopkinson, J. M. (1966). Studies on the expansion of the leaf surface. VI. Senescence and the usefulness of old leaves. *J. Exp. Bot.* **17,** 762–770.
125. Hylmo, B. (1953). Transpiration and ion absorption. *Physiol. Plant.* **6,** 383–405.
126. Hylmo, B. (1955). Passive components in the ion absorption of the plant. I. The zonal ion and water absorption in Brouwer's experiments. *Physiol. Plant.* **8,** 433–441.
127. Hylmo, B. (1958). Passive components in the ion absorption of the plant. II. The zonal water flow, ion passage and pore size in roots of *Vicia faba. Physiol. Plant.* **11,** 382–400.
128. Jackson, J. E., and Weatherley, P. E. (1962). The effect of hydrostatic pressure gradients on the movement of potassium across the root cortex. *J. Exp. Bot.* **13,** 128–143.
129. Jackson, J. E., and Weatherley, P. E. (1962). The effect of hydrostatic pressure gradients on the movement of sodium and calcium across the root cortex. *J. Exp. Bot.* **13,** 404–413.
130. Jacoby, B. (1967). Effect of roots on calcium ascent in bean stems. *Ann. Bot. (London)* [N.S.] **31,** 727–730.
131. Jefferies, R. L. (1973). The ionic relations of seedlings of *Triglochin maritima* L. *In* "Ion Transport in Plants" (W. P. Anderson, ed.), pp. 297–321. Academic Press, London.
132. Jefferies, R. L. (1980). The role of organic solutes in osmo-regulation in halophytic higher plants. *In* "Genetic Engineering of Osmoregulation. Impact on Plant Productivity for Food, Chemicals and Energy" (D. W. Rains, R. C. Valentine, and A. Hollaender, eds.), pp. 133–154. Plenum, New York.
133. Jennings, D. H. (1968). Halophytes, succulence and sodium in plants—a unified theory. *New Phytol.* **67,** 899–911.

134. Jenny, H., and Overstreet, R. (1939). Surface migration of ions and contact exchange. *J. Phys. Chem.* **43,** 1185–1196.
135. Jeschke, W. J. (1982). Shoot dependent regulation of sodium and potassium fluxes in roots of whole barley seedlings. *J. Exp. Bot.* **33,** 601–618.
136. Jones, L. P. H., and Handreck, K. A. (1965). Studies on silica in the oat plant. VI. Uptake of silica from soils by the plant. *Plant Soil* **23,** 79–96.
137. Katsuta, M. (1961). The breakdown of reserve protein in pine seeds during germination. *J. Jpn. For. Sci.* **43,** 241–244.
138. Kelday, I. S., and Bowling, D. J. F. (1975). The effect of cycloheximide on uptake and transport of ions by sunflower roots. *Ann. Bot. (London)* [N.S.] **39,** 1023–1027.
139. Kelday, I. S., and Bowling, D. J. F. (1980). Profiles of chloride concentration and PD in the root of *Commelina communis* L. *J. Exp. Bot.* **31,** 1347–1365.
140. Kissel, D. E., and Ragland, J. C. (1967). Redistribution of nutrient elements in corn (*Zea mays* L.). 1. NPK Ca and Mg redistribution in the absence of nutrient accumulation after silking. *Soil Sci. Soc. Am. Proc.* **31,** 227–230.
141. Klepper, B., and Greenway, H. (1968). Effects of water stress in phosphorus transport to the xylem. *Planta* **80,** 142–146.
142. Klepper, B., and Kaufmann, M. R. (1966). Removal of salt from xylem sap by leaves and stems of guttating plants. *Plant Physiol.* **41,** 1743–1747.
143. Kramer, P. J. (1956). Relative amounts of mineral absorption through various regions of roots. *U.S. At. Energy Comm. Rep.* **TID 7512,** 287–295.
144. Kramer, P. J., and Wiebe, M. M. (1954). Longitudinal gradients of ^{32}P absorption in roots. *Plant Physiol.* **29,** 229–234.
145. Kramer, P. J., and Wilbur, K. M. (1949). Absorption of radioactive phosphorus mycorrhizal roots of pine. *Science* **110,** 8–9.
146. Langston, R. (1956). Studies on marginal movement of cobalt 60 in cabbage. *Proc. Am. Soc. Hortic. Sci.* **68,** 366–369.
147. Lanyi, J. K. (1974). Salt dependent properties of proteins from extremely halophilic bacteria. *Bacteriol. Rev.* **38,** 272–290.
148. Larkum, A. W. D., and Hill, A. E. (1970). Ion and water transport in *Limonium.* V. The ionic status of chloroplasts in the leaf of *Limonium vulgare* in relation to the activity of salt glands. *Biochim. Biophys. Acta* **203,** 133–138.
149. Laties, G. C. (1969). Dual mechanisms of salt uptake in relation to compartmentation and long distance transport. *Annu. Rev. Plant Physiol.* **20,** 89–116.
150. Laties, G. C., and Budd, K. (1964). The development of differential permeability in isolated studies of corn roots. *Proc. Natl. Acad. Sci. U.S.A.* **52,** 462–469.
151. Läuchli, A. (1972). Translocation of inorganic solutes. *Annu. Rev. Plant Physiol.* **23,** 197–218.
152. Läuchli, A. (1973). Investigations of ion transport in plants by electron probe analysis: Principles and perspectives. *In* "Ion Transport in Plants" (W. P. Anderson, ed.), pp. 1–10. Academic Press, London.
153. Läuchli, A. (1976). Symplasmic transport and ion release to the xylem. *In* "Transport and Transfer Processes in Plants" (I. F. Wardlaw and J. B. Passioura, eds.), pp. 101–112. Academic Press, London and New York.
154. Läuchli, A. (1976). Genotypic variation in transport. *In* "Encyclopedia of Plant Physiology, New Series" (U. Lüttge and M. G. Pitman, eds.), Vol. IIB, pp. 372–393. Springer-Verlag, Berlin and New York.
155. Läuchli, A., Kramer, D., Pitman, M. G., and Lüttge, U. (1974). Ultrastructure of xylem parenchyma cells of barley roots in relation to ion transport in the xylem. *Planta* **119,** 85–99.

156. Läuchli, A., Lüttge, U., and Pitman, M. G. (1973). Ion uptake and transport through barley seedlings. Differential effects of cycloheximide. *Z. Naturforsch., B: Anorg. Chem., Org. Chem.* **28B,** 431–434.
157. Läuchli, A., Pitman, M. G., Lüttge, U., Kramer, D., and Ball, E. (1978). Are developing xylem vessels the site of an exudation from root to shoot? *Plant, Cell Environ.* **1,** 217–223.
158. Läuchli, A., Spurr, A. R., and Epstein, E. (1971). Lateral transport of ions into the xylem of corn roots. II. Evaluation of a stelar pump. *Plant Physiol.* **48,** 118–124.
159. Lazaroff, N., and Pitman, M. J. (1966). Calcium and magnesium uptake by barley seedlings. *Aust. J. Biol. Sci.* **19,** 991–1005.
160. Lee, R. B. (1982). Selectivity and kinetics of ion uptake by barley plants following nutrient deficiency. *Ann. Bot. (London)* [N.S.] **50,** 429–449.
161. Lepp, N. W., and Fairfax, J. A. W. (1976). The role of acid rain as a regulator of foliar nutrient uptake and loss. *In* "Microbiology of Aerial Plant Surfaces" (C. H. Dickinson and T. F. Preece, eds.), pp. 107–118. Academic Press, London and New York.
162. Linck, A. J. (1955). Studies on the distribution of phosphorus in *Pisum sativum* in relation to fruit development. Ph.D. Dissertation, Ohio State University, Columbus.
163. Loneragan, J. F. (1968). Nutrient requirements of plants. *Nature (London)* **220,** 1307–1308.
164. Loneragan, J. F., and Asher, C. J. (1967). Responses of plants to phosphate concentration in solution culture. II. Rate of phosphate absorption and its relation to growth. *Soil Sci.* **103,** 311–318.
165. Loneragan, J. F., and Snowball, K. (1967). Rate of calcium absorption by plant roots and its relation to growth. *Aust. J. Agric. Res.* **20,** 479–499.
166. Loneragan, J. F., and Snowball, K. (1969). Calcium requirements of plants. *Aust. J. Agric. Res.* **20,** 465–478.
167. Lopushinsky, W. (1964). Effect of water movement on ion movement into the xylem of tomato roots. *Plant Physiol.* **39,** 494–501.
168. Loughman, B. C. (1969). The uptake of phosphate and its transport within the plant. *In* "Ecological Aspects of the Mineral Nutrition of Plants" (I. R. Rorison, ed.), pp. 309–322. Blackwell, Oxford.
169. Loughman, B. C., and Russell, R. S. (1957). The absorption and utilisation of phosphate by young barley plants. *J. Exp. Bot.* **8,** 280–293.
170. Lundegårdh, H. (1955). Mechanisms of absorption, transport, accumulation and secretion of ions. *Annu. Rev. Plant Physiol.* **6,** 1–24.
171. Lüttge, U. (1971). Structure and function of plant glands. *Annu. Rev. Plant Physiol.* **22,** 23–44.
172. Lüttge, U. (1975). Salt glands. *In* "Ion Transport in Plant Cells and Tissues" (D. A. Baker and J. L. Hall, eds.), pp. 335–376. North-Holland Publ., Amsterdam.
173. Lüttge, U., and Higinbotham, N. (1979). "Transport in Plants." Springer-Verlag, Berlin and New York.
174. Macklon, A. E. S. (1975). Cortical cell fluxes and transport to the stele in excised root segments of *Allium cepa* L. *Planta* **122,** 109–130.
175. Mason, T. G., and Maskell, E. J. (1931). Further studies on transport in the cotton plant. I. Preliminary observations on the transport of phosphorus, potassium and calcium. *Ann. Bot. (London)* [N.S.] **45,** 125–174.
176. Mason, T. G., Maskell, E. J., and Phillis, E. (1936). Further studies on transport in the cotton plant. III. Concerning the independence of solute movement in the phloem. *Ann. Bot. (London)* [N.S.] **50,** 23–58.
177. Meiri, A., and Anderson, W. P. (1970). Observations on the effects of pressure differ-

ences between the bathing medium and the exudates of excised barley roots. *J. Exp. Bot.* **21**, 899–907.

178. Meiri, A., and Anderson, W. P. (1970). Observations on the exchange of salt between the xylem and neighbouring cells in *Zea mays* primary roots. *J. Exp. Bot.* **21**, 908–914.
179. Mer, C. L., Dixon, P. F., Diamond, B. C., and Drake, C. F. (1969). The dominant influence of nitrogen on growth correlation in etiolated oat seedlings. *Ann. Bot. (London)* [N.S.] **27**, 693–721.
180. Millikan, C. R. (1971). Mid-season movement of ^{45}Ca in apple trees. *Aust. J. Agric. Res.* **22**, 923–930.
181. Millikan, C. R., Bjarnsson, E. N., Osborn, R. L., and Harger, F. C. (1971). Calcium concentration in tomato fruits in relation to the incidence of blossom end rot. *Aust. J. Exp. Agric. Anim. Husb.* **11**, 570–575.
182. Mitchell, J. W., and Martin, W. E. (1937). Effects of indolyl acetic acid on growth and chemical composition of etiolated bean plants. *Bot. Gaz. (Chicago)* **90**, 171–183.
183. Mitsui, S., and Takatoh, H. (1963). Nutritional study of silica in Graminaceous crops. *Soil Sci. Plant Nutr.* **9**, 54–65.
184. Mohamed, G. E., and Marshall, C. (1979). The pattern of distribution of phosphorus and dry matter with time in spring wheat. *Ann. Bot. (London)* [N.S.] **44**, 721–730.
185. Moorby, J. (1977). Integration and regulation of translocation in the whole plant. *Symp. Soc. Exp. Biol.* **31**, 425–454.
186. Moorby, J. (1981). "Transport Systems in Plants," Integrated Themes in Biology Series. Longmans, Green, London and New York.
187. Mothes, K. (1961). Aktiver Transport als regulatives Prinzip für gerichtete stoffverteilung in hoheren Pflanzen. *In* "Biochemie des Aktiven Transport." pp. 72–81. Springer-Verlag, Berlin and New York.
188. Muenscher, W. C. (1922). Effect of transpiration on the absorption of salt by intact plants. *Am. J. Bot.* **9**, 311–330.
189. Neales, T. F., Anderson, M. J., and Wardlaw, I. F. (1963). The role of the leaves in the accumulation of nitrogen by wheat during ear development. *Aust. J. Agric. Res.* **14**, 725–736.
190. Okamoto, H. (1962). Transport of cations from cotyledons to seedling of the embryonic plants of *Vigna sesquipedalis*. *Plant Cell Physiol.* **3**, 83–94.
191. Osmond, C. B. (1968). Ion absorption in *Atriplex* leaf tissues. 1. Absorption by mesophyll cells. *Aust. J. Biol. Sci.* **21**, 1119–1130.
192. Osmond, C. B., and Greenway, M. (1972). Salt responses of carboxylation enzymes from species differing in salt tolerance. *Plant Physiol.* **49**, 260–263.
193. Osmond, C. B., Lüttge, U., West, K. R., Pallaghy, C. K., and Schachar-Hill, B. (1969). Ion absorption in *Atriplex* leaf tissue. II. Secretion of ions to epidermal bladders. *Aust. J. Biol. Sci.* **22**, 797–814.

193a. Pate, J. S. (1975). Exchange of solutes between phloem and xylem and circulation in the whole plant. *In* Encyclopedia of Plant Physiology, New Series" (M. H. Zimmermann and J. A. Milburn, eds.), Vol. I, pp. 451–473. Springer-Verlag, Berlin and New York.

194. Pate, J. S. (1976). Nutrients and metabolites of fluids recovered from the soybean and phloem: Significance in relation to long-distance transport in plants. *In* "Transport and Transfer Processes in Plants" (I. F. Wardlaw and J. B. Passioura, eds.), pp. 253–281. Academic Press, London and New York.
195. Pate, J. S., Atkins, C. A., Hamel, K., McNeil, D. L., and Layzell, D. B. (1979). Transport of organic solutes in phloem and xylem of a nodulated legume. *Plant Physiol.* **63**, 1082–1088.

196. Pate, J. S., Sharkey, P. J., and Atkins, C. A. (1977). Nutrition of a developing legume fruit. Functional economy in terms of carbon, nitrogen and water. *Plant Physiol.* **59,** 506–510.
197. Pate, J. S., Sharkey, P. J., and Lewis, O. A. M. (1974). Phloem bleeding from legume fruits—a technique for study of plant nutrition. *Planta* **120,** 229–243.
198. Pate, J. S., Sharkey, P. J., and Lewis, O. A. M. (1975). Xylem to phloem transfer of solutes in fruiting shoots of a legume, studied by a phloem bleeding technique. *Planta* **122,** 11–16.
199. Perry, M. W., and Greenway, H. (1973). Permeation of uncharged molecules and water through tomato roots. *Ann. Bot. (London)* [N.S.] **37,** 225–232.
200. Pettersson, S. (1966). Artificially induced water and sulphate transport through sunflower roots. *Physiol. Plant.* **19,** 581–601.
201. Pitman, M. G. (1965). Sodium and potassium uptake by seedlings of *Hordeum vulgare. Aust. J. Biol. Sci.* **18,** 10–24.
202. Pitman, M. G. (1966). Uptake of potassium and sodium by seedlings of *Sinapis alba. Aust. J. Biol. Sci.* **19,** 257–269.
203. Pitman, M. G. (1972). Uptake and transport of ions in barley seedlings. II. Evidence for two active stages in transport to the shoot. *Aust. J. Biol. Sci.* **25,** 243–257.
204. Pitman, M. G. (1972). Uptake and transport of ions in barley seedlings. III. Correlation of potassium transport to the shoot with plant growth. *Aust. J. Biol. Sci.* **25,** 905–919.
205. Pitman, M. G. (1975). Whole plants. *In* "Ion Transport in Plant Cells and Tissues" (D. A. Baker and J. L. Hall, eds.), pp. 267–308. North-Holland Publ., Amsterdam.
206. Pitman, M. G. (1976). Nutrient uptake by roots and transport to the xylem: uptake processes. *In* "Transport and Transfer Processes in Plants" (I. F. Wardlaw and J. B. Passioura, eds.), pp. 85–98. Academic Press, London and New York.
207. Pitman, M. G. (1977). Ion transport in the xylem. *Annu. Rev. Plant Physiol.* **28,** 71–89.
208. Pitman, M. G., and Cram, W. J. (1973). Regulation of inorganic transport in plants. *In* "Ion Transport in Plants" (W. P. Anderson, ed.), pp. 465–448. Academic Press, London.
209. Pitman, M. G., Lüttge, U., Kramer, D., and Ball, E. (1974). Free space characteristics of barley leaf slices. *Aust. J. Plant Physiol.* **1,** 65–75.
210. Pitman, M. G., Lüttge, U., Lauchli, A., and Ball, E. (1974). Action of abscisic acid on ion transport as affected by root temperature and nutrient status. *J. Exp. Bot.* **25,** 147–155.
211. Pitman, M. G., and Wellfare, D. (1978). Inhibition of ion transport in excised barley roots by abscisic acid; relative to water permeability. *J. Exp. Bot.* **29,** 1125–1138.
212. Pollak, G., and Waisel, Y. (1970). Salt secretion in *Aeluropus litoralis* (Willd.) Part. *Ann. Bot. (London)* [N.S.] **34,** 879–888.
213. Priestley, J. H., and Wormall, A. (1925). On the solutes exuded by root pressure from vines. *New Phytol.* **24,** 24–38.
214. Rains, D. W., Valentine, R. C., and Hollaender, A. (1980). "Genetic Engineering of Osmoregulation. Impact on Food, Chemicals and Energy." Plenum, New York.
215. Riga, A. J., and Bukovac, M. J. (1961). Distribution du ^{32}P, du ^{45}Ca and du ^{65}Zn chez le haricot (*Phaseolus vulgaris* L.) après absorption radiculaire. Redistribution de ces éléments au cours de la germination de la graine et du développement de la jeune plantule. *Bull. Inst. Agron. Stn. Rech. Gembloux* **29,** 165–196.
216. Ringoet, A., Rechenmann, R. V., and Veen, H. (1967). Calcium movement in oat leaves measured by semi-conductor detectors. *Radiat. Biol.* **7,** 81–90.
217. Robards, A. W., and Clarkson, D. T. (1976). The role of plasmodesmata in the transport of water and nutrients across roots. *In* "Intercellular Communication in Plants.

Studies on Plasmodesmata" (B. E. Gunning and A. W. Robards, eds.), pp. 181–202. Springer-Verlag, Berlin and New York.

218. Robards, A. W., and Jackson, S. M. (1976). Root structure and function—an integrated approach. *In* "Perspectives in Experimental Biology" (N. Sunderland, ed.), Vol. 2, pp. 413–422. Pergamon, Oxford.

219. Routien, J. B., and Dawson, R. F. (1943). Some interrelationships of growth, salt absorption, respiration and mycorrhizal development in *Pinus echinata*. *Am. J. Bot.* **30,** 440–451.

220. Rozema, J., Gude, H., and Pollak, G. (1981). An ecophysiological study of the salt secretion of four halophytes. *New Phytol.* **89,** 201–218.

221. Russell, R. S. (1977). "Plant Root Systems." McGraw-Hill, London.

222. Russell, R. S., and Clarkson, D. T. (1976). Ion transport in root systems. *In* "Perspectives in Experimental Biology" (N. Sunderland, ed.), Vol. 2, pp. 401–411. Pergamon, Oxford.

223. Russell, R. S., and Martin, R. P. (1953). A study of the absorption and utilisation of phosphate by young barley plants. 1. The effect of external concentration on the distribution of absorbed phosphate between roots and shoots. *J. Exp. Bot.* **4,** 108–127.

224. Russell, R. S., and Sanderson, J. (1967). Nutrient uptake by different parts of the intact roots of plants. *J. Exp. Bot.* **18,** 491–508.

225. Russell, R. S., and Shorrocks, V. M. (1959). The relationship between transpiration and the absorption of inorganic ions by intact plants. *J. Exp. Bot.* **10,** 301–316.

226. Saeed, A. F. M. (1975). The distribution of mineral elements in *Xanthium pennsylvanicum* in relation to growth. D.Phil. Thesis, University of Sussex, U.K.

227. Scott, F. M. (1949). Plasmodesmata in xylem elements. *Bot. Gaz. (Chicago)* **110,** 492–495.

228. Seth, A. K., and Wareing, P. F. (1967). Hormone-directed transport of metabolism and its possible role in plant senescence. *J. Exp. Bot.* **18,** 65–77.

229. Shimony, C., Fahn, A., and Reinhold, L. (1973). Ultrastructure and ion gradients in the salt gland of *Avicennia marina*. *New Phytol.* **72,** 27–36.

230. Skelton, B. J., and Shear, G. M. (1971). Calcium translocation in the pea nut *Arachis hypogaea* L. *Agron. J.* **63,** 409–412.

231. Steward, F. C., and Millar, F. K. (1954). Salt accumulation in plants. A reconsideration of the role of growth and metabolism. *Symp. Soc. Exp. Biol.* **8,** 367–406.

232. Steward, F. C., and Sutcliffe, J. F. (1959). Plants in relation to inorganic salts. *In* "Plant Physiology: A Treatise" (F. C. Steward, ed.), Vol. 2, Chapter 4, pp. 253–478. Academic Press, New York.

233. Stewart, G. R., and Lee, J. A. (1974). The role of proline accumulation in halophytes. *Planta* **120,** 279–289.

234. Storey, R., and Wyn Jones, R. G. (1975). Betaine and choline levels in plants and their relationship to sodium chloride stress. *Plant Sci. Lett.* **4,** 161–168.

235. Stout, P. R., and Hoagland, D. R. (1939). Upward and lateral movement of salt in certain plants as indicated by radioactive isotopes of potassium sodium and phosphorus absorbed by roots. *Am. J. Bot.* **26,** 320–324.

236. Sudia, T. W., and Green, D. G. (1972). The translocation of ^{65}Zn and ^{134}Co between seed generations in soybean (*Glycine max* (L) Mer). *Plant Soil* **37,** 695–697.

237. Sutcliffe, J. F. (1957). The selective uptake of alkali cations by storage tissues and intact barley plants. *Potassium Symp. Annu. Meet. Board Tech. Advis. Int. Potash Inst., Berne,* pp. 1–11.

238. Sutcliffe, J. F. (1958). Ion secretion in plants. *Int. Rev. Cytol.* **11,** 179–200.

239. Sutcliffe, J. F. (1959). Salt uptake in plants. *Biol. Rev. Cambridge Philos. Soc.* **34,** 159–220.
240. Sutcliffe, J. F. (1962). "Mineral Salts Absorption in Plants." Pergamon, Oxford.
241. Sutcliffe, J. F. (1969). Some relationships between growth and ion absorption in plant root cells. *Bull. Soc. Fr. Physiol. Veg.* **15,** 115–124.
242. Sutcliffe, J. F. (1971). Trace elements in plants—uptake and translocation. *Tech. Bull.—Minist. Agric., Fish. Food (G.B.)* **21,** 35–40.
243. Sutcliffe, J. F. (1973). The role of protein synthesis in ion transport. *In* "Ion Transport in Plants" (W. P. Anderson, ed.), pp. 399–406. Academic Press, London.
244. Sutcliffe, J. F. (1975). Regulation of ion transport in the whole plant. *In* "Perspectives in Experimental Biology" (N. Sunderland, ed.), Vol. 2, pp. 433–444. Pergamon, Oxford.
245. Sutcliffe, J. F. (1976). Regulation in the whole plant. *In* "Encyclopedia of Plant Physiology, New Series" (U. Lüttge and M. G. Pitman, eds.), Vol. IIB, pp. 394–417. Springer-Verlag, Berlin and New York.
246. Sutcliffe, J. F., and Baset, Q. A. (1973). Control of hydrolysis of reserve materials in the endosperm of germinating oat (*Avena sativa* L.) grains. *Plant Sci. Lett.* **1,** 15–20.
247. Sutcliffe, J. F., and Bryant, J. A. (1977). Germination and early seedling growth. *In* "The Physiology of the Garden Pea" (J. F. Sutcliffe and J. S. Pate, eds.), pp. 45–82 Academic Press, London.
248. Sutcliffe, J. F., and Counter, E. R. (1959). Absorption of alkali cations by plant tissue cultures. *Nature (London)* **183,** 1513–1514.
249. Tal, M. (1971). Salt tolerance in the wild relatives of the cultivated tomato: Responses of *Lycopersicon esculentum, L. peruvianum* and *L. esculentum minor* to sodium chloride solution. *Aust. J. Agric. Res.* **24,** 353–361.
250. Treichel, S. P. (1975). Der Einfluss Van NaCl auf die Prolin Konzentration Vershiedener Halophyten. *Z. Pflanzenphysiol.* **76,** 56–68.
251. Treichel, S. P., Kirst, G. O., and van Willert, D. J. (1974). Verinderung der Activitat der phosphoenol pyruvate-carboxylase durch NaCl bei Halophyten verschiedner Biotypes. *Z. Pflanzenphysiol.* **71,** 437–449.
252. Tukey, H. B., Jr. (1970). The leaching of substances from plants. *Annu. Rev. Plant Physiol.* **21,** 305–324.
253. Tyree, M. T. (1970). The symplast theory: A general theory of symplastic transport according to the principles of irreversible thermodynamics. *J. Theor. Biol.* **26,** 181–214.
254. van den Honert, T. H., Hooymans, J. J. M., and Volkers, W. S. (1955). On the relation between water absorption and mineral uptake by plant roots. *Acta Bot. Neerl.* **4,** 139–155.
255. van Steveninck, R. F. M. (1976). Effects of hormones and related substances on ion transport. *In* "Encyclopedia of Plant Physiology, New Series" (U. Lüttge and M. G. Pitman, eds.), Vol. IIB, pp. 307–342. Springer-Verlag, Berlin and New York.
256. Varner, J. E. (1964). Gibberellic acid controlled synthesis of amylase in barley endosperm. *Plant Physiol.* **39,** 413–415.
257. Waisel, Y. (1972). "Biology of Halophytes." Academic Press, New York.
258. Wardlaw, I. F., and Moncur, L. (1976). Source, sink and hormonal control of translocation in wheat. *Planta* **128,** 93–100.
259. Weatherley, P. E. (1963). The pathway of water movement across the root cortex and leaf mesophyll of transpiring plants. *In* "The Water Relations of Plants" (A. J. Rutter and F. M. Whitehead, eds.), pp. 85–100. Blackwell, Oxford.
260. Weatherley, P. E. (1969). Ion movement within the plant and its integration with other

physiological processes. *In* "Ecological Aspects of the Mineral Nutrition of Plants" (I. M. Rorison, ed.), pp. 323–340. Blackwell, Oxford.

261. Weatherley, P. E. (1974). The hydraulic resistance of the root system. *In* "Structure and Function of Primary Root Tissue" (J. Kolek, ed.), pp. 297–308. Czech Acad. Sci., Bratislava.
262. Weatherley, P. E. (1975). Water relations of the root system. *In* "The Development and Function of Roots" (J. G. Torrey and D. T. Clarkson, eds.), pp. 397–413. Academic Press, London and New York.
263. Widders, I. E., and Lorenz, O. A. (1983). Effects of leaf age and position on the shoot apex on potassium absorption by tomato leaf slices. *Ann. Bot. (London)* [N.S.] **52,** 489–498.
264. Widders, I. E., and Lorenz, O. A. (1983). Effects of leaf age on potassium efflux and net flux in tomato leaf slices. *Ann. Bot. (London)* [N.S.] **52,** 499–506.
265. Wiebe, M. H., and Kramer, P. J. (1954). Translocation of radioactive isotypes from various regions of roots of barley seedlings. *Plant Physiol.* **29,** 342–348.
266. Wildes, R. A., Pitman, M. G., and Schaefer, N. (1975). Comparison of isomers of fluorophenyl alanine as inhibitors of ion transport across barley roots. *Aust. J. Plant Physiol.* **2,** 659–661.
267. Wildes, R. A., Pitman, M. G., and Schaefer, N. (1976). Inhibition of ion uptake to barley roots by cycloheximide. *Planta* **128,** 35–40.
268. Williams, R. F. (1948). The effect of phosphorus supply on the rates of intake of phosphorus and nitrogen and upon certain aspects of phosphorus metabolism in graminaceous plants. *Aust. J. Sci. Res., Ser. B* **1,** 333–361.
269. Williams, R. F. (1955). Redistribution of mineral elements during development. *Annu. Rev. Plant Physiol.* **6,** 25–43.
270. Wyn Jones, R. G. (1980). An assessment of quaternary ammonium and related compounds as osmotic effectors in crop plants. *In* "Genetic Engineering of Osmoregulation: Impact on Plant Productivity for Food, Chemicals and Energy" (D. W. Rains, R. C. Valentine, and A. Hollaender, eds.), pp. 155–170. Plenum, New York.
271. Yagi, M. I. A. (1972). Relationships between the distribution of mineral elements and the growth of bean plant. D.Phil. Thesis, University of Sussex, U.K.
272. Yeo, A. R. (1974). Salt tolerance in the halophyte *Suaeda maritima* (L) Dum. D.Phil. Thesis, University of Sussex, U.K.
273. Yeo, A. R. (1983). Salinity resistance: Physiological prices. *Physiol. Plant.* **58,** 214–222.
274. Yomo, H., and Varner, J. E. (1971). Hormonal control of a secretory tissue. *Curr. Top. Dev. Biol.* **6,** 111–144.
275. Yu, G. H., and Kramer, P. J. (1969). Radial transport of ions in roots. *Plant Physiol.* **44,** 1095–1100.
276. Zeroni, M., and Hall, M. A. (1980). Molecular effects of hormone treatment on tissue. *In* "Encyclopedia of Plant Physiology, New Series" (J. Macmillan, ed.), Vol. IX, pp. 511–586. Springer-Verlag, Berlin and New York.
277. Ziegler, H., and Lüttge, U. (1967). Die salzdrusen von *Limonium vulgare* II Mitteilung. Die Lokalisierung des chlorids. *Planta* **74,** 1–17.
278. Zimmermann, U. (1977). Cell turgor pressure regulation and turgor pressure-mediated transport processes. *Symp. Soc. Exp. Biol.* **31,** 117–154.

CHAPTER SIX

Phloem Transport

J. E. Dale and J. F. Sutcliffe

I. Introduction

Long-distance transport in the phloem has engaged the attention of plant physiologists for many years, and with an enthusiasm which may sometimes seem to border on masochism they are continuing to attack this complicated problem. Since the accounts given by Biddulph (20) and Swanson (275) in Volume II of this treatise were published, factual knowledge of the subject has increased considerably, mainly as a result of the application of radioactive isotopes to trace the movement of materials, but the increase in understanding of the underlying mechanism of translocation has been much less. Critical experiments which would enable us to decide between several conflicting hypotheses have not so far been devised.

Among recent publications of major importance to which reference must be made immediately are the edited volumes by Zimmermann and

Plant Physiology
A Treatise
Vol. IX: Water and Solutes in Plants

Milburn (329), Aronoff *et al.* (8), Wardlaw and Passioura (302), and Eschrich (74), which summarize the views of leading workers up to 1980. The former superseded the treatment of phloem transport in Volume XIII of Ruhland's *Encyclopedia of Plant Physiology* (251) and the three latter volumes are the proceedings of symposia held specifically to discuss the problem in all its aspects. Comprehensive monographs by Crafts and Crisp (43), Canny (32), and Peel (237) appeared in the early 1970s, with a more recent one by Moorby (210); there have also been important reviews by Cronshaw (48), Eschrich (72), Evert (79), Giaquinta (114), MacRobbie (186), Milthorpe and Moorby (205), Wardlaw (299, 300), and Weatherley and Johnson (309). Faced with an enormous literature, including a plethora of research papers presenting many conflicting views, students have some difficulty in acquiring an adequate overview of the subject in the limited time at their disposal.

Our aim in this chapter is to present to them, and their teachers, a concise and yet coherent account of phloem transport as we see it, indicating those areas in which significant progress has been made during the past twenty years and assessing prospects for the future. To keep this chapter to an acceptable length much of the detailed evidence upon which current ideas are based has had to be omitted; for this information the works listed in the preceding paragraph or the literature cited at the end of the text may be consulted. At the outset, the approach adopted is explained.

The simplest model of translocation is one in which a substance is transported from one place, traditionally called a "source," through a channel, believed to be the sieve elements, to a site, or "sink," where it is either utilized or stored (Fig. 1). While this model is almost certainly a gross oversimplification, as explained in the next section, it is a useful starting point for the analysis of what is an integrated system in which loading of translocates at the source and unloading at the sink are at least as important as the mechanism of transport along the channel per se. Indeed, from the time of Münch (218) it has been realized that the supplying and receiving organs in a plant have an important role in determining the rate and direction of movement of materials in the phloem.

FIG. 1. The source–sink concept. Arrow indicates direction of flow.

Starting from the source–sink concept, the structural and physiological characteristics of each component of the system will be described. Then the various mechanisms that have been proposed to account for translocation through the sieve tubes will be evaluated. Although most attention will be paid to the transport of carbohydrates, notably sucrose, which as the substance transported in largest quantities has received most intensive investigation, the movement of other substances will also be referred to. The regulation of ion transport in the whole plant is discussed in more detail in Chapter 5 and the interrelationships between transport of materials and their utilization are dealt with by Pate in Volume VIII of this treatise.

II. Sources and Sinks

A. General Concepts

The terms "sources" and "sinks" were coined in the 1920s by Mason and Maskell (193, 194), and since that time the source–sink model has been widely used in descriptions of phloem transport. A source can be defined as a site which supplies specified materials to the transport system and conversely a sink is a region in which substances are being received (305). Well-known examples of sources are mature leaves and storage organs, such as tubers, rhizomes, and cotyledons. Sinks include shoot apices, roots, and developing storage organs. It should be noted that an organ which is a sink for some substances may be at the same time a source of others. An example of this is a shoot apex, which is usually a sink for carbohydrates and a source of auxins.

Although the terms were originally applied to organs, sources and sinks can also be identified at the cellular and subcellular levels (209). Generally cells behave as sinks during their growth phase, but later they are often converted to sources at least for carbohydrates, either by becoming photosynthetic or by releasing previously stored materials, e.g., starch. The behavior of an organ as a whole depends on the summation of the effects of the numerous sources and sinks represented by the individual cells which compose it. Nongreen cells, such as those in the epidermis of a leaf or in the chlorotic zones of a variegated leaf, continue to act as sinks for carbohydrates after the organ as a whole has become a net source. Whether a particular organ or cell acts as a source or sink may depend on environmental conditions, such as temperature or light intensity. A darkened leaf, for example, may be a sink for carbohydrates, whereas in the light it is a source (137–139, 161). An interesting situation occurs in guard cells, which are sinks for potassium ions in the light and sources in the dark (see Chapter 3). The phloem in a source organ may be looked on as a sink for materials transported from photosynthetic or storage cells, while in a sink the sieve elements constitute a source from which substances are released.

Sources and sinks can also be identified at the subcellular level. Some organelles, e.g., chloroplasts, may be acting as sources of carbohydrates while others, e.g., mitochondria in the same cell, are behaving as sinks. Amyloplasts and vacuoles may function either as sources or sinks depending on whether they are storing or releasing materials at a particular time. Their behavior will be related to temporal changes in enzyme activities and membrane permeability. Biochemically, a reaction or sequence of reactions in which a product accumulates at a particular site leads to the creation of a source, actual or potential, and conversely reactions in which a substance is consumed causes development of a sink. The simplest source–sink system imaginable is one in which the product of a reaction at one site diffuses to another site at which the reaction is reversed.

The ability of sources and sinks to supply or utilize material may be quantified on the basis of their metabolic activity. Warren-Wilson (305) has pointed out that source strength measured in terms of the amount of a substance exported in unit time depends on its size or capacity, which may be expressed in terms of mass, volume, or, in the case of a leaf, surface area, and its activity, that is, the rate per unit of size at which materials are being produced either by photosynthesis or by mobilization of stored reserves. Thus,

$$\begin{array}{cccccc} \text{source strength} & = & \text{source size} & \times & \text{source activity} \\ g\,t^{-1} & & \mathrm{g} & & g\,\mathrm{g}^{-1}\,t^{-1} \\ & & \mathrm{cm}^3 & & g\,\mathrm{cm}^{-3}\,t^{-1} \\ & & \mathrm{cm}^2 & & g\,\mathrm{cm}^{-2}\,t^{-1} \end{array}$$

where t is time, g is mass in grams, and cm^3 and cm^2 are units of volume and surface area, respectively.

This relationship resembles that between relative growth rate (RGR), leaf area ratio (LAR), and net assimilation rate (NAR) in classical growth analysis (76, 148).

$$\begin{array}{ccccc} \text{RGR} & = & \text{LAR} & \times & \text{NAR} \\ g\,\mathrm{g}^{-1}\,t^{-1} & & \mathrm{cm}^2\,\mathrm{g}^{-1} & & g\,\mathrm{cm}^{-2}\,t^{-1} \end{array}$$

Sink strength may similarly be described in terms of size and activity; in this case activity is measured by the ability of the sink to utilize or store the materials supplied to it (see Section II,D).

B. THE LEAF AS A SOURCE OF ASSIMILATES

1. *Influence of Leaf Size*

A leaf in the primordial stage imports the materials required for growth from elsewhere. The source of carbohydrates are usually adjacent mature leaves, but in some cases, such as sprouting tubers and germinating seeds, storage organs may make an important contribution. During bud burst in the spring, the roots may supply sugars through the xylem in certain trees, as in sugar maple. Recirculation of organic nitrogen and mineral nutrients originating from senescing leaves may supply part of the needs of growing leaf primordia, but a major source is the root system. When leaf growth is reduced, for example, by limitation of nutrients, or by other causes (e.g., in etiolated seedlings), transition of a leaf from sink to source is delayed.

The phase of early leaf growth in which cell division is a prominent feature is followed normally by lamina expansion, and when it is exposed to light, a leaf changes rapidly from a heterotrophic to an autotrophic mode of nutrition. Leaf expansion is characterized anatomically by an increase in both cell number and cell size (55), and at the same time there is extensive development of the vascular system. The main veins of a leaf

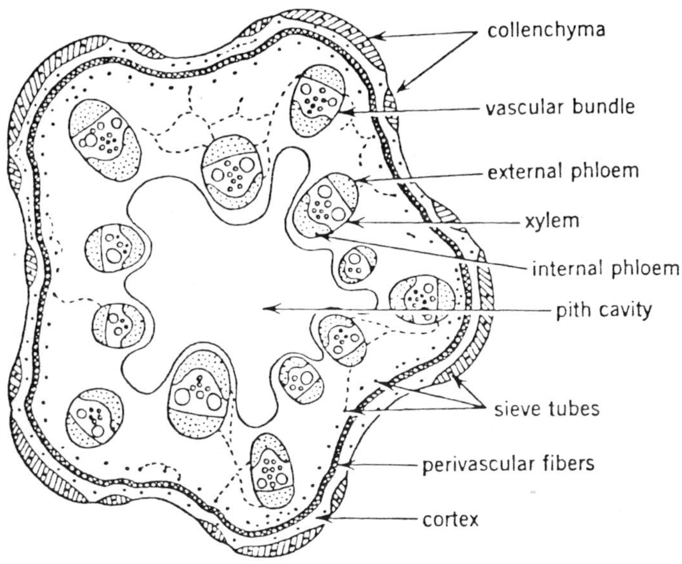

FIG. 2. The arrangement of bicollateral bundles seen in transverse section of a stem of *Cucurbita* (×8). Reproduced from Esau (64).

differentiate acropetally as the lamina expands, whereas the minor veins which appear later usually develop first at the leaf tip and gradually extend downward. The basipetal maturation of a leaf is particularly marked in many monocotyledons such as grasses in which growth of the lamina is prolonged by the presence of a basal meristem (162, 180). Maturation of the minor veins seems to coincide with or just precede the onset of export of assimilates, suggesting that they are of particular importance in this connection (81, 152, 286, 287). In species of Cucurbitaceae (e.g., squash) and Solanaceae (e.g., tomato) in which the vascular bundles are bicollateral (Fig. 2), there is evidence that the adaxial sieve elements of the main veins mature first and are associated particularly with the import of materials, whereas the later-formed abaxial elements function mainly as export channels (142). What is not yet clear is whether the initial import and subsequent export can occur in the same sieve tube or whether transport is strictly unidirectional throughout the life of an individual element. In species with a single phloem strand in each vascular bundle, net movement in opposite directions may occur at different stages of development, but this does not necessarily imply a reversal in the direction of flow in individual sieve tubes because maturation of new elements occurs after export begins. The crucial question of the possibility of simultaneous bidirectional flow in the same sieve tube is discussed in Section III,D below.

2. *Photosynthetic Activity*

Coincidentally with anatomical maturation of a leaf there is progressive development of photosynthetic activity from the tip downward (55, 181). It follows that the tip of a leaf ceases to be a sink for assimilates some time before the base. For leaves of squash it has been shown by ^{14}C feeding experiments that the tip stops importing carbon compounds when the leaf has reached about 10% of its final size, while the base continues to import until the blade is about 45% expanded (286, 287). There is evidence that carbon fixed by the leaf tip may be supplied to the basal region as long as the latter acts as a sink. In a similar way photosynthesis in the terminal leaflets of a tomato leaf provides assimilates to the still-expanding basal leaflets (142). However, in many species lamina expansion proceeds rapidly and the duration of the phase in which there is redistribution of carbon between sources and sinks in the same leaf or of simultaneous export and import may be quite short.

Import of carbohydrates by a leaf falls quickly as photosynthetic activity develops and data for sugar beet (Table I) show that a leaf becomes a net exporter of photosynthates by the time it is about 50% of its final length. Photosynthetic activity expressed as carbon fixed per unit of leaf area reaches a maximum, depending on species, by the time that the leaf is from

TABLE I

EVENTS OBSERVED DURING DEVELOPMENT OF THE SEVENTH LEAF OF SUGAR BEET[a]

Event	Percentage of final laminar length
Marked increase in net photosynthesis (dm^{-2})	10
Area of leaf exhibiting phloem loading begins to increase	22
Maximum observed import of ^{14}C translocate	25
Peak of net photosynthesis (dm^{-2})	35–40
Export initially observed from leaf	35
Phloem loading observed in all orders of veins across lamina	45–50
Entire leaf appears to be a source	50

[a] Data from Fellows and Geiger (81).

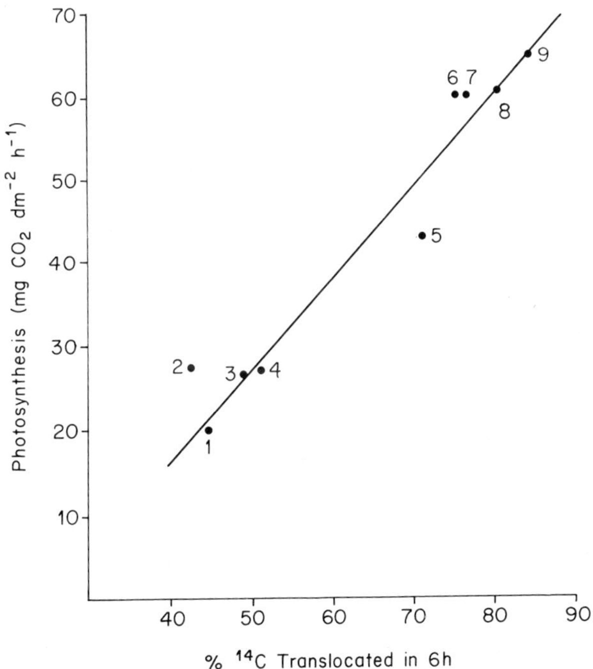

FIG. 3. The relationship between maximum rates of photosynthesis and the percentage of assimilated ^{14}C that is exported within 6 h from leaves of different species: 1, tomato; 2, tobacco; 3, castor bean; 4, soybean; 5, sunflower; 6, sorghum; 7, millet; 8, sugarcane; 9, corn. Adapted from Hofstra and Nelson (144).

one-third to two-thirds expanded, and then it may gradually decline. For most species, carbon fixation capacity per leaf continues to increase until lamina expansion nears completion; i.e., source strength reaches a maximum at around the time, or just before, the leaf is fully grown.

The relationship between photosynthetic carbon fixation and transport of assimilates from a leaf is far from simple. This complexity stems from the fact that some of the carbon fixed is utilized in leaf respiration and in a wide variety of synthetic processes, while, on the other hand, some of the carbon exported may have been released from temporary stores. Nevertheless, Hofstra and Nelson (144) found that there was a highly significant correlation between the rate of photosynthesis and the percentage of the fixed carbon which is translocated in a 6-h period in a number of different plant species (Fig. 3). They found, and later workers have confirmed (185, 212), that C_4 plants (e.g., *Zea mays,* sugarcane) have high rates of carbon fixation and a correspondingly high percentage of the fixed carbon is

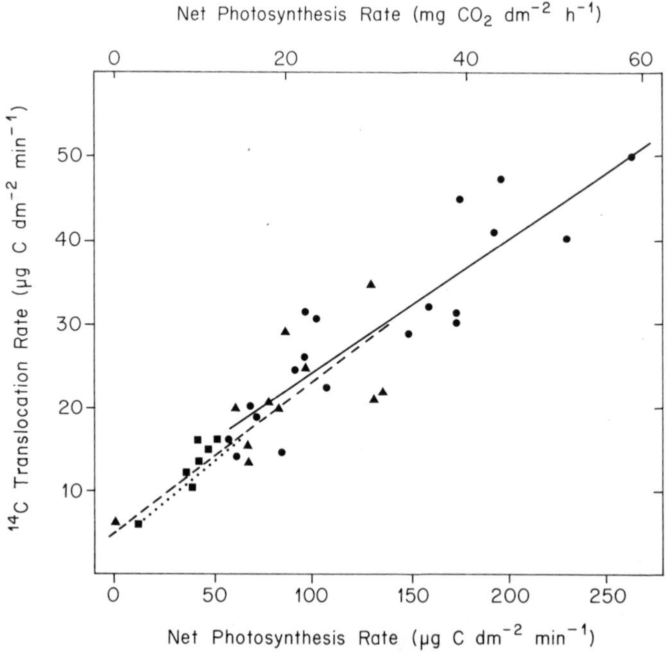

FIG. 4. The relationship between net photosynthesis rate in a range of light intensities (7200 fc, ——●; 3700 fc, ------▲; 2000 fc, ······ ■) and translocation rate measured as import at a sink leaf. Data are for sugar beet. Note: It is not possible to convert foot-candles to SI units in this work. Adapted from Servaites and Geiger (253).

exported. In sugar beet, a C_3 plant, a linear relationship between the rate of net photosynthesis of a source leaf and import of carbon by an adjoining sink leaf has been found over a wide range of light intensities (Fig. 4). Servaites and Geiger (253) concluded from these observations that translocation is limited by the rate of photosynthesis. However, only about 20% of the ^{14}C fixed by the plants in these experiments was exported during the period of observation, indicating that there was appreciable storage in the leaf. The export of carbon from chloroplasts may be 30–50% less than the rate of fixation as a result of the accumulation of starch within the organelles (37). In such circumstances, partitioning of carbon between export and storage pools may be of critical importance. Support for this view comes from the work of Ho (140), who studied the relationship between net photosynthesis and carbon transport from a mature tomato leaf. He found that when the rate of carbon fixation was 2 mg dm^{-2} h^{-1} or more,[1] export was proportional to fixation, but when the amount of carbon fixed fell below 1 mg dm^{-2} h^{-1} export was maintained at this level for a time, presumably by hydrolysis of starch.

The existence of pools in a leaf from which carbon is exported at different rates is indicated by the time course of loss of label from a leaf given a short pulse of $^{14}CO_2$ (Fig. 5). The cumulative loss of carbon follows two and possibly three exponential phases, the rate of loss from each being an order of magnitude less than from the preceding one. This pattern is interpreted as indicating three pools into which fixed carbon may enter (Fig. 6). The first of these, termed the export pool, is relatively small and short-lived in the sense that a carbon atom spends on average less than 0.5 h in it. It may be thought of as comprising, in the broadest sense, all the carbon within a leaf that is en route from the site of fixation to the sieve elements; in the narrower sense of Outlaw *et al.* (221), it is the sucrose in the companion cells of the minor veins. Ho (140) has demonstrated that there is a highly significant correlation between the level of sucrose in a tomato leaf and the rate of export of carbon. The second pool is a medium-term store in which it is calculated that a carbon atom spends on average about 16 h (13, 212). The carbon in this pool occurs mainly as starch in the plastids of the mesophyll cells and its turnover rate will depend on the rates of synthesis and hydrolysis of starch. The third pool is that from which fixed carbon is released at a very slow rate from stable molecules such as cellulose which do not turn over rapidly.

The mechanism which controls partition of fixed carbon between these pools is not known, but it is presumably related to the activities of enzyme systems and their accessibility to the photosynthetic products. The fact

[1] The preferred SI units are mg m^{-2} s^{-1}, 10 mg dm^{-2} h^{-1} ≡ 0.28 mg m^{-2} s^{-1}.

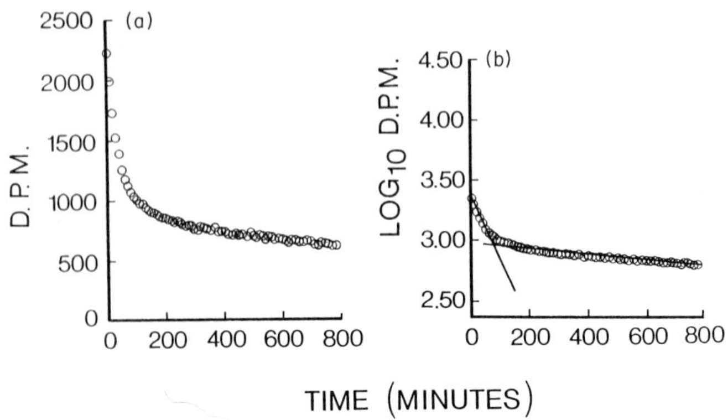

FIG. 5. The time course of loss of radioactivity from an area of wheat leaf fed with $^{14}CO_2$ expressed on arithmetic (a) and logarithmic (b) scales. The two straight lines in (b) were fitted by eye. Redrawn from Dale *et al.* (54).

that both storage and export pools exist raises the important question as to whether there is competition between them for the available photosynthate or whether the export pool takes precedence and only excess carbon, which cannot immediately be loaded into the phloem, is converted to storage products. Chatterton and Silvius (37) observed that accumulation of starch in soybean leaves is inversely related to the length of the photosynthetic period. Plants grown in a 14-h photoperiod converted about 60% of the carbon fixed during this time into starch, whereas those grown

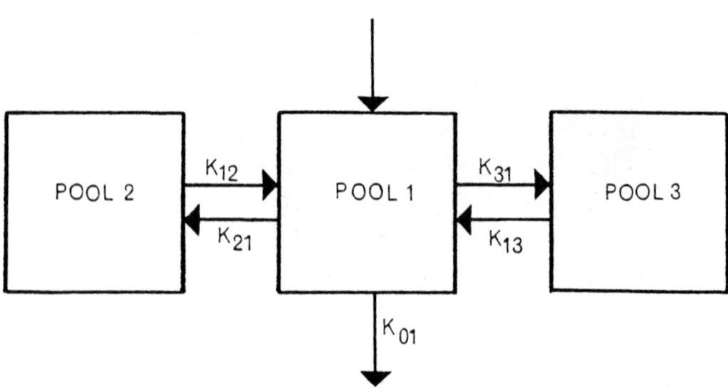

FIG. 6. A three-compartment model for the partition of carbon in a leaf. Carbon enters pool 1 from which it is exported or exchanged for storage in pools 2 and 3. The rate constants K_{01}, K_{12}, K_{21}, K_{31}, K_{13} are derived from compartmental analysis.

in a 7-h photoperiod converted about 90% to starch. The difference was attributed to a reduction in the proportion of carbon used in the synthesis of structural carbohydrates, such as cellulose, in the shorter photoperiod. Leaves grown in a 7-h photoperiod were thinner and photosynthetically more efficient in terms of carbon fixed per unit of dry mass than were leaves exposed to a 14-h photoperiod. It appears that starch accumulation is controlled independently of CO_2 fixation and is modified by the intensity of other processes utilizing the same substrates, including sucrose synthesis and translocation (258).

Although translocation rates are often correlated with rates of photosynthesis, translocation continues in the dark after photosynthesis stops. Since there is such a rapid turnover of carbon in the export pool, it seems likely that it is the medium-term pool, and in particular starch, which provides the carbon translocated at night. It may be that in darkness there is an increase in activity of enzymes responsible for starch degradation (starch phosphorylase) and for sucrose synthesis (sucrose synthase, sucrose-phosphate synthase). Such changes in enzyme activities would be further evidence for control of translocation through biochemical mechanisms. The buffering of translocation through storage pools, notably starch, so that large temporal fluctuations are avoided is important and may also involve temporary storage in stems (315). Unfortunately, we know little of the control mechanisms involved, except in the case of sugar cane (117, 134) in which stems are specialized for long-term storage of sucrose.

3. The Source–Channel Interface

a. The Anatomical Background. Carbon assimilates from photosynthesis are loaded into the phloem of the minor veins of the leaf. During development repeated branching of the vein network leads to the production of high-order veinlets which mature basipetally. As a result of this extensive branching all mesophyll cells are relatively close to a minor vein (102, 319). In sugar beet minor veins of the fifth order are the main sites of assimilate loading and ramify so extensively (Table II, Fig. 7) that the average mesophyll cell is only about 65 μm (two to three average cell diameters) from such a vein when the leaf is about 60% expanded. A length of minor vein equal to the average mesophyll cell diameter receives assimilate from about 30 mesophyll cells situated on average between two and three cell diameters away (104).

The minor veins are anatomically complex including parenchymatous cells and tracheids as well as sieve tube elements and companion cells (Fig. 8). In C_4 plants (52, 176) the vein is invested by a chlorenchymatous bundle sheath in which sugars are synthesized from carboxylic acids produced in

TABLE II

Extent of the Minor Veins in a Sugar Beet Leaf Whose Length Is 60% of Its Final Size[a]

Characteristic	Data
Extent of minor veins	70 cm cm^{-2} blade
Volume of sieve element–cell complex in minor veins	0.15 mm^3 cm^{-2} blade
Proportion of volume occupied by sieve element–companion cell complex	0.6%
Surface area of sieve element–companion cell complex in minor veins	88 mm^2 cm^{-2} blade
Maximum distance from mesophyll cell to nearest minor vein	100 μm

[a] Data of Geiger and Cataldo (104) and Sovonick *et al.* (263).

the mesophyll (125, 133, 135). In several species of C_3 plants the companion cells of the minor veins are of greater diameter than the sieve element with which they are associated; in major veins the reverse is true and companion cells are normally substantially smaller than the sieve tubes, mainly because of the larger overall size of the latter in major as compared with minor veins (104, 214).

Distribution of plasmodesmata in the mesophyll and minor veins is important in any consideration of the pathway of loading and has been studied in detail in sugar beet leaves (102, 103, 105). There are frequent plasmodesmata connections between sieve tube and companion cells, and they are also common between phloem parenchyma and mesophyll cells (Table III). On the other hand, connections between one sieve tube and

TABLE III

Frequency of Plasmodesmata between Cells in the Mesophyll and in the Vascular Bundles of Sugar Beet[a]

Components	Occurrence
Mesophyll cells and other mesophyll cells	Rare
And phloem parenchyma	Common
Phloem parenchyma and other phloem parenchyma	Common
And companion cells	Abundant
And sieve elements	Rare
Companion cells and other companion cells	Rare
And sieve elements	Abundant
Sieve elements and other sieve elements	Rare

[a] Data from Geiger *et al.* (105).

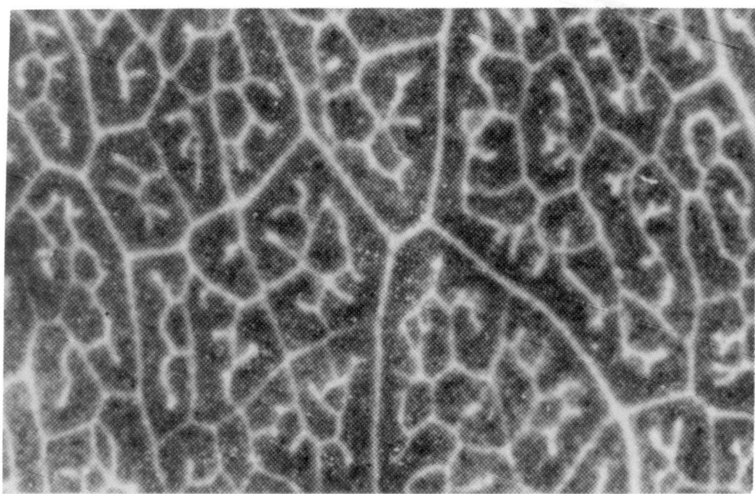

FIG. 7. Autoradiograph of sugar beet leaf supplied with [^{14}C]sucrose. The minor vein network shows clearly as white regions on a dark field and contains the highest amounts of ^{14}C. Reproduced from Geiger *et al.* (107).

another are as rare as are those between adjoining mesophyll cells, the latter presumably reflecting the physical separation of these cells in the fully expanded leaf with consequent reduction in the surface areas of confluent walls.

The parenchymata of the minor veins may also show the characteristics of transfer cells (see Section III and the chapter by J. L. Hall in Volume VII of this treatise) in which the area of the plasma membrane is increased by

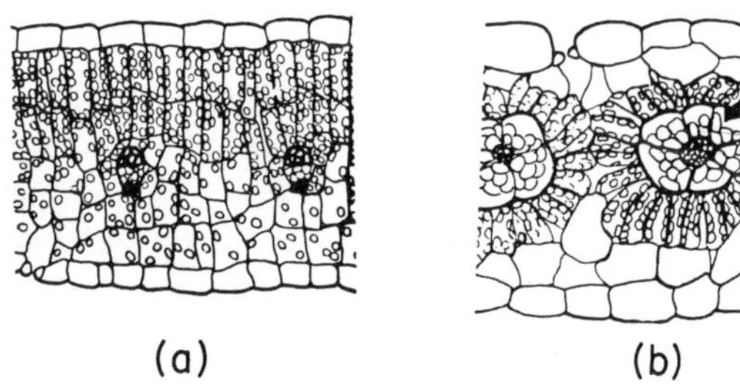

FIG. 8. Drawings of transverse sections of leaves of (a) *Atriplex patula* (C_3) and (b) *A. rosea* (C_4). Note the differences in location of photosynthetic tissues and the bundle sheath (arrowed) containing many large chloroplasts in *A. rosea*. Adapted from Boynton *et al.* (28).

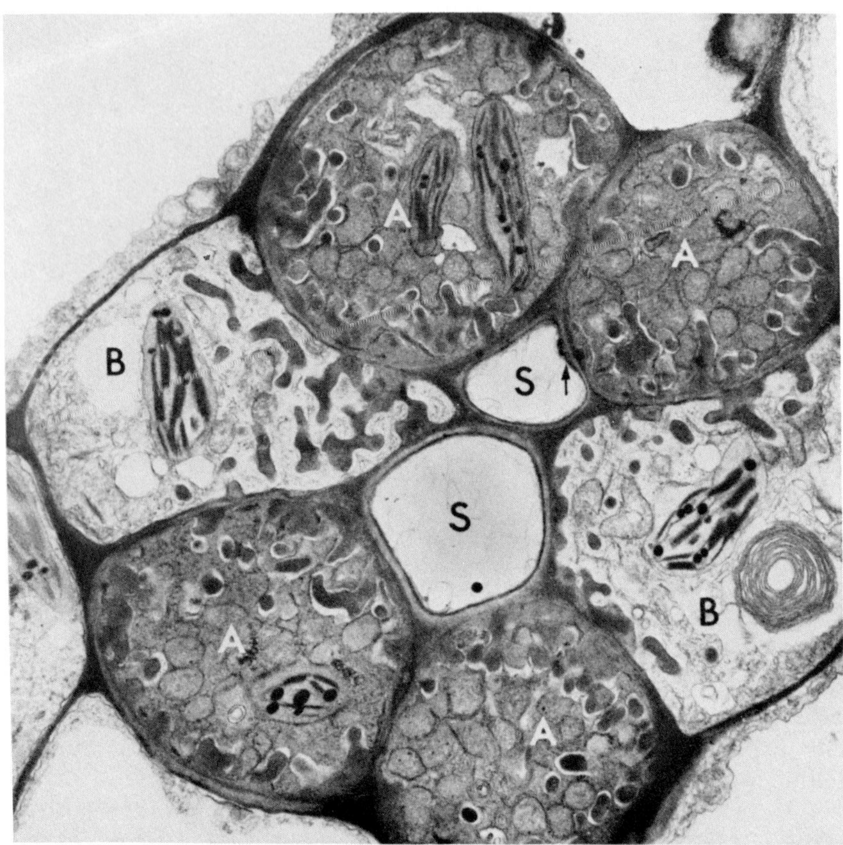

FIG. 9. Minor vein of *Senecio vulgaris* showing sieve elements (S) and A- and B-type transfer cells. The arrow in the upper sieve element indicates a plasmodesmatal connection. Reproduced from Pate and Gunning (227).

ingrowth of the cell wall (Fig. 9; see Section III,B). If the plasma membrane is the site of an active transport mechanism, the increased membrane area could facilitate greater fluxes across it (124, 227). Unfortunately, there is little direct evidence to support this idea and despite the existence of transfer cells in the minor veins, where solutes are moved at rapid rates, a fully convincing demonstration of their functional significance in loading has yet to be made.

b. Phloem Loading. Loading may be defined as the process by which translocated molecules are accumulated and concentrated in the sieve elements. Where this occurs in the minor veins of the leaf, it is termed

TABLE IV

TIME INTERVAL BETWEEN THE SUPPLY OF LABELED CO_2 AND THE APPEARANCE OF LABEL IN THE TRANSLOCATION PATH (LOADING DELAY)

Species	Loading delay (minutes)	Reference
Maize C_4	3–5	Troughton *et al.* (285)
Wheat C_3	2–3	A. Bauermeister (unpublished)
Soybean C_3	5	Moorby *et al.* (211)
Tomato C_3	4–9	Moorby and Jarman (212)
Sunflower C_3	6–9	Williams *et al.* (315)

primary loading, with the term secondary loading used to mean re-entry of translocate into sieve elements from temporary storage sites in organs other than leaves (see Section II,C). It is primary loading that is considered here. The subject has recently been reviewed by Giaquinta (114).

The spatial separation of the site of carbon fixation, the chloroplast, and the phloem means that there is a delay before assimilated carbon is moved out of an exporting leaf. Nevertheless radioactive carbon has been detected in the petiole of a leaf within minutes of supplying labeled CO_2 to the lamina (Table IV). Part of the observed delay is due to the metabolic interconversions that occur, leading to the formation of sucrose in the mesophyll cells or elsewhere. In C_4 plants (e.g., corn) the position is further complicated by the fact that the cells containing plastids in which carbon fixation occurs are separate from those in which carbon reduction is completed (176). However, the contribution made by metabolic processes to loading delay is likely to be of the order of seconds, and the major cause of the delay is the time taken for assimilates to move from a photosynthesizing cell to the sieve tube.

If it is assumed that assimilate travels by diffusion, then application of Fick's second law of diffusion (220) gives the equation

$$x^2 = 4Dt$$

where D is the coefficient of diffusion of the molecule in question in water and t is the time taken to diffuse along a path of length x. On the further assumption that assimilate diffuses as sucrose, for which D is 0.52×10^{-5} $cm^2\ s^{-1}$ at 25°C, and that the diffusion path length is unlikely to be more than four times the average distance between mesophyll cells and minor vein, we have

$$t = (240 \times 10^{-4})^2/(4 \times 0.52 \times 10^{-5})$$
$$= 27\ s$$

This value is substantially less than the observed delays (Table IV) and indicates that the rate at which sucrose reaches the sieve tube from the chloroplast is unlikely to be limited by diffusion.

A number of plasmolytic studies (102, 248) have shown that the osmotic potential of the cells of the sieve element–companion cell complex is around −3.0 MPa compared with values of −1.3 MPa for the mesophyll cells and about −0.8 MPa for the phloem parenchyma. Other methods including the use of aphid stylets (60, 163, 207, 238, 239) and negative staining (95) have indicated that this difference is due mainly to the high sucrose concentration in the sieve elements and companion cells. This indicates that movement into sieve tubes is not likely to be by diffusion. Autoradiographic studies which demonstrate that after a short pulse labeling with $^{14}CO_2$, label initially present only in the mesophyll is found almost exclusively in the minor veins after 5 to 10 min (103) also indicate a nondiffusional loading mechanism. At least for sucrose, therefore, loading occurs uphill, against a concentration gradient, and involves an active process; it cannot be explained in terms of diffusion alone.

Two questions now arise. These concern, first, the pathway of movement of assimilate from chloroplast to the sieve element and, second, the site of the active loading process. It is possible that the pathway could be entirely through the symplast. This requires the presence of plasmodesmata between cells along the path, through which solutes move (289), and a symplastically sited barrier and pump at some point close to, or at, the minor vein. Such a route from mesophyll cells to phloem parenchyma and via the companion cells to sieve elements is possible in sugar beet (Table III) and in wheat (171). However ". . . the symplastic path . . . cannot continue from cell to cell in the same way to the sieve elements, for the very feature of easy diffusion that distinguishes the plasmodesmata for cell-to-cell movement denies them the role of a barrier and pumping site, unless some of them are functionally quite different from the rest" (171).

The problem of identifying a credible loading site within the symplast has led workers to consider the pathway to include an apoplastic element with the loading pump located at the apoplast–symplast boundary at the minor vein. Kursanov and Brovchenko (173–175) washed disks of sugar beet leaves in distilled water and found that within 30 min between 20 and 30% of soluble sugars in the leaf had been lost by leakage. Using a pulse of $^{14}CO_2$ they demonstrated the appearance of labeled compounds in the free space within 10 min of feeding, and by 120 min the free space contained about 40% of newly labeled compounds. When portions of the mesophyll tissues and of vascular bundles were incubated in sucrose solutions, uptake of sugar into the latter was found to be significantly greater than that into the mesophyll. Movement of labeled assimilate into the free space of leaves has been confirmed by other workers (107), and it has also been found that when labeled sucrose is fed to the free space of abraded leaves, it is translocated at a rate comparable with that of photosynthetically produced assimilate. The assumption that sucrose is loaded unchanged and not following hydrolysis seems to be justified (108 – 110). Using an elegant isotope trapping method, Geiger and co-workers (105, 107) supplied leaves of sugar beet with $^{14}CO_2$ and followed accumulation of label into unlabeled sucrose solution applied to abraded surfaces of the same leaves (Figs. 10,11). The fact that label was trapped at all clearly indicates passage into the free space without necessarily confirming the labeled compound to be sucrose. Cor-

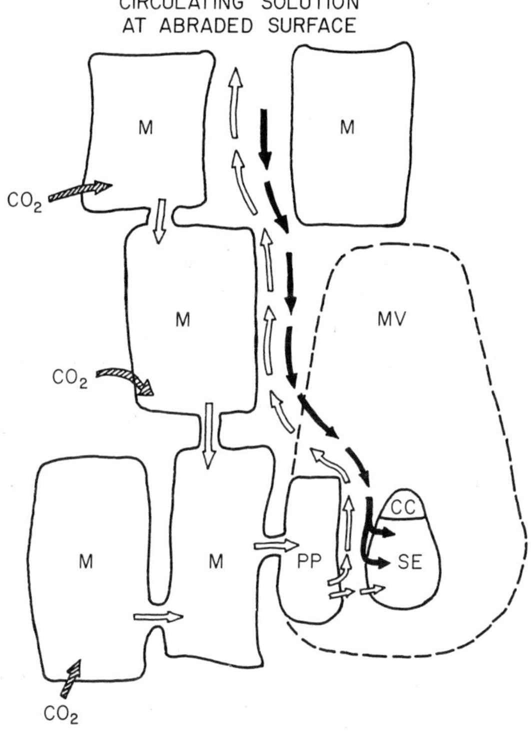

FIG. 10. Diagram of Geiger's isotope trapping technique. Unlabeled sucrose is supplied to the abraded surface of leaves, which are also fed with $^{14}CO_2$, which is fixed in the mesophyll cells (M). The added sucrose (solid arrows) increases the rate of translocation of sucrose from the minor vein (MV). Labeled sucrose (unfilled arrows) is thought to be loaded by the sieve element – companion cell complex (SE – CC) from the free space, but some will enter the pool at the leaf surface from which it is sampled. Redrawn from Geiger *et al.* (107).

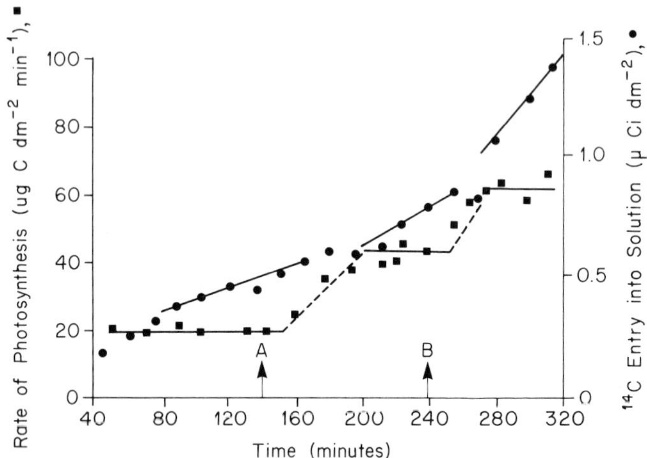

FIG. 11. The effect of increasing light intensity on photosynthetic rate (■) and on the exit of labeled carbon into free space (●) of sugar beet leaves. At point A, light intensity was increased from 22,500 to 32,250 lx, and at point B, it was increased further to 45,200 lx. Adapted from Geiger *et al.* (107).

relative evidence indicates that conditions which increase photosynthetic carbon fixation, such as increasing irradiance, also increase translocation and the amount of label trapped.

In sugar beet leaves it is believed that sucrose enters the free space from the phloem parenchyma cells (61), movement to these cells being symplastic. Once in the apoplast, movement will be by diffusion along a concentration gradient which can be made steeper either by enhanced efflux of sucrose or by more rapid removal by the loading process. The site of loading is the point at which assimilate is moved from apoplast to symplast across a boundary membrane and is thought to be at the boundary of the sieve element–companion cell complex (103, 111, 124, 171), although unequivocal evidence for this is scanty.

Numerous studies support the idea that loading is an active process. It can be stimulated by the addition of ATP to the leaf (173) and inhibited by compounds such as DNP and cyanide (103, 106, 115, 131, 256). Other evidence for an active mechanism involved in loading comes from the high fluxes calculated for leaves of sugar beet (263) in which for a translocation rate of 0.95 μg sucrose min^{-1} cm^{-2} leaf blade a loading flux across the plasma membrane of 3.2×10^{-9} mol min^{-1} cm^{-2} is required. Such high rates are indicative of an active transport system.

A very large number of substances can be identified in the sieve tube contents (see Table VIII), but it is believed that only comparatively few compounds are actively loaded, the rest moving into the sieve tube by diffusion; presumably for many compounds the concentration in the sieve tube is lower than outside and diffusion is thermodynamically feasible (147). Those compounds for which active loading appears to occur include sucrose and the related galacto sugars, sorbitol, some amino acids, and several inorganic ions (124, 201, 254). The sugars are of particular interest in that they appear to be readily transported across membranes and yet are not equally readily broken down in the transport path. Bieleski (23) considers these sugars to constitute a class of "transfer carbohydrates" which are metabolically inert in the sieve element.

The specificity of the loading mechanism for sugars is somewhat uncertain. Fondy and Geiger (97) found that when a variety of sugars including sucrose, fructose, stachyose, and mannitol were fed to sugar beet leaves, only sucrose was loaded rapidly and concentrated in the phloem. This is compatible with the view that a single sugar, or a limited range of sugars, can be actively loaded. However, species differ and Bieleski (23) found evidence for active uptake of sorbitol (the major translocated sugar in many woody Rosaceous species) and glucose in slices of pear leaves. Fructose, mannose, and mannitol had slight but quantitatively different effects on uptake of these sugars, and it was concluded that their uptake involved different transfer mechanisms. However, it cannot be said for certain that these results indicate two carrier systems at the surface of the sieve element–companion cell complex since movement to that site could have been affected.

In squash, in which galacto sugars are important transport carbohydrates, stachyose is loaded into the phloem following synthesis elsewhere in the leaf, probably in the mesophyll (136). It remains unknown whether the loading system in this species will accommodate galactose and sucrose with one or several carriers. It may be noted that in a number of species there is evidence for active loading of the amino acid leucine by a carrier different from that involved in sucrose loading (9, 254, 294).

A satisfactory explanation of the mechanism by which sucrose is loaded into the sieve tube element is now available. In early work on this topic Kursanov (172) proposed a carrier system to transport the phosphorylated products of sucrose hydrolysis across the sieve element membrane. However, the hydrolysis of sucrose, as part of the loading process, seems not to occur (109) and carrier-based mechanisms requiring such hydrolysis and resynthesis are not now in favor.

Recently, several authors (58, 109, 111, 113, 114, 149, 168, 169, 188, 189, 192) have proposed schemes based on proton cotransport of sugars at the plasma membrane of the sieve element. These involve an ATP-driven pump giving a proton efflux and potassium influx and the movement of sugar, and in some work amino acids, by proton cotransport from the apoplast into the sieve element (Fig. 12). Evidence supporting this hypothesis comes from studies on a number of species. In work on *Ricinus* (188, 189) ^{14}C-labeled sugars were injected in potassium or sodium-based buffer solutions at pH 5 or 8 into the hollow petiole. Phloem sap was collected from a stem cut 10 to 20 mm from the petiole and exudate was found to be much more heavily labeled when the buffer solution contained potassium and was at pH 5. Such a result would be expected on the scheme in Fig. 12, which also accounts for the well-known facts of high pH of phloem sap and its high potassium content. The pH dependence of loading has also been shown for sugar beet leaf tissue (103). Feeding the inhibitor *p*-chloromercuribenzoate sulfonate (PCMBS) to the free space of sugar beet leaves did not affect photosynthetic rate but slowed down the translocation rate (113). This compound reacts with sulfhydryl groups and may affect entry of sucrose into minor veins by altering the outer surface of the plasma membrane of the sieve element–companion cell complex across which loading occurs. Giaquinta (113, 114) suggests that PCMBS inhibits plasmalemma ATPase, although Delrot *et al.* (59) argue for a direct effect on the sugar carrier complex itself. In contrast the fungal toxin, fusicoccin

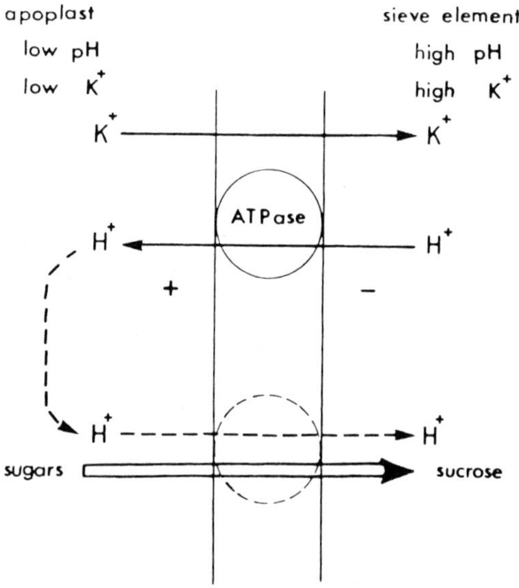

FIG. 12. A scheme for loading sucrose into the sieve element by means of proton cotransport. An ATP-energized pump extrudes protons into the apoplast, while a carrier transports sucrose and protons together from apoplast to the symplast of the sieve element. From a diagram supplied by D. A. Baker.

which is a potent activator of H^+-K^+ exchange, and of ATPase activity in many systems (191), stimulates both sucrose uptake and proton efflux in sugar beet and *Ricinus* (113, 114, 189).

Kinetic studies on sucrose-specific proton cotransport have been attempted by a number of workers and values of K_m and V_{max} derived (58, 254). Interestingly, it has been claimed that a second loading mechanism may operate in abraded leaf disks of sugar beet supplied with concentrations of sucrose sufficient to saturate the proton cotransport mechanism (200).

C. Storage Organs as Sources

As well as storage compartments within leaves there are also other storage compartments elsewhere in the plants. Stems and roots can be regarded as both sinks and sources because of their roles in storage of metabolites. For example, in cereals small but significant quantities of dry matter are translocated from the adjacent regions of the stem to the ear during grain filling in maize (4) and wheat (303). In other species stem storage can be very considerable. In Jerusalem artichoke (151) the stem is a major temporary store for dry matter prior to tuber formation. Tuberization begins when the leaves become senescent, and stem dry weight declines by up to 70% in 80 days, this loss being more than sufficient to account for tuber growth (Fig. 13). Incoll and Neales (151) suggest that the hormonal mechanisms involved in tuber formation may be associated with the establishment of dry matter flow to the tuber from the stem storage centres, perhaps through effects on the activity of hydrolytic enzymes there.

It has also been suggested (315) that in sunflower stem tissue may provide temporary short-term storage sites which buffer the transport of assimilate. During the day material may be unloaded from the phloem, only to be reloaded (secondary loading) at night when the assimilate supply from the leaves falls.

Biochemical interactions between sink and source are seen when specialized storage organs, such as potato tubers or cereal grains, are considered. In the latter case the interaction between embryonic axis and endosperm mediated through gibberellins is well known (158, 320) and activity of the source is dependent initially on a signal from the sink. Amylases are only one group of hydrolytic enzymes involved, and control over peptidase and glucanase activity is also important in modulating the flow of metabolites from the endosperm to the growing axis of the young seedling.

Baset and Sutcliffe (12) demonstrated a close correlation between the rates of change in amylase and protease activities in extracts from oat grain

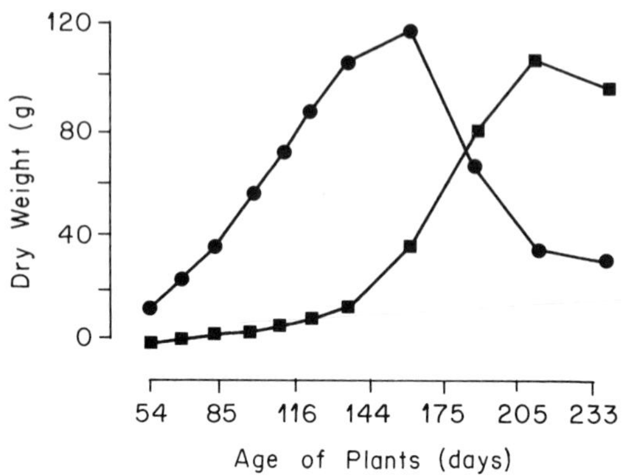

FIG. 13. Changes in dry weight of the stem of *Helianthus tuberosus* as the tubers fill at the end of the growing period; ●, stem dry weight; ■, tuber dry weight. Adapted from Incoll and Neales (151).

endosperm during the first 7 days of germination and the rate of transport of hydrolysis products of starch and protein into the developing axis.

The mechanism controlling mobilization and transport of stored materials from the cotyledons of dicot seedlings is not so clear (121, 122, 274). Yomo and Varner (321) found that when pea seedlings were grown in vermiculite soaked in a solution containing casein hydrolysate, there was only about half as much protease activity in cotyledon extracts as there was when seedlings were fed with water. They also observed that only about half as much nitrogen had left the cotyledons of casein-fed plants and they concluded that transport of stored nitrogen is controlled by protease activity which in turn is regulated by a feedback inhibition through the accumulation of soluble nitrogen compounds in the cotyledons. Guardiola and Sutcliffe (121, 122) reached a different conclusion from a detailed study of the changes in protease activity, protein and soluble nitrogen contents, in excised and attached pea cotyledons. They argued that protease activity and hence mobilization and transport of nitrogen is more likely to be controlled by the axis through growth-regulating substances, despite their inability to demonstrate this directly.

D. SINKS

1. *General Considerations*

Some of the problems of defining sinks have already been outlined (Section II,A). For convenience sinks are considered to be regions in which translocated substances are being received. Meristems and actively growing organs constitute the most important sinks in a plant.

Three major questions can be asked about sinks:

1. What governs the amount of translocated material flowing into a particular sink?
2. Is there competition between sinks and if so what governs this?
3. How is unloading at a sink achieved and controlled?

These will be considered in turn.

2. The Flow into a Sink

The size of a sink and its location with respect to the source might be expected to influence assimilate flow to it. By analogy with an earlier definition of source strength [Section II,A (305)], sink strength can be defined as

$$\underset{g\,t^{-1}}{\text{sink strength}} = \underset{\text{g}}{\text{sink size}} \times \underset{g\,\text{g}^{-1}\,t^{-1}}{\text{sink activity}}$$

It is of interest to know what happens if sink size is varied while sink activity is kept constant, and vice versa; it is also informative to follow changes in sink strength since size and activity vary concurrently.

The developing ear of a cereal offers a convenient system for examining the first point for sink size can readily be altered by removing some of the grains, and in short-term experiments the complicating effects on sink activity might be expected to be slight (153). Wardlaw and Moncur (301) showed that removal of half to a third of the grains on the developing ear of wheat reduced import of ^{14}C-labeled assimilate fixed in the flag leaf by between about 30 and 65%, indicating that where sink size is large, import is greater than when it is smaller. Removal of grains had a very rapid effect on translocate movement, a reduction in the amount of assimilate reaching the base of the peduncle being shown within 10 min of the removal treatment. Speed of translocation was also related to sink size (Fig. 14). When sink demand, based on grain number, was doubled, speed of translocation was more than doubled, implying that there was a fall in the concentration of translocate in the conducting elements. Treatment of the developing ear with the inhibitor DCMU, which prevented photosynthesis in the awns and other structures of the ear, also increased the speed of translocation as the demand for assimilate increased.

A quite different approach in which sink size was varied but activity kept constant was employed by Peel and Ho (239), who established colonies containing either 4 or 5 or 15–20 aphids *(Tuberolachnus salignus)* on stems of willow. A leafy shoot acting as the source was fed with $^{14}CO_2$ and the rate of honeydew secretion by the colonies, and specific activity (cpm mg^{-1}) of

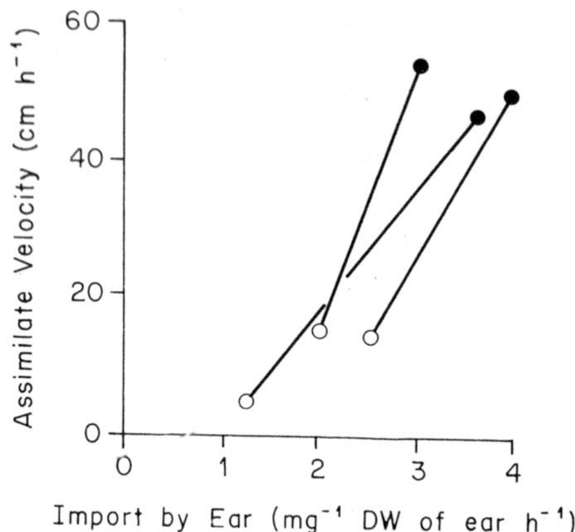

FIG. 14. The relationship between sink size and the speed of assimilate movement through the peduncle in wheat. In three separate experiments sink size was varied by removing between one-half and one-third of grains in the ear (○) and the treated plants were compared with controls in which no grains were removed (●). Adapted from Wardlaw and Moncur (301).

the honeydew was determined (Table V). Averaging the results of nine experiments, the rate of honeydew production from the larger colony was about 3.5 times that from the smaller. On the face of it, these data indicate the rate of honeydew production and the number of aphids in the colony to be in approximately similar ratio. However, in all experiments the specific activity of the honeydew obtained from the large colony was greater than that from the smaller by a factor varying from 1.5 to 3300. That is, more assimilate was secreted as honeydew by the larger colony than would be expected on the basis of size alone. This was interpreted as

TABLE V

HONEYDEW PRODUCTION AND SPECIFIC ACTIVITY BY LARGE AND SMALL APHID COLONIES ON LEAFY SHOOTS OF WILLOW SUPPLIED WITH $^{14}CO_2$[a,b]

Property	Colony size: Large	Colony size: Small	Ratio (large : small)
Rate of honeydew production (mg h^{-1})	1.11 ± 0.38	0.32 ± 0.09	3.47
Specific activity (cpm mg^{-1})	129,900	36,900	3.52

[a] Data from Peel and Ho (239).
[b] Values are means from nine experiments.

meaning that a much greater length of stem contributed to the honeydew from the larger colony, the larger sink, than to that from the smaller. In other words, there was also an increase in size of the source.

The cases so far considered are atypical in that they involve arbitrary manipulation of sink size in the case of wheat and establishment of wholly artificial sinks in the case of the aphid studies. The intact growing fruit represents a more natural system in which one might expect increases in size, measured as fruit dry mass, to be increasingly offset by an ontogenetic decline in sink activity, expressed by rate of increase of dry mass per unit size, as the fruit reaches maximum size and sink strength falls to zero. Data for developing tomato fruits bear this out. Studying import of carbon by fruits on plants grown under standard conditions of defoliation and truss removal, Walker and Ho (296) found that, as fruit size increased, sink activity decreased; in consequence sink strength, measured as rate of import of carbon, also fell since the increase in sink size was accompanied by a proportionately greater fall in activity (Table VI). Calculations based on their data show that sink activity fell from 0.0065 to 0.00089 mg mg^{-1} h^{-1} as the fruit increased from 20 to 90% of its final volume.

In general, for sinks of the same kind and having the same or similar activity, sink strength is proportional to size. This relationship does not hold for sinks whose growth is determinate and finite, where ontogenetic changes in activity can effectively counteract size increases. In the case of meristems of indeterminate growth, such as those of root or stem apices, and cambia, increases in size, though often slight, is not offset by decrease in sink activity; if this were to happen, growth would decline and eventually cease altogether.

A number of lines of evidence indicate that sink location with respect to the source is important. In cereals, local photosynthesis in the awns and

TABLE VI

INFLUENCE OF SINK SIZE ON CARBON IMPORT BY TOMATO FRUITS[a]

Fruit size (% final volume)	Pedicel phloem area (mm^2)	Carbon import: Per fruit (mg h^{-1})	Carbon import: Per unit phloem area (mg mm^{-2} h^{-1})
20	0.87 ± 0.05	5.87 ± 0.39	6.75 ± 0.59
30	0.90 ± 0.04	5.28 ± 0.27	5.87 ± 0.40
50	1.07 ± 0.07	3.35 ± 0.08	3.13 ± 0.22
90	1.24 ± 0.08	3.11 ± 0.14	2.51 ± 0.20

[a] Data from Walker and Ho (296).

other structures of the ear makes an important contribution to the carbon economy of developing grain, although significant amounts of carbon are also imported from the adjacent flag leaf (30, 36). Using wheat plants treated so that only two tillers remained, Cook and Evans (41) showed that assimilate from labeled carbon fed to the flag leaf on one tiller was preferentially supplied to the adjacent ear rather than to the ear 140 cm distant on the other stem, irrespective of whether it was larger or smaller in terms of grain number; the sink's location was more important than its size.

Other data show that developing fruits tend to be supplied with assimilate from adjacent rather than distant leaves. This is so for tomato (164) and for cotton, in which the developing boll is supplied with assimilate mainly from the leafy bracts which surround it and from the leaf carried at the same sympodial node as the fruit itself (29, 145).

Results such as these have led to the general conclusion that sinks tend to be supplied by the nearest source (32). Such a system is clearly efficient in terms of the distance and time taken to move assimilate. Support for the view that this relationship between source and sink is anatomical comes from numerous observations indicating that labeled carbon fixed in one leaf is translocated exclusively, or at least preferentially, to younger leaves or even to a part of such a leaf in direct anatomical connection with it (Fig. 15)(11, 159, 243). In *Phaseolus* each primary leaf normally supplies assimilated carbon to one of the lateral leaflets of the developing first trifoliate

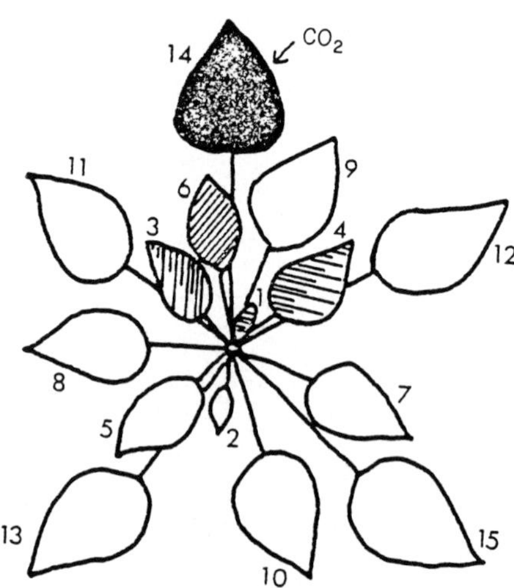

FIG. 15. Distribution of radioactivity in leaves around the crown of sugar beet plants becoming radioactive after exposure of leaf 14 to $^{14}CO_2$. Adapted from Joy (159).

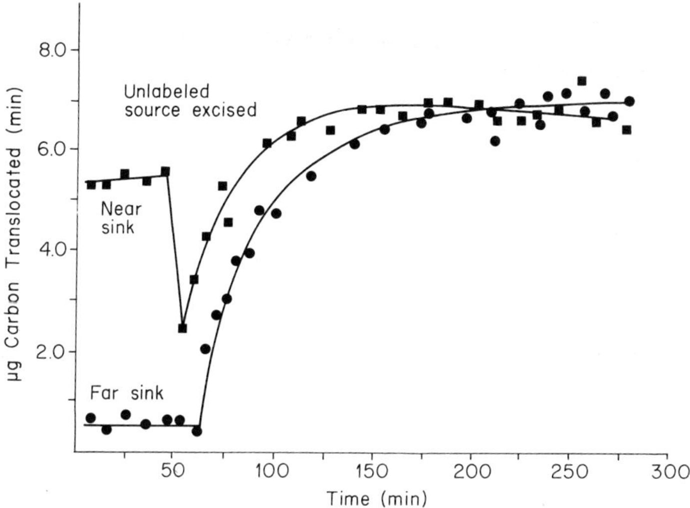

FIG. 16. Translocation of labeled carbon from a primary leaf of *Phaseolus vulgaris* to the near (■) and far (●) leaflet of the first trifoliate leaf, before and after excision of the other primary leaf. Adapted from Borchers-Zampini *et al.* (26).

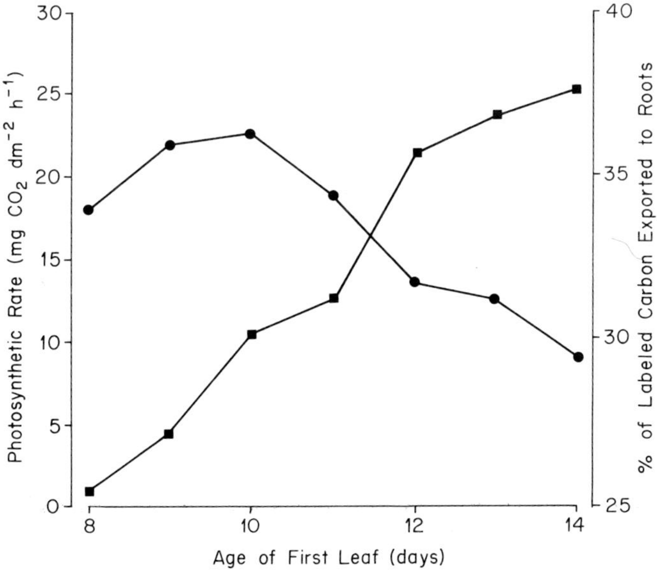

FIG. 17. Changes with time in the photosynthetic rate of the first leaf of barley (●) and of the percentage of labeled carbon exported to the roots in a six hour period (■).

leaf (216). However, if one of the primary leaves is removed, the other supplies both leaflets, and very soon, within 30 min (Fig. 16), does so at the same rate as before (26, 278). Despite the increased rate of export from the remaining primary leaf there is no immediate effect on net photosynthesis; this suggests that in this case, treatment directly effects the partition of fixed carbon between storage and export pools (see Section II,C).

Proximity of source and sink does not necessarily govern the movement of metabolites other than carbon assimilates, for which, as in the case of nitrogen compounds, the sources may be the roots or some other distant part of the plant. The roots themselves tend to be supplied with assimilate from older leaves (Fig. 17); young leaves export only a small proportion, often less than 30%, of fixed carbon to the roots, the bulk being moved in an acropetal direction to younger leaves, primordia, and stem apices.

3. Competition between Sinks

When a large and a small sink are anatomically equidistant from a source which can supply both, the larger sink will be favored over the smaller. Such a situation can be considered to represent competition between sinks. Inevitably, where more than one sink is supplied by a single source, as in the case of grains in a cereal ear or a truss of fruit on a tomato, some competition between sinks must exist and is resolved in favor of the larger or more active sink. Evans (77) has described this as the "Matthew effect": "Unto those that have shall be given" (New English Bible, Matt. 25: 29). However, the fact that individual sinks are often supplied from nearby sources tends to limit competition among them, at least for assimilates; proximity of source and sink must constitute an important part of the mechanism for regulating and coordinating growth of the plant and avoiding competition. Extreme competition between sinks often results from experimental manipulation of a system in which competition is normally much less marked.

4. Unloading

Metabolites which are transported in the phloem eventually leave that tissue to be used for growth in meristematic sinks or to be stored in specialized cells, storage sinks. The term unloading is used here to denote movement of translocated material out of the sieve elements; it carries no implication about the nature, active or passive, of the process. Since sinks are more or less distant from the phloem, exit from the sieve elements is followed by postunloading movement.

The idea that many translocated compounds enter the sieve elements by diffusion and that comparatively few compounds, including transfer carbohydrates, are actively loaded has already been mentioned (Section

II,B,3,b). If this view is correct, then a reasonable corollary is that most metabolites leave the phloem by diffusion and that only the transfer carbohydrates such as sucrose present any difficulties in unloading. Because of a general lack of knowledge of unloading, this view is accepted as a working hypothesis and we now consider the unloading and postunloading movement of sucrose for which limited data are available.

Keener *et al.* (160) have discussed a number of possible unloading mechanisms for sucrose. The first of these involves simple diffusion out of the sieve tube and diffusion along a concentration gradient to the sink, a view also adopted by Gifford and Evans (116) in a perceptive review; a related mechanism postulates a passive unloading facilitated by a permease-enhancing movement in both directions across the sieve element membrane depending upon concentration gradients and metabolic activities involving sucrose. For both these mechanisms unloading will continue if sink demand falls until sucrose concentrations inside and outside the sieve element come to equilibrium. There is no evidence that this occurs.

Another possible mechanism involves active unloading of sucrose followed by its hydrolysis by invertase (75). A variant of this mechanism

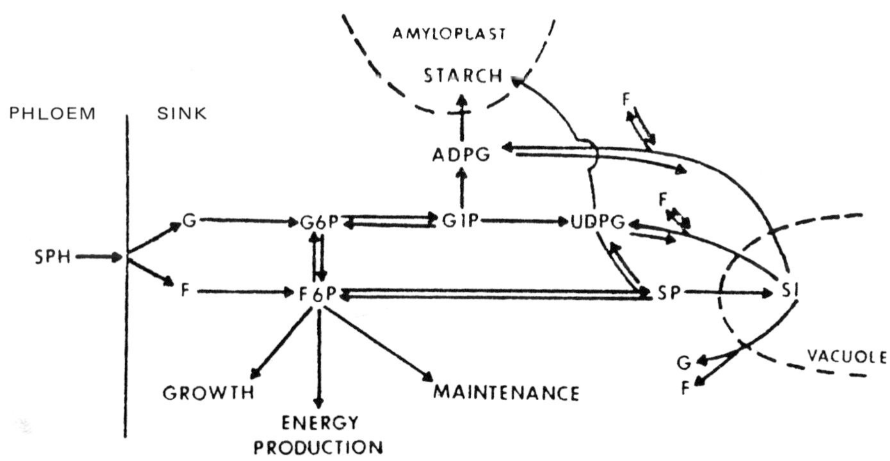

FIG. 18. A generalized sink model showing metabolic pathways and intermediates involved in movement of sucrose from the phloem. ATP, adenosine triphosphate; ADP, adenosine diphosphate; ADPG, adenosine diphosphate glucose; F, fructose; F6P, fructose-6-phosphate; G, glucose; G1P, glucose-1-phosphate; G6P, glucose-6-phosphate; M^+, restored maintainer; M^-, depleted maintainer; SI, sucrose in the vacuole; SO, sucrose in the cytoplasm; SP, sucrose phosphate; Sph, sucrose in the phloem; UDPG, uridine diphosphate glucose. Adapted from Keener *et al.* (160).

favored by Keener *et al.* (160) is that sucrose movement out of the sieve element is governed by activity of invertase at the sieve element membrane, the association being so tight that sucrose cannot move out of the phloem without being split into glucose and fructose (Fig. 18). There is evidence from both sugarcane and sugar beet which is relevant to these ideas.

In sugarcane stems, sucrose is translocated in the phloem and enters the free space, where it is hydrolyzed by a cell wall-bound invertase (117, 134). The resulting hexoses are actively accumulated in the metabolic compartment of storage parenchyma cells and resynthesized into sucrose phosphate, which is then stored in the vacuole as sucrose (Fig. 19). Major points of control in the processes could be through the levels and activities of invertase and sucrose-phosphate synthase.

Giaquinta (112) found that immature roots of sugar beet contained high activities of an invertase but that this disappeared prior to the commence-

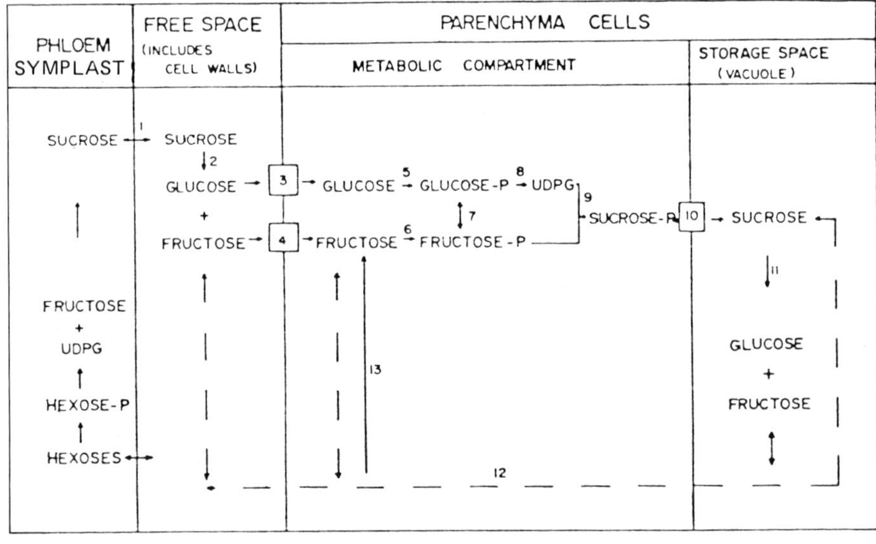

FIG. 19. The sugar cycle in sugarcane storage: (1) movement of sugars between phloem and free space (which includes cell walls), (2) hydrolysis of sucrose in free space by acid invertase, (3) carrier-mediated transfer of fructose into metabolic compartment, (5)–(7) hexose phosphorylation and interconversion, (8) synthesis of uridine diphosphate-glucose, (9) synthesis of sucrose phosphate, (10) transfer of sucrose moiety of sucrose phosphate into storage, (11) hydrolysis of stored sucrose by acid invertase (immature tissue only), (12) diffusional movement of sugars in the direction of the prevailing gradient, and (13) hydrolysis of sucrose by neutral invertase. Redrawn from Glasziou and Gayler (117).

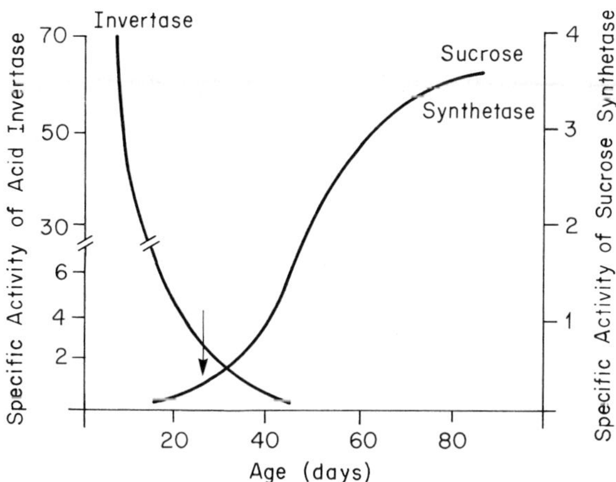

FIG. 20. Changes in the activities of invertase and sucrose synthase during sugar beet root development. Sucrose concentration begins to increase markedly after about day 25 (arrow). Redrawn from Giaquinta (112).

ment of the sucrose-storage phase at which time sucrose synthase activity markedly increased (Fig. 20). Giaquinta (112) argues that sucrose synthase, which is a reversible enzyme, hydrolyzes sucrose in the growing root and that this enzyme plays an important role in the partition of sucrose between storage on the one hand and metabolic activities on the other. In the mature beet root the evidence suggests that sucrose storage occurs without hydrolysis in the apoplast as is found in sugarcane (117). In the root the postunloading pathway could well be symplastic and by movement through plasmodesmata, whereas in sugarcane apoplastic movement occurs linked to an active uptake mechanism in the storage parenchyma.

The supply of assimilate to a sink may thus involve movement in apoplast or symplast, or both, with the participation of active transport processes as well as diffusion. The concentration gradient between phloem and sink appears to be an important feature of the mechanism controlling exit from the phloem (116). If uptake and utilization of a given translocate at the sink are rapid, then the gradient will be steep, whereas if the translocated substance is not being utilized rapidly, the gradient between channel and sink boundary may be a shallow one.

Data indicating the significance of concentration gradient come from work of Walker and Ho (296) on tomato fruit. They observed that the specific mass transfer (SMT) of carbon, i.e., carbon translocated in unit time per unit of cross-sectional area of the phloem, fell as the fruit on a standard truss grew. For fruits which were about 20% expanded the value of SMT was 6.75 mg carbon mm^{-2} phloem h^{-1}, falling to a value of 2.51 when the fruits reached 90% of final size. Walker and Ho (296) found an increase in sucrose concentration in larger fruits for which SMT was falling, suggesting that a feedback from the sink itself may inhibit transport to it. The observation that sucrose is exported from cooled fruits in which its concentration is raised is further support for this idea (297). Some of the metabolic pathways for sugars in the developing tomato fruit have been explored by Walker *et al.* (298).

There is no need to postulate a direct effect of the sink on the transport system characteristics, although the possibility exists that the sink could directly affect sieve tube permeability, perhaps by a hormonal mechanism. There are other implications of the gradient in solute potential between sieve elements and the sink. Without prejudging subsequent discussion in Section IV, it is now widely accepted that some pressure flow, in the sense envisaged in the Münch hypothesis (218), is inevitable, given the characteristics of the sieve tube system. It follows that an increase in osmotic potential (i.e., a reduction in solute concentration) at the sink end of the channel will lead to further supply of solutes to that region by the pressure flow mechanism. This can proceed independently of any import of solute into the channel at a source, although in such circumstances there would be a slow and progressive increase in the osmotic potential of the phloem sap with a corresponding decrease in the pressure flow itself. There would thus be an element of global as well as local control over flow to a sink.

5. *Hormone-Directed Transport*

The possibility that the flow of translocate is under hormonal control originating at the sink has attracted considerable attention (25, 56, 184, 217, 229–231, 304). The idea of such a control is based on observation and experiment. Growing tissues, which are active sinks, are frequently found to be rich sources of growth-regulating substances, especially auxins, but also gibberellins and cytokinins (119). Observations of this kind are suggestive but cannot be taken as an unequivocal indication of causality between high hormone level and sink activity.

More direct evidence comes from experiments using system in which local application of growth substance leads to a local accumulation of translocate at the treated region, hence the term "hormone-directed transport." An example of the approach used is taken from the work of two staunch proponents of the idea of hormone-directed transport, Patrick and Wareing (232). In a typical experiment plants of *Phaseolus vulgaris,* French bean, were decapitated below the first fully expanded trifoliate leaf, leaving the rest of the plant intact. Lanolin paste containing 0.1% IAA, argued by Patrick and Woolley (235) to be equivalent to application of an aqueous solution of 10^{-3} to $10^{-5}M$, was applied to the cut stumps. Then $^{14}CO_2$ was fed to one of the primary leaves 0, 3, 6, or 9 h after treatment; and 3 h after feeding, the distal 1.5 cm of the internode adjacent was harvested and radioactivity in it determined. The results (Fig. 21) showed significantly greater amounts of label in IAA-treated than in lanolin-treated regions when treatments were made 3 h or more before $^{14}CO_2$ feeding. These data indicate that treatment with IAA changes the pattern of distribution of labeled assimilate within decapitated plants, and such results can be regularly found.

The experimental approach using *Phaseolus* has been extended to study the effects of compounds believed to inhibit the polar movement of IAA. Patrick and Wareing (234) found that application of TIBA between the site of IAA application and the source leaf reduced basipetal movement of IAA and inhibited acropetal movement of assimilate, although the movement basipetally was not affected. Similar effects were found using other

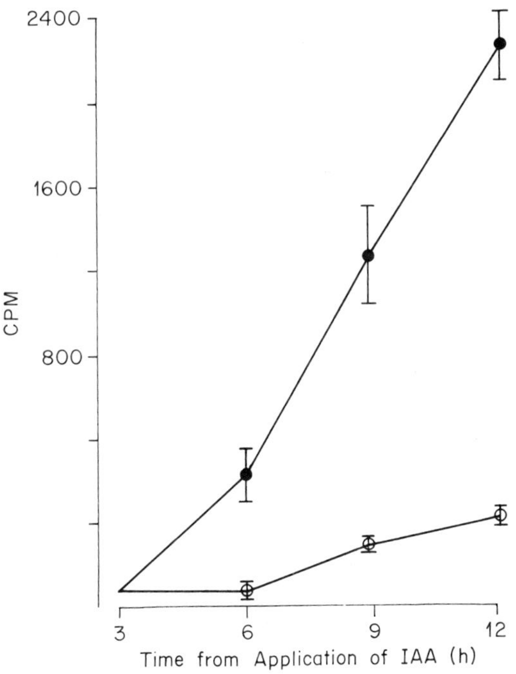

FIG. 21. Time course for transport of labeled assimilate out of primary leaf of decapitated plants of *Phaseolus vulgaris*. Stem stumps were treated with IAA (●) or with lanolin alone (○). Adapted from Patrick and Wareing (232).

inhibitors of IAA transport, eosin and 1-(2-carboxyphenyl)-3-phenylpropane-1,3-dione. Patrick (230, 231) argues that since movement of IAA is necessary for metabolite transport to the treated region to be promoted, these must be an effect of IAA remote from the site of application, and he suggests that this occurs along the length of the transport pathway.

Treatment of decapitated stumps of *Phaseolus vulgaris* with kinetin (288) and with gibberellic acid (215) also results in local accumulation of labeled assimilate at the treated area, but in the case of these growth substances there was no evidence to suggest that they acted other than at the site of application. Sprays of gibberellic acid or cytokinins have also been found to enhance transport of ^{14}C-labeled assimilates to treated fruits or shoot tips (170, 255), but few other data are available from intact systems, although effects of treatments on growth over long time spans have been demonstrated many times.

A major difficulty in studies on hormone-directed transport is that of deciding whether the effect of growth substances treatment is directly on the transport system or whether it is indirect and mediated via growth processes. The problem arises from the fact that in the majority of studies a substantial interval of at least several hours is allowed to elapse between hormone treatment and determination of transport to the treated region. This is because, as Fig. 21 shows, the response is seldom rapid, although effects on translocation rate in experiments lasting only 1 h have been found for decapitated soybean plants and plants of sunflower treated with aqueous solutions of gibberellic acid (50 ppm) or IAA (5 ppm) (137). In experiments of longer duration, it is possible that the effect of hormone treatment is to promote growth which leads to the establishment of a new sink. On this argument, cogently discussed by Phillips (242), enhanced transport to the treated region is an indirect effect which follows hormone-induced growth. The fact that gibberellins and cytokinins, as well as IAA, are active in hormone-directed transport is support for the idea that growth responses are important, since it is unlikely that all of these classes of compound would have direct effects on the translocation process. Indeed, as already pointed out, the sites of action of different growth substances appear to differ.

However, accepting that some responses to hormones are mediated through growth, the possibility of direct effects remains. Patrick (229) has listed a number of possible effects of hormone treatment: (1) hormone-induced changes at the source affecting capacity to synthesize translocated metabolites, (2) effects on sink strength, altering the capacity to accumulate translocated metabolites, (3) effects on the relative ability of sinks to compete against each other for available translocate, (4) direct effects on the sieve tube loading or unloading mechanisms, and (5) direct effects on the process of longitudinal transport within the sieve tube.

There is little evidence that hormone treatment affects photosynthetic activity at all rapidly (137), unless there are effects on stomatal closure, and for hormones to be effective in altering source strength the response would have to be specific to the extent that a particular sink affected only the activity of the source supplying that sink; an effect on the global production or supply of assimilate would be of less significance to any particular sink. Effects of hormones on sink strength are more difficult to assess, although if hormone treatment increases growth, it must also increase the rate of utilization of supplied metabolites and hence steepen the concentration gradient between channel and sink. With regard to point (3) competition between sinks has already been discussed; unequivocal information on the way in which hormones can affect one sink relative to another is difficult to obtain since treatment is almost inevitably confounded with differences in sink size and location.

There is little support for the idea that plant growth substances affect the loading of assimilate in leaves (13, 54, 232). In studies using wheat, it was found that treatment with IAA, GA_3, or benzyladenine had no effect on the kinetics of export of ^{14}C- or ^{11}C-labeled assimilate out of the treated area of leaf (Table VII). As regards effects on unloading, little is known. It has been suggested that in *Phaseolus* IAA treatment may affect unloading at the sink and that it may affect longitudinal transfer in the sieve tubes throughout the transport pathway (230, 233). Data for intact wheat plants (13) showed no effect of treatment with IAA on either velocity or direction of transport of ^{11}C-labeled assimilate, nor was the kinetic profile of trans-

TABLE VII

RATE CONSTANTS[b] FOR PARTITION OF LABELED CARBON IN LEAVES OF HERON WHEAT TREATED WITH 10^{-4} M IAA, GA_3, OR BENZYLADENINE AND FED WITH $^{14}CO_2$[a]

Treatment	Rate constants ($min^{-1} \times 10^4$)		
	k_{01}	k_{12}	k_{21}
IAA-treated plants			
Control	226 ± 13	11.8 ± 1.6	93 ± 10.6
+IAA	234 ± 10	13.1 ± 1.2	95 ± 6.0
GA_3-treated plants			
Control	189 ± 29	13.3 ± 1.5	103 ± 12.1
+GA_3	188 ± 15	12.5 ± 1.3	93 ± 5.0
BAP-treated plants			
Control	244 ± 16	13.9 ± 1.0	110 ± 11.7
+BAP	224 ± 11	12.7 ± 1.4	122 ± 7.6

[a] Data from Dale *et al.* (54).

[b] See Fig. 6.

location in leaves affected. These results indicate that there is no general effect of growth regulators on the translocation process, although absence of effects in the leaf does not preclude effects on transport in phloem of stem tissue. The suggestion that IAA has to be present along the length of the transport path for the promotion of assimilate movement to be obtained remains an intriguing one. Whether the presence of IAA affects the exchange of materials between phloem and adjacent parenchyma (see Section II,C) is not proved. If it does so, this would be evidence for hormone effects on the temporary storage pools in the stem and indicates effects on secondary loading. It would also indicate another way by which sinks could regulate their own metabolite supply, apart from local effects on the concentration gradient between unloading site and site of utilization.

E. ENVIRONMENTAL EFFECTS

The effects of environmental factors on translocation should now be considered. Because factors such as temperature, light, and water stress often have large effects on both sink and source strengths, identification of direct effects on translocation is often difficult. For example, in studies on plants of the C_4 grass *Pennisetum,* Pearson and Derrick (236) found that at low temperature leaves had lower photosynthetic rates and higher sucrose concentrations, as well as retaining a greater proportion of fixed carbon which was exported more slowly, than leaves at higher temperatures; distinguishing source and channel effects here is far from easy.

Mention will be made subsequently of the effects of local application of low temperatures (0 – 2°C) on translocation. In general, results using localized applications of less extreme temperatures have shown translocation to increase with temperature up to a maximum rate at between 20 and 30°C (276). This might be expected for two reasons, first that viscosity of the sieve tube content falls as temperature increases so that the effective resistance of the path to bulk flow will fall and second because at high temperatures enhanced respiration rates will generate larger quantities of ATP, which may be used for phloem-loading processes.

With regard to light, again the difficulty of separating the large effects on source and sink from any on translocation is obvious. There are a number of reports suggesting that light increases translocation (126, 213, 284). In experiments feeding corn leaves with $^{11}CO_2$, Troughton *et al.* (284) have shown substantial increases in velocity of translocation on illumination and equally large decreases in velocity on shading (Fig. 22).

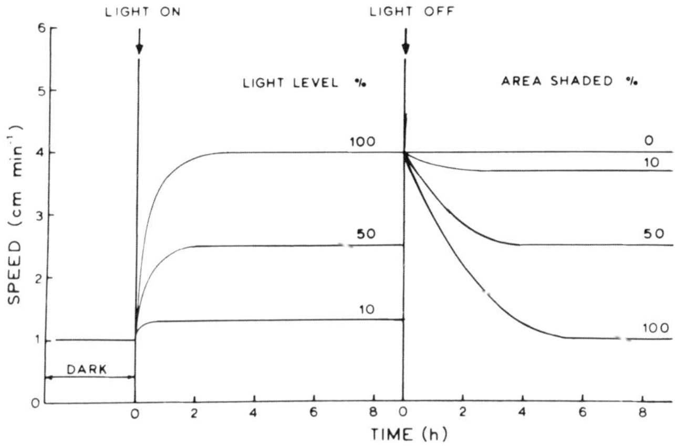

Fig. 22. The effect of light level and darkness on the velocity of translocation in leaves of corn. 100% light was 240 W m^{-2}. From Troughton *et al.* (284).

However, analysis of the kinetic profiles led these workers to conclude that light exerted its effect on the loading process.

Almost invariably water stress leads to a reduction in leaf photosynthesis, but this does not always affect the relative amounts of assimilate translocated. Indeed, phloem transport in many species seems to be relatively insensitive to water stress (130, 143, 157). Thus in tuber-forming plants of potato, Munns and Pearson (219) found moderate water stress to reduce photosynthesis by about 20%, but the relative amount of assimilate exported from the leaf remained at about 63% of the total fixed for both control and stressed plants. Similarly in studies on plants of cotton and sorghum, Sung and Krieg (272) found that even at leaf water potentials as low as −2.7 to −3.3 MPa relative translocation rates were still around 60% of those for unstressed controls (Fig. 23). Overall then, the effect of water stress seems to be initially greater on photosynthesis than on translocation, implying that the transport system may be at least temporarily buffered by stored compounds (see Fig. 6).

However, there are some indications that in model systems phloem transport can be affected by water status. In *Ricinus*, for example, incision of the shoot leads to exudation from the punctured phloem. If derooted shoots are removed from water, there is an almost immediate fall in the rate of exudation, although restoration of the water supply leads to a resumption of exudation at the initial rate (128). This implies a rapid effect of water stress on phloem physiology in this material.

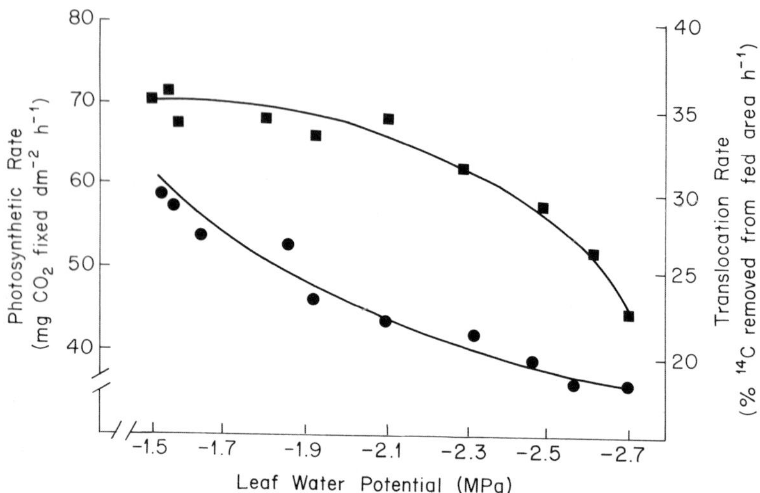

FIG. 23. The effect of increasing leaf water stress in sorghum on photosynthesis (●) and translocation (■) rates. Adapted from Sung and Krieg (272).

III. The Channel

A. BACKGROUND CONSIDERATIONS

So far it has been assumed that the transport process occurs in the phloem. Swanson (275) in Vol. II of this treatise sets out clearly the evidence for this belief, and he and Canny (32) give detailed accounts of the classic studies of Mason, Maskell, and Phillis and others (193–199) which founded present-day understanding.

The use of radioactive isotopes as tracers including ^{36}S and ^{32}P (19) as well as isotopes of carbon has greatly facilitated the localization of transport pathways in plants. The distribution of labeled assimilates in a stem or leaf petiole after feeding $^{14}CO_2$ to a photosynthesizing leaf can be determined by placing sections of the plant material in contact with X-ray film which is sensitive to ionizing radiation. The development of fine-grain photographic emulsions and improvements in techniques of fixation, resin embedding, and sectioning have made it possible to achieve resolutions of better than 1 μm under favorable conditions and so to localize materials within individual sieve tubes. However, since assimilates are water soluble, special precautions have to be taken to avoid movement of labeled compounds within, or out of, the tissue during its processing. A number of effective microautoradiographic techniques have been developed for this purpose (93, 100) and principles and problems of the methods have been discussed in detail by Bonnemain (24) and Fritz (99).

Developments in techniques of electron microscopy which have coincidentally led to improvements in microautoradiography have also led to a clearer understanding of phloem development and structure, although at certain key points our knowledge is tantalizingly imprecise.

The difficulties implict in any fine structural study of the sieve tube are many and have been discussed by several authors (48, 79, 156, 309). Basically, because the contents of the functional sieve tube are in motion and fixation cannot be instantaneous, some movement of the contents is inevitable before the tissue is killed. To compound the problem, the contents of the sieve tube are also under considerable pressure, and unless very careful methods are used, mechanical damage results in a pressure surge which completely disrupts the natural distribution of materials in the intact sieve tube (see Fig. 25). Consequently our knowledge of sieve plant structure and of the distribution of P-protein can only be tentative.

B. STRUCTURE AND DEVELOPMENT OF SIEVE ELEMENTS

The sieve elements are believed to be the channels for long-distance transport of materials in the phloem. They are associated with various kinds of parenchyma cells (Fig. 9), some of which are believed to have a role in the loading and unloading of the conducting elements. In addition, the phloem may contain thick-walled sclereids and fibers (sclerenchyma) which have a mechanical function but are not directly involved in the transport process.

The sieve elements are so-called because of the presence of numerous perforations ("pores") in the cell walls in specific areas (the sieve areas). The contents of adjoining sieve elements are contiguous through these pores. In most angiosperms, longitudinally oriented files of sieve elements form continuous sieve tubes which may extend from the leaves to the roots of a plant, which in the case of trees may be many meters. The sieve tubes are interrupted at intervals by transverse or oblique cross walls which are perforated by a number of prominent pores and are termed "sieve plates" (Fig. 24). In gymnosperms and other vascular plants, the sieve elements are usually spindle-shaped cells, the so-called sieve cells, with sieve areas on the adjoining walls of two overlapping cells.

The length of individual sieve elements varies widely, ranging from about 1400 to 1500 μm in the secondary phloem of conifers, 75 to

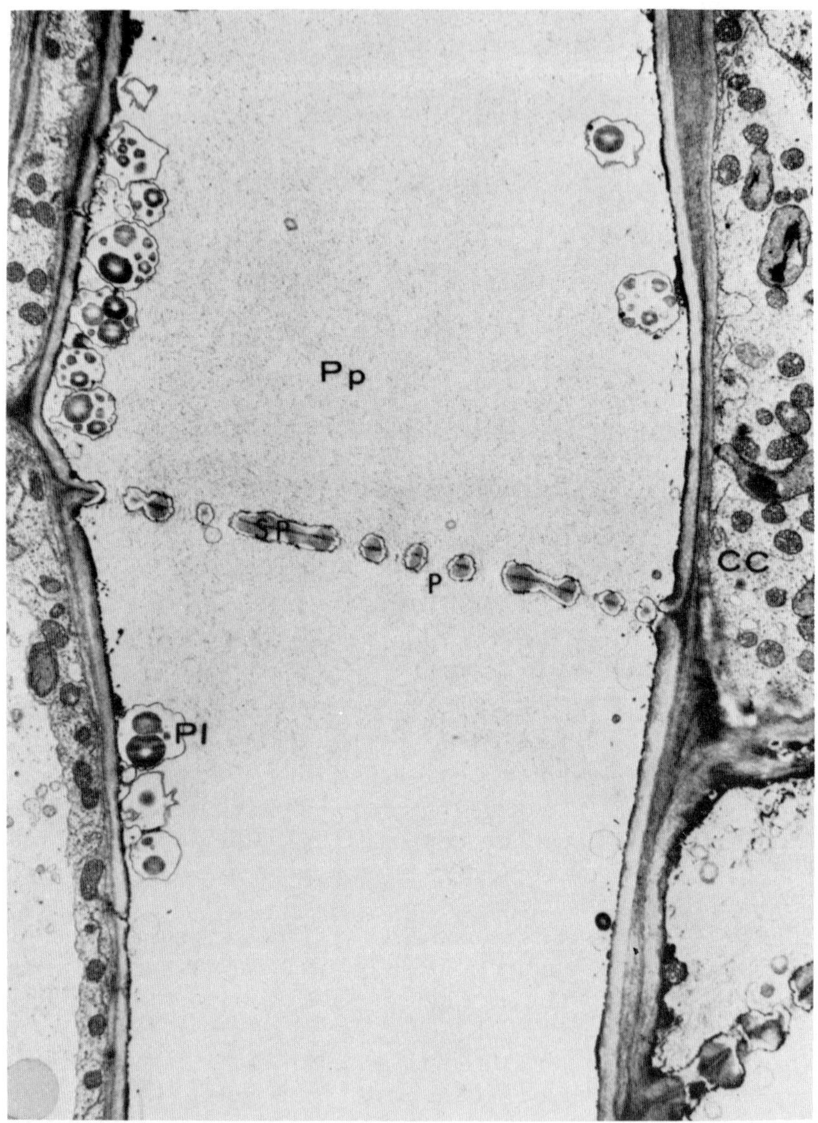

FIG. 24. Electron micrograph (×4400) of sieve elements and adjacent cells of *Nicotiana tabacum.* The plant was wilted before preparation of the specimen and the pores (P) in the sieve plate (SP) are more or less unblocked except for a fine reticulum of P protein (Pp), which also occupies all of the upper sieve element and the upper portion of the lower. A number of plastids (Pl) are seen at the cell margins. Note also the dense contents of the adjacent companion cells (CC). From Anderson and Cronshaw (6).

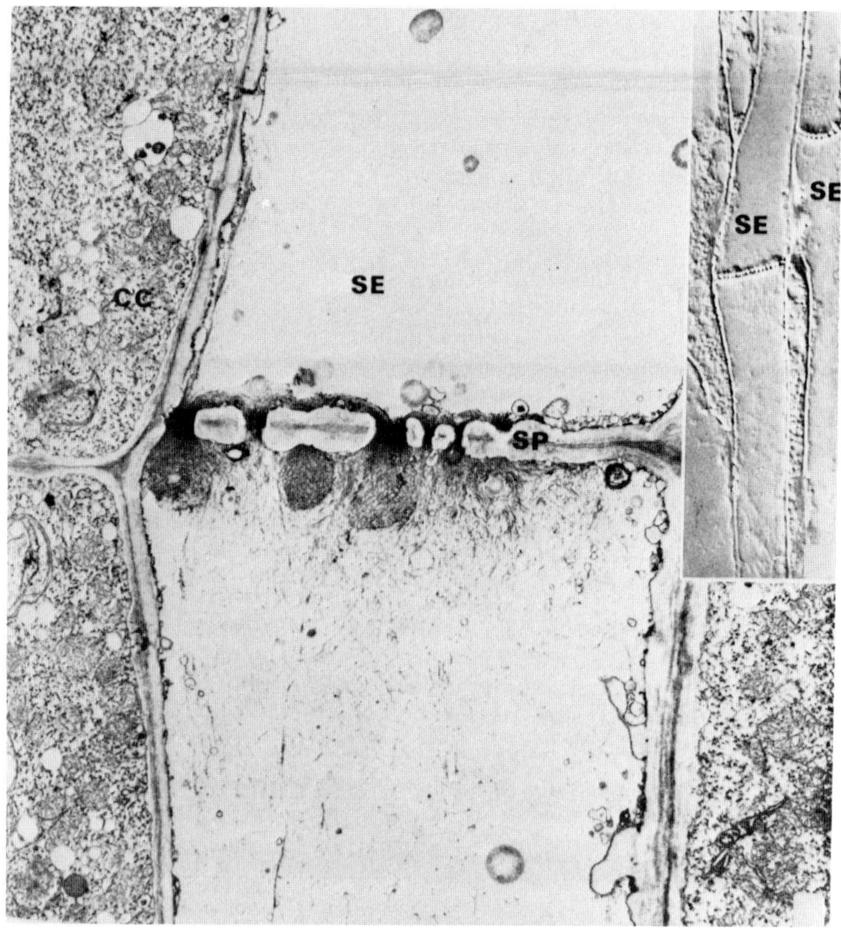

FIG. 25. Electron micrograph (×5412) of sieve elements (SE) and adjacent cells of *Nicotiana tabacum* killed in liquid nitrogen. The pores in the sieve plate (SP) are blocked from the lower side, probably as a result of pressure surge during killing. The inset, top right, shows similar sieve elements and companion cells photographed under Nomarski interference microscopy (×541). From Anderson and Cronshaw (5).

1300 μm in dicotyledons, and 500 to 5000 μm in palms. Diameters range normally from about 10 to 60 μm, but sieve tubes can be as wide as 400 μm in some palms (177, 224).

Sieve tubes usually have one or more files of modified parenchyma cells closely associated with them; these are the companion cells (Fig. 25, inset). Similar cells associated with the sieve elements in gymnosperms are usually termed albuminous cells, but the use of a different name may not be justified (270). A companion cell is usually, but not invariably, formed by an unequal longitudinal division of the sieve element mother cell which results in the formation of a narrow companion cell and a wider sieve tube element. These relative sizes are reversed in the minor veins of some leaves as noted earlier. Although companion cells and sieve elements remain structurally, and it is believed functionally, related throughout their lives, they differentiate very differently. Whereas sieve elements undergo the dramatic changes in ultrastructure described below which transform them into functional conducting channels, the companion cells remain relatively unaltered with prominent vacuoles, numerous well-developed mitochondria, and a prominent nucleus which is sometimes extremely lobed. There are abundant protoplasmic connections between the companion cells and sieve elements, and they are commonly branched on the companion cell side. The range of structure and arrangement of companion cells in phloem has been reviewed by Esau (66) and Behnke (15).

Recent anatomical studies of phloem in the minor veins of leaves have led to the rediscovery of the "Ubergangszellen" described by Fischer (92) and now named "transfer cells" (123). They are characterized by protuberances of the cell wall which may extend for a considerable distance into the cell, increasing the surface area of the plasma membrane (Fig. 9). The type A transfer cells adjoin sieve elements and have numerous protoplasmic connections with them, which lead Pate and Gunning (227) to the conclusion that they are modified companion cells. Type B transfer cells, which in structure resemble more nearly parenchyma cells, do not always abut onto sieve elements, but when they do, the convoluted wall is most prominent on the sieve element side and there are few, if any, protoplasmic connections (Fig. 9).

The differentiation of sieve elements to form sieve tubes has been described from electron microscopy studies by Cronshaw (45, 47), Parthasarathy (225), Srivastava (271), Kollman (167), and others. At first, the cell which gives rise to a sieve element appears to be a normal parenchyma cell with a prominent nucleus, abundant mitochondria, plastids, dictyosomes, and a well-developed endoplasmic reticulum with numerous ribosomes. There is a clearly distinguishable central vacuole which is bounded by a distinct membrane and which will accumulate dyes, such as neutral red.

The first indication of sieve element differentiation is often the unequal cell division which leads to formation of the associated companion cell. The wall of the young sieve element then begins to thicken, and it acquires a characteristic appearance under the light microscope which was described as nacreous (lustrous) by nineteenth-century microscopists. In electron micrographs it can be seen as being more electron opaque than the walls of surrounding cells. The plasma membrane has the usual triple-layered structure with two electron-opaque layers separated by a lipid-rich layer which is less electron opaque. Concomitant with early changes in the cell wall there is an increase in the thickness of the electron-opaque layers, especially the outer layer, and this might be related to increased cellulose synthetic activity at the cell surface. Assemblies of microtubules and abundant dictyosome vesicles in the peripheral cytoplasm are also indications of active cell wall synthesis.

Another early sign of differentiation in a prospective sieve tube element is the appearance of prominent plasmodesmata, especially in those regions of the cell walls that will later become the sieve plates. Sieve pores develop by the gradual breakdown of cell wall materials around individual plasmodesmata. At the same time callose, a polysaccharide containing β1-3-linked glucan residues (66, 69, 73), is deposited on the developing sieve plate (Fig. 26). It appears first as a cylindrical jacket surrounding each plasmodesmata and later spreads over the cellulose wall between the sieve pores.

At this stage while the sieve element is still increasing in size ultrastructural changes occur inside the cell. Cisternae of the endoplasmic reticulum begin to form stacks and in some cases fibers of a dense proteinaceous material (P-protein) are formed among them (50, 51). As development of the sieve element proceeds, the accumulation of P-protein leads to the formation of large structures (protein bodies) consisting of densely packed tubules or fibrils (44). The sieve elements in most species of dicotyledon usually have one major protein body per cell, but some, such as those of *Cucurbita,* have a large number which are of two different types. The larger of the two contains the bulk of the P-protein in the form of fine fibrils, while the second consists of tubular elements (44, 48).

Formation of the protein bodies is followed rapidly by dramatic changes in the protoplast. The nucleus breaks down and chromatin is dispersed in the surrounding cytoplasm; the vacuole, ribosomes, and dictyosomes disappear; and the mitochondria and plastids migrate to a peripheral position in the cell. The protein bodies are disassembled and P-protein fibers and tubules appear dispersed as a fine reticulum throughout the cell (44, 80) or in a peripheral position close to the plasma membrane which remains intact. P-protein in one form or another has been found in the sieve elements of all dicotyledons that have been examined with the electron microscope and in some monocotyledons. It is reported to be absent from palms and from some of the Gramineae and does not apparently occur in gymnosperms (49, 225). The function of P-protein is still unknown, and its implications for the translocation process are discussed later (Section IV).

From the point of view of the mechanism of translocation, considerable interest centers on the structure of the sieve plate in the mature functional sieve element. In the later stages of sieve tube development, the sieve plate pores become widened by enlargement of the plasmodesmatal canal (65,

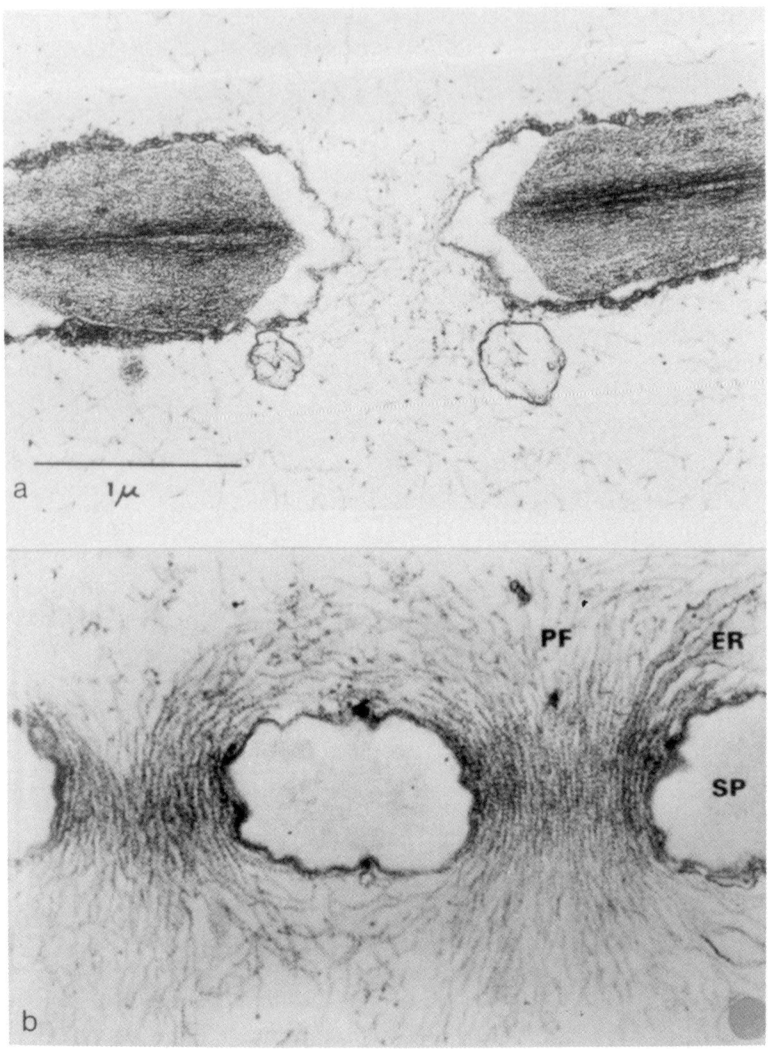

FIG. 26. Electron micrographs to show the pores of *Nicotiana tabacum* [(a) ×28,000] and *Aristolochia brasiliensis* [(b) ×43,000]. Note the callose lining to the pores in (a) and its absence in (b). Note also the abundance of plasmatic filaments, sometimes called microfilaments, in (b). From Cronshaw (46) (a) and from Behnke (14) (b).

69). The diameter of the pores in mature sieve tubes of dicotyledons ranges from less than 1 μm to about 15 μm (67). The nature of the contents of the pores is still uncertain. Almost all investigators agree that the pores are lined with plasma membrane and varying amounts of callose may be present. Callose is deposited and may block the pores when sieve elements are wounded (62, 73, 281) or when the tissue becomes dormant (63, 66). The callose can disappear and the sieve tube become functional again in the following season (see also Section IV).

Some investigators believe that the sieve plate pores are blocked or partially occluded in their normal state by P-protein (see 267). Others (5, 6, 224) think that the blocked condition is abnormal and that it is caused by a sudden release of pressure in sieve tubes when the tissue is damaged as in preparation for microscopy.

The view that the sieve plate pores are unobstructed ("open"), or nearly so, in functioning sieve tubes (Figs. 24, 26) is supported by the observation that various pathogenic organisms, including viruses and mycoplasmas of dimensions comparable to the diameter of the pores, travel rapidly through sieve tubes (17, 79). The trypanosomelike organism *Phytomonas leptocasarum* which causes phloem necrosis in coffee has a diameter of 0.4 to 0.6 mm, which is considerably greater than that of the sieve pores. Nevertheless it moves rapidly through the sieve tubes and it is difficult to visualize this happening if the pores are filled with solid material.

Knowledge of the state of the pores in functioning sieve tubes would help to distinguish between various mechanisms of phloem transport that have been proposed (see Section IV). However, the structure is so labile and the technical difficulties so great that electron microscopy may not provide unequivocal evidence in the near future. Johnson (156) has pointed out that during the time taken to fix sieve tubes chemically, or even by fast freezing, filaments of the size of P-protein fibers could be so displaced by Brownian movement as to render their observed position in electron micrographs quite misleading. He had urged electron microscopists not to waste their time on an unresolvable problem but to concentrate on such questions as the structure of the P-protein and its behavior under various conditions of flow in the sieve tubes. Its distribution in the sieve tube lumen may be crucial in determining the resistance of the channel to bulk flow (307).

C. SUBSTANCES TRANSPORTED

A wide variety of substances is translocated in the phloem, 80% or more of the material being carbohydrate (Table VIII). Chromatographic analy-

TABLE VIII

COMPOSITION AND SOME PHYSICAL PROPERTIES OF PHLOEM EXUDATE OBTAINED FROM PLANTS OF *Ricinus communis*[a]

Component	Concentration (mg cm^{-3})
Sucrose	80–106
Reducing sugars	Absent
Protein	1.45–2.20
Amino acids (as glutamic acid)	5.2
Keto acids (as malic acid)	2.0–3.2
Phosphate	0.35–0.55
Sulfate	0.024–0.048
Chloride	0.355–0.675
Nitrate	Absent
Bicarbonate	0.010
Potassium	2.3–4.4
Sodium	0.046–0.276
Calcium	0.020–0.092
Magnesium	0.109–0.122
Ammonium	0.029
ATP	0.24–0.36
Indoleacetic acid	10.5×10^{-6}
Gibberellin (as GA_3)	2.3×10^{-6}
Cytokinin	10.8×10^{-6}
Abscisic acid	105.7×10^{-6}
Total dry matter	100–125
pH	8.0–8.2
Osmotic potential	−1.4 to −1.5 MPa
Viscosity	1.34 N s m^{-2} at 20°C

[a] Data from Hall and Baker (127).

ses of the phloem exudates of 16 tree species by Zimmermann (325) have confirmed the earlier evidence mentioned (Section II,B,3,b) that sucrose is the major carbohydrate translocated. The transfer carbohydrates may also include substantial amounts of galacto sugars such as raffinose, stachyose, and verbascose in some species such as *Fraxinus*. Sorbitol is a major translocate in some Rosaceous trees; in cherry about half of the total dry weight of phloem exudate consists of sorbitol, the rest being mainly sucrose. The total sugar concentration of sieve tube sap was found by Zimmermann (325, 326) to range between 10 and 25% in different trees, depending on the species, time of day, and season.

The hexoses glucose and fructose are not found in phloem exudates, although these sugars are present in the phloem cells. This suggests that these monosaccharides are not translocated and that they probably occur only in nonconducting cells of the phloem.[2] Swanson and El-Shishiny (277) analyzed sections of grape vine at increasing distances from a leaf supplied with $^{14}CO_2$. They found that the largest amount of radioactivity was in the sucrose fraction and that the ratio of labeled hexoses to labeled sucrose decreased with increasing distance from the source. This, coupled with the 1 : 1 ratio of glucose to fructose, suggests that these monosaccharides are the secondary breakdown products of sucrose hydrolysis and not translocatory sugars. It is interesting to note that the translocated sugars are always nonreducing and weight for weight these sugars contain more available energy because some is released on hydrolysis of the glycosidic bond.

The nature of the nitrogenous compounds in the phloem has received attention. (See Chapter 4 of Volume VII.) It is commonly accepted that the amides asparagine and glutamine are the forms in which nitrogen is transported, for example, from the cotyledons of germinating seeds. Mittler (207) detected a number of amino acids as well as the amides in exudate from aphid stylets attached to willow stems. He found that the concentration of nitrogenous substances in the exudate varied with the season, being highest during the rapid leaf expansion, decreasing to a low value during the summer, and increasing again when the leaves became senescent.

The concentration of organic nitrogenous compounds in the phloem sap of a number of plants has been found to be low in comparison with carbohydrate, and this probably reflects the small amount of nitrogen exported from photosynthesizing leaves. Zimmermann (325) found that the total concentration of amino acids and amides in the sieve tube exudate from the white ash, *Fraxinus excelsior,* was usually less than 0.001 *M* but increased prior to leaf fall when some of the nitrogen was mobilized and returned to the stem.

A number of other substances have been found in phloem exudate in small amounts. Ziegler (322) found that inorganic phosphate occurs in the exudates from a number of trees at concentrations of less than 1 mg cm^{-3} and organic phosphorus compounds at an even lower level (0.01 mg cm^{-3}). Sieve tube sap is relatively rich in potassium but low in calcium (Table VIII).

The impression is sometimes given that the sieve tubes are metabolically inert. This seems to be quite incorrect. Ziegler (323) has listed a formidable array of reactions that can occur (Fig. 27) in the sieve tube sap of *Robinia pseudoacacia,* and as might be expected from interest in the

[2] Alternatively the monosaccharides may be transported so efficiently that they do not accumulate in the channels of transport and appear only in the accumulating cells.

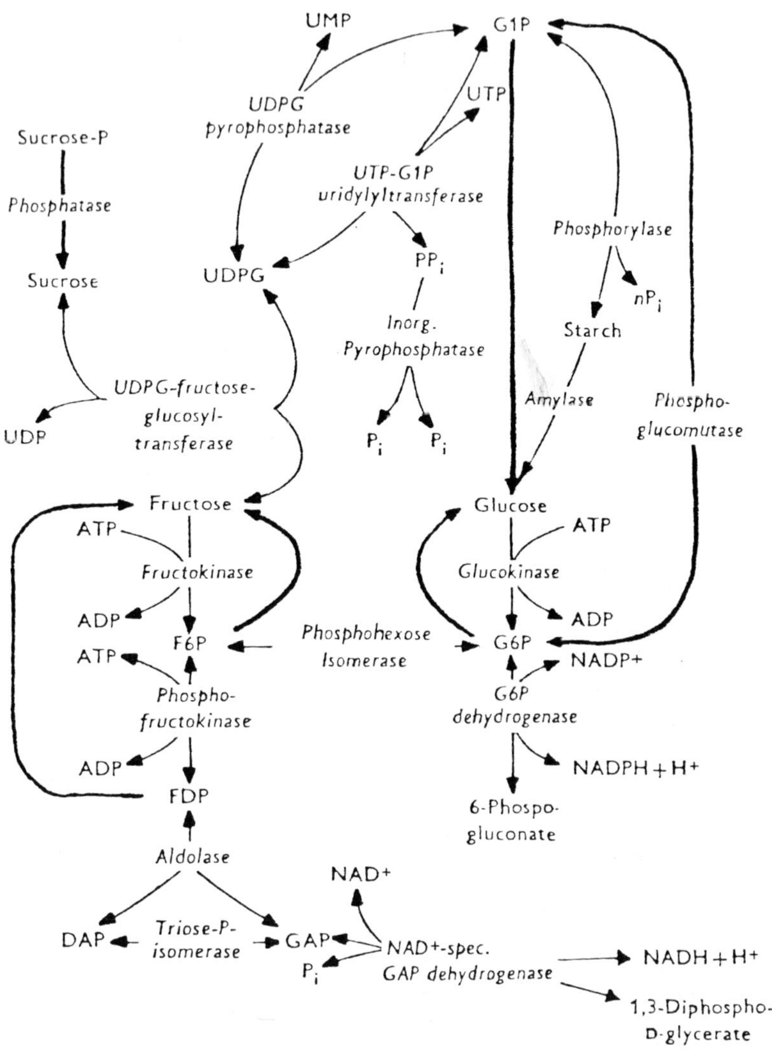

FIG. 27. Enzymes of carbohydrate metabolism known to be present in the sieve tube sap of *Robinia pseudoacacia* and the reactions they catalyze. After Ziegler (323).

phloem-loading mechanisms (Section II,B,3,b) there have been a large number of attempts to localize ATP and ATPase activity in sieve tubes (see 18, 48). The picture that emerges from these studies is of cells with a very considerable metabolic potential. How far this potential is realized in the functioning sieve tube is uncertain.

D. TRANSLOCATION VARIABLES

1. *Direction of Transport*

The study of the movement of sugars from a leaf has been greatly facilitated by the use of labeled carbon, which when supplied as $^{14}CO_2$, or with the short-lived isotope ^{11}C as $^{11}CO_2$, is rapidly converted to radioactive sugar that is exported to various parts of the plant, where it can be detected by autoradiography. In one of the earliest tracer experiments using $^{13}CO_2$ since ^{14}C was not then available, Rabideau and Burr (246) demonstrated transport of labeled sugar down the petiole of a bean leaf and upward and downward in the stem from the point of leaf insertion. Subsequent workers have confirmed that movement occurs in the stem both upward into the growing apex of shoots and downward to roots (see 32, 275). Most of the sugar moves into young growing leaves and root apices and very little is detected in mature leaves.

Although movement of materials in the phloem follows a predominantly longitudinal path through the stem, considerable lateral transfer of sugars is known to occur in some plants. Biddulph and Cory (21) observed that some 25% of labeled translocates were transported radially from the phloem of bean plants, and they suggested that this occurs mainly through the vascular rays to the xylem. In some species it is possible to induce lateral transport by defoliation. Joy (159) observed that when leaves from one side of a beet plant were removed and $^{14}CO_2$ was fed to the other side, radioactivity was later present on both sides of the stem (see also Fig. 15). As Joy pointed out, the complex anastomoses of the vascular system in beet may facilitate transfer between vascular bundles. In contrast, Zimmermann (327) defoliated one of the two main branches of a Y-shaped tree and found that less than 1% of the photosynthates produced moved into the defoliated branch. Lack of tangential movement in this experiment may have been due to failure of the defoliated branch to act as a powerful sink through lack of growth.

Because of the possibility of radial transfer from phloem to xylem, some upward movement of photosynthates in the stem may be due to movement in the transpiration stream. However, this possibility has been excluded in a number of experiments by separating the xylem and phloem over short distances of the stem and inserting waxed paper between them. There seems to be clear evidence from such experiments that substances can be transported in either direction through the stem of plants via the phloem. The bidirectional movement of substances simultaneously through the same section of stem or petiole has also been established. As long ago as 1938, Palmquist (223) demonstrated that the dye fluorescein moved through the phloem from the terminal to a lateral leaflet in a trifoliate leaf of bean *(Phaseolus vulgaris)* while movement of carbohydrate occurred in the opposite direction. Whether such bidirectional movement can occur simultaneously in the same sieve tube is a matter of some controversy. Biddulph and Cory (22) demonstrated that two-way movement of ^{14}C-labeled assimilates in young phloem tissue of *Phaseolus* occurred in separate vascular bundles, some conducting in one direction and others in the other.

An ingenious attack on the problem was made by Eschrich (71) using *Vicia faba.* He supplied one leaf with fluorescein and the leaf below with $^{14}CO_2$; on the stem between the two leaves he established aphids and analyzed the honeydew from single insects. He found that the solution obtained could contain fluorescein or ^{14}C-labeled compounds or both. Since the aphid stylet samples only a single sieve element, the latter result

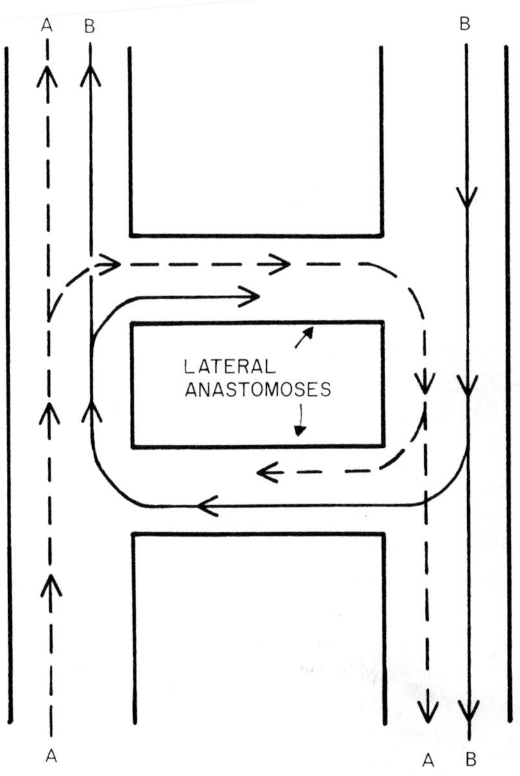

FIG. 28. The possibility of lateral transfer between adjacent sieve elements translocating in opposite directions. After Eschrich (71).

suggests transport bidirectionally in the same sieve element. However, this cannot be accepted as conclusive evidence of bidirectional flow for it remains possible that the aphid constitutes a sink, drawing translocate to itself from both directions; it is also possible that some mixing occurs through lateral exchange (Fig. 28) between elements which each translocate in one direction only.

A rather similar approach using autoradiography was made by Trip and Gorham (283), who supplied tritiated glucose to a lower leaf and $^{14}CO_2$ to an upper leaf of a soybean plant. When grain counts were made on autoradiographs of phloem from an intermediate region of the stem, it was found that both ^3H and ^{14}C were present in a number of sieve tubes. The fact that the ratio of ^3H to ^{14}C was the same in each tube was taken as evidence that a loop path was not involved, although this is scarcely convincing.

Despite the possibility of bidirectional movement in the same sieve tube, which is of considerable theoretical importance in relation to the mechanism of transport, one of the most characteristic features of phloem transport is that different substances tend to move simultaneously in the *same* direction. Experiments in which ^{32}P-labeled phosphate has been fed to leaves (40) indicate that export of phosphate only occurs when sugar is also being transported from the leaf. Likewise, the movement of growth regulators, such as 2,4-D (249), and of viruses (17) seems to be linked directionally with transport of carbohydrates.

2. *Mass Transfer and Velocity*

Translocation involves movement of dry matter and the most important measure of this is the amount of dry matter transported per unit time per unit area of the transport system; this is termed specific mass transfer (SMT) and can be based either on cross-sectional area of phloem as a whole or on that of the sieve tube thought to be involved in transport, although such values are difficult to obtain with any accuracy (see 32, 34). Estimates of SMT usually give values less than 5 g h^{-1} cm^{-2} phloem, although much higher estimates have been made (226). It is possible to partition SMT components in at least two ways. In the first, which assumes that translocation proceeds by a mechanism analogous to diffusion,

$$\underset{\text{g h}^{-1}\text{ cm}^{-2}}{\text{SMT}} = \underset{\text{g cm}^{-3}\text{ cm}^{-1}}{\text{solute concentration gradient}} \times \underset{\text{cm}^{2}\text{ h}^{-1}}{\text{constant}}$$

and in the second, which assumes bulk flow of solution,

$$\underset{\text{g h}^{-1}\text{ cm}^{-2}}{\text{SMT}} = \underset{\text{g cm}^{-3}}{\text{solute concentration}} \times \underset{\text{cm h}^{-1}}{\text{velocity}}$$

Grange and Peel (120) attempted to distinguish between these models experimentally by measuring solute concentration and gradient, and velocity and SMT, for willow plants labeled with $^{14}CO_2$ and "sampled" by aphids at two points 40 cm apart. Their data, although not conclusive, favor the second model and suggest that solute concentration and velocity of flow may be useful parameters in quantitative studies of translocation. It may be noted that Swanson (275) and Weatherley (306) quote examples based on the second model for the derivation of translocation velocities. However, gradients in solute concentration do exist in the phloem and have been demonstrated repeatedly since the classic studies of Mason and Maskell (193–199). Furthermore, whether osmotically driven pressure flow (see Section IV,D) or diffusionlike processes operate in translocation, the plot of SMT against distance would be expected to be hyperbolic (Fig. 29) with values of SMT being low when distance between the site of origin of translocate and the site of utilization is long (33, 34). This is an important point when considering translocation in tall trees and vines. In such cases, substantial storage of translocate in ray and other parenchyma may occur, so that it is from these cells, along the length of the stem, that transport occurs to sites where the demand is high. The restriction on SMT imposed by distance may thus be alleviated.

Solute concentration estimates are comparatively few. Grange and Peel (120) found values ranging from 1 to 25% in their studies on *Salix,* with most values within the range of 5 to 15%. These values appear to be typical. As far as velocity is concerned, estimates of this parameter vary widely from less than 1 to more than 1000 cm h^{-1}. Canny (32) and Ross

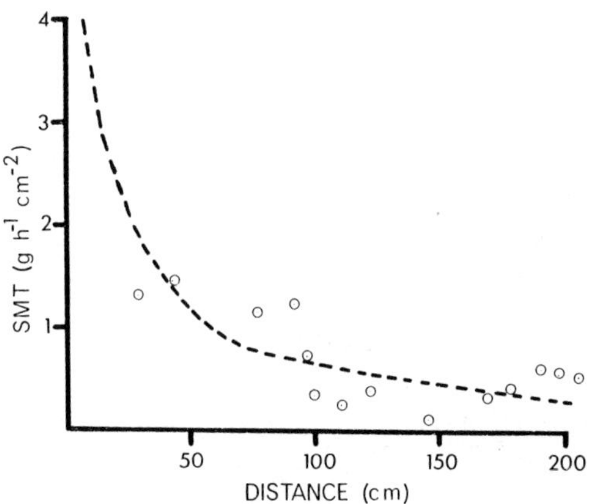

FIG. 29. The variation of translocation, expressed as specific mass transfer (SMT), with distance in fruit stalks of the sausage tree, *Kigelia africana.* Redrawn from Canny (32).

and Tyree (250) discuss some of the difficulties associated with velocity measurement, especially where radioisotopes have been used. Recent estimates of velocity of translocation in sunflower using ^{11}C give values ranging from 54 to 180 cm h^{-1} (0.15–0.5 mm s^{-1}) by one method and values of up to 400 cm h^{-1} by another (87, 315). As well as depending upon the method used, velocity measurements obviously also depend upon the sensitivity of the instruments used to detect the moving marker. With more sensitive detectors, velocity estimates will be higher since the marker, radioactive label, or fluorescent dyestuff will be "seen" earlier and will appear to have traveled faster than in cases in which larger amounts of marker have to reach the detector to be recorded.

3. Translocation Profiles

If a leaf is fed with labeled CO_2, a proportion of the label will appear in the sieve elements. If the periods of uptake, fixation, and loading of carbon into the phloem are very short, then a plot of the time course for counts detected at a particular point in the transport system will be of the general type shown in Fig. 30. The profile, called a kinetic profile (94), contains information about both loading and transport processes, and even with short-term pulse labeling can never be rectilinear since the loading of label is a relatively slow process. The exact shape of the profile therefore de-

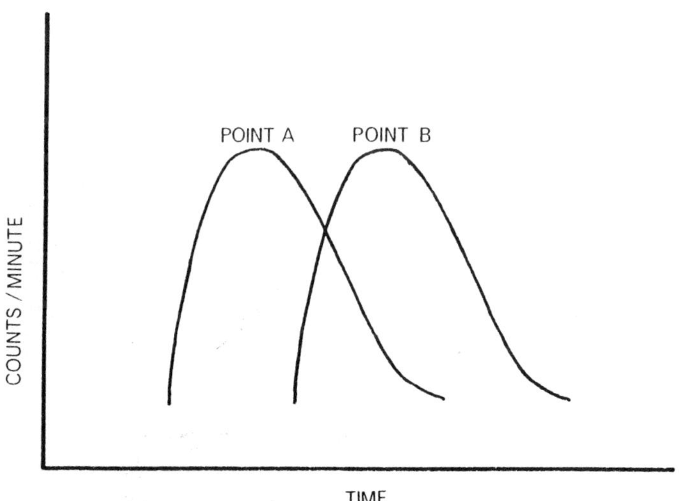

FIG. 30. Idealized kinetic profiles of translocation recorded at two points along a channel through which a probe of label is being transported.

pends upon the characteristics of the leaf, including the area of lamina labeled, the rate of fixation of CO_2, the distance from sites of fixation to sieve elements, and the kinetics of loading; Fischer (94) refers to these factors as kinetic size and source pool kinetics. Once in the transport system, label will move in a preferred direction and if movement is in a single sieve element of uniform dimensions, with a constant velocity. Then, and with no loss of label to surrounding tissues, the profile seen in Fig. 30 might be expected to be repeated at successive points in the system. But it is extremely unlikely that transport will occur in only one sieve element in a bundle and that velocity in different elements will be identical or even within an element constant with time. In consequence the profile will change, tending to spread and flatten at successive positions. That is to say, uniform "slug" or "plug" flow does not occur, even after pulse labeling. An example of this is seen (Fig. 31) from work of Christy and Fisher (39). They used morning glory vines supplied with a pulse of $^{14}CO_2$ and interpreted the change in shape of the profiles at successive points along the stem as due to a range of velocities in the many sieve elements carrying translocate and to exchange between sieve elements. Figure 31 also demonstrates that SMT for label is less for the more distant leaf.

Using ^{11}C it is possible to determine the movement of labeled assimilate along quite thick stems directly since the position emission of this short-

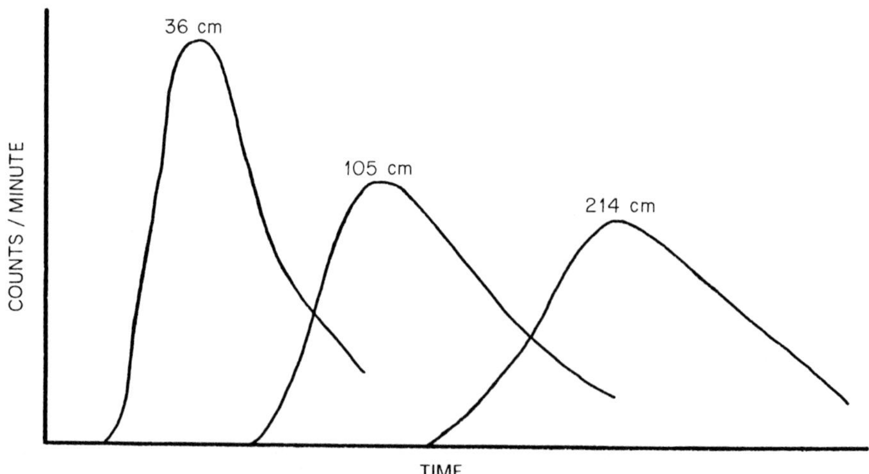

FIG. 31. Kinetic profiles for the rate of arrival of ^{14}C-labeled compounds at sink leaves of *Ipomoea* at various points along the stem, following supply of $^{14}CO_2$ to a mature leaf. Adapted from Christy and Fisher (39).

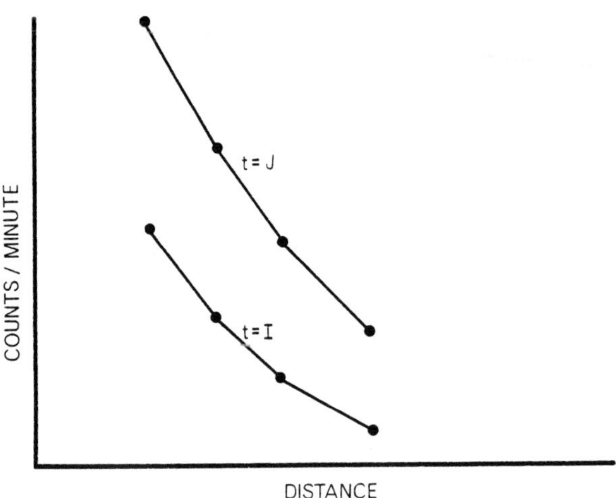

FIG. 32. Idealized spatial profiles of translocation record at two times (J and I) for a channel through which a pulse of label is being transported (cf. Fig. 30).

lived isotope is relatively energetic and can be counted by using Geiger tubes or scintillation spectrometers. In experiments feeding $^{11}CO_2$ to sunflower leaves and following movement of labeled assimilate in the plant, Williams *et al.* (315) found no evidence of a "slug" type of profile for movement along petiole and stem. Indeed, in many cases there was no evidence for a fall in counts of label with time at particular counting sites, while at others there was a prolonged slow increase in label indicating accumulation along the transport path.

Minchin (206) has analyzed kinetic profiles using simple mathematical models. His approach enables properties of the transport system to be examined without the complications introduced by the loading factor. In analyses of data for corn he also found evidence for significant loss of assimilate by phloem leakage, leading to retention of label at specific sites. This additional complication which distorts the kinetic profile can also be seen in data of Moorby *et al.* (211) in one of the pioneer runs using $^{11}CO_2$ fed to soybean plants.

It is also possible to present data in an alternative fashion, as a spatial profile, in which counts are plotted against distance from the source of label at specified times (Fig. 32). Early studies with tracers showed the profile to be linear when plotted on a semilog scale, but more recently with more data available nonlinear plots have been found; the significance of this and of profile shape in general are discussed by Horwitz (146), Canny (32), Fisher (94), and Ross and Tyree (250).

IV. Mechanisms of Translocation

A. INTRODUCTION

Most accounts of mechanisms of phloem transport have concentrated on movement in the sieve tubes, relegating problems of loading and efflux to subordinate status. Without subscribing to this unduly restrictive view, we shall follow the traditional approach, stressing the significance of input and output to the transport channel wherever appropriate. The basic problem may be stated in simple terms. How are many large molecules, present in widely differing concentrations, moved at high speeds along a channel whose ultrastructural characteristics seem to be incompatible with the observed rates of flow (309)? It is against this background that a number of proposals are considered and a widely held view is noted that more than one transport mechanism may be involved (7, 77, 84, 268).

Before considering hypotheses of translocation in detail, it is desirable to contrast the physical bases for solute movement and to compare diffusion, surface flow, and bulk flow, sometimes termed mass flow.

1. Diffusion

It has long been realized that the thermal agitation and movement of molecules by diffusion is far too slow a process to be involved in long-distance transport. Tyree and Dainty (291) have pointed out that where bulk flow of fluid is rapid and the product of solute concentration and mean velocity greatly exceeds (say by 100 times) the product of diffusion coefficient (usually of the order of 10^{-5} cm^{-2} s^{-1}) and concentration gradient, then the contribution of diffusion can be ignored. But diffusion may be significant in two other ways.

First, the dimensions of the sieve tube and the order of velocity of movement along it are such that because of lateral diffusion solute concentration across the sieve tube radius will be uniform (291) and local gradients can therefore be ignored.

Diffusion is also important in transport up to and away from the sieve tube. No matter what the mechanism of transport along the sieve tube or the velocity, ultimately the supply of translocate to cells can be limited by diffusion through the apoplastic or symplastic pathway. Clearly, the shorter the diffusion path, the less likely such a limitation is to be important, and the fact that there is little or no evidence for sink activity being restricted by translocation suggests that vascular development at the sink may serve to limit the length of the diffusion path.

2. *Surface Flow*

The application of the principle of surface flow to explain translocation originally introduced by Mangham (190) and Van den Honert (295) has been recently discussed by a number of workers (3, 83, 311). The phenomenon involves the spreading of surface-active molecules along a surface to form a monolayer. The rate of this movement is governed by the surface tension of the liquid phase over which spreading occurs and that of the monolayer. In contrast to diffusion of a solute in liquid which is very slow, surface flow may be very rapid and even approach the rate for diffusion of a gas. Nevertheless, Aikman and Wildon (3) have calculated that the maximum velocity for surface flow along fibrils of the size found in the sieve tube is of the order of 72×10^{-6} mm h^{-1}, which is about seven orders of magnitude less than the speed of translocation. It is possible that a surface-active molecule such as a sugar phosphate or amino acid might be absorbed onto the P-protein surface and move rapidly along it. A small amount of a rapidly flowing component could drag a larger amount of molecules by bulk flow.

While surface flow may well be inevitable between cellular surfaces and molecules which absorb or are surface active with respect to them, sucrose seems an unlikely molecule for such a role. Calculations (83, 183) of the area of membrane surface (plasmalemma and P-protein filaments) in relation to the amount of sucrose in the sieve tube suggest that the surface is too small to accommodate more than one-sixth of the sucrose present. Thus, the evidence for surface flow is equivocal, at least for long-distance transport. Within the cell such movement may well occur, but its significance is difficult to assess.

3. *Bulk Flow*

Since diffusion is far too slow and surface flow appears to be inadequate, most of current theories of translocation assume bulk flow of the sieve tube contents. By this is meant the movement of solution in the sieve tubes by a nonselective fluid flow in which all molecules moving do so with the same velocity. The introduction of alien but transportable substances such as carbon particles (10) is followed by their spread and movement at a speed comparable with that of native molecules in the system. Swanson (275) and Weatherley (308) have summarized the evidence favoring bulk flow of solution in translocation.

The motive force for bulk flow can originate in a number of ways: (1) through electrochemical effects, (2) by mechanical means involving pumping or wafting, and (3) as a result of osmotic and hydrostatic gradients. Each of these possibilities is associated with at least one theory of translocation; thus (1) is linked with the electroosmotic theory (264, 266, 268), (2) with numerous streaming and contractile protein hypotheses, and (3) with the well-known pressure flow hypothesis first proposed by Münch (218). Of these hypotheses, all except that of Münch postulate the motive force to be located within the channel itself, rather than at the source and sink.

B. ELECTROOSMOSIS

Most biological membranes are charged, with a pd across them of −2 to −100 mV, and provided that they have a porous structure the possibility of electroosmotic movement across them exists. The physical basis for the phenomenon has been described by Bennet-Clark (16) and Spanner (266) and is summarized as follows: At the interface between a solid and liquid phase containing electrolytes, an electrical double layer is established as a result of fixed charges, often negative and arising from absorption of OH^- ions on to the solid phase surface, or in the case of membranes from ionization of fixed groups such as carboxyl groups. As a result of this, pores in membranes, if they are so charged, will tend to have a predominance of cations within them and application of an electrical potential across the membrane will lead to migration of anions and cations in opposite directions but with cation flow predominating (Fig. 33). This flow will exert a drag, by friction, on the fluid and the other solutes it contains, and bulk flow in the direction of migration of the cations will occur. This is electroosmosis.

Spanner (264, 266) originally proposed that such a mechanism, operating at the pores of the sieve plate, is involved in the translocation process. He suggested that the pores of the sieve plates are blocked with a meshwork of negatively charged P-protein, the dimensions of which allow movement of ions in response to a gradient in electrical potential across the plate which is established and maintained by movement of K^+ across it. A further feature of the proposal is that K^+ is recirculated either through the adjacent companion cell or adjacent sieve elements or even apoplastically through sieve element walls or the sieve plate itself. Active uptake of K^+ into the sieve element upstream from the plate and passive leakage out of the sieve element downstream are also suggested. The scheme is summarized in Fig. 34.

The key features of the proposal are the existence of a site for electroosmotic activity within the pores of the sieve plate, involving P-protein, the continuous polarization of the plate by circulation of K^+ across it, and the drag of sieve tube contents across the plate as a result of the K^+ flux.

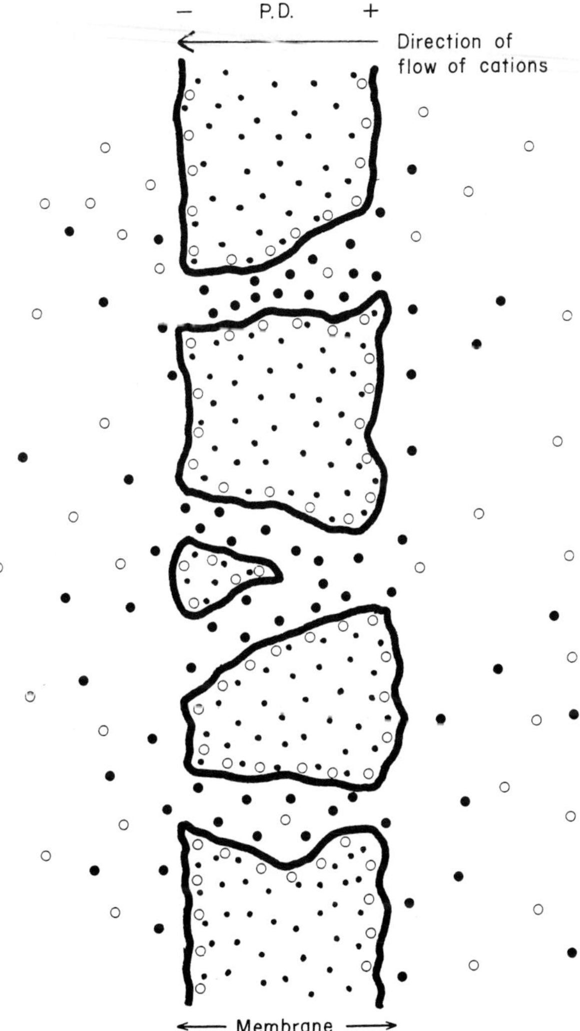

Fig. 33. A scheme for electroosmosis whereby cations (●) flow across a membrane when an electrical potential is applied across it. (○, anions). After Spanner (266).

Current uncertainty about pore structure prevents a final decision whether the structural requirement is met, although Spanner (267) believes that there is adequate evidence that correctly fixed functioning sieve tubes have sieve pores filled with P-protein. That there is a gradient in

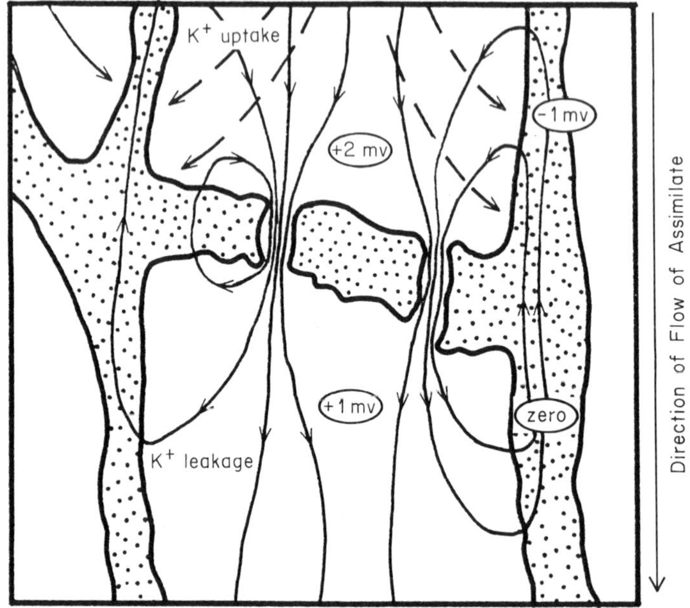

FIG. 34. Spanner's scheme for electroosmosis based on the circulation of K^+ ions (solid lines) and the supply of high-energy compounds (broken lines) involved in the generation of electrical potential difference across the sieve plate. The values of potential indicated in the round boxes are hypothetical. Redrawn from Spanner (266).

electrical potential across the plate in the direction required for electroosmosis is suggested by data of Bowling (27), who used microelectrodes to demonstrate a mean gradient of 19 mV across the sieve plates of *Vitis vinifera*. Unfortunately, as MacRobbie (186) has pointed out, it is not certain that these data relate to actively translocating and intact sieve elements. This familiar dilemma seems incapable of solution at the present time. If Bowling's data are accepted, then it seems reasonable, in the light of the known high concentration of potassium in the sieve tube (127), to ascribe the generation of this potential to K^+ movement. However, demonstration of the potassium pumps required by the theory has not been achieved. Slight potassium deficiency is known to reduce translocation (132, 202), but a large number of other metabolic processes will also be affected and it is uncertain whether the effect of deficiency is direct and on the transport process itself, or indirect on the supply of metabolites for loading or on the utilization of metabolites at the sink.

The electroosmotic theory in this form has been subjected to detailed criticism on a number of grounds by MacRobbie (186). The most fundamental of her objections are (1) that for electroosmosis to be effective the charge must be carried by ions of one sign only; in consequence, an efficient system is incapable of transporting both cations and anions, although both species of ion are moved in the phloem; (2) that the theory demands a pattern of differentiation in the sieve element whereby it is leaky to K^+ downstream from the sieve plate but shows an active K^+ influx upstream, above the plate; (3) that the required transmembrane K^+ flux through the sieve plate (Table IX) is excessively large compared with values for K^+ influx into guard cells of *Commelina communis* (241) and in *Nitella* (269); and (4) that the energy required to prime the K^+ fluxes is too large in relation to that available from respiratory ATP production unless the sieve elements are substantially longer than 100 μm, which often is not the case. MacRobbie also noted that two-way transport in the same sieve element is incompatible with electroosmosis, as indeed it is with most other proposed mechanisms.

Spanner (268) substantially modified his earlier hypothesis by including a proton extrusion pump to replace the potassium pump originally involved. His new interpretation is that Münch-type bulk flow leads to accumulation of P-protein at the sieve plate pore. This will be negatively charged due to the high pH of the sieve tube contents and therefore interferes with the flow of ions across the plate, so that a large potential difference builds up there. Eventually this reaches a value sufficient to hyperpolarize the plasmalemma and allow massive inflow of protons, by "punchthrough" (42) from the adjacent apoplast. The pH rises, and the negative charge on the P-protein will be neutralized so that the potential across the plate is reversed; in consequence, potassium ions above the plate are driven through it by electroosmotic forces, taking with them other transported molecules. The proton extrusion pump now acts to remove H^+ ions so that the pH rises, the P-protein regains its negative charge, and the cycle begins again. Spanner (268) discusses the problem presented by movement of anions, the energetics of such a mechanism, and the case in *Cucurbita,* where P-protein is basic; he considers that all these points can be

TABLE IX

Transmembrane K^+ Fluxes in Plants

Plant	Flux (pmol cm^{-2} s^{-1})	Reference
Nitella	0.2–0.85	Spanswick and Williams (269)
Commelina guard cells	$1.5 \times 10^2 - 1.9 \times 10^2$	Penny and Bowling (241)
Required for electroosmosis	$0.25 \times 10^6 - 2.0 \times 10^6$	MacRobbie (186)

accommodated in the revised scheme (268). It will be recalled that proton extrusion pumps have also been invoked in phloem-loading mechanisms. It is not known whether punchthrough occurs, and at present there is no unequivocal evidence that the membranes involved are excitable. There is clearly a great need for more detailed analysis of sieve tube potentials.

C. Protoplasmic Streaming and Related Hypotheses

1. *Protoplasmic Streaming*

Swanson (275) outlined three main objections to protoplasmic streaming hypotheses of translocation. These were, first, that streaming had not been observed in mature sieve elements; second, that streaming velocities in other systems such as internodal cells of *Chara* and *Nitella* were of an order of magnitude too low in comparison with velocities of translocation; and, third, that streaming is usually associated with an active metabolic condition, whereas he considered the sieve element to be relatively inactive metabolically. This last objection is probably invalid (323) and perhaps significantly there is a large content of ATP in sieve tubes (101, 324). A further point to note is that rotational streaming within an individual sieve element does not lead to net transport in one direction; for this to take place streaming, or some other means of transport, must occur across the sieve plate and it has been calculated (265) that the energy required to prime such a flow (2.5 g sucrose ml^{-1} sieve tube sap day^{-1}) is improbably high.

There is still almost general agreement with the view that protoplasmic streaming does not occur in mature sieve elements. However, Thaine (279), using living preparations of phloem of *Primula obconica*, filmed what were considered to be protoplasmic strands, 1 – 7 μm in diameter and running across the sieve plate. He observed the movement of particles, including mitochondria, along these transcellular strands at speeds of from 3 to 5 cm h^{-1} and proposed that such streaming was associated with translocation. It is agreed that living sieve elements are difficult to identify unequivocally, and Esau *et al.* (70) considered Thaine's observations to be of streaming in phloem parenchyma rather than in sieve elements. While there seems little doubt that protein microfilaments (~ 25 nm in diameter) penetrate the sieve plate pore of specimens prepared for electron microscopy, there are no reports of strands of the size reported by Thaine despite many attempts to find them. This is also significant in relation to Thaine's more recent model (280) in which transcellular strands (5 μm in diameter) are considered to be membrane bound and to possess a detailed

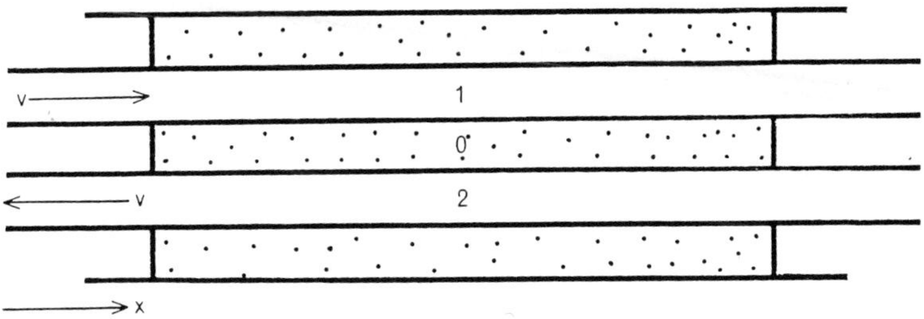

Fig. 35. Canny's scheme for translocation based on streaming in strands. Two such strands (1 and 2) are shown passing through a reservoir of sap (0), with the flow in each being in opposite directions but with similar velocity, v and $-v$. The direction of translocation is indicated by x. Sugar concentration in the reservoir in the next sieve tube in the right-hand direction will be less than that in the reservoir 0. Redrawn from Canny (32).

fine structure which allows a peristaltic pumping to arise from circumferential contraction of the strand-bounding membrane system. Although such a model appears to be energetically feasible (2), absence of any confirmatory evidence on fine structure (154, 155) makes it unlikely.

A rather different kind of protoplasmic streaming hypothesis has been proposed by Canny (31). In this model (Fig. 35) it is considered that the sieve element cytoplasm consists of two phases, one of which is strandlike, with restricted ability to translocate and which traverses the lumina and sieve plates of adjacent elements. Within the strands streaming occurs in opposite directions, whereas the outer cytoplasm, which is not continuous from element to element, is stationary. Canny postulates that there is a continuing and passive exchange of contents between the moving strand and the surrounding reservoir. With such a system differences in translocate concentrations will tend to be evened out so that the additional sucrose, say, resulting from loading at the effective source, will be moved toward the sink as a result of successive equilibration between adjacent sieve elements. The mathematical basis for this model, considered by Canny and Phillips (35), has been criticized (186, 291) and the structural requirements, as demanding in their way as those for Thaine's hypothesis (280), have not been shown to have been met.

2. *Contractile Protein Mechanisms*

Increasing evidence for the ubiquity of actin and myosin-like proteins in plant cells and of their involvement in streaming in several systems including *Nitella, Chara,* and *Plasmodium* (317), coupled with the demonstration of substantial amounts of protein in the sieve element, has led to suggestions that contractile proteins of the actin–myosin type may be involved in translocation. Mention has already been made of the idea that material may be transported by peristaltic movement of trancellular strands (280). Fensom (82) suggested that protein fibrils in the sieve element are aggregated to form channels within which material moves, again by peristalsis. The structures suggested by Fensom are of a diameter and order of magnitude less than those postulated by Thaine. In a later development, Fensom and Williams (86) proposed that the protein microfibrillar material is structured to include short lateral branches which by contractile oscillating movements propel material along by wafting (Fig. 36). To maintain directional flow very high rates of oscillation of the microfibrillar material would be necessary.

In considering contractile mechanisms two requisites are obvious; first, the energy available must be adequate for the functioning of the contractile proteins, and second, the existence of such proteins must be demonstrated. On the first of these points, calculations (2) suggest that in general contractile mechanisms are both energetically feasible and economical, although the power requirement for the wafting strand model is probably excessively high (1).

Evidence for the existence of actin and myosin in the sieve tube elements is less encouraging (282). A standard, and specific, method for identifying actin in biological systems is to decorate it with heavy meromyosin subfraction I derived from rabbit plasma; where this is done, interaction between the two proteins yields a characteristic arrowhead structure plainly visible

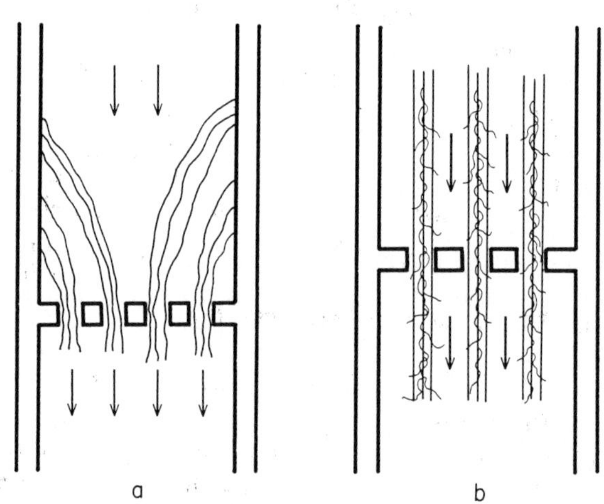

FIG. 36. A scheme for mass flow in sieve tubes activated by swishing microfilament material attached either to the sieve tube wall (a) or to fibrils running axially through the sieve plate (b). After Fensom and Williams (86).

under the electron microscope. Attempts to demonstrate the existence of actin in intact sieve elements and in phloem protein preparations have either failed completely or resulted in less than convincing pictures (150, 222, 252, 316).

There is no evidence for the major amounts of actin and myosin that would be necessary to drive a contractile mechanism within the sieve elements. That is not to say that a mechanism based on other nonactinlike proteins may not exist, although so far there is no evidence for it. In considering mechanisms based on protein in the sieve element, it must also be remembered that P-protein is not found in a substantial number of species including many monocotyledons and gymnosperms (48).

It has been suggested that the existence of a contractile mechanism might be demonstrated by following the movement of labeled assimilate along the translocation path and identifying pulses, i.e., discrete packets of label, moving along the sieve element. However, there are substantial technical difficulties in identifying such pulses at the subcellular level, and these are compounded by the short time intervals that are likely between successive pulses (1). Discontinuities in the spatial profile for ^{14}C and ^{42}K shown by Fensom (82) for *Helianthus* have been suggested as being caused by the method of introducing the label (77). It must be concluded that good evidence for the involvement of contractile proteins in translocation does not exist. We are therefore left with the pressure flow hypothesis by a process of elimination.

D. Pressure Flow Hypothesis

The term mass flow which is sometimes used to describe pressure flow should be avoided, since this is equally a feature of streaming or electroosmotic mechanisms.

1. *General Features*

Since its original proposal by Münch (218), this hypothesis of the mechanism of translocation has attracted more attention than any other, and at the present time, despite its apparent limitations, it is the most widely accepted. Detailed descriptions of the Münch model will be found in most elementary textbooks of plant physiology, and a number of critical accounts have been published recently (84, 203, 237, 308).

Because the hypothesis is so well known, the following description is necessarily brief. In essence, the Münch hypothesis requires a gradient of solute concentration along the transport channel, concentration being greatest at the source and least at the sink. This gradient, dominated by sucrose, is maintained by the addition of assimilate at the sieve element–companion cell complex in the minor veins of the leaf and by loss of solutes at the cells of the sink. As a result of the semipermeable nature of the system, water enters the channel at the source, thus generating a hydrostatic pressure, of osmotic origin, which leads to a pressure-driven bulk flow of solution, which is passive, along the channel down a gradient of both concentration and hydrostatic pressure, toward the sink. The water and solute relations of source, sink, xylem, and phloem suggested in the pressure flow scheme are summarized in Fig. 37.

The simplicity of the Münch model is not the least of its attractions, and it is widely recognized as inevitable that some osmotically primed pressure flow will occur, given the existence of source, sink, and semipermeable transport system (7, 84, 186). The question is whether such flow is entirely adequate to explain observed translocation fluxes.

The physical model, involving two linked osmometers, frequently used to introduce the pressure flow hypothesis suggests a single source and a

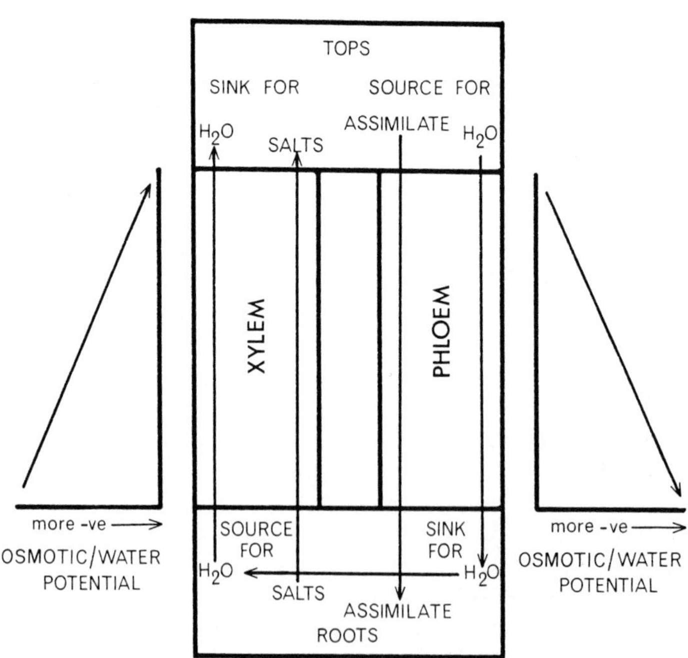

FIG. 37. A summary of the source–sink relationships suggested for the pressure flow hypothesis.

single sink. The whole-plant model in which the leaf is substituted for one osmometer, and say, the root system for the second, although conceptually attractive, is also misleading. This is because different sieve elements may be linked to a variety of sources and sinks making it possible, for example, for sieve tubes in different bundles, or even in the same bundle, to supply different sink regions and to be loaded from different sources. Evidence that this must be the case comes from the fact that assimilate from a given leaf may be transported to the stem apex and younger leaves and also to the root system. This feature has important implications which will be discussed later.

Because on this hypothesis solution moves by bulk flow, it follows that all solutes will move in the same direction in a single sieve element. The direction of flow is determined by the osmotic potential and the hydrostatic pressure gradient brought about by the pressure of high concentrations of sucrose or some other transfer carbohydrate. Thus it is possible for any particular compound except the transfer carbohydrate to be moved against its concentration gradient.

Pressure-driven flow is energetically efficient since it does not involve the addition of energy to prime protein contraction or cytoplasmic streaming or to promote the fluxes of potassium or protons suggested in the electroosmotic hypothesis. This being so, the effects of cold treatment, narcotics, and metabolic inhibitors, all of which impair translocation to a greater or lesser extent, have to be explained, as does the question of why translocation occurs in living tissue at all, for pressure flow could operate equally well in a nonliving channel.

2. *Evidence for Pressure Flow*

A requirement for the pressure flow hypothesis is that water should enter the channel by osmosis, at least at the source. The sieve element should therefore have a semipermeable membrane and be plasmolyzable. Currier *et al.* (53) demonstrated repeated cycles of plasmolysis and deplasmolysis of sieve elements of *Vitis,* especially for material taken from dormant vines whose sieve plates show characteristic heavy callose deposition, and where it is possible that the plasmalemma is less permeable than in actively translocating sieve elements.

The pressure flow hypothesis also requires a gradient in hydrostatic pressure within the sieve tube from source (high) to sink (low). The demonstration of such gradients has proved difficult, perhaps because it has been assumed that all phloem at a given point in a stem will be translocating in the same direction from a common source to a common sink, whereas this is not necessarily the case. What is required is the demonstration of a pressure gradient along a single sieve element series and this is difficult if not impossible to achieve. Hammel (129) inserted small manometers directly into the phloem of oak trees at distances 5 m apart. Despite substantial variation his figures suggest a small gradient in pressure along the direction required. Rather similar findings have been reported by Milburn (204), who also discusses the technical difficulties in this kind of work. A related approach using micromanometers attached to aphid stylets was used by Wright and Fisher (318), who reported values of sieve tube turgor to vary from 0.51 to 0.93 MPa, although gradients were not measured.

An alternative approach to demonstrate turgor differences between leaf and root sufficient to meet the requirements of the pressure flow hypothesis has also been made by Fisher (96). In this work, sucrose concentrations in sieve elements of petioles and roots were estimated by a negative staining method (95). Briefly, this involves the freeze substitution of tissue with acetone and embedding in epoxy resin containing the dye Sudan B; the presence of water-soluble solutes in cells leads to the exclusion of the dye so that sieve elements stain more lightly than surrounding cells, which contain fewer solutes (Fig. 38). This approach has technical difficulties, notably in calibration, but the data indicate sucrose concentration to be substantially higher (11.5%) in the petiole than in the root sieve elements (3.5%). Measurements of water potential in the tissues enabled a turgor difference of about 0.39 to 0.44 MPa to be calculated between source and sink.

From the early work of Mason and Maskell (see 199) it has been known that concentration gradients exist between source and sink with respect to sucrose. This is to be expected on the pressure flow hypothesis, but caution is necessary in interpreting results where there is failure to demonstrate such a gradient since this may reflect the sampling of large numbers of sieve elements not all of which are transporting in the same direction from the same source.

On the pressure flow hypothesis, water is recirculated via the xylem, although in reality, some at least could be used in growth of the sink itself. Affecting the water potential of the xylem should also affect movement in the phloem. Weatherley *et al.* (310) lowered the xylem water potential of willow stem segments by allowing uptake of mannitol or sucrose solution. As expected, reduction in the water potential was associated with a reduction in the volume of exudate from stylets of aphids feeding on the segments. The converse of this would be to cause an increase in exudation rate by an increase in water potential in the xylem. Peel and Weatherley (240) applied a hydrostatic pressure of 0.4 MPa to the end of willow stem segments and observed an increase of about 40% in the exudation rate from stylets. However, despite the greater flux, the concentration of sucrose in the exudate did not fall over a period of 3 h, indicating a continuing supply

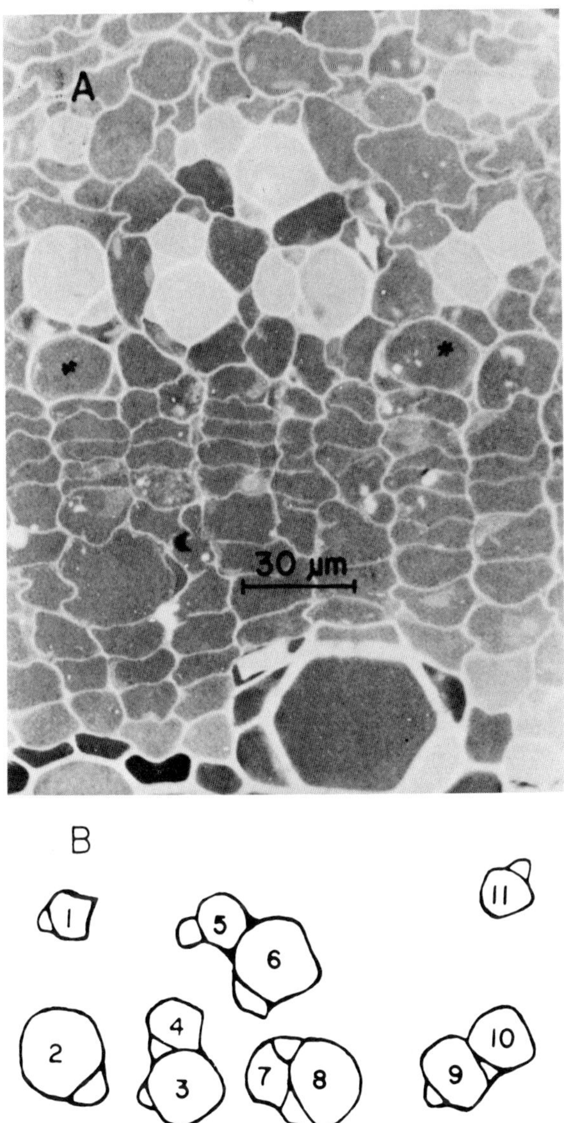

FIG. 38. (A) Transverse section through phloem of stem of *Phaseolus vulgaris* processed by a negative staining method. The sieve element–companion cell pairs (in B) show as clear cells indicating a high solute (sugar) content. Reproduced from Fisher (95).

of sucrose to the sieve elements. Whether this was the case for all sieve elements or only for those tapped by the stylets is not known, but the possibility remains of a feedback mechanism whereby rapid efflux of solute to a sink, in this case the stylet, is compensated by an increased solute supply to that part of the channel concerned. The control of phloem sap composition has been discussed by Smith and Milburn (260–262), who conclude that the high concentration of potassium in sieve tube sap may have an osmoregulatory significance in the maintenance of sieve tube turgor.

3. Evidence against Pressure Flow

Just as there is little direct and unequivocal experimental evidence for the pressure flow hypothesis, so evidence against it is also diffuse and seldom clear-cut. The most important objections are centered around the structure of the sieve element, which on the face of it seems ill adapted for pressure flow. Before we consider this in detail, two other aspects will be mentioned.

The composition of sieve tube sap shows certain curious but consistent features, namely, a high pH ranging from 7.4 to 8.6, a high ratio of $K^+ : Na^+$, and a low ratio of $Ca^{2+} : Mg^{2+}$ (Table VIII). There are no obvious reasons why a purely passive mechanism such as pressure flow should be associated with such values. This has led to the suggestion that they may be related to the requirement of other mechanisms such as electroosmosis (high $K^+ : Na^+$) or actin-based streaming (low $Ca^{2+} : Mg^{2+}$) rather than pressure flow. Such indirect argument is weakened by the fact that sap composition could equally well be a consequence of the loading or efflux mechanisms rather than of that concerned with transport. Alternatively, it might be related to the mechanism responsible for maintaining structural integrity of the path, an idea explored later.

a. Inhibitor Studies, Anoxia, and Low-Temperature Effects. Inhibitor studies offer a very good example of the difficulties of obtaining unequivocal data on the mechanism of transport (106, 314). Because the pressure flow model involves a passive mechanism, it should be unaffected by metabolic inhibitors. But such compounds as cyanide and DNP do affect transport, and it might appear as though metabolic activity is involved in translocation. However, for this to be the case, it is necessary to rule out any effect of the inhibitor on loading or unloading, and since it may be moved within the experimental system, this is not always easy. Even if it is clearly demonstrated that compounds affect the path of transport, it must be shown that they do this by affecting metabolic processes without altering the physical characteristics of the pathway. Again, such a distinction may be difficult to make. As an example we may consider the effects of cyanide which has long been known to inhibit translocation in a number of systems (115, 141, 313). Qureshi and Spanner (245) found cyanide, in solution or in gaseous form, to inhibit movement of ^{14}C assimilates, and ^{137}Cs, along the long, thin stolons of *Saxifraga stolonifera* (Fig. 39). They argued that their results are explicable if cyanide treatment interfered with the energy supply necessary for the transport mechanism and that their data are inimical to the pressure flow hypothesis. The objection that can be raised here, that the inhibitor may affect loading or unloading processes along the stem, cannot be made for the work of Giaquinta and Geiger (115), who produced evidence that cyanide inhibited transport in *Phaseolus vulgaris* petioles by causing structural damage and blockage of the sieve pores, rather than by impairing metabolism. The mechanism causing such blockage is not clear, but callose formation is probably involved (292, 293). Very high concentrations of cyanide are necessary to cause an effect that is in any case reversible. Of other inhibitors used, the respiratory uncoupler DNP appears to act on the loading process in soybean (131) and valinomycin, which is a K^+-binding ionophore, may also affect translocation (see 84) by affecting loading.

The evidence advanced against pressure flow on the basic of inhibitor studies is thus equivocal. This is true also for studies on the effects of

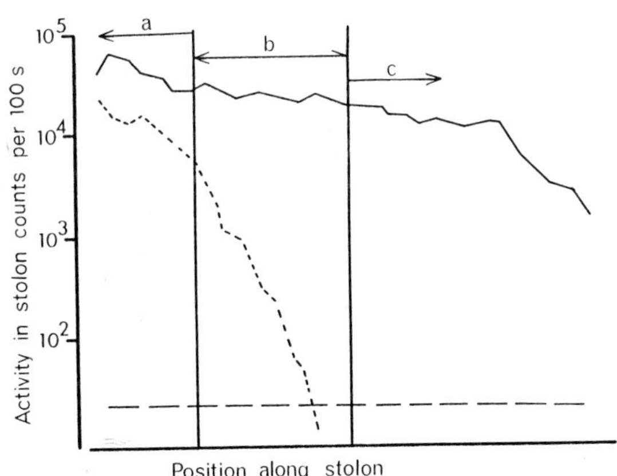

FIG. 39. The effects of cyanide on translocation in a stolon of *Saxifraga stolonifera*. $^{14}CO_2$ was fed to a leaf on the parent plant (a) and assimilate was allowed to move toward the daughter plant (c). Region b was exposed to HCN gas. Plants were harvested and counted after 4 h. (——, control; ------, HCN treated; – – –, background level of radioactivity.) Redrawn from Qureshi and Spanner (245).

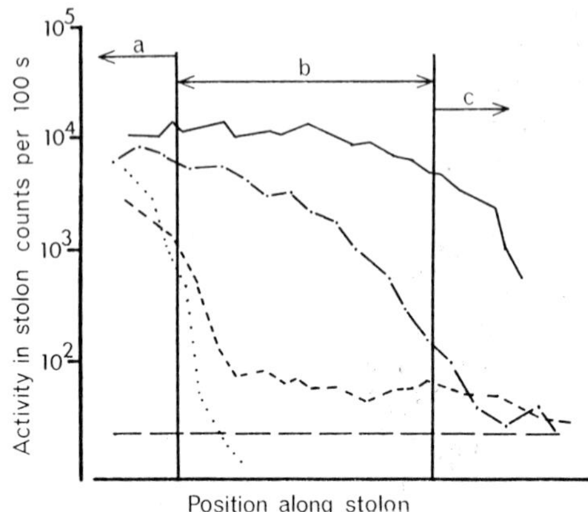

FIG. 40. The effect of nitrogen treatment on translocation in stolons of *Saxifraga stolonifera* and recovery from anoxia. $^{14}CO_2$ was fed to a leaf on the parent plant (a) and assimilate was allowed to move toward the daughter plant (c) over a 4-h period. Region b was subjected to treatment with nitrogen gas. (······, $^{14}CO_2$ fed 1.5 h after start of nitrogen treatment; ------, $^{14}CO_2$ fed 2 h after start of nitrogen treatment; —·—, $^{14}CO_2$ fed 3 h after start of nitrogen treatment; ——, $^{14}CO_2$ fed 5 h after start of nitrogen treatment; – – –, background level of radioactivity.) Redrawn from Qureshi and Spanner (244).

nitrogen-induced anoxia. Mason and Phillis (198) found that when oxygen was carefully excluded from a 20-cm length of cotton stem, transport to the terminal fruit was greatly inhibited over prolonged periods. Similar effects of anoxia have been found by using stolons of *S. stolonifera* (Fig. 40). Such results suggest that either there is an inhibition of metabolic processes which are essential to transport or treatment leads to structural disruption or change which increases path resistance and thus reduces the rate of transport. Demonstration of such effects on structure depends upon electron microscopy using an appropriate technology, and it is by no means certain that existing methods are adequate. It also has to be pointed out that a number of workers have not been able to demonstrate significant and persistent reduction of transport following nitrogen treatment (257, 312).

Low-temperature treatment of the transport path also inhibits translocation (106) and this would not be expected on the pressure flow hypothesis unless treatment affects viscosity of the translocated sap or changes the resistance of the path by altering the structure. In fact, both of these possibilities seem to be realized, and the effects of low temperature on sieve element integrity are especially interesting. Plants appear to fall into two categories: chill-sensitive plants show a threshold at 12°C and at lower temperatures substantial disruption of sieve element fine structure occurs; for chill-insensitive plants this threshold is much lower at 0°C. Recovery from structural damage is slow and occurs over many hours in chill-sensitive species but is more rapid in insensitive plants, occurring even when the path is kept chilled during the recovery period. During chilling, sieve pores appear to become blocked with cytoplasm in the chilled zone, and increased deposition of callose also results. Such effects increasing path resistance must inhibit pressure flow, and it is unnecessary to assume an action of low temperature on metabolism of the path to explain these results.

b. Sieve Tube Fine Structure. It has been repeatedly argued that the structure of the sieve tube is incompatible with the pressure flow hypothesis (267, 307, 309). This seems especially true of gymnosperms in which very complicated sieve plate structure can be encountered (225). It is also a relevant objection in cases in which it can be demonstrated that the sieve plate is blocked with callose, cytoplasm, P-protein, or whatever. Furthermore, the inclusion of substances such as P-protein in the sieve tube lumen of many angiosperms, but not gymnosperms, presents an additional complication for any explanation based on pressure flow.

The starting point for quantitative analysis of the relation between structure and function of the sieve tube is the assumption that flow through the sieve elements is laminar and that the Poiseuille equation relating flow through narrow tubes applies. This equation in its simplest form states that

$$P = (8nv/r^2) \times 10^{-4} \text{ MPa m}^{-1}$$

where P is the pressure drop along a tube of radius r (centimeters) and containing a fluid of viscosity n moving with a velocity of v (centimeters per second). Crafts and Crisp (43) discuss variants of the equation to take account of the geometry of sieve plate pores, and Weatherley (307) considers a modification designed to take account of occlusion of pores with filamentous material, assuming this to be regularly packed (85). Specimen values based on accounts by Weatherley and Johnson (309) and Weatherley (307) are given. Assuming that sieve tube diameter is 24 μm; viscosity of contents is 1.5×10^{-2} Ns m^{-2}; velocity is 100 cm h^{-1}; sieve pores account for 50% of area; pore diameter is 5 μm; plate frequency is 20 per cm; and plate thickness is 1 μm and on the assumption that all the pores are open, the pressure drop is calculated at about 0.025 MPa m^{-1} for the tube alone with an extra 0.002 MPa m^{-1} for the sieve plates giving a total value around 0.027 MPa m^{-1}. On the other hand, if it is assumed that the pores are occluded by regularly packed parallel filaments of 100-nm diameter

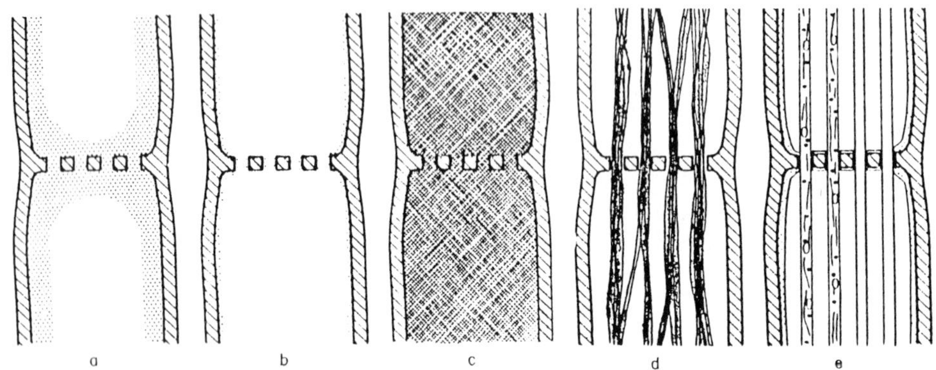

FIG. 41. Possible arrangements of the contents of translocating sieve elements. (a) Pores blocked by cytoplasm. (b) Pores not blocked and empty lumina of sieve elements connected. (c) Pores and lumina blocked by a network of microfibrillar material. (d) Bundles of filaments blocking pores and passing through lumina. (e) Membrane-bound transcellular strands. Adapted from Weatherley and Johnson (309).

and 2000 nm apart, application of the formula of Fensom and Spanner (85) indicates a pressure drop of 0.014 MPa for each sieve plate, equivalent to 27.6 MPa m^{-1}, the resistance of the tube itself being negligible in comparison. Whereas the lower figure is clearly feasible, the higher most certainly is not, and sieve tubes would burst if subjected to it. Weatherley and Johnson (309) considered five possible structures for sieve elements (Fig. 41) and concluded that pressure flow was possible only in one case in which both lumen and sieve plate pores were open. This conclusion has been disputed by Crafts and Crisp (43), who calculate a lower pressure requirement. In an experimental attack on the problem using soybean, Fisher (96) showed a lower sucrose concentration in the sieve elements at the sink than at the source sufficient to give a turgor difference of 0.4 MPa, which was adequate to drive a pressure flow mechanism with a velocity of 48 to 54 cm h^{-1} even if the sieve plate pores were only 70% open. More direct evidence of this type is urgently required, especially for tall species.

4. *Mathematical Models of Pressure Flow*

Since the work of Horwitz (146) modeling the profile of tracer movement along the phloem, there have been many recent attempts to develop mathematical models of the Münch pressure flow hypothesis (38, 88–91, 118, 187, 247, 290). These go beyond the simple use of the Poiseuille equation and have generated additional controversy, particularly with respect to the nature of unloading (90, 118, 187). Space does not allow

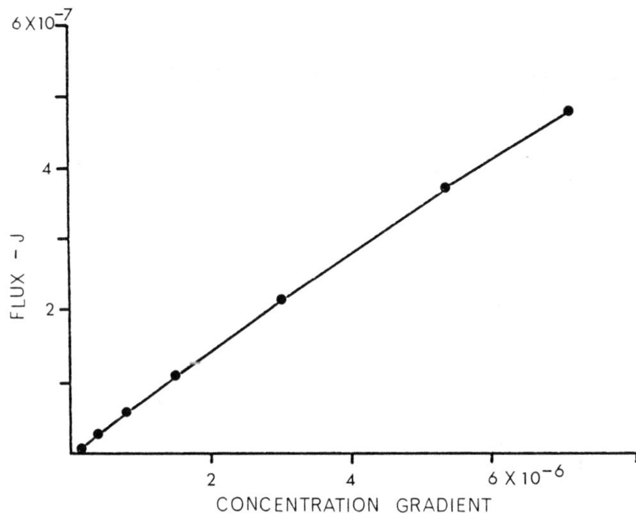

FIG. 42. Plot of computer simulation of sugar flux in a sieve tube J (mol s^{-1} cm^{-2}, equivalent to SMT) against concentration gradient dc/dx (mol cm^{-3} cm^{-1}). The hydraulic conductivity of the sieve tube was set at the low values of 5.5×10^{-4} cm^2 MPa s^{-1} equivalent to that of tubes with substantial blockage of sieve plates. Redrawn from Ross and Tyree (250).

comprehensive analysis of these approaches, but one is of special interest.

Mason and Maskell (193–199) in their studies on cotton repeatedly stressed that the movement of sugar in translocation showed important similarities to a diffusion process. Thus sugar moves from a source at high concentration to a sink at lower concentration, with a flux J (moles per second square centimeter phloem) showing a remarkably constant proportionality to the concentration gradient dc/dx (moles per cubic centimeter centimeter). Ross and Tyree (250) used a steady-state model of the pressure flow system to examine the relationship between J and dc/dx in the phloem, assuming that loading and unloading can be ignored. They found that the plot of J against dc/dx was close to linearity (Fig. 42), with the slope K varying from 0.073 to 0.062 cm^2 s^{-1}. These values were obtained by assuming a low hydraulic conductivity of the sieve tube lumen and pores, equivalent to substantial blockage. Where hydraulic conductivities equivalent to unblocked pores were used, values of K rose to between 7 and 70 cm^2 s^{-1}. Calculations based on data of Zimmermann (326, 328) suggested that for *Fraxinus,* values of K were around 42 to 70 cm^2 s^{-1}, compatible with the existence of unplugged pores as required for long-distance pressure flow in tall species. This modeling approach shows that there need be no incompatibility between the requirement of the pressure flow hypothesis and the classic experimental observations and that movement by presure flow can show similarities to a diffusion process.

E. Hybrid Mechanisms

The inevitability of pressure flow, given the properties of the translocating system, has already been remarked upon and even proponents of alternative mechanisms accept this (84, 268). Recently, there have been attempts to take account of this in considering hybrid mechanisms which rely on pressure flow in combination with one or other transport mechanism. For example, Spanner (268) postulates that pressure flow leads to the accumulation of P-protein at the seive plate and that this sets up the required conditions for the electroosmotic pumping mechanism considered in Section IV,B; it may be noted that Spanner concedes that pressure flow alone is adequate to explain translocation in *Zea,* a species which lacks P-protein and in which the transport path for metabolites is short and seldom more than 2 to 3 m.

A relay mechanism which combines pressure flow either with electroosmosis or with a membrane transport pump has been proposed by Lang (179). The model (Fig. 43) shows a number of overlapping units in the transport path with pumping sites arranged in parallel at the top and bottom of each overlapping set. At the top of any one unit, sucrose is pumped in from the bottom of the adjacent array against a concentration gradient. Along the unit there is also a concentration gradient, and material moves within it by pressure flow. At the base of the unit sucrose is pumped out into the next array. In a tree 50 m in height Lang suggests that there could be, say, 10 of these units each 6 m long and overlapping by as much as 1 m. The energy required for such a mechanism would account for about 20% of the sugar input at the leaf, but specific mass transfer would be greatly increased over that from pressure flow alone.

By including a metabolically based component, hybrid mechanisms such as this and others (84) allow for direct control of the distribution of metabolites by action along the transport path. So far though, a rigorous mathematical analysis of Lang's model has not been made. However, growing evidence of exchange of assimilate between sieve tube and adjacent cells is accompanied by the belief that phloem-loading mechanisms probably exist along the length of the stem, at least in some herbaceous species (304). If, as Smith and Milburn (262) have suggested, phloem loading is turgor dependent, increasing in response to a fall in sieve tube turgor, then the basic requirements for a hybrid pressure flow–metabolic pump mecha-

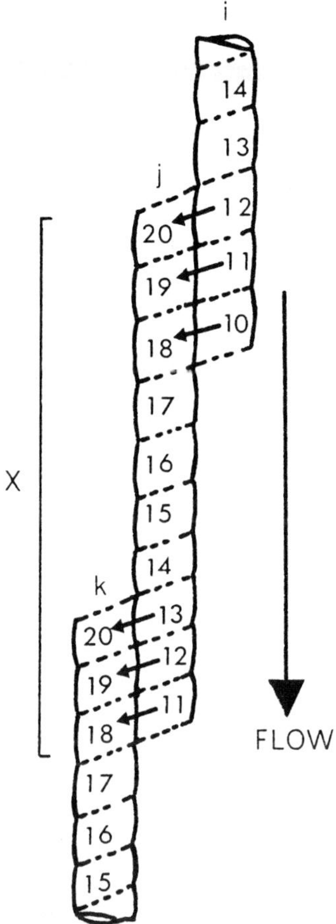

FIG. 43. Diagram of the hybrid mechanism suggested by Lang. Three lengths of adjacent sieve tubes are shown i, j, and k. Over the length marked X sieve tube j is loaded by an active mechanism at its distal end from sieve tube i, while at its proximal end material in j is loaded into sieve tube k. The loading mechanism operates in parallel across adjacent sieve elements and movement within the sieve tube is by pressure flow. The numbers in each sieve element represent relative concentrations of sugar. Redrawn from Lang (179).

nism are met, without necessarily requiring any fixed relay sites as suggested by Lang. It is likely that within the next decade there will be substantial progress in elucidating the importance of transfer of metabolites between the sieve elements and temporary stores in adjacent tissues. Such movement may prove to be highly significant in explaining long-distance transport in tall plants.

V. Regulation of Translocation

Uncertainty over mechanisms should not prevent attempts at understanding the overall control and regulation of translocation in the intact plant, despite the fact that this is extremely complex. Part of this complexity comes from the organization of the plant in nutritional terms, and again the fundamental point must be made that the plant contains a multiplicity of sinks served by a multiplicity of sources via an extremely complicated network of channels. Ideally, we need to know and understand the regulation of translocation between individual sources and individual sinks, but this is likely to remain an unachievable goal. Instead, a statistical type of model is often adopted in which it is assumed that all sinks, all sources, and all channels have similar properties and interrelationships with one another. The inherent dangers in such an approach should be obvious; it could be argued that our failure to resolve the details of the mechanisms of translocation comes from this approach, which yields results hiding important real differences in rate, or even direction, of transport, which are not recognized but which are of fundamental importance to each specific plant.

Regulation of translocation may be brought about at the channel or at source or sink or at all these sites. It is clear that the extent and pattern of development of phloem must exert a regulatory influence at the level of the whole plant or tissue. Specific mass transfer depends upon sieve element connections between source and sink, and in the absence of such connections rates are unlikely to be high. It is significant that phloem elements occur very close to apical and other meristems, differentiating from the procambial strand in advance of the associated xylem in stem, root, and leaf. A feature that may be important in woody species is the production of secondary phloem. While much is known about xylem differentiation, considerably less is known about the factors controlling the extent and rate of differentiation of secondary sieve elements (98, 182, 259).

In an interesting study on a number of wild species and modern cultivars of wheat, Evans *et al.* (78) showed a general linear relationship between maximum rate of import of assimilate by developing ears and cross-sectional area of phloem in the rachis (Fig. 44). It could be argued that phloem area limited assimilate input, but this view has been rejected (116) in the light of observations by Wardlaw and Moncur (301) that severing up to half of the phloem connections in the peduncle of wheat increased the speed of translocation. Gifford and Evans (116) argue that vascular connections develop in anticipation of carbon flow in the wheat ear, and they share the view of Milthorpe and Moorby (205) that vascular transport does not usually exert any significant control over sink growth.

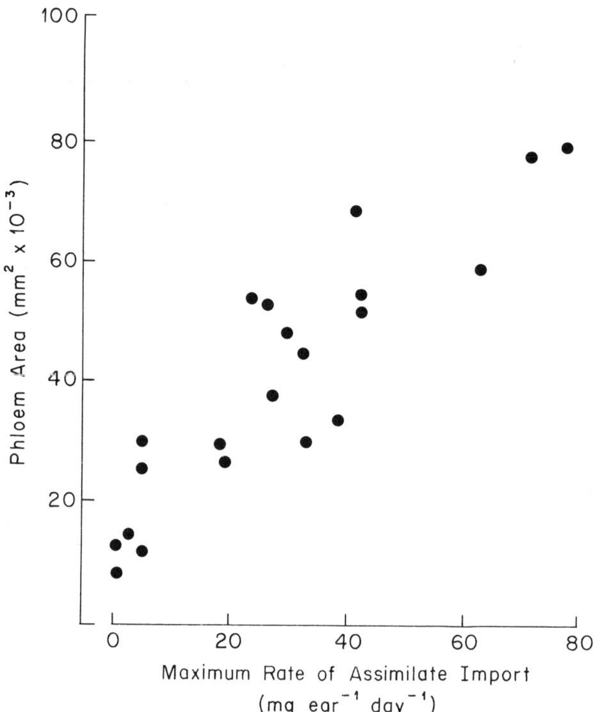

FIG. 44. The relationship between area of phloem of the stem and estimated maximum rate of import of carbohydrates by ears of a range of cultivated and "wild" wheats. Redrawn from Evans *et al.* (78).

One possible exception to this conclusion is the case in which in species such as *Vitis* there is seasonal formation and disappearance of callose at the sieve plate (57, 63). Here there is long-term regulation of transport by alteration in the resistance of the channel. The possibility of short-term regulation by occlusion of sieve plate pores by callose or P-protein cannot be ruled out—it certainly occurs to seal off damaged phloem tissues (62, 73).

The interrelationships between source and sink have been discussed in Section II. That there is some form of feedback whereby sink strength affects source activity is widely believed (116, 165, 178, 208), but the nature of the modulating mechanism is not clear. In the case of the germinating grain of cereals and other species, the signal that relates source activity to sink demand is hormonal (158, 320). The signal may be hormonal in other systems, too, but alternative mechanisms are possible, including one in which the sieve tube itself acts as a sensor to link source

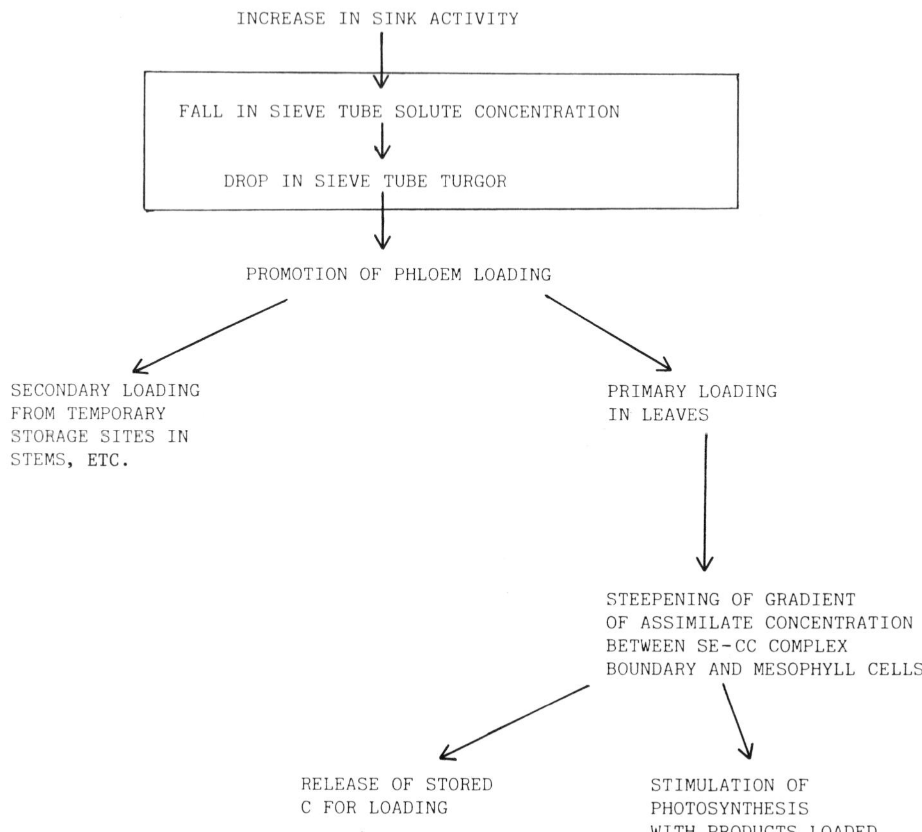

FIG. 45. A scheme to relate sink and source activities.

and sink as shown in Fig. 45. Here, it is suggested that a fall in sieve tube turgor, resulting from sink activity, leads to a stimulation of phloem loading (260–262), either from temporary storage sites in the stem or as primary loading from the leaves. The extent to which photosynthesis will be affected directly depends on the relative importance of primary and secondary loading and the extent to which primary loading proceeds using currently produced assimilate as opposed to material released from temporary storage in the leaf (e.g., starch; see also Fig. 6).

In many ways this model highlights the major advances since Swanson dealt with this topic in 1959. Our knowledge and awareness of the importance of the loading process constitutes an important step forward helping to integrate our understanding of translocation as a "whole-plant" phenomenon. The next 25 years may bring new insights into the regulation of translocation and its central role in the coordination of the activities of the growing plant and may also lead to an understanding of the mechanisms of translocation which have proved and still are so puzzling.

References

1. Aikman, D. P. (1980). Contractile proteins and hypotheses concerning their role in phloem transport. *Can. J. Bot.* **58,** 826–832.
2. Aikman, D. P., and Anderson, W. P. (1971). A quantitative investigation of a peristaltic model for phloem translocation. *Ann. Bot. (London)* [N.S.] **35,** 761–772.
3. Aikman, D. P., and Wildon, D. C. (1978). Phloem transport: The surface flow hypothesis. *J. Exp. Bot.* **29,** 387–393.
4. Allison, J. C. J., and Watson, D. J. (1966). The production and distribution of dry matter in maize after flowering. *Ann. Bot. (London)* [N.S.] **30,** 365–381.
5. Anderson, R., and Cronshaw, J. (1969). The effect of pressure release on the sieve plate pores of *Nicotiana. J. Ultrastruct. Res.* **29,** 50–59.
6. Anderson, R., and Cronshaw, J. (1970). Sieve plate pores in tobacco and bean. *Planta* **91,** 173–180.
7. Anderson, W. P. (1974). The mechanism of phloem translocation. *Symp. Soc. Exp. Biol.* **28,** 63–85.
8. Aronoff, S., Dainty, J., Gorham, P. R., Srivastava, L. M., and Swanson, C. A., eds. (1975). "Phloem Transport," NATO Adv. Study Inst. Ser., Ser. A, Life Sci., Vol. 4. Plenum, New York.
9. Baker, D. A., Malek, F., and Denvar, F. D. (1980). Phloem loading of amino acids from the petioles of *Ricinus* leaves. *Ber. Dtsch. Bot. Ges.* **93,** 203–209.
10. Barclay, G. F., and Fensom, D. S. (1973). Passage of carbon black through sieve plates of unexcised *Heracleum sphondylium* after micro-injection. *Acta Bot. Neerl.* **22,** 228–232.
11. Barlow, H. W. B. (1979). Sectorial patterns on leaves of fruit tree shoots produced by radioactive assimilates and solutions. *Ann. Bot. (London)* [N.S.] **43,** 595–602.
12. Baset, Q. A., and Sutcliffe, J. F. (1975). Regulation of the export of potassium, nitrogen, phosphorus, magnesium and dry matter from the endosperm of etiolated oat seedlings (*Avena sativa,* cv Victory). *Ann. Bot. (London)* [N.S.] **39,** 31–41.
13. Bauermeister, A., Dale, J. E., Williams, E., and Scobie, J. (1980). Movement of ^{14}C and ^{11}C-labeled assimilate in wheat leaves: The effect of IAA. *J. Exp. Bot.* **31,** 1199–1209.
14. Behnke, H. D. (1971). The control of the sieve plate pores in *Aristolochia. J. Ultrastruct. Res.* **36,** 493–498.
15. Behnke, H. D. (1975). Companion cells and transfer cells. *In* "Phloem Transport" (S. Aronoff *et al.,* eds.), pp. 153–175. Plenum, New York.
16. Bennet-Clark, T. A. (1959). Water relations of cells. *In* "Plant Physiology: A Treatise" (F. C. Steward, ed.), Vol. 2, pp. 105–191. Academic Press, New York.
17. Bennett, C. W. (1940). The relation of virus to plant tissue. *Bot. Rev.* **6,** 427–473.
18. Bentwood, B. J., and Cronshaw, J. (1978). Cytochemical localisation of adenosine triphosphatase in the phloem of *Pisum sativum* and its relation to the function of transfer cells. *Planta* **140,** 111–120.
19. Biddulph, O. (1956). Visual indications of ^{36}S and ^{32}P translocation in the phloem. *Am. J. Bot.* **43,** 143–148.

20. Biddulph, O. (1959). Translocation of inorganic solutes. *In* "Plant Physiology: A Treatise" (F. C. Steward, ed.) Vol. 2, pp. 553–603. Academic Press, New York.
21. Biddulph, O., and Cory, R. (1957). An analysis of translocation in the phloem of the bean plant using THO, P^{32} and $C^{14}02$. *Plant Physiol.* **32,** 608–619.
22. Biddulph, O., and Cory, R. (1965). Translocation of C^{14} metabolites in the phloem of the bean plant. *Plant Physiol.* **40,** 119–129.
23. Bieleski, R. L. (1976). Transport of sorbitol in pear leaf slices. *In* "Transport and Transfer Processes in Plants" (I. F. Wardlaw and J. B. Passioura, eds.), pp. 185–190. Academic Press, New York.
24. Bonnemain, J. L. (1980). Microautoradiography as a tool for the recognition of phloem transport. *Ber. Dtsch. Bot. Ges.* **93,** 99–107.
25. Booth, A., Moorby, J., Davies, C. R., Jones, H., and Wareing, P. F. (1962). Effects of indolyl-3-acetic acid on the movement of nutrients in plants. *Nature (London)* **194,** 204–205.
26. Borchers-Zampini, C., Glamm, A. B., Hoddinott, J., and Swanson, C. A. (1980). Alterations in source-sink patterns by modifications of source strength. *Plant Physiol.* **65,** 1116–1120.
27. Bowling, D. J. F. (1968). Measurement of the potential across the sieve plate. *Planta* **89,** 108–125.
28. Boynton, J. E., Nobs, M. A., Björkman, O., and Pearcy, R. W. (1971). Hybrids between *Atriplex* species with and without *β*-carboxylation photosynthesis: Leaf anatomy and ultrastructure. *Year Book—Carnegie Inst. Washington* **69,** 629–632.
29. Brown, K. J. (1968). Translocation of carbohydrate in cotton: Movement to the fruiting bodies. *Ann. Bot. (London)* [N.S.] **32,** 703–713.
30. Buttrose, M. S. (1962). Physiology of the cereal grain. III. Photosynthesis in the wheat ear during grain development. *Aust. J. Biol. Sci.* **15,** 611–618.
31. Canny, M. J. (1962). The mechanism of translocation. *Ann. Bot. (London)* [N.S.] **26,** 603–617.
32. Canny, M. J. (1973). "Phloem Translocation." Cambridge Univ. Press, London and New York.
33. Canny, M. J. (1973). Translocation and distance. 1. Growth of the fruit of the Sausage Tree, *Kigelia pinnata*. *New Phytol.* **72,** 1269–1280.
34. Canny, M. J. (1975). Mass transfer. *In* "Encyclopedia of Plant Physiology, New Series" (M. H. Zimmermann and J. A. Milburn, eds.), Vol. I, pp. 139–153. Springer-Verlag, Berlin and New York.
35. Canny, M. J., and Phillips, O. M. (1963). Quantitative aspects of a theory of translocation. *Ann. Bot. (London)* [N.S.] **27,** 379–402.
36. Carr, D. J., and Wardlaw, I. F. (1965). The supply of photosynthetic assimilates to the grain from the flag leaf and ear of wheat. *Aust. J. Biol. Sci.* **18,** 711–719.
37. Chatterton, N. J., and Silvius, J. E. (1979). Photosynthate partitioning into starch in soybean leaves. 1. Effects of photoperiod versus photosynthetic period duration. *Plant Physiol.* **64,** 749–753.
38. Christy, A. L., and Ferrier, J. M. (1973). A mathematical treatment of Münch's pressure flow hypothesis of phloem translocation. *Plant Physiol.* **52,** 531–538.
39. Christy, A. L., and Fisher, D. B. (1978). Kinetics of ^{14}C photosynthate translocation in Morning Glory vines. *Plant Physiol.* **61,** 283–290.
40. Colwell, R. N. (1942). The use of radioactive phosphorus in translocation studies. *Am. J. Bot.* **29,** 798–807.
41. Cook, M. K., and Evans, L. T. (1976). Effects of sink size, geometry and distance from source on the distribution of assimilates in wheat. *In* "Transport and Transfer Pro-

cesses in Plants" (I. F. Wardlaw and J. B. Passioura, eds.), pp. 139–153. Academic Press, New York.

42. Coster, M. G. L. (1965). A quantitative analysis of the voltage-current relationship of fixed-charge membranes and the associated property of 'punch through.' *Biophys. J.* **5,** 669–686.
43. Crafts, A. S., and Crisp, C. E. (1971). "Phloem Transport in Plants." Freeman, San Francisco, California.
44. Cronshaw, J. (1971). The P protein compounds of sieve elements. *Proc. Int. Conf. Electron Microsc. 7th, 1970,* pp. 429–430.
45. Cronshaw, J. (1974). Phloem differentiation and development. *In* "Dynamic Aspects of Plant Ultrastructure" (A. W. Robards, ed.), pp. 391–413. McGraw-Hill, London.
46. Cronshaw, J. (1975). Sieve element cell walls. *In* "Phloem Transport" (S. Aronoff *et al.,* eds.), pp. 129–147. Plenum, New York.
47. Cronshaw, J. (1980). Histochemical localisation of enzymes in the phloem. *Ber. Dtsch. Bot. Ges.* **93,** 123–139.
48. Cronshaw, J. (1981). Phloem structure and function. *Annu. Rev. Plant Physiol.* **32,** 465–484.
49. Cronshaw, J., and Esau, K. (1967). Tubular and fibrillar components of mature and differentiating sieve elements. *J. Cell Biol.* **34,** 801–816.
50. Cronshaw, J., and Esau, K. (1968). P protein in the phloem of *Cucurbita.* I. The development of P protein bodies. *J. Cell Biol.* **38,** 25–39.
51. Cronshaw, J., and Esau, K. (1968). P protein in the phloem of *Cucurbita.* II. The P protein of mature sieve elements. *J. Cell Biol.* **38,** 292–303.
52. Crookston, R. K. (1980). The structure and function of C_4 vascular tissue—some unanswered questions. *Ber. Dtsch. Bot. Ges.* **93,** 71–78.
53. Currier, H. B., Esau, K., and Cheadle, V. I. (1955). Plasmolytic studies of phloem. *Am. J. Bot.* **42,** 68–81.
54. Dale, J. E., Bauermeister, A., and Williams, E. J. (1981). The use of compartmental analysis to examine effects of plant growth regulating substances on transport of assimilate in wheat leaves. *In* "Mathematics and Plant Physiology" (D. A. Rose and D. A. Charles-Edwards, eds.), pp. 79–90. Academic Press, London and New York.
55. Dale, J. E., and Milthorpe, F. L. (1983). General features of the production and growth of leaves. *In* "The Growth and Functioning of Leaves" (J. E. Dale and F. L. Milthorpe, eds.), pp. 151–178. Cambridge Univ. Press, London and New York.
56. Davies, C. R., and Wareing, P. F. (1965). Auxin-directed transport of radiophosphorus in stems. *Planta* **65,** 139–156.
57. Davis, J. D., and Evert, R. F. (1970). Seasonal cycle of development in woody vines. *Bot. Gaz. (Chicago)* **131,** 128–138.
58. Delrot, S., and Bonnemain, J. L. (1981). Involvement of protons as a substrate for the sucrose carrier during phloem loading in *Vicia faba* leaves. *Plant Physiol.* **67,** 560–564.
59. Delrot, S., Despeghel, J. P., and Bonnemain, J. L. (1980). Phloem loading in *Vicia faba* leaves: Effect of N-ethylmaleimide and para-chloromercuribenzenesulphonic acid on H^+ extrusion, K^+ and sucrose uptake. *Planta* **149,** 144–148.
60. Dixon, A. F. G. (1975). Aphids and translocation. *In* "Encyclopedia of Plant Physiology, New Series" (M. M. Zimmermann and J. A. Milburn, eds.), Vol. I, pp. 154–170. Springer-Verlag, Berlin and New York.
61. Doman, D. C., and Geiger, D. R. (1979). Effect of exogenously supplied foliar potassium on phloem loading in *Beta vulgaris* L. *Plant Physiol.* **64,** 528–533.
62. Engleman, E. (1965). Sieve elements of *Impatiens sultanii.* 1. Wound reaction. *Ann. Bot. (London)* [N.S.] **29,** 83–101.

63. Esau, K. (1948). Phloem structure in the grapevine and its seasonal changes. *Hilgardia* **18,** 217–296.
64. Esau, K. (1965). "Plant Anatomy," 2nd ed. Wiley, New York.
65. Esau, K. (1965). Anatomy and cytology of *Vitis* phloem. *Hilgardia* **37,** 17–72.
66. Esau, K. (1969). The phloem. *In* "Handbuch der Pflanzenanatomie" (W. Zimmermann, P. Ozenda, and M. D. Wulfe, ed.), Vol. 5, Part 2. Borntraeger, Berlin.
67. Esau, K., and Cheadle, V. I. (1959). Size of pores and the controls in sieve element of dicotyledons. *Proc. Natl. Acad. Sci. U.S.A.* **45,** 156–162.
68. Esau, K., and Cheadle, V. I. (1965). Cytologic studies on phloem. *Univ. Calif. Publ. Bot.* **36,** 253–344.
69. Esau, K., Cheadle, V. I., and Risley, E. D. (1962). Development of the sieve plate pores. *Bot. Gaz. (Chicago)* **123,** 233–243.
70. Esau, K., Englemann, E. M., and Bisalputra, T. (1963). What are transcellular strands? *Planta* **59,** 617–623.
71. Eschrich, W. (1967). Bidirektionelle Translokation in Siebröhren. *Planta* **73,** 37–49.
72. Eschrich, W. (1970). Biochemistry and fine structure of phloem in relation to transport. *Annu. Rev. Plant Physiol.* **21,** 193–214.
73. Eschrich, W. (1975). Sealing systems in phloem. *In* "Encyclopedia of Plant Physiology, New Series" (M. H. Zimmerman and J. A. Milburn, eds.), Vol. I, pp. 39–56. Springer-Verlag, Berlin and New York.
74. Eschrich, W. (1980). Phloem loading and related processes. *Ber. Dtsch. Bot. Ges.* **93,** 1–378.
75. Eschrich, W. (1980). Free space invertase, its possible role in phloem unloading. *Ber. Dtsch. Bot. Ges.* **93,** 363–378.
76. Evans, G. C. (1972). "The Quantitative Analysis of Plant Growth." Blackwell, Oxford.
77. Evans, L. T. (1976). Transport and distribution in plants. *In* "Transport and Transfer Processes in Plants" (I. F. Wardlaw and J. B. Passioura, eds.), pp. 1–13. Academic Press, New York.
78. Evans, L. T., Dunstone, R. L., Rawson, H. M., and Williams, R. F. (1970). The phloem of the wheat stem in relation to requirements for assimilate by the ear. *Aust. J. Biol. Sci.* **23,** 743–752.
79. Evert, R. F. (1977). Phloem structure and histochemistry. *Annu. Rev. Plant Physiol.* **28,** 199–222.
80. Evert, R. F., Eschrich, W., and Eichhorn, S. E. (1973). P protein distribution in mature sieve elements of *Cucurbita maxima. Planta* **109,** 193–210.
81. Fellows, R. J., and Geiger, D. R. (1974). Structural and physiological changes in sugar beet leaves during sink to source conversion. *Plant Physiol.* **64,** 877–885.
82. Fensom, D. S. (1972). A theory of translocation in phloem of *Heracleum* by contractile protein microfibrillar material. *Can. J. Bot.* **50,** 479–497.
83. Fensom, D. S. (1975). Possible mechanisms of phloem transport: Other possible mechanisms. *In* "Encyclopedia of Plant Physiology, New Series" (M. H. Zimmermann and J. A. Milburn, eds.), Vol. I, pp. 354–366. Springer-Verlag, Berlin and New York.
84. Fensom, D. S. (1980). Problems arising from a Münch-type pressure flow mechanism of sugar transport in phloem. *Can. J. Bot.* **59,** 425–432.
85. Fensom, D. S., and Spanner, D. C. (1969). Electro-osmotic and biopotential measurements of phloem strands of *Nymphoides. Planta* **88,** 321–331.
86. Fensom, D. S., and Williams, E. J. (1974). A note on Allen's suggestion for long distance translocation in the phloem of plants. *Nature (London)* **250,** 490–492.
87. Fensom, D. S., Williams, E. J., Aikman, D., Dale, J. E., Scobie, J., Ledingham, K. W. O., Drinkwater, A., and Moorby, J. (1977). Translocation of ^{11}C from leaves of *Helianthus:* Preliminary results. *Can. J. Bot.* **55,** 1787–1793.

88. Ferrier, J. M. (1976). An approximate analytical equation for sugar concentration waves in Münch phloem translocation systems. *Can. J. Bot.* **54,** 2130–2132.
89. Ferrier, J. M. (1978). Further theorctical analysis of concentration-pressure-flux waves in phloem transport systems. *Can. J. Bot.* **56,** 1086–1090.
90. Ferrier, J. M., and Christy, A. L. (1977). Role of concentration-dependent unloading in mathematical models of Münch transport. *Plant Physiol.* **60,** 173–174.
91. Ferrier, J. M., Tyree, M. T., and Christy, A. L. (1975). The theoretical time-dependent behaviour of a Münch pressure-flow system: The effect of sinusoidal time variation in sucrose loading and water potential. *Can. J. Bot.* **53,** 1120–1127.
92. Fischer, A. (1884). "Untersuchungen uber das Siebrohren-System der Cucurbitaceen." Borntraeger, Berlin.
93. Fisher, D. B. (1972). Artifacts in the embedment of water-soluble compounds for light microscopy. *Plant Physiol.* **49,** 161–165.
94. Fisher, D. B. (1975). Translocation kinetics of photosynthesis. *In* "Phloem Transport" (S. Aronoff *et al.,* eds.), pp. 327–358. Plenum, New York.
95. Fisher, D. B. (1978). The estimation of sugar concentration in individual sieve-tube elements by negative staining. *Planta* **139,** 19–24.
96. Fisher, D. B. (1978). An evaluation of the Münch hypothesis for phloem transport in soybean. *Planta* **139,** 25–28.
97. Fondy, B. R., and Geiger, D. R. (1977). Sugar selectivity and other characteristics of phloem loading in *Beta vulgaris* L. *Plant Physiol.* **59,** 953–960.
98. Ford, E. D., Robards, A. W., and Piney, M. D. (1978). Influence of environmental factors on cell production and differentiation in the early wood of *Picea sitchensis. Ann. Bot. (London)* [N.S.] **42,** 683–692.
99. Fritz, E. (1980). Microautoradiographic localisation of assimilates in phloem: Problems and new methods. *Ber. Dtsch. Bot. Ges.* **93,** 109–121.
100. Fritz, E., and Eschrich, W. (1970). ^{14}C-mikroautoradiographie wasserlüslicher substanzen in phloem. *Planta* **92,** 267–281.
101. Gardner, D. C. J., and Peel, A. J. (1969). ATP in the sieve tube sap from willow. *Nature (London)* **222,** 774.
102. Geiger, D. R. (1975). Phloem loading. *In* "Encyclopedia of Plant Physiology, New Series" (M. H. Zimmermann and J. A. Milburn, eds.), Vol. I, pp. 395–431. Springer-Verlag, Berlin and New York.
103. Geiger, D. R. (1976). Phloem loading in source leaves. *In* "Transport and Transfer Processes in Plants" (I. F. Wardlaw and J. B. Passioura, eds.), pp. 167–184. Academic Press, New York.
104. Geiger, D. R., and Cataldo, D. A. (1969). Leaf structure and translocation in sugar beet. *Plant Physiol.* **44,** 45–54.
105. Geiger, D. R., Giaquinta, R. T., Sovonick, S. A., and Fellows, R. J. (1973). Solute distribution in sugar beet leaves in relation to phloem loading and translocation. *Plant Physiol.* **52,** 585–589.
106. Geiger, D. R., and Sovonich, S. A. (1975). Effects of temperature, anoxia and other metabolic inhibitors on translocation. *In* "Encyclopedia of Plant Physiology, New Series" (M. H. Zimmermann and J. A. Milburn, eds.), Vol. I, pp. 256–286. Springer-Verlag, Berlin and New York.
107. Geiger, D. R., Sovonick, S. A., Shock, T. L., and Fellows, R. J. (1974). Role of free space in translocation in sugar beet. *Plant Physiol.* **54,** 892–898.
108. Giaquinta, R. T. (1977). Phloem loading of sucrose: pH dependence and selectivity. *Plant Physiol.* **59,** 750–755.
109. Giaquinta, R. T. (1977). Sucrose hydrolysis in relation to phloem translocation in *Beta vulgaris. Plant Physiol.* **60,** 339–343.

110. Giaquinta, R. T. (1978). Source and sink leaf metabolism in relation to phloem translocation, carbon partitioning and enzymology. *Plant Physiol.* **61,** 380–383.
111. Giaquinta, R. T. (1979). Phloem loading of sucrose. Involvement of membrane ATPase and proton transport. *Plant Physiol.* **63,** 744–748.
112. Giaquinta, R. T. (1979). Sucrose translocation and storage in the sugar beet. *Plant Physiol.* **63,** 828–832.
113. Giaquinta, R. (1980). Mechanism and control of phloem loading of sucrose. *Ber. Dtsch. Bot. Ges.* **93,** 187–201.
114. Giaquinta, R. T. (1983). Phloem loading of sucrose. *Annu. Rev. Plant Physiol.* **34,** 347–387.
115. Giaquinta, R. T., and Geiger, D. R. (1977). Mechanism of cyanide inhibition of phloem translocation. *Plant Physiol.* **59,** 178–180.
116. Gifford, R. M., and Evans, L. T. (1981). Photosynthesis, carbon partitioning and yield. *Annu. Rev. Plant Physiol.* **32,** 485–509.
117. Glasziou, K. T., and Gayler, K. R. (1972). Storage of sugars in stalks of sugar cane. *Bot. Rev.* **38,** 471–490.
118. Goeschl, J. D., Magnuson, C. E., DeMichele, D. W., and Sharpe, P. J. M. (1976). Concentration-dependent unloading as a necessary assumption for a closed form mathematical model of osmotically driven pressure flow in phloem. *Plant Physiol.* **58,** 556–562.
119. Goodwin, P. B. (1978). Phytohormones and growth and development of organs of the vegetative plant. *In* "Phytohormones and Related Compounds—A Comprehensive Treatise" (D. S. Letham, P. B. Goodwin, and T. J. V. Higgins, eds.), pp. 37–174. Elsevier North-Holland, Amsterdam.
120. Grange, R. I., and Peel, A. J. (1978). Evidence for solution flow in the phloem of willow. *Planta* **138,** 15–23.
121. Guardiola, J. L., and Sutcliffe, J. F. (1971). Control of protein hydrolysis in the cotyledons of germinating pea (*Pisum sativum* L.) seeds. *Ann. Bot. (London)* [N.S.] **35,** 791–808.
122. Guardiola, J. L., Sutcliffe, J. F. (1972). Transport of materials from the cotyledons during germination of the seeds of the garden pea (*Pisum sativum* L.). *J. Exp. Bot.* **23,** 322–337.
123. Gunning, B. E. S., Pate, J. S., and Briarty, L. (1968). Specialised transfer cells in minor veins of leaves and their possible significance in phloem translocation. *J. Cell Biol.* **37,** C7–12.
124. Gunning, B. E. S., Pate, J. S., Minchin, F. R., and Marks, J. (1974). Quantitative aspects of transfer cell structure in relation to vein loading in leaves and solute transport in legume nodules. *Symp. Soc. Exp. Biol.* **28,** 87–126.
125. Gutierrez, M., Gracen, V. E., and Edwards, G. E. (1974). Biochemical and cytological relationships in C_4 plants. *Planta* **119,** 279–300.
126. Habeshaw, D. (1969). The effect of light on the translocation from sugar beet leaves. *J. Exp. Bot.* **20,** 64–71.
127. Hall, S. M., and Baker, D. A. (1972). The chemical composition of *Ricinus* phloem exudate. *Planta* **106,** 131–140.
128. Hall, S. M., and Milburn, J. A. (1973). Phloem transport in *Ricinus:* Its dependence on the water balance of the tissues. *Planta* **109,** 1–10.
129. Hammel, H. T. (1968). Measurement of turgor pressure and its gradient in the phloem of oak. *Plant Physiol.* **43,** 1042–1048.
130. Hanson, A. D., and Tully, R. E. (1979). Amino acids translocated from turgid and water-stressed barley leaves. II. Studies with ^{13}N and ^{14}C. *Plant Physiol.* **64,** 467–471.

131. Harel, S., and Reinhold, L. (1966). The effect of 2,4-dinitrophenol on translocation in the phloem. *Physiol. Plant.* **19,** 634–643.
132. Hartt, C. (1969). Effect of potassium deficiency upon translocation of ^{14}C in attached blades and entire plants of sugar cane. *Plant Physiol.* **44,** 1461–1469.
133. Hatch, M. D., and Osmond, C. B. (1976). Compartmentation and transport in C_4 photosynthesis. *In* "Encyclopedia of Plant Physiology, New Series" (C. R. Stocking and U. Heber, eds.), Vol. 3, pp. 144–184. Springer-Verlag, Berlin and New York.
134. Hatch, M. D., Sacher, J. A., and Glasziou, K. T. (1963). Sugar accumulation cycle in sugar cane I Studies on enzymes of the cycle. *Plant Physiol.* **38,** 338–343.
135. Hattersley, P. W., Watson, L., and Osmond, C. B. (1976). Metabolite transport in leaves of C_4 plants: Specification and speculation. *In* "Transport and Transfer Processes in Plants" (I. F. Wardlaw and J. B. Passioura, eds.), pp. 191–201. Academic Press, New York.
136. Hendrix, J. E. (1977). Phloem loading in squash. *Plant Physiol.* **60,** 567–569.
137. Hew, C. S., Nelson, C. D., and Krotkov, G. (1967). Hormonal control of translocation of photosynthetically assimilated ^{14}C in young soybean plants. *Am. J. Bot.* **54,** 252–256.
138. Heyser, W., Heyser, R., Eschrich, W., and Fritz, E. (1977). The influence of externally-applied sucrose on phloem transport in the maize leaf strip. *Planta* **137,** 145–151.
139. Heyser, W., Leonard, O. A., Heyser, R., Fritz, E., and Eschrich, W. (1975). The influence of light, darkness and lack of CO_2 on phloem translocation in detached maize leaves. *Planta* **122,** 143–154.
140. Ho, L. C. (1976). The relationship between the rates of carbon transport and of photosynthesis in tomato leaves. *J. Exp. Bot.* **27,** 87–97.
141. Ho, L. C., and Mortimer, D. C. (1971). The site of cyanide inhibition of sugar translocation in sugar beet leaf. *Can. J. Bot.* **49,** 1769–1775.
142. Ho, L. C., and Shaw, A. F. (1977). Carbon economy and translocation of ^{14}C in leaflets of the seventh leaf of tomato during leaf expansion. *Ann. Bot. (London)* [N.S.] **41,** 833–848.
143. Hoddinott, J., Ehret, D. L., and Gorham, P. R. (1979). Rapid influences of water stress on photosynthesis and translocation in *Phaseolus vulgaris. Can. J. Bot.* **57,** 768–776.
144. Hofstra, G., and Nelson, C. D. (1969). A comparative study of translocation of assimilated ^{14}C from leaves of different species. *Planta* **88,** 103–112.
145. Horrocks, R. D., Kerby, T. A., and Buxton, D. R. (1978). Carbon source for developing bolls in normal and superokra leaf cotton. *New Phytol.* **80,** 335–340.
146. Horwitz, L. (1958). Some simplified mathematical treatment of translocation in plants. *Plant Physiol.* **33,** 81–93.
147. Houseley, T. L., Peterson, D. M., and Schrader, L. E. (1977). Long distance translocation of sucrose, serine, leucine, lysine and CO_2 assimilates. I. Soybean. *Plant Physiol.* **59,** 217–220.
148. Hunt, R. (1978). "Plant Growth Analysis," Stud. Biol. No. 96. Arnold, London.
149. Hutchings, V. M. (1978). Sucrose and proton co-transport in *Ricinus* cotyledons. *Planta* **138,** 229–235.
150. Ilker, R., and Currier, M. B. (1974). Heavy meomyosin complexing filaments in the phloem of *Vicia faba* and *Xylosma congestum. Planta* **120,** 311–316.
151. Incoll, L. D., and Neales, T. F. (1970). The stem as a temporary sink before tuberization in *Helianthus tuberosus. J. Exp. Bot.* **21,** 469–476.
152. Isebrands, J. G., and Larson, P. R. (1973). Anatomical changes during leaf ontogeny of *Populus deltoides. Am. J. Bot.* **60,** 199–208.
153. Jenner, C. F. (1980). Effects of shading or removing spikelets in wheat: Testing assumptions. *Aust. J. Plant Physiol.* **7,** 113–122.

154. Johnson, R. P. C. (1968). Microfilaments in pores between frozen-etched sieve elements. *Planta* **81**, 314–332.
155. Johnson, R. P. C. (1973). Filaments but no membranous transcellular strands in sieve pores in freeze-etched, translocating phloem. *Nature (London)* **244**, 464–465.
156. Johnson, R. P. C. (1978). The microscopy of P protein filaments in freeze-etched sieve tube pores. *Planta* **143**, 191–205.
157. Johnson, R. R., and Moss, D. N. (1976). Effects of water stress on $^{14}CO_2$ fixation translocation in wheat during grain filling. *Crop Sci.* **16**, 697–701.
158. Jones, R. L. (1971). Gibberellic acid-enhanced release of β-1,3-glucanase from barley aleurone cells. *Plant Physiol.* **47**, 412–416.
159. Joy, K. W. (1964). Translocation in sugar beet. 1. Assimilation of $^{14}CO_2$ and distribution of material from the leaves. *J. Exp. Bot.* **15**, 485–494.
160. Keener, M. E., DeMichele, D. W., and Sharpe, P. J. M. (1979). Sink metabolism: A conceptual framework for analysis. *Ann. Bot. (London)* [N.S.] **44**, 659–670.
161. Keith, D. (1979). Mass transfer and ^{14}C translocation is detached maize leaves. *Can. J. Bot.* **57**, 657–665.
162. Kemp, D. R. (1980). The location and size of the extension zone of emerging wheat leaves. *New Phytol.* **84**, 729–737.
163. Kennedy, J. S., and Mittler, T. E. (1953). A method of obtaining phloem sap via the mouth parts of aphids. *Nature (London)* **171**, 528.
164. Khan, A. A., and Sagar, G. R. (1967). Translocation in tomato: The distribution of the products of photosynthesis of the leaves of a tomato plant during the phase of fruit production. *Hortic. Res.* **7**, 61–69.
165. King, R. W., Wardlaw, I. F., and Evans, L. T. (1967). Effect of assimilate utilisation on photosynthetic rate in wheat. *Planta* **77**, 261–276.
166. Knight, B. K., Mitton, G. D., Davidson, H. R., and Fensom, D. S. (1974). Micro-injection of ^{14}C sucrose and other tracers into isolated phloem strands of *Heracleum. Can J. Bot.* **52**, 1491–1499.
167. Killman, R. (1975). Sieve element structure in relation to function. *In* "Phloem Transport" (S. Aronoff *et al.*, eds.), pp. 225–242. Plenum, New York.
168. Komor, E., Rotter, M., and Tanner, W. (1977). A proton-cotransport system in a higher plant: Sucrose transport in *Ricinus communis. Plant Sci. Lett.* **9**, 153–162.
169. Komor, E., Rotter, M., Waldhauser, J., Martin, E., and Cho, B. H. (1980). Sucrose proton symport for phloem loading in the *Ricinus* seedling. *Ber. Dtsch. Bot. Ges.* **93**, 211–219.
170. Kriedeman, P. E. (1968). An effect of kinetin on the translocation of ^{14}C-labeled photosynthate in citrus. *Aust. J. Biol. Sci.* **21**, 569–571.
171. Kuo, J., O'Brien, J. P., and Canny, M. J. (1974). Pit-field distribution, plasmodesmatal frequency, and assimilate flux in the mestome sheath cells of wheat leaves. *Plant* **121**, 97–118.
172. Kursanov, A. L. (1963). Metabolism and the transport of organic substances in the phloem. *Adv. Bot. Res.* **1**, 209–274.
173. Kursanov, A. L., and Brovchenko, M. I. (1961). Effect of ATP on the entry of assimilates into the conducting system of sugar beet. *Sov. Plant Physiol. (Engl. Transl.)* **8**, 211–217.
174. Kursanov, A. L., and Brovchenko, M. I. (1969). Free space as an intermediate zone between photosynthesising and conducting cells of leaves. *Sov. Plant Physiol. (Engl. Transl.)* **16**, 965–972.
175. Kursanov, A. L., and Brovchenko, M. I. (1970). Sugars in the free space of leaves: Their origin and possible involvement in transport. *Can. J. Bot.* **48**, 1243–1250.

176. Laetsch, W. M. (1974). The C-4 syndrome: A structural analysis. *Annu. Rev. Plant Physiol.* **25,** 27–52.
177. Lamoureux, C. M. (1975). Phloem tissue in angiosperms and gymnosperms. *In* "Phloem Transport (S. Aronoff *et al.*, eds.), pp. 1–20. Plenum, New York.
178. Lang, A. (1978). Interactions between source, path and sink in determining phloem translocation rate. *Aust. J. Plant Physiol.* **5,** 665–674.
179. Lang, A. (1979). A relay mechanism for phloem translocation. *Ann. Bot. (London)* [N.S.] **44,** 141–145.
180. Langer, R. H. M. (1979). "How Grasses Grow," 2nd ed., Stud. Biol. No. 34. Arnold, London.
181. Larsen, P. R., Isebrands, J. G., and Dickson, R. E. (1972). Fixation patterns of ^{14}C within developing leaves of Eastern Cottonwood. *Planta* **107,** 301–314.
182. Lawton, J. R. (1976). Seasonal variation in the secondary phloem from the main trunks of willow and sycamore trees. *New Phytol.* **77,** 761–771.
183. Lee, D. R. (1972). The possible significance of filaments in sieve elements. *Nature (London)* **235,** 286.
184. Lepp, N. W., and Peel, A. J. (1970). Some effects of IAA and kinetin upon the movement of sugars in the phloem of willow. *Planta* **90,** 230–235.
185. Lush, W. M., and Evans, L. T. (1974). Longitudinal translocation of ^{14}C-labeled assimilates in leaf blades of *Lolium temulentum. Aust. J. Plant Physiol.* **1,** 433–443.
186. MacRobbie, E. A. C. (1971). Phloem translocation. Facts and mechanisms: A comparative survey. *Biol. Rev. Cambridge Philos. Soc.* **46,** 429–481.
187. Magnuson, C. E., Goeschl, J. D., Sharpe, P. J. H., and DeMichele, D. W. (1979). Consequences of insufficient equations in models of the Münch hypothesis of phloem transport. *Plant, Cell Environ.* **2,** 181–188.
188. Malek, F., and Baker, D. A. (1977). Proton co-transport of sugars in phloem loading. *Planta* **135,** 297–299.
189. Malek, F., and Baker, D. A. (1978). Effect of fusicoccin on proton cotransport of sugars in the phloem loading of *Ricinus communis* L. *Plant Sci. Lett.* **11,** 233–239.
190. Mangham, S. (1917). On the mechanism of translocation in plant tissues. An hypothesis with special reference to sugar conduction in sieve tubes. *Ann. Bot. (London)* [OS.] **31,** 293–311.
191. Marré, E. (1977). Effects of fusicoccin and hormones on plant cell membrane activities: Observation and hypotheses. *In* "Regulation of Cell Membrane Activities in Plants" (E. Marré and O. Ciferri, eds.), pp. 185–202. Elsevier/North-Holland, Amsterdam.
192. Martin, E., and Komor, E. (1980). Role of phloem in sucrose transport in *Ricinus* cotyledons. *Planta* **148,** 367–373.
193. Mason, T. G., and Maskell, E. J. (1928). Studies on the translocation of carbohydrates in the cotton plant. 1. A study of diurnal variation in the carbohydrates of leaf bark and wood and of the effects of ringing. *Ann. Bot. (London)* [O.S.] **42,** 189–253.
194. Mason, T. G., and Maskell, E. J. (1928). Studies on the transport of carbohydrates in the cotton plant. II. The factors determining the rate and direction of movement of sugars. *Ann. Bot. (London)* [O.S.] **42,** 571–636.
195. Mason, T. G., and Maskell, E. J. (1931). Further studies on transport in the cotton plant. 1. Preliminary observations on the transport of phosphorus, potassium and calcium. *Ann. Bot. (London)* [O.S.] **45,** 125–173.
196. Mason, T. G., Maskell, E. J., and Phillis, E. (1936). Further studies on transport in the cotton plant. IV. Concerning the independence of solute movement in the phloem. *Ann. Bot. (London)* [O.S.] **50,** 23–58.
197. Mason, T. G., Maskell, E. J., and Phillis, E. (1936). Further studies on transport in the

cotton plant. IV. On the simultaneous movement of solutes in opposite directions through the phloem. *Ann. Bot. (London)* [O.S.] **50,** 167–174.
198. Mason, T. G., and Phillis, E. (1936). Further studies on transport in the cotton plant. V. Oxygen supply and the activation of diffusion. *Ann. Bot. (London)* [O.S.] **50,** 455–499.
199. Mason, T. G., and Phillis, E. (1937). The migration of solutes. *Bot. Rev.* **3,** 47–71.
200. Maynard, J. W., and Lucas, W. J. (1982). A re-analysis of the two component phloem loading systems in *Beta vulgaris. Plant Physiol.* **69,** 734–739.
201. McNeil, D. L. (1979). The kinetics of phloem loading of valine in the shoot of a nodulated legume (*Lupinus albus* L. cv. Ultra). *J. Exp. Bot.* **30,** 1003–1012.
202. Mengel, K., and Haeder, H. E. (1977). Effect of potassium supply on the role of phloem sap exudation and the composition of phloem sap in *Ricinus communis. Plant Physiol.* **59,** 282–284.
203. Milburn, J. A. (1975). Pressure flow. *In* "Encyclopedia of Plant Physiology, New Series" (M. H. Zimmermann and J. A. Milburn, eds.), Vol. I, pp. 328–353. Springer-Verlag, Berlin and New York.
204. Milburn, J. A. (1980). The measurement of turgor pressure in sieve tubes. *Ber. Dtsch. Bot. Ges.* **93,** 153–166.
205. Milthorpe, F. L., and Moorby, J. (1969). Vascular transport and its significance in plant growth. *Annu. Rev. Plant Physiol.* **20,** 117–138.
206. Minchin, P. E. H. (1978). Analysis of tracer profiles with applications to phloem transport. *J. Exp. Bot.* **29,** 1441–1450.
207. Mittler, T. E. (1957). Studies on the feeding and nutrition of *Tuberolachnus salignus* Gmelin (Homophera Aphididae). 1. The uptake of phloem sap. *J. Exp. Bot.* **34,** 334–345.
208. Mondal, M. H., Brun, W. A., and Brenner, M. L. (1978). Effect of sink removal on photosynthesis and senescence in leaves of soybean (*Glycine max* L.) plants. *Plant Physiol.* **61,** 394–397.
209. Moorby, J. (1977). Integration and regulation of translocation within the whole plants. *Symp. Soc. Exp. Biol.* **31,** 425–454.
210. Moorby, J. (1981). "Transport Systems in Plants," Integrated Themes in Biology Series. Longmans, Green, London and New York.
211. Moorby, J., Ebert, M., and Evans, W. T. S. (1963). The translocation of ^{11}C-labeled assimilate in the soybean. *J. Exp. Bot.* **14,** 210–220.
212. Moorby, J., and Jarman, P. D. (1975). The use of compartmental analysis in the study of the movement of carbon through leaves. *Planta* **122,** 155–168.
213. Moorby, J., Troughton, J. H., and Currie, B. G. (1974). Investigations of carbon transport in plants. I. *J. Exp. Bot.* **25,** 937–944.
214. Morretes, B. L. de (1962). Terminal phloem in vascular bundles of leaves of *Capsicum annuum* and *Phaseolus vulgaris. Am. J. Bot.* **49,** 560–567.
215. Mulligan, D. R., and Patrick, J. W. (1979). Gibberellic-acid-promoted transport of assimilates in stems of *Phaseolus vulgaris.* L. *Planta* **145,** 233–238.
216. Mullins, M. G. (1970). Transport of ^{14}C assimilates in seedlings of *Phaseolus vulgaris* L. in relation to vascular anatomy. *Ann. Bot. (London)* [N.S.] **34,** 889–896.
217. Mullins, M. G. (1970). Hormone-directed transport of assimilates in decapitated internodes of *Phaseolus vulgaris. Ann. Bot. (London)* [N.S.] **34,** 897–909.
218. Münch, E. (1930). "Die Stoffbewegungen in der Pflanze." Fischer, Jena.
219. Munns, R., and Pearson, C. J. (1974). Effect of water deficit on translocation of carbohydrate in *Solanum tuberosum. Aust. J. Plant Physiol.* **1,** 529–537.
220. Nobel, P. S. (1974). "Introduction to Biophysical Plant Physiology." Freeman, San Francisco, California.

221. Outlaw, W. H., Fisher, D. B., and Christy, A. L. (1975). Compartmentation in *Vicia faba* leaves. IV. Kinetics of ^{14}C sucrose redistribution among individual tissues following pulse labeling. *Plant Physiol.* **55,** 704–711.
222. Palevitz, B. A., and Hepler, P. K. (1975). Is P-protein actin-like?—Not yet. *Planta* **125,** 261–271.
223. Palmquist, E. M. (1938). The simultaneous movement of carbohydrates and fluorescein in opposite directions in the phloem. *Am. J. Bot.* **25,** 97–105.
224. Parthasarathy, M. V. (1974). Ultrastructure of phloem in palms. III. Mature phloem. *Protoplasma* **79,** 265–315.
225. Parthasarathy, M. V. (1975). Sieve element structure. *In* "Encyclopedia of Plant Physiology, New Series" (M. H. Zimmermann and J. A. Milburn, eds.), Vol. I, pp. 3–57. Springer-Verlag, Berlin and New York.
226. Passioura, J. B., and Ashford, A. E. (1974). Rapid translocation in the phloem of wheat roots. *Aust. J. Plant Physiol.* **1,** 521–527.
227. Pate, J. S., and Gunning, B. E. S. (1969). Vascular transfer cells in angiosperm leaves: A taxonomic and morphological survey. *Protoplasma* **68,** 135–156.
228. Pate, J. S., Sharkey, P. J., and Lewis, O. A. M. (1974). Phloem bleeding from legume fruits—a technique for study of fruit nutrition. *Planta* **120,** 229–243.
229. Patrick, J. W. (1976). Hormone-directed of metabolites. *In* "Transport and Transfer Processes in Plants" (I. F. Wardlaw and J. B. Passioura, eds.), pp. 433–446. Academic Press, New York.
230. Patrick, J. W. (1979). Auxin-promoted transport of metabolites in stems of *Phaseolus vulgaris* L. Further studies of effects remote from the site of hormone application. *J. Exp. Bot.* **30,** 1–13.
231. Patrick, J. W. (1979). An assessment of auxin-promoted transport in whole and decapitated stems of *Phaseolus vulgaris* L. *Planta* **146,** 107–112.
232. Patrick, J. W., and Wareing, P. F. (1973). Auxin-promoted transport of metabolites in stems of *Phaseolus vulgaris.* L. Some characteristics of the experimental transport system. *J. Exp. Bot.* **24,** 1158–1171.
233. Patrick, J. W., and Wareing, P. F. (1976). Auxin-promoted transport of metabolites in stems of *Phaseolus vulgaris* L.; Effects at the site of hormone application. *J. Exp. Bot.* **27,** 969–982.
234. Patrick, J. W., and Wareing, P. F. (1978). Auxin-promoted transport of metabolites in stems of *Phaseolus vulgaris* L. Effects remote from the site of hormone application. *J. Exp. Bot.* **29,** 359–366.
235. Patrick, J. W., and Woolley, D. J. (1973). Auxin physiology of decapitated stems of *Phaseolus vulgaris* L. treated with indol-3yl-acetic acid. *J. Exp. Bot.* **24,** 949–957.
236. Pearson, C. J., and Derrick, G. A. (1977). Thermal adaptation of *Pennisetum:* Leaf photosynthesis and photosynthate translocation. *Aust. J. Plant Physiol.* **4,** 763–769.
237. Peel, A. J. (1974). "Transport of Nutrients in Plants." Butterworth, London.
238. Peel, A. J. (1975). Investigations with applied stylets into the physiology of the sieve tube. *In* "Encyclopedia of Plant Physiology, New Series" (M. H. Zimmermann and J. A. Milburn, eds.), Vol. I, pp. 171–195. Springer-Verlag, Berlin and New York.
239. Peel, A. J., and Ho, L. C. (1970). Colony size of *Tuberolachnus salignus* (Gmelin) in relation to mass transport of ^{14}C-labeled assimilates from the leaves of willow. *Physiol. Plant.* **23,** 1033–1038.
240. Peel, A. J., and Weatherley, P. E. (1963). Studies in sieve tube exudation through aphid mouthparts. II. The effects of pressure gradients in the wood and metabolic inhibitors. *Ann. Bot. (London)* [N.S.] **27,** 197–211.
241. Penny, M. G., and Bowling, D. J. F. (1975). A study of potassium gradients in the

epidermis of intact leaves of *Commelina communis* L. in relation to stomatal opening. *Planta* **119,** 17–25.
242. Phillips, I. D. J. (1975). Apical dominance. *Annu. Rev. Plant Physiol.* **26,** 341–367.
243. Porter, H. K. (1966). Leaves as collecting and distributing agents of carbon. *Aust. J. Sci.* **29,** 31–40.
244. Qureshi, F. A., and Spanner, D. C. (1973). The effect of nitrogen on the movement of tracers down the stolon of *Saxifraga sarmentosa,* with some observations on the influence of light. *Planta* **110,** 131–144.
245. Qureshi, F. A., and Spanner, D. C. (1973). Cyanide inhibition of phloem transport along the stolons of *Saxifraga sarmentosa* L. *J. Exp. Bot.* **24,** 751–762.
246. Rabideau, G. S., and Burr, G. O. (1945). The use of ^{13}C isotope as a tracer for transport studies in plants. *Am. J. Bot.* **32,** 349–359.
247. Rand, R. M., and Cooke, J. R. (1978). Fluid dynamics of phloem flow: An axisymmetric model. *Trans. ASAE* **21,** 898–906.
248. Roeckl, B. (1949). Nachweis eines Konzentrationshubs zwischen Palisadenzeller und Siebrohren. *Planta* **36,** 530–550.
249. Rohrbaugh, L. M., and Rice, E. L. (1949). Effects of application of sugar on the translocation of sodium 2,4-dichlorophenoxyacetate by bean plants in the dark. *Bot. Gaz. (Chicago)* **110,** 85–89.
250. Ross, S. M., and Tyree, M. T. (1979). Mason and Maskell's diffusion analogue reconsidered with a translocation theory. *Ann. Bot. (London)* [N.S.] **44,** 637–639.
251. Ruhland, W., ed. (1967). "Handbuch der Pflanzen physiologie," Vol. 18. Springer-Verlag, Berlin and New York.
252. Sabnis, D. D., and Hart, J. W. (1974). Studies on the possible occurrence of actomyosin-like proteins in phloem. *Planta* **118,** 271–281.
253. Servaites, J. C., and Geiger, D. R. (1974). Effects of light intensity and oxygen on photosynthesis and translocation in sugar beet. *Plant Physiol.* **54,** 575–578.
254. Servaites, J. C., Schrader, L. E., and Jung, D. M. (1979). Energy-dependent loading of amino acids and sucrose into the phloem of soybean. *Plant Physiol.* **64,** 546–550.
255. Shindy, W., and Weaver, R. J. (1967). Plant regulators alter translocation of photosynthetic products. *Nature (London)* **214,** 1024–1025.
256. Shiroya, M. (1968). Comparison of upward and downward translocation of ^{14}C from a single leaf of sunflower. *Plant Physiol.* **43,** 1605–1610.
257. Sij, J. W., and Swanson, C. A. (1973). Effect of petiole anoxia on phloem transport in squash. *Plant Physiol.* **51,** 368–371.
258. Silvius, J. E., Chatterton, N. J., and Kremer, D. F. (1979). Photosynthate partitioning in soybean leaves at two irradiance levels. Comparative responses of acclimated and unacclimated leaves. *Plant Physiol.* **64,** 872–875.
259. Skene, D. S. (1972). The kinetics of tracheid development in *Tsuga canadensis* Carr. and its relation to tree vigour. *Ann. Bot. (London)* [N.S.] **36,** 179–187.
260. Smith, J. A. C., and Milburn, J. A. (1980). Osmoregulation and the control of phloem sap composition in *Ricinus communis* L. *Planta* **148,** 28–34.
261. Smith, J. A. C., and Milburn, J. A. (1980). Phloem transport, solute flux and the kinetics of sap exudation in *Ricinus communis* L. *Planta* **148,** 35–41.
262. Smith, J. A. C., and Milburn, J. A. (1980). Phloem turgor and the regulation of sucrose loading in *Ricinus communis* L. *Planta* **148,** 42–48.
263. Sovonick, S. A., Geiger, D. R., and Fellows, R. J. (1974). Evidence for active phloem loading in the minor veins of sugar beet. *Plant Physiol.* **54,** 886–891.
264. Spanner, D. C. (1958). The translocation of sugar in sieve tubes. *J. Exp. Bot.* **9,** 332–342.

265. Spanner, D. C. (1962). A note on the velocity and energy requirement of translocation. *Ann. Bot. (London)* [N.S.] **26,** 511–516.
266. Spanner, D. C. (1975). Electro-osmotic flow. *In* "Encyclopedia of Plant Physiology, New Series" (M. H. Zimmermann and J. A. Milburn, eds.), Vol. I, pp. 301–327. Springer-Verlag, Berlin and New York.
267. Spanner, D. C. (1978). Sieve-plate pores, open or occluded? A critical review. *Plant, Cell Environ.* **1,** 7–20.
268. Spanner, D. C. (1979). The electroosmotic theory of phloem transport: A final restatement. *Plant, Cell Environ.* **2,** 107–121.
269. Spanswick, R. M., and Williams, E. J. (1964). Electrical potentials and Na, K, and Cl concentrations in the vacuole and cytoplasm of *Nitella translucens. J. Exp. Bot.* **15,** 193–200.
270. Sristava, L. M. (1970). The secondary phloem of *Austrobaileya scandens. Can. J. Bot.* **48,** 341–359.
271. Srivastava, L. M. (1975). Structure and differentiation of sieve elements in angiosperms and gymnosperms. *In* "Phloem Transport" (S. Aronoff *et al.*, eds.), pp. 33–62. Plenum, New York.
272. Sung, F. J. M., and Krieg, D. R. (1979). Relative sensitivities of photosynthetic assimilation and translocation of ^{14}carbon to water stress. *Plant Physiol.* **64,** 852–856.
273. Sutcliffe, J. F. (1976). Regulation of ion transport in the whole plant. *In* "Perspectives in Experimental Biology" (N. Sunderland, ed.), Vol. 2, pp. 433–444. Pergamon, Oxford.
274. Sutcliffe, J. F., and Baset, Q. A. (1973). Control of hydrolysis of reserve materials in the endosperm of germinating oat grains. *Plant Sci. Lett.* **1,** 15–20.
275. Swanson, C. A. (1959). Translocation of organic solutes. *In* "Plant Physiology: A Treatise" (F. C. Steward, ed.), Vol. 2, pp. 481–551. Academic Press, New York.
276. Swanson, C. A., and Böhning, R. H. (1951). The effect of petiole temperature on the translocation of carbohydrates from bean leaves. *Plant Physiol.* **26,** 557–564.
277. Swanson, C. A., and El-Shishiny, E. D. H. (1958). Translocation of sugars in grape. *Plant Physiol.* **33,** 33–37.
278. Swanson, C. A., Hoddinott, J., and Sij, J. W. (1976). The effects of selected sink leaf parameters on translocation rates. *In* "Transport and Transfer Processes in Plants" (I. F. Wardlaw and J. B. Passioura, eds.), pp. 347–356. Academic Press, New York.
279. Thaine, R. (1962). A translocation hypothesis based on the structure of plant cytoplasm. *J. Exp. Bot.* **13,** 152–160.
280. Thaine, R. (1969). Movement of sugars through plants by cytoplasmic pumping. *Nature (London)* **222,** 873–875.
281. Thomas, B., and Hall, M. A. (1979). The control of wound callose formation in willow phloem. *J. Exp. Bot.* **30,** 449–458.
282. Thompson, R. G. (1980). Are there contractile proteins present in phloem? *Can. J. Bot.* **58,** 821–825.
283. Trip, P., and Gorham, P. R. (1968). Bi-directional translocation of sugars in sieve tubes of squash plants. *Plant Physiol.* **43,** 877–882.
284. Troughton, J. M., Currie, B. G., and Chang, F. H. (1977). Relations between light level, sucrose concentration and translocation of carbon 11 in *Zea mays* leaves. *Plant Physiol.* **59,** 808–820.
285. Troughton, J. M., Moorby, J., and Currie, B. G. (1974). Investigations of carbon transport in plants. 1. *J. Exp. Bot.* **25,** 684–694.
286. Turgeon, R., and Webb, J. A. (1973). Leaf development and phloem transport in *Cucurbita pepo;* transition from import and export. *Planta* **113,** 179–191.

287. Turgeon, R., and Webb, J. A. (1975). Physiological and structural ontogeny of the source leaf. *In* "Phloem Transport" (S. Aronoff *et al.*, eds.), pp. 297–314. Plenum, New York.
288. Turvey, P. M., and Patrick, J. W. (1979). Kinetin-promoted transport of assimilates in stem of *Phaseolus vulgaris*. *Planta* **147,** 151–155.
289. Tyree, M. T. (1970). The symplast concept; a general theory of symplastic transport according to the thermodynamics of irreversible processes. *J. Theor. Biol.* **26,** 181–214.
290. Tyree, M. T., Christy, A. L., and Ferrier, J. M. (1974). A simple iterative steady state solution of Münch pressure flow systems applied to long and short translocation paths. *Plant Physiol.* **54,** 589–560.
291. Tyree, M. T., and Dainty, J. (1975). Theoretical considerations. *In* "Encyclopedia of Plant Physiology, New Series" (M. H. Zimmermann and J. A. Milburn, eds.), Vol. I, pp. 367–392. Springer-Verlag, Berlin and New York.
292. Ullrich, W. (1961). Zur sauerstoffabhängigkeit des transportes in den Siebrohren. *Planta* **57,** 402–429.
293. Ullrich, W. (1961). Über die Bildung von Kallose bei einer Hemmung des Transportes in den Siebröhen durch Cyanid. *Planta* **59,** 387–390.
294. Van Bel, A. J., and Ammerlaan, A. (1981). Light promoted diffusional amino acid efflux from *Commelina* leaf disks. *Planta* **152,** 115–123.
295. Van den Honert, J. M. (1932). On the mechanism of the transport of organic materials in plants. *Proc. Ned. Akad. Wet.* **35,** 1104–1111.
296. Walker, A. J., and Ho, L. C. (1977). Carbon translocation in the tomato: Carbon import and fruit growth. *Ann. Bot. (London)* [N.S.] **41,** 813–823.
297. Walker, A. J., and Ho, L. C. (1977). Carbon translocation in the tomato: Effects of fruit temperature on carbon metabolism and the rate of translocation. *Ann. Bot. (London)* [N.S.] **41,** 825–832.
298. Walker, A. J., Ho, L. C., and Baker, D. A. (1978). Carbon translocation in the tomato: Pathways of carbon metabolism in the fruit. *Ann. Bot. (London)* [N.S.] **42,** 901–909.
299. Wardlaw, I. F. (1968). The control and pattern of movement of carbohydrate in plants. *Bot. Rev.* **34,** 79–105.
300. Wardlaw, I. F. (1974). Phloem transport: Physical, chemical or impossible. *Annu. Rev. Plant Physiol.* **25,** 515–539.
301. Wardlaw, I. F., and Moncur, L. (1976). Source, sink and hormonal control of translocation in wheat. *Planta* **128,** 93–100.
302. Wardlaw, I. F., and Passioura, J. B., eds. (1976). "Transport and Transfer Processes in Plants." Academic Press, New York.
303. Wardlaw, I. F., and Porter, H. K. (1967). The redistribution of stem sugars in wheat during grain development. *Aust. J. Biol. Sci.* **20,** 309–318.
304. Wareing, P. F. (1978). Hormonal regulation of assimilate movement. *Monogr.—Br. Crop Prot. Counc.* **21,** 175–180.
305. Warren-Wilson, J. (1972). Control of crop processes. *In* "Crop Processes in Controlled Environment" (A. R. Rees, K. E. Cockshull, D. W. Hand, and R. G. Hurd, eds.), pp. 7–30. Academic Press, London and New York.
306. Weatherley, P. E. (1962). The mechanism of sieve tube translocation: Observation, experiment and theory. *Adv. Sci.* **57,** 571–577.
307. Weatherley, P. E. (1972). Translocation in sieve tubes. Some thoughts on structure and mechanism. *Physiol. Veg.* **10,** 731–742.
308. Weatherley, P. E. (1975). Some aspects of the Münch hypothesis. *In* "Phloem Transport" (S. Aronoff *et al.*, eds.), pp. 535–555. Plenum, New York.

309. Weatherley, P. E., and Johnson, R. P. C. (1968). The form and function of the sieve tube: A problem in reconciliation. *Int. Rev. Cytol.* **24,** 149–192.
310. Weatherley, P. E., Peel, A. J., and Hill, G. P. (1959). The physiology of the sieve tube. Preliminary investigations using applied mouth parts. *J. Exp. Bot.* **10,** 1–16.
311. Whittle, C. M. (1971). The behaviour of ^{14}C profiles in *Helianthus* seedlings. *Planta* **98,** 136–149.
312. Willenbrink, J. (1957). Über die Hemming des Stofftransporte in der Siebrohren durch lokale Inaktivierung verschiedener Atmongenzyme. *Planta* **48,** 269–342.
313. Willenbrink, J. (1966). Transport ^{14}C markierter assimilate in phloem von *Pelargonium zonale* und *Phaseolus vulgaris. Planta* **71,** 171–183.
314. Willenbrink, J. (1980). Aspects arising from the use of inhibitors in phloem transport studies. *Can. J. Bot.* **58,** 816–820.
315. Williams, E. J., Dale, J. F., Moorby, J., and Scobie, J. (1979). Variations in translocation during the photoperiod: Experiments feeding $^{11}CO_2$ to sunflower. *J. Exp Bot.* **30,** 727–738.
316. Williamson, R. E. (1972). An investigation of the contractile protein hypothesis of phloem translocation. *Planta* **106,** 149–157.
317. Williamson, R. E. (1980). Actin in motile and other processes in plant cells. *Can. J. Bot.* **58,** 766–772.
318. Wright, J. P., and Fisher, D. B. (1980). Direct measurement of sieve tube turgor pressure using severed aphid stylets. *Plant Physiol.* **65,** 1133–1135.
319. Wylie, R. B. (1939). Relations between tissue organisation and vein distribution in dicotyledon leaves. *Am. J. Bot.* **26,** 219–225.
320. Yomo, H., and Varner, J. E. (1971). Hormonal control of a secretory tissue. *Curr. Top. Dev. Biol.* **6,** 111–144.
321. Yomo, H., and Varner, J. E. (1973). Control of the formation of amylases and proteases in the cotyledons of germinating peas. *Plant Physiol.* **51,** 708–713.
322. Ziegler, H. (1956). Unterzuchungen über die Leitung und Sekretion der Assimilate. *Planta* **47,** 447–500.
232. Ziegler, H. (1974). Biochemical aspects of phloem transport. *Symp. Soc. Exp. Biol.* **28,** 43–62.
324. Ziegler, H. (1975). Nature of transported substances. *In* "Encyclopedia of Plant Physiology, New Series" (M. H. Zimmermann and J. A. Milburn, eds.), Vol. I, pp. 59–100. Springer-Verlag, Berlin and New York.
325. Zimmermann, M. H. (1957). Translocation of organic substances in trees. I. The nature of the sugars in the sieve exudate of trees. *Plant Physiol.* **32,** 288–291.
326. Zimmermann, M. H. (1957). On the translocation mechanism in the phloem of white ash. *Plant Physiol.* **32,** 399–404.
327. Zimmermann, M. H. (1960). Transport in the phloem. *Annu. Rev. Plant Physiol.* **11,** 167–190.
328. Zimmermann, M. H. (1969). Translocation velocity and specific mass transfer in sieve tubes of *Fraxinus americana. Planta* **84,** 272–278.
329. Zimmermann, M. H., and Milburn, J. A., eds. (1975). "Encyclopedia of Plant Physiology, New Series," Vol. I. Springer-Verlag, Berlin and New York.

CHAPTER SEVEN

Solutes in Cells: Their Responses during Growth and Development

F. C. STEWARD

I. Introduction

From light and air, water and familiar solutes, nature fabricates and maintains the life of autotrophic cells and plants. Some deceptively simple questions might be posed. What do plants and their cells do with all the essential water and solutes from the environment? Why does so little dry matter maintain such great disparity between endogenous and exogenous fluids? The responses to such questions are, however, not simple and direct. Nevertheless the formation and maintenance of the adult plant body must surely harness the properties of water and its solutes.

Angiosperms begin their sexual life cycles as fertilized eggs in embryo sacs and are at first nourished heterotrophically in the ovule by elaborated nutrients from the parent sporophyte. Eventually, however, after embryos have developed shoot- and root-growing regions, they lead an independent existence when they receive solutes from their environments. The ultimate aim, therefore, should be to interpret the solute composition of the cells as it changes throughout development.

Plant Physiology
A Treatise
Vol. IX: Water and Solutes in Plants

Following a modern trend, the chapter by Jennings dealt with the problems of salts and ions, not only at the cellular level, but also essentially at membranes and surfaces where electrogenic phenomena occur.

Ideas of membrane structure and function have become increasingly sophisticated. They derive from the static pictures of electron microscopy (Chapter 1 of Volume VII), from the electrogenic properties they are required to fulfill (Chapter 4 of this volume) and the molecular structures in terms of bilayers which now incorporate other molecules (proteins, enzymes) that endow them with specific properties (16). Specificity in membrane properties may not, however, be due to position on the surface for it may vary in time. Some membranes have even been conceived to have "gates" which open or close. Among all this postulated detail, properties of membranes are freely invoked to explain active transfers of solutes at the molecular level (2, 6, 8, 9). Be this as it may, the overall exchanges between plants and their environments need to be interpreted at the level of cells, organs, and the whole plant body and, inevitably so, as they grow and develop. In doing this, however, some elements of precision must yield to the dictates of practicality. Some concentrations of ions or molecules, as at specific locations in organelles or cytoplasm, become included in the total content of cells. Nevertheless this approach which accepts cells as integrated metabolizing units and organs as interrelated functional members of the plant body is a necessary feature of an overall synthesis.

In the early 1920s vacuolated, developed cells were still being regarded as osmotic systems that separated their intracellular fluids from the external media by membranes already endowed with semipermeable properties. The main focus was then on the retention of cellular (vacuolar) contents and of an osmotic potential to retain turgor. The evident requirement for solutes, especially inorganic ions, to traverse cellular membranes was attributed to varying degrees of permeability of their membranes. Not, however, until late in the 1920s and early 1930s was the active intervention of metabolic energy, whether mediated indirectly via photosynthesis or more specifically and directly via respiratory metabolism, seen as a means to drive a nonequilibrium system and so enable specific ions or solutes to move against diffusion gradients and to achieve higher internal concentrations. Thus vacuolated plant cells came to be seen as capable of osmotic work. Thereafter, a chapter in Volume II of this treatise, as written in 1959 (19), summarized work on various systems and linked respiration as the ultimate source of energy with the ability of cells, compatible with their stage of development and growth, to accumulate and retain ions at concentrations very much greater than in their ambient media. The cellular mechanisms of absorption and accumulation as they were seen in 1959 were briefly summarized in Volume II, Chapter 4.

The later literature on bioenergetics (Volume VII, Chapter 2), on intermediary metabolism (Volume VII, Chapters 2 and 3, and Volume VIII, Chapter 2), on protein synthesis and metabolism (Volume VIII, Chapter 2, pp. 126 ff., and Chapter 3), and on the fine structure of cells, organelles, and membranes (Volume VII, Chapter 1) has transformed the background against which the cellular mechanisms of the absorption and accumulation of ions and molecules needs to be interpreted. One quotation from 1959, however, is still apt (Volume VII, p. 420) for it stated that ". . . no single, detailed and specific mechanism of ion accumulation in all cells can be regarded as universally and finally established." In view of the nature of the process which is so intimately related to the organization of a living cell as a whole, this is not surprising. Nevertheless the message of Volume II, Chapter 3, 25 years later, is that one should now look to the electrogenic properties of molecules at membranes for the unifying principles that will relate to cells in general and to plant cells in particular. Indeed a comprehensive discussion of the binding and transport of anions in living tissues (8) includes communications that deal specifically with chloride transport mechanisms in plants, especially as they are exemplified by guard cells of stomata (9), with electrogenic effects of the members of *Acetabularia* (6), with *Chara* (2), and *Limonium* (15). These are but isolated examples of a trend that is now widespread.

But the earlier chapter (25) did not rest solely upon the cellular machinery of salt accumulation for (25; see Section III, p. 425 therein) it considered the salt relations of intact plants in soil (25, Section III,A) and aquatic environments, dealt with transfers of solutes from roots to aerial shoots, and dealt (25, Section III,B) with the interrelations of salt accumulation with growth and development of the plant body. An epilog (25, Section III,C) set the stage for the future. It recognized that the problem of how plants absorb salts is first a problem of their cells and how they invoke their metabolic machinery, which has "different characteristic features according as the cells are growing, dividing or elongating." Also as cells "form part of the integrated whole which constitutes the plant body, their activity is coordinated and regulated by mechanisms, still incompletely understood, which preserve balance and integration in the plant body. . . . The fact that this complete pattern is so controlled and integrated is evident, the method by which this is accomplished is, however, totally unknown." Although we now know more about how a given cell absorbs solutes, specially ions, from dilute solutions in the first place, we know little about the stimuli that prompt that cell to part with those solutes so that they may be directed to even more strongly accumulating cells elsewhere in the plant body.

A current response to the preceding quotations is the theme of this chapter. It may, however, seem somewhat premature since it anticipates considerations of growth and development of cells that properly relate to a later volume. But even to refer here to all the relevant features of Volume VII is not possible. Nevertheless the following trends may be noted.

Since membranes are to be seen as in control of the movements of solutes and water to and fro in cells and their organelles, they are vital components in any conceptual cellular framework. They act as barriers between compartments but also furnish structural frameworks on which multienzymes may exist and function. To do this they must house their complex component parts in proximity so that the often reversible steps they facilitate in energy transfers and syntheses permit the wheels of metabolism to turn. Thus membranes are no longer to be seen only as barriers to be surmounted, for they may also incorporate into their structures the catalytic agents to make specific biochemistry work. Robertson (16) epitomizes this aspect under the title "The Lively Membranes."

The physical structures that obtain at liquid surfaces and their bearing on concepts of membranes are still fascinating aspects of physical chemistry which has had a profound influence on cell physiology. This started with work on oriented unimolecular films pioneered by N. K. Adam, using the Adam – Langmuir trough. Adam distributed insoluble substances, like fatty acids, on the surface of clean water and studied precisely their physical chemistry as they, being insoluble, reside at the water surface. One can still recall the excitement with which one heard in the mid-1920s Adam describe (see 1) how the fatty acids orient themselves at a water – air surface to form films which, when dispersed, behave as two-dimensional gases and, when more condensed, as liquids that may "vaporize" or "solidify" and the historic analogy with the gas laws emerged. Although Adam did not apply himself to the cell physiology applications of his remarkable work, this was later developed by his associate Danielli (4). However, the membranes *in situ* in cells are more than the physical surfaces of contact between immiscible fluids because they have a morphology of their own.

Robertson (16), following Davson and Danielli (4), builds not only on the oriented molecular thin films (with their more water-soluble "polar" heads and their more lipophilic "nonpolar" hydrocarbon chains), but on the complex bilayers which more realistically represent membranes *in situ*. Robertson, in an imaginative work (16) for which we are indebted, now describes membranes in different situations in cells and skillfully assigns in successive figures molecular configurations to them in accordance with their biochemistry on the one hand and their physiological functions on the other.

But whether the solutes of cells are approached as attributes of organelles and their metabolism, as constituents of cytoplasmic inclusions, or as governed by the properties of boundary surface membranes, it is ultimately necessary to recognize them as they exist in cells which have grown.

The topics selected are all from work that developed in the author's laboratory over many years. This arbitrary choice has the merit that it covers the entire range of behavior from tissue slices to cultured explants and to free cells as they respond either to dilute solutions of single salts or, by contrast, to fully heterotrophic media in which they may grow and develop. Furthermore, certain exogenous stimuli that induce rapid growth when resolved into their partial systems also interact with inorganic nutrients (trace elements) in ways that affect the uptake of water and of salts as cells grow. But this plan of presentation owes much to the foreknowledge that free somatic cells, as of carrot, may also behave as somatic embryos. Thus the behavior of cells that originate from tissue explants and which can be studied in solutions may be linked to their behavior *in situ,* even to their responses during development.

II. Solutes in Relation to Cells in Culture

A. From Thin Tissue Disks to Aseptically Cultured Explants

1. Thin Disks in Dilute Salt Solutions in Retrospect

The study of salt accumulations in the cells of slices of storage organs was well documented in Volume II, Chapter 4. The conditions which allow slices of potato tuber (which initially had acquired concentrations of potassium in the growing tuber greater than in the soil solutions) to accumulate the bromide ion were determined as early as 1929. The essential conditions were (1) the use of very thin tissue slices (0.66 mm), thin enough so that no central cells remain inactive as in the tuber; (2) the use of batches of washed slices in a large excess of very dilute, usually single, salt solutions; and (3) the maintenance of the cells under such conditions that they were not limited by access to oxygen and with appropriate constant temperature.

Under all these conditions, and after a short lag period, thin potato slices absorbed anions (Br) and cations (K or Rb) at linear rates with time affected by temperature and oxygen tension for relatively long periods. This focused attention upon continuing changes in the tissue, which were quite inconsistent with speedy attainment of equilibria, and these were that (1) there was some gain in fresh weight, starch hydrolysis, accumulation of sugars, and high rates of carbon dioxide evolution related to oxygen partial pressure and to temperature; (2) concomitantly, the previously stored non-protein-N compounds formed alcohol insoluble (protein-N) compounds; and (3) in short, the tissue (previously quiescent) moved toward renewed activity compatible with the ability of potato cells to divide as this may be demonstrated in moist air.

These responses sufficed to sketch out the biochemical background of salt accumulation as it occurs in potato cells (Volume, II, Chapter 4, pp. 335 ff.) and as they responded to various external factors that regulated aspects of their metabolism indicative of that renewed growth referred to above. The most significant criterion of the activity of the cells in this respect was held to be the synthesis of protein which is in turn linked to enhanced aerobic respiration. An early investigation (30) exploited several salient characteristics of potato tubers and their cut disks. The relevant observations, which still stand, were as follows.

Thin disks cut from tubers stored at temperatures of the order of 12°C remain normally viable and accumulate KBr from dilute, aerated solutions, while, concurrently, they exhibit enhanced respiration by cells that show starch hydrolysis, accumulation of sugars, and the ability to synthesize protein from their stored soluble-N reserves. Prolonged storage of tubers at low temperatures (2°C) is associated with progressive starch hydrolysis and accumulation of sugars and the eventual loss of the ability of cells in cut disks to accumulate salts, to synthesize protein, and to divide when in contact with moist air. These observations focused attention, even at that early date (1943), upon the needed ability of the cells to grow and to synthesize protein for their accumulation of K and Br. But to study the active participation of the processes of growth the system as used from 1929 to 1939 was limited in ways that were not overcome until research was resumed in Rochester, N.Y., after a six-year gap.

2. *The Developed Use of Standard Tissue Explants and Completely Aseptic Conditions*

It was considered that any alternative system should be capable of being exploited under aseptic conditions and, as necessary, use complete heterotrophic nutrient media to permit the contrast between quiescent and actively growing cells to be sharply drawn. Also it should be more versatile in the concomitant study of salt uptake, protein metabolism, and growth than were the thin potato disks.

These requirements were happily met by the aseptic culture system which exploited small standard (2–3-mg) secondary phloem explants (free of cambium) from the storage root of carrot. These were used in specially designed tubes (26, see Plate VI for use of the tubes therein; also Fig. 4.8 from Ref. 21) or flasks (21, see Fig. 6.2 and Fig. 3 of Ref. 18) and in media that provided for their maintenance in the quiescent nongrowing state or, when suitably supplemented with exogenous growth factors, for their ability to grow by random proliferation (19, see Figs. 7 and 8 in color therein). Meanwhile the study of nitrogen metabolism in this system was being advanced greatly by the use of the chromatographic procedures then being developed (Volume VIII, Chapter 2, Section IV).

However, as the intake and accumulations of ions, metabolites, and water by intact cells are increasingly seen to be dependent upon variables that determine their growth and metabolic activity, the terms of reference change. Although, hitherto, actual concentrations in internal fluids—extracted or expressed—were useful in assessing internal concentrations, greater use was now made of the total content of water and solutes in tissue and cells as they grow. To this end methods of maceration were employed to determine cell number and average cell size; thus, throughout the work with tissue explants, data could be expressed per cell or per unit of water at will. The important point here is that many phenomena which others might wish to isolate and deal with as at membrane surfaces or in internal cytoplasmic inclusions and organelles are here inevitably included in the overall behavior of cells. This is the price paid for the holistic approach which nevertheless has given access to data and concepts that otherwise could not have been obtained.

It was not realized at the outset that the small aspetic growing explants may give rise in culture to free cells that behave as somatic embryos. Thus the behavior of proliferated explants in culture may not be linked to that of pre-embryonic cells and later work on somatic embryogenesis in carrot may be brought to bear on the interpretation of its salt and water relations. These aspects were re-examined in 1970 by Steward and Mott (22) long after the 1959 review by Steward and Sutcliffe. The 1970 review summarized an emerging trend made possible by the aseptic culture procedures. This trend was developed in five papers in the *Annals. of Botany* (3, 10–13) with R. L. Mott in which the attention was focused upon solute accumulations in the cultured cells with special reference to ions. In another series of five papers with K. V. N. Rao *et al.* the attention was focused (13, 23, 24, 28, 29) especially on metabolites in the cultured explants and the effect of trace elements on their growth and composition. Sections II,B; III,A,B; and IV now present the special points that emerged.

B. Quiescent versus Growing Cells: Exogenous Electrolytes versus Nonelectrolytes

1. Perspectives

The aseptic cultured carrot system ranges from tissue explants that may remain quiescent for long periods and are unable to grow by cell multiplication (though not irreversibly so) to similar (clonal) explants in the state of prolonged active growth. These different levels of activity are determined by the composition of the ambient media.

To keep the tissue healthy but quiescent, a dilute calcium chloride solution is used and then, in the presence of KCl, NaCl, or both, the ability of the cells in this state to absorb these ions may be studied. With the addition of sugar (sucrose) to the medium some increase in cell number occurs, but relatively more in cell size, with consequential effects on the uptake of salts. But a complement of inorganic and organic nutrients (i.e., a basal medium) supplemented by growth factors (conveniently supplied in coconut milk or water) produces a great stimulus to growth which may be recorded in fresh weight, and number of cells; the cells, however, remain smaller as long as rapid cell division persists.

The study of the time course of treatments such as those indicated in the preceding paragraph, coupled with the analysis of the cultured tissue for K^+, Na^+, Cl^- sugars and osmotic value (10), sheds light on the interactions between electrolytes and nonelectrolytes in the maintenance of osmotic value and on the circumstances in which potassium absorption predominates most heavily over sodium and in which stoichiometric balance between cations (K^+ and Na^+) and anions (Cl^-) is or is not encountered. The point of view and short answers to these questions may be set out first as follows (10, see pp. 634 ff.):

The problems of ion accumulation have usually been approached as though they relate to specific ions separately, with specific electrochemical mechanisms for each ion. Nevertheless the point of view that emerges here is different. Essentially this is that solutes (charged or uncharged) are accumulated in the cells but that the ultimate driving force for their uptake and regulation derives from the development of the cells and is subject to the ensuing control of their osmotic value.

A salient feature is that cells in different stages of their development have quite different characteristics with respect to salts and solutes and to water. At the outset when growth by cell multiplication is the main feature, with all that this entails in metabolic terms, the cells have special relations to solutes. These relationships depend on potassium as the key ion, for it may be—and usually is—then accumulated *without* its accompanying anion (Cl) but is balanced internally with organic anions, albeit the degree of ion accumulation is at this stage relatively low.

Later, as replication of organelles and cell division subside the solute relations change (10, see pp. 636 ff.). The predominantly enlarging cells now become less dependent upon potassium, or even salt concentrations, to maintain their osmotic value as their volume and water content increases in the nutrient-rich medium. However, the osmotic values of these more mature and fully vacuolated cells, which do not divide, may now be sustained *either* by inorganic salts (balanced as between anions and cations) and with less need for potassium as a specific requirement *or* by organic solutes built up from exogenous sources of sugar.

In this context the controls do not operate through specific electrochemical properties of ions considered singly. Salts *or* sugars here act as osmotically active components in a system in which the stability of the cells is controlled through the concentration of its internal fluids. In fact, at this stage, *reversible* replacement of sugars by salts or vice versa is possible by controlling the medium. Thus, the emphasis is transferred from the specific properties of ions, or even individual solutes, to the colligative properties of solutions which are mediated by the effects of total solutes upon the properties of water.

2. *Findings*

a. Mature Enlarged Cells in Explants That Do Not Grow. These conditions prevailed in aseptic cultures of standard explants exposed to a calcium chloride solution and without exogenous nutrients. Over 20 days the explants were maintained in a healthy condition though with virtually no increase of cell number or total fresh weight or of average cell size (10, see Fig. 2a, p. 629). However, with added KCl and NaCl in the medium this tissue maintained, after some 4 to 6 days, steady levels of Cl, K, and Na above those in the initial explants (10, see Fig. 2b). In this strictly nongrowing tissue maintained solely by its endogenous nutrients any increments of K^+, Na^+, and Cl^- that occurred in response to their external concentrations to achieve new internal levels were in strictly stoichiometric balance as between ions K^+ plus Na^+ and Cl^- absorbed, with K^+ being in excess of Na^+ (10, see Fig. 2 and its explanatory note).

b. Cells after Growth Induction (10, see Fig. 3). If, however, explants were exposed for a similar period (20 days) to a complete basal medium supplemented with a total complement of growth factors, they grew rapidly (after a lag period of about 4 days) and increased greatly in fresh weight and in the number of their cells. Up to about 12 days cells under these conditions divided faster than they enlarged so that their average cell size declined, but thereafter it increased even above that of the cells in the initial explant. In this rapidly growing tissue system the uptake of potassium greatly exceeded that of sodium, and it was also in stoichiometrical excess of that of Cl^-; this held whether the basis of expression is micromoles per unit fresh weight, per explant, or per million cells. The general implications of these and other data not here recorded are as follows.

Quiescent, mature cells (in explants in $CaCl_2$ solution) can nevertheless absorb salts from dilute solutions and maintain their vitality for sustained periods; these cells rely upon endogenous sugars and organic solutes for much of their osmotic value. This uptake, without growth, in dilute salt solutions without nutrients, or exogenous sugar, emphasized cations balanced by anions ($K^+ + Na^+ = Cl^-$) although K^+ intake exceeded that of Na^+ (10, see Fig. 2).

The uptake of potassium (relatively unaccompanied by sodium and greatly in excess of chloride) that occurred during growth in (B + CM) contrasted sharply with the nongrowing system (10, see Figs. 2 and 3). The presence of sucrose alone in a salt solution greatly reduced the concentration of salts that otherwise would have obtained in the cells in its absence. To the extent that these explants received their first stimulus to grow, they relied endogenously upon sugars and organic solutes for much of their osmotic value, but as cells mature and growth subsides and if sugar is unavailable to the cells, they may acquire K^+ plus Na^+ and Cl^- in more nearly stoichiometric proportions.

Thus, as the cells of explants are stimulated by endogenous and exogenous stimuli to grow and develop, they need to build up internally an osmotic value. Inasmuch as exogenous solutes (organic and inorganic) can participate in this they respond to the overall drive for their uptake and regulation operate concomitantly with the *de novo* development and control of osmotic value in the cells. This approach does not treat each ion separately, nor does it seek to erect specific explanations for the osmotic work performed on each. Mature cells present in cultured carrot explants may be induced to replace, reversibly, preabsorbed sugars and organic solutes with exogenous salts and vice versa and in so doing they preserve their osmotic value. This emphasizes that what the cells actually control is their osmotic value. In doing this, organic solutes on the one hand and inorganic salts and charged ions on the other are replaceable alternatives (10, see Fig. 6 and its explanatory note).

c. The Progressive Uptake of Ions and Nonelectrolytes in Explants as They Grow. The situation may be summarized as follows. The system as used focused upon the most rapidly growing cultures which were dependent for their growth on the exogenous stimuli (as in CM) over and above the nutrients and sucrose in the basal medium. These cultures contained cells which individually passed through their cycle of division and enlargement and collectively traced out a sigmoidal curve of growth for the explant as a whole. Different phases of that time course of growth I, II, and III are associated with changes in the solute uptake and content of the cells and also with concomitant metabolic characteristics, notably in protein synthesis.

Early in the experimental growth of cultured explants the emphasis is on cells that multiply by division (Fig. 1). Organic solutes absorbed from the ambient medium are used to create form and complex polymeric substances, and the cells then have a specific requirement for potassium (selectively preferred to sodium), and in the tissue it is first balanced by organic anions, not by chloride. Moreover, at this stage the degree of "accumulation" of potassium over its content in the ambient medium is relatively low. Later as cells develop and enlarge, these relationships change; the emphasis is then upon the maintenance of osmotic value in the cells within their enlarging vacuoles. The developing vacuoles store organic solutes preferentially at first, but later these solutes may be replaced by neutral salts (KCl, NaCl). This is especially so when organic supplies in the medium are depleted, or previously stored solutes may be withdrawn or used in anabolism. This later accumulation of salts, however, does not show as markedly the disparity between $K > Na$ or $K^+ + Na^+ \gg Cl^-$, which was so evident in the cell multiplication phase of growth. Moreover, the later accumulations of individual ions over the ambient medium are also characteristically greater as cells enlarge and mature. Thus to interpret the relations of salts to cells in culture should involve their growth and development as also concomitant changes in their organic solutes.

Figure 1 shows diagrammatically the typical responses of carrot explants, isolated from the carrot root and exposed to the complete basal medium under the standard conditions adopted. However, the explants only responded to a limited extent unless they also received the supplementary stimuli most conveniently supplied in coconut milk. Also the minimal response on the basal medium alone and to the complete growth induction stimulus varied somewhat from clone to clone.

The time course of growth in the basal medium plus coconut milk (B + CM) has been contrasted with that on the basal medium alone for different clones of explants with different affinities for potassium and sodium using various criteria of growth and solute content. (For this purpose the control basal medium was enriched with potassium to allow for the potassium content of the added coconut milk.) Published figures show the time course of growth in fresh weight in milligrams per explant and salt uptake in concentrations of K^+, Na^+, and Cl^- micromoles per gram fresh weight, or content in terms of micromoles per explant (22, see Fig. 6). One such clone [22, see Fig. 6a(i)] reached only a low content of potassium and sodium when on the basal medium ($K + Na = 98$ μmol g fresh weight^{-1}), although its K was very prominent over Na (K : Na ratio at 18 days 2.3); another clone in parallel cultures reached 222 μmol of K + Na per gram fresh weight and its sodium was relatively more conspicuous (K : Na ratio 1.1). A greater affinity of explants on the basal medium alone for sodium tended to be associated with a longer lag period in their response to the

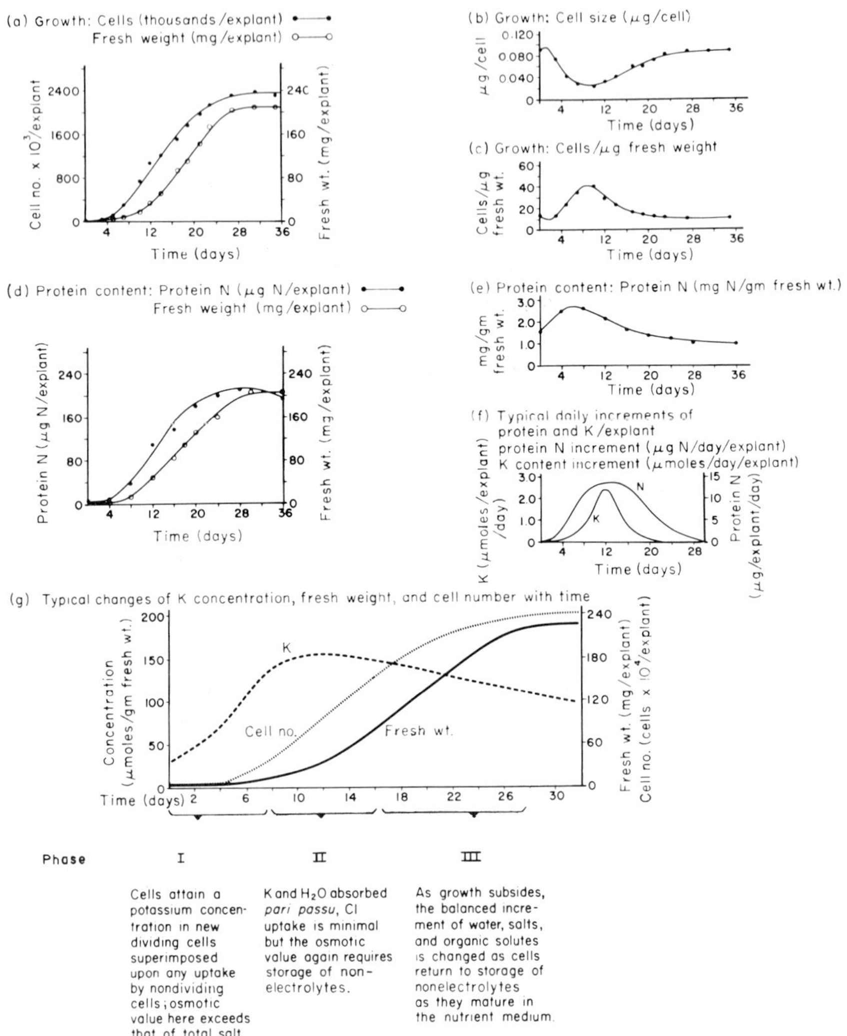

FIG. 1. Time courses in the growth and composition of cultured carrot phloem explants, at 23°C in diffuse light. (a) Growth, cells (thousands) ●——● and fresh weight (milligrams) ○——○ per explant; (b) growth, cell size (micrograms/cell); (c) growth, cells per microgram fresh weight; (d) protein content, protein nitrogen (micrograms N) ●——● and fresh weight (milligrams) ○——○ per explant; (e) protein content, protein nitrogen (milligrams N per gram fresh weight); typical daily increments of protein (microgram N) and K (micromoles) per day per explant; (g) typical changes of K concentration, fresh weight, and cell number with time.

complete growth factors and a slower total growth rate; by contrast a greater affinity for potassium from the basal medium alone was associated with a shorter lag period and a more rapid response to the complete growth factors (see also Refs. 10, 11 for other similar examples).

Thus in some as yet obscure fashion the ability of explants to respond to the complete exogenous growth factors which control cell multiplication and enlargement is related to the avidity with which the tissue acquires potassium; a high potassium uptake from the basal medium can usually be associated with a large response to the growth induction stimuli in B + CM. Conversely, a richer preference for sodium is correlated with a relatively weak growth response to B + CM.

d. Idiosyncracies of Different Clones of Explants. Clones were arbitrarily numbered 1 to 19 in descending order of their "growth" (i.e., increase in fresh weight) on the basal medium alone during the standard conditions of these experiments [Fig. 2a(i)]. On radiating axes similarly numbered [Fig. 2b(i)], the salt content (μmoles per explant) and salt concentrations [Fig. 2c(i)] in μmoles per gram fresh weight are shown on diagrams that also indicate the levels of their total cation ($K^+ + Na^+$), their (K^+) and (Na^+) as on [Figs. 2b(i) and c(i)]. Figures 2a(ii), b(ii), and c(ii) use the same designs to show the responses of similar explants in the medium (B + CM) that promotes maximum growth.

As shown, otherwise standard carrot explants which receive identical culture media and environmental conditions display differences in their responses. When the data from 19 different clones (all within the same stock) are arranged in the descending order of their ability to "grow" in the basal medium alone, the resulting polygonal diagram, Fig. 2a(i) (from Ref. 8, see p. 650), reflects the innate abilities of the explants to absorb water during the time period in question. However, as the cells absorb water and the explants increase in weight, they also take in salts (K and Na) from the medium. If in their responses the cells of the different clones did not discriminate between water and salts, one could expect the form of the diagrams constructed in a similar manner [Fig. 2b(i) and c(i)] to follow the pattern of the diagram [Fig. 2a(i)] that relates to their responses in fresh weight. Moreover, if the great stimulus to growth, together with cell division and enlargement in the medium B + CM should act impartially upon water and salts and merely accentuate the responses evident in the basal medium alone, then one would expect the series of diagrams in Figs. 2a(ii), b(ii), and c(ii) to follow the pattern set by Figs. 2a(i), b(i), and c(i). Briefly, the conclusions are the following.

Carrot explants, even in a full basal medium (B), but lacking the exogenous stimuli (CM) which release their full capacity for growth, absorb

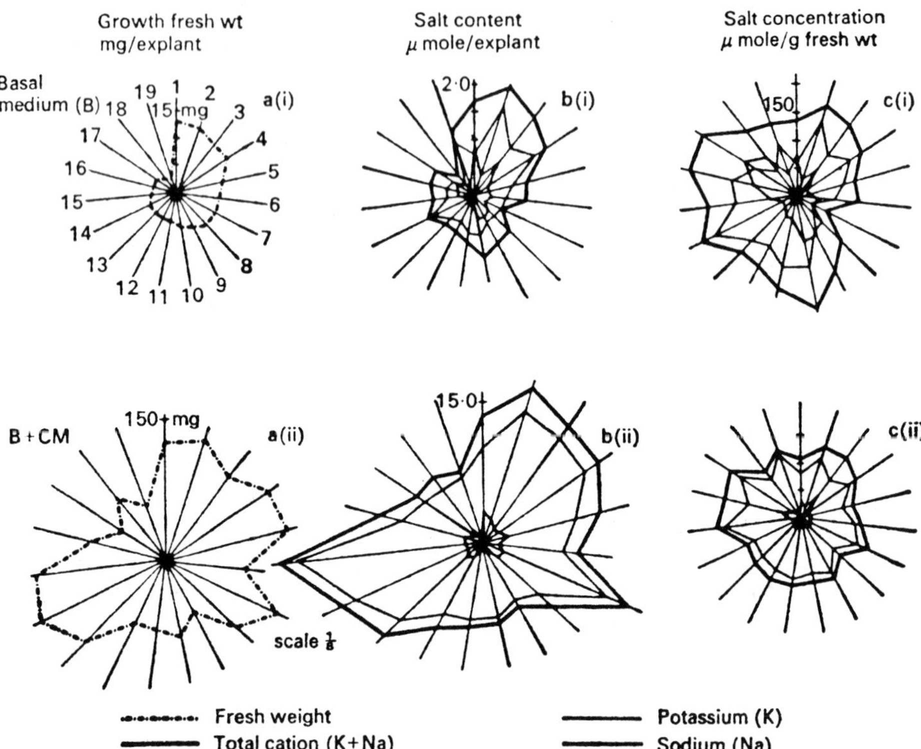

FIG. 2. Salt content and growth of carrot explants: idosyncrasies of 19 clones. Note: The scales in [a(ii)] and [b(ii)] are smaller by (1 : 6) than those in [a(i)] and [b(i)] to accommodate the large growth due to the coconut milk (CM) in the medium. This does *not* apply to [c(ii)] relative to [c(i)] for here the parameters plotted are concentrations. From Mott and Steward (11, see p. 650).

water as limited by their endogenous characteristics [Fig. 2a(i)]. When there limitations are relieved by the provision of the coconut milk stimuli to growth as in Fig. 2(ii), the figure that reflects their relative uptake of water is not merely an enlarged replica of Fig. 2a(i), but it deviates from this in ways which reflect clonal differences in the uptake of water by the tissue as dictated by the exogenous stimuli. The corresponding diagrams for salt (K + Na) and ion (K^+, Na^+) concentrations, i.e., Figs. 2b(i) and c(i), 2b(ii) and c(ii), show the extent to which the tissue discriminates between water and salts [cf. 2a(i) and 2b(i)] and the extent to which, under the influence of the growth stimuli, its relative responses in terms of water, salts, and ions are changed [cf. 2a(i), b(i), and c(i) and 2a(ii), b(ii), and c(ii)].

The diagrams show that the explants which had responded to the growth stimulus had the *lower* salt concentrations [cf. 2c(ii) with 2c(i)] and in these the disparity between the K + Na content was greatest when they grew [cf. 2b(ii) with 2b(i)]. Also, inasmuch as the external medium (B + CM) was effective in overcoming the endogenous limitations of explants to grow, it also tended to eliminate the clonal differences between explants in their discrimination between water, potassium, and sodium for the diagrams of 2c(ii) are more nearly circular than those of 2c(i). The interpretation here is that the more intensely growing, growth-factor-stimulated cells tend to have *all* their endogenous limitations more fully satisfied. Thus they respond more uniformly to water, salts, and ions, and clonal differences are thereby submerged.

These interrelationships indicate again that uptake of water and salts and problems of internal secretion of ions from the external environment and of the balance between nonelectrolytes and electrolytes in the production and maintenance of the osmotica of cells should be considered not in isolation, but as part of a more general synthesis that involves the ability of cells to grow as they both multiply and enlarge. Although the implication is that a rich content of potassium is associated with the greatest ability to grow in response to the specific stimulants of growth induction, the potassium *alone* will not bring this about. Neither do the events at a membrane surface that are involved in the actual entry of K^+ alone determine the final outcome.

The experiments and interpretations described lead to the short conclusion that "actively growing cells with small developing vacuoles first acquire their appropriate levels of potassium unbalanced by equivalent chloride, and their osmotic value is furnished as necessary by organic solutes from sugar. Larger, highly vacuolated non-dividing cells, however, may exchange and build up larger concentrations of salts which then replace organic solutes in the osmoticum. These replacements, however, require the expenditure of metabolic energy to facilitate the uptake of specific ions. Thus, it is the osmotic value of the internal solutions, *not* the concentrations of particular solutes which is first regulated during the progressive uptake of solutes in cells and explants as they grow and mature" (11, p. 651).

One should therefore recognize the limitations of investigations and theoretical formulations that relate solely to the so-called active transport, for this represents but one important step in an absorption process that should properly relate to the behavior of the cell system as a whole. The use of preformed cells or subcellular preparations in as restricted a state as possible and observed over the minimum time span may be conductive to knowledge of, and precision in the description of specific ion pumps, membrane potentials and electrogenic interactions of facilitated ion exchanges, fluxes, permeases, etc. Such observations derive from, and apply to, preformed, delimited systems with prescribed characteristics. Hence they shed virtually no light upon the means by which the system came about *de novo* or its progressive march toward a steady state, for the usually mature cells or preparations investigated have been deliberately kept in as static a state as possible. Moreover, insofar as solutes or ions have been studied singly, there has been little incentive to understand their interactions or their part in the concerted action of all solutes which are subject to overriding controls that operate in cells, as they develop and grow (11, see p. 638). It is when one considers cells in relation to their ontogeny that one should ask what mechanism initiates the absorption process, *de novo,* and controls its kind and rate, and finally brings it to a stop.

III. Procedures and Media That Affect Growth and Solutes

A. Salt Relations of Carrot Explants as They Grow in Different Media

The discussion in Section II,B. linked the time course of growth of cultured carrot explants to characteristics in the absorption of salts by their cells. It also pointed to relations between exogenous neutral salts and nonelectrolytes (sucrose), showing that they may make mutually replaceable contributions to the internal osmotic value of the cell contents. With this in mind other experiments using methods already described (10) have explored variables that affect the growth and the solutes absorbed and accumulated.

1. Minimal Growth on $CaCl_2$

The cells of carrot explants cultured only in dilute calcium chloride solution, but otherwise under the conditions conducive to their growth, remain viable for long periods (in excess of 20 days) under aseptic conditions (10) and retain their low concentrations of solutes (osmotic value of about 300 mOsm) against an almost infinite outwardly directed concentration gradient. Supplied with other exogenous salts (KCl, NaCl) in addition to the dilute calcium chloride such explants absorbed salts (K + Na = Cl) to high levels by 6 days and retained them for long periods. During this period the constituent cells showed virtually no cell division and only very limited enlargement and their osmotic value (now almost completely due to inorganic salts) could attain levels (500 mOsm) which were comparable to carrot root cells *in situ* or to the maximum osmotic value for a given clone of carrot explants in culture.

2. In a Basal Medium Only

Depending upon the characteristics of a given clone the cells of carrot explants may grow to some extent in a complete basal nutrient medium (B), even though it is not supplemented by specific growth factors, over a protracted period (20 days); this growth involved some cell division but mainly cell enlargement (12, see Fig. 2, p. 659). The content of solutes in such enlarging cells tended to increase linearly with time, along with the intake of water as the cells expanded. The maximum osmotic value so built up was largely due to salts, i.e., K and Na accompanied by Cl^- (in a given clone 400–500 mOsm, of which 54% was due to salts).

3. Growth in a Basal Medium (B) plus Growth Factors (B + CM)

If the basal medium (B) is supplemented by growth factors (B + CM), rapid growth ensues and the explants then trace out a sigmoidal time course of growth (Fig. 1) which may be treated in phases each of which has special characteristics in terms of the balance between cell division and enlargement on the one hand and the uptake of water and solutes on the other (11). Cells predominantly in the dividing state (as at 6 to 10 days of culture) had relatively low osmotic value (~200 mOsm) composed almost completely of potassium salts of *organic* anions; this is a distinctive feature of the dividing cells. The later enlargement (12 days et seq.) of the preformed cells, with their cytoplasm now being spread in thin parietal layers, was distinguished by the development of high solute concentrations and osmotic values. Such a time course in a medium that built up high osmotic values produced explants that could have osmotic values of the order of 500 mOsm (of which in a given case 29% was attributable to salts and 11% of that was due to K^+). Calculating in terms of total amounts of solutes per average cell, the cells of a given clone reached 42×10^{-6} μmol (42 pmol) of sugar, nitrogen compounds, etc., and approximately 18×10^{-6} μmol of K + Na as chloride or organic acid anions. These concentrations were acquired during the time course of the growth of the cells and should not be attributed solely to a brief physicochemical event expressible in terms which ignore the reality of that growth. However, other work has also shown that when the growth of cells is completed, the still viable, mature cells may alternate reversibly between internal solutions rich in salts or in organic solutes according to the state of supply in their ambient media.

This situation recalls the historic one in barley roots (see Steward and

Sutcliffe, Volume II, pp. 367 ff.) which when given only limited access to nutrients exported this supply to their growing shoots and received carbohydrate in return. This exchange of mineral nutrients from the roots for sugars from the shoots produced the "low salt–high sugar" barley roots which, when excised and in aerated dilute neutral salt (KBr) solutions, could rapidly replace their stored carbohydrate and organic acids with salts from the ambient medium (7).

4. *Conditions which Affect the Final Salt Content of Cultured Cells*

After the growth of the cells of carrot explants subsides, they have a final cell size, solute content per cell, osmotic value, and solute composition which is the result of the very different time course and routes by which their final steady state was achieved. Cells in explants which were denied access to nutrients, absorbable salts, and growth stimuli (as in a $CaCl_2$ medium only) reached a minimum solute content of 9×10^{-6} μmol cell^{-1}. At the other extreme, cells of explants which had passed through a protracted phase of rapid division and eventual cell enlargement had solute contents of the order of 60×10^{-6} μmol, while the smaller ones lacking such access to nutrients and stimuli had contents of the order of 30×10^{-6} μmol.

Thus the osmotic concentration, the total solute content, and the composition of the matured vacuolated carrot cells resulted from the conditions which promoted or restricted their growth by cell multiplication and cell enlargement and also from the concurrent supply of absorbable nutrients and solutes which were available.

5. *The Arrest and Restoration of Growth and Absorption*

Carrot explants could be placed under conditions that predictably regulated their growth with observable effects on the uptake of salts from the ambient medium (12, see the informative summary, pp. 668–670). For these purposes the time course of growth of carrot was divided into three phases; I, 0–6 days, the phase of growth induction; II, 6–10 days, when explants normally multiply their cells faster than they enlarge and average cell size may be at a minimum; and III, 10–20 days, when explants and cells pursue their time course along the typical sigmoidal curve of growth (Fig. 1). The experimental design exposed explants to alternatives of different degrees of growth-promoting stimuli and nutrition. These varied between the extremes of a full basal medium with a complete growth factor supplement (B + CM) that stimulated great growth in 20 days and, at the other extreme, a salt solution containing in a calcium chloride medium KCl and NaCl to the total concentration of the basal medium as supplemented by the coconut milk. The status of the explants (fresh weight, cell number, and average cell size) and their solute status (total micromoles per explant, per gram fresh weight, or per million cells and their K, Na, and Cl contents) were determined at the end of each phase. Sequential experimental treatments were designed to arrest and then restore growth and absorption along its time course.

6. *Resume*

Briefly, when cells are interrupted along their time course of growth, they maintain the osmotic values of their contents with the solutes then available (salts and/or sugars and their nonelectrolyte derivatives). But when and if growth is restored by the degree of stimuli and salts resupplied, both the osmotic value and the solute composition readjust to the status that is appropriate for the growth by division and enlargement that ensues. The absorptive behavior of previously arrested cells is, therefore, predictably interpreted in terms of their renewed growth and solute supply.

Throughout Sections II and III,A. the theme is that the ultimate regulation in the cultured carrot explants is one of control of total internal solute concentrations and not primarily of individual solutes. As they divide, grow, and enlarge, the cells provide the internal solute space; then, granted a sufficient degree of metabolic activity, they can "take in their stride" the internal secretion of given solutes, even to high accumulation ratios from dilute solutions. The absorbed solutes in effect represent the internally reduced activity of water that cells create as they grow and to be physically stable retain as long as they are viable. Since the economy of the cells operates in terms of total solutes, effects may be (and have been) dramatized by studying single solutes (especially ions) supplied at low concentration. Granted an adequate total solute supply, the efficiency of cells as absorbing systems has been expressed in terms of their absorption or accumulation ratios for that one solute—ratios that may obviously approach infinity under circumstances that may be contrived.

B. Effects Due to Nitrogenous Supplements

Sections II,A,B and III,A have focused attention on inorganic salts (KCl and/or NaCl) as absorbable solutes that may accumulate as ions in the cellular fluids and to which reference may be made when active transfers are in question. Since reference needs to be made to total solutes in cells, the respective roles of sucrose and of sources of nitrogen (other than the nitrate of the basal medium) to the cells arise.

Hitherto exogenous nitrogen has either been absent (as in $CaCl_2$ solutions), present only in minimal nutritional amounts as nitrate (as in the basal medium), or present in low amounts even in the supplemented medium (B + CM). However, it is also well known that growth and the medium B + CM is accentuated by added casein hydrolysate, depending on the source of the clone of explants (5).

1. *Effects of* NO_3, NH_4, *and* NH_4NO_3: *The Plan*

Since nitrate is not normally accumulated as such in the initial carrot explants but often appears in reduced forms (as glutamine or arginine or even as protein synthesized), the use of additional nitrogen from nitrate or ammonium makes demands also upon organic acids derived from sugar.

Explants were cultured in the normal basal medium (B), which contains both K and Na, with a full complement of growth factors (CM). This medium was then supplemented by salts (KNO_3, NH_4Cl, NH_4NO_3), each salt being used at three concentration levels (6, 12, and 18 m*M*). All these cultures were sampled for analysis at 18 days. The extensive data as published (3) may be summarized as follows (see also Fig. 3).

2. *Summary of Results*

1. The main contrasts were between *all* the culture media that contained NO_3 and all those that contained NH_4 (Fig. 3).

2. All the nitrate-containing media produced explants high in K^+, far in excess of their Cl^- or NO_3^-, but with organic acid anions of the order required to balance the excess of K + Na over inorganic anions (3, see Table II).

3. All the ammonium-containing media, whether accompanied by Cl^- or NO_3^-, produced cultures with much lower K^+ levels than those solely on NO_3 media, and with correspondingly lower levels of organic acid ions. In the ammonium-containing cultures the normal disparity between K^+ and Na^+ in the final tissue was also much reduced (3, Table I).

4. Osmotic values milliosmolar or micromoles per gram fresh weight were markedly greater in the NO_3 than in the NH_4 cultures (3, see Table II); this was attributable in part to their greater content of sugars, both reducing and nonreducing. But the apparent disparity between the osmotic values (in the NO_3 and the NH_4 series) was narrowed because the cultures in the NH_4 series stored more soluble, non-protein-N (amino acids and amides) than those in the NO_3 series, and this occurred even though the NO_3 media were also more conductive to protein synthesis.

5. A curious but unexpected feature was that the distinctive effects of NO_3 on the composition of the extractable solutes were so relatively little affected by the threefold external concentration range of the salts in the ambient media. Only when NH_4 was accompanied by nitrate was there a very evident and apparently competitive reduction in potassium content at increasing concentrations of NH_4NO_3. In other words, after the nature of

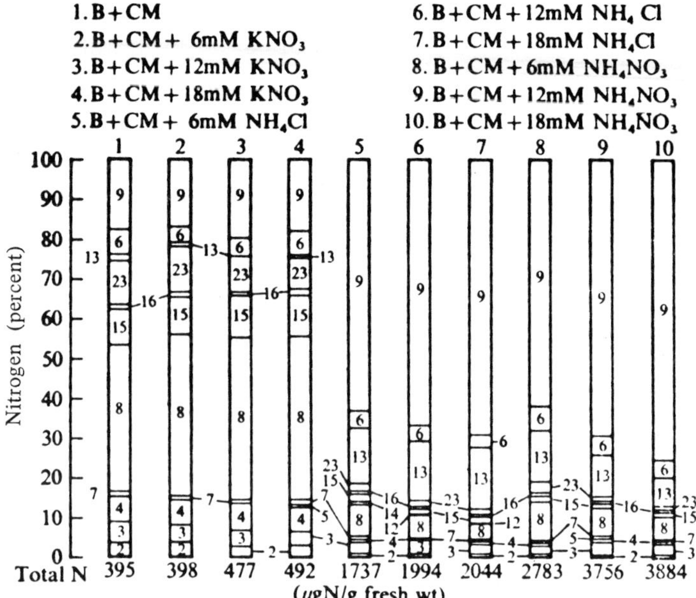

Key to amino acids: 2. Aspartic; 3. Glutamic; 4. Serine; 5. Glycine; 6. Asparagine; 7. Threonine; 8. Alanine; 9. Glutamine; 12. Lysine; 13. Arginine; 14. Methionine; 15. Proline; 16. Valine; 23. γ-Aminobutyric

FIG. 3. Effects of supplementary salts and their concentrations on levels of soluble nitrogen compounds, in carrot explants (clone 940A) cultured in basal medium (B) plus coconut milk (CM). From Craven *et al.* (3).

the external solution had determined what the relative composition of the internal solution would be with respect to its major components, the remaining effects if any were exerted directly on the total growth of the explants, more than selectively on the concentrations of the cations (K) accumulated.

With increasing concentration of NH_4NO_3 in the ambient medium the resultant explants accounted for much more total-N absorbed (per gram fresh weight) than in the corresponding NH_4Cl series (compare 8, 9, 10 with 5, 6, 7 in Fig. 3), but the increase was mainly evident in the additional storage of nitrogen in the soluble nonprotein form. Parenthetically, it should be noted that these increases in soluble-N had as their counterparts *decreases* in salts, especially of K, and organic acids. However, the figures of total soluble-N do not tell the whole story because the complement of soluble-N compounds was so very different in the cultures grown on nitrate or ammonium sources.

Figure 3 presents the relative composition of the soluble-N moiety (i.e., N of each compound expressed as a percentage of the total non-protein-N recovered) as affected by the different salts and their concentrations in the external medium. Figure 3 shows that all the ammonium-nourished cultures (numbers 6–10) contained predominantly glutamine with some arginine, very little alanine and asparagine. All the nitrate-nourished cultures contained predominantly alanine with much less glutamine, very little arginine but with more of γ-aminobutyric acid.

3. Conclusions: Responses of Carrot Explants to Supplementary Sources of Nitrogen

The general conclusions to be drawn are most conveniently restated from Craven *et al.* (3) as follows. It is clear that satisfying interpretations (of the solute content of cultured carrot cells) should not be sought in terms of single solutes or ions since each such response is but one in a network of changes which affect a wide range of solutes, organic and inorganic, electrolytes and nonelectrolytes in the cells as they grow.

Such plants, heterotrophically nourished (i.e., on B + CM plus nitrate or ammonium) store organic solutes as their vacuoles form and cell enlargement ensues. However, the complement of osmotically active solutes changes with time and the maturity of the cells (11) and with the nature and availability of salts in the medium (12) and with the source of nitrogen supplied. In general the osmotic value of the cell contents in the full culture medium (B + CM) increased with the average size of the cells. Sugars (supplied in the basal medium) contribute to the solute content of explants increasingly with time. The absorbed nitrate is converted to protein and to soluble organic forms in the carrot cultures for nitrate is not appreciably stored as such, even at increased exogenous levels of the potassium or ammonium salt. The source of nitrogen (NO_3 or NH_4) and whether it is accompanied by nonmetabolizable cation (K) or anion (Cl) greatly influences the growth and consequently the osmotic value of the solutes in the cells.

When growing cells receive their nitrogen as ammonium, they absorb it over and above their ability to form protein; the resultant high concentration of soluble-N compounds partially replaces sugars as osmotica as sugars and organic acids are activated to form organic nitrogen-rich compounds (i.e., amide and arginine-N). Thus, glutamine incorporates the metabolizable cation ammonium and in effect replaces stronger organic acids (e.g., malate) and their dissociated nonmetabolizable cation (K). Ammonium, however, affects the concentrations of all the solutes in the system, a result reminiscent of an earlier one obtained on potato slices (which was described in Volume II, Chapter 4, pp. 332–333). In that early experiment the penetration of ammonium from ammonium chloride affected the concentrations of all other ions in the cells.

It is as if the growth factors and the endogenous properties of the explants determined the new spaces created for solutes as the explants grew, but thereafter the metabolism of the cells as influenced by ammonium or nitrate dictated how they should be filled.

The data emphasize that the accumulation of solutes in cells as they grow is not primarily determined by an independent flux of inorganic ions. When cells are well supplied with carbohydrate as they grow (endogenously by photosynthesis, exogenously as in culture, or by translocation *in situ* in the plant body) and have free access to reduced nitrogen compounds, their need for large "luxury" accumulation of nonmetabolizable alkali cations (K) and for nonmetabolizable anions (Cl^- or Br^-) as alkali halides, in near anion–cation balance, may not arise. However, such salts may be and are accumulated if and when the cells are placed under such stress that their endogenous organic solutes are drawn upon by translocation, or they are directly used to maintain the cells when external nutrient supplies are restricted.

The intervention of ammonium into the pattern of solute accumulation in cells as they grow shows the great importance to be attached to the organic molecules in the solute complement of the cells and the extent to which the accumulation of alkali cations (K) or even their halides, which are often prominent in mature cells may be both dispensible and replaceable. As ammonium alters the relative composition of the solutes in cells, it affects the content of virtually all the solutes, organic and inorganic, which the cells normally store.

Finally, all factors which control the whole complement of solutes in cells as they respond to nutritive, developmental, and environmental considerations will affect the milieux in which the salt uptake and ion accumulation also occur. Interactions between growth factors and trace elements now to be discussed have prominent consequences to this end.

IV. Other Salient Factors That Affect the Complement of Solutes in Cells

A. OUTLOOK

Although angiosperms are notable as autotrophic green plants, their zygotes and developing embryos are nevertheless usually dependent upon the parent sporophyte for a highly developed heterotrophic mode of nutrition. In fact, their later autotrophy is more a result of their organization and "division of labor" as between organs than an unequivocal characteristic of their cells. Therefore, a salient problem of development is the responsiveness of cells *in situ* to signals they may receive and that may produce differences in composition between mature cells which are all derivatives of the same zygote. This is apparent as between the living cells in different tissues and organs of the plant body. It is, therefore, pertinent to ask how far the now known exogenous stimuli that regulate the composition of cells as they grow in isolation could explain the differences in composition of cells *in situ*. Posing the problem in this way is an alternative to the technically very different task of following the development of zygotic derivatives throughout ontogeny. Furthermore it is now more reasonable because the cells of cultured carrot have shown through their somatic embryogenesis their ability to develop as embryos and to form plants.

B. Interactions between Growth Factors and Trace Elements

The stimulus of coconut milk restores resting, quiescent cells of carrot to the degree of activity comparable to the "built-in capacity to grow" in zygotes nourished by the contents of embryo sac and ovule. This stimulus has been extensively fractionated and the effect of its components, singly and in combinations, upon carrot cells investigated. This has been done with respect to the effects imposed upon growth of cells, on metabolism, and upon the composition of the pool of solutes, organic and inorganic, in the activated cells.

While the element potassium plays a key role in both the growth and composition of the growing cultures, attention also needs to be paid to certain trace elements of which the most important are iron, without which the growth factors of coconut milk do not act, and manganese and molybdenum that further modulate the system after it is activated by iron (14). Thus the growth induced by the factors found naturally in such fluids as those from coconut and the modulating effect of the trace elements on the carrot explants provide a system in which growth metabolism and composition of the cells responds to an array of interacting factors. To investigate this requires that experiments involve standardized conditions and many parallel treatments for which the carrot culture system is peculiarly well adapted.

1. *The Design of Comprehensive Experiments*

The cells of standard carrot explants as grown routinely may vary in their soluble components over a very wide range. This applies especially to the total organic solutes. To demonstrate this the combination of growth-promoting stimuli, similar to those that have produced embryos from cells, have been studied in relation to the trace elements to which they are especially responsive. To demonstrate this, combinations of an essentially trace-element-free basal medium, which when not supplemented limited the growth of explants to their endogenous contents, were made with trace elements singly and in combination with the whole coconut milk system and its component parts, singly and in combination. To distinguish the effects of the growth-promoting complex (CM) from any possible content of trace elements, it was prepared in a trace-element-free form and its growth induction stimulus tested in its interaction with trace elements, especially Fe, Mn, and Mo.

Furthermore the component parts of CM, designated systems I and II (which are modulated by inositol and IAA, respectively), could also be related to trace elements. In doing this the objective was to describe the response of growing explants to growth factors and trace elements in terms of cell growth and also in terms of the composition of the cells as grown.

Organic solutes of the carrot cells known to be very responsive to the growth induction stimuli were the alcohol-soluble nonprotein nitrogen compounds, and available methods sufficed when appropriately standardized for their extraction, separation, and determination. Thus changes in the soluble nitrogen compounds could be related to the changes in protein–nitrogen compounds of the tissue as it responded to the applied culture media. Subsequently, the analysis was also extended to the principal inorganic ions K, Na, and Cl and the total osmotic pressure of all solutes.

2. *Effects on Total Solutes and Salts Accumulated*

Only representative data and figures from a more complete study (29) are presented here. The experiment in question involved two parallel clones, grown for 18 days, under four trace-element regimes (i–iv), each combined with nine growth factor regimes (numbers 0–8). To assess the effect of each treatment upon the growth and solutes of the tissue as grown, enough replicate cultures were necessary to provide for the measurements that described growth (fresh weight, cell number in thousands, average cell size in micrograms per cell) and solute content (K, Na, and Cl and total solutes) of the cultured tissue harvested. For the 36 combinations

TABLE I

EFFECTS OF TRACE ELEMENTS, GROWTH-PROMOTING SYSTEMS, AND THEIR COMPONENT PARTS ON THE GROWTH AND THE SOLUTE CONTENT OF CARROT EXPLANTS (CLONE 886-A) CULTURED 18 DAYS AT 21 °C[a]

Series	Medium	Trace element regime[b]	Growth factor regime[c]	Growth fresh wt. (mg explant^{-1})	Cell number (thousands)	Cell size (μg cell^{-1})	Solute concentrations (μmol g^{-1} fresh wt.) K	Na	Cl	Total solutes[d]
a	B modified according to trace elements only	(i)	0	18	105	0.173	75	24	44	334
		(ii)	0	9	52	0.170	103	69	63	793
		(iii)	0	8	63	0.114	115	72	68	703
		(iv)	0	8	43	0.143	106	88	53	735
b	B as above plus all growth factors	(i)	8	176	1333	0.132	96	7	33	358
		(ii)	8	152	1347	0.115	85	25	12	334
		(iii)	8	184	1534	0.117	96	31	24	367
		(iv)	8	157	1507	0.104	81	91	33	342
c	B with different trace elements + system I	(i)	3	24	487	0.050	44	19	10	208
		(ii)	3	19	384	0.037	39	43	11	354
		(iii)	3	20	400	0.045	44	39	11	343
		(iv)	3	18	457	0.039	42	43	19	216
d	B with different trace elements + system II (1AA + Z)	(i)	6	25	453	0.058	60	10	9	227
		(ii)	6	21	494	0.041	57	12	10	316
		(iii)	6	18	512	0.029	73	11	7	325
		(iv)	6	19	457	0.032	68	9	10	316

e	B with different trace elements + Z	(i)	5	30	239	0.126	61	103	16	385
		(ii)	5	21	158	0.116	65	148	34	622
		(iii)	5	21	143	0.137	118	135	31	575
		(iv)	5	23	147	0.117	124	158	49	606
f	B with different trace elements + systems I and II	(i)	7	35	817	0.043	86	12	7	284
		(ii)	7	21	268	0.077	52	60	23	699
		(iii)	7	14	147	0.083	59	37	19	475
		(iv)	7	12	382	0.034	88	32	9	353

[a] From Table 1, Steward *et al.* (29).

[b] *Key to trace element regimes:*

(i) B, containing Fe and all five trace elements.

(ii) Basal medium (B**)[b] designates a trace-element-free basal medium to which Fe and other specified trace elements may be added; see (iii) and (iv).

(iii) B** with Fe, without Mo, with Mn and other trace elements.

(iv) B** with Fe, without Mo, without Mn, and with other trace elements.

[c] *Key to growth-factor regimes:*

0. Basal medium (B)

1. B + *myo*-inositol (25 mg/l)
2. B + AF^{c}_{aesc} (2 mg/l)
3. B + inositol + AF_{aesc} (System I)
4. B + IAA (0.5 mg/l)
5. B + zeatin (0.1 mg/l)
6. B + IAA + zeatin (System II)
7. Systems I + II
8. B + 10% untreated CM, or B** + trace-element-free CM (CM**) for trace element regimes ii, iii, and iv.

AF_{aesc} designates an ethyl acetate-soluble active fraction from immature *Aesculus* fruits which interacts with *myo*-inositol.

[d] Total solutes estimated from freezing-point depression of hot water extracts of tissue; milliosmoles/kg water = μmoles/g fresh weight tissue.

of treatments, this required for each clone a large number of replicated cultures. In turn these demanded the technical help of assistants experienced in the various procedures of culture, sampling, and analysis. The two clones yielded compatible sets of data, so that the results obtained on one clone suffice here. Table I presents data for clone 866-A for six of the nine growth factor regimes arranged in series (a – f). This table emphasizes the great range of response open to carrot cells in culture as shown by their growth and solute contents. The regimes of Table I series a limited the growth of which the explants were capable to that fostered only by their endogenous organic content and to the basal nutrient medium which did not contain any of the special growth-promoting agents as in coconut milk, but as in a(i), it did contain 5 trace elements. Under these conditions the cells, somewhat modified by the trace element regimes (i – iv) enlarged to 0.170 μg cell^{-1} and acquired their highest total solute concentrations [793 μmol g fresh weight^{-1}, as in Table I a(ii)] accompanied by high K (103 μmol g fresh weight^{-1}). In media with a complete growth factor regime (number 8) and a range of added trace elements the tissue in culture produced the greatest growth in fresh weight (184 mg explant^{-1}) and number of cells (1534 $\times$ 10^3 explant^{-1}) with a substantial K^+ concentration (96 μmol g fresh weight^{-1}), but a much lower Na^+ (31 μmol g fresh weight^{-1}). Along with the great increase in cell number the cells under regime 8 were smaller and their total solutes (334 μmol g fresh weight^{-1}) and their K^+ and Na^+ concentrations were lower than when the cells enlarged without much cell multiplication (as under treatments 0).

When the explants received in lieu of coconut milk one or other other of two potential growth factor systems (I, mediated by inositol; II, mediated by IAA), they responded differentially in terms of cell number and size and in terms of solute contents according as they received one or the other growth factor system (cf. Table I series c and d). Series c invoked the partial growth factor system I in lieu of CM, i.e., growth factor regime 3, which is mediated by inositol and an active fraction from *Aesculus* (AFaesc). Series d invoked the comparable partial system II as in growth factor regime 6, which is mediated by IAA and zeatin as the active fraction (AF_{11}). Series e with growth factor regime 5 invoked the zeatin moiety of system II *without* its IAA cofactor, and series f with growth factor regime 7 utilized in lieu of coconut milk the components of I and II in combination. Series e revealed a striking effect (repeatedly encountered) in which the use of zeatin, unbalanced by IAA, caused the cells to accumulate more Na^+ than K^+, and series f showed also that I and II *together* do not equal CM (suggesting some still unidentified component of CM—probably a coupling factor between I and II).

In other words, one has now to visualize a network of factors which act over and above the basal nutrient medium and which singly or in combination and along with trace elements activate the cells to grow by multiplication and enlargement and, in so doing their water absorption and solute contents all respond to changes in the balances of the stimuli that the cells receive.

To develop the concepts implicit in Table I use has been made of symmetrically planned experiments designed to reveal the range of the responses to the component parts of each growth factor system (I or II) as they act separately or in combination. To this end growth (fresh weight in milligrams per explant, cell number, and average cell size) and solutes (total solutes, K and Na in micromoles per million cells) were all determined. To make it possible to scan all these interrelationships, use was made of circular diagrams that permit responses to many different variables to be graphically presented in one set of diagrams. Each diagram has sectors, each of which relates to a growth factor combination and within each sector radiating axes (i–iv) relate to different trace element regimes; along these axes plots of the solute content and concentrations or growth criteria may be made (29, see Fig. 3a–d). The value of these formulations is that they show how the growth and total solute contents of the tissue as grown in culture in the basal medium respond to each of four trace element regimes and also how the cultured tissue as so grown also responded to each growth factor system (I or II) acting intact or with its two component parts independently. Without reproducing the actual data and figures here, it can be stated that they portray a network of effects brought about in the growing cultures by the different stimuli applied; these stimuli are over and above the nutrients of the basal medium, and they affect the growth and solutes of the cultures concomitantly.

3. Effects of Growth-Promoting Substances and Trace Elements on the Soluble Nitrogen Compounds of Carrot Explants

The analysis of the soluble-N content of cultured carrot explants complements the studies made on total solutes and salts. A useful device to enable a large amount of data to be scanned is to construct histograms to represent the nitrogen content of the cultured tissue after its response to different culture regimes. The histograms show the nitrogen of each compound as a percentage of the total soluble (nonprotein) nitrogen analyzed. Other data that characterize the cultured tissue as grown are shown at the base of the histograms, namely, the total soluble and protein-N in micrograms of nitrogen per gram fresh weight, the number of cells in millions per gram fresh weight, as a measure of average cell size. In this way the composition of carrot explants can be represented with respect to many soluble-N compounds, using a numbered code for each compound and a consistent arrangement on the histograms. The responses of the cultured tissue to combinations of growth factor and trace element regimes may be represented. One such experiment, symmetrically designed, involved many treatments and cultures (24, see Fig. 5). The resultant histograms were arranged in 12 blocks of four which when scanned cover a very wide range of treatments and responses of soluble-N compounds together with data that also record the concomitant effects on the growth of the cells.

Some clear conclusions emerged. Whenever the cultured tissue responds to the complete growth-promoting complex (CM), it overrides responses otherwise due to its partial components and the soluble-N in the tissue as grown is then dominated by alanine. When the growth stimulus is mediated by the partial system I or II, the soluble N is dominated by glutamine, but, even so, the soluble N of the cultured tissue is also responsive throughout to the modulating effects of trace element regimes that involve Fe and Mo.

It is, therefore, impressive to scan the various blocks of histograms and to sense that as carrot cells grow in response to an array of exogenous growth factors and trace elements, which act upon the innate genetic and metabolic resources of the cells. They also regulate their total solutes, their soluble-N compounds, as well as their inorganic salts and ions.

4. Perspectives on the Growth and Composition of Cultured Carrot Cells: An Epitome

Carrot cells in aseptic media with and without the stimulants to growth can in 18 to 20 days display a very wide range in the number, average size, and composition of cells with respect to ions (K^+, Na^+, Cl^-), total solutes, and their nonprotein nitrogen compounds.

Cells in explants on a basal medium without any supplements that stimulate renewed cell multiplication enlarge (regime 0 of series a, Table I). When modulated by the trace element regimes i–iv, these cells acquired average sizes that ranged between 0.114 and 0.173 μg cell^{-1}. Cells in cultures subjected to the partial growth-promoting system I, as modulated by the Fe/Mo trace element regimes (i–iv) remained small and ranged from 0.037 to 0.050 μg cell^{-1} (Table I, series c).

Under the complete system (CM as in regime 8 of Table I) cells both multiplied and enlarged and, subject to the trace element regimes (i–iv), achieved very large numbers of cells (1333–1507 $\times 10^3$) per explant that ranged in their average size from 0.132 to 0.117 μg per cell according to the conditions (Table I, series b).

The partial growth-promoting systems I and II (regimes 3 and 6 of Table I, series c and d) produced fewer cells than the complete system

(CM), and since these were modulated by the trace element regimes, their average sizes ranged from 0.029 to 0.058.

Within these ranges of cell multiplication and average cell size due to the growth factor and trace element regimes, the cells varied greatly in their total solute content (micromoles per gram) and in their content of K^+, Na^+, and Cl^-). For total solutes the range encompassed was from 793 to 208 μmol g^{-1}; for K^+ it was from 115 to 39 μmol g^{-1}; for Na^+ it was 158 to 7 μmol g^{-1}; and for Cl^- it was from 68 to 7 μmol g^{-1}.

Under almost all treatments K^+ predominated over Na^+ and in combination (K^+ and Na^+) they predominated over Cl^-, the balance being represented by organic acid anions. But an unexpected and dramatic shift occurred whenever the growth factor system II was unbalanced in the supply of zeatin unaccompanied by IAA (regime 5 of Table I). In all such circumstances the Na^+ in the cells as grown predominated over K^+ (see series e). Nevertheless, whenever zeatin acted in the presence of IAA the sodium in the cells was again less than the K^+ (e.g., regime 6 of Table I, series d).

The soluble (nonprotein) nitrogen compounds also provide striking examples of the effects of the conditions that predetermine cell multiplication and enlargement on the soluble compounds stored. Cells grown in a basal medium with nitrate sugar and minerals, but lacking all the special stimuli to growth, store their nonprotein nitrogen mainly as the dicarboxylic acids, chiefly glutamic acid. A complete complement of trace elements (TE) increased the total nitrogen content per gram over an otherwise trace-element-free basal medium (B**) and the increase was in the content of other compounds, particularly the amides, although glutamic acid still predominated. The active fraction of system II (zeatin) acting alone or with the cofactor (IAA) (23, Fig. 1, series E) along with Fe and Mo in the medium developed cells in which glutamine predominated; the comparable combinations for system I (AFaesc and inositol) also produced cells in which glutamine was the richest nitrogen component. However, the use of the complete system (CM) restored the prominence of alanine.

The unusual effect of zeatin, i.e., unaccompanied by IAA (which especially emphasized glutamine over other nitrogen compounds), also produced the unusual accumulation in the cells as grown of sodium over potassium.

Finally, the conditions that regulate the multiplication, enlargement, and maturity of the cultured cells also control their content of total solutes. They also control the distribution of nitrogen among the different compounds stored and within the content of inorganic salts, its allocation to K^+, Na^+, and Cl^- during growth. Published diagrams render these complex interrelationships evident; they leave the mechanisms of regulation as problems to be solved.

V. Morphogenetic and Environmental Effects on the Solutes of Cells

This topic was introduced in Volume II, Chapter 4, Part III of this treatise in which special attention was piad to the absorption and redistribution of bromide ions throughout the plant body of *Cucurbita pepo* plants, to the absorption of rubidium bromide throughout the roots of barley and its preferential transfer to the shoot, and to the entry of bromide into the developing buds of *Populus.* The activated cells of the cambial region in the spring accumulate the ion laterally from within and supply it to the developing buds. Cesium-137 was also used to study the effect of light and darkness on its subsequent movement into leaves after it was applied directly to the exposed cambial surface of *Acer.* The same isotope was also used in studics of absorption and distribution from roots throughout the developing plant body of *Narcissus.*

The treatment here draws heavily on a "case history" of the potato plant (27, and references therein cited) and upon Volume VIII, Chapter 2, Section IV in this treatise. The former stresses how responsive the potato plant is to environmental factors that control the development, form, and composition of its organs; the latter stresses how the great range of metabolism that is genetically feasible to a given genome is nevertheless modulated by responses to environments that also control development.

GROWTH AND COMPOSITION OF POTATO PLANTS: A CASE HISTORY

1. Responses to Environmental Factors during Growth

Plants from the same clone were grown from standardized cuttings in water culture and exposed to environments throughout their growth to bring out their responses to combinations of environmental factors. The experiments were performed in controlled environmental chambers in which parallel sets of plants could be exposed during their growth to regimes of short and long days combined with high- or low-temperature regimes to be exploited concurrently. The very great morphological responses to these variables were recorded during growth and the definitive effects observed on the development of the organs were traced to responses which originate in the growing regions, especially shoot apices (27, Figs. 11 – 13). After the effects of temperature, particularly low temperature, upon the responses of the plants were noted, the same techniques were exploited to show the distinctive temperature effects according to whether the low or high temperatures were administered by day or by night.

The data have been summarized and discussed (27, see pp. 32–35 and pp. 35–43); it should suffice to say that the *durations* of the day and night environments in question interact with effects of temperature, particularly night temperature, to determine the form and branching habits of shoots, forms of leaves, the development of stolons and tubers, and whether the shoot apex is transformed into a floral shoot. Similarly, shoot apices in the axils of leaves, especially those low on the axis, may give rise to branches that are very different; they may grow erect and more or less recapitulate the main shoot or prostrate as stolons that either persist as such or form tubers. Since such dramatic effects on form were induced in, and traceable to, the growing points per se, it is not surprising that equally dramatic responses which affect the ultimate composition of the fluids in the cells of

TABLE II

EFFECTS OF ENVIRONMENTS AND NITROGEN SOURCES ON THE OSMOTIC VALUE AND METABOLITES OF CARROT LEAVES (CV. 'DANVERS') AFTER 60 DAYS[a]

				Sugars[e]		N-compounds		
No.	Environment[b]	N source[c]	Osmotic value[d]	Reducing	Non-reducing	Free	Combined	Potassium[f]
I–1	SD: HDT/LNT	FN	487	28	30	175	3115	203
I–2	SD: HDT/LNT	NO_3-N	473	25	28	209	2825	206
I–3	SD: HDT/LNT	NH_4-N	749	18	24	395	3026	308
II–1	SD: LDT/HNT	FN	457	14	37	295	3565	168
II–2	SD: LDT/HNT	NO_3-N	420	12	38	244	3268	155
II–3	SD: LDT/HNT	NH_4-N	650	19	39	447	3449	245
III–1	LD: HDT/LNT	FN	459	21	22	162	2993	182
III–2	LD: HDT/LNT	NO_3-N	443	32	27	186	2949	171
III–3	LD: HDT/LNT	NH_4-N	786	17	20	422	2407	268
IV–1	LD: LDT/HNT	FN	440	11	33	424	3236	161
IV–2	LD: LDT/HNT	NO_3-N	433	12	35	402	3179	156
IV–3	LD: LDT/HNT	NH_4-N	886	20	31	443	2787	313

[a] From Craven *et al.* (3). Table 7, Ref. 27. These data are included because they complement the data on potato (see Table III).

[b] SD, day length 10 h; LD, day length 14 h; HDT, day temperature 24°C; LDT, day temperature 12°C; HNT, night temperature 24°C; LNT, night temperature 12°C.

[c] FN, Hoagland's No. 2 medium; NO_3-N, Modified Hoagland's No. 2 medium with N as NO_3; NH_4-N, Modified Hoagland's No. 2 medium with N as NH_4.

[d] In mOsm [for methods, see Mott and Steward (10)].

[e] In μmol g^{-1} fresh wt.

[f] In μg N g^{-1} fresh wt.

the mature organs also occur. These effects on form and composition have been determined with special reference to the nitrogen compounds of leaves and tubers. Polygonal diagrams (27, Figs. 7a,b, 10a,b, with their explanatory notes) illustrate how contrasting combinations of environmental factors influenced the growth and form of potato plants which was also shown photographically (27, Figs. 3, 5, and 6 and 8 and 9). The published results (27, Figs. 14, 15, and 16) also show that the impact of the environmental factors is not only upon growth and form but extends to the content of soluble N and protein in the different organs and to the relative proportions of the compounds in the soluble fraction.

Whether the effects of interacting environmental variables are sought in the balance of total soluble and protein-N (27, p. 34, Table 5) or in the content of particular solutes such as asparagine and glutamine (27, p. 34, Table 6), the ultimate message is clear. The final content of the solutes of the cells in leaf or tuber, respectively, is the outcome of their developmental history. This in turn results from the impact of the environmental

TABLE III

EFFECT OF ENVIRONMENT ON THE CONTENT OF IONS AND TOTAL SOLUTES OF "KB-165" POTATO PLANTS[a,b]

		μmol g^{-1} fresh wt.[c]			
Organ	Environment	K	Na	Cl	Total solutes
Older leaves	SD: HDT/LNT	173	2.4	31	332
Older leaves	SD: LDT/LNT	107	0.7	22	219
Older leaves	SD: HDT/HNT	74	1.6	15	207
Older leaves	LD: LDT/LNT	77	0.7	7	228
Tubers	SD: HDT/HNT	71	0.5	5	152
Tubers	SD: LDT/LNT	72	1.1	2	164
Tubers	LD: HDT/HNT	86	0.6	4	143

[a] Reproduced from Table 8, Ref. 27.

[b] SD: HDT/HNT = short days (10 h), high day temperature (24°C), high night temperature (24°C). SD: LDT/LNT = short days (10 h), low day temperature (12°C), low night temperature (12°C). LD: HDT/HNT = long days (14 h), high day temperature (24°C), high night temperature (24°C). LD/LDT/LNT = long days (14 h), low day temperature (12°C), low night temperature (12°C).

[c] For all methods of determination see Mott and Steward (10). Potassium is the principal cation of potato leaves and tubers and it is largely balanced by organic anions. Salt concentrations, and the changes induced by environments, were greater in leaves than in tubers. Growth at high temperatures produced higher total solutes and K under short than long days, but growth at lowered temperatures benefited only potassium and not total solutes under short days.

factors in question first upon the form of the apices (27, pp. 8, 24–28, Figs. 2, 11–13) for this sets in train their respective developmental histories.

Table II shows that a range of interacting environmental factors that affected the growth and form of potato plants also affected the composition and osmotic value of carrot leaves during their growth. Table III shows the effects of environments on the contents, in terms of ions (K^+, Na^+, and Cl^-) and total solutes, of potato leaves and tubers.

Reference should be made, however, to the original source (27, pp. 39–43) for a discussion of the problems of regulation of form and composition during growth by interacting environmental variables.

The relevant point here is a general one; it is that the environmental factors that interact to determine growth and form, as in the special case of leaves and tubers of potato, also regulate the composition of their solutes, both organic and inorganic. Theories that purport to explain how plant cells acquire their specific complements of solutes should, therefore, be compatible with these demonstrable facts.

VI. Resume, Reflections, and Concluding Remarks

This chapter has developed its main theme with reference to cells of angiosperms, that is, the cells of the most highly organized land plants which encounter in nature the most complex environments. The earlier chapter (25) properly gave much attention (25, see pp. 287–309) to the uses made of aquatic plants in investigations on cell physiology, especially in studies which capitalized on special features of their morphology. As, for example, the so-called giant cells of members of the Siphonales and the large internodal cells of *Chara* and *Nitella.* Interesting as these organisms were, and are, the special features that make them attractive experimental objects also render them, from the standpoint of this chapter, atypical and not representative of the plant kingdom as a whole.

The large coenocytic vesicles (such as members of the Valoniaceae) are not really analogous to the parenchyma of angiosperms. They have retained their attraction over the years as structures suitable for electrophysiology, but the accessible fluids in the large central cavities of the vesicles should not be regarded as strictly homologous with the vacuoles of angiosperm cells. Moreover, as algae, these plants lack growing regions homologous with shoot and root apices, and they do not present in their growth and morphology the equivalent problems of tissue differentiation and organogenesis. However, interesting as these organisms were, and are, a retrospective comment may be appropriate.

Direct acquaintance with the *Valonias* in their habitats at Tortugas (20, see Fig. 8, p. 288, and references therein cited) in the 1930s emphasized how unrealistic was the simplistic homology between these multinucleate coenocytic vesicles (especially when isolated under aquarium conditions) with angiosperm cells. Nevertheless the *Valonias* at Tortugas presented the author (20) with a first encounter with the effects of complex environmental factors on the solute composition of plants as they grow. Moreover the aplanospores of *V. ventricosa* provided a first encounter with naked protoplasts capable of being grown into plants (17, see Figs. 3–15 and 19; 17, see Fig. 8). In fact, it was an accident of war that abruptly shifted what was to have been an emphasis on the uptake of inorganic ions of the alkali metals into *Valonia* sporelings as they grew on marble blocks in sea water in London to, years later, the use of tissue from storage organs, not in slices, but in aseptic explants and later as free totipotent cells.

Therefore, this chapter has presented an overview of the content of solutes, inorganic and organic, of certain angiosperm cells. By using the special features of cells in aseptic culture, attention was focused upon the solute content of cells as they multiply, enlarge, and develop. Attention is addressed not only to electrolytes and to ions that may be conspicuously accumulated to concentrations greater than in ambient media, but also and necessarily to organic solutes and nonelectrolytes that contribute so greatly to the solute content of cells. The data and their interpretations are in Sections I–IV.

Even when mature cells are supplied in culture with a basal medium, ostensibly with the organic and inorganic nutrients which are the normal requirements for growth, the ability of the explanted cells to respond may be minimal unless they are activated by special exogenous stimuli that promote growth by cell multiplication and enlargement. The most rapidly growing cultures in media that promote cell division contain cells which individually pass through their cycle of division and enlargement and trace out a sigmoidal curve of growth for the explants as a whole. In the early experimental phase of growth, when the emphasis is on cell division, the average size of cells is small, the cells develop a specific requirement for potassium, selectively preferred to sodium, and balanced by organic solutes rather than halide. These relationships change as cells develop, and with enlarging vacuoles the emphasis is then on the maintenance of osmotic value as their vacuoles enlarge. The developing cells first store, preferentially, organic solutes but later these are replaceable by neutral salts (KCl, NaCl) when, and if, their sugars are depleted. Later, as this occurs the first disparities between K and Na, and between K plus Na and Cl typical of the cell multiplication phase become less evident. As the cells of the explants grow they seem to control, primarily, the internal activity of their water by their total solute concentrations.

Cells which multiply require energy to create new structures, and they do not then emphasize the accumulation of solutes in bulk. However, when cells enlarge, energy is used in the storage of solutes (organic and inorganic) and their osmotic value supports their cytoplasm which is being "spread out thin." These events involve more than the properties of membranes or their relations to individual ions or molecules for they also require an appreciation of cells as compartmented metabolic and osmotic machines and of their various energy relationships.

A range of factors will in the outcome determine whether isolated cells merely survive on the one hand or grow by cell multiplication and enlargement on the other. The agents that convey this message to cells in effect put the idling machinery of growth into (or out of) gear and in doing this they regulate the use of nutrients and accessible solutes.

The recognizable complex exogenous agents that cause the growth induction and which are of natural origin [as obtained from the liquid extractable from coconuts (CM) or from immature *Aesculus* fruits] achieve these ends as they affect metabolism and stimulate protein synthesis from previously stored or external sources of nitrogen, and, concomitantly, they affect the water relations of the cells as they cause them to acquire characteristic complements of soluble compounds (organic and inorganic) with colligative properties.

Mott and Steward (13) treated solute accumulations in plant cells as an aspect of nutrition and development. In so doing they presented data on the range of solute contents that carrot cells in culture may display in response to the growth that is stimulated by different exogenous treatments (see Table IV). But where, and how, are the controlling master switches activated for cells *in situ* in the plant body?

The responses of cells to exogenous stimuli when they are in culture must surely reflect similar responses of cells to endogenous stimuli during their development *in situ.* One might select from the maze of effects which have been described isolated examples in which a single solute could seem to be regulated by a single causal factor. However, such an exercise would not lead to a mature understanding of the whole system in which electrolytes and nonelectrolytes respond collectively during the growth and development of the cells. As cells multiply, enlarge, and mature there is a built-in program in the solutes they acquire at different stages along this route in response to the culture and environmental conditions. *In situ* the stimuli which control these relationships are endogenous. The study of the isolated cells in culture has enabled them to be investigated in response to exogenous stimuli which substitute for those to which cells may respond *in situ.*

The chemical stimuli to which the carrot cells respond in culture derive from the normal environments of developing embryos, and it is a curious

TABLE IV

RANGE OF SOLUTE CONTENT AND COMPOSITION IN CELLS OF CARROT IN RELATION TO THE GROWTH STIMULATED IN CULTURED EXPLANTS[a]

Treatment		Growth		Solute contents				
No.	Culture medium[b]	Fresh weight increase	Average cell size (μg cell^{-1})	K (μmol g^{-1} fresh weight)	Na	Cl	Total solutes[c]	Organic solutes as % of total[d]
1.	Explants fresh from root	×1	0.065	77	17	34	402	53
2.	Explants as inoculated	×1	0.065	43	12	26	175	37
3.	$CaCl_2$ (0.5 mM)	×1	0.124	25	8	36	180	62
4.	$CaCl_2$ + Sucrose (2%)	×1	0.114	18	6	23	452	90
5.	$CaCl_2$ + KCl (1 mM) + NaCl (1 mM)	×1	0.104	151	36	183	420	13
6.	$CaCl_2$ + Sucrose + KCl (1 mM) + NaCl (1 mM)	×1	0.162	48	5	46	510	79
7.	B + CM (at 6 days)	×4	0.051	96	17	10	246	8
8.	B + CM (at 18 days)	×70	0.117	59	16	24	370	59
9.	B** + CM** + T.E.[e]	×15	0.053	98	41	14	371	25
10.	B + CM + NH_4NO_3	×47	0.110	40	2	20	247	66
11.	B + CM + NH_4Cl	×27	0.047	37	12	12	202	51
12.	Basal medium (B)	×2	0.126	124	15	38	389	29
13.	Basal medium (B**) + T.E.[e]	×2	0.170	88	44	35	538	51
14.	B + INOS	×2	0.140	111	8	22	441	46
15.	B + AF_1	×8	0.034	106	38	20	310	7
16.	B + INOS + AF_1 (system I)	×12	0.045	37	53	28	210	14
17.	B + IAA	×11	0.095	54	37	20	358	49
18.	B + Z	×10	0.126	61	103	16	385	15
19.	B + IAA + Z (system II)	×8	0.058	60	10	9	227	38
20.	B + System I + system II	×8	0.041	78	19	8	234	17

[a] Data from Mott & Steward (13).

[b] Explants, initially 2.5 to 3 mg fresh weight, were cultured 18 days (except where otherwise noted) in the media specified, where the abbreviations represent: B, a basal medium after White (see 11, p. 624); CM, coconut milk at 10% v/v; (INOS), *myo*-inositol at 25 mg/L; IAA, indoleacetic acid at 0.5 mg/L; AF, an active fraction from immature *Aesculus* fruits at 2 mg/L; Z, zeatin at 0.1 mg/L; NH_4NO_3 at 18 mM; or $CaCl_2$ at 0.5 mM.

[c] Total solutes estimated from freezing-point depression of hot-water extracts of tissue or by summation of individual solutes and expressed as micromoles of ideal solute.

[d] The contribution of organic nonelectrolytes (e.g., sugars, amino acids, and amides) was obtained by reducing the total solutes by the content of inorganic cations ($K^+ + Na^+$) and their accompanying anions, whether organic or inorganic.

[e] These data are from other unpublished work with K. V. N. Rao and the authors and show the responses of explants cultured in media which supplied trace elements except Mn and Mo. CM**, coconut milk freed from trace elements; T.E., appropriate supplies of trace elements but lacking Mn and Mo.

but significant fact that the similar stimuli and conditions promote somatic embryogenesis. Hence it is reasonable to equate the observed physiology of the cells in culture with that of cells *in situ* during development of the plant body.

However, particular attention should be paid to the consequences of the normal need of land plants to acquire nitrogen from nitrate in soils and carbon via photosynthesis in leaves. The introduction of reduced nitrogen compounds in these circumstances curtails the need for a balanced supply of the nonmetabolizable ions of alkali halides to meet the total solute requirements. Thus the solutes of the cells and organs of angiosperms not less than their form should also be evaluated inasmuch as they are affected by the environmental variables which they encounter as plants grown in their habitats.

This general approach to the factors that affect the solutes of cells as they grow complements the conventional emphasis on observations on mature cells in near steady states and under conditions that preclude active growth and development and when attention is commonly focused on a single ion or molecule followed over a short period of time. While this approach may illuminate how specific molecules traverse complex membranes, this is only one step toward understanding how cells acquire and maintain their solutes as they develop.

However there is yet another broad area to be considered in the responses of genetically identical cells to the environmental factors that operate during the development and organogenesis of land plants. These environmental factors also operate during development on cells which are responsive in culture to a network of interacting stimuli that exert effects as they act singly or in combinations.

Thus the theme is that the solutes of cells *in situ* are essentially determined by the morphological setting in which they occur and by the stimuli, over and above their genetically imparted information, that control their metabolism and vital activity. In short, instead of isolating cells from the complications of growth metabolism and development to investigate them remote from these complications, the reverse is now desirable. By the use of the aseptic culture methods as described, the salt and water relations have been investigated under actual circumstances in which the cells can grow, develop, and mature. Moreover the facts of embryogenesis permit that the cultured carrot cells may be regarded as totipotent, that is, inherently capable of all that development entails. Of necessity, however, this has required not only the full and familiar requirements that autotrophic plants need, but also a range of supplementary conditions necessary to stimulate and maintain their activity. Thus a facile, mechanistic simplicity does not emerge.

The long-acknowledged ability of cells to harness metabolic energy to useful work, as in the active transfer of solutes (especially ions) against concentration gradients (amplified by the later developments described in this volume), is still the basis on which such movements, where they occur, as recognized as biologically motivated; they are not mere passive aspects of a physical chemistry that is remote from vital activity. But the theme referred to goes further than this for it enters areas and modes of thought where insight and perception cannot easily be translated into physicochemical terms.

The capacity to grow—so prominent in totipotent cells which first multiply and then develop—is itself difficult to define, although its consequences are obvious enough when encountered. This is especially so when endogenous stimuli and exogenous conditions need to be in subtle balance and involve the interactions among growth-promoting substances and their cofactors, not to speak of trace elements over and above the macronutrients and vitamins. Cells that utilize all this complexity of nutrition and metabolism and growth are the setting for the ongoing uptake of water, salts, and solutes. But there are also morphological criteria to be considered for the cells in organs have used their unique genetic information, nutrition, and metabolism to create (and recreate) internal complexity with the need to acquire and store solutes in spaces to be filled. In the maintained composition of the cellular fluids, such solutes control the internal activity of water that imparts to the cells and organs of the plant body their physical stability.

Although it is satisfying to know that the cells harness their vital activity and complexity of structure to perform such useful physicochemical work, it is still philosophically frustrating that the physicochemical machinery to bring it about is so inseparable from the system by which it was produced and which it serves.

Moreover, it is not a simplification to focus upon one solute molecule or ion. The clear indication now is that once self-duplication of cells is achieved the ultimate regulatory controls operate upon the total complement of solutes in cells. Some decades of focusing attention upon indicator ions or molecules may have diverted attention from considerations of total solute concentrations which reduce the internal activity of water and which may often override the importance of even the dramatic accumulation of a particular ion or solute over the external medium.

Angiosperm cells that multiply maintain at first a relatively low concentration of solutes. The first soluble components acquired are prominently organic. The principal ion they first acquire is K^+, usually in marked (even total) excess over Na^+ not balanced by Cl^-. As cell enlargement ensues and cytoplasm is more spread out over cell surfaces (that increase more than cell volume), the physical stability of the system is maintained by the reduced activity of its internal water due to the colligative properties of *all* its solutes. From this point on the control is exercised more over water through total solutes than over particular solutes. As cell expansion proceeds, and if organic solutes become limiting, total solute levels may be maintained by salts and nonelectrolytes that may now behave as alternative osmotically active solutes, depending upon their respective supplies. It is in these circumstances that ions from the environments may be accumulated to high degree over the ambient media and the initial disparity between inorganic cations and anions ($K^+ + Na^+$ and Cl^-) may be reduced and the great preference for K^+ over Na^+ declines. However, the regulation of the soluble constituents of cells baffles simplistic interpretations. The regulatory factors present an array or network (growth factors; their cofactors, trace elements, environments), and they regulate a complex array of solutes that change in composition when cells develop *in situ* or in culture.

But there is a further complexity to be faced. Who could have expected, in the early days of the osmotic view of cells, that their ultimate fate in angiosperms would begin with their antecedents in the shoot-growing points? So much so that the physiology of cells would be molded, over and above genetic factors or nutrition, by the interaction of day length with night length and with day and night temperatures to determine the final setting of cells in organs of perennation, as in the contrasts between cells of potato tubers or leaves? And these interactions insofar as they regulate the solutes of cells *in situ* are as subtle (or more so) as those that regulate the behavior of isolated cells such as those of carrot as they grow.

Thus this chapter ends, not with final solutions, but with a dilemma. This is that to understand how cells develop their complements of solutes is to comprehend the problems of embryogenesis and morphogenesis, not only in terms of form, but in terms of cell contents. Furthermore, if all these events are to be regarded as "orchestrated," then while the "script" is inherited via the DNA, it is to be seen as "interpreted" through the properties of water (the universal solvent and source of stability) and through the bioenergetics involving phosphate bond energy as the "currency" through which all cellular transactions are negotiated.

In other words, it is not only by membranes or permeases or ion fluxes that cells can accumulate and regulate their contents. Cells are improbable morphological entities and their development and operation *in toto* as physical systems seems to need description and definition in terms of entropies and free energies in ways which allow their contribution to the organism as a stable physical system in its environment to be comprehended. This is the combined challenge that morphogenesis, genetics, biochemistry, and bioenergetics are still required to meet before we know fully how plants absorb and distribute their water salts and solutes.

Acknowledgments

These acknowledgments are now made by the writer in his dual role as author of the chapter and editor of Volume IX. It would be ungracious to overlook the many special circumstances that have contributed to the diversity of this account. The circumstances which now call for special thanks concern the many awards from granting agencies and foundations which have given the work the stability that it has enjoyed over the years. The level of continuing support has permitted different institutions to physically house the various aspects of the research which has required the skills and application of a diverse group of investigators and assistants. But to be comprehended the results of all this work have needed to be analyzed and made available and for their cooperation with this task the editors and publishers of the various journals cited are to be thanked. This is especially so for the comprehensive and numbered series of publications now available for citation in this chapter from the *Annals of Botany* and *Planta.* Although special recognition is given in the citations by name to the principal investigators and authors of the published works, this should not obscure the debt which all the authors owe to the technicians and assistants whose special skills have contributed so crucially to the success of the work as planned on the scale required. The author has deemed it both a privilege and a duty to present this summarized account. But in doing this, liberal use has had to be made of material now to be found in the *Annals of Botany* and which has also drawn upon the texts as published in *Planta.* For all of this help these collective acknowledgments are now made.

References

1. Adam, N. K. (1921). The properties and molecular structure of thin films of palmitic acid on water. Part 1. *Proc. R. Soc. London, Ser. A* **99,** 336. (This is the first of 21 papers on thin films in the *Proc. R. Soc. London, Ser. A.* For full bibliography, classified and an account of work with the Langmuir-Adam trough as developed by Adam and interpreted by him, see N. Carrington *et al.*, on N. K. Adam, *In* "Biographical Memoirs of the Royal Society," Vol. 20, pp. 1–26, esp pp. 13–18.)
2. Bielby, M. J. (1982). Cl^- channels in *Chara. Philos. Trans. R. Soc. London, Ser. B* **299,** 435–445.
3. Craven, G. H., Mott, R. L., and Steward, F. C. (1972). Solute accumulation in plant cells. IV. Effects of ammonium ions on growth and solute content. *Ann. Bot. (London)* [N.S] **36,** 897–914.
4. Davson, H., and Danielli, J. F. L. (1952). "The Permeability of Natural Membranes." Cambridge Univ. Press, London and New York.
5. Degani, N., and Steward, F. C. (1969). The effect of various media on the growth responses of different clones of carrot explants. *Ann. Bot. (London)* [N.S.] **33,** 483–504.
6. Gradman, D., Fritton, J., and Goldfarb, T. (1982). Electrogenic Cl^- pump in *Acetabularia. Philos. Trans. R. Soc. London, Ser. B* **299,** 447–457.
7. Hoagland, D. R., and Broyer, T. C. (1936). General nature of the process of salt accumulation by roots with description of experimental methods. *Plant Physiol.* **11,** 471–507.
8. Keynes, R. D., and Ellroy, J. C. (1982). The binding and transport of anions in living tissues. *Philos. Trans. R. Soc. London, Ser. B* **299,** 365–607.
9. MacRobbie, E. A. B. (1982). Chloride transport in stomatal guard cells. *Philos. Trans. R. Soc. London, Ser. B* **299,** 469–481.

10. Mott, R. L., and Steward, F. C. (1972). Solute accumulation in plant cells. I. Reciprocal relations between electrolytes and non-electrolytes. *Ann. Bot. (London)* [N.S.] **36,** 621–639.
11. Mott, R. L., and Steward, F. C. (1972). Solute accumulations in plant cells. II. The progressive uptake of non-electrolytes and ions in carrot explants as they grow. *Ann. Bot. (London)* [N.S.] **36,** 641–653.
12. Mott, R. L., and Steward, F. C. (1972). Solute accumulation in plant cells. III. Treatments which arrest and restore the course of absorption and growth in cultured carrot explants. *Ann. Bot. (London)* [N.S.] **36,** 655–670.
13. Mott, R. L., and Steward, F. C. (1972). Solute accumulation in plant cells. V. An aspect of nutrition and development. *Ann. Bot. (London)* [N.S.] **36,** 915–937.
14. Neumann, K. H., and Steward, F. C. (1968). Investigation on the growth and metabolism of cultured explants of *Daucus carota.* I. Effects of iron, molybdenum and maganese on growth. *Planta* **81,** 333–350.
15. Parr, A. J., and Hanke, D. E. (1982). Chloride stimulated ATPase activity in *Limonium vulgare* Mill. *Philos. Trans. R. Soc. London, Ser. B* **299,** 459–468.
16. Robertson, R. N. (1983). "The Lively Membranes." Cambridge Univ. Press, London and New York.
17. Steward, F. C. (1968). "Growth and Organization in Plants." Addison-Wesley, Reading, Massachusetts.
18. Steward, F. C. (1970). From cultured cells to whole plants: The induction and control of their growth and morphogenesis. *Proc. R. Soc. London, Ser. B* **175,** 1–30.
19. Steward, F. C. (1970). Totipotency, variation and clonal development of cultured cells. *Endeavour* **29,** 117–124.
20. Steward, F. C., and Martin, J. C. (1937). The distribution and physiology of *Valonia* at the Dry Tortugas with special reference to the problems of salt accumulation in plants. *Pap. Tortugas Lab.* **31,** 89–170. (*Carnegie Inst. Washington Publ.* **475**).
21. Steward, F. C., and Krikorian, A. D. (1971). "Plants, Chemicals, and Growth." Academic Press, New York and London.
22. Steward, F. C., and Mott, R. L. (1970). Cells, solutes and growth: Salt accumulation re-examined. *Int. Rev. Cytol.* **28,** 275–369.
23. Steward, F. C., and Rao, K. V. N. (1970). Investigation on the growth and metabolism of cultured explants of *Daucus carota.* III. The range of responses induced in carrot explants by exogenous growth factors and by trace elements. *Planta* **91,** 129–145.
24. Steward, F. C., and Rao, K. V. N. (1971). Investigation on the growth and metabolism of cultured explants of *Daucus carota.* IV. Effects of iron, molybdenum and the components of growth promoting systems and their interactions. *Planta* **99,** 240–264.
25. Steward, F. C., and Sutcliffe, J. F. (1959). Plants in relation to inorganic salts. *In* "Plant Physiology: A Treatise" (F. C. Steward, ed.), Vol. 2, Chapter 4, pp. 253–478. Academic Press, New York.
26. Steward, F. C., Caplin, S. M., and Millar, F. K. (1952). Investigations on growth and metabolism, nutrition and growth in undifferentiated cells. *Ann. Bot. (London)* [N.S.] **16,** 57–77.
27. Steward, F. C., Moreno, U., and Roca, W. M. (1951). Growth, form and composition of potato plants as affected by environment. *Ann. Bot. (London)* **48,** Suppl. 2, 1–45.
28. Steward, F. C., Neumann, K. H., and Rao, K. V. N. (1968). Investigations on the growth and metabolism of cultured explants of *Daucus carota.* II. Effects of iron, molybdenum and manganese on metabolism. *Planta* **81,** 351–371.
29. Steward, F. C., Mott, R. L., and Rao, K. V. N. (1973). Investigations on the growth and

metabolism of cultured explants of *Daucus carota.* V. Effects of trace elements and growth factors on the solutes accumulated. *Planta* **111,** 219–273.
30. Steward, F. C., Berry, W. E., Preston, C., and Ramamurti, T. K. (1943). The absorption and accumulation of solutes by living plant cells. X. Time and temperature effects on salt uptake by potato discs and the influence of the storage conditions of the tubers on metabolism and other properties. *Ann. Bot. (London)* [N.S.] **7,** 221–260.

Index to Plant Names

Numbers in this index designate the pages on which reference is made in the text to the plant in question. No reference is made in the index to plant names included in the titles that appear in the reference lists. In general, when a plant has been referred to in the text sometimes by its common name, sometimes by its scientific name, all such references are listed in the index after the scientific name; cross reference is made, under the common name, to this scientific name. However, in a few instances in which a common name as used cannot be referred with certainty to a particular species, the page numbers follow the common name.

Z

Subject Index

D

Subject Index

D

X

Y

Z